MATHEMATICS —
The Core Course for A-level

L. Bostock, B.Sc.

Formerly Senior Mathematics Lecturer
Southgate Technical College

S. Chandler, B.Sc.

Formerly of the Godolphin and Latymer School

Stanley Thornes (Publishers) Ltd.

First published in 1981 by Stanley Thornes (Publishers) Ltd.,
Old Station Drive, Leckhampton, CHELTENHAM GL53 0DN

Reprinted 1982
Reprinted 1983
Reprinted 1984 with minor corrections
Reprinted 1985
Reprinted 1986
Reprinted 1987
Reprinted 1988

British Library Cataloguing in Publication Data

Bostock, L.
    Mathematics — the core course for A-level.
    1. Mathematics — 1961 —
    I. Title   II. Chandler, S.
    510   QA39.2

    ISBN 0–85950–306–2

Printed and bound in Great Britain at The Bath Press, Avon.

PREFACE

What knowledge and skills should be acquired during an A-level Mathematics course?

This question has been studied closely by various interested organisations such as the SCUE, CNAA and the Schools' Council. After looking at the problems caused by the diversity of mathematical backgrounds possessed by students, all of whom had passed A-level Mathematics, the conclusion reached was that all students who intend to follow higher education courses in which mathematics is used, should start with a common core of mathematical knowledge and ability. The core syllabus which they recommended is already being adopted fairly closely by Examining Boards, for example the University of London Syllabus B, and this book covers the whole of this common core syllabus.

Much of the material in this book is taken from our Pure Mathematics books, and the new work follows the same format with many worked examples and exercises to illustrate each development of a topic. At the end of each chapter there is a miscellaneous exercise including some questions of sufficient difficulty to appeal to the most able students. Also, where appropriate, there is a multiple choice exercise.

With the varied nature of O-level courses, the starting point for an A-level text presents problems. We have started most topics from the beginning, assuming only a very basic mathematical knowledge.

The order in which topics are presented is another problem. We have chosen an order which we feel forms a logical sequence and is suitable for a student working on his own. However the course teacher will find considerable flexibility.

We are grateful to the following Examination Boards for permission to reproduce questions from their past examination papers (part questions are indicated by the suffix p):

University of London (U of L)
Joint Matriculation Board (JMB)
University of Cambridge Local Examinations Syndicate (C)
Oxford Delegacy of Local Examinations (O)
The Associated Examining Board (AEB)

1981

L. Bostock
S. Chandler

v

CONTENTS

Preface   v

Notes on the use of the book   xi

Chapter 1   **Algebraic Relationships**   1

Nature of equations, identities, inequalities, functions.
Partial fractions. Quadratic equations.

Chapter 2   **Algebraic Topics**   23

Indices. Surds. Logarithms. Remainder theorem. Pascal's
triangle.

Chapter 3   **Functions**   43

Mapping. Functions. Domain and range. Graphs.
Quadratic functions. Inequalities. Exponential
functions. Rational functions. Logarithmic functions.
Inverse functions.

Chapter 4   **Coordinate Geometry I**   69

Cartesian coordinates. Coordinate geometry of the
straight line. The meaning of equations.

Chapter 5   **Differentiation I**   106

Gradient of a curve. Differentiation of polynomials.
Tangents and normals. Stationary values.

Chapter 6   **Trigonometric Functions**   137

Radians. Arc and sector. Trigonometric ratios. Sine
rule. Cosine rule. Inverse trigonometric functions.
Trigonometric equations. General solutions.

Chapter 7   **Trigonometric Identities**   199

Further solution of equations. The expression
$a \cos \theta + b \sin \theta$. Small angles. Further properties of triangles.

Chapter 8   **Differentiation II**   255

Differentiation of trig functions, exponential functions,
logarithmic functions. Differentiation of compound
functions. Implicit functions. Logarithmic
differentiation. Parametric equations.

Chapter 9   **Integration**   299

Indefinite integrals. Integration as the reverse of
differentiation. Integration of products, fractions,
trig functions. Change of variable. Differential
equations. Definite integration as summation. Area
by integration.

Chapter 10   **Coordinate Geometry II**   360

Coordinate geometry of the circle and parabola.
Intersection of lines and curves. Cubic curves.
Parametric coordinates. Loci.

Chapter 11   **Curve Sketching**   420

Even, odd, continuous and periodic functions. General
methods. $y = 1/f(x)$. Modulus notation. Further
inequalities.

Chapter 12   **Vectors and Three Dimensional Coordinate Geometry**   455

Vectors and scalars. Addition. Position vectors. Base
vectors. Cartesian components. Equations of a straight
line. Scalar product. Equation of a plane.

Chapter 13   **Complex Numbers**   532

Imaginary numbers. The algebra of complex numbers.
Complex roots of quadratic equations. Cube roots of
unity. Argand diagram. Modulus and argument and their
properties.

Chapter 14   **Permutations and Combinations**   564

Chapter 15   **Series**   586

Sequences. Σ notation. Arithmetic progressions.
Geometric progressions. Convergence. Arithmetic and
geometric mean. Binomial theorem. Approximations.
Number series. Summation using partial fractions and
method of differences. Proof by induction.

Chapter 16   **Solution of Equations**   641

Polynomial equations: exact solutions in special cases,
approximate location of the roots, estimation of the
number of real roots. Graphical methods and iterative
methods for the approximate solution of equations.

Chapter 17   **Numerical Applications**   674

Reduction of a relationship to linear form. Definite
integration used to find areas, volumes of revolution,
centroids. Approximate integration using the
Trapezium rule and Simpson's rule. Small increments.
Rates of change.

Answers   718

Index   749

NOTES ON USE OF THE BOOK

Notation

$=$	is equal to	$:$	such that
$\equiv$	is identical to	$\mathbb{N}$	the natural numbers
$\simeq$	is approximately equal to*	$\mathbb{Z}$	the integers
$>$	is greater than	$\mathbb{Q}$	the rational numbers
$\geqslant$	is greater than or equal to	$\mathbb{R}$	the real numbers
$<$	is less than	$\mathbb{R}^+$	the positive real numbers
$\leqslant$	is less than or equal to		excluding zero
∞	infinitely large	$\mathbb{C}$	the complex numbers
$\Rightarrow$	implies	$[a, b]$	the interval $\{x : a \leqslant x \leqslant b\}$
$\Leftarrow$	is implied by	$(a, b]$	the interval $\{x : a < x \leqslant b\}$
$\Longleftrightarrow$	implies and is implied by	(a, b)	the interval $\{x : a < x < b\}$
$\in$	is a member of		

$|x|$ the modulus of x, $\begin{cases} |x| = x & \text{for } x \geqslant 0 \\ |x| = -x & \text{for } x < 0 \end{cases}$

$\binom{n}{r}$ the binomial coefficient $\dfrac{n!}{r!(n-r)!}$

A stroke through a symbol negates it. i.e. $\neq$ means 'is not equal to'.

Abbreviations

$\parallel$	parallel
$+$ ve	positive
$-$ve	negative
w.r.t.	with respect to

*Practical problems rarely have exact answers. Where numerical answers are given they are correct to two or three decimal places depending on their context, e.g. π is 3.142 correct to 3 d.p. and although we write $\pi = 3.142$ it is understood that this is not an exact value. We reserve the symbol $\simeq$ for those cases where the approximation being made is part of the method used.

Instructions for answering Multiple Choice Exercises

These exercises are at the end of most chapters. The questions are set in groups, each group representing one of the variations that may arise in examination papers. The answering techniques are different for each type of question and are classified as follows:

TYPE I

These questions consist of a problem followed by several alternative answers, only *one* of which is correct.

Write down the letter corresponding to the correct answer.

TYPE II

In this type of question some information is given and is followed by a num-number of possible responses. *One or more* of the suggested responses follow(s) directly and necessarily from the information given.

Write down the letter(s) corresponding to the correct response(s).
e.g. PQR is a triangle

(a) $\hat{P} + \hat{Q} + \hat{R} = 180°$

(b) PQ + QR is less than PR

(c) if $\hat{P}$ is obtuse, $\hat{Q}$ and $\hat{R}$ must both be acute.

(d) $\hat{P} = 90°$, $\hat{Q} = 45°$, $\hat{R} = 45°$.

The correct responses are (a) and (c).
(b) is definitely incorrect and (d) may or may not be true of triangle PQR, i.e. it does not follow directly and necessarily from the information given. Responses of this kind should not be regarded as correct.

TYPE III

Each problem contains two independent statements (a) and (b).
1) If (a) always implies (b) but (b) does not always imply (a)   write A.
2) If (b) always implies (a) but (a) does not always imply (b)   write B.
3) If (a) always implies (b) *and* (b) always implies (a)   write C.
4) If (a) denies (b) and (b) denies (a)   write D.
5) If none of the first four relationships apply   write E.

TYPE IV

A problem is introduced and followed by a number of pieces of information. You are not required to solve the problem but to decide whether:
1) the given information is *all* needed to solve the problem. In this case write A;

2) the total amount of information is insufficient to solve the problem. If so write I;

3) the problem can be solved without using one or more of the given pieces of information. In this case write down the letter(s) corresponding to the items not needed.

TYPE V

A single statement is made. Write T if it is true and F if it is false.

CHAPTER 1

ALGEBRAIC RELATIONSHIPS

THE NATURE OF ALGEBRAIC EXPRESSIONS

Consider the expressions $(x + 2)^2 = 2x + 7$ [1]

$$(x + 2)^2 = x^2 + 4x + 4 \qquad [2]$$

$$(x - 2) > 1 \qquad [3]$$

$$(x + 2)^2 \qquad [4]$$

By inspection we see that these four expressions are all different in nature and by investigating each of them we shall try to identify these differences.

(1) $(x + 2)^2 = 2x + 7$

Substituting 1 for x in both the left hand side (LHS) and right hand side (RHS) separately we find

$$\text{LHS} = (1 + 2)^2 = 3^2 = 9$$

$$\text{RHS} = 2 + 7 \quad = 9$$

i.e.   LHS = RHS   when $x = 1$.

However substituting 2 for x in a similar way we find

$$\text{LHS} = (2 + 2)^2 = 16$$

$$\text{RHS} = 4 + 7 \quad = 11$$

i.e.   LHS $\neq$ RHS   when $x = 2$.

Now rearranging the original expression gives

$$x^2 + 4x + 4 = 2x + 7$$

$$\Rightarrow \quad x^2 + 2x - 3 = 0$$

i.e. $(x + 3)(x - 1) = 0$

from which we see that   LHS = RHS   if and only if

either $x + 3 = 0$ i.e. $x = -3$

or $x - 1 = 0$ i.e. $x = 1$

Thus $(x + 2)^2 = 2x + 7$ only if $x = -3$ or if $x = 1$ and the equality is not true for any other value of x.

Expressions of this type are called equations and the equality is true only for a number of distinct values of the unknown quantity (or quantities).

The process of finding these values is referred to as solving the equation.

(2) $(x + 2)^2 = x^2 + 4x + 4$

If we substitute 1 for x in both sides of this expression we find

$$\text{LHS} = (1 + 2)^2 \qquad = 9$$

$$\text{RHS} = 1^2 + 4.1 + 4 = 9$$

i.e.   LHS = RHS   when $x = 1$
If we substitute -1 for x as before we find

$$\text{LHS} = (-1 + 2)^2 = 1$$

$$\text{RHS} = (-1)^2 + (4)(-1) + 4 = 1$$

i.e.   LHS = RHS   when $x = -1$.
Whatever other numerical value we substitute for x we find that   LHS = RHS, so it appears (but is not proved) that   LHS = RHS   for all values of x.
The following geometric illustration confirms that this suspicion is correct.
Consider a square of side $x + 2$ units

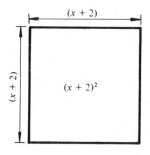
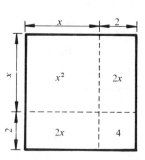

Since these two squares are identical, their areas are identical.
Hence $(x + 2)^2 = x^2 + 4x + 4$ for all values of x and we say that
$(x + 2)^2$ *is identical to* $x^2 + 4x + 4$.
Using the symbol ≡ for 'is identical to' the relationship is written

$$(x + 2)^2 \equiv x^2 + 4x + 4.$$

Relationships of this type are called identities and both sides are equal for any value of the unknown quantity. They are, in fact, two forms for the same expression, and we shall use the identity symbol whenever we are dealing with an identity relationship and we strongly recommend that the reader does the same.

(3) $(x-2)>1$

Reading from left to right the symbol $>$ means greater than (and $<$ means less than). This relationship is obviously different from the first two and it is called an inequality.

By inspection we see that for $x-2$ to have a value greater than one then x must have a value greater than three.

i.e. if $$x-2>1$$

then $$x>3$$

Consider a line as being made up of adjacent points. We can represent all the real values that x, the variable, can have by the positions of points on a line, (known as a *number line*).

negative
values of x
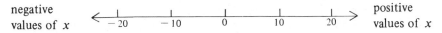
positive
values of x

The position of a point to the *left* of a second point corresponds to a value of x *less than* the value of x at the second point.

i.e. $$-99<1, \quad -10<-5 \quad \ldots \text{etc.}$$

The values of x given by the statement $x>3$ can then be represented by a section of this line.

i.e.

From this we see that *not* all values of x satisfy the inequality but that there is an infinite set of values that do, i.e. the solution of an inequality is a range (or ranges) of values of the variable involved.

Note that $x=3$ is *not* included in the range and this is indicated by an open circle at $x=3$. For $x \geqslant 3$, which means x is greater than or equal to 3, the value $x=3$ *is* included in the range. This is indicated on a number line by a solid circle.

i.e.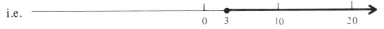

(4) $(x+2)^2$

This expression is not related to any other expression and can take different values depending on the value given to x. For example, if we give x the value 1, then the value of $(x+2)^2$ is $(1+2)^2 = 9$.

Such an expression is called a function (of x in this case) and, considering the values given to x as input and the corresponding values of the function as output, we denote this idea by

$$f: x \to (x+2)^2$$

Thus $f: r \to 2r+1$ can be read as 'the function which, when we input a value for r, gives us, as output, the value of $2r+1$'.

Using f as the symbol for function, we use $f(x)$ for a function of x, $f(r)$ for a function of $r, \ldots$, etc., so we may also write the statements above in the forms $f(x) \equiv (x+2)^2$ and $f(r) \equiv 2r+1$.

To represent the value of a function when $x = 1$, say, we write $f(1)$.

So, for the function $f: x \to (x+2)^2$ we have

$$f(1) = (1+2)^2 = 9$$

$$f(3) = (3+2)^2 = 25$$

$$f(-2) = (-2+2)^2 = 0$$

Note that the values both of x and $f(x)$ are variable but whereas x can be given *any* value, the value of $f(x)$ depends on that of x, so x *is referred to as the independent variable* and $f(x)$ *as the dependent variable*.

EXERCISE 1a

1) State which of the following are equations and which are identities

(a) $x^2 - 3 = 2$ (b) $x^2 - 9 = (x-3)(x+3)$ (c) $\dfrac{1}{x} - \dfrac{1}{x+1} = \dfrac{1}{x^2+x}$

(d) $\dfrac{1}{x-1} + \dfrac{1}{x+1} = \dfrac{2}{x^2-1}$ (e) $p^2 + 2p - 3 = 3 - 2p - p^2$

(f) $(x+y)(x-y) = x^2 - y^2$ (g) $y - 1 = \dfrac{1}{y}$ (h) $\dfrac{2q}{q^2-1} = \dfrac{1}{q-1} + \dfrac{1}{q+1}$

2) Find the range of values of x for which the following inequalities are true and illustrate the range on a number line.

(a) $x - 5 > 0$ (b) $x + 1 \leqslant -1$ (c) $0 \geqslant x - 4$ (d) $3 < 4 - x$

3) Find the values of $f(0), f(1), f(-2), f(5)$ where the function f is defined by

(a) $f: x \to x^2 - 3x$ (b) $f: x \to \dfrac{1}{x-2}$ (c) $f: x \to (x-7)(x+2)$

(d) $f: x \to \dfrac{1}{x+1} - \dfrac{2}{2x-3}$

POLYNOMIALS AND FRACTIONAL FUNCTIONS

If each individual term of a function is of the form ax^n, where a is a constant (i.e. has a fixed numerical value) and n is a positive integer, the function is called a *polynomial*. Thus $x^2 - 2x$, $3x^6 - 7x^4 + 6$, $(x-4)^2$ are polynomials but $\sqrt{x}$, $\dfrac{1}{x}$, $\sqrt{x^2 - 2}$ are not.

The highest power of x that occurs in a polynomial defines *the degree or order of the polynomial*.

Thus $5x^6 - 7x^3 + 6x$ is a polynomial of degree 6.

A fractional function of the form $\dfrac{3x^2 - 7}{x^3 + 1}$ where both the numerator and denominator are polynomials, is referred to as a 'proper' fraction if the degree of the numerator is less than the degree of the denominator. However if the degree of the numerator is greater than, or equal to, the degree of the denominator the fraction is referred to as 'improper'.

An improper numerical fraction such as $\dfrac{9}{7}$ may be written as $\dfrac{7+2}{7} = 1 + \dfrac{2}{7}$

Similarly an improper algebraic fraction such as $\dfrac{x^2 - 1}{x^2 + 1}$ may be written

$$\frac{x^2 + 1 - 2}{x^2 + 1} \equiv \frac{x^2 + 1}{x^2 + 1} - \frac{2}{x^2 + 1} \equiv 1 - \frac{2}{x^2 + 1}$$

PARTIAL FRACTIONS

Consider first a function such as $f(x) \equiv \dfrac{2}{x+1} + \dfrac{x}{x^2 + 1}$.

$f(x)$ may be expressed as a single fraction with a common denominator thus

$$f(x) \equiv \frac{2}{x+1} + \frac{x}{x^2+1} \equiv \frac{2(x^2+1) + x(x+1)}{(x+1)(x^2+1)} \equiv \frac{3x^2 + x + 2}{(x+1)(x^2+1)}$$

It is often useful to be able to reverse this operation, that is to take a function such as $f(x) \equiv \dfrac{x-2}{(x+3)(x-4)}$ and express $f(x)$ as the sum of two (or in some cases more) separate fractions.

This process is called expressing, or decomposing, $f(x)$ in partial fractions.

If the original fraction is 'proper' then the separate (or partial) fractions will also be 'proper'.

Thus $\dfrac{x+3}{(x-2)(x+4)}$ can be expressed as $\dfrac{A}{x-2} + \dfrac{B}{x+4}$

and $\dfrac{x+3}{(x-2)(x^2+4)}$ can be expressed as $\dfrac{A}{x-2} + \dfrac{Bx+C}{x^2+4}$ where A, B and C are constants to be determined. The method for evaluating these constants depends to some extent on the factors in the denominator.

EXAMPLES 1b

1) Decompose $\dfrac{x+3}{(x-2)(x+4)}$ into partial fractions.

This example is a proper fraction with *linear* (of degree one) *factors* only and so its partial fractions are also proper. As these partial fractions have linear denominators their numerators contain only one constant.

$$\frac{x+3}{(x-2)(x+4)} \equiv \frac{A}{x-2} + \frac{B}{x+4}$$

so

$$\frac{x+3}{(x-2)(x+4)} \equiv \frac{A(x+4) + B(x-2)}{(x-2)(x+4)}$$

As the denominators are obviously identical, the numerators must also be identical,

i.e.

$$x+3 \equiv A(x+4) + B(x-2)$$

Now   LHS = RHS   for any value of x.
Choosing to substitute 2 for x (to eliminate B) gives

$$2+3 = A(2+4) + B(0)$$

$\Rightarrow$

$$A = \frac{5}{6}$$

Substituting -4 for x (to eliminate A) gives

$$-4+3 = B(-4-2)$$

$\Rightarrow$

$$B = \frac{1}{6}$$

Therefore

$$\frac{x+3}{(x-2)(x+4)} \equiv \frac{5}{6(x-2)} + \frac{1}{6(x+4)}$$

2) Express $\dfrac{x^2-3}{(x-1)(x^2+1)}$ in partial fractions.

This example contains a *quadratic factor* in the denominator,

therefore

$$\frac{x^2-3}{(x-1)(x^2+1)} \equiv \frac{A}{x-1} + \frac{Bx+C}{x^2+1}$$

each numerator on RHS being chosen so that each partial fraction is proper,

$\Rightarrow$

$$\frac{x^2-3}{(x-1)(x^2+1)} \equiv \frac{A(x^2+1) + (Bx+C)(x-1)}{(x-1)(x^2+1)}$$

therefore

$$x^2-3 \equiv A(x^2+1) + (Bx+C)(x-1) \qquad [1]$$

Substituting 1 for x (so eliminating B and C) gives

$$1^2 - 3 = A(1^2 + 1)$$

$\Rightarrow$ $$A = -1$$

There is no value which we can substitute for x to eliminate A (as there is no value of x for which $x^2 + 1 = 0$).
But substituting 0 for x will eliminate B giving

$$-3 = A(1) + C(-1)$$

As $A = -1$, $$-3 = -1 - C$$

$\Rightarrow$ $$C = 2$$

Any other value can now be substituted for x to find B, a small value being sensible.
Substituting 2 for x gives

$$(2)^2 - 3 = A(2^2 + 1) + (2B + C)(2 - 1)$$

$$1 = 5A + 2B + C$$

But $A = -1$ and $C = 2$ so $B = 2$

Therefore $$\frac{x^2 - 3}{(x - 1)(x^2 + 1)} \equiv \frac{-1}{x - 1} + \frac{2x + 2}{x^2 + 1}$$

Alternatively the value of B may be found as follows.
Expanding the RHS of [1] gives

$$x^2 - 3 \equiv Ax^2 + A + Bx^2 - Bx + Cx - C$$

or $$x^2 - 3 \equiv (A + B)x^2 + (C - B)x + A - C \qquad [2]$$

As this is an identity, the coefficients (quantity) of x^2 on the LHS and RHS of [2] must be equal.

Therefore $$1 = A + B$$

(this is referred to as comparing the coefficients of x^2).
As $A = -1$, $B = 2$.
In practice a mixture of these two methods will give a simple solution.

3) Express $\dfrac{x - 1}{(x + 1)(x - 2)^2}$ in partial fractions.

As $(x - 2)^2 \equiv (x - 2)(x - 2)$, it is called a *repeated factor*, but it is also quadratic so we may initially think of $\dfrac{x - 1}{(x + 1)(x - 2)^2}$ as $\dfrac{A}{x + 1} + \dfrac{Bx + C}{(x - 2)^2}$.
But this is not the simplest partial fraction form, as we shall see.

Considering just the fraction $\dfrac{Bx + C}{(x - 2)^2}$ and letting $C = -2B + D$

then
$$\frac{Bx + C}{(x - 2)^2} \equiv \frac{Bx - 2B + D}{(x - 2)^2}$$

$$\equiv \frac{B(x - 2)}{(x - 2)^2} + \frac{D}{(x - 2)^2}$$

$$\equiv \frac{B}{x - 2} + \frac{D}{(x - 2)^2}$$

In general any repeated factor of the form $(ax + b)^2$ in a denominator will give rise to two partial fractions of the form $\dfrac{A}{ax + b}$ and $\dfrac{B}{(ax + b)^2}$.

Similarly a repeated factor $(ax + b)^3$ gives rise to three partial fractions of form $\dfrac{A}{ax + b}$, $\dfrac{B}{(ax + b)^2}$ and $\dfrac{C}{(ax + b)^3}$.

Returning to the original problem we have

$$\frac{x - 1}{(x + 1)(x - 2)^2} \equiv \frac{A}{x + 1} + \frac{B}{x - 2} + \frac{D}{(x - 2)^2}$$

Therefore $x - 1 \equiv A(x - 2)^2 + B(x + 1)(x - 2) + D(x + 1)$

substituting 2 for x gives

$$1 = 3D \quad \text{so} \quad D = \frac{1}{3}$$

substituting -1 for x gives

$$-2 = A(-3)^2 \quad \text{so} \quad A = -\tfrac{2}{9}$$

comparing the coefficients of x^2 gives

$$0 = A + B \quad \text{so} \quad B = \tfrac{2}{9}$$

therefore $\dfrac{x - 1}{(x + 1)(x - 2)^2} \equiv -\dfrac{2}{9(x + 1)} + \dfrac{2}{9(x - 2)} + \dfrac{1}{3(x - 2)^2}$

4) Express $\dfrac{x^3}{(x + 1)(x - 3)}$ in partial fractions.

This function is an *improper fraction* and it is necessary first to divide the denominator into the numerator to obtain a mixed fraction.

$$
\begin{array}{r}
x + 2 \\
x^2 - 2x - 3 \overline{\smash{\big)}\ x^3 } \\
\underline{x^3 - 2x^2 - 3x} \\
2x^2 + 3x \\
\underline{2x^2 - 4x - 6} \\
7x + 6
\end{array}
$$

It is not necessary to go any further as the remainder at this stage is of degree one (less than the degree of the divisor) so we can say

$$\frac{x^3}{(x+1)(x-3)} \equiv x + 2 + \frac{\text{Remainder}}{(x+1)(x-3)}$$

$$\equiv x + 2 + \frac{A}{x+1} + \frac{B}{x-3}$$

therefore $x^3 \equiv (x+2)(x+1)(x-3) + A(x-3) + B(x+1)$

Substituting 3 for x gives $27 = 4B \implies B = \frac{27}{4}$

Substituting -1 for x gives $-1 = -4A \implies A = \frac{1}{4}$

Therefore

$$\frac{x^3}{(x+1)(x-3)} \equiv x + 2 + \frac{1}{4(x+1)} + \frac{27}{4(x-3)}$$

EXERCISE 1b

Express in partial fractions

1) $\dfrac{3}{(x+1)(x-1)}$

2) $\dfrac{x}{(x-4)(x-1)}$

3) $\dfrac{x-1}{(x+2)(x-2)}$

4) $\dfrac{2}{(2x-1)(x-2)}$

5) $\dfrac{x+3}{x(x+1)}$

6) $\dfrac{2x-1}{(x+1)(3x+2)}$

7) $\dfrac{3x}{(x-1)(x-2)(x-3)}$

8) $\dfrac{x^2-2x+4}{2x(x-3)(x+1)}$

9) $\dfrac{2x-1}{(3x-1)(2x+1)}$

10) $\dfrac{5x-x^2}{x(x-1)(2x+1)}$

11) $\dfrac{1-3x}{(2x-1)(x+2)}$

12) $\dfrac{x(x+1)}{(x-1)(x-2)}$

13) $\dfrac{x^2-4}{x(x+1)(x+3)}$

14) $\dfrac{6}{(x+1)(x-1)(x-4)^2}$

15) $\dfrac{3x-1}{(x^2-9)(x^2-1)}$

16) $\dfrac{(x-2)(2x+3)}{(x-1)(x^2-9)}$

17) $\dfrac{2}{(x-1)(x^2+1)}$

18) $\dfrac{x-3}{(x+4)(x^2-2)}$

19) $\dfrac{x^2+3}{x(x^2+2)}$

20) $\dfrac{2x^2+x+1}{(x-3)(2x^2+1)}$

21) $\dfrac{x^3-1}{(x+2)(2x+1)(x^2+1)}$

22) $\dfrac{x^2+1}{x(2x^2-1)(x-1)}$

23) $\dfrac{x}{(x-1)(x-2)^2}$

24) $\dfrac{x^2-1}{x^2(2x+1)}$

25) $\dfrac{3}{x(3x-1)^2}$

26) $\dfrac{x^2+x+1}{(x^2-1)(x^2+1)}$

27) $\dfrac{x^2}{(x-1)(x+1)}$

28) $\dfrac{x^2-2}{(x+3)(x-1)}$

29) $\dfrac{x^3+3}{(x+1)(x-1)}$

QUADRATIC EQUATIONS

Any equation of the form

$$ax^2 + bx + c = 0$$

is called a quadratic equation.

Solution of Quadratic Equations which Factorise

Consider the equation $\quad 2x^2 - 7x + 3 = 0$

the LHS factorises and the equation becomes

$$(2x - 1)(x - 3) = 0$$

from which we see that either

$$2x - 1 = 0 \quad \Rightarrow \quad x = \tfrac{1}{2}$$

or $\qquad\qquad\qquad\qquad x - 3 = 0 \quad \Rightarrow \quad x = 3$

Solutions to equations can sometimes be 'lost' if care is not taken with apparently straightforward equations. Consider, for example, the equation $2x^2 - 14x = 0$ and the following two solutions.

(1)	$2x^2 - 14x = 0$	(2)	$2x^2 - 14x = 0$
$\div 2$	$x^2 - 7x = 0$	factorising,	$2x(x - 7) = 0$
$\div x$	$x - 7 = 0$	therefore either	$2x = 0$ or $x - 7 = 0$
therefore	$x = 7$	giving	$x = 0$ or $x = 7$

This shows that the first solution results in the loss of the answer $x = 0$ and this is because the equation was divided by the common factor x.

Therefore although it is correct (and desirable) to divide by constant factors, division by a common factor containing the unknown quantity will lead to a loss of some solutions. This must be remembered when solving any equation, quadratic or otherwise.

Now consider the equation $\quad t(t - 3) = t^2 - 4\quad$ and the following two solutions.

(1) $t(t - 3) = t^2 - 4$

$\qquad t^2 - 3t = t^2 - 4$

$\qquad 4 - 3t = 0$

$\qquad\qquad t = \tfrac{4}{3}$

(2) Substituting $\dfrac{1}{m}$ for t gives

$$\frac{1}{m}\left(\frac{1}{m}-3\right) = \frac{1}{m^2}-4$$

$$\frac{1}{m^2}-\frac{3}{m} = \frac{1}{m^2}-4$$

$$\Rightarrow \quad 1-3m = 1-4m^2$$

$$4m^2-3m = 0$$

$$m(4m-3) = 0$$

$$m = 0 \quad \text{or} \quad m = \tfrac{3}{4}$$

i.e. $\quad \dfrac{1}{t}=0 \quad$ or $\quad \dfrac{1}{t}=\dfrac{3}{4}$

therefore $\quad t=\infty \quad$ or $\quad t=\tfrac{4}{3}$

Note that the symbol ∞ means infinitely large.
This shows that the first solution resulted in the loss of the answer $\quad t=\infty$
This is because the t^2 terms on the LHS and RHS were equal and were cancelled.
In most problems an infinite solution would have no practical meaning but in some
cases it would be applicable. For example if t represents $\tan\theta$, then $\quad t=\infty$
gives the solution $\quad \theta = 90°$.

So if, in a quadratic equation, the squared term apparently disappears, remember to
consider the infinite solution.

EXERCISE 1c

Solve the following quadratic equations.

1) $x^2 + 5x - 6 = 0$ 2) $3x^2 - 7x = 0$ 3) $4x^2 - 8x + 4 = 0$

4) $x(x-3) = x^2 - 6$ 5) $3 - x - 2x^2 = 0$ 6) $x^2 - 2ax + a^2 = 0$

7) $\dfrac{1}{x^2} - 1 = \dfrac{1}{x} - 1$ 8) $x(1-x) = x(2x-1)$

9) $(x-2)(x+3) = (x-2)(4-x)$ 10) $\dfrac{1}{x+1} + \dfrac{2}{x+2} = 1$

Solution of Quadratic Equations which do not Factorise

Consider the equation $\quad 2x^2 - 5x + 1 = 0$

Dividing by 2 gives

$$x^2 - \frac{5}{2}x + \frac{1}{2} = 0$$

$$x^2 - \frac{5}{2}x = -\frac{1}{2}$$

Adding $\left(\frac{1}{2} \times \frac{5}{2}\right)^2$ to both sides to make the LHS a perfect square gives

$$x^2 - \frac{5}{2}x + \left(\frac{5}{4}\right)^2 = -\frac{1}{2} + \left(\frac{5}{4}\right)^2$$

i.e.

$$\left(x - \frac{5}{4}\right)^2 = \frac{17}{16}$$

therefore

$$x - \frac{5}{4} = \frac{\pm\sqrt{17}}{4}$$

$$x = \frac{5}{4} \pm \frac{4.123}{4}$$

therefore either $x = 2.28$ or $x = 0.22$.

This method is called 'completing the square' and when applied to the general quadratic equation, viz. $ax^2 + bx + c = 0$,   results in the familiar formula

$$x = \frac{-b \pm \sqrt{b^2 - 4ac}}{2a}$$

Nature of the Roots of a Quadratic Equation

Using the formula to solve the equation $ax^2 + bx + c = 0$ we see that

either $x = \dfrac{-b + \sqrt{b^2 - 4ac}}{2a}$ or $x = \dfrac{-b - \sqrt{b^2 - 4ac}}{2a}$

Therefore in general a quadratic equation has two solutions (called roots).

If $b^2 - 4ac$ is positive, $\sqrt{b^2 - 4ac}$ can be evaluated and the equation will have two real and distinct (i.e. different) roots.

If $b^2 - 4ac$ is zero the equation is satisfied by only one value of x $\left(x = -\dfrac{b}{2a}\right)$ and we say that it has a repeated root or equal roots.

If $b^2 - 4ac$ is negative, $\sqrt{b^2 - 4ac}$ has no real value so the equation has no real roots.

To summarise, the equation $ax^2 + bx + c = 0$

has two real distinct roots   if $b^2 - 4ac > 0$

has equal roots   if $b^2 - 4ac = 0$

has no real roots   if $b^2 - 4ac < 0$

and $b^2 - 4ac$ is called the discriminant.

EXAMPLES 1d

1) Determine the nature of the roots of the equations

(a) $4x^2 - 7x + 3 = 0$ (b) $x^2 + ax + a^2 = 0$ (c) $x^2 - px - q^2 = 0$

(a) $4x^2 - 7x + 3 = 0$

therefore $b^2 - 4ac = (-7)^2 - 4(4)(3) = 1$

i.e. $b^2 - 4ac > 0$

So the equation has two distinct real roots.

(b) $x^2 + ax + a^2 = 0$

therefore   '$b^2 - 4ac$' $= (a)^2 - 4(1)(a^2) = -3a^2$

As a^2 is positive irrespective of the value of a, '$b^2 - 4ac$' < 0

So the equation has no real roots.

(c) $x^2 - px - q^2 = 0$

$$b^2 - 4ac = (-p)^2 - 4(1)(-q^2) = p^2 + 4q^2$$

As p^2 and q^2 are both positive, $b^2 - 4ac > 0$

therefore the equation has two real distinct roots.

2) Find the value of k if $2x^2 - kx + 8 = 0$ has equal roots.

For the roots of $2x^2 - kx + 8 = 0$ to be equal,

$$b^2 - 4ac = 0$$

i.e. $(-k)^2 - 4(2)(8) = 0$

$$k^2 = 64$$

therefore $k = \pm 8$

EXERCISE 1d

Solve the following quadratic equations by completing the square.

1) $2x^2 - 6x + 4 = 0$ 2) $x^2 + 4x - 8 = 0$ 3) $2x^2 + 7x + 3 = 0$

4) $x^2 - 2x + a = 0$ 5) $x^2 - 2ax + b = 0$ 6) $ax^2 + bx + c = 0$

Determine the nature of the roots of the following equations but do not solve the equations.

7) $x^2 - 6x + 9 = 0$ 8) $x^2 - 6x + 10 = 0$ 9) $2x^2 - 5x + 3 = 0$

10) $3x^2 + 4x + 2 = 0$ 11) $4x^2 - 12x + 9 = 0$ 12) $4x^2 - 12x - 9 = 0$

13) For what values of k is $9x^2 + kx + 16$ a perfect square?

14) The roots of $3x^2 + kx + 12 = 0$ are equal. Find k.

15) Find a if $x^2 - 5x + a = 0$ has equal roots.

16) Prove that $kx^2 + 2x - (k - 2) = 0$ has real roots for any value of k.

17) Show that the roots of $ax^2 + (a + b)x + b = 0$ are real for all values of a and b.

18) Find a relationship between p and q if the roots of $px^2 + qx + 1 = 0$ are equal.

19) If x is real and $p = \dfrac{3(x^2 + 1)}{2x - 1}$, prove that $p^2 - 3(p + 3) \geqslant 0$.

Relationships Between the Roots and Coefficients of a Quadratic Equation

Let α and β be the roots of the equation $ax^2 + bx + c = 0$

i.e. $\qquad (x - \alpha)(x - \beta) = 0 \qquad$ [1] $\Big\}$ have the same solution

and $\qquad ax^2 + bx + c = 0 \qquad$ [2]

but $\qquad\qquad (x - \alpha)(x - \beta) \equiv x^2 - (\alpha + \beta)x + \alpha\beta$

and dividing $ax^2 + bx + c = 0$ by a gives $x^2 + \dfrac{b}{a}x + \dfrac{c}{a} = 0$

therefore $\quad x^2 - (\alpha + \beta)x + \alpha\beta = 0 \qquad$ [3] $\Big\}$ have the same solution.

and $\qquad\qquad x^2 + \dfrac{b}{a}x + \dfrac{c}{a} = 0 \qquad$ [4]

As the LHS of [3] and [4] have the same coefficient of x^2 it follows that the coefficients of x and the constant terms are also equal

i.e. $\qquad\qquad x^2 - (\alpha + \beta)x + \alpha\beta \equiv x^2 + \dfrac{b}{a}x + \dfrac{c}{a}$

(**Note** that the LHS of [2] and [3] are not identical unless $a = 1$.)

Therefore $\qquad\qquad\qquad \alpha + \beta = -\dfrac{b}{a}$

$$\alpha\beta = \dfrac{c}{a}$$

and the equation may be written

$$x^2 - (\text{sum of roots})\, x + (\text{product of roots}) = 0$$

So if $2x^2 - 3x + 6 = 0$ has roots α and β,

the sum of its roots $(\alpha + \beta)$ is $\quad -\left(-\dfrac{3}{2}\right) = \dfrac{3}{2}$

and the product of its roots $(\alpha\beta)$ is $\quad \dfrac{6}{2} = 3.$

Also if a quadratic equation has roots whose sum is 7 and whose product is 10 the equation can be written as $x^2 - 7x + 10 = 0$.

EXAMPLES 1e

1) The roots of the equation $2x^2 - 7x + 4 = 0$ are α and β. Find the values of $\dfrac{1}{\alpha} + \dfrac{1}{\beta}$ and $\dfrac{1}{\alpha\beta}$

Hence write down the equation whose roots are $\dfrac{1}{\alpha}$ and $\dfrac{1}{\beta}$.

Given $2x^2 - 7x + 4 = 0$ we see that $\alpha + \beta = -\left(-\dfrac{7}{2}\right) = \dfrac{7}{2}$

and $$\alpha\beta = \dfrac{4}{2} = 2$$

To evaluate $\dfrac{1}{\alpha} + \dfrac{1}{\beta}$ we need to express it in terms of $\alpha + \beta$ and $\alpha\beta$ whose values are known.

Expressing $\dfrac{1}{\alpha} + \dfrac{1}{\beta}$ as a single fraction gives

$$\frac{1}{\alpha} + \frac{1}{\beta} = \frac{\alpha + \beta}{\alpha\beta} = \frac{\frac{7}{2}}{2} = \frac{7}{4}$$

and $$\frac{1}{\alpha\beta} = \frac{1}{2}$$

Therefore the required equation has roots whose sum $\left(\dfrac{1}{\alpha} + \dfrac{1}{\beta}\right)$ is $\dfrac{7}{4}$ and whose product $\dfrac{1}{\alpha\beta}$ is $\dfrac{1}{2}$,

so the equation is $$x^2 - \tfrac{7}{4}x + \tfrac{1}{2} = 0$$

or $$4x^2 - 7x + 2 = 0$$

Alternatively, for the given equation $2x^2 - 7x + 4 = 0, \quad x = \alpha, \beta$

for the required equation $$X = \frac{1}{\alpha}, \frac{1}{\beta}$$

therefore $$X = \frac{1}{x} \Rightarrow x = \frac{1}{X}$$

Substituting $\dfrac{1}{X}$ for x in the given equation we get

$$2\left(\frac{1}{X}\right)^2 - 7\left(\frac{1}{X}\right) + 4 = 0$$

i.e. $$4X^2 - 7X + 2 = 0$$

and this is the required equation.

Note. This method can be used only if each new root depends in the same way on each original root. For example, if the given equation has roots α, β and the required equation has roots α^2, β^2 it can be used, but if the required equation has roots $\alpha + \beta, \alpha - \beta$ it cannot.

2) If α and β are the roots of $x^2 + 3x - 2 = 0$ find the values of $\alpha^3 + \beta^3$ and $\alpha^3\beta^3$. Write down the equation whose roots are α^3 and β^3.

From $x^2 + 3x - 2 = 0$ we see that $\alpha + \beta = -3$

and $$\alpha\beta = -2$$

To express $\alpha^3 + \beta^3$ in terms of $\alpha + \beta$ and $\alpha\beta$ we can use

$$(\alpha + \beta)^3 \equiv \alpha^3 + 3\alpha^2\beta + 3\alpha\beta^2 + \beta^3$$

$$\equiv \alpha^3 + \beta^3 + 3\alpha\beta(\alpha + \beta)$$

therefore $$\alpha^3 + \beta^3 \equiv (\alpha + \beta)^3 - 3\alpha\beta(\alpha + \beta)$$

$$= (-3)^3 - 3(-2)(-3)$$

$$= -45$$

$$\alpha^3\beta^3 \equiv (\alpha\beta)^3 = (-2)^3 = -8$$

As the required equation has roots α^3 and β^3 the sum of its roots is $\alpha^3 + \beta^3 = -45$ and the product of its roots is $\alpha^3\beta^3 = -8$.

Therefore the required equation is $x^2 - (-45)x + (-8) = 0$

i.e. $$x^2 + 45x - 8 = 0$$

Note that although the alternative method could be used in this example, as $X = x^3$,   it is not recommended, because the resulting equation $(\sqrt[3]{X})^2 + 3(\sqrt[3]{X}) + 4 = 0$ is not easy to simplify.

3) Find the range of values of k for which the equation $x^2 - 2x - k = 0$ has real roots. If the roots of this equation differ by one, find the value of k.

If $x^2 - 2x - k = 0$ has real roots $b^2 - 4ac$ is greater than or equal to zero (this includes the case of equal roots)

i.e. $(-2)^2 - 4(1)(-k) \geqslant 0$

i.e. $4 + 4k \geqslant 0$

or $1 + k \geqslant 0$

$$k \geqslant -1$$

Let one root of the equation be α, then the other is $\alpha + 1$.

Sum of the roots is $\qquad\qquad 2\alpha + 1 = -(-2)$

$$2\alpha = 1$$

$$\alpha = \tfrac{1}{2}$$

Product of the roots is $\qquad \alpha(\alpha + 1) = -k$

Therefore $\qquad\qquad\qquad k = -\tfrac{3}{4}$

EXERCISE 1e

1) Write down the sums and products of the roots of the following equations:

(a) $x^2 - 3x + 2 = 0$ (b) $4x^2 + 7x - 3 = 0$ (c) $x(x - 3) = x + 4$

(d) $\dfrac{x-1}{2} = \dfrac{3}{x+2}$ (e) $x^2 - kx + k^2 = 0$ (f) $ax^2 - x(a + 2) - a = 0$

2) Write down the equation, the sum and product of whose roots are:

(a) $3, 4$ (b) $-2, \tfrac{1}{2}$ (c) $\tfrac{1}{3}, -\tfrac{2}{5}$ (d) $-\tfrac{1}{4}, 0$

(e) a, a^2 (f) $-(k + 1), k^2 - 3$ (g) $\dfrac{b}{a}, \dfrac{c^2}{b}$

3) The roots of the equation $2x^2 - 4x + 5 = 0$ are α and β. Find the value of:

(a) $\dfrac{1}{\alpha} + \dfrac{1}{\beta}$ (b) $(\alpha + 1)(\beta + 1)$ (c) $\alpha^2 + \beta^2$ (d) $\alpha^2\beta + \alpha\beta^2$

(e) $(\alpha - \beta)^2$ (f) $\dfrac{\alpha}{\beta} + \dfrac{\beta}{\alpha}$ (g) $\dfrac{1}{\alpha + 1} + \dfrac{1}{\beta + 1}$ (h) $\dfrac{1}{2\alpha + \beta} + \dfrac{1}{\alpha + 2\beta}$

(i) $\dfrac{1}{\alpha^2 + 1} + \dfrac{1}{\beta^2 + 1}$

4) The roots of $x^2 - 2x + 3 = 0$ are α and β. Find the equation whose roots are:

(a) $\alpha + 2, \beta + 2$ (b) $\dfrac{1}{\alpha}, \dfrac{1}{\beta}$ (c) α^2, β^2 (d) $\dfrac{\alpha}{\beta}, \dfrac{\beta}{\alpha}$

(e) $\alpha - \beta, \beta - \alpha$

5) *Write down* and simplify the equation whose roots are the reciprocals of the roots of $3x^2 + 2x - 1 = 0$, without solving the given equation.

6) *Write down* and simplify the equation whose roots are double those of $4x^2 - 5x - 2 = 0$, without solving the given equation.

7) *Write down* and simplify the equation whose roots are one less than those of $5x^2 + 3x - 1 = 0$.

8) Write down the equation whose roots are minus those of $2x^2 - 3x - 1 = 0$.

9) Find the value of k if the roots of $3x^2 + 5x - k = 0$ differ by two.

10) Find the value of p if one root of $x^2 + px + 8 = 0$ is the square of the other.

11) If α and β are the roots of $ax^2 + bx + c = 0$, find the equation whose roots are $\dfrac{1}{\alpha}, \dfrac{1}{\beta}$.

12) Find a relationship between a and c if the roots of $ax^2 + bx + c = 0$ are the reciprocal of each other.

13) If one root of $ax^2 + bx + c = 0$ is treble the other prove that $3b^2 - 16ac = 0$.

14) For what value of k are the roots of $3x^2 + (k-1)x - 2 = 0$ equal and opposite?

SUMMARY

Partial Fractions

1. Linear factors:

$$\frac{3x - 4}{(2x - 3)(x + 5)} \equiv \frac{A}{2x - 3} + \frac{B}{x + 5}$$

2. Quadratic factors:

$$\frac{3x - 4}{(2x - 3)(x^2 + 5)} \equiv \frac{A}{2x - 3} + \frac{Bx + C}{x^2 + 5}$$

3. Repeated factors:

$$\frac{3x - 4}{(2x - 3)(x + 5)^2} \equiv \frac{A}{2x - 3} + \frac{B}{x + 5} + \frac{C}{(x + 5)^2}$$

Quadratic Equations

The equation $\quad ax^2 + bx + c = 0 \quad$ has two roots given by

$$x = \frac{-b \pm \sqrt{b^2 - 4ac}}{2a}$$

These roots are real and distinct if $\quad b^2 - 4ac > 0$

real and equal if $\quad b^2 - 4ac = 0$

not real if $\quad b^2 - 4ac < 0$

If these roots are α and β

$$\alpha + \beta = -\frac{b}{a}$$

$$\alpha\beta = \frac{c}{a}$$

MULTIPLE CHOICE EXERCISE 1

(Instructions for answering these questions are given on page xii.)

TYPE I

1) If $\dfrac{x + p}{(x - 1)(x - 3)} \equiv \dfrac{q}{x - 1} + \dfrac{2}{x - 3}$, the values of p and q are:

(a) $p = -2, q = 1$ (b) $p = 2, q = 1$ (c) $p = 1, q = -2$
(d) $p = 1, q = 1$ (e) $p = 1, q = -1$.

2) If $x^2 + px + 6 = 0$ has equal roots and $p > 0$, p is:
(a) $\sqrt{48}$ (b) 0 (c) $\sqrt{6}$ (d) 3 (e) $\sqrt{24}$.

3) If $x^2 + 4x + p \equiv (x + q)^2 + 1$, the values of p and q are:
(a) $p = 5, q = 2$ (b) $p = 1, q = 2$ (c) $p = 2, q = 5$
(d) $p = -1, q = 5$ (e) $p = 0, q = -1$.

4) If the equation $2x^2 + 3x + 1 = 0$ has roots α, β the equation whose roots are $\dfrac{1}{\alpha}, \dfrac{1}{\beta}$ is:

(a) $3x^2 + 2x + 1 = 0$ (b) $x^2 + 3x + 2 = 0$ (c) $2x^2 + x + 3 = 0$
(d) $x^2 - 3x + 2 = 0$ (e) none of these.

5) If α and β are the roots of the equation $x^2 - px + q = 0$ the value of $\alpha^2 + \beta^2$ is:

(a) $p - q$ (b) $p^2 + 2q$ (c) p^2 (d) $p^2 - 2q$ (e) $-p^2 - 2q$.

6) $x - 3 > 2$ corresponds to:
(a) $x > 3$ (b) $x > 5$ (c) $x > 1$ (d) $x < 3$ (e) $x > 0$.

7) $f(x) \equiv x^2 + \dfrac{1}{x} + 1$ corresponds to:

(a) $f(1) = 1$ (b) $f(-1) = 3$ (c) $f(0) = 1$ (d) $f(1) = 3$
(e) $f(-1) = -1$.

8) $\dfrac{2}{(x + 1)(x - 1)} \equiv \dfrac{A}{x + 1} + \dfrac{B}{x - 1}$ corresponds to:

(a) $A = 1, B = 1$ (b) $A = -1, B = 1$ (c) $A = x, B = 1$
(d) $A = 0, B = 2$ (e) $A = x - 1, B = x + 1$.

TYPE II

9) $2x^2 - 7x + 4 = 0$ has roots α and β.
(a) $\alpha\beta = -2$.
(b) $\alpha + \beta = 3\frac{1}{2}$.
(c) α and β are real and unequal.

10) $f(x) \equiv \dfrac{2x}{(x + 2)(x - 2)}$.

(a) $f(x)$ is an improper fraction.

(b) $f(0) = 2$.

(c) $f(x) \equiv \dfrac{1}{x - 2} + \dfrac{1}{x + 2}$.

11) $\dfrac{2}{(x - 1)(x^2 + 1)} \equiv \dfrac{A}{x - 1} + \dfrac{Bx + C}{x^2 + 1}$.

(a) $A = 1$.

(b) $2 \equiv A(x^2 + 1) + (Bx + C)(x - 1)$.

(c) $A + B = 0$.

12) $f(x) \equiv x^2 - 2x + 2$.

(a) $f(x) \equiv (x - 1)^2 + 1$.

(b) $f(1) = 0$.

(c) $f(x) = 0$ has equal roots.

TYPE III

13) (a) $x(x - 2) \equiv x^2 - 2x$.

     (b) $x = 2$.

14) (a) $x + 1 > 2$.

     (b) $x < 0$.

15) (a) $f(3) = -5$.

     (b) $f(x) \equiv x^2 - 6x + 4$.

16) (a) $x^2 - 4x + 2 = 0$ has roots α and β.

     (b) $2x^2 - 4x + 1 = 0$ has roots $\dfrac{1}{\alpha}$ and $\dfrac{1}{\beta}$.

TYPE IV

17) Solve the equation $ax^2 + bx + c = 0$.

(a) $a = 1$.

(b) One root is twice the other root.

(c) $c = 2$.

18) Write down the value of $f(2)$.

(a) $f(X)$ is a polynomial of degree 1.

(b) $f(0) = 1$.

(c) $f(1) = 2$.

19) Express $f(x)$ in partial fractions.

(a) $f(x)$ is a proper fraction.

(b) $f(0) = 1$.

(c) $f(x) \equiv \dfrac{A}{(x-1)^2(x+1)}$.

TYPE V

20) A quadratic equation always has two real solutions.

21) The relationship $x(x^2 + 4) = x^2 + 4x$ is an identity.

22) The expression $\dfrac{x-2}{(x-1)(x+1)^2}$ can be expressed as three separate fractions.

23) The inequality $x + 5 > 3$ is satisfied by a finite number of values of x.

24) The equation $ax^2 + bx + c = 0$ has two roots.

25) If $f(x) \equiv \dfrac{x}{(x^2+1)(x-1)}$ then $f(x) \equiv \dfrac{A}{x^2+1} + \dfrac{B}{x+1} + \dfrac{C}{x-1}$.

MISCELLANEOUS EXERCISE 1

1) Express $\dfrac{9x}{(2x+1)^2(1-x)}$ as a sum of partial fractions with constant

numerators. (U of L)p

2) Resolve the expression $\dfrac{(x-2)}{(x^2+1)(x-1)^2}$ into its simplest partial fractions.

(U of L)p

3) Express the function $\dfrac{7x+4}{(x-3)(x+2)^2}$ as the sum of three partial fractions

with numerators independent of x. (JMB)p

4) If α, β are the roots of the equation

$$ax^2 - bx + c = 0$$

form the equation whose roots are $\alpha + \dfrac{1}{\alpha}, \beta + \dfrac{1}{\beta}$. (U of L)p

5) If the roots of $x^2 + px + q = 0$ are α and β, where α and β are non-

zero, form the equation whose roots are $\dfrac{2}{\alpha}, \dfrac{2}{\beta}$. (U of L)p

6) If α and β are the roots of the equation $ax^2 + bx + c = 0$ show that the

roots of the equation $acx^2 - (b^2 - 2ac)x + ac = 0$ are $\dfrac{\alpha}{\beta}$ and $\dfrac{\beta}{\alpha}$.

(AEB)'71p

7) The roots of the quadratic equation $x^2 - px + q = 0$ are α and β. Form, in terms of p and q, the quadratic equation whose roots are $\alpha^3 - p\alpha^2$ and $\beta^3 - p\beta^2$.   (AEB)'75p

8) Form a quadratic equation with roots which exceed by 2 the roots of the quadratic equation $3x^2 - (p - 4)x - (2p + 1) = 0$.
Find the values of p for which the given equation has equal roots.

(U of L)p

9) Given that $g(x) \equiv \dfrac{5 - x}{(1 + x^2)(1 - x)}$, express $g(x)$ in partial fractions.

(U of L)p

10) Find the constants A and B in the identity

$$\frac{x + 7}{(2x - 1)(x + 2)} \equiv \frac{A}{(2x - 1)} + \frac{B}{(x + 2)}$$ (U of L)

11) The equation $ax^2 + bx + c = 0$ has roots α, β.
Express $(\alpha + 1)(\beta + 1)$ in terms of a, b and c.   (U of L)

12) Given that the roots of the equation $ax^2 + bx + c = 0$ are β and $n\beta$, show that $(n + 1)^2 ac = nb^2$ (U of L)

13) Express $\dfrac{(x - 1)}{x(x + 1)}$ in partial fractions.   (U of L)

CHAPTER 2

ALGEBRAIC TOPICS

INDICES

If $a^3 \equiv a \times a \times a$ then $a^2 \times a^3 \equiv (a \times a) \times (a \times a \times a) \equiv a^5$.

From similar illustrations it can be seen that

$$a^n \times a^m \equiv a^{n+m} \qquad [1]$$

and

$$a^n \div a^m \equiv a^{n-m} \qquad [2]$$

and

$$(a^n)^m \equiv a^{nm} \qquad [3]$$

These three identities form the basic laws of indices and if they are to hold for values of m and n other than positive integral values, we must attach some meaning to zero, negative and fractional indices.

From [2], $\qquad a^3 \div a^5 \equiv a^{-2}$

But $\qquad a^3 \div a^5 \equiv \dfrac{a \times a \times a}{a \times a \times a \times a \times a} \equiv \dfrac{1}{a^2}$

therefore $\qquad a^{-2}$ means $\dfrac{1}{a^2}$

In general $\qquad a^{-n} \equiv \dfrac{1}{a^n}$

Also from [2] $\qquad a^3 \div a^3 \equiv a^0$

but $\qquad \dfrac{a^3}{a^3} \equiv 1$

therefore the meaning of a^0 is that a fraction has cancelled completely to unity,

i.e. $\qquad a^0 \equiv 1.$

Using [1], a^1 may be written as $a^{\frac{1}{2}} \times a^{\frac{1}{2}}$

i.e. $a^1 \equiv a^{\frac{1}{2}} \times a^{\frac{1}{2}}$,   so $a^{\frac{1}{2}} \equiv \sqrt{a}$

similarly $a^1 \equiv a^{\frac{1}{5}} \times a^{\frac{1}{5}} \times a^{\frac{1}{5}} \times a^{\frac{1}{5}} \times a^{\frac{1}{5}} \Rightarrow a^{\frac{1}{5}} \equiv \sqrt[5]{a}$

In general $a^{\frac{1}{n}} \equiv \sqrt[n]{a}$

Also $a^{\frac{3}{4}} \equiv a^{\frac{1}{4}} \times a^{\frac{1}{4}} \times a^{\frac{1}{4}}$

i.e. $a^{\frac{3}{4}}$ means the cube of the fourth root of a

In general $a^{n/m} \equiv (\sqrt[m]{a})^n$ or $\sqrt[m]{a^n}$

EXAMPLES 2a

1) Simplify $\left(\dfrac{125}{27}\right)^{-\frac{2}{3}}$

$\left(\dfrac{125}{27}\right)^{-\frac{2}{3}}$ means the reciprocal of $\left(\dfrac{125}{27}\right)^{\frac{2}{3}}$

i.e. $\left(\dfrac{125}{27}\right)^{-\frac{2}{3}} = \left(\dfrac{27}{125}\right)^{\frac{2}{3}} = \left(\sqrt[3]{\dfrac{27}{125}}\right)^2 = \left(\dfrac{3}{5}\right)^2 = \dfrac{9}{25}$

2) Simplify $\dfrac{t^{\frac{1}{2}} + t^{-\frac{1}{2}}}{t^{\frac{3}{2}}}$

Multiplying top and bottom by $t^{\frac{1}{2}}$ to eliminate the negative index gives

$$\frac{t^{\frac{1}{2}} + t^{-\frac{1}{2}}}{t^{\frac{3}{2}}} \times \frac{t^{\frac{1}{2}}}{t^{-\frac{1}{2}}} \equiv \frac{t^1 + t^0}{t^2}$$

$$\equiv \frac{t+1}{t^2}$$

EXERCISE 2a

Evaluate

1) $\left(\dfrac{1}{2}\right)^{-2}$

2) $\dfrac{1}{2^{-1}}$

3) $27^{-\frac{1}{3}}$

4) $\left(\dfrac{16}{49}\right)^{\frac{1}{2}}$

5) $(125)^{-\frac{1}{3}}$

6) $(121)^{\frac{3}{2}}$

7) $\left(\dfrac{1}{4}\right)^{\frac{5}{2}}$

8) $\left(\dfrac{1}{9}\right)^{-\frac{3}{2}}$

9) $\left(\dfrac{100}{9}\right)^{-\frac{3}{2}}$

10) $\left(\dfrac{27}{8}\right)^{\frac{2}{3}}$

11) $\left(-\dfrac{1}{7}\right)^{-2}$

12) $(0.36)^{\frac{1}{2}}$

13) $(0.04)^{-2}$ 14) $(2.56)^{-\frac{1}{2}}$ 15) $(2\frac{7}{9})^{\frac{3}{2}}$ 16) $\left(\dfrac{125}{27}\right)^{0}$

17) $3^{-1} . 2^{2} . 4^{0}$ 18) $12^{\frac{1}{2}} . 3^{\frac{1}{2}}$ 19) $27^{\frac{1}{4}} . 3^{\frac{1}{4}}$ 20) $32^{\frac{1}{2}} . 2^{-\frac{1}{2}}$

21) $\dfrac{9^{\frac{1}{2}} . 8^{\frac{1}{2}}}{2^{\frac{1}{2}}}$ 22) $\dfrac{5^{\frac{1}{3}} . 5^{0} . 25^{\frac{1}{3}}}{125^{\frac{1}{3}}}$ 23) $\dfrac{8^{\frac{1}{3}} . 16^{\frac{1}{3}}}{32^{-\frac{1}{3}}}$ 24) $\dfrac{9^{\frac{1}{3}} . 27^{-\frac{1}{2}}}{3^{-\frac{1}{6}} . 3^{-\frac{2}{3}}}$

Simplify

25) $\dfrac{y^{\frac{2}{6}} . y^{-\frac{2}{3}}}{y^{\frac{1}{4}}}$ 26) $\dfrac{p^{\frac{1}{2}} . p^{-\frac{3}{4}}}{p^{-\frac{1}{4}}}$ 27) $\dfrac{\sqrt{x} . \sqrt{x^3}}{x^{-3}}$ 28) $\dfrac{(\sqrt{t})^{3} . t^{2}}{\sqrt{(t^{5})}}$

29) $\dfrac{x^{2} + x^{\frac{5}{2}}}{x^{-\frac{1}{2}}}$ 30) $\dfrac{y^{\frac{1}{2}} + y^{-\frac{1}{4}}}{y^{-\frac{3}{4}}}$

31) $\dfrac{(x-1)^{\frac{1}{2}} + (x-1)^{-\frac{1}{2}}}{(x-1)^{\frac{1}{2}}}$ 32) $\dfrac{m^{\frac{3}{2}} - m^{-\frac{1}{2}}}{m^{\frac{1}{2}} + m^{-\frac{1}{2}}}$

SURDS

Expressions such as $\sqrt{4}$, $\sqrt{25}$ have exact numerical values, viz. $\sqrt{4} = 2$, $\sqrt{25} = 5$.

But expressions such as $\sqrt{2}, \sqrt{3}, \sqrt{5}, \ldots$ cannot be written as numerically exact quantities.

For example we might say that $\sqrt{2} = 1.4$ correct to 2 s.f.
or that $\sqrt{2} = 1.4142136$ correct to 8 s.f.
but we can never find an exact quantity equal to $\sqrt{2}$.

Such numbers are called irrational and it is often convenient to leave them in the form $\sqrt{2}$ (or $\sqrt{3}$ ...) and they are then called surds.

(**Note**. $\sqrt{2}$ means the positive square root of 2, so although the solution to $x^2 = 4$ is $x = \pm 2$, $\sqrt{4} = 2$)

As surds occur frequently in solutions it is useful to be able to simplify them.

EXAMPLES 2b

1) Express $\sqrt{48}$ as the simplest possible surd.

$$\sqrt{48} = \sqrt{(16 \times 3)} = \sqrt{16} . \sqrt{3} = 4\sqrt{3}$$

2) Expand and simplify:
(a) $(2 - 3\sqrt{3})(3 + 2\sqrt{3})$ (b) $(5 - 2\sqrt{7})(5 + 2\sqrt{7})$.

(a) $(2 - 3\sqrt{3})(3 + 2\sqrt{3}) = 6 - 9\sqrt{3} + 4\sqrt{3} - 6(\sqrt{3})^2$

$$= 6 - 5\sqrt{3} - 6 \times 3$$

$$= -12 - 5\sqrt{3}$$

(b) $(5 - 2\sqrt{7})(5 + 2\sqrt{7}) = 25 - 10\sqrt{7} + 10\sqrt{7} - 4(\sqrt{7})^2$

$$= 25 - 28$$

$$= -3$$

When the solution to a problem results in an answer containing surds it is accepted practice to leave that answer (simplified as far as possible) in surd form unless an approximation is asked for (e.g. give your answer correct to three significant figures). Simplification of a fractional answer can often be achieved by removing surds from the denominator and this process is called *rationalising the denominator*.

3) Rationalise the denominator of $\dfrac{3}{\sqrt{2}}$.

$$\frac{3}{\sqrt{2}} = \frac{3}{\sqrt{2}} \cdot \frac{\sqrt{2}}{\sqrt{2}} = \frac{3\sqrt{2}}{2}$$

4) Simplify $\dfrac{3 - \sqrt{5}}{1 + 3\sqrt{5}}$.

Referring back to Example 2(b) we see that expanding brackets of the form $(a + b)(a - b)$ gives $a^2 - b^2$ thus eliminating any surds contained in either a or b. So to rationalise the denominator of $\dfrac{3 - \sqrt{5}}{1 + 3\sqrt{5}}$ we multiply numerator and denominator by $1 - 3\sqrt{5}$.

i.e.
$$\frac{3 - \sqrt{5}}{1 + 3\sqrt{5}} = \frac{(3 - \sqrt{5})(1 - 3\sqrt{5})}{(1 + 3\sqrt{5})(1 - 3\sqrt{5})} = \frac{18 - 10\sqrt{5}}{1^2 - (3\sqrt{5})^2}$$

$$= \frac{9 - 5\sqrt{5}}{-22} = \frac{5\sqrt{5} - 9}{22}$$

EXERCISE 2b

1) Express in terms of the simplest possible surds:
(a) $\sqrt{8}$ (b) $\sqrt{12}$ (c) $\sqrt{50}$ (d) $\sqrt{18}$ (e) $\sqrt{200}$
(f) $\sqrt{72}$ (g) $\sqrt{125}$ (h) $\sqrt{288}$ (i) $\sqrt{450}$ (j) $\sqrt{2000}$.

2) Simplify:
(a) $\sqrt{2}(3 - \sqrt{2})$ (b) $\sqrt{2}(3 - 2\sqrt{2})$ (c) $\sqrt{3}(\sqrt{27} - 1)$
(d) $(\sqrt{2} - 1)(\sqrt{2} + 1)$ (e) $(\sqrt{3} - 2)(\sqrt{3} - 1)$ (f) $(2\sqrt{2} + 1)(\sqrt{2} - 2)$
(g) $(3\sqrt{3} - 2)(3\sqrt{3} + 2)$ (h) $(2\sqrt{5} + 3)(3\sqrt{5} - 2)$ (i) $(\sqrt{3} - 1)(\sqrt{2} + 1)$
(j) $(2\sqrt{6} - 3)^2$

3) Rationalise the denominators and simplify:

(a) $\dfrac{1}{\sqrt{3}}$ (b) $\dfrac{1}{\sqrt{8}}$ (c) $\dfrac{2}{\sqrt{32}}$ (d) $\dfrac{1}{\sqrt{2}+1}$

(e) $\dfrac{1}{\sqrt{3}-1}$ (f) $\dfrac{3}{2-\sqrt{3}}$ (g) $\dfrac{5}{2+\sqrt{5}}$ (h) $\dfrac{2}{2\sqrt{3}-3}$

(i) $\dfrac{1}{3+2\sqrt{5}}$ (j) $\dfrac{1}{\sqrt{6}-\sqrt{5}}$ (k) $\dfrac{1}{2\sqrt{3}+\sqrt{2}}$ (l) $\dfrac{1}{\sqrt{2}+1}+\dfrac{1}{\sqrt{2}-1}$

LOGARITHMS

Logarithm is another word for an index or power.

Now $2^3 = 8$

i.e.   3 is the power to which the base 2 must be raised to obtain 8

or   *3 is the logarithm which, with a base 2, gives 8.*

This is written simply as $3 = \log_2 8$

Similarly $3^2 = 9$

i.e.   2 is the logarithm which, with a base 3, gives 9

or $2 = \log_3 9$

Also $(\frac{1}{5})^{-2} = 25$

i.e. -2 is the logarithm which, with a base $\frac{1}{5}$, gives 25

or $-2 = \log_{\frac{1}{5}} 25$

So we see that the base of a logarithm may be any number. The tables of common logarithms, which are usually used for calculations, have a base 10. From these tables (or from a calculator) it is found that the logarithm of 5 is 0.6990.

i.e. $10^{0.6990} = 5$ or $\log_{10} 5 = 0.6990$

However it is usual to omit the base 10 and to write simply

$$\log 5 = 0.6990 \quad \text{or} \quad \lg 5 = 0.6990$$

but any base other than 10 must be stated.

Therefore $\lg 100 = 2$ (i.e. $100 = 10^2$)

and $\log_{0.1} 100 = -2$ (i.e. $100 = 0.1^{-2}$)

In general $\log_a b = c \iff b = a^c$

EXERCISE 2c

1) Express in logarithmic form:

(a) $5^3 = 125$ (b) $7^2 = 49$ (c) $4096 = 8^4$ (d) $4^{\frac{5}{2}} = 32$

(e) $1331 = (121)^{\frac{3}{2}}$ (f) $10^{-2} = 0.01$ (g) $5^0 = 1$ (h) $8^{-\frac{1}{3}} = \frac{1}{2}$

(i) $125 = (\frac{1}{5})^{-3}$ (j) $9^{-\frac{3}{2}} = \frac{1}{27}$ (k) $1 = a^0$ (l) $\pi^2 = 9.8696$

(m) $p = q^2$ (n) $a^c = b$ (o) $x^y = 2$

2) Express in index form:

(a) $\log_5 625 = 4$ (b) $\log 1000 = 3$ (c) $\log_3 27 = 3$ (d) $\log 1 = 0$

(e) $\log_{\frac{1}{2}} 4 = -2$ (f) $\log_{25} 5 = \frac{1}{2}$ (g) $\log_a 1 = 0$ (h) $\log_x y = 2$

(i) $\log_4 p = q$ (j) $\log_a 5 = b$ (k) $\log_{xy} = z$ (l) $p = \log_q r$

3) Evaluate

(a) $\log_4 64$ (b) $\lg 10\,000$ (c) $\log_9 3$ (d) $\log_{\frac{1}{2}} 4$

(e) $\log_{0.1} 10$ (f) $\log_{125} 25$ (g) $\log_{121} 11$ (h) $\log_{81} 3$

(i) $\lg 0.1$ (j) $\log_8 0.5$ (k) $\log_9 9^{\frac{1}{2}}$ (l) $\log_2 2^3$

(m) $\lg 1$ (n) $\log_a 1$ (o) $\log_a a^3$ (p) $\log_a a^b$

(q) $10^{\lg 10}$ (r) $5^{\log_5 25}$

The Laws of Logarithms

Let $\log_a b \equiv x$ and $\log_a c \equiv y$

therefore $a^x \equiv b$ and $a^y \equiv c$

so $(a^x)(a^y) \equiv bc$

or $a^{x+y} \equiv bc$

therefore $x + y \equiv \log_a bc$

i.e. $\log_a b + \log_a c \equiv \log_a bc$ [1]

Similarly $a^x \div a^y \equiv \dfrac{b}{c}$

or $a^{x-y} \equiv \dfrac{b}{c}$

therefore $x - y \equiv \log_a \dfrac{b}{c}$

i.e. $\log_a b - \log_a c \equiv \log_a \dfrac{b}{c}$ [2]

Let $\log_a b^n \equiv z$

therefore $\qquad a^z \equiv b^n$

so $\qquad a^{\frac{z}{n}} \equiv b$

therefore $\qquad \log_a b \equiv \dfrac{z}{n}$

or $\qquad n \log_a b \equiv z$

i.e. $\qquad n \log_a b \equiv \log_a b^n$ $\qquad\qquad$ [3]

The identities [1], [2], [3] are known as the three basic laws of logarithms, i.e.

$$\log_a x + \log_a y \equiv \log_a xy$$

$$\log_a x - \log_a y \equiv \log_a \frac{x}{y}$$

$$\log_a x^y \equiv y \log_a x$$

Using these three laws an expression such as $\log\left(\dfrac{a^2}{b^3}\right)$ may be written in terms of $\log a$ and $\log b$

i.e. $\qquad \log \dfrac{a^2}{b^3} \equiv \log a^2 - \log b^3$

$$\equiv 2 \log a - 3 \log b$$

Conversely an expression such as $\log 100 - 2 \log 50$ may be written as the logarithm of a single number

i.e. $\qquad \log 100 - 2 \log 50 = \log 100 - \log 50^2$

$$= \log 100 - \log 2500$$

$$= \log\left(\frac{100}{2500}\right)$$

$$= \log \frac{1}{25}$$

Changing the Base of a Logarithm

Tables are not readily available which list the values of expressions such as $\log_7 2$.

However if $\qquad \log_7 2 = x$

then $\qquad 7^x = 2$

so $\qquad x \lg 7 = \lg 2$

or $\qquad x = \dfrac{\lg 2}{\lg 7}$

i.e. we have changed the base from 7 to 10 and can now use our ordinary tables or calculator to evaluate $\log_7 2$.

Therefore
$$\log_7 2 = \frac{\lg 2}{\lg 7} = \frac{0.3010}{0.8451} = 0.3562$$

A general formula for changing from base a to base b can be derived in the same way.

If
$$\log_a c \equiv x$$

then
$$c \equiv a^x$$

so
$$\log_b c \equiv x \log_b a$$

$\Rightarrow$
$$x \equiv \frac{\log_b c}{\log_b a}$$

or
$$\log_a c \equiv \frac{\log_b c}{\log_b a}$$

In the special case when $c = b$, this identity becomes

$$\log_a b \equiv \frac{\log_b b}{\log_b a}$$

or
$$\log_a b \equiv \frac{1}{\log_b a}$$

EXPONENTIAL EQUATIONS (i.e. where the variable is an index)

To solve an equation like $5^x = 10$ we can make use of the third law of logarithms to 'bring down' the power. Taking logs of both sides gives

$$\lg 5^x = \lg 10$$

$\Rightarrow$
$$x \lg 5 = \lg 10$$

$\Rightarrow$
$$x = \frac{\lg 10}{\lg 5} = \frac{1}{0.6990} = 1.43 \text{ to } 3 \text{ s.f.}$$

Now consider an equation of the type

$$2^{2x} + 3(2^x) - 4 = 0$$

Taking logs is no help this time as $2^{2x} + 3(2^x)$ cannot be combined into a single expression, i.e. $\log [2^{2x} + 3(2^x)]$ cannot be simplified. But if y is substituted for 2^x, the equation becomes quadratic in y

i.e.
$$y^2 + 3y - 4 = 0$$

$$(y + 4)(y - 1) = 0$$

$$y = -4 \quad \text{or} \quad y = 1$$

i.e. $$2^x = -4 \quad \text{or} \quad 2^x = 1$$

There are no real values of x for which $2^x = -4$.

But if $2^x = 1$ then $x = 0$, which is the only solution to this equation.

EXAMPLES 2d

1) Solve for x and y the equations

$$xy = 80$$
$$\lg x - 2\lg y = 1$$

$$xy = 80 \qquad\qquad\qquad [1]$$

$$\lg x - 2\lg y = 1 \qquad\qquad\qquad [2]$$

Using the laws of logarithms, [2] becomes

$$\lg \frac{x}{y^2} = 1 \qquad \text{i.e.} \qquad \frac{x}{y^2} = 10 \quad \text{or} \quad x = 10y^2$$

Substituting $10y^2$ for x in [1] we get

$$10y^3 = 80 \Rightarrow y^3 = 8$$

Therefore $$y = 2 \text{ and } x = 40$$

2) Solve the equation $\log_3 x - 4\log_x 3 + 3 = 0$.

Using the identity $\log_a b \equiv \dfrac{1}{\log_b a}$, $\log_x 3$ may be replaced by $\dfrac{1}{\log_3 x}$

So the given equation becomes $\log_3 x - \dfrac{4}{\log_3 x} + 3 = 0$

$\Rightarrow$ $$(\log_3 x)^2 - 4 + 3\log_3 x = 0$$

Substituting y for $\log_3 x$ gives

$$y^2 + 3y - 4 = 0$$
$$(y + 4)(y - 1) = 0$$

therefore $$y = -4 \text{ or } 1$$

i.e. either $$\log_3 x = -4 \qquad \text{or} \quad \log_3 x = 1$$

so $$x = 3^{-4} = \tfrac{1}{81} \quad \text{or} \qquad x = 3$$

EXERCISE 2d

1) Express in terms of $\log a$, $\log b$ and $\log c$:

(a) $\log ab$ (b) $\log abc$ (c) $\log \dfrac{a}{b}$ (d) $\log \dfrac{ab}{c}$ (e) $\log \dfrac{a}{bc}$

(f) $\log \dfrac{1}{a}$ (g) $\log a^2 b$ (h) $\log \sqrt{\dfrac{a}{b}}$ (i) $\log \dfrac{a^2}{b}$ (j) $\log \dfrac{a^3}{10}$

(k) $\log \dfrac{1}{10a}$ (l) $\log \dfrac{100}{\sqrt{b}}$.

2) Simplify:
(a) $\log 3 + \log 4$ (b) $\log 6 - \log 2$ (c) $\log 2 + \log 6 - \log 4$
(d) $2 \log 3 + \log 2$ (e) $\frac{1}{2} \log 4 - \log 6$ (f) $2 - 2 \log 5$

(g) $\frac{1}{2} \log 9 + 1$ (h) $\frac{1}{2} \log 25 - 2 \log 3 + 2 \log 6$ (i) $\dfrac{\log 9}{\log 3}$

(j) $\log (x + 1) - \log (x^2 - 1)$ (k) $\dfrac{\log 125}{\log 5}$

(l) $2 \log_a 5 + \log_a 4 - 2 \log_a 10$.

3) Solve the equations:
(a) $3^x = 6$ (b) $5^x = 4$ (c) $2^{2x} = 5$ (d) $3^{x-1} = 7$
(e) $4^{2x+1} = 3$ (f) $(5^x)(5^{x-1}) = 10$.

4) Evaluate:
(a) $\log_2 10$ (b) $\log_7 5$ (c) $\log_{20} 2$ (d) $\log_{0.3} 5$.

5) Solve the equations:
(a) $2(2^{2x}) - 5(2^x) + 2 = 0$ (b) $3^{2x+1} - 26(3^x) - 9 = 0$
(c) $4^x - 6(2^x) - 16 = 0$.

6) If $\log_2 x + \log_x 2 = 2$,   find x.

7) Solve the simultaneous equations $\log_x y = 2$ $xy = 8$.

8) Solve simultaneously $2 \lg y = \lg 2 + \lg x$ and $2^y = 4^x$.

9) Solve the simultaneous equations $\log_3 x = y = \log_9 (2x - 1)$.

10) Find the positive value of x that satisfies the equation
$\log_2 x = \log_4 (x + 6)$.

11) Show that $(\log_a b^2) \times (\log_b a^3) \equiv 6$.

12) Solve the simultaneous equations $\log (x + y) = 0$
and $2 \log x = \log(y + 1)$.

THE REMAINDER THEOREM

If the remainder is required when a polynomial is divided by a linear function it can be found by long division as follows:

$$\begin{array}{r}
x^2 - 5x - 4 \\
x - 2 \overline{\smash)x^3 - 7x^2 + 6x - 2} \\
\underline{x^3 - 2x^2} \\
-5x^2 + 6x - 2 \\
\underline{-5x^2 + 10x} \\
-4x - 2 \\
\underline{-4x + 8} \\
-10
\end{array}$$

Thus when $f(x) \equiv x^3 - 7x^2 + 6x - 2$ is divided by $x - 2$ there is a remainder of -10 and a quotient $x^2 - 5x - 4$.
This result can be written in the form

$$f(x) \equiv x^3 - 7x^2 + 6x - 2 \equiv (x-2)(x^2 - 5x - 4) - 10$$

If 2 is substituted for x in this identity so that $x - 2 = 0$, the quotient is eliminated giving $f(2) = -10$.

In general if $f(x)$ is a polynomial function of x which, when divided by $x - a$, gives a quotient $Q(x)$ and a remainder R, the relationship between these expressions is

$$f(x) \equiv (x-a)Q(x) + R$$

Substituting a for x in this identity gives

$$f(a) = R$$

This result is known as the remainder theorem and can be summarised as follows.

When $f(x)$ is divided by $x - a$ the remainder is $f(a)$.

Note. This theorem gives a (simple) method for evaluating the remainder only. If the quotient is required, long division must be used.

So, for example, if we want the remainder when $x^3 - 2x^2 + 6$ is divided by $x + 3$ we write

$$f(x) \equiv x^3 - 2x^2 + 6$$

and the remainder, R, is given by

$$R = f(-3) = (-3)^3 - 2(-3)^2 + 6 = -39$$

And if we want the remainder when $6x^2 - 7x + 2$ is divided by $2x - 1$ then as

$$f(x) \equiv 6x^2 - 7x + 2 \equiv (2x - 1)Q(x) + R$$

$$R = f(\tfrac{1}{2})$$

THE FACTOR THEOREM

The remainder theorem states that when $f(x)$ is divided by $(x-a)$ the remainder, R, is $f(a)$.

Now if $x-a$ is a factor of $f(x)$ there will be no remainder,

i.e. $$R = 0 \quad \text{and} \quad f(a) = 0.$$

This property is known as the factor theorem and is defined as follows:

If, for a given polynomial function $f(x)$, $f(a) = 0$,
then $x - a$ is a factor of $f(x)$.

This theorem is very useful when factorising polynomials of degree greater than 2.

EXAMPLE 2e

Factorise completely $x^4 - 3x^3 + 4x^2 - 8$.

Let $$f(x) \equiv x^4 - 3x^3 + 4x^2 - 8$$

then $$f(1) = 1 - 3 + 4 - 8 \neq 0$$

therefore $x - 1$ is *not* a factor of $f(x)$.

$$f(-1) = 1 + 3 + 4 - 8 = 0$$

therefore $x + 1$ *is* a factor of $f(x)$.

Now that one factor has been found it should be 'taken out', either by inspection or by long division.

By inspection $x^4 - 3x^3 + 4x^2 - 8 \equiv (x+1)(x^3 - 4x^2 + 8x - 8)$

Now let $$g(x) \equiv x^3 - 4x^2 + 8x - 8$$

$$g(-1) = -1 - 4 - 8 - 8 \neq 0$$

Therefore $x + 1$ is *not* a factor of $g(x)$

$$g(2) = 8 - 16 + 16 - 8 = 0$$

Therefore $x - 2$ *is* a factor of $g(x)$.

By inspection $x^3 - 4x^2 + 8x - 8 \equiv (x-2)(x^2 - 2x + 4)$

and $x^2 - 2x + 4$ has no linear factors.

Therefore $$x^4 - 3x^3 + 4x^2 - 8 \equiv (x+1)(x-2)(x^2 - 2x + 4)$$

Note (1) The factors of 8 are $1, 2, 4, 8$ so the values we choose for a must belong to the set $\{\pm 1, \pm 2, \pm 4, \pm 8\}$.

(2) After having taken out the first factor from $f(x)$ it must be tried again as a possible factor of $g(x)$ since repeated factors occur frequently.

The Factors of $a^3 - b^3$ **and** $a^3 + b^3$

$$a^3 - b^3 = 0 \quad \text{when} \quad a = b$$

therefore $a - b$ is a factor of $a^3 - b^3$.

Hence
$$a^3 - b^3 \equiv (a - b)(a^2 + ab + b^2)$$

$$a^3 + b^3 = 0 \quad \text{when} \quad a = -b$$

so $a + b$ is a factor of $a^3 + b^3$

giving
$$a^3 + b^3 \equiv (a + b)(a^2 - ab + b^2)$$

These two identities are very useful and should be memorised.

EXERCISE 2e

1) Find the remainder when the following functions are divided by the linear factors indicated.

(a) $x^3 - 2x + 4, x - 1$ (b) $x^3 + 3x^2 - 6x + 2, x + 2$

(c) $2x^3 - x^2 + 2, x - 3$ (d) $x^4 - 3x^3 + 5x, 2x - 1$

(e) $9x^5 - 5x^2, 3x + 1$ (f) $x^3 - 2x^2 + 6, x - a$

(g) $x^2 + ax + b, x + c$

2) Determine whether the following linear functions are factors of the given polynomials.

(a) $x^3 - 7x + 6, x - 1$ (b) $2x^2 + 3x - 4, x + 1$

(c) $x^3 - 6x^2 + 6x - 2, x - 2$ (d) $x^3 - 27, x - 3$

(e) $2x^4 - x^3 - 6x^2 + 5x - 1, 2x - 1$

(f) $x^3 + ax^2 - a^2x - a^3, x + a$

3) Factorise the following functions as far as possible.

(a) $x^3 + 2x^2 - x - 2$ (b) $x^3 - x^2 - x - 2$ (c) $x^4 - 1$

(d) $x^4 + x^3 - 3x^2 - 4x - 4$ (e) $2x^3 - x^2 + 2x - 1$

(f) $27x^3 - 1$ (g) $x^3 + a^3$ (h) $x^3 - y^3$

4) If $x^2 - 7x + a$ has a remainder 1 when divided by $x + 1$, find a.

5) If $x - 2$ is a factor of $ax^2 - 12x + 4$ find a.

6) One solution of the equation $x^2 + ax + 2 = 0$ is $x = 1$, find a.

7) One root of the equation $x^2 - 3x + a = 0$ is 2. Find the other root.

BINOMIAL EXPRESSIONS

A binomial is the sum (or difference) of two terms.

e.g. $a + b$, $2x + 3y$, $p^2 - r$ are binomials.

It is often necessary to expand (i.e. multiply out) a power of a binomial, e.g.

$$(x + y)^3 \equiv (x + y)^2(x + y) \equiv (x^2 + 2xy + y^2)(x + y) \equiv x^3 + 3x^2y + 3xy^2 + y^3$$

This multiplication is a tedious process for powers of 3 and over so we describe below a far quicker way of obtaining such expansions.

Pascal's Triangle

Consider the following expansions:

$$(1 + x)^0 \equiv 1$$

$$(1 + x)^1 \equiv 1 + x \qquad\qquad [1]$$

$$(1 + x)^2 \equiv 1 + 2x + x^2 \qquad\qquad [2]$$

$$(1 + x)^3 \equiv 1 + 3x + 3x^2 + x^3 \qquad\qquad [3]$$

$$(1 + x)^4 \equiv 1 + 4x + 6x^2 + 4x^3 + x^4 \qquad\qquad [4]$$

Closer inspection of [4], i.e.

$$(1 + x)^4 \equiv (\boxed{1} + \boxed{3x} + \boxed{3x^2} + \boxed{x^3})(1 + x)$$

$$\equiv 1 + x + 3x + 3x^2 + 3x^2 + 3x^3 + x^3 + x^4$$

$$\equiv \boxed{1} + \boxed{(1 + 3)x} + \boxed{(3 + 3)x^2} + \boxed{(3 + 1)x^3} + \boxed{x^4}$$

$$\equiv 1 + \quad 4x + \quad 6x^2 + \quad 4x^3 + x^4$$

shows that the coefficient of any one power of x is the sum of the coefficients of the same and preceding power of x in the previous expansion.

For the expansions [1], [2], [3] and [4], writing the coefficients only as a triangular array gives

$$1 \quad 1$$
$$1 \quad 2 \quad 1$$
$$1 \quad \boxed{3 \quad + \quad 3} \quad 1$$
$$1 \quad 4 \quad \boxed{6} \quad 4 \quad 1$$

Knowing that any one number is obtained by adding together the two numbers in the row above as shown, we can add as many further rows as we wish, e.g.

$$
\begin{array}{ccccccccc}
 & & & & 1 & \ 1 & & & \\
 & & & 1 & 2 & 1 & & & \\
 & & 1 & 3 & & 3 & 1 & & \\
 & 1 & 4 & 6 & & 4 & 1 & & \\
 1 & 5 & 10 & & 10 & 5 & 1 & & \\
 1 & 6 & 15 & 20 & 15 & 6 & 1 & & \\
 1 & 7 & 21 & 35 & 35 & 21 & 7 & 1 &
\end{array}
$$

The numbers in row five give the coefficients in the expansion of $(1 + x)^5$, and so on for the subsequent rows.

This triangular array is known as Pascal's Triangle.

Using this triangle we may write

$$(1 + x)^5 \equiv 1 + 5x + 10x^2 + 10x^3 + 5x^4 + x^5$$

EXAMPLES 2f

1) Expand $(1 + 3y)^3$.

From Pascal's Triangle

$$(1 + x)^3 \equiv 1 + 3x + 3x^2 + x^3$$

replacing x by $3y$ we have

$$(1 + 3y)^3 \equiv 1 + 3(3y) + 3(3y)^2 + (3y)^3$$
$$\equiv 1 + 9y + 27y^2 + 27y^3$$

2) Expand $(a + b)^4$ and $(a + b)^6$.

From Pascal's Triangle

$$(1 + x)^4 \equiv 1 + 4x + 6x^2 + 4x^3 + x^4$$

Writing $\quad (a + b)^4 \equiv a^4 \left(1 + \dfrac{b}{a}\right)^4$

and replacing x by $\dfrac{b}{a}$ we have

$$(a + b)^4 \equiv a^4 \left[1 + 4\left(\frac{b}{a}\right) + 6\left(\frac{b^2}{a^2}\right) + 4\left(\frac{b^3}{a^3}\right) + \frac{b^4}{a^4}\right]$$
$$\equiv a^4 + 4a^3 b + 6a^2 b^2 + 4ab^3 + b^4$$

Note that (a) the sum of the powers of a and b is always four in each term,
(b) that as the powers of a decrease, the powers of b increase,
(c) the numerical coefficients are those for the expansion of $(1 + x)^4$.

From these observations we can write out the expansion of $(a + b)^6$ directly

$$(a + b)^6 \equiv a^6 + 6a^5b + 15a^4b^2 + 20a^3b^3 + 15a^2b^4 + 6ab^5 + b^6$$

3) Expand $(2x - 3y)^3$.

Using $(a + b)^3 \equiv a^3 + 3a^2b + 3ab^2 + b^3$

and replacing a by $2x$, b by $-3y$, we have:

$$(2x - 3y)^3 \equiv (2x)^3 + 3(2x)^2(-3y) + 3(2x)(-3y)^2 + (-3y)^3$$
$$\equiv 8x^3 - 36x^2y + 54xy^2 - 27y^3$$

EXERCISE 2f

Expand the following binominal expressions.

1) $(1 + 2x)^4$ 2) $(1 - x)^3$ 3) $(x - 1)^3$

4) $(1 - 3y)^4$ 5) $(1 + x)^7$ 6) $(2x - 1)^3$

7) $(2x + y)^3$ 8) $\left(x + \dfrac{1}{x}\right)^4$ 9) $(p - 2q)^3$

10) $(x^2 - y)^5$ 11) $\left(x - \dfrac{1}{x}\right)^3$ 12) $(a - b)^3(a + b)^3$

Use Pascal's Triangle to simplify:

13) $(1 + \sqrt{2})^3$ 14) $(\sqrt{3} - \sqrt{2})^4$ 15) $(1 + \sqrt{3})^3 + (1 - \sqrt{3})^3$

16) By expanding $(1 + 0.01)^3$ evaluate $(1.01)^3$ without using tables.

17) Find, without using tables, the value of $(2.1)^4$.

SUMMARY

Indices:

$$a^{-n} \equiv \frac{1}{a^n}$$

$$a^0 \equiv 1$$

$$a^{\frac{1}{n}} \equiv \sqrt[n]{a}$$

$$a^{\frac{n}{m}} \equiv \sqrt[m]{a^n}$$

Logarithms:

$$\log_a b = c \iff b = a^c$$

$$\log_a x + \log_a y \equiv \log_a xy$$

$$\log_a x - \log_a y \equiv \log_a \frac{x}{y}$$

$$\log_a x^y \equiv y \log_a x$$

$$\log_a c \equiv \frac{\log_b c}{\log_b a}$$

$$\log_a b \equiv \frac{1}{\log_b a}$$

Remainder Theorem: When $f(x)$ is divided by $x - a$ the remainder is $f(a)$.

Factor Theorem: If $f(a) = 0$, $x - a$ is a factor of $f(x)$.

$$a^3 + b^3 \equiv (a + b)(a^2 - ab + b^2)$$

$$a^3 - b^3 \equiv (a - b)(a^2 + ab + b^2)$$

MULTIPLE CHOICE EXERCISE 2

(Instructions for answering these questions are given on page xii.)

TYPE I

1) If $\log_x y = 2$:
(a) $x = 2y$ (b) $x = y^2$ (c) $x^2 = y$ (d) $y = 2x$ (e) $y = \sqrt{x}$.

2) $x^3 - 3x^2 + 6x - 2$ has remainder 2 when divided by:

(a) $x - 1$ (b) $x + 1$ (c) x (d) $x + 2$ (e) $2x - 1$.

3) $\log 5 - 2 \log 2 + \frac{3}{2} \log 16$ is equal to:
(a) $\log 80$ (b) 10 (c) 0 (d) $2 \log 12$ (e) 1.

4) $x^3 - 3x^2 + 2x - 6$ has a factor:
(a) $x - 3$ (b) $x - 2$ (c) $x - 4$ (d) $x + 3$ (e) $x + 2$.

5) If $2^{2x+1} - 6(2^x) = 0$ then x is:
(a) 1.5 (b) $\log_2 3$ (c) $\log 3$ (d) $\log_3 2$ (e) 3.

6) $\dfrac{p^{-\frac{1}{2}} \times p^{\frac{3}{4}}}{p^{-\frac{1}{4}}}$ simplifies to:

(a) 1 (b) $p^{-\frac{1}{2}}$ (c) $p^{\frac{3}{4}}$ (d) p (e) $p^{\frac{1}{2}}$.

7) In the expansion of $(a - 2b)^3$ the coefficient of b^2 is:
(a) $-2a^2$ (b) $-8a$ (c) $12a$ (d) $-4a$ (e) -12.

TYPE II

8) $f(x) \equiv x^3 - 7x^2 + 3x - 1$.
(a) $f(x)$ has a remainder -15 when divided by $x - 2$,
(b) $f(x)$ has no linear factors with integral coefficients,
(c) $f(x)$ is a polynominal of degree 3.

9) $\frac{1}{2} \log 16 - 1$:
(a) can be expressed as a single logarithm,
(b) has an exact numerical value,
(c) is equal to $\log 7$.

10) $f(x) \equiv 2x^2 + 3x - 2$.
(a) $f(x)$ can be expressed as the sum of two partial fractions,
(b) the equation $f(x) = 0$ has two real distinct roots,
(c) $x + 2$ is a factor of $f(x)$.

11) $\dfrac{2\sqrt{3} - 2}{2\sqrt{3} + 2}$:

(a) can be expressed as a fraction with a rational denominator,
(b) is an irrational number,
(c) is equal to -1.

12) $f(x) \equiv (3 - 5x)^4$.
(a) $f(x)$ has a remainder 16 when divided by $x - 1$,
(b) the expansion of $f(x)$ contains four terms,
(c) the equation $f(x) = 0$ is satisfied by only one value of x.

TYPE III

13) (a) $a = \log x$.
    (b) $a \log_x 10 = 1$.

14) (a) $f(x) \equiv x^2 + 2x + 1$.
    (b) $f(x)$ has a remainder 1 when divided by x.

15) (a) $\log_b a = c$.
    (b) $a = b^c$.

16) (a) $a = \log c + \log b$.
    (b) $a = \log (c + b)$.

TYPE V

17) If $f(x)$ is divided by $ax - 1$ the remainder is $f\left(\dfrac{1}{a}\right)$.

18) If $x - a$ is a factor of $x^2 + px + q$, the equation $x^2 + px + q = 0$ has a root equal to a.

19) $3 \log x + 1 = \log 10x^3$ is an equation.

20) In the expansion of $(1 + x)^6$ the coefficient of x is 6.

MISCELLANEOUS EXERCISE 2

1) Find the real values of x for which $\log_3 x - 2\log_x 3 = 1$.   (U of L)p

2) Solve for x, correct to two significant figures, the equations:
   (a) $4^{2x+1} \cdot 5^{x-2} = 6^{1-x}$
   (b) $4^x - 2^{x+1} - 3 = 0$.   (AEB)'71p

3) Without using tables, write down the values of:
   (a) $\log_4 8\sqrt{2}$
   (b) $(\log_5 49) \times (\log_7 125)$.   (U of L)p

4) If a and b are both positive and unequal, and $\log_a b + \log_b a^2 = 3$ find b in terms of a.   (U of L)p

5) If $x = \log_a b$, write down an expression for b in terms of a and x. Hence prove that
$$\log_s t = \frac{\log_r t}{\log_r s}.$$
Given that $\log_3 6 = m$ and $\log_6 5 = n$, express $\log_3 10$ in terms of m and n.   (JMB)p

6) (i) Solve for real x the equation $4(3^{2x+1}) + 17(3^x) - 7 = 0$.
   (ii) If s and t are positive numbers other than 1, prove that:
   (a) $\log_s t + \log_{1/s} t = 0$
   (b) $\log_s t = \dfrac{1}{\log_t s}$.   (U of L)

7) Solve the simultaneous equations $\log_y x = 2$, $5y = x + 12\log_x y$.   (U of L)p

8) Solve the simultaneous equations $\log_2 x + \log_2 y = 3$, $\log_y x = 2$.   (AEB)'76p

9) Solve the simultaneous equations $\log(x-2) + \log 2 = 2\log y$
$$\log(x - 3y + 3) = 0.$$ (U of L)p

10) Show that $\log_{16}(xy) = \frac{1}{2}\log_4 x + \frac{1}{2}\log_4 y$. Hence, or otherwise, solve the simultaneous equations
$$\log_{16}(xy) = 3\tfrac{1}{2}$$
$$\frac{(\log_4 x)}{(\log_4 y)} = -8$$ (AEB)'75p

11) If $P(x) \equiv p_n x^n + p_{n-1} x^{n-1} + \ldots + p_0$ is divided by $(x-a)$, show that the remainder is $P(a)$.
If $Q(x) \equiv x^4 + hx^3 + gx^2 - 16x - 12$ has factors $(x+1)$ and $(x-2)$, find the constants h, g and the remaining factors.   (U of L)p

12) If $f(x) \equiv x^6 - 5x^4 - 10x^2 + k,$ find the value of k for which $x - 1$ is a factor of $f(x)$.
When k has this value, find another factor of $f(x)$, of the form $x + a$, where a is a constant. (C)p

13) If $f(x) \equiv ax^2 + bx + c$ leaves remainders $1, 25, 1$ on division by $x - 1$, $x + 1, x - 2$ respectively, show that $f(x)$ is a perfect square.

(U of L)p

14) Without using tables, solve each of the following equations for x, expressing your answers as simply as possible:
(a) $9 \log_x 5 = \log_5 x$,

(b) $\log_8 \dfrac{x}{2} = \dfrac{\log_8 x}{\log_8 2}$. (JMB)

15) Prove that when a polynomial $f(x)$ is divided by $ax + b$, where $a \neq 0$, the remainder is $f(-b/a)$.
Find the polynomial in x of the third degree, which vanishes when $x = -1$ and $x = 2$, has the value 8 when $x = 0$ and leaves the remainder $16/3$ when divided by $3x + 2$. (JMB)

16) Express $\log_9 xy$ in terms of $\log_3 x$ and $\log_3 y$.
Without using tables, solve for x and y the simultaneous equations

$$\log_9 xy = \frac{5}{2}$$

$$\log_3 x \log_3 y = -6$$

expressing your answers as simply as possible. (JMB)

CHAPTER 3

FUNCTIONS

MAPPING

Consider the function f defined by $f: x \to 2x + 1$.

If we input the value 2 for x, f gives 5 as output and we say that this function maps 2 to 5, or $2 \to 5$.

Similarly f maps -2 to -3, -1 to -1, 0 to 1,....

This mapping can be illustrated as follows.

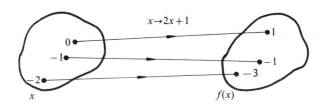

In this example, one value of the independent variable x maps to just one unique value of the dependent variable $f(x)$. Such a function is said to define a one-one mapping.

Now consider the function f given by $f: x \to x^2 - 2x + 6$.

This function maps -1 to 9, 0 to 6, 1 to 5, 2 to 6, 3 to 9,....

i.e.

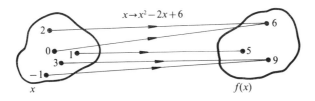

43

In this example one value of x again gives rise to just one value of $f(x)$ but this value of $f(x)$ is not necessarily uniquely obtained. For example, both 0 and 2 map to 6. Thus there are at least some values of $f(x)$ which arise from more than one value of x. Such a function is called a many–one mapping.

Hence we may regard a function as a 'rule' for mapping a number a to a number b. However the strict mathematical definition of the word function is restricted to those relationships for which one input value gives rise to *just one* output value.

So the two relationships considered above, viz. $x \rightarrow 2x + 1$ and $x \rightarrow x^2 - 2x + 6$ are functions under this definition. But if we consider the relationship $x \rightarrow \pm\sqrt{x}$, we find that if we input the value 4, say, for x we get as output the two values 2 and -2.

Thus $x \rightarrow \pm\sqrt{x}$ is not a function, although it does define a perfectly good mapping for positive values of x.

GRAPHICAL REPRESENTATION OF FUNCTIONS (CURVE SKETCHING)

Consider again the function $f(x) \equiv x^2 - 2x + 6$.

For any arbitrarily chosen value of x, the corresponding value of $f(x)$ can be calculated, e.g.

x	-3	-2	-1	0	1	2	3	4
$f(x)$	21	14	9	6	5	6	9	14

Having arranged the values of $f(x)$ in the order of ascending values of x, we can see that there is some pattern to the values obtained for $f(x)$.

This pattern is easier to visualise if the results are displayed graphically, i.e. if the values of $f(x)$ on a vertical number line are plotted against the corresponding values of x on a horizontal number line.

The following graph has been drawn using only eight values of x, separated by intervals of one unit. We could choose values of x from a larger range and use smaller intervals. The corresponding values of $f(x)$ so obtained would all be found to lie on the smooth curve we have already drawn.

Therefore this curve is a graphical representation of the varying values of $f(x)$ obtained from varying values of x and all possible related values of $f(x)$ and x lie on the curve.

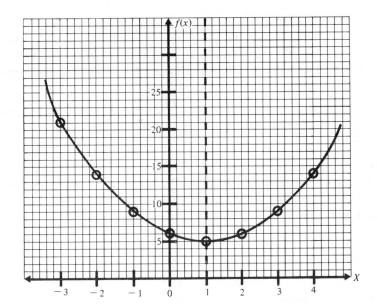

We can now use the graph to deduce some properties of the function.

(a) The lowest point on the curve is where $x = 1$ and $f(x) = 5$ so although x, as the independent variable, can be given any real value, the corresponding value of $f(x)$ is always greater than 5, or equal to 5 when $x = 1$. We say that the function has a least value of 5.

(b) The curve is symmetrical about the line $x = 1$, as two values of x equidistant from $x = 1$, $(1 + a$ and $1 - a)$, give the same value for $f(x)$.

(c) Conversely, for any one value of $f(x)$, there are two corresponding values for x of the form $1 + a$ and $1 - a$.

All these properties can be confirmed algebraically as follows:

completing the square on R.H.S gives

$$f(x) \equiv x^2 - 2x + 6 \equiv (x - 1)^2 + 5$$

(a) Whatever value we substitute for x, $(x - 1)^2$ is always positive as it is a squared quantity. Therefore the least value of $(x - 1)^2$ is zero and this occurs when $x = 1$. Hence the lowest value that $f(x)$ can have is 5, i.e. $f(x)$ *has a least value of 5 when* $x = 1$.

(b) Consider two values of x symmetrically placed on either side of $x = 1$, i.e. $x = 1 + a$ and $x = 1 - a$

$$f(1 + a) = a^2 + 5$$
$$f(1 - a) = a^2 + 5$$

Therefore values of x that are symmetrical about $x = 1$ give the same value for $f(x)$.

(c) Conversely, for any given values of $f(x)$, c say, the corresponding values of x are given by the solution of the equation

$$c = x^2 - 2x + 6$$

or $\qquad\qquad x^2 - 2x + 6 - c = 0$

From the formula, the roots of this equation are

$$x = 1 + \sqrt{c-5} \quad\text{and}\quad x = 1 - \sqrt{c-5}$$

Thus for x to have real values, $c \geqslant 5$, i.e. $f(x)$ has a least value of 5 and these two real values of x, corresponding to one value of $f(x)$, are symmetrical about $x = 1$.

DOMAIN AND RANGE

For the function $f: x \rightarrow x^2 - 2x + 6$ we have assumed that we can input any real value of x. However for a full definition of a function, the set of permissible input values must be stated.

So the function illustrated above is fully defined as

$$f: x \rightarrow x^2 - 2x + 6, \quad x \in \mathbb{R}$$

(in this context $x \in \mathbb{R}$ means that x can be any member of the set of real numbers).

The set of input values for which any function is defined is called the *domain of the function*, and the domain is not always the complete set of real numbers as we shall see in the following examples.

The analysis of the function $f(x) \equiv x^2 - 2x + 6$ also showed that, for the domain $x \in \mathbb{R}$, the output values were limited, i.e. $f(x) \geqslant 5$. Once the domain of any function is defined there is a corresponding set of output values. This set of output values is called the *range of the function* (or the *image-set*). Hence the range of f where $f: x \rightarrow x^2 - 2x + 6$, $x \in \mathbb{R}$ is

$$f(x) \in \mathbb{R}, \quad f(x) \geqslant 5$$

EXAMPLES 3a

1) Draw a sketch graph of the function defined by $f: x \rightarrow 2x + 1$, $x \in \mathbb{R}$ and state its range.

The following arbitrary set of values
for x gives

x	0	2	4
$f(x)$	1	5	9

from which we get the sketch graph
on the right.

As x can be any real number, this
line is infinite in both directions so
the range of f also is the complete
set of real numbers.

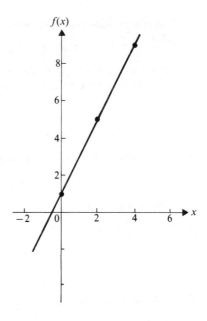

2) For the function illustrated in Example 1, A is the point on the line for
which $x = -1$ and B is the point for which $x = 2$. Define the function
represented by the line segment AB and state its range.

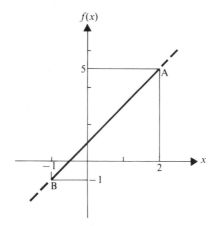

The mapping represented by the line segment AB is $x \rightarrow 2x + 1$ but only
for values of x from -1 to 2.

So the function represented by AB is $f : x \rightarrow 2x + 1$, $x \in \mathbb{R}$, $-1 \leqslant x \leqslant 2$

The range of this function is the set of values of $f(x)$ corresponding to
$-1 \leqslant x \leqslant 2$

i.e. the range of f is $-1 \leqslant f(x) \leqslant 5$.

3) State the image-set of the function defined as

$$f(x) \equiv 2x + 1, \quad x \in \{0, 1, 2, 3, 4\}$$

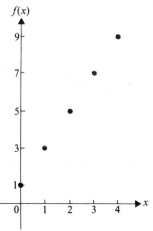

For this function there are just 5 input values. The corresponding output values form the image-set. So the image-set is

$$\{1, 3, 5, 7, 9\}$$

The graphical representation of this function is the set of five points shown on the right.

Note that when the domain of a function is not stated it is assumed to be $x \in \mathbb{R}$.

EXERCISE 3a

1) Given $f: x \to x^2$ for $x \in \mathbb{R}$,

(a) draw the graph of f and state its range,

(b) if the domain is redefined as $x \in \mathbb{R}^+$, sketch the graph that now represents f.

2) The domain of each of the following mappings is the set of real numbers. State which of these mappings are functions.

(a) $a \to a^2 + 5$,   (b) $b \to \pm\sqrt{b}$,   (c) $c \to \sqrt{c}$

3) In Utopia, income tax on earnings is calculated as follows:
The first £10 000 is tax free, the next £10 000 is taxed at 5% and the remaining income is taxed at 10%. Taking income as 'input' and tax payable as 'output', state whether these rules for calculating tax constitute a function. If they do, state the implied domain and range.

4) If in Question 3 £I is income and £T is the tax payable, express the mapping $I \to T$ as algebraic formulae of the form $T \equiv f(I)$, stating the values of I for which they are valid.

THE QUADRATIC FUNCTION

Functions of similar form usually have properties in common, so their graphs are usually similar in shape. A knowledge of the common characteristics

of such functions will allow us to draw a sketch of the graph for any one
particular function without calculating a series of corresponding values of $f(x)$
and x. The function $\quad f(x) \equiv x^2 - 2x + 6 \quad$ analysed above is a quadratic
function and any function whose general form is

$$f(x) \equiv ax^2 + bx + c$$

where a, b and c are constants, is called a quadratic function.

Completing the square on the R.H.S of this expression

gives
$$f(x) \equiv a\left[x^2 + \frac{b}{a}x + \frac{c}{a}\right]$$

$$\equiv \left\{\frac{4ac - b^2}{4a}\right\} + a\left(x + \frac{b}{2a}\right)^2$$

Now whatever value x takes, $\left\{\dfrac{4ac - b^2}{4a}\right\}$ is constant and equal to K say

and $\left(x + \dfrac{b}{2a}\right)^2 \geqslant 0$ as it is a squared quantity.

Therefore the function has the general form

$$f(x) \equiv K + a \text{ (zero or a + ve quantity)}.$$

So if a is *positive* i.e. $a > 0$, $f(x)$ is at least equal to K

i.e. $f(x)$ has a *least* value of $\dfrac{4ac - b^2}{4a}$, occuring when $x = -\dfrac{b}{2a}$

When a is *negative*, i.e. $a < 0$, $f(x)$ can never be greater than K

i.e. $f(x)$ has a *greatest* value of $\dfrac{4ac - b^2}{4a}$, when $x = -\dfrac{b}{2a}$

Now $x = -\dfrac{b}{2a}$ is the value of x corresponding to the greatest or least value

of $f(x)$.

Taking two values of x that are symmetrical about $-\dfrac{b}{2a}$, i.e. $x = -\dfrac{b}{2a} \pm k$,

gives
$$f\left(-\frac{b}{2a} + k\right) = f\left(-\frac{b}{2a} - k\right) = ak^2 + \frac{4ac - b^2}{4a}$$

i.e. input values of x that are symmetrical about $x = -\dfrac{b}{2a}$ give as output

the same value of $f(x)$.

From this analysis we deduce that when $f(x) \equiv ax^2 + bx + c$

$f(x)$ has $\begin{cases} \text{a least value of } f\left(-\dfrac{b}{2a}\right) \text{ if } a > 0 \\[2ex] \text{a greatest value of } f\left(-\dfrac{b}{2a}\right) \text{ if } a < 0 \end{cases}$

and is symmetrical in shape about the line $x = -\dfrac{b}{2a}$ which is called the *axis* of the curve.

The diagrams below represent the two alternative graphs of a quadratic function.

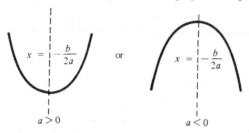

The curve representing any particular quadratic function can now be sketched using this information. So, for example, to sketch the curve representing

$f(x) \equiv 2x^2 - 7x - 4$ we proceed as follows:

$$b = -7, \quad a = 2$$

therefore $\qquad f\left(-\dfrac{b}{2a}\right) = f(\tfrac{7}{4}) = -\tfrac{81}{8}$

i.e. $f(x)$ has a least value of $-\tfrac{81}{8}$ when $x = \tfrac{7}{4}$.

To locate the curve accurately on the axes we need one more pair of corresponding values for x and $f(x)$.
$f(0)$ is easy to find, i.e. $f(0) = -4$.

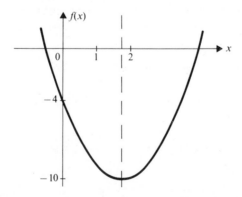

Alternatively a quick sketch of a quadratic function can be obtained as follows

$$f(x) \equiv 2x^2 - 7x - 4 \equiv (2x + 1)(x - 4)$$

The coefficient of x^2 is positive, $(a > 0)$, so $f(x)$ has a least value.
When $f(x) = 0$ the corresponding values of x are roots of the quadratic
equation

$$(2x + 1)(x - 4) = 0$$

i.e. $\qquad\qquad\qquad x = -\tfrac{1}{2}$ and $x = 4$

The average of these values $(\tfrac{7}{4})$ gives the value of x about which the curve is
symmetrical.

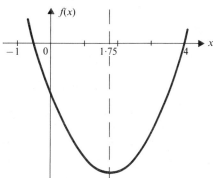

Note. This method is suitable only when the function factorises.

EXERCISE 3b

Find the greatest or least value of the following functions.

1) $x^2 - 3x + 5$ 2) $2x^2 - 4x + 5$ 3) $3 - 2x - x^2$

4) $7 + x - x^2$ 5) $x^2 - 2$ 6) $2x - x^2$

Sketch the graphs of the following quadratic functions, showing clearly the
greatest or least value of $f(x)$ and the value of x at which it occurs, where
$f(x)$ is

7) $x^2 - 2x + 5$ 8) $x^2 + 4x - 8$ 9) $2x^2 - 6x + 3$

10) $4 - 7x - x^2$ 11) $x^2 - 10$ 12) $2 - 5x - 3x^2$

Draw a quick sketch of the graph of each of the following functions, showing
the axis of symmetry clearly.

13) $(x - 1)(x - 3)$ 14) $(x + 2)(x - 4)$ 15) $(2x - 1)(x - 3)$

16) $(1 + x)(2 - x)$ 17) $x^2 - 9$ 18) x^2

19) $4 - x^2$ 20) $3 - 7x - 6x^2$

INEQUALITIES

Consider the real numbers 5 and 2.

Now $$5 > 2.$$
The introduction of an extra term on both sides leaves the inequality sign unchanged

e.g. $$5 + 4 > 2 + 4 \quad \text{and} \quad 5 - 4 > 2 - 4$$

i.e. $$9 > 6 \qquad\qquad 1 > -2$$

Multiplication of both sides by a positive number also leaves the inequality sign unchanged.

e.g. $$5 \times 2 > 2 \times 2 \quad \text{i.e.} \quad 10 > 4$$

However if we multiply both sides by a negative number, the inequality is no longer true. For example, multiplying by -2, we find that
the L.H.S. becomes -10 and the R.H.S. becomes -4,

i.e. $$\text{L.H.S.} < \text{R.H.S.} \quad \text{since} \quad -10 < -4$$

This illustrates the general fact that multiplication or division of both sides by a negative number reverses the inequality sign.

To summarise, if a and b are real numbers such that
$$a > b$$

then $\quad a + k > b + k \quad$ for *all* real values of k

and $\qquad ak > bk \qquad$ for positive values of k

but $\qquad ak < bk \qquad$ for negative values of k

EXAMPLE 3c

Find the range of values of x satisfying the inequality $\quad x - 3 < 2x + 5$

$$x - 3 < 2x + 5$$

$\Rightarrow \qquad x < 2x + 8 \quad$ (adding 3 to both sides)

$\Rightarrow \qquad -x < 8 \qquad$ (subtracting $2x$ from both sides)

$\Rightarrow \qquad x > -8 \quad$ (multiplying by -1)

Therefore the range of values of x that satisfies the inequality is

$$x > -8$$

QUADRATIC INEQUALITIES

Any inequality relationship that involves a quadratic function is called a quadratic inequality. For example

$$(x-2)(2x+1)>0$$

The range of values of x satisfying this inequality can be found graphically as follows:

Let $f(x) \equiv (x-2)(2x+1)$.
The diagram on the right shows
a sketch of $f(x)$ from which
we see that $f(x)>0$
(i.e. the curve is above the x-axis)
for values of x greater than 2
and less than $-\frac{1}{2}$.

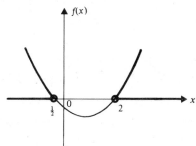

Therefore the ranges of values of x that satisfy

$$(x-2)(2x+1)>0 \quad \text{are} \quad x<-\tfrac{1}{2} \quad \text{and} \quad x>2$$

EXERCISE 3c

Find the range (or ranges) of values of x that satisfy the following inequalities.

1) $x+2>4-x$
2) $2x-1<x-4$
3) $x>5x-2$

4) $3x-5<4$
5) $2(x-1)>3(x-1)$

6) $2(x-1)<2(x+1)$
7) $(x-1)(x-2)>0$

8) $(x+1)(x-2)>0$
9) $(x-3)(x-5)<0$

10) $(2x-1)(x+1)<0$
11) $x^2-4x>5$
12) $4x^2<1$

13) $5x^2>3x+2$
14) $(2-x)(x+4)<0$

15) $(3-2x)(x+5)>0$
16) $3x>x^2+2$
17) $(x-1)^2>4x^2$

18) $(x-1)(x+2)<x(4-x)$

Problems involving Quadratic Inequalities

EXAMPLES 3d

1) Find the range of values of k for which the equation

$$x^2-kx+(k+3)=0 \quad \text{has real roots.}$$

For $x^2-kx+(k+3)=0$ to have real roots

$$b^2 - 4ac \geqslant 0$$

i.e. $(-k)^2 - 4(k+3) \geqslant 0$

$$k^2 - 4k - 12 \geqslant 0$$

Let $f(k) \equiv k^2 - 4k - 12$

$$\equiv (k-6)(k+2)$$

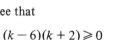

From the sketch of $f(k)$ we see that

$$(k-6)(k+2) \geqslant 0$$

for $k \leqslant -2$ and $k \geqslant 6$.

2) Find the set of values of p for which

$$f(x) \equiv x^2 + 3px + p$$

is greater than zero for all real values of x.

$f(x) \equiv x^2 + 3px + p$ is a quadratic function of x and, as the coefficient of x^2 is positive, $f(x)$ has a least value.

So if $f(x) > 0$ for all x, the least value of $f(x)$ has to be greater than zero, i.e. $f(x)$ has a graph of the type in the sketch. Completing the square on the R.H.S. gives

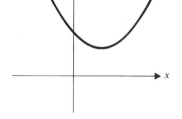

$$f(x) \equiv \left(x + \frac{3p}{2}\right)^2 + p - \frac{9p^2}{4}$$

Therefore the least value of $f(x)$ is $p - \dfrac{9p^2}{4}$

So for $f(x) > 0$ for all x,

$$p - \frac{9p^2}{4} > 0$$

$$4p - 9p^2 > 0$$

$$p(4 - 9p) > 0$$

Let $g(p) \equiv p(4 - 9p)$.

From the sketch of $g(p)$ we see that

$$p(4-9p) > 0 \quad \text{for} \quad p > 0 \quad \text{and} \quad p < \tfrac{4}{9}$$

$$\text{or} \quad 0 < p < \tfrac{4}{9}$$

Therefore $f(x) > 0$ for all real x, for the set of values of p given by $0 < p < \tfrac{4}{9}$.

3) Find the ranges of values of x for which

$$2x - 1 < x^2 - 4 < 12$$

There are two inequality relationships here

viz: (a) $2x - 1 < x^2 - 4$ and   (b) $x^2 - 4 < 12$

and we are looking for the ranges of values of x that satisfy *both* of these inequalities.

(a) $2x - 1 < x^2 - 4$ (b) $x^2 - 4 < 12$

    $-x^2 + 2x + 3 < 0$ $x^2 - 16 < 0$

        $x^2 - 2x - 3 > 0$ If $g(x) \equiv x^2 - 16$

If $f(x) \equiv x^2 - 2x - 3$ $\equiv (x - 4)(x + 4)$

            $\equiv (x - 3)(x + 1)$ the graph of $g(x)$ is

the graph of $f(x)$ is

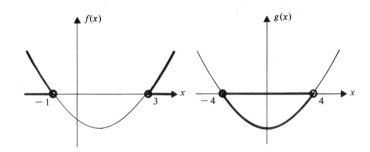

For $f(x) > 0$,   For $g(x) < 0$

    $x < -1$ and $x > 3$ $-4 < x < 4$

Illustrating these ranges on a number line

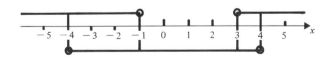

we see that the ranges of values of x that satisfy both (a) and (b), i.e. where the lines overlap, are

$$-4 < x < -1 \qquad \text{and} \qquad 3 < x < 4$$

EXERCISE 3d

1) Find the range of values of x for which $\quad 3-x < 4(x-2)$.

2) Find the range of values of x for which $\quad x-5 > 3(2-x)$.

3) Find the ranges of values of k for which the equation

$$x^2 + (k-3)x + k = 0$$

has (a) real distinct roots, (b) roots of the same sign.   (JMB)p

4) If x is real and $\quad x^2 + (2-k)x + 1 - 2k = 0 \quad$ show that k cannot lie between certain limits, and find these limits.   (JMB)p

5) Find the limitations required on the values of the real number c in order that the equation $\quad x^2 + 2cx - c + 2 = 0 \quad$ shall have real roots.   (JMB)p

6) Prove that, if $\quad x^2 > k(x+1) \quad$ for all real x, then $\quad -4 < k < 0$.   (C)p

7) Find the condition that must be satisfied by k in order that the expression

$$2x^2 + 6x + 1 + k(x^2 + 2)$$

may be positive for all real values of x.   (JMB)p

8) Find the range of values of x for which

$$(x-4) < x(x-4) \leqslant 5 \qquad \text{(U of L)p}$$

9) (a) If $\quad x = 2 \quad$ is a root of the equation

$$\alpha^2 x^2 + 2(2\alpha - 5)x + 8 = 0$$

find the possible value (or values) of α and the corresponding value (or values) of the other root.

(b) Find the range (or ranges) of possible values of the real number α if

$$\alpha^2 x^2 + 2(2\alpha - 5)x + 8 > 0$$

for all real values of x.   (C)

10) Determine for each of the three expressions $f(x)$, $g(x)$ and $h(x)$ the range (or ranges) of values of x for which it is positive. Give your answers correct to two places of decimals. Explain, briefly, the reasons for your answers.

(a) $f(x) \equiv x^2 + 4x - 6$ (b) $g(x) \equiv -x^2 - 8x + 2$ (C)p

11) (a) State the range of values of x for which $2x^2 + 5x - 12$ is negative.

(b) The value of the constant a is such that the quadratic function
$f(x) \equiv x^2 + 4x + a + 3$ is never negative. Determine the nature of
the roots of the equation $af(x) = (x^2 + 2)(a - 1)$. Deduce the
value of a for which this equation has equal roots. (AEB) '73p

12) By eliminating x and y from the equations

$$\frac{1}{x} + \frac{1}{y} = 1, \quad x + y = a, \quad \frac{y}{x} = m$$

where $a \neq 0$, obtain a relation between m and a. Given that a is real,
determine the ranges of values of a for which m is real. (JMB)p

13) If $a > 0$, prove that the quadratic expression $ax^2 + bx + c$ is positive
for all real values of x when $b^2 < 4ac$. Hence find the range of values of p
for which the quadratic function of x

$$f(x) \equiv 4x^2 + 4px - (3p^2 + 4p - 3)$$

is positive for all real values of x.
Illustrate your result by making sketch graphs of $f(x)$ for each of the cases
$p = 0$ and $p = 1$. (U of L)

14) (a) If a is a positive constant, find the set of values of x for which
$a(x^2 + 2x - 8)$ is negative. Find the value of a if this function has a
minimum value of -27.
(b) Find two quadratic functions of x which are zero at $x = 1$, which
take the value 10 when $x = 0$ and which have a maximum value
of 18. Sketch the graphs of these two functions. (U of L)

15) Find the set of values of k for which $f(x) \equiv 3x^2 - 5x - k$ is greater
than unity for all real values of x.
Show that, for all k, the minimum value of $f(x)$ occurs when $x = \frac{5}{6}$.
Find k if this minimum value is zero. (U of L)p

16) Prove that $3x^2 - 4x + 2 > 0$ for all real values of x.
(U of L)p

17) Find the value of $k(\neq 1)$ such that the quadratic function of x

$$k(x + 2)^2 - (x - 1)(x - 2)$$

is equal to zero for only one value of x.
Find also (a) the range of values of k for which the function possesses a minimum value, (b) the range of values of k for which the value of the function
never exceeds 12.5.
Sketch the graph of the function for $k = \frac{1}{2}$ and for $k = 2\frac{1}{2}$. (U of L)

18) Find the set of values of x for which

$$2x^2 + 3x + 2 > 4$$

(U of L)

19) The roots of the equation

$$9x^2 + 6x + 1 = 4kx$$

where k is a real constant, are denoted by α and β.

(a) Show that the equation whose roots are $1/\alpha$ and $1/\beta$ is

$$x^2 + 6x + 9 = 4kx$$

(b) Find the set of values of k for which α and β are real.

(c) Find also the set of values of k for which α and β are real and positive.

(U of L)

20) The function f is given by $f: x \to x^2 - 3x - 4$, where $x \in \mathbb{R}$.
Find the range of f and the values of x for which $f(x) = 0$. (C)p

SOME OTHER SIMPLE FUNCTIONS

We will now consider the graphical representation of some other simple functions. A full analysis of the general form of these functions is not possible at this stage, but enough information can be deduced to give a rough idea of the shapes of the curves that represent them.

Exponential Functions

An exponential function is one where the variable appears as an exponent, (i.e. an index)

e.g. $2^x, \ 3^{-x}, \ 10^{x+1}, \ 5^{-3x} + 2$

are all exponential functions of x.
Consider the function $f(x) \equiv 2^x$ for which the following table shows corresponding values of x and $f(x)$.

x	...	-10	...	-5	-4	-3	-2	-1	0	1	2	3	4	5	...	10	...
$f(x) \equiv 2^x$	...	$\frac{1}{1024}$	...	$\frac{1}{32}$	$\frac{1}{16}$	$\frac{1}{8}$	$\frac{1}{4}$	$\frac{1}{2}$	1	2	4	8	16	32	...	1024	...

From this table we see that:

1. $f(x) > 0$ for all real values of x,

2. as x increases $f(x)$ increases at a rapidly accelerating rate,

3. $f(x) = 1$ when $x = 0$

4. as x decreases (i.e. $x = -10, -100, \ldots$) $f(x)$ quickly becomes numerically smaller and we say that, as x approaches minus infinity, $f(x)$ approaches the value zero. This is written,

$$x \to -\infty, \quad f(x) \to 0$$

From these observations the sketch of $f(x) \equiv 2^x$ is drawn:

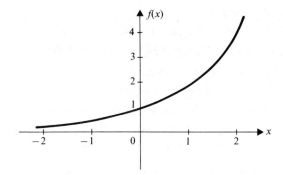

The curve approaches the negative x-axis but never actually touches it or crosses it and we say that the x-axis is an *asymptote* to the curve.

Another way of expressing the behaviour of $f(x)$ for negative values of x is based on the fact that

$f(x)$ approaches a limiting value (or limit) of zero as x approaches minus infinity.

i.e. the limit of $f(x)$ as $x \to -\infty$ is zero.

This property is written

$$\lim_{x \to -\infty} [f(x)] = 0$$

Any function of the form a^x, where $a > 1$, is represented by a curve similar to that deduced for 2^x.

Rational Functions

A function where both numerator and denominator are polynomials is called a rational function.

e.g. $$\frac{1}{x}, \quad \frac{x}{x^2 - 1}, \quad \frac{2x^2 - 7x}{x + 1}$$

are rational functions of x.

Consider the function $f(x) \equiv \dfrac{1}{x}$ and the following table of corresponding

values for x and $f(x)$

x	...	-3	-2	-1	0	1	2	3	4	...	10	...
$f(x) \equiv \dfrac{1}{x}$	...	$-\dfrac{1}{3}$	$-\dfrac{1}{2}$	-1	?	1	$\dfrac{1}{2}$	$\dfrac{1}{3}$	$\dfrac{1}{4}$	...	$\dfrac{1}{10}$	...

From this table we see that:

1. For $x > 0$, $f(x) > 0$ and as $x \to \infty$, $f(x) \to 0$.

2. For $x < 0$, $f(x) < 0$ and as $x \to -\infty$, $f(x) \to 0$
   i.e. x and $f(x)$ have the same sign.

3. For $x = 0$, $f(0) = \dfrac{1}{0}$ which has no finite value and is said to be *undefined*. In such circumstances we investigate the behaviour of $f(x)$ as x approaches zero.

Now x can approach zero in two ways

i.e. decrease from positive values towards zero (approach zero from above)

or increase from negative values towards zero (approach zero from below)

From (1) and (2) above we see that $f(x)$ and x have the same sign. As x decreases, $f(x)$ increases.

From the following table

x	-1	...	$-\frac{1}{10}$	...	$-\frac{1}{100}$	$\rightarrow$..	0	..$\leftarrow$	$\frac{1}{100}$	...	$\frac{1}{10}$	...	1	..
$f(x)$	-1	...	-10	...	-100	$\rightarrow$..		..$\leftarrow$	100	...	10	...	1	..

we see that

as x decreases to zero, $\quad f(x) \rightarrow \infty$

and as x increases to zero, $\quad f(x) \rightarrow -\infty$.

From these observations we can draw the following sketch representing $f(x) \equiv \dfrac{1}{x}$.

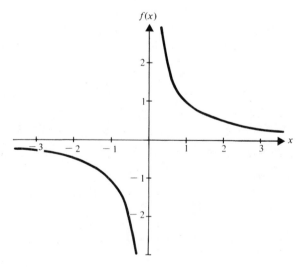

This curve has two asymptotes, the horizontal and vertical axes.

All the curves looked at so far have been unbroken or *continuous*, but this curve has a 'break' or *discontinuity* at the point where $x = 0$ and $f(0)$ is *undefined*.

In general, if $f(x)$ is any function of x and if $f(x)$ is undefined for a finite value of x, $x = a$ say, then the curve representing $f(x)$ has a discontinuity where $x = a$.

Also, we see that

$$\lim_{x \to \infty} \left[\frac{1}{x}\right] = 0 \qquad \text{and} \qquad \lim_{x \to -\infty} \left[\frac{1}{x}\right] = 0$$

but $\lim_{x \to 0} \left[\frac{1}{x}\right]$ does not have a unique value as $\frac{1}{x}$ approaches ∞ or $-\infty$

depending on whether x approaches zero from above or below. In this case we

say that $\lim_{x \to 0} \left[\frac{1}{x}\right]$ does not exist.

In general, for $\lim_{x \to a} [f(x)]$ to exist with a value k, say,

$$\left. \begin{array}{l} \text{then as } x \to a \text{ from above} \\ \textit{and} \text{ as } x \to a \text{ from below} \end{array} \right\} \quad f(x) \to k$$

LOGARITHMIC FUNCTIONS

For values of $a > 0$, any function of the form $\log_a x$, $\log_a(2x+1), \ldots$
is a logarithmic function.

Consider the function $f(x) \equiv \log_2 x$ and the following table of corresponding values for x and $f(x)$.

x	$\ldots -1$	0	1	2	$\ldots 4$	$\ldots 8$
$f(x) \equiv \log_2 x$	$\ldots$ does not exist	?	0	1	$\ldots 2$	$\ldots 3$

From this table we see that:

1. when $x = -1$, $f(-1) = \log_2(-1) = b$ say. But there is no real value of b for which $2^b = -1$.   So $f(-1)$ does not exist, and all negative values of x lead to the same conclusion,

i.e. $f(x) = \log_a x$ does not exist for negative values of x,

2. for $x > 1$, $f(x) > 0$ and as $x \to \infty$, $f(x) \to \infty$,

3. for $x = 0$, $f(0)$ is undefined so we investigate the behaviour of $f(x)$ as $x \to 0$ from above,

x	0	$\ldots \frac{1}{64}$	$\ldots \frac{1}{16}$	$\ldots \frac{1}{4}$	$\ldots \frac{1}{2}$	$\ldots 1$
$f(x)$		$\ldots -6$	$\ldots -4$	$\ldots -2$	$\ldots -1$	$\ldots 0$

From the table we see that as x decreases to zero, $f(x) \to -\infty$ and that for $0 < x < 1$, $f(x) < 0$.

From these observations we can now sketch the graphical representation of $f(x) \equiv \log_2 x$.

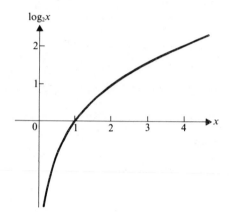

Any function of the form $\log_a x$ will have a graph of similar shape.

Note that, while $\log_a x$ does not exist for negative values of x, the value of $\log_a x$ *can* be negative.

SIMPLE VARIATIONS OF FUNCTIONS

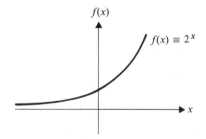

The curve on the right is one with which we are now familiar and represents the function $f(x) \equiv 2^x$.

We will now consider how to obtain the graphs of some simple variations of this function.

1) $g(x) \equiv 2^x + 1$.

Comparing $f(x) \equiv 2^x$ and $g(x) \equiv 2^x + 1$ we see that, for any one value of x, $g(x)$ is one unit greater than $f(x)$.

So we deduce that the curve representing $g(x)$ is the same shape as that representing $f(x)$ but is raised vertically by one unit:

i.e.

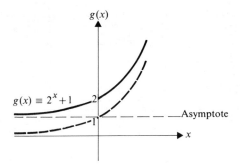

In general, if the curve representing any function $f(x)$ is known, the curve representing $f(x)+c$ is the same shape but is raised vertically by c units.

2) $g(x) \equiv 2^{-x}$.

Comparing $g(x) \equiv 2^{-x}$ with $f(x) \equiv 2^x$ we see that for $g(x)$, an input $x = a$ gives an output value of 2^{-a} while for $f(x)$, an input $x = -a$ gives the same output value of 2^{-a}

i.e. $\qquad g(a) \equiv f(-a) \quad$ or $\quad g(x) \equiv f(-x) \quad$ for $\quad x \in \mathbb{R}$

So the curve representing $g(x)$ is the same as that representing $f(x)$ except that the negative and positive values of x are transposed:

i.e.

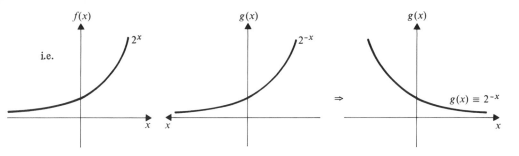

In general, for any function $f(x)$, the curve representing $f(-x)$ is the reflection in the vertical axis of the graph of $f(x)$.

3) $g(x) \equiv -2^x$.

Comparing $g(x) \equiv -2^x$ with $f(x) \equiv 2^x$ we see that $g(x) \equiv -f(x)$ so the curve representing $g(x)$ is the same shape as that for $f(x)$ with the positive and negative values of $f(x)$ transposed. The vertical axis is returned to its conventional position by a reflection in the horizontal axis giving

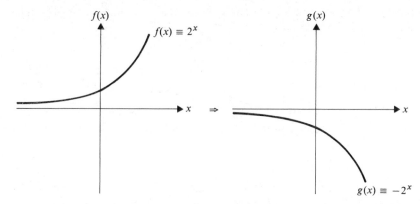

In general the curve representing $-f(x)$ is the reflection in the horizontal axis of the curve representing $f(x)$.

EXERCISE 3e

1) Write down the values of $f(x) \equiv (\frac{1}{2})^x$ corresponding to $x = 2, 4, 6$ and to $x = -2, -4, -6$. From these values deduce the behaviour of $f(x)$ as $x \to \infty$ and as $x \to -\infty$. Sketch the graph of $f(x) \equiv (\frac{1}{2})^x$, marking any asymptotes.

2) Draw sketch graphs of the following functions:

(a) 3^x (b) 2^{x-1} (c) 3^{-x} (d) 2^{2x} (e) $1 + 2^x$ (f) $-(2^x)$

showing clearly, in each case, any asymptotes.

3) For what value of x is $f(x) \equiv \dfrac{1}{1-x}$ undefined? Describe the behaviour of $f(x)$ as x approaches this value from above and from below.

Write down also, $\displaystyle\lim_{x \to \infty} [f(x)]$ and $\displaystyle\lim_{x \to -\infty} [f(x)]$.

Use this information to sketch the graph of $f(x) \equiv \dfrac{1}{1-x}$ marking the asymptotes clearly.

4) By following a procedure similar to that indicated in (3), or otherwise, draw sketch graphs of the following functions:

(a) $-\dfrac{1}{x}$ (b) $\dfrac{1}{x-2}$ (c) $\dfrac{2}{x+1}$ (d) $1 + \dfrac{1}{x}$ (e) $\dfrac{x-1}{x}$

(f) $\dfrac{x}{x+1}$

5) Sketch the curve representing $f(x) \equiv x^2 - 4$. On the same axes sketch the graphs of the following functions

(a) $x^2 - 2$ (b) $x^2 + 2$ (c) $4 - x^2$

6) Sketch the graph of the function $f(x) \equiv \lg x$. Also make sketch graphs of the following functions

(a) $\lg(-x)$ (b) $-\lg x$ (c) $1 + \lg x$ (d) $1 - \lg x$

7) Sketch the graph of the function $f(x) \equiv 2 - 3x$ and on the same set of axes sketch the graph of $3x - 2$.

8) On the same set of axes draw sketch graphs of the functions $f(x) \equiv \lg x$ and $g(x) \equiv 10^x$. Describe how the second graph can be obtained from the first graph.

9) Repeat Question 8 for the functions $f(x) \equiv 2x - 1$ and $g(x) \equiv \frac{1}{2}(x + 1)$.

INVERSE FUNCTIONS

Consider the mapping $f: x \rightarrow 2x$, $x \in \{2, 3, 4\}$.

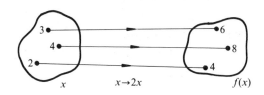

Under this function the domain $\{2, 3, 4\}$ maps to the image-set $\{4, 6, 8\}$. It is possible to reverse this mapping, i.e. we can map each member of the image-set $\{4, 6, 8\}$ back to the corresponding member of the domain by halving each member of the image-set,

i.e. $4 \rightarrow 2$, $6 \rightarrow 3$, $8 \rightarrow 4$

We can also express this procedure as an algebraic relationship,

i.e.   if $x \in \{4, 6, 8\}$ then $x \rightarrow \frac{1}{2}x$ maps $4 \rightarrow 2$, $6 \rightarrow 3$, $8 \rightarrow 4$

This reverse mapping is a one–one mapping so it is a function in its own right and it is called the inverse function of f where $f: x \rightarrow 2x$. Denoting this inverse function by f^{-1} we see that $f^{-1}: x \rightarrow \frac{1}{2}x$, $x \in \{4, 6, 8\}$ reverses the mapping $f: x \rightarrow 2x$, $x \in \{2, 3, 4\}$.
In fact for all real values of x, $f: x \rightarrow 2x$ can be reversed. So we can say that if f is the function defined by $f: x \rightarrow 2x$, $x \in \mathbb{R}$ then f^{-1} is the function which reverses this mapping and which is defined by $f^{-1}: x \rightarrow \frac{1}{2}x$, $x \in \mathbb{R}$.

The graphs of f and f^{-1} are shown below.

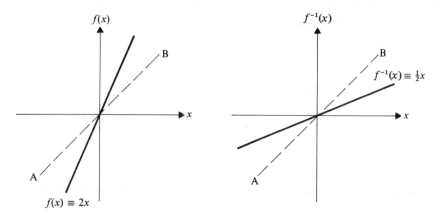

Now consider the function $f: x \to x^2$, $x \in \mathbb{R}$.
This is a many–one mapping, i.e. there are two
values of x which map to one value of $f(x)$.
For example both 2 and -2 map to 4.

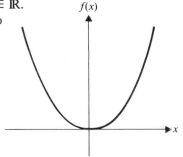

We can reverse this mapping by taking the positive and negative square root of
each member of the image-set. Expressing this in algebraic form gives the
mapping $x \to \pm\sqrt{x}$. However this is a one–many mapping so it is *not* a
function and we say that $f: x \to x^2$ for $x \in \mathbb{R}$ does not have an inverse
function.

If, however, we restrict the domain of f to
$x \in \mathbb{R}^+$, i.e. we redefine our function as
$f: x \to x^2$, $x \in \mathbb{R}^+$ then this becomes a
one–one mapping.

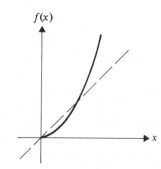

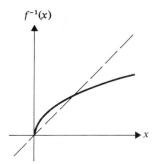

We can now express the reverse mapping as
$x \to \sqrt{x}$ which is a one–one mapping.
So the function $f: x \to x^2,\quad x \in \mathbb{R}^+$
does have an inverse, viz.

$$f^{-1}: x \to \sqrt{x},\quad x \in \mathbb{R}^+$$

To summarise, a given function f maps the domain of f to the image-set of f.
If the reverse mapping of the image-set of f to the domain of f is a one–one
mapping it is called the inverse function and is denoted by f^{-1}.

Note that if f defines a one–one mapping then f^{-1} usually exists, but if f
defines a many–one mapping then f^{-1} does not exist.

THE GRAPHS OF FUNCTIONS AND THEIR INVERSES

The graphs of both $f(x) \equiv 2x$ and $f^{-1}(x) \equiv \tfrac{1}{2}x$ for $x \in \mathbb{R}$,

and   both $f(x) \equiv x^2$ and $f^{-1}(x) \equiv x$ for $x \in \mathbb{R}^+$

have been sketched above. Observing these we see that in each case the graph of
$f^{-1}(x)$ is the reflection in the line AB of the graph of $f(x)$.
This is true for the graph of any function f and its inverse, f^{-1}.
(The line AB is the graph of $f : x \to x$ and this function is its own inverse, i.e.
$f(x) \equiv f^{-1}(x)$ in this case.)
So if the graph representing a function is known, the graph representing its
inverse can be sketched. It must be appreciated, however, that even when a
function f does not possess an inverse, the curve representing f can be reflected
in the line AB but in this case the reflected graph does *not* correspond to f^{-1}.

EXAMPLE 3f

Given the function $f(x) \equiv 2^x,\quad x \in \mathbb{R},$ find f^{-1} as a function of x and
sketch the graph of f^{-1}.

$f(x) \equiv 2^x$ maps x to 2^x.
To find f^{-1} we have to reverse this process, i.e. map values of 2^x back to
values of x.
If $2^x \equiv w$ say, then taking logarithms to base 2 on each side gives

$$x \equiv \log_2 w$$

Hence the relationship $\qquad\qquad 2^x \to x$

can be expressed as $\qquad\qquad\qquad w \to \log_2 w$

The relationship $\quad w \to \log_2 w \quad$ is a one–one mapping for $\quad w \in \mathbb{R} \quad$ so it is a function and it reverses the mapping $\quad x \to 2^x$.

Replacing the variable w by the variable x we have

$$f(x) \equiv 2^x \quad \Rightarrow \quad f^{-1}(x) \equiv \log_2 x$$

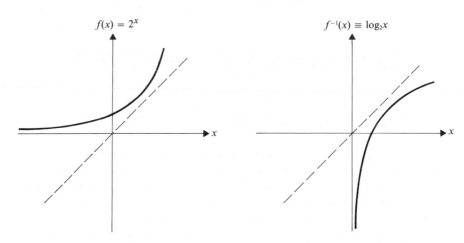

$f(x) = 2^x$

$f^{-1}(x) \equiv \log_2 x$

EXERCISE 3f

1) Determine which of the following functions have an inverse for $\quad x \in \mathbb{R}$. If f^{-1} exists, express it as a function of x.

(a) $3x$ (b) $2x+1$ (c) $x^2 - 4$ (d) $\log_5 x$

2) On the same set of axes draw a sketch graph of each function given in Question 1 with, where it exists, the inverse function.

3) State the inverse function, with its domain, of each of the functions given below.

(a) $f: x \to \frac{1}{2}x - 3, \quad x \in \mathbb{R}$ (b) $f: x \to 5x, \quad x \in \{-1, 0, 1\}$
(c) $f: x \to 5x, \quad x \in \mathbb{N}$
(d) $f: x \to (x-1)(x-3), \quad x \in \mathbb{R}, \quad x \geqslant 2$
(e) $f: x \to (x-2)(x+2), \quad x \in \mathbb{R}^+$
(f) $f: x \to 10^x, \quad x \in \mathbb{R}^+$

CHAPTER 4

COORDINATE GEOMETRY I

LOCATION OF A POINT IN A PLANE

Graphical methods lend themselves particularly well to the investigation of the geometric properties of many kinds of curves and surfaces. At this stage we will restrict ourselves to plane figures (i.e. those that can be described fully using only two dimensions). To represent any figure on a graph we need, as a start, a simple and unambiguous way of describing the position of a point.
Consider the problem of describing the location of a town, Birmingham say. There are many ways in which this can be done, but all require reference to at least one known place and known directions (called a system, or frame, of reference). Within this frame of reference, two measurements (or coordinates) will be needed to locate the town precisely.

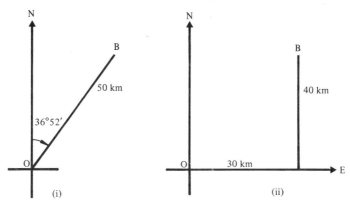

The position of B is described in two alternative ways in the diagrams above.

In (i) the system of reference is the fixed point O and the direction due N from O.

The coordinates of B are 50 km from O on a bearing 36°52′.

In (ii) the system of reference is the directions due E and due N from a fixed
point O.
The coordinates of B are 30 km E of O and 40 km N of O.

The two systems most frequently used for mathematical analysis are basically
similar to the two practical systems described above.

POLAR COORDINATES

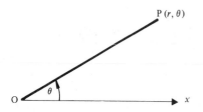

The system of reference is a fixed
point O, called the pole and a fixed
direction from O, the line Ox, called
the initial line.

The coordinates of a point P are the distance of P from O and the angle OP
makes with Ox, measured in an anticlockwise sense from Ox.
These coordinates are written as an ordered pair (r, θ), i.e. (distance, angle).

Using this system the diagram
represents the position of the point
whose coordinates are $(2, 30°)$

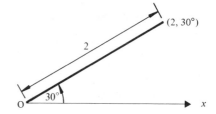

CARTESIAN COORDINATES

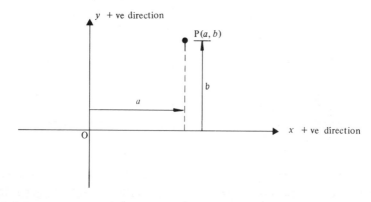

The system of reference is a fixed point O, the origin, and a pair of
perpendicular lines through O.

It is usual to draw these lines as shown in the diagram. The horizontal line is called the x-axis, the vertical line the y-axis.

The coordinates of a point P are the directed distances of P from O parallel to the axes. A positive coordinate is a distance measured in the positive direction of the axis and a negative coordinate is a distance in the opposite direction.

The coordinates are given as an ordered pair (a, b) with the x *coordinate* or *abscissa* first and the y *coordinate* or *ordinate* second.

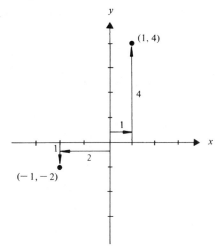

Taking the positive direction of each axis as shown, the diagram represents the points whose Cartesian coordinates are

$$(1, 4) \text{ and } (-2, -1),$$

referred to in future as the points $(1, 4)$ and $(-2, -1)$.

EXERCISE 4a

Represent on a diagram points whose polar coordinates are:

1) $(1, 45°)$ 2) $(3, 90°)$ 3) $(2, 60°)$ 4) $(1, 150°)$ 5) $(1, 270°)$

6) $(2, 200°)$ 7) $(3, 300°)$

Represent on a diagram points whose Cartesian coordinates are:

8) $(1, 1)$ 9) $(4, 2)$ 10) $(2, 4)$ 11) $(-1, 5)$ 12) $(3, -1)$ 13) $(0, 3)$

14) $(-2, -5)$ 15) $(0, 0)$

COORDINATE GEOMETRY

Coordinate geometry is the name given to the analysis, using graphical methods, of geometric properties. The properties of straight lines and many curves are most simply found using Cartesian coordinates and so this system of reference is used more frequently than any other. For this analysis we need to refer to three types of points:

(a) fixed points whose coordinates are known, e.g. the point $(4, 5)$,

(b) fixed points whose coordinates are not known numerically. These are referred to as the points $(x_1, y_1), (x_2, y_2) \ldots$ etc. or (a, b) etc.

(c) points which are not fixed (general points). A general point is referred to as the point (x, y).

It is conventional to use the letters $P, Q, R \ldots$ etc. for general points and the letters $A, B, C \ldots$ for fixed points.

In order to avoid distorting the shape of a curve when drawing it on a Cartesian plane, the two axes are graduated using identical scales.

The Length of a Line Joining Two Points

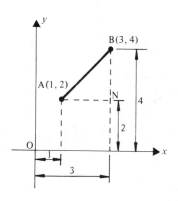

From the diagram we see that the length of the line joining $A(1, 2)$ and $B(3, 4)$ can be found using Pythagoras' Theorem where

$$AB^2 = AN^2 + BN^2$$
$$= (3 - 1)^2 + (4 - 2)^2$$
$$= 8$$

Therefore $AB = \sqrt{8} = 2\sqrt{2}$.

In general, if $A(x_1, y_1)$ and $B(x_2, y_2)$ are any two points

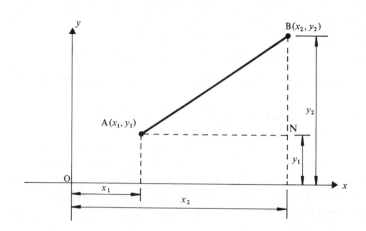

we see from the diagram, and using Pythagoras' Theorem, that

$$AB^2 = AN^2 + BN^2$$

$$= (x_2 - x_1)^2 + (y_2 - y_1)^2$$

Therefore $\quad\quad AB = \sqrt{(x_2 - x_1)^2 + (y_2 - y_1)^2}$

Note that this formula still holds when some, or all, of the coordinates are negative, e.g.

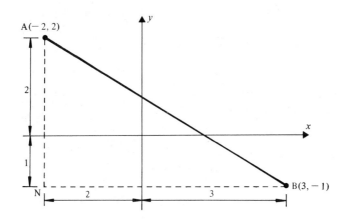

$$AN = 2 + 1 = 2 - (-1) = (y_1 - y_2)$$

Therefore $\quad\quad AN^2 = (y_1 - y_2)^2 = (y_2 - y_1)^2$

Similarly $\quad\quad NB = 2 + 3 = 3 - (-2) = (x_2 - x_1)$

Therefore $\quad\quad NB^2 = (x_2 - x_1)^2$

Therefore $\quad\quad AB^2 = (x_2 - x_1)^2 + (y_2 - y_1)^2$

Therefore *the length of the line joining* $A(x_1, y_1)$ *to* $B(x_2, y_2)$ *is given by*

$$AB = \sqrt{(x_2 - x_1)^2 + (y_2 - y_1)^2}$$

The Midpoint of the Straight Line Joining Two Given Points

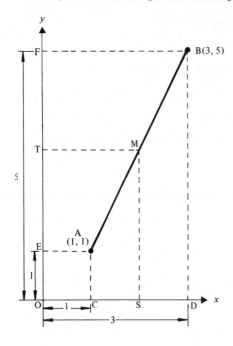

From the diagram we see that if M is the midpoint of AB, then S is the midpoint of CD, Therefore the x coordinate of M is given by OS where

$$OS = OC + \tfrac{1}{2}CD$$
$$= 1 + \tfrac{1}{2}(3 - 1)$$
$$= \tfrac{1}{2}(3 + 1) = 2$$

Similarly T is the midpoint of EF, so the y coordinate of M is given by OT, where

$$OT = OE + \tfrac{1}{2}EF$$
$$= 1 + \tfrac{1}{2}(5 - 1)$$
$$= \tfrac{1}{2}(5 + 1) = 3$$

Therefore M is the point $(2, 3)$.

In general, if M is the midpoint of the line joining $A(x_1, y_1)$ and $B(x_2, y_2)$ then we see from the diagram that

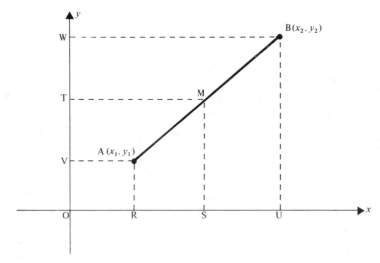

the x coordinate of M is OS, where

$$OS = OR + \tfrac{1}{2}RU$$

$$= x_1 + \tfrac{1}{2}(x_2 - x_1)$$

$$= \tfrac{1}{2}(x_1 + x_2)$$

and the y coordinate of M is OT, where

$$OT = OV + \tfrac{1}{2}VW$$

$$= y_1 + \tfrac{1}{2}(y_2 - y_1)$$

$$= \tfrac{1}{2}(y_1 + y_2)$$

Note that this formula also holds when some, or all, of the coordinates are negative:

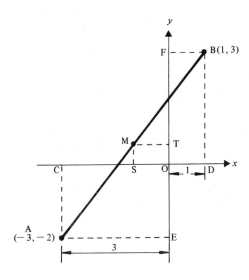

If M is the midpoint of AB, then S is the midpoint of CD. Therefore the x coordinate of M is $-$ OS where

$$OS = OC - CS$$

$$= OC - \tfrac{1}{2}CD$$

$$= 3 - \tfrac{1}{2}(3 + 1)$$

$$= \tfrac{1}{2}(3 - 1)$$

Therefore $- OS = \tfrac{1}{2}(1 - 3)$
i.e. the x coordinate of M is the arithmetic mean of the x coordinates of A and B.

Similarly the y coordinate of M is OT, where

$$OT = OF - \tfrac{1}{2}FE$$

$$= 3 - \tfrac{1}{2}(3 + 2)$$

$$= \tfrac{1}{2}(3 - 2)$$

i.e. the y coordinate of M is the arithmetic mean of the y coordinates of A and B.
Therefore M is the point $(-1, \tfrac{1}{2})$

Therefore, *if M is the midpoint of the line joining $A(x_1, y_1)$ and $B(x_2, y_2)$, the coordinates of M are*

$$\left[\tfrac{1}{2}(x_1 + x_2), \tfrac{1}{2}(y_1 + y_2) \right]$$

EXERCISE 4b

1) Find the length of the line joining the following pairs of points:
(a) $(1, 2), (4, 6)$ (b) $(3, 1), (2, 0)$ (c) $(4, 2), (2, 5)$
(d) $(-1, 4), (2, 6)$ (e) $(0, 0), (-1, -2)$ (f) $(-1, -4), (-3, -2)$

2) Find the coordinates of the midpoints of the lines joining the pairs of points given in 1.

3) Find the length of the line from the origin to the point $(7, 4)$.

4) Show, using Pythagoras' Theorem, that the lines joining $A(1, 6)$, $B(-1, 4)$ and $C(2, 1)$ form a right angled triangle.

5) Show that $\triangle ABC$ is isosceles where A, B and C are the points $(7, 3)$, $(-4, 1)$ and $(-3, -2)$.

6) Find the midpoint of the base of $\triangle ABC$ in No. 5. Hence find the area of $\triangle ABC$.

7) Prove that the lines OA and OB are perpendicular where A, B are the points $(4, 3)$, $(3, -4)$ respectively.

8) In the triangle ABC, A, B, and C are the points $(0, 2)$, $(1, 5)$ and $(-1, 4)$. Find the coordinates of the point D such that AD is a median and find the length of this median.

9) A, B and M are three points such that M is the midpoint of AB. The coordinates of A and M are $(5, 7)$ and $(0, 2)$ respectively. Find the coordinates of B.

GRADIENT

The gradient of a straight line is a measure of its slope with respect to the *x*-axis, and is defined as *the increase in the y coordinate divided by the increase in the x coordinate between one point on the line and another point on the line.*

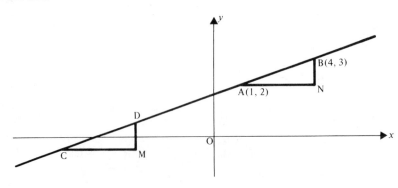

Consider the straight line passing through the points A(1, 2) and B(4, 3).
From A to B

the increase in the y coordinate is 1

the increase in the x coordinate is 3

Therefore the gradient of AB is $\frac{1}{3}$.
Note that NB measures the increase in the y coordinate
and AN measures the increase in the x coordinate

so the gradient of $AB = \dfrac{NB}{AN}$.

If C and D are any two other points on the same line, then as $\triangle ABN$ is
similar to $\triangle CDM$,

$$\frac{NB}{AN} = \frac{MD}{CM} = \frac{1}{3}$$

i.e. the gradient of a line may be found from *any* two points on the line.

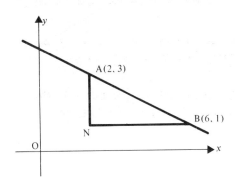

Consider the line passing through the
points A(2, 3) and B(6, 1)

From A to B,

the y coordinate *decreases by* 2, i.e. *increases by* -2

the x coordinate *increases by* 4

Therefore the gradient of the line AB is $\dfrac{-2}{4} = -\dfrac{1}{2}$.

Note that from B to A the gradient is

$$\frac{\text{increase in } y\ (B \to A)}{\text{increase in } x\ (B \to A)} = \frac{2}{-4} = -\frac{1}{2}$$

i.e. it does not matter in which order the two points are considered, provided
they are considered in the *same* order when calculating the increases in both
x and y.

From the two examples considered we see that the gradient of a line may be positive or negative.

From the first example we see that a *positive gradient* indicates an 'uphill' slope with respect to the positive direction of the x-axis, i.e. a line which makes an acute angle with the positive sense of the x-axis.

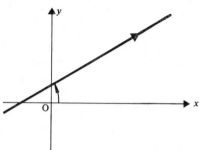

From the second example we see that a *negative gradient* indicates a 'down-hill' slope with respect to the positive direction of the x-axis, i.e. a line which makes an obtuse angle with the positive sense of the x-axis.

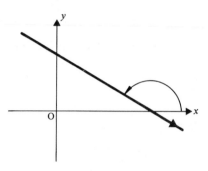

Note that in both cases:

(a)

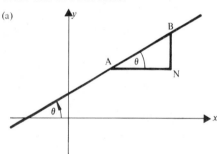

$$\text{gradient of } AB = \frac{BN}{AN} = \tan \theta$$

(b)

$$\text{gradient of } AB = \frac{-AN}{BN} = -\tan \alpha$$

$$= \tan \theta$$

Therefore the *gradient of a line* may also be defined as *the tangent of the angle that it makes with the positive direction of the x-axis*, where the angle is measured in an anticlockwise sense.
In general

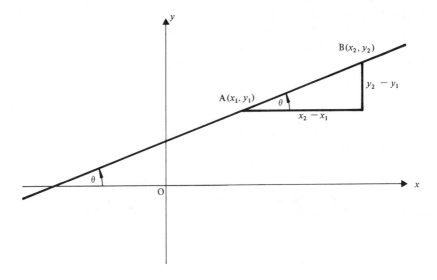

The gradient of the line passing through $A(x_1, y_1)$ and $B(x_2, y_2)$ is

$$\frac{\text{the increase in the } y \text{ coordinate}}{\text{the increase in the } x \text{ coordinate}} = \frac{y_2 - y_1}{x_2 - x_1} = \tan \theta$$

As the gradient of a straight line is the increase in y divided by the increase in x from one point to another point on the line, *gradient measures the increase in y per unit increase in x, i.e. the rate of increase of y with respect to x.*

Parallel Lines

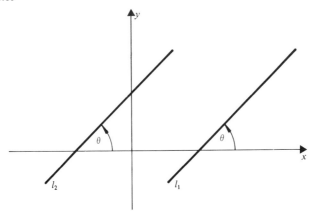

If l_1 and l_2 are parallel lines, they are equally inclined to the positive direction of the x-axis (corresponding angles), so $\tan \theta$ is the gradient of both l_1 and l_2, i.e. *parallel lines have equal gradients.*

Perpendicular Lines

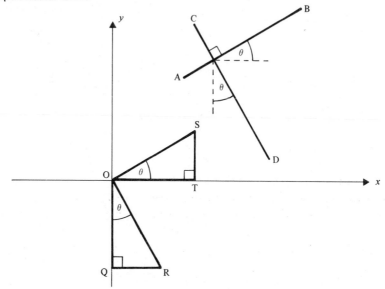

Consider the perpendicular lines AB and CD whose gradients are m_1 and m_2 respectively.

If AB makes an angle θ with the x-axis
then CD makes an angle θ with the y-axis.
If OS is drawn parallel to AB and OR is drawn parallel to CD then triangles OST and ORQ are similar,

therefore
$$\frac{ST}{OT} = \frac{QR}{OQ}$$

But gradient of $OS = \dfrac{ST}{OT} = $ gradient of $AB = m_1$

and gradient of $OR = -\dfrac{OQ}{QR} = $ gradient of $CD = m_2$.

Therefore $\quad m_1 = -\dfrac{1}{m_2} \quad$ or $\quad m_1 m_2 = -1$

i.e. *the product of the gradients of perpendicular lines is* -1, or, if one line has a gradient m, the gradient of any line perpendicular to it is $-\dfrac{1}{m}$.

EXAMPLE 4c

Show that the point $(-\frac{6}{7}, 0)$ is on the median through A of triangle ABC

where A, B, C are the points $(2, 4), (-2, 3), (1, -2)$. If, also, the point (a, b) is on this median, find a relationship between a and b.

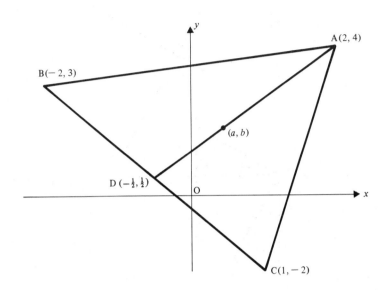

If AD is the median through A, D is the midpoint of BC. i.e. D is the point $(-\frac{1}{2}, \frac{1}{2})$.
If $E(-\frac{6}{7}, 0)$ is on AD then

gradient AD and gradient AE should be equal.

Now gradient $AD = \dfrac{4 - \frac{1}{2}}{2 + \frac{1}{2}} = \dfrac{7}{5}$

and gradient $AE = \dfrac{4 - 0}{2 + \frac{6}{7}} = \dfrac{7}{5}$

Therefore E is on the median AD.

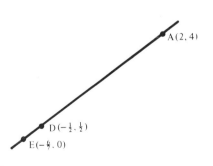

A condition that $P(a, b)$ should be on AD is
gradient AP = gradient AD.

Gradient AP $= \dfrac{4-b}{2-a}$.

Therefore P lies on AD if

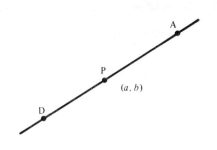

$$\dfrac{4-b}{2-a} = \dfrac{7}{5}$$

$\Rightarrow \qquad 20 - 5b = 14 - 7a$

$\Rightarrow \quad 7a - 5b + 6 = 0$

EXERCISE 4c

1) Find the gradients of the lines passing through the following pairs of points:
(a) $(0,0),(1,3)$ (b) $(1,4),(3,7)$ (c) $(5,4),(2,3)$
(d) $(-1,4),(3,7)$ (e) $(-1,-3),(-2,1)$ (f) $(-1,-6),(0,0)$
(g) $(-2,5),(1,-2)$ (h) $(3,-2),(-1,4)$ (i) $(h,k),(0,0)$

2) Write down the gradients of the lines which are inclined at the following angles to the positive direction of the x-axis.
(a) 45° (b) 135° (c) 0 (d) 90°

3) Determine, by comparing gradients, whether the three points whose coordinates are given, are collinear (i.e. lie on the same straight line).
(a) $(0,-1),\quad (1,1),\quad (2,3)$
(b) $(0,2),\qquad (2,5),\quad (3,7)$
(c) $(-1,4),\quad (2,1),\quad (-2,5)$

4) Determine whether AB is parallel or perpendicular to CD where:
(a) $A(0,-1),\quad B(1,1),\qquad C(1,5),\qquad D(-1,1)$
(b) $A(1,1),\qquad B(3,2),\qquad C(-1,1),\qquad D(0,-1)$
(c) $A(3,3),\qquad B(-3,1),\quad C(-1,-1),\quad D(1,-7)$
(d) $A(2,6),\qquad B(-1,-9),\quad C(2,11),\qquad D(0,1)$

Questions 5–13 are miscellaneous problems on coordinate geometry. A clear, reasonably accurate diagram showing all the given information will usually suggest the most direct method for answering a particular problem.

5) $A(1,3),\ B(5,7),\ C(4,8),\ D(a,b)$ form a rectangle ABCD. Find a and b.

6) The points $A(1,5),\ B(4,-1)$ and $C(-2,-4)$ form triangle ABC. Prove that the triangle is right angled, and find its area.

7) ABCD is a quadrilateral where A, B, C and D are the points $(3,-1)$, $(6,0),(7,3)$ and $(4,2)$. Prove that the diagonals bisect each other at right angles and hence find the area of ABCD.

8) The vertices of a $\triangle$ABC are at the points $A(a, 0), B(0, b), C(c, d)$. If $\angle B = 90°$, find a relationship between a, b, c and d.

9) A circle, radius two units, centre the origin, cuts the x-axis at A and B and cuts the positive y-axis at C. Prove that AB subtends a right angle at C.

10) In No. 9, if $D(a, b)$ is a point on the circumference of the circle, find a relationship between a and b.

11) Prove that the point $(5, -1)$ is on the perpendicular bisector of the line joining $A(1, -3)$ to $B(3, 3)$. If $D(h, k)$ is another point on this perpendicular bisector, find a relationship between h and k.

12) The point $A(5, 0)$ lies on a circle, centre the origin. Find the radius of this circle. Prove that the points $B(4, -3), C(-3, 4)$ are also on the circumference of this circle. If this circle cuts the negative x-axis at D, find the area of the quadrilateral ABDC.

13) A point $P(a, b)$ is equidistant from the y-axis and from the point $(4, 0)$. Find a relationship between a and b.

THE MEANING OF EQUATIONS

The Cartesian system of reference provides a means of defining the position of any point in a plane. This plane is called the xy plane.

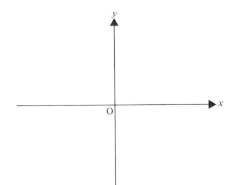

In general, x and y are independent variables (i.e. they can each take any value, independently of the value of the other) unless some restriction is placed on them.

Consider the set of points for which $x = 2$.

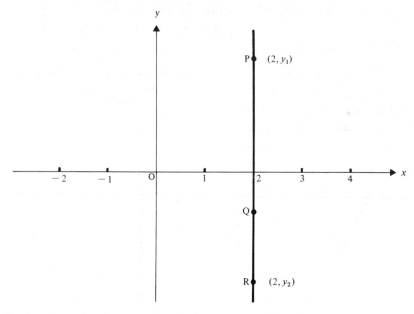

As the value of y is not restricted, these points all lie on the line parallel to the y-axis, passing through P, Q and R as shown.

So the equation $x = 2$ defines the line through P, Q, R in the xy plane. $x = 2$ is called the equation of the line through P, Q, R which is briefly referred to as 'the line $x = 2$'.

Now consider the set of points for which $x > 2$.

All points to the right of the line $x = 2$ have an x coordinate which is greater than 2.

So the inequality $x > 2$ defines the shaded region of the xy plane, shown in Fig. 1.

Similarly the inequality $x < 2$ defines the region of the xy plane left unshaded in Fig. 1.

Note. The region defined by $x > 2$ does not include the line $x = 2$. When a region does not include points on the boundary lines these are drawn as broken lines. When a region does include points on the boundary lines these are drawn as solid lines.

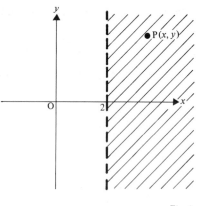

Fig. 1

Now consider the function

$$f(x) \equiv (x - 3)(x + 1)$$

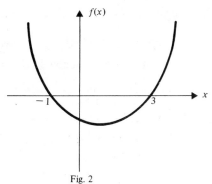

Fig. 2 shows the curve representing this function and Fig. 3 shows the same curve drawn on the xy plane.

Fig. 2

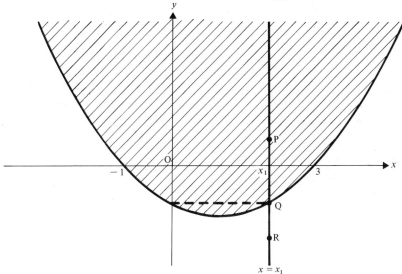

$x = x_1$

If P, Q and R are points on the line $x = x_1$ as shown, the y coordinate of Q is $f(x_1)$,

i.e. at Q, $\qquad\qquad y = (x_1 - 3)(x_1 + 1)$

the y coordinate of P is greater than the y coordinate of Q,

i.e. at P, $\qquad\qquad y > (x_1 - 3)(x_1 + 1)$

the y coordinate of R is less than the y coordinate of Q,

i.e. at R, $\qquad\qquad y < (x_1 - 3)(x_1 + 1)$

This argument applies for all values of x.
Therefore the inequality $y > (x - 3)(x + 1)$ defines the set of points contained in the shaded region of the xy plane,
and the inequality $y < (x - 3)(x + 1)$ defines the unshaded region of the xy plane.

Whereas *only for points on the curve is* $y = (x - 3)(x + 1)$

$$y = (x - 3)(x + 1) \quad \text{is called the equation of the curve}$$

which is often referred to simply as *the curve* $y = (x - 3)(x + 1)$.

Note. An equation such as $x^2 - 7x + 3 = 0$ contains only one variable and its solution comprises a finite set of values of x. Equations containing two variables, such as $y = (x - 3)(x + 1)$, have as their solution, an infinite set of ordered pairs (x, y) and the elements of such a solution set are definable as follows.

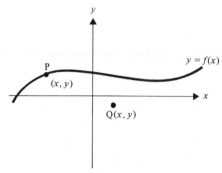

If A is the solution set of the equation $y = f(x)$, the elements of A are the coordinates, (x, y), of all points on the curve $y = f(x)$. Conversely the coordinates (x, y) of points *not* on the curve are *not* elements of A.

Thus $P(x, y) \in A$, but $Q(x, y) \notin A$.

In general if $f(x)$ is any function of x, then in the xy plane

$$y = f(x) \quad \text{defines a curve}$$

and is called the equation of that curve.

$$y > f(x) \quad \text{and} \quad y < f(x) \quad \text{define regions of the plane}$$

EXAMPLES 4d

1) Determine whether the points $(5, 11)$ and $(-2, -20)$ are on the curve $y = (x - 4)(x + 6)$.

Substituting 11 for y in L.H.S. of the equation of the curve gives

$$\text{L.H.S} = 11$$

Substituting 5 for x in R.H.S. gives

$$\text{R.H.S} = (5 - 4)(5 + 6) = 11$$

Therefore L.H.S = R.H.S when $x = 5$ and $y = 11$.
Therefore $(5, 11)$ is a member of the solution set of $y = (x - 4)(x + 6)$ and is therefore on the curve.
Substituting -20 for y in L.H.S. gives

$$\text{L.H.S} = -20$$

Substituting -2 for x in R.H.S. gives

$$\text{R.H.S} = (-2-4)(-2+6) = -24$$

So L.H.S $>$ R.H.S   when $x = -2$ and $y = -20$.
Therefore $(-2, -20)$ is not a point on the curve.

2) Draw a sketch to show the region of the xy plane defined by the following inequalities $0 \leqslant x \leqslant 2$, $y \geqslant 0$, $y \leqslant x^2$.

The relationship $0 \leqslant x \leqslant 2$ contains two inequalities, viz. $x \geqslant 0$, $x \leqslant 2$, which must be considered separately,

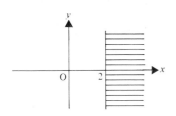 $x \geqslant 0$ indicates the line $x = 0$ (i.e. the y-axis) and the region to the right of the y-axis, $(x > 0)$.   We eliminate the region to the left of the y-axis by shading it out, as it does not satisfy $x \geqslant 0$,   i.e. we shade the *unwanted* region.

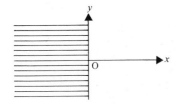 $x \leqslant 2$ indicates the line $x = 2$ and the region to the left of this line, so we shade out the region to the right of this line that does not satisfy $x \leqslant 2$.

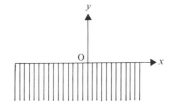 $y \geqslant 0$ is the line $y = 0$ and the region above the x-axis, so we shade out the region below the x-axis.

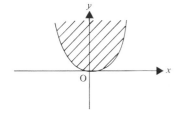 $y \leqslant x^2$ is the curve $y = x^2$ and the region below it, so we shade out the region above $y = x^2$.

Combining these four diagrams gives the diagram below

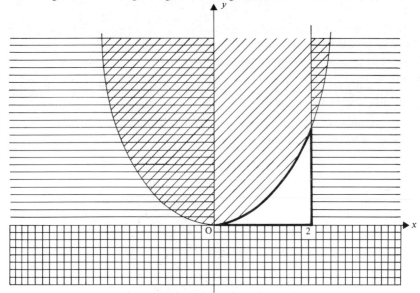

Therefore the unshaded region, including the boundary lines, is the set of points that satisfies all the given inequalities.

EXERCISE 4d

Draw a sketch of the curve whose equation is:

1) $y = x^2 - 3x + 4$ 2) $y = 1/x$ 3) $y = x(1-x)$ 4) $y = 2^{-x}$

Determine whether the given point lies on the given curve:

5) $(0, 1)$, $y = x^2 - 2$ 6) $(2, -3)$, $y = 3 - x - x^2$

7) $(-4, -0.2)$, $y = \dfrac{1}{x}$ 8) $(3, 2)$, $y^2 = x - 1$

9) $(-3, -6)$, $y + x^2 = 3$ 10) $(1, -2)$, $\dfrac{1}{x} + \dfrac{1}{y} = \dfrac{1}{2}$

Draw a sketch showing the region of the xy plane defined by the following inequalities:

11) $x > 3$ 12) $y < 2$ 13) $x < 0$ 14) $y > (x + 3)(x - 2)$

15) $0 < x < 2$ 16) $-1 < y < 3$

17) $y < 2$ and $y > (x - 2)(x + 2)$ 18) $x \geqslant 2$, $y \leqslant 4$

19) $1 \leqslant x \leqslant 3$, $y \leqslant \dfrac{1}{x}$ 20) $2 \geqslant x \geqslant 0$, $y \geqslant -7$, $y \leqslant (x - 2)(x + 5)$

21) $y \geqslant 0$, $1 \leqslant x \leqslant 4$, $y \leqslant \dfrac{1}{x}$

THE EQUATION OF A PARTICULAR CURVE

So far in our work on functions and graphs we have begun with a function
and have deduced from its properties the curve that represents it.
The reverse process in which we begin with a curve, given geometrically, and
deduce its equation, will now be examined.
Consider, for example, the circle whose centre is at the point $(4, 2)$ and whose
radius is 2.

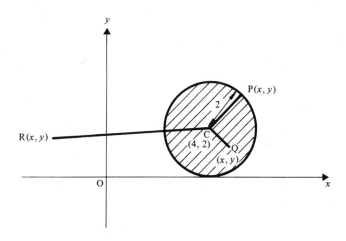

Any point P on the circumference of this circle is such that $PC = 2$.
Any point Q inside the circle satisfies the inequality $CQ < 2$.
Any point R outside the circle satisfies the inequality $CR > 2$.
The distance of any point (x, y) from C is given by

$$\sqrt{(x - 4)^2 + (y - 2)^2}$$

Therefore the coordinates (x, y) of P must satisfy the equation

$$\sqrt{(x - 4)^2 + (y - 2)^2} = 2$$

or $$(x - 4)^2 + (y - 2)^2 = 4$$

Therefore this equation defines the set of points on the circumference of the
circle and so is the equation of the circle.
Similarly the coordinates (x, y) of Q satisfy the inequality

$$(x - 4)^2 + (y - 2)^2 < 4$$

So this inequality defines the region inside the circle
and the inequality

$$(x - 4)^2 + (y - 2)^2 > 4$$

defines the region outside the circle.

EQUATION OF A STRAIGHT LINE

Straight lines play an important part in any geometric analysis and we will now concentrate our attention on these. We will return to the problem of finding the equations of particular curves later in the book.

A straight line may be defined in many ways, e.g.:

1) a line which passes through the origin and has a gradient of $\frac{1}{2}$,

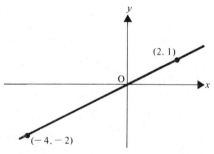

2) a line which passes through the points $(2, 1)$ and $(-4, -2)$.

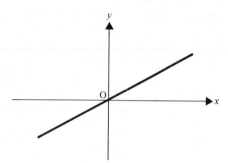

and there are many other ways of defining a straight line.

1) The equation of this line can be found as follows:

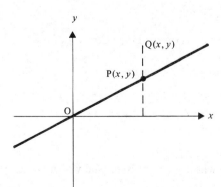

If $P(x, y)$ is any point on the line, the gradient of $OP = \frac{1}{2}$.

The gradient of OP is given by $\dfrac{y-0}{x-0} = \dfrac{y}{x}$

Therefore the coordinates of P satisfy the equation

$$\frac{y}{x} = \frac{1}{2} \quad \text{or} \quad 2y = x$$

Therefore $2y = x$ is the equation of the line.

Note. For any point $Q(x, y)$ above P, $y > \frac{1}{2}x \Rightarrow 2y > x$.
Therefore the inequality $2y > x$ defines the region above the line.
Similarly $2y < x$ defines the region below the line.

2) To obtain the equation of this line we know that
$P(x, y)$ is a point on the line $\iff$ gradient of PA (or PB) = gradient of AB.

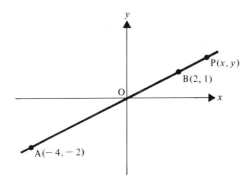

Gradient of PA $= \dfrac{y-1}{x-2}$

Gradient of AB $= \dfrac{1-(-2)}{2-(-4)} = \dfrac{1}{2}$

Therefore the coordinates of P satisfy the equation

$$\frac{y-1}{x-2} = \frac{1}{2}$$

or $\qquad\qquad\qquad\qquad 2y = x$

Note. It is conventional to use integers for coefficients whenever possible.
Note also that these apparently different definitions give the same line, and although there are other ways of defining a straight line we will concentrate on these two, which are the commonest.

Consider the more general case of the line whose gradient is m and which passes through the origin.

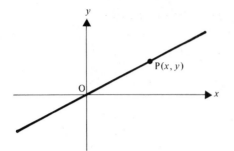

If $P(x,y)$ is any point on this line, then the gradient of OP is m.

Therefore the coordinates of P satisfy the equation

$$\frac{y}{x} = m$$

or $$y = mx$$

Generalising even further to cover any straight line, consider the line whose gradient is m and which cuts the y-axis at a directed distance c from the origin.

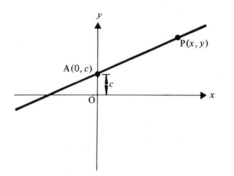

Note: c is called the *intercept* on the y-axis

$P(x, y)$ is any point on the line $\Longleftrightarrow$ gradient of AP is m.
Therefore the coordinates of P satisfy the equation

$$\frac{y - c}{x - 0} = m$$

or $$y = mx + c$$

The equation above is called the standard form for the equation of a straight line. It follows that

(a) an equation of the form

$$y = mx + c$$

represents a straight line with gradient m and intercept c on the y-axis.

(b) any equation involving a linear relationship between x and y is the equation of a straight line, i.e.

$$ax + by + c = 0$$

where a, b and c are constants, is the equation of a straight line.

EXAMPLES 4e

1) Write down the gradient of the line $3x - 4y + 2 = 0$ and find the equation of the line through the origin which is perpendicular to the given line.

Writing $3x - 4y + 2 = 0$ in standard form gives

$$y = \frac{3}{4}x + \frac{1}{2}$$

Comparing with $\qquad\qquad y = mx + c$

we see that the gradient, m, of the line is $\frac{3}{4}$
So the gradient of the perpendicular line is $-\frac{4}{3}$ $(= -1/m)$
The required line passes through the origin (i.e. has zero intercept on the y-axis).
Therefore its equation is

$$y = -\frac{4}{3}x + 0 \qquad (y = mx + c)$$

$\Rightarrow \qquad\qquad 3y + 4x = 0$

2) Sketch the line $x - 2y + 3 = 0$.

This line can be located accurately in the xy plane when we know two points on the line. As the intercepts on the axes can be found by inspection (i.e. $x = 0 \Rightarrow y = \frac{3}{2}$ and $y = 0 \Rightarrow x = -3$), we shall use these to place the line on the xy plane.

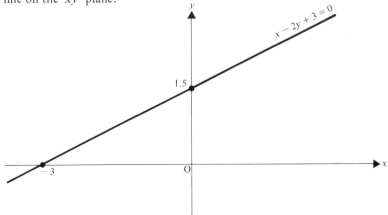

EXERCISE 4e

1) Write down the equation of the line passing through the origin and with gradient

(a) 2   (b) -1 (c) $\frac{1}{3}$ (d) $-\frac{1}{4}$ (e) 0   (f) ∞

Draw a sketch showing these lines on the same pair of axes.

2) Write down the equation of the line passing through the given point and with the given gradient.

(a) $(0, 1), \frac{1}{2}$ (b) $(0, 0), \frac{1}{2}$ (c) $(-1, -4), \frac{1}{2}$

Sketch these lines on the same pair of axes.

3) Write down the equation of the line passing through the given points.

(a) $(0, 0), (2, 1)$ (b) $(1, 4), (3, 0)$ (c) $(2, 0), (0, 4)$

(d) $(-1, 3), (-4, -3)$

4) Write down the inequality which defines the region:

(a) above the line through the origin with gradient 1,

(b) below the line through $(1, 2)$ and $(0, 4)$,

(c) above the line $x + y - 2 = 0$,

(d) below the line $2x - y + 4 = 0$.

5) Write down the equation of the line passing through the origin and perpendicular to:

(a) $3x + 2y - 4 = 0$ (b) $x - 2y + 3 = 0$

6) Write down the equation of the line passing through the given point and perpendicular to the given line.

(a) $(2, 1), 3x + y - 2 = 0$ (b) $(-1, -2), 2x - 3y + 6 = 0$

7) Write down the equation of the line passing through $(3, -2)$ and parallel to:

(a) $5x - y + 3 = 0$ (b) $x + 7y - 5 = 0$

The Equation of a Line with Gradient m passing through the Point (x_1, y_1)

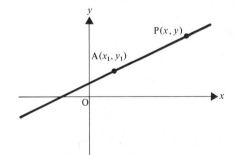

If $P(x, y)$ is any point on the line, the gradient of AP is m.

Therefore the coordinates of P satisfy the equation

$$\frac{y - y_1}{x - x_1} = m$$

$\Rightarrow$
$$y - y_1 = m(x - x_1) \qquad\qquad\qquad [1]$$

EXAMPLES 4f

1) Find the equation of the line with gradient $-\frac{1}{3}$, passing through $(2, -1)$.

Substituting $-\frac{1}{3}$ for m, 2 for x_1 and -1 for y_1 in [1] gives the equation of the line as

$$y - (-1) = -\tfrac{1}{3}(x - 2)$$

or
$$x + 3y + 1 = 0$$

Alternatively the equation can be found as follows:
As any straight line has an equation $y = mx + c$,
the equation of this line can be written

$$y = -\tfrac{1}{3}x + c$$

As the point $(2, -1)$ lies on this line, its coordinates satisfy the equation.

i.e.
$$-1 = -\tfrac{1}{3}(2) + c$$

$\Rightarrow$
$$c = -\tfrac{1}{3}$$

Therefore the equation is
$$y = -\tfrac{1}{3}x - \tfrac{1}{3}$$

or
$$x + 3y + 1 = 0$$

Note. The worked examples in the text must necessarily contain a lot of explanation but this should not mislead the reader into thinking that his solutions should be equally long. The temptation to 'overwork' a problem should be avoided, particularly in the case of coordinate geometry problems which are basically simple. The reader will find that, with a little practice, either of the methods illustrated above will enable the required equation of a line to be written down directly.

The Equation of the Line passing through (x_1, y_1) and (x_2, y_2)

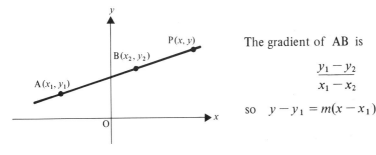

The gradient of AB is

$$\frac{y_1 - y_2}{x_1 - x_2}$$

so $y - y_1 = m(x - x_1)$

gives

$$y - y_1 = \frac{y_1 - y_2}{x_1 - x_2}(x - x_1)$$

e.g. the line through $(1, -2)$ and $(3, 5)$ has equation

$$y - 5 = \frac{5 - (-2)}{3 - 1}(x - 3) \quad \Rightarrow \quad 7x - 2y - 11 = 0$$

INTERSECTION

If two curves cut at a point A, A is called a point of intersection.

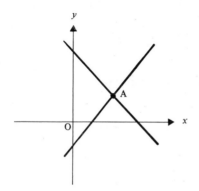

If A is the point of intersection of the lines

$$y - 3x + 1 = 0 \qquad [1]$$

and $\quad y + x - 2 = 0 \qquad [2]$

then the coordinates of A satisfy both equations [1] and [2]. So A can be found by solving these equations simultaneously.

$$[2] - [1] \quad \Rightarrow \quad 4x - 3 = 0 \quad \Rightarrow \quad x = \tfrac{3}{4}, \ y = \tfrac{5}{4}$$

Therefore $(\tfrac{3}{4}, \tfrac{5}{4})$ is the point of intersection.

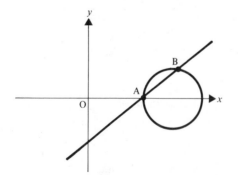

If A and B are the points of intersection of the circle

$$x^2 + y^2 - 3x + 2 = 0 \quad [1]$$

and the line $\qquad y = x - 1 \quad [2]$

then the coordinates of both A and B will satisfy both equations [1] and [2].

Solving these equations simultaneously by substituting $x - 1$ for y in [1] we have

$$x^2 + (x - 1)^2 - 3x + 2 = 0$$

$$\Rightarrow \qquad 2x^2 - 5x + 3 = 0$$

$$\Rightarrow \qquad (2x - 3)(x - 1) = 0$$

$$\Rightarrow \qquad x = \tfrac{3}{2} \quad \text{or} \quad 1.$$

Substituting $\tfrac{3}{2}$ and 1 for x in [2] gives $y = \tfrac{1}{2}$ and 0.
Therefore A and B are the points $(\tfrac{3}{2}, \tfrac{1}{2}), (1, 0)$.

In general the coordinates of the points of intersection of two curves $y = f(x)$ and $y = g(x)$ can be found from the simultaneous solution of the equations $y = f(x)$ and $y = g(x)$.

EXAMPLES 4f continued

2) Find the equation of the line through $(1, 2)$ which is perpendicular to the line $3x - 7y + 2 = 0$.

Writing $3x - 7y + 2 = 0$ in standard form gives $y = \tfrac{3}{7}x + \tfrac{1}{2}$
showing that the given line has a gradient of $\tfrac{3}{7}$.
So the required line has gradient $-\tfrac{7}{3}$ and passes through $(1, 2)$.
Using $y - y_1 = m(x - x_1)$ gives its equation as

$$y - 2 = -\tfrac{7}{3}(x - 1)$$

or $\qquad\qquad\qquad\qquad 7x + 3y - 13 = 0$

Note that the line perpendicular to $\qquad 3x - 7y + 2 = 0$

has an equation $\qquad\qquad\qquad\qquad 7x + 3y - 13 = 0$

i.e. the coefficients of x and y have been transposed and the sign between the x and y terms has changed.

In fact, given the line l with equation $ax + by + c = 0$, any line perpendicular to l has equation $bx - ay + k = 0$ and this property of perpendicular lines can be used to shorten the working of problems.

3) A, B and C are the points $(0, 4), (2, 3)$ and $(-2, -1)$.
Find the circumcentre of triangle ABC.

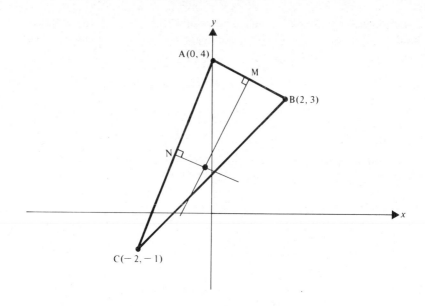

The circumcentre of a triangle is the point of intersection of the perpendicular bisectors of its sides.

AC has gradient $$\frac{4-(-1)}{0-(-2)} = \frac{5}{2}$$

and its midpoint is $$\left(\frac{0-2}{2}, \frac{4-1}{2}\right) \Rightarrow \left(-1, \frac{3}{2}\right)$$

Therefore the perpendicular bisector of AC has gradient $-\frac{2}{5}$ and passes through $(-1, \frac{3}{2})$

Hence its equation is $$y = -\tfrac{2}{5}x + \tfrac{11}{10}$$

i.e. $$4x + 10y - 11 = 0 \qquad\qquad [1]$$

Similarly the gradient of AB is $\dfrac{4-3}{0-2} = -\dfrac{1}{2}$ and its midpoint is $\left(1, \dfrac{7}{2}\right)$.

Therefore the perpendicular bisector of AB has gradient $+2$ and passes through $(1, \frac{7}{2})$

and its equation is $$y = 2x + \tfrac{3}{2}$$

i.e. $$4x - 2y + 3 = 0 \qquad\qquad [2]$$

Solving equations [1] and [2] simultaneously gives

$$12y - 14 = 0 \;\Rightarrow\; y = \tfrac{7}{6}, \;\; x = -\tfrac{1}{6}$$

Therefore the circumcentre of triangle ABC is the point $(-\frac{1}{6}, \frac{7}{6})$.

EXERCISE 4f

1) Find the equation of the line perpendicular to $x + 2y + 3 = 0$ and through the point $(5, 2)$.

2) Find the equation of the line through the origin which is parallel to $x - y + 2 = 0$.

3) Find the equation of the line joining the points:
(a) $(0, 1), (2, 4)$ (b) $(-1, 2), (1, 5)$ (c) $(3, -1), (3, 2)$

4) Determine which of the following pairs of lines are perpendicular:
(a) $x - 2y + 4 = 0$, $2x + y - 3 = 0$
(b) $x + 3y - 6 = 0$, $3x + y + 2 = 0$
(c) $x + 3y - 2 = 0$, $y = 3x + 2$
(d) $y + 2x + 1 = 0$, $x = 2y - 4$

5) Find the coordinates of the points of intersection of the pairs of lines given in Question 4.

6) Find the equations of the perpendicular bisectors of the lines joining:
(a) $(0, 0), (2, 4)$ (b) $(3, -1), (-5, 2)$ (c) $(5, -1), (0, 7)$

7) The line $4x - 5y + 20 = 0$ cuts the x-axis at A and the y-axis at B. Find the equation of the median through O of triangle OAB.

8) Find the equation of the altitude through O of triangle OAB defined in Question 7.

9) The sides of a triangle are the lines $y = 0$, $x - 3y + 5 = 0$ and $2x + y - 7 = 0$. Find the coordinates of the vertices of the triangle.

10) Find the equation of the perpendicular from the point $A(5, 3)$ to the line $2x - y + 4 = 0$. Hence find the distance of A from the line.

SUMMARY

If A and B are the points (x_1, y_1) and (x_2, y_2)

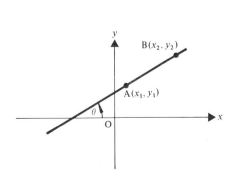

The *length* of AB is
$$\sqrt{(x_1 - x_2)^2 + (y_1 - y_2)^2}$$

The *midpoint* of AB is the point
$$\frac{x_1 + x_2}{2}, \frac{y_1 + y_2}{2}$$

The *gradient* of AB is
$$\frac{y_1 - y_2}{x_1 - x_2} = \tan \theta$$

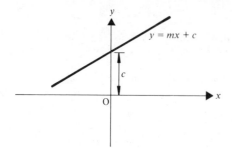

The *equation* $y = mx + c$ defines the straight line with gradient m and intercept c on the y-axis.

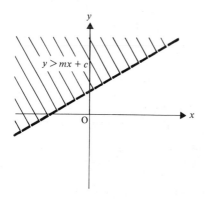

The inequality $y > mx + c$ defines the region of the xy plane above the line $y = mx + c$.

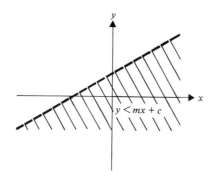

The inequality $y < mx + c$ defines the region below the line $y = mx + c$.

If lines l_1, l_2 have equations $y = m_1 x + c_1$, $y = m_2 x + c_2$ then
l_1 and l_2 are *parallel* if $m_1 = m_2$
l_1 and l_2 are *perpendicular* if $m_1 m_2 = -1$

The equation of any line perpendicular to $ax + by + c = 0$ is of the form $bx - ay + k = 0$.

MULTIPLE CHOICE EXERCISE 4

(Instructions for answering these questions are given on p. xii)

TYPE I

1) The length of the line joining $(3, -4)$ to $(-7, 2)$ is:
(a) $2\sqrt{13}$ (b) 16 (c) $2\sqrt{34}$ (d) $2\sqrt{5}$ (e) 6.

2) The midpoint of the line joining $(-1, -3)$ to $(3, -5)$ is:
(a) $(1, 1)$ (b) $(0, 0)$ (c) $(2, -8)$ (d) $(1, -4)$ (e) $(1, -1)$.

3) The gradient of the line joining $(1, 4)$ and $(-2, 5)$ is:
(a) $\frac{1}{3}$ (b) $-\frac{1}{3}$ (c) 3 (d) -3 (e) 1.3.

4) The gradient of the line perpendicular to the join of $(-1, 5)$ and $(2, -3)$ is:
(a) $\frac{3}{8}$ (b) $-2\frac{2}{3}$ (c) $\frac{1}{2}$ (d) 2 (e) $2\frac{2}{3}$.

5) The line joining $(1, 3)$ to (a, b) has unit gradient.
(a) $b - a = 2$ (b) $a - b = 2$ (c) $a + b = 2$ (d) $b - a = 4$
(e) $a - b = 4$.

6) The equation of the line through the origin and perpendicular to $3x - 2y + 4 = 0$ is:
(a) $3x + 2y = 0$ (b) $2x + 3y + 1 = 0$ (c) $2x + 3y = 0$
(d) $2x - 3y - 1 = 0$ (e) $3x - 2y = 0$.

7) The equation of the line with gradient 1 passing through the point (h, k) is:
(a) $y = x + k - h$ (b) $y = \dfrac{k}{h}x + 1$ (c) $y = x + h - k$
(d) $ky = hx - 1$ (e) $y + x = k - h$.

8) The two lines $x + y = 0$ and $2x - y + 3 = 0$ intersect at the point:
(a) $(-\frac{1}{3}, \frac{1}{3})$ (b) $(1, -1)$ (c) $(-3, 3)$
(d) $(-1, 1)$ (e) $(3, -3)$.

9) The shaded region defined by $y > 0$, $y > x(x - 1)$ is:

a)

b)

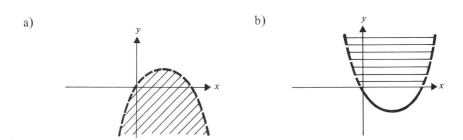

c)

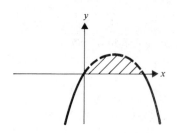

d)

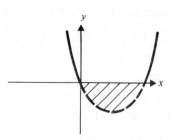

e)

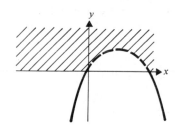

10) The curves $y = x^2$, $y = x(2 - x)$ intersect at:
(a) $(0, 0), (1, 1)$ (b) $(2, 4)$ (c) $(0, 0), (2, 4)$
(d) $(0, 0), (-1, 1)$ (e) $(1, 1)$.

TYPE II

11) A and B are two points with coordinates $(3, 4)$, $(-1, 6)$.
(a) Gradient of AB is $-\frac{1}{2}$.
(b) Midpoint of AB is the point $(2, 5)$.
(c) Length of AB is $2\sqrt{5}$.

12) A, B and C are the points $(5, 0), (-5, 0), (2, 3)$.
(a) AB and BC are perpendicular.
(b) Area of triangle ABC is 15 square units.
(c) A, B and C are collinear.

13) A, B, C are the points $(0, 13), (0, -13), (5, -12)$.
(a) A, B, C lie on the circumference of a circle, centre the origin.
(b) The equation of AC is $5x + y - 13 = 0$.
(c) The midpoint of BC is the origin.

14) The equation of a line l is $y = 2x - 1$.
(a) The line through the origin perpendicular to l is $y + 2x = 0$.
(b) The line through $(1, 2)$ parallel to l is $y = 2x - 3$.
(c) l passes through $(1, 1)$.

15) The equation of a line l is $7x - 2y + 4 = 0$.
(a) l has a gradient of $3\frac{1}{2}$.
(b) l is parallel to $7x + 2y - 3 = 0$.
(c) l is perpendicular to $2x + 7y - 5 = 0$.

TYPE III

16) (a) Two lines l_1 and l_2 have gradients m and $\dfrac{1}{m}$ respectively.

(b) Two lines l_1 and l_2 are perpendicular.

17) A, B and C are three points where:
(a) C is equidistant from A and B.
(b) The coordinates of A, B and C are $(1, 2)$, $(3, 6)$ and $(2, 4)$.

18) (a) A straight line has equation $x + y - 1 = 0$.
(b) A straight line has intercepts 1 and 1 on the x and y axes respectively.

TYPE IV

19) Find the equation of a line.
(a) The line is perpendicular to $x + y = 0$.
(b) The line cuts the y-axis at the point $(0, 3)$.
(c) The area of the triangle enclosed by the line and the coordinate axes is $4\frac{1}{2}$ square units.

20) Find the equation of the median through A of triangle ABC.
(a) A is the point $(5, -1)$.
(b) The equation of BC is $y = 3x - 4$.
(c) AC is parallel to $x + 2y - 3 = 0$.

21) Find the point of intersection of the two lines l_1 and l_2.
(a) l_1 passes through $(2, 5)$ and $(7, 3)$.
(b) l_2 is parallel to $y = 3x - 4$.
(c) The intercept of l_2 on the x-axis is 1.

22) Find the point of intersection of the curve $y = f(x)$ and the line l.
(a) l is perpendicular to $y = 3x - 4$.
(b) $f(x) \equiv (x - 2)(x + 5)$.
(c) l cuts $y = f(x)$ in two points.

23) Find the coordinates of the circumcentre of triangle ABC.
(a) The equation of AC is $x - y + 3 = 0$.
(b) The equation of BC is $x + 2y - 4 = 0$.
(c) The equation of AB is $y = 5$.

TYPE V

24) The line joining $(0, 0)$ and $(1, 3)$ is equal in length to the line joining $(0, 1)$ and $(3, 0)$.

25) If a line has gradient m and intercept d on the x-axis, its equation is $y = mx - md$.

26) The line $2x - y + 5 = 0$ has intercepts 5 and 2 on the y-axis and x-axis respectively.

27) The line passing through $(3, 1)$ and $(-2, 5)$ is perpendicular to the line $4y = 5x - 3$.

28) The line $3y = 7x - 2$ has an intercept of -2 on the y-axis.

MISCELLANEOUS EXERCISE 4

1) Show that the triangle whose vertices are $(1, 1)$, $(3, 2)$, $(2, -1)$ is isosceles.

2) Find the area of the triangular region defined by $y \geqslant 2x - 1$, $x \geqslant 0$, $y \leqslant 0$.

3) Find the coordinates of the vertices of the triangular region defined by $y \geqslant 0$, $y \leqslant x + 5$, $y \leqslant -\frac{1}{2}x + 3$.

4) Write down the equation of the line which goes through the point $(7, 3)$ and which is inclined at $45°$ to the positive direction of the x-axis.
Find the area enclosed by this line and the coordinate axes.

5) Find the coordinates of the points of intersection of $y = x^2 - 9$ and $y = x - 3$. Find the length of the line joining these two points.

6) Find the equation of the line through $A(5, 2)$ which is perpendicular to the the line $y = 3x - 5$. Hence find the coordinates of the foot of the perpendicular from A to the line.

7) The coordinates of a point P are $(t + 1, 2t - 1)$.
Plot the position of P when $t = -1, 0, 1, 2$. Show that these four points are collinear and find the equation of the line on which they lie.

8) Write down the equation of the perpendicular bisector of the line joining $(a, b), (2a, -3b)$.

9) A circle has a radius 4 and centre at the point $(2, 0)$. If $P(x, y)$ is any point inside the circumference of this circle, write down the condition that must be satisfied by the coordinates of P.

10) Write down the equation of the circumference of the circle defined in Question 9.

11) Find the coordinates of the points of intersection of $y = x^2$ and $y = x(4 - x)$. Draw a sketch showing the region defined by
$$x(4 - x) > y > x^2$$

12) The equation of a circle is $(x - 1)^2 + (y - 1)^2 = 4$.
Find the coordinates of A and B, the points of intersection of the line $x + y = 2$ and the circle.

13) Show that the point $C(1, 3)$ is on the circumference of the circle whose equation is given in Question 12. Find the midpoint M of AB and show that $MC = MA = MB$. What can you deduce about the line AB?

14) The equations of two adjacent sides of a rhombus are $y = 2x + 4$, $y = -\frac{1}{3}x + 4$. If $(12, 0)$ is one vertex and all vertices have positive coordinates, find the coordinates of the other three vertices.

15) A line is drawn through the point $A(1, 2)$ to cut the line $2y = 3x - 5$ in P and the line $x + y = 12$ in Q.
If $AQ = 2AP$, find the coordinates of P and Q.   (U of L)p

16) One side of a rhombus lies along the line $5x + 7y = 1$ and one of the vertices is $(3, -2)$. One diagonal of the rhombus is the line $3y = x + 1$. Find the coordinates of the other vertices and the equations of the three remaining sides.   (U of L)

CHAPTER 5

DIFFERENTIATION I

TANGENTS AND NORMALS

If A, B are two points on a curve (any curve) then

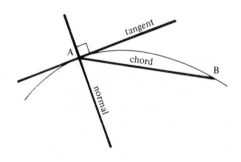

the line joining A and B is called a *chord*,

the line touching the curve at A is called the *tangent* to the curve at A,

the line perpendicular to the tangent at A is called the *normal* to the curve at A

GRADIENT OF A CURVE

The gradient of a curve, which is a measure of its slope, changes continually as one moves along it.

Suppose, when moving along the curve in the diagram (in the sense $B \to A$), that at the point A the gradient stops changing and remains constant. One would then move along the straight line AT, i.e. along the tangent at A.

So the gradient of the curve at the point A is the same as the gradient of the tangent at A.

The gradient of a curve at any point is defined as the gradient of the tangent to the curve at that point and measures the rate of increase of y with respect to x.

An *approximate* value for the gradient of a curve at a point can be found by plotting the curve, drawing the tangent by eye and measuring its gradient. This method has to be used for a curve when the coordinates of a finite number of points are known, but its equation is not, e.g. data from an experiment. When the equation of a curve is known, an accurate method for determining gradients is necessary so that we can further our analysis of curves and functions.
Consider first the problem of finding the gradient of a curve at a given point A.

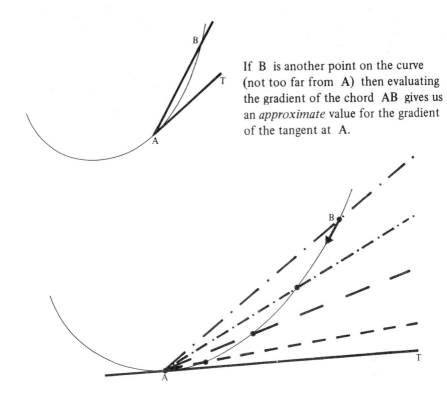

If B is another point on the curve (not too far from A) then evaluating the gradient of the chord AB gives us an *approximate* value for the gradient of the tangent at A.

The closer B is to A, the better is the approximation.

i.e.   as B → A

           gradient of chord AB → gradient of tangent AT

or   limit {gradient of chord AB} = gradient of tangent AT
           as B→A

Let us now consider an example where we can use this definition to find the gradient of a curve at a particular point.

Find the gradient of the curve $y = x(2x - 1)$ at the point on the curve where $x = 1$.

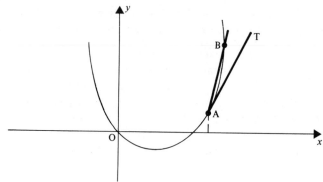

From the definition above

$$\text{Gradient of } AT = \lim_{B \to A} \{\text{gradient of chord } AB\}$$

So we must first calculate the gradient of a chord AB and observe what happens to this gradient as B progressively approaches A.

One method is to take a succession of points $B_1, B_2, B_3 \ldots$ where $x = 1.5, 1.25, 1.125 \ldots$, (i.e. halving the remaining difference between the x coordinates of A and B) and calculating the gradients of the chords $AB_1, AB_2 \ldots$ etc.

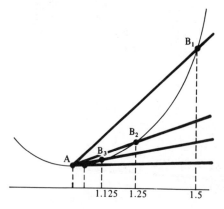

grad AB_1 = 4

grad AB_2 = 3.5

grad AB_3 = 3.25

grad AB_4 = 3.125

grad AB_5 = 3.0625

From this sequence of values we observe that

as B → A, gradient of chord AB → 3,

i.e. we deduce that the gradient of the curve at A is 3.
However this method is unsatisfactory, not least because of the large amount of numerical calculation involved. So instead of placing B at particular positions we will introduce a variable quantity for the difference between the x coordinates of A and B.

The Delta Prefix

A variable quantity, prefixed by δ, means a small increase in that quantity:

i.e.

δx is a small increase in x

δy is a small increase in y

δv is a small increase in v

(**Note** δ is only a prefix; it cannot be treated as a factor.)
If δx is the increase in the x coordinate in moving from A to B, then the x coordinate of B is $1 + \delta x$.
For all points on the curve, $y = x(2x - 1)$.
So at A, $x = 1$, $y = 1(2 \times 1 - 1) = 1$
     at B, $x = 1 + \delta x$, $y = (1 + \delta x)[2(1 + \delta x) - 1] = (1 + \delta x)(2\delta x + 1)$

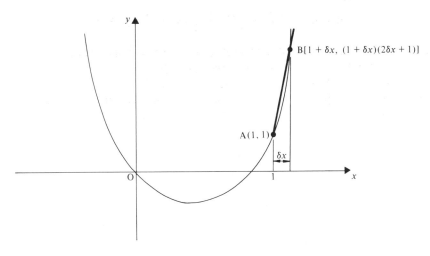

Therefore the gradient of AB which is $\dfrac{\text{increase in } y}{\text{increase in } x}$

$$= \frac{(1 + \delta x)(2\delta x + 1) - 1}{(1 + \delta x) - 1}$$

$$= \frac{2(\delta x)^2 + 3\delta x}{\delta x}$$

$$= 2\delta x + 3$$

As B approaches A, the difference between their x coordinates approaches zero, i.e. $\delta x \to 0$.

Therefore   as $B \to A$, gradient of $AB \to 3$,

or   gradient at $A = \lim_{B \to A} \{\text{gradient of chord } AB\}$

$$= \lim_{\delta x \to 0} \{2\delta x + 3\}$$

$$= 3$$

EXAMPLE 5a

Using the method described above find the gradient of the curve $y = \dfrac{1}{x}$ at the point where $x = 2$.

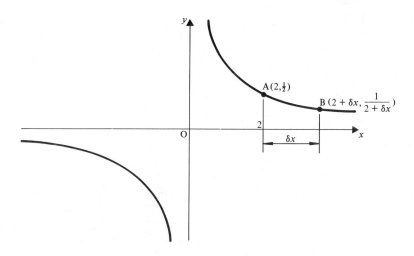

Let δx be the increase in moving from $A(2, \tfrac{1}{2})$ to a near point B on the curve so that the coordinates of B are $\left(2 + \delta x, \dfrac{1}{2 + \delta x}\right)$

The gradient of chord $AB = \left(\dfrac{1}{2 + \delta x} - \dfrac{1}{2}\right) \Big/ \delta x$

$$= \frac{-\delta x}{2\delta x(2 + \delta x)}$$

$$= \frac{-1}{2(2 + \delta x)} = -\frac{1}{4 + 2\delta x}$$

Therefore gradient at $A = \lim_{\delta x \to 0} \left(-\frac{1}{4 + 2\delta x} \right)$

$$= -\frac{1}{4}$$

EXERCISE 5a

Use the method of Example 5a to find the gradient of the given curve at the point indicated.

1) $y = x^2 + 1$ where $x = 3$ 2) $y = (x + 1)(x - 1)$ where $x = 2$

3) $y = 2x(x - 4)$ where $x = 0$ 4) $y = (x + 2)(x - 1)$ where $x = -3$

5) $y = x^3 - 3$ where $x = 1$ 6) $y = \dfrac{1}{x^2}$ where $x = 1$

GRADIENT FUNCTION

Earlier in this chapter we found that the gradient of the curve $y = x(2x - 1)$ is   3   at the point on the curve where $x = 1$.   We will now derive a function for the gradient at *any* point on the curve. Then we can find the gradient at a particular point by substitution into this derived function. Instead of taking a fixed point on the curve we will take   A   as any point (x, y) on the curve. Its y coordinate can then be written as $x(2x - 1)$ since both coordinates satisfy the equation $y = x(2x - 1)$.

Let   B   be another point on the curve such that the increase in the x coordinate in moving from   A   to   B   is δx. Then   B   is the point $[x + \delta x, (x + \delta x)(2x + 2\delta x - 1)]$.

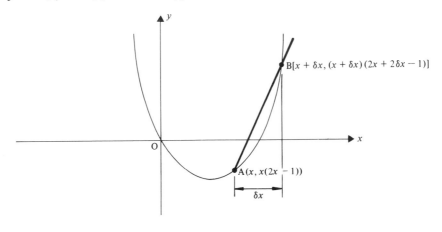

gradient of chord $AB = \dfrac{(x + \delta x)(2x + 2\delta x - 1) - x(2x - 1)}{\delta x}$

$$= \dfrac{2x^2 + 4x\delta x + 2(\delta x)^2 - \delta x - x - 2x^2 + x}{\delta x}$$

$$= \dfrac{4x\delta x - \delta x + 2(\delta x)^2}{\delta x}$$

$$= 4x - 1 + 2\delta x$$

Then the gradient at any point A on the curve

$$= \lim_{\delta x \to 0} \{4x - 1 + 2\delta x\}$$

$$= 4x - 1$$

So the function $4x - 1$ gives the gradient at any point on the curve $y = x(2x - 1)$.

We can find the gradient of the curve (i.e. the rate of increase of y with respect to x) at a particular point on $y = x(2x - 1)$ by substituting the x co-ordinate of that point into the function $4x - 1$.

Thus $4x - 1$ is called the *gradient function* of $y = x(2x - 1)$ and the process of deriving it is called *differentiation with respect to* x.

Now $4x - 1$ was derived from the function $x(2x - 1)$ so $4x - 1$ is called the *derivative*, or *derived function*, of $x(2x - 1)$.

Using $\dfrac{d}{dx}$ as a symbol to denote 'derivative w.r.t. x of'

we may write $\dfrac{d}{dx}[x(2x - 1)] = 4x - 1$,

or, when $y = x(2x - 1)$, $\dfrac{dy}{dx} = 4x - 1$

where $\dfrac{dy}{dx}$ means the *derivative* of y w.r.t. x and is sometimes called the *differential coefficient* of y.

As an alternative notation we can use the symbol D for derivative, or derived function, of

i.e. $D[x(2x - 1)] = 4x - 1$

where D is referred to as the differential operator.

Now $4x - 1$ represents the rate of increase of y w.r.t. x

therefore $\dfrac{dy}{dx}$ represents also the rate of increase of y w.r.t. x.

Similarly $\dfrac{dv}{dt}$ means the derivative of v w.r.t. t or the rate of increase of v

w.r.t. t etc.

EXAMPLE 5b

Differentiate $y = x^3 + 3$ with respect to x.

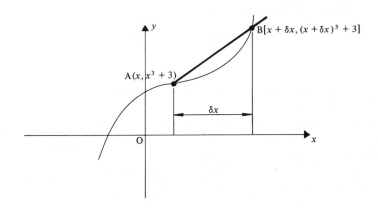

Let A be the point $(x, x^3 + 3)$ and B the point $[x + \delta x, (x + \delta x)^3 + 3]$ on $y = x^3 + 3$.

$$\text{Gradient of AB} = \frac{(x + \delta x)^3 + 3 - (x^3 + 3)}{\delta x}$$

which simplifies to $3x^2 + 3x\delta x + (\delta x)^2$.

$$\text{Gradient at A} = \lim_{\delta x \to 0} \{3x^2 + 3x\delta x + (\delta x)^2\}$$

$$= 3x^2$$

i.e.
$$\frac{dy}{dx} = 3x^2$$

EXERCISE 5b

Use the method of Example 5b to differentiate the following with respect to x.

1) $y = x^2$ 2) $y = x^3$ 3) $y = x^4$ 4) $y = x$ 5) $y = 3x^2$

6) $y = 5x^3$ 7) $y = \dfrac{1}{x^2}$ 8) $y = x^2 + 3x$ 9) $y = x^2 - 2x + 1$

GENERAL DIFFERENTIATION

Consider any curve $y = f(x)$.

Let A$[x, f(x)]$ be any point on $y = f(x)$.

Let δx be the increase in the x coordinate in moving from A to another point B on the same curve. So B is the point $[x + \delta x, f(x + \delta x)]$.

Let δy be the corresponding increase in the y coordinate in moving from A to B,

i.e. $\delta y = f(x + \delta x) - f(x)$.

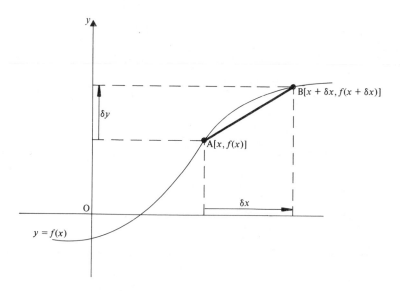

$$\text{Gradient of AB} = \frac{\delta y}{\delta x} = \frac{f(x + \delta x) - f(x)}{\delta x}$$

Therefore the gradient at A is

$$\lim_{\delta x \to 0} \left\{ \frac{\delta y}{\delta x} \right\} = \lim_{\delta x \to 0} \left\{ \frac{f(x + \delta x) - f(x)}{\delta x} \right\}$$

i.e. $\dfrac{dy}{dx} = \lim\limits_{\delta x \to 0} \left\{ \dfrac{\delta y}{\delta x} \right\}$ or $\dfrac{d}{dx} f(x) = \lim\limits_{\delta x \to 0} \left\{ \dfrac{f(x + \delta x) - f(x)}{\delta x} \right\}$

This formal definition of differentiation can be used to differentiate any function that we have not yet met. This process is called differentiating from first principles.

Fortunately it is not always necessary to go back to first principles because certain categories of functions can be differentiated by rules, some of which we will now find.

Differentiation of a Constant

Consider the equation $y = c$. This represents a straight line parallel to the x-axis and so it has zero gradient.

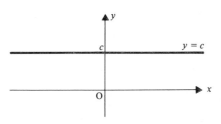

i.e. $$\frac{dy}{dx} = 0 \qquad \text{or} \qquad \frac{d}{dx}(c) = 0.$$

Differentiation of ax where a is a Constant

Consider the equation $y = ax$. This represents a straight line with gradient a,

i.e. $$\frac{dy}{dx} = a \qquad \text{or} \qquad \frac{d}{dx}(ax) = a$$

Differentiation of x^n

The table below summarises the results from some of the questions in Exercise 5b.

$f(x)$	x^2	x^3	x^4	$x^{-2}\left(=\dfrac{1}{x^2}\right)$
$\dfrac{d}{dx}f(x)$	$2x$	$3x^2$	$4x^3$	$-2x^{-3}$

From this table it appears that to differentiate a power of x we multiply by that power and then subtract one from the power, i.e.

$$\frac{d}{dx}(x^n) = nx^{n-1} \qquad [1]$$

This result, deduced from a few examples, is in fact valid for all powers of x including fractional and negative powers, although proof is not possible at this stage. (This is one example of many such 'rules' whose validity, for the time being, we must take on trust.)

Accepting that rule [1] above can be used for any power of x, the following examples illustrate its application,

$$\frac{d}{dx}(x^9) = 9x^8$$

$$\frac{d}{dx}(x^{-4}) = -4x^{-4-1} = -4x^{-5}$$

$$\frac{d}{dx}(\sqrt{x}) = \frac{d}{dx}(x^{\frac{1}{2}}) = \frac{1}{2}x^{\frac{1}{2}-1} = \frac{1}{2}x^{-\frac{1}{2}}$$

EXERCISE 5c

Differentiate by rule the following functions with respect to x.

1) x^5 2) x^{10} 3) x^{-3} 4) $x^{\frac{1}{3}}$ 5) x^{-1} 6) $x^{-\frac{1}{2}}$

7) $\dfrac{1}{x^4}$ 8) $\sqrt{x^3}$ 9) $2x$ 10) $\dfrac{1}{x^5}$ 11) $\dfrac{1}{\sqrt{x}}$ 12) $\sqrt[4]{x}$

13) 5 14) $-4x$ 15) $\dfrac{1}{x^{-\frac{1}{2}}}$ 16) $\sqrt[3]{(x^2)}$

Differentiation of ax^n

Selecting some more results from Exercise 5b we see that

$$\frac{d}{dx}(3x^2) = 3(2x)$$

and

$$\frac{d}{dx}(5x^3) = 5(3x^2)$$

These results suggest that

$$\frac{d}{dx}(ax^n) = a\frac{d}{dx}(x^n) = anx^{n-1} \qquad [2]$$

where a is a constant.
Again selecting from Exercise 5b we have

$$\frac{d}{dx}(x^2 + 3x) = 2x + 3 = \frac{d}{dx}(x^2) + \frac{d}{dx}(3x)$$

and

$$\frac{d}{dx}(x^2 - 2x + 1) = 2x - 2 = \frac{d}{dx}(x^2) - \frac{d}{dx}(2x) + \frac{d}{dx}(1)$$

This suggests that the operation 'differentiate' is distributive across addition and subtraction of functions: i.e.

$$\frac{d}{dx}\left[f(x) + g(x)\right] = \frac{d}{dx}f(x) + \frac{d}{dx}g(x) \qquad [3]$$

Therefore $\qquad\dfrac{d}{dx}(3x^2 - 7x) = 3(2x) - 7 = 6x - 7$

and $\qquad\dfrac{d}{dx}\left(x + \dfrac{1}{x}\right) = \dfrac{d}{dx}(x + x^{-1}) = 1 - x^{-2}$

Note that differentiation is *not* distributive across multiplication and division:

e.g. $\qquad\dfrac{d}{dx}\left[x(x-3)\right] = \dfrac{d}{dx}(x^2 - 3x) = 2x - 3$

but $\qquad\left[\dfrac{d}{dx}(x)\right] \times \left[\dfrac{d}{dx}(x-3)\right] = (1) \times (1) = 1$

i.e. $\qquad\dfrac{d}{dx}\left[x(x-3)\right] \neq \dfrac{d}{dx}(x) \times \dfrac{d}{dx}(x-3)$

In order to differentiate at this stage, therefore, any product must be expanded and any quotient must be divided out to give terms which are added or subtracted.

EXAMPLES 5d

1) Differentiate $f(x) \equiv \dfrac{4x^2 + x - 1}{2x}$.

$$f(x) \equiv \dfrac{4x^2 + x - 1}{2x} \equiv \dfrac{4x^2}{2x} + \dfrac{x}{2x} - \dfrac{1}{2x}$$

therefore $\qquad f(x) \equiv 2x + \tfrac{1}{2} - \tfrac{1}{2}x^{-1}$

therefore $\qquad\dfrac{d}{dx}f(x) = 2 + 0 - \tfrac{1}{2}(-x^{-2}) = 2 + \tfrac{1}{2}x^{-2} = 2 + \dfrac{1}{2x^2}$

2) Find the gradient of the curve $y = (x - 3)(x^2 + 2)$ at the point on the curve where $x = 1$.

$$y = (x - 3)(x^2 + 2) = x^3 - 3x^2 + 2x - 6$$

therefore $\qquad\dfrac{dy}{dx} = 3x^2 - 6x + 2.$

When $x = 1$, $\qquad\dfrac{dy}{dx} = 3(1)^2 - 6(1) + 2 = -1$

therefore the gradient of $y = (x - 3)(x^2 + 2)$ is -1 at the point where $x = 1$.

3) Find the coordinates of the point on the curve $y = \dfrac{2}{x^2}$ at which its gradient is $\tfrac{1}{2}$.

$$y = \frac{2}{x^2} = 2x^{-2}$$

therefore $\qquad\qquad\qquad \dfrac{dy}{dx} = -4x^{-3}$

when $\quad \dfrac{dy}{dx} = \frac{1}{2}, \qquad\qquad -4x^{-3} = \frac{1}{2}$

$\Rightarrow \qquad\qquad\qquad\qquad -\dfrac{4}{x^3} = \frac{1}{2}$

$\Rightarrow \qquad\qquad\qquad\qquad x^3 = -8$

$\Rightarrow \qquad\qquad\qquad\qquad x = -2$

At the point on $\quad y = \dfrac{2}{x^2}\quad$ where the gradient is $\frac{1}{2}$, the x coordinate is -2

and when $\quad x = -2, \quad y = \frac{1}{2}$

Therefore the gradient of $\quad y = \dfrac{2}{x^2}\quad$ is $\frac{1}{2}$ at the point $(-2, \frac{1}{2})$.

EXERCISE 5d

Differentiate with respect to x.

1) $3x^2 - 7x$ 2) $x^4 - 9x^3 + 6$ 3) $(x-3)(2x+5)$ 4) $(x-4)^2$

5) $(7x+1)(7x-1)(x-1)$ 6) $\dfrac{(x^3+2)}{x}$ 7) $x^{-2}(1+x)$

8) $\dfrac{x^2 - 7x + 4}{x^3}$ 9) $\sqrt{x}(x-1)$ 10) $\dfrac{(x-1)}{\sqrt{x}}$

11) $(2x+1)^3$ 12) $1 + \dfrac{1}{x} + \dfrac{1}{x^2} + \dfrac{1}{x^3}$

Find the gradient of the given curve at the given point on the curve.

13) $y = x^2 - 3$ where $x = 1$ 14) $y = 3x^2 - 2$ where $x = 3$

15) $y = \sqrt{x}$ where $x = 2$ 16) $y = (x-3)(x+4)$ where $x = -1$

17) $y = \dfrac{1}{x}$ where $x = 3$ 18) $y = (2x-3)(x+1)$ where $x = 0$

19) $y = 1 - \dfrac{1}{x}$ where $x = 3$ 20) $y = x^3 + 7x - 4$ where $x = -3$

21) $y = \dfrac{(3x - 1)}{x}$ where $x = \frac{1}{2}$ 22) $y = 2\sqrt{x}(1 - \sqrt{x})$ where $x = 4$

23) $y = x^2 + \dfrac{3}{x}$ where $x = -1$ 24) $y = \dfrac{(\sqrt{x} - 1)}{\sqrt{x}}$ where $x = 9$

Find the coordinates of the point(s) on the given curve at which its gradient has the given value.

25) $y = x^2 - x + 3$,   1   26) $y = 5 + 3x - 2x^2$, -3

27) $y = (x - 1)(x + 1)$,   2   28) $y = (x + 3)(x - 5)$,   0

29) $y = x^3 + x^2$,   1   30) $y = \frac{1}{3}x^3 + 3x^2 + 7x + 1$, -1

31) $y = x + \dfrac{1}{x}$,   2   32) $y = \sqrt{x}$,   2

33) $y = 1 - \dfrac{1}{x}$,   4   34) $y = \dfrac{1}{x^2}$, $\dfrac{1}{4}$

35) $y = \dfrac{1}{x^3}$, -3 36) $y = x^3$, -1

EQUATIONS OF TANGENTS AND NORMALS

Now that we know how to find the gradient of a curve at a given point on the curve, we can find the equation of the tangent or normal to the curve at that point.

EXAMPLES 5e

1) Find the equation of the tangent to the curve $y = x^2 - 3x + 2$ at the point where it cuts the y-axis.

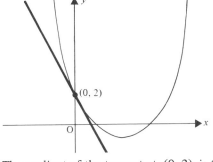

(0, 2)

$y = x^2 - 3x + 2$ cuts the y-axis where $x = 0$, $y = 2$.

The gradient of the tangent at $(0, 2)$ is the value of $\dfrac{dy}{dx}$ when $x = 0$.

As $y = x^2 - 3x + 2$, $\dfrac{dy}{dx} = 2x - 3$.

When $x = 0$ the gradient of the curve is -3.
Therefore the tangent has gradient -3 and passes through $(0, 2)$.
So its equation is $y = -3x + 2$,
i.e. $3x + y - 2 = 0$.

2) Find the equation of the normal to the curve $y = \dfrac{1}{x}$ at the point on the

curve where $x = 2$.   Find the coordinates of the point where this normal cuts
the curve again.

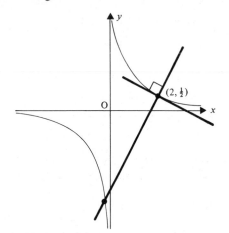

As $y = \dfrac{1}{x}$, $\dfrac{dy}{dx} = -\dfrac{1}{x^2}$

So when $x = 2$,

$y = \dfrac{1}{2}$ and $\dfrac{dy}{dx} = -\dfrac{1}{4}$

Therefore the tangent at $(2, \tfrac{1}{2})$ has gradient $-\tfrac{1}{4}$
and the normal (i.e. the line perpendicular to the tangent) at $(2, \tfrac{1}{2})$ has
gradient 4.
The equation of this normal is $y - \tfrac{1}{2} = 4(x - 2)$ or $8x - 2y - 15 = 0$.
The points of intersection of the normal and the curve are given by solving
simultaneously

$$\begin{cases} 8x - 2y - 15 = 0 & \qquad [1] \\[2mm] \qquad\qquad y = \dfrac{1}{x} & \qquad [2] \end{cases}$$

i.e. $\qquad\qquad\qquad\qquad 8x - \dfrac{2}{x} - 15 = 0$

$\qquad\qquad\qquad\qquad 8x^2 - 15x - 2 = 0 \qquad\qquad\qquad [3]$

We already know that [1] and [2] cut where $x = 2$,   so $x - 2$ is a factor
of [3]

Hence $\qquad\qquad\qquad\qquad (x - 2)(8x + 1) = 0.$

Therefore [1] and [2] cut again where $x = -\tfrac{1}{8}$ and $y = -8$ (from [1])
i.e. at the point $(-\tfrac{1}{8}, -8)$.

EXERCISE 5e

Find the equation of the tangent to the given curve at the given point on the curve.

1) $y = x^2 - 2$ where $x = 1$

2) $y = x^2 + 3x - 1$ where $x = 0$

3) $y = \dfrac{1}{x}$ where $x = -1$

4) $y = (x - 2)(x^2 + 1)$ where $x = -1$

5) $y = x^2 - 5x + 2$ where $x = 3$

6) $y = x^2 - 2$ where $x = 0$

7) Find the equations of the normals to the curves in Questions 1–6 at the given points.

8) Find the equation of the normal to the curve $y = x^2 + 3x - 2$ at the point where the curve cuts the y-axis.

9) Find the equation of the tangent to the curve $y = x^2 + 5x - 2$ at the point where this curve cuts the line $x = 4$.

10) Find the equations of the tangents to the curve $y = (2x - 1)(x + 1)$ at the points where the curve cuts the x-axis. Find the point of intersection of these tangents.

11) Find the equations of the normals to the curve $y = x^2 - 5x + 6$ at the points where the curve cuts the x-axis.

12) Find the equations of the tangents to the curve $y = 3x^2 + 5x - 1$ at the points of intersection of the curve and the line $y = x - 1$.

13) Find the coordinates of the point on $y = x^2$ at which the gradient is 2. Hence find the equation of the tangent to $y = x^2$ whose gradient is 2.

14) Find the coordinates of the point on $y = x^2 - 5$ at which the gradient is 3.
Hence find the value of c for which the line $y = 3x + c$ is a tangent to $y = x^2 - 5$.

15) Find the equation of the normal to $y = x^2 - 3x + 2$ which has a gradient of $\frac{1}{2}$.

16) Find the equation of the tangent to $y = 2x^2 - 3x$ which has a gradient of 1.

17) Find the value of k for which $y = 2x + k$ is a normal to $y = 2x^2 - 3$.

18) Find the equation of the tangent to $y = (x - 5)(2x + 1)$ which is parallel to the x-axis.

STATIONARY VALUES

A stationary value of a function $f(x)$ is any value of $f(x)$ at which its rate of change with respect to x is zero,

i.e. stationary values of $f(x)$ occur when $\dfrac{d}{dx}\left[f(x)\right] = 0.$

For a graphical representation of a stationary value, consider the curve whose equation is $y = f(x)$.

At a stationary value of $f(x)$, $\dfrac{dy}{dx} = 0$

i.e. the gradient of the curve is zero
i.e. the tangent to the curve is parallel to the x-axis.

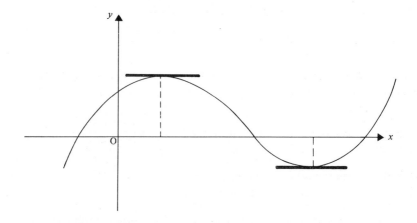

Therefore stationary values of $f(x)$ are the values of the y coordinates of points on the curve $y = f(x)$ at which the tangent is parallel to the x-axis.

EXAMPLE 5f

Find the stationary values of $x^3 - 3x^2 + 2$.

If $$f(x) \equiv x^3 - 3x^2 + 2$$

$$\frac{d}{dx}f(x) \equiv 3x^2 - 6x$$

At stationary values of $f(x)$, $\dfrac{d}{dx}f(x) = 0$

$\Rightarrow$ $$3x^2 - 6x = 0$$

$$x(x - 2) = 0$$

i.e. stationary values of $x^3 - 3x^2 + 2$ occur when $x = 0$ and when $x = 2$. These stationary values are:

$$f(0) = 2 \quad \text{and} \quad f(2) = 2^3 - 3(2)^2 + 2 = -2$$

Note that for the curve $y = x^3 - 3x^2 + 2$,

$$\frac{dy}{dx} = 3x^2 - 6x = 0 \quad \text{when} \quad x = 0 \text{ or } 2$$

so the gradient of $y = f(x)$ is zero at the points $(0, 2)$ and $(2, -2)$.

EXERCISE 5f

Find the value(s) of x at which the following functions have stationary values:

1) $x^2 - 3$ 2) $x^2 - 5x + 1$ 3) $x^3 - 12x + 1$ 4) $2x^4 - 2x^3 - x^2$

Find the stationary value(s) of the following functions:

5) $x^2 - 6x + 3$ 6) $x + \dfrac{1}{x}$ 7) $3x^3 - 4x + 2$

Find the coordinates of the points on the given curves at which the gradient is zero.

8) $y = (x - 2)(x + 3)$ 9) $y = x^3 - 5x^2 + 3x - 2$ 10) $y = \dfrac{x^2 - 2}{x}$

TURNING POINTS

The gradient of a curve can be zero at several points. The shape of the curve in the immediate neighbourhood of one of these points belongs to one of the three categories shown below.

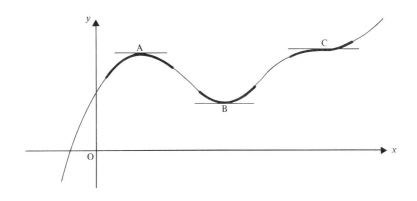

Moving along the curve in the positive direction of the x-axis:

(a) In the neighbourhood of A, the gradient changes from positive, through zero at A, to negative.
A is called a *maximum turning point*.
The y coordinate of A is called a *maximum value of y* [or of $f(x)$ where $y = f(x)$].

(b) In the neighbourhood of B, the gradient changes from negative, through zero at B, to positive.
B is called a *minimum turning point*.
The y coordinate of B is called a *minimum value of y* [or of $f(x)$].

Note that the gradient at a maximum or minimum turning point *must* be zero.

Note also that the terms maximum and minimum values are not synonymous with greatest and least values. Maxima and minima apply to the behaviour of a function in the *immediate neighbourhood only* of its stationary values.

(c) The curve does not turn at C, i.e. although the gradient is zero at C, it does not change sign when moving through C. However the sense in which the curve is turning does change (from clockwise to anticlockwise).

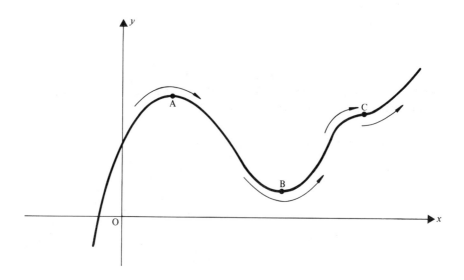

Any point on a curve at which the sense of turning changes is called a *point of inflexion*. Apart from C, there are two other points of inflexion in the diagram, one between A and B and another between B and C. Thus the gradient at a point of inflexion is not necessarily zero.

INVESTIGATING THE NATURE OF STATIONARY VALUES

We know how to find the coordinates of points on $y = f(x)$ at which $f(x)$ has stationary values, but we need to investigate further so that we can distinguish between them, and there are several ways in which this can be done.

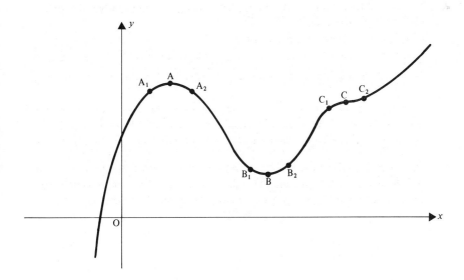

Consider the points A_1 and A_2, B_1 and B_2, C_1 and C_2 which are left and right respectively of A, B, C and *close* to them.

(1) Consider the values of y.

For A (a maximum value)

$$y \text{ at } A_1 < y \text{ at } A$$
$$y \text{ at } A_2 < y \text{ at } A$$

For B (a minimum value)

$$y \text{ at } B_1 > y \text{ at } B$$
$$y \text{ at } B_2 > y \text{ at } B$$

For C (a point of inflexion)

$$y \text{ at } C_1 < y \text{ at } C$$
$$y \text{ at } C_2 > y \text{ at } C$$

i.e.

	Maximum	Minimum	Inflexion
Values of y either side of stationary value	Both smaller	Both larger	One smaller and one larger

(2) Now consider the behaviour of the gradient $\dfrac{dy}{dx}$.

For A   at A_1, $\dfrac{dy}{dx}$ is $+$ ve

                       at A, $\dfrac{dy}{dx}$ is zero

                       at A_2, $\dfrac{dy}{dx}$ is $-$ ve

For B   at B_1, $\dfrac{dy}{dx}$ is $-$ ve

                       at B, $\dfrac{dy}{dx}$ is zero

                       at B_2, $\dfrac{dy}{dx}$ is $+$ ve

For C   at C_1, $\dfrac{dy}{dx}$ is $+$ ve

                       at C, $\dfrac{dy}{dx}$ is zero

                       at C_2, $\dfrac{dy}{dx}$ is $+$ ve

i.e.

	Maximum	Minimum	Inflexion
Sign of $\dfrac{dy}{dx}$ when moving through a stationary value	$+\ \ 0\ \ -$ ╱ — ╲	$-\ \ 0\ \ +$ ╲ — ╱	$+0+\ \ -0-$ or ╱–╱ ╲–╲

(3) When passing through A, $\dfrac{dy}{dx}$ changes from positive to negative.

i.e. $\dfrac{dy}{dx}$ decreases as x increases

or the rate of increase w.r.t. x of $\dfrac{dy}{dx}$ is negative

i.e. $\dfrac{d}{dx}\left(\dfrac{dy}{dx}\right)$ is negative.

This clumsy notation for the rate of increase w.r.t. x of $\dfrac{dy}{dx}$ is condensed to $\dfrac{d^2y}{dx^2}$.

Similarly when passing through **B**, $\dfrac{dy}{dx}$ changes from negative to positive

i.e. $\dfrac{dy}{dx}$ increases as x increases

or $\dfrac{d^2y}{dx^2}$ is positive.

Points of inflexion are not so easily dealt with by this method. It is true to say that $\dfrac{d^2y}{dx^2} = 0$ at such points but $\dfrac{d^2y}{dx^2}$ can also be zero at maxima and minima.

A more detailed analysis of points of inflexion will be carried out later.

i.e.

	Maximum	Minimum
Sign of $\dfrac{d^2y}{dx^2}$	negative (or zero)	positive (or zero)

The three tables above summarise the three alternative methods for determining the nature of stationary values. The third method fails if, at a stationary value of y, $\dfrac{d^2y}{dx^2}$ is found to be zero. In this case either of the first two methods has to be used.

EXAMPLES 5g

1) Find the points on $y = x^4 + 4x^3 - 6$ at which the gradient is zero and determine the nature of these points.

If

$$y = x^4 + 4x^3 - 6 \qquad\qquad [1]$$

$$\frac{dy}{dx} = 4x^3 + 12x^2 \qquad\qquad [2]$$

and $$\frac{d^2y}{dx^2} = 12x^2 + 24x$$ [3]

The gradient of [1] is zero when $\frac{dy}{dx} = 0$

i.e. when $4x^3 + 12x^2 = 0$

$\Rightarrow$ $4x^2(x + 3) = 0$

$\Rightarrow$ $x = -3$ and $x = 0$

So when $x = -3$, $\frac{dy}{dx} = 0$

and from [3] $\frac{d^2y}{dx^2} = 12(-3)^2 + 24(-3) = 36$ (>0)

therefore y has a minimum value here.

From [1] $y = (-3)^4 + 4(-3)^3 - 6 = -33$

therefore $(-3, -33)$ is a minimum turning point.

When $x = 0$, $\frac{dy}{dx} = 0$, $y = -6$

and $\frac{d^2y}{dx^2} = 0$ which is inconclusive

Let us now see what the sign of $\frac{dy}{dx}$ is on either side of the point where $x = 0$.

x	-1	0	$+1$
$\dfrac{dy}{dx}$	$+$	0	$+$
slope	╱	—	╱

From this table we see that $(0, -6)$ is a point of inflexion.

2) Sketch the curve $y = 2x^3 + x^2 - 4x + 1$.

Finding the maximum and minimum turning points will help to give a general idea of the shape and position of this curve.

If $y = 2x^3 + x^2 - 4x + 1$

$\frac{dy}{dx} = 6x^2 + 2x - 4$

$\frac{d^2y}{dx^2} = 12x + 2$

At turning points $\dfrac{dy}{dx} = 0$,   i.e. $6x^2 + 2x - 4 = 0$

$$(3x - 2)(x + 1) = 0$$

$$x = \tfrac{2}{3} \quad \text{and} \quad x = -1.$$

When $x = \tfrac{2}{3}$, $\dfrac{dy}{dx} = 0$

$$\dfrac{d^2 y}{dx^2} = 12(\tfrac{2}{3}) + 2 > 0$$

$$y = 2(\tfrac{2}{3})^3 + (\tfrac{2}{3})^2 - 4(\tfrac{2}{3}) + 1 = -\tfrac{17}{27}$$

Therefore $(\tfrac{2}{3}, -\tfrac{17}{27})$ is a minimum turning point.   [1]

When $x = -1$, $\dfrac{dy}{dx} = 0$

$$\dfrac{d^2 y}{dx^2} = (12)(-1) + 2 < 0$$

$$y = 2(-1)^3 + (-1)^2 - 4(-1) + 1 = 4$$

Therefore $(-1, 4)$ is a maximum turning point.   [2]

As $y = 2x^3 + x^2 - 4x + 1$,   the curve cuts the y-axis at $(0, 1)$.   [3]
From results [1], [2] and [3] we can sketch the curve.

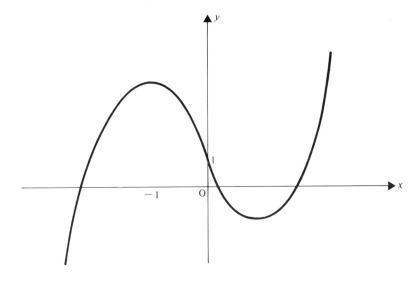

Note that in some cases, the intercepts on the x-axis can be found. It is not always easy to solve the equation which gives these values however, e.g. in the example above, $y = 0 \Rightarrow 2x^3 + x^2 - 4x + 1 = 0$.

3) A farmer has an adjustable electric fence that is 100 m long. He uses this fence to enclose a rectangular grazing area on three sides, the fourth side being a fixed hedge. Find the maximum area he can enclose.

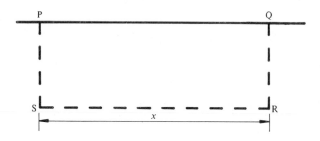

The farmer can vary the length (up to 100 m) of his enclosure and the width and area are then dependent on this chosen length.
Let x m be the length SR.

Then as $PS + SR + RQ = 100$

$$PS = \tfrac{1}{2}(100 - x)$$

Therefore the area, A, of the enclosure is given by

$$A = x[\tfrac{1}{2}(100 - x)] = \tfrac{1}{2}x(100 - x)$$

Now A is a quadratic function of x.

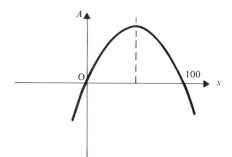

From the sketch and our knowledge of quadratic functions we can see that A is greatest when $x = 50$.

Therefore the maximum area that can be enclosed under the given conditions is

$$50 \times 25 \text{ m}^2 = 1250 \text{ m}^2.$$

Note that the maximum value of the function $\tfrac{1}{2}x(100 - x)$ can also be found as follows:

If $\qquad A = 50x - \tfrac{1}{2}x^2, \qquad \dfrac{\mathrm{d}A}{\mathrm{d}x} = 50 - x$

and A has a stationary value when $\dfrac{\mathrm{d}A}{\mathrm{d}x} = 0,$ i.e. when $x = 50.$

Now $\dfrac{\mathrm{d}^2 A}{\mathrm{d}x^2} = -1.$

So when $x = 50,$ $\dfrac{\mathrm{d}A}{\mathrm{d}x} = 0,$ $\dfrac{\mathrm{d}^2 A}{\mathrm{d}x^2} < 0,$ hence A is maximum.

The maximum value is $25 \times 50\,\mathrm{m}^2 = 1250\,\mathrm{m}^2.$

This second method, using differentiation, is necessary when finding maximum or minimum values of functions that are not quadratic. However the first method if preferable for quadratic functions when knowledge of their properties allows their maximum or minimum value to be found by inspection.

Note that in this case greatest value and maximum value are the same.

i.e. if $f(x) \equiv ax^2 + bx + c$

then the curve $y = ax^2 + bx + c$ crosses the x-axis where $ax^2 + bx + c = 0.$

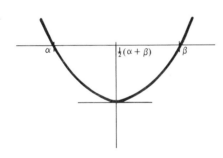

If α and β are the roots of this equation then the turning point of the curve has an x coordinate $\tfrac{1}{2}(\alpha + \beta).$

But $\tfrac{1}{2}(\alpha + \beta) = -\dfrac{b}{2a}.$

Therefore $f(x) \equiv ax^2 + bx + c$ has a stationary value when $x = -\dfrac{b}{2a}.$

EXERCISE 5g

Find the stationary values of the following functions and investigate their nature:

1) $x^3 - 3x$ 2) $x + \dfrac{4}{x}$ 3) $x - \dfrac{4}{x^2}$ 4) $x^2(3 - x)$

5) $(x - 3)(2x + 1)$ 6) x^3 7) x^4

8) $x^2(x^2 - 2)$ 9) $x^2(3 + 2x - 3x^2)$

Find the coordinates of the turning points of the following curves and sketch the curves.

10) $y = x^2 - x^3$ 11) $y = 1 - x^4$ 12) $y = x + \dfrac{1}{x}$

13) $y = x^2 - 4$ 14) $y = 3 - x + x^2$ 15) $y = 3 + 24x - 21x^2 - 4x^3$

16) $y = x^3 + 3$ 17) $y = 2x^4$ 18) $y = x^5 - 5x$

MULTIPLE CHOICE EXERCISE 5

(Instructions for answering these questions are given on page xii)

TYPE I

1) The function $x^3 - 12x + 5$ has a stationary value when:
(a) $x = \sqrt{6}$ (b) $x = -2$ (c) $x = 0$ (d) $x = 4$ (e) $x = 1$.

2) When $x = 1$ the function $x^3 - 3x^2 + 7$ is:
(a) stationary   (b) increasing   (c) maximum   (d) decreasing   (e) minimum.

3) The rate of increase of the function $x^2 - \dfrac{1}{x^2}$ w.r.t. x is:

(a) $2x + \dfrac{2}{x^3}$ (b) $2x - \dfrac{1}{2x}$ (c) $2x - \dfrac{2}{x^3}$ (d) $2x + \dfrac{3}{x^3}$ (e) $2x + \dfrac{1}{2x}$.

4) When $x = 0$ the function $x^3 - 2$ is:
(a) stationary   (b) maximum   (c) minimum   (d) increasing   (e) decreasing.

5) The function $\dfrac{1}{x}$ has a stationary value when:

(a) $x = 1$ (b) $x = 0$ (c) $x = -1$ (d) $x = 2$

(e) there is no real finite value of x for which $\dfrac{1}{x}$ is stationary.

6) The gradient function of $y = (x - 3)(x^2 + 2)$ is:
(a) $2x$ (b) $2x - 3$ (c) $3x^2 - 6x + 2$ (d) $-3(2x + 2)$
(e) $x^3 - 3x^2 + 2x - 6$.

7) The point $(1, 1)$ on the curve $y = x^3 - 3x^2 + 3x$ is:
(a) a maximum point   (b) a point on inflexion
(c) a minimum point   (d) none of these.

TYPE II

8) $f(x) \equiv x + \dfrac{1}{x}$.

(a) $f(x)$ is stationary when $x = -1$.

(b) $\dfrac{d}{dx} f(x) = 1 - \dfrac{1}{x^2}$.

(c) $y = f(x)$ has no turning points.

9) $y = x^3 - 4x + 5$.

(a) $\dfrac{d^2 y}{dx^2} = 9x$.

(b) The curve has two turning points.
(c) y is increasing when $x = 2$.

10) $y = x^4$.
(a) y is decreasing when $x = 1$.
(b) x^4 has only one stationary value.

(c) $\dfrac{dy}{dx} = 4x^3$.

TYPE III

11) (a) $y = f(x)$ is maximum when $x = 2$.

     (b) $\dfrac{d^2 y}{dx^2} < 0$ when $x = 2$.

12) (a) $y = f(x)$ and $\dfrac{dy}{dx} = 0$ and $\dfrac{d^2 y}{dx^2} > 0$ when $x = a$.

     (b) $y = f(x)$ and $f(a)$ is a minimum value of y.

13) When $x = 2$ (a) $\dfrac{dy}{dx} = 0$.

                      (b) y is stationary.

14) When $x = a$ (a) $\dfrac{dy}{dx} > 0$.

                      (b) $\dfrac{d^2 y}{dx^2} > 0$.

TYPE IV

15) Find the turning points on $y = f(x)$.

(a) $\dfrac{d^2 y}{dx^2} = 3x - 2$.

(b) $\dfrac{dy}{dx}$ is a quadratic function of x.

(c) The equation $\dfrac{dy}{dx} = 0$ has two real roots, $x = 0$, $x = 1$.

16) Find the values of x which correspond to stationary values of $f(x)$.
(a) $f(x)$ is a polynomial of degree 4.

(b) $\dfrac{d}{dx} f(x) = g(x)$.

(c) $g(x) \equiv x(x-1)(x-2)$.

17) A beer can is made from sheet metal. Find the radius of the base.
(a) The can holds 0.5 litre.
(b) The can is a cylinder.
(c) The surface area is to be as small as possible.

18) Sketch the curve $y = f(x)$.
(a) The curve has two stationary values.
(b) There are no discontinuities.
(c) The curve passes through $(2, 0)$ and $(0, 1)$.

MISCELLANEOUS EXERCISE 5

1) If $x^3 y = 2$, find the value of $\dfrac{d^2 y}{dx^2}$ when $x = 1$.

2) If $vt = 7$, find the value of $\dfrac{dv}{dt}$ when $v = 3$.

3) Find the derivatives of the following functions:
(a) $t^2 - 5t$ (b) $(y-1)^2$ (c) $(v - 2v^3)/v$.

4) Differentiate the function $f(x) \equiv x + \dfrac{1}{x}$ from first principles.

5) Plot the graph of $y = \sqrt{(1+x)}$ for integral values of x between -1 and 4.
Use your graph to estimate the value of $\dfrac{dy}{dx}$ when $x = 2$.

6) Find the coordinates of the point on $y = x^2 - 7x + 3$ at which the gradient is 2. Hence find the equation of the normal to $y = x^2 - 7x + 3$ which is parallel to $x + 2y - 1 = 0$.

7) Find the equations of the tangents to $y = x^3 + 3x$ which are parallel to the line $y = 15x + 2$.

8) Find the equation of the normal to $y = x^2 - 4$ which is parallel to $x + 3y - 1 = 0$.

9) Find the turning points on $y = 3x^4 + 4x^3 - 12x^2$. Give a rough sketch of the curve.

10) Find the stationary values of $f(x) \equiv x + \dfrac{1}{x}$ and use these to sketch the curve $y = x + \dfrac{1}{x}$.

11) Sketch the curve $y = (x - 3)^3$.

12) A closed cylindrical can has height h and base radius r. The volume is $0.01 \, \text{m}^3$. Show that $h = \dfrac{1}{100\pi r^2}$.

Show further that S, the surface area, is given by $S = 2\pi r^2 + \dfrac{1}{50r}$.

Hence find the value of r for which S is minimum.

13) The gradient function of $y = ax^2 + bx + c$ is $4x + 2$. The function has a minimum value of 1. Find the values of a, b and c.

14) An open rectangular box is made from a square sheet of cardboard by removing a square from each corner and joining the cut edges. If the cardboard is of edge $0.5 \, \text{m}$, find the maximum volume of the box.

15) A cylinder is cut from a solid sphere of radius $5 \, \text{cm}$. If the height of the cylinder is $2h$, show that the volume of the cylinder is $2\pi h(25 - h^2)$, assuming that the curved edges of the cylinder reach the surface of the sphere. Find the maximum volume of such a cylinder.

16) If a piece of string of fixed length is made to enclose a rectangle, show that the enclosed area is greatest when the rectangle is a square.

17) A point P moves along a straight line such that, after a time t seconds, its displacement s metres from a fixed point on the line is given by
$$s = 5t^2 + 1$$
What does $\dfrac{ds}{dt}$ represent?

If velocity is the rate of increase of displacement w.r.t. time, find the velocity of P when $t = 20$.

18) Find the maximum displacement of a particle from a point O, if its displacement s metres from O after time t seconds is given by
$$s = 2 + 3t - t^2$$

19) The table below shows the results of an experiment in which some hot liquid was left to cool and its temperature measured at intervals of one minute.

Time (minutes)	0	1	2	3	4	5
Temperature ($^\circ$C)	100°	93°	87°	83°	80°	78°

By drawing a graph of these results, estimate the rate of decrease of the temperature after three minutes.

20) In an experiment a small ball bearing was catapulted into a tank of viscous liquid. The table below records the penetration of the ball at second intervals.

Time (seconds)	0	1	2	3	4	5	6
Penetration (cm)	0	20	30	35	37	38	38.2

Plot these results on a graph and use your graph to estimate the rate of penetration after four seconds.

21) Show that the point $(p, 4p^2)$ lies on the curve $y = 4x^2$ for all real values of p. Find the equation of the tangent to $y = 4x^2$ at $(p, 4p^2)$.

22) Show that $\left(t, \dfrac{1}{t}\right)$ lies on the curve $y = \dfrac{1}{x}$ for all values of t. Find the equation of the tangent at $\left(t, \dfrac{1}{t}\right)$ to $y = \dfrac{1}{x}$. Find the area of the triangle enclosed by this tangent and the coordinate axes.

CHAPTER 6

TRIGONOMETRIC FUNCTIONS

CIRCULAR FUNCTIONS

Measurement of Rotation

When a line OP is pivoted at O and rotates from its initial position OP_0 to a new position OP_1, the angle P_0OP_1 is a measure of the rotation of OP.

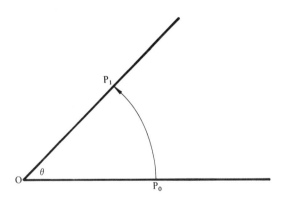

The angle θ is usually measured in one of two units.

The Degree

The ancient Babylonian Mathematicians, because they thought that the solar year was 360 days long, divided one complete revolution into 360 equal parts, each part now being known as one degree $(1°)$. Using the degree as the unit of rotation, half a revolution corresponds to $180°$ and quarter of a revolution, i.e. a right angle, corresponds to $90°$.

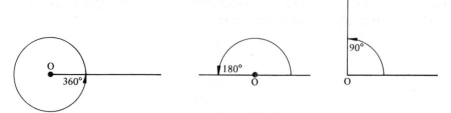

Angles smaller than a degree are usually given as decimal parts, e.g. half a degree is $0.5°$, but in some fields (e.g. navigation) a degree may be divided into 60 minutes ($60'$) and each minute into 60 seconds ($60''$).

The Radian

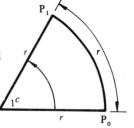

If an arc P_0P_1 of a circle is drawn so that it is equal in length to the radius of the circle, then the angle P_0OP_1 is called one *radian* (1^c)

The number of radians in one complete revolution is therefore given by the ratio $\dfrac{\text{circumference}}{\text{radius}}$.

But the circumference of a circle of radius r is of length $2\pi r$.

Thus there are $\dfrac{2\pi r}{r} = 2\pi$ radians in one revolution.

i.e. 2π radians = 360 degrees

Further, half a revolution is π radians = 180 degrees

and one right angle is $\dfrac{\pi}{2}$ radians = 90 degrees

When an angle is quoted in terms of π it is normal to omit the radian symbol. Thus we would write $180° = \pi$ (not π^c)

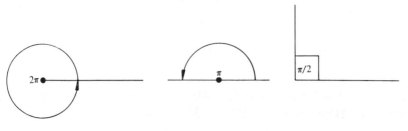

Other angles which are simple fractions of $180°$ can easily be expressed in radians in terms of π, using the relationship $180° = \pi$.

e.g.
$$60° = 60 \times \frac{\pi}{180} = \frac{\pi}{3}$$

$$135° = 135 \times \frac{\pi}{180} = \frac{3\pi}{4}$$

and conversely
$$\frac{7\pi}{6} = \frac{7\pi}{6} \times \frac{180°}{\pi} = 210°$$

$$\frac{5\pi}{3} = \frac{5\pi}{3} \times \frac{180°}{\pi} = 300°$$

Unit conversion carried out in this way is not convenient for angles which are not simple fractions of a revolution. It would not, for instance, be easy to express $47° \, 34'$ as a multiple of π. We would first have to express $34'$ as a decimal part of a degree (i.e. $34' = 0.567°$) and then use

$$47.567° = 47.567 \times \frac{\pi}{180} = 0.830^c$$

Calculations such as this are tedious and can be avoided either by using conversion tables or by using an electronic calculator which offers this facility. In order to visualise the size of an angle of 1 radian, it helps to remember that

$$\pi \text{ radians} = 180°$$

and
$$\pi = 3.14 \text{ (to 2 d.p.)}$$

so
$$1 \text{ radian} = \frac{180°}{3.14} = 57.3° \text{ (to 3 s.f.)}$$

Thus one radian is a little less than $60°$.

EXERCISE 6a

1) Without using tables, express the following angles in radians, giving your answer in terms of π:

$30°; \quad 270°; \quad 120°; \quad 45°; \quad 240°; \quad 150°; \quad 20°; \quad 300°; \quad 22.5°; \quad 80°$

2) Without using tables, express the following angles in degrees:

$$\frac{3\pi}{4}; \quad \frac{5\pi}{6}; \quad \frac{\pi}{10}; \quad \frac{3\pi}{2}; \quad \frac{11\pi}{6}; \quad \frac{\pi}{3}; \quad \frac{4\pi}{3}; \quad \frac{\pi}{12}; \quad \frac{\pi}{8}; \quad \frac{4\pi}{9}$$

3) Use tables to express the following angles in radians:

$35° \, 30'; \quad 78° \, 12'; \quad 54° \, 45'; \quad 70°; \quad 16°$

4) Use tables to express the following angles in degrees:

$2.86^c; \quad 4.21^c; \quad 1^c; \quad 3.47^c; \quad 3^c$

MENSURATION OF A CIRCLE

The reader will already be familiar with the formulae for the area and the circumference of a circle of radius r.

$$\text{Circumference} = 2\pi r \qquad \text{Area} = \pi r^2$$

These formulae can be used to derive further results.

Length of an Arc

Consider an arc which subtends an angle θ at the centre of the circle, *where θ is measured in radians*.

From the definition of a radian, the arc which subtends 1 radian at the centre is of length r. Hence an arc which subtends θ radians at the centre is of length $r\theta$.

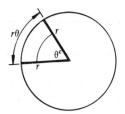

Area of a Sector

The area of a sector containing an angle of θ radians at the centre can be found by considering the sector as a fraction of a circle. The ratio of the area of the sector to the area of the circle is equal to the ratio of the angle θ contained in the sector to the angle 2π contained in the circle.

i.e.

$$\frac{\text{area of sector}}{\pi r^2} = \frac{\theta}{2\pi}$$

$\Rightarrow$

$$\text{area of sector} = \tfrac{1}{2}r^2\theta$$

Hence if an arc AB subtends an angle θ radians at the centre, O, of a circle of radius r

> Length of arc   AB $= r\theta$
>
> Area of sector   AOB $= \tfrac{1}{2}r^2\theta$

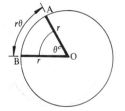
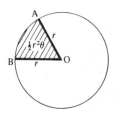

EXAMPLES 6b (π is taken as 3.142)

1) A railway line changes direction by $20°$ when passing round a circular arc of
length 500 m. What is the radius of the arc?

Length of arc $= r\theta$
(where θ is in radians)

and $\qquad 20° = \dfrac{20}{180} \times \pi = \dfrac{\pi}{9}$ radians

Hence $\qquad\qquad 500 = r\left(\dfrac{\pi}{9}\right)$

i.e. $\qquad\qquad r = \dfrac{4500}{\pi}$

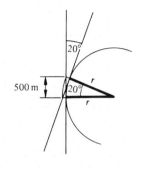

The radius of the track is therefore 1432 m.

2) A chord AB divides a circle of radius 2 m into two segments. If AB
subtends an angle of $60°$ at the centre of the circle, find the area of the minor
segment.

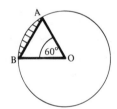

$$60° = \frac{\pi}{3} \text{ radians}$$

Area of sector $AOB = \frac{1}{2}r^2\theta = \frac{1}{2} \times 2^2 \times \dfrac{\pi}{3} = \dfrac{2\pi}{3} = 2.094 \text{ m}^2$

Area of triangle $AOB = \frac{1}{2}r^2 \sin\theta = \frac{1}{2} \times 2^2 \sin 60° = 1.732 \text{ m}^2$

Area of minor segment (shaded in the diagram) $= (2.094 - 1.732)\text{ m}^2$

$\qquad\qquad\qquad\qquad\qquad\qquad\qquad\qquad\qquad = 0.362 \text{ m}^2$

3) Two discs of radii 3 cm and 4 cm are laid on a table with their centres
5 cm apart. Find the perimeter of the 'figure-eight' shape so formed.

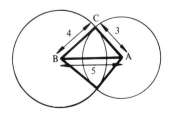 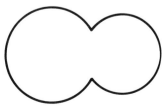

Let the discs intersect at C and have their centres at A and B.
Triangle ABC is right angled at C $(5^2 = 3^2 + 4^2)$

Thus $\tan \alpha = \frac{4}{3}$ and $\alpha = 53.13° = 0.927^c$

Hence $\beta = 90° - \alpha = 36.87°$ $= 0.644^c$

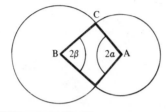 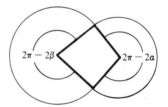

The perimeter is made up of
an arc subtending $(2\pi - 2\alpha)^c$ in the circle of radius 3 cm and
an arc subtending $(2\pi - 2\beta)^c$ in the circle of radius 4 cm.
Hence, using arc length $= r\theta$, we have

$$\text{perimeter} = 3(6.284 - 1.854) + 4(6.284 - 1.288) \text{ cm}$$

$$= 33.3 \text{ cm.}$$

EXERCISE 6b

1) If d is the length of an arc of a circle of radius r and θ is the angle
subtended by the arc at the centre of the circle, complete the following table:

d (cm)		9	2		5	
r (cm)	3		4	2		10
θ	30°	$\dfrac{2\pi}{3}$		$x°$	1.3^c	47° 18′

2) The moon subtends an angle of $31'$ at the earth and its distance from the
earth is 382 100 km. Find, in kilometres, the diameter of the moon.

3) Calculate, in degrees and minutes, the angle subtended at the centre of a
circle of radius 3.12 cm by an arc of length 6.14 cm.

4) If a is the area of a sector of a circle of radius r and θ is the angle
contained in the sector at the centre of the circle, complete the following table:

a (m²)		16	8		6	
r (m)	4		3	7		5
θ	30°	$\dfrac{3\pi}{4}$		$x°$	2.13^c	64° 28′

5) Calculate in degrees and minutes the angle at the centre of a circle of radius 5.29 cm contained in a sector of area 4.13 cm².

6) An arc **AB** of length 5 cm is marked on a circle of radius 3 cm. Find the area of the sector bounded by this arc and the radii from **A** and **B**.

7) A chord **AB** of length 4 cm divides a circle of radius 3.3 cm into two segments. Find the area of each segment.

8) A chord **AB** of length 5.2 cm subtends an angle of 120° at the centre of a circle. Calculate:
(a) the length of the arc **AB**,
(b) the area of the sector containing the angle 120°,
(c) the area of the minor segment cut off by **AB**.

9) Two discs, each of radius 11.6 cm, are laid on a table with their centres 14 cm apart. Find:
(a) the length of their common chord,
(b) the area common to the two discs.

10) The area of a sector of a circle, diameter 10.23 cm, is 22.86 cm². What is the length of the arc of the sector?

CIRCULAR FUNCTIONS

For any acute angle θ there are six trigonometric ratios, each of which is defined by referring to a right angled triangle containing θ.

$$\frac{AB}{OB} = \text{sine } \theta \quad \text{usually written} \quad \sin \theta$$

$$\frac{OA}{OB} = \text{cosine } \theta \qquad \text{or} \quad \cos \theta$$

$$\frac{AB}{OA} = \text{tangent } \theta \qquad \text{or} \quad \tan \theta$$

$$\frac{OB}{AB} = \text{cosecant } \theta \qquad \text{or} \quad \text{cosec } \theta \quad \left(\equiv \frac{1}{\sin \theta} \right)$$

$$\frac{OB}{OA} = \text{secant } \theta \qquad \text{or} \quad \sec \theta \qquad \left(\equiv \frac{1}{\cos \theta} \right)$$

$$\frac{OA}{AB} = \text{cotangent } \theta \qquad \text{or} \quad \cot \theta \qquad \left(\equiv \frac{1}{\tan \theta} \right)$$

Each of the above ratios has a unique value for any one acute angle and these values are available in tables or from electronic calculators.

Since we are now regarding an angle as the measure of rotation from a given position of a straight line about a fixed point, it is clear that the size of an angle is unlimited, as the line can keep on rotating indefinitely. The meaning of the six trigonometric ratios is, as yet, restricted to acute angles, since the definition used so far for each ratio refers to an angle in a right angled triangle.

If we wish to extend the application of trigonometric ratios to angles of any size, they must be defined in a more general way.

TRIGONOMETRIC RATIOS FOR A GENERAL ANGLE

The system of reference in which a general angle is measured is very similar to that used for polar coordinates (see, p. 70). The point about which the line OP rotates is the pole or origin, O, and the position from which the angle is measured is the initial line or x axis. An angle formed when the line rotates anticlockwise is taken as positive, while clockwise rotation corresponds to negative angles.

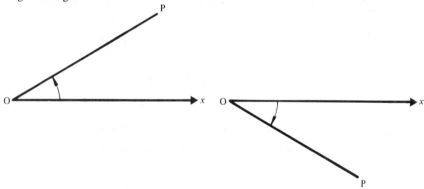

The pair of Cartesian axes divides the plane of rotation into four quadrants numbered 1, 2, 3, 4 as shown.

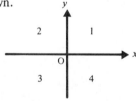

As the line OP rotates, P moves round the first quadrant and in this region its coordinates are both positive. As OP moves into the second quadrant its x coordinate becomes negative. In the third quadrant both the x and y coordinates of P are negative. Finally in the fourth quadrant we have a positive x coordinate and a negative y coordinate. The length, r, of OP (the radius vector) is taken always to be positive.

Denoting the angle through which OP has rotated as θ, the trigonometric ratios of θ are always defined in the following way

$$\sin \theta = \frac{y}{r}$$

$$\cos \theta = \frac{x}{r}$$

$$\tan \theta = \frac{y}{x}$$

The numerical values of these ratios can be found as follows.

From any position of P, a vertical line drawn from P to meet the x-axis at Q, forms a right angled triangle OPQ.

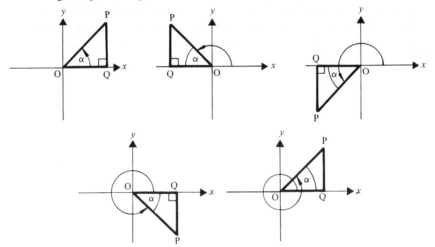

The angle POQ so formed is always acute regardless of the value of θ, and is called the associated acute angle, α. The value of α for a particular value of θ is the difference between θ and π (180°) or 2π (360°) or further multiples of π for larger angles.

e.g. when $\theta = 160°$ $\alpha = 180° - 160° = 20°$

when $\quad \theta = \dfrac{7\pi}{6} \qquad \alpha = \dfrac{7\pi}{6} - \pi = \dfrac{\pi}{6}$

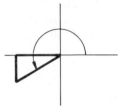

when $\quad \theta = 275° \qquad \alpha = 360° - 275° = 85°$

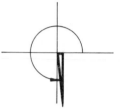

when $\quad \theta = 517° \qquad \alpha = 540° - 517° = 23°$

The signs of the trigonometric ratios depend upon the signs of x and y, the coordinates of the rotating point P, i.e. they depend upon the quadrant into which P has rotated.

1st Quadrant

All six ratios are positive and, since θ is acute, their numerical values can be obtained directly from tables.

If OP has rotated through more than a complete revolution.

$\alpha = \theta - 360°$

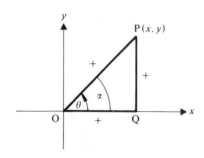

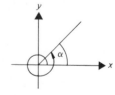

2nd Quadrant

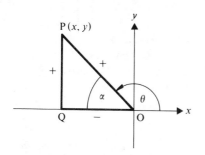

The associated acute angle, α, is $180° - \theta$.

Using the signs of the coordinates as shown, we see that:

the sine ratio, $\dfrac{y}{r}$, is positive,

the cosine ratio, $\dfrac{x}{r}$, is negative,

the tangent ratio, $\dfrac{y}{x}$, is negative.

The trigonometric ratios of θ when P is in the second quadrant are therefore defined as follows:

$$\left.\begin{array}{l} \sin\theta = +\sin\alpha \\ \cos\theta = -\cos\alpha \\ \tan\theta = -\tan\alpha \end{array}\right\} \text{ where } \alpha = 180° - \theta$$

3rd Quadrant

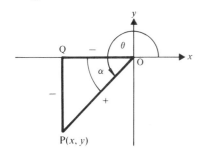

Here $\alpha = \theta - 180°$, and

the tangent ratio, $\dfrac{y}{x}$, is positive

while the sine ratio, $\dfrac{y}{r}$,

and the cosine ratio, $\dfrac{x}{r}$,

are both negative.

The trigonometric ratios of θ when P is in the third quadrant are therefore defined as follows:

$$\left.\begin{array}{l} \sin\theta = -\sin\alpha \\ \cos\theta = -\cos\alpha \\ \tan\theta = +\tan\alpha \end{array}\right\} \text{ where } \alpha = \theta - 180°$$

4th Quadrant

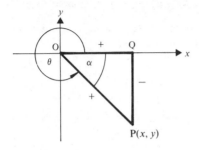

In this quadrant $\alpha = 360° - \theta$ and we see that the cosine ratio is positive but the tangent and sine ratios are negative.

The trigonometric ratios of θ when P is in the fourth quadrant are therefore defined as follows:

$$\left. \begin{array}{l} \sin \theta \; = \; - \sin \alpha \\ \cos \theta \; = \; + \cos \alpha \\ \tan \theta \; = \; - \tan \alpha \end{array} \right\} \quad \text{when} \quad \alpha \; = \; 360° - \theta$$

These results can be summarised in a quadrant diagram as shown.

$s+$	$s+$		s
$c-$	$c+$	s	c
$t-$	$t+$		t
$s-$	$s-$		
$c-$	$c+$	t	c
$t+$	$t-$		
(i)		or	(ii)

In diagram (ii) the ratios shown are those which are positive in each quadrant. The quadrant rule together with the value of α, the associated acute angle, enable us to find the value of any trig ratio of any angle.

Negative Angles

If OP rotates clockwise so that P moves through the quadrants in reverse order, i.e. 4th, 3rd, 2nd, 1st, θ is taken to be negative,

e.g.

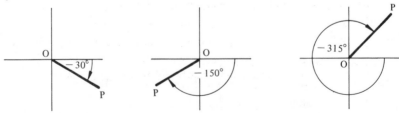

Every position of OP could be reached either by anticlockwise or by clockwise rotation and therefore corresponds to two different values of θ, one positive and one negative.

e.g.

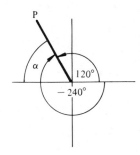

$\theta = + 120°$ or $\theta = -240°$
In both cases $\alpha = 60°$, hence the angle $+ 120°$ and the angle $-240°$ have the same trig ratios.

P is in the second quadrant where only the sine ratio is positive.

Hence

$$\sin 120° = \sin (- 240°) = + \sin 60°$$
$$\cos 120° = \cos (- 240°) = - \cos 60°$$
$$\tan 120° = \tan (- 240°) = - \tan 60°$$

EXAMPLES 6c

1) Find the sine, cosine and tangent of $243°$.

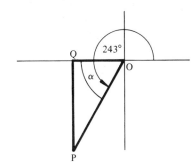

$\theta = 243°$

$\alpha = \theta - 180° = 63°$

P is in the third quadrant where only the tangent ratio is positive.

Hence

$$\sin 243° = - \sin 63° = - 0.8910$$
$$\cos 243° = - \cos 63° = - 0.4540$$
$$\tan 243° = + \tan 63° = 1.9626$$

2) If $\cos \theta = 0.866$ and $\tan \theta$ is negative, find $\sin \theta$.

θ is in a quadrant where the cosine ratio is positive and the tangent ratio is negative,
i.e. in the fourth quadrant where $\alpha = 360° - \theta$.

Now $$\cos \alpha = 0.866$$

$$\Rightarrow \qquad\qquad \alpha = 30°$$

Hence $$\theta = 330°$$

In the fourth quadrant the sine ratio is negative.

Therefore $$\sin \theta = -\sin \alpha = -\sin 30°$$

i.e. $$\sin \theta = -0.5$$

3) Given that $\tan \theta = 1$ and $-2\pi < \theta < 2\pi$, give four possible values for θ.

The tangent ratio is positive in the first and third quadrants.

When $\tan \alpha = 1$, $\alpha = \dfrac{\pi}{4}$ (or $45°$).

In this problem the range of values of θ is specified in radians so the solution should also be given in radians. Thus we use $\alpha = \dfrac{\pi}{4}$.

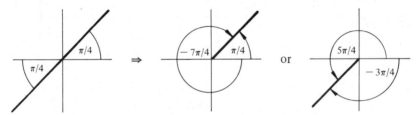

Hence $\theta = \dfrac{\pi}{4}$ or $-\dfrac{7\pi}{4}$, $\dfrac{5\pi}{4}$ or $-\dfrac{3\pi}{4}$.

In this example we are solving a simple trig equation.

4) An angle θ has an associated acute angle of $53°$. Find $\sin \theta$.

Only α is given in this case, so θ could be in any of the four quadrants.

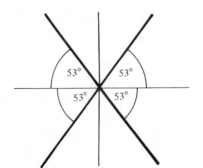

In quadrants 1 and 2,
$\sin \theta = \sin 53° = 0.7986$.

In quadrants 3 and 4,
$\sin \theta = -\sin 53° = -0.7986$.

Therefore $\sin \theta = \pm 0.7986$.

EXERCISE 6c

1) Write down the associated acute angle when θ is:

$93°$; $\dfrac{3\pi}{4}$; $308°$; $\dfrac{8\pi}{3}$; $22°$; $864°$; $\dfrac{7\pi}{6}$; $\dfrac{11\pi}{4}$; $-\dfrac{5\pi}{4}$; $-176°$

2) Complete the following table

	sin θ	cos θ	tan θ	α	θ
(a)	0.3907		-0.4245		
(b)		-0.7071	1.0000		
(c)					$\dfrac{5\pi}{6}$
(d)					$300°$
(e)					$170°$
(f)	0.5000	-0.8660			
(g)		0.5000			
(h)				$45°$	

3) Within the range $-360° \leqslant \theta \leqslant 360°$, give all values of θ for which:
(a) $\sin\theta = 0.4$ (b) $\cos\theta = -0.5$ (c) $\tan\theta = 1.2$
(d) $\operatorname{cosec}\theta = -1.5$ (e) $\sec\theta = 2.5$ (f) $\cot\theta = -2$

4) Find the smallest (positive or negative) angle for which:
(a) $\cos\theta = 0.8$ and $\sin\theta$ is positive,
(b) $\sin\theta = -0.6$ and $\tan\theta$ is negative,
(c) $\cos\theta$ is positive and $\tan\theta = \sin 30°$.

NUMERICAL TRIGONOMETRY

Triangles are involved in many practical measurements (e.g. in surveying). A triangle has three sides and three angles and is specified by the values of any three members of the set {3 sides, 2 angles}. (The sum of the three angles of a triangle is $180°$, so if any two angles are known the third angle can be calculated directly and is therefore not an independent item.) From any set of data sufficient to define a triangle, the remaining sides and angles can be calculated. This is called *solving* the triangle and requires the use of one of the various formulae that relate the sides and angles of a triangle.
The two relationships that are used most frequently are the sine rule and the cosine rule.

The Sine Rule

In a triangle ABC in which A, B, C are used to denote the angles at the vertices A, B, C and a, b, c are used to denote the sides opposite to these vertices respectively,

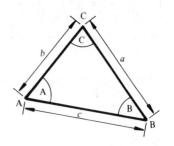

$$\frac{a}{\sin A} = \frac{b}{\sin B} = \frac{c}{\sin C}$$

Proof

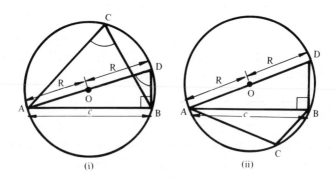

(i)   (ii)

Taking O and R as the centre and radius of the circle through A, B and C, draw the diameter AD and join DB.

Then $\angle ABD = 90°$ (angle in a semicircle).

Thus $c = 2R \sin D$

But, in diagram (i) $C = D$ (angles in same segment)

and in diagram (ii) $C = 180° - D$ (opposite angles of a cyclic quadrilateral).

Thus, in both diagrams, $\sin C = \sin D$.

Hence $c = 2R \sin C \Rightarrow \dfrac{c}{\sin C} = 2R$

Similarly $\dfrac{b}{\sin B} = 2R$ and $\dfrac{a}{\sin A} = 2R$

$\Rightarrow$ $\dfrac{a}{\sin A} = \dfrac{b}{\sin B} = \dfrac{c}{\sin C} = 2R$

Any pair of these three equal ratios comprises an equation containing two sides and two angles.

Thus the sine rule can be used to solve a triangle in which we know either two sides and one angle

or   two angles and one side

provided that one given side is opposite to a given angle.

For instance, if in a triangle ABC it is known that $A = 73°$, $B = 49°$ and $a = 12.2$ cm   then, as a, A and B are known, b can be calculated using

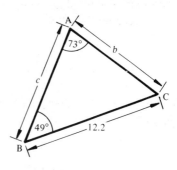

$$\frac{a}{\sin A} = \frac{b}{\sin B} \Rightarrow b = \frac{12.2 \sin 49°}{\sin 73°}$$

$$= 9.63 \text{ cm}$$

Also, since $C = 180° - 73° - 49° = 58°$,

c can be calculated using

$$\frac{a}{\sin A} = \frac{c}{\sin C} \Rightarrow c = \frac{12.2 \sin 58°}{\sin 73°}$$

$$= 10.82 \text{ cm}$$

In this example the given data defines one and only one triangle.

Sometimes, when two sides and one angle are specified, two different triangles can be found from the given data.

The Ambiguous Case

Suppose that we have to solve the triangle in which $A = 24°$, $c = 2.6$ cm and $a = 1.1$ cm.

Knowing a, A and c we can find C using

$$\frac{a}{\sin A} = \frac{c}{\sin C} \Rightarrow \sin C = \frac{2.6}{1.1} \sin 24°$$

i.e. $\sin C = 0.9614$

Now in a triangle, $\sin C = 0.9614 \Rightarrow C = 74°$ or $106°$.

But we must check whether these angles are possible values for C, by considering:

(a) if $C = 74°$ and $A = 24°$ (given)

$$74° + 24° = 98° \text{ which is less than } 180°.$$

So $74°$ is a possible value for C corresponding to $B = 82°$.

(b) if $C = 106°$ and $A = 24°$

$$106° + 24° = 130° \text{ which is less than } 180°.$$

So $106°$ is *also* a possible value for C corresponding to $B = 50°$.

Thus we see that there are two possible triangles including the given data.

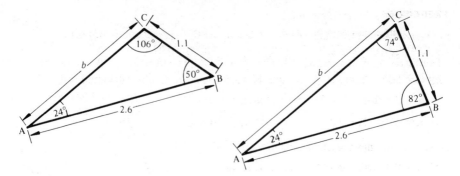

This is known as the *ambiguous case* and it can easily be understood if an attempt is made to *construct* the triangle from the given specification.

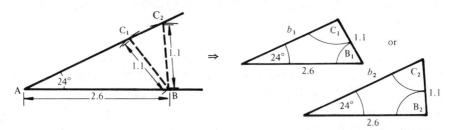

The constructions show that there are two possible positions for C, each corresponding to a different pair of values for B and b.

In each case the solution of the triangle can be completed using $\dfrac{a}{\sin A} = \dfrac{b}{\sin B}$

Note. It must not be thought that there are *always* two possible triangles when one angle and two sides are given. The following example shows that this is not necessarily so.

If $A = 37°$, $a = 4.59$ cm   and $c = 2.1$ cm   show that there is only one possible triangle ABC and find the remaining angles.

Using $\dfrac{a}{\sin A} = \dfrac{c}{\sin C} \Rightarrow \sin C = \dfrac{2.1}{4.59} \sin 37°$

i.e. $\sin C = 0.2753$

Now, in a triangle, $\sin C = 0.2753 \Rightarrow C = 16°$ or $164°$.

If $C = 16°$ and $A = 37°$, $A + C = 53°$ $(< 180°)$.
So $16°$ is a possible value for C corresponding to $B = 127°$.

If $C = 164°$ and $A = 37°$, $A + C = 201°$ $(> 180°)$.
So $164°$ is *not* a possible value for C.

Hence there is only one triangle defined by the given data and its other angles are $C = 16°$ and $B = 127°$

EXERCISE 6d

In Questions 1–7, the data refers to triangle ABC, using standard notation.

1) $B = 35°$, $a = 2.7\,cm$, $b = 5.1\,cm$; find A.

2) $C = 52°$, $B = 49°$, $c = 8.62\,cm$; find b.

3) $b = 3.8\,cm$, $A = 25°$, $a = 1.8\,cm$; find C.

4) $A = 87°$, $B = 63°$, $c = 3.2\,cm$; find a.

Solve the triangles specified in Questions 5–7.

5) $a = 9.12\,cm$, $b = 4.87\,cm$, $A = 63°$.

6) $c = 5.73\,cm$, $B = 19.3°$, $b = 2.16\,cm$.

7) $B = 83.2°$, $C = 43.7°$, $a = 19.86\,cm$.

8) In a triangle PQR, angle $PQR = \theta$ and angle $QPR = 2\theta$. Prove that
$$\cos \theta = \frac{p}{2q}.$$

9) Prove the sine rule for a triangle ABC by drawing AD perpendicular to BC, meeting BC at D. Consider two cases (a) B acute, (b) B obtuse.

10) From a point P on the same level as the base of a tower, the angle of elevation of the top of the tower is 24.8°. From a point Q, 5 m vertically above P, the angle of elevation of the top of the tower is 18.7°. Find the height of the tower.

The Cosine Rule

We have seen that the sine rule can be applied only when we know an angle opposite to a given side. It cannot be used, for instance, when two sides and the included angle are given. Such cases can usually be solved using the cosine rule

$$a^2 = b^2 + c^2 - 2bc \cos A$$

A reminder of the proof of this formula is given below.

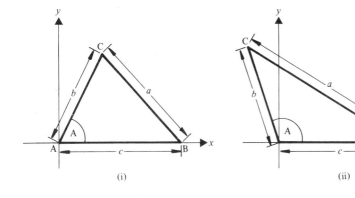

(i)   (ii)

Placing the triangle ABC on Cartesian axes as shown, we see that, in diagram (i) the coordinates of B and C respectively are

$$(c, 0) \quad \text{and} \quad (b \cos A, b \sin A)$$

in diagram (ii) the coordinates of B and C respectively are

$$(c, 0) \quad \text{and} \quad (-b \cos \{\pi - A\}, b \sin A) \Rightarrow (b \cos A, b \sin A)$$

Thus
$$a^2 = (b \sin A)^2 + (b \cos A - c)^2$$
$$= b^2 \sin^2 A + b^2 \cos^2 A - 2bc \cos A + c^2$$

i.e.
$$a^2 = b^2 + c^2 - 2bc \cos A$$

Similarly it can be shown that

$$b^2 = c^2 + a^2 - 2ca \cos B$$

and
$$c^2 = a^2 + b^2 - 2ab \cos C$$

The calculation involved in using this formula is less straightforward than that required in using the sine formula. Consequently the cosine rule is used only when the sine rule is inapplicable. For instance, suppose that we have a triangle ABC in which $a = 17.5$ cm, $b = 8.4$ cm and $c = 11.9$ cm and we are asked to find the largest angle.

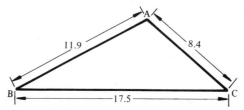

The longest side is opposite to the largest angle so we must find A.
But as a, b, c are given, we must use the cosine rule (rearranged so that A can be found conveniently), i.e.

$$\cos A = \frac{b^2 + c^2 - a^2}{2bc}$$

$$= \frac{(8.4)^2 + (11.9)^2 - (17.5)^2}{2(8.4)(11.9)}$$

$$= -0.4706$$

Hence $A = 118.1°$ (obtuse because $\cos A$ is negative)

When the solution of a triangle requires initial use of the cosine rule, the remaining sides or angles can be found using the sine rule. For example, to solve the triangle PQR given that $p = 12.1$ cm, $q = 7.3$ cm and $R = 37.5°$ we first use

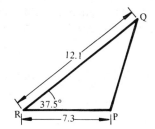

$r^2 = p^2 + q^2 - 2pq \cos R$

$\Rightarrow \quad r = 7.72 \, \text{cm}$

Now the sine rule can be used since
we know r, R, p and q.

Thus $\qquad \dfrac{r}{\sin R} = \dfrac{q}{\sin Q} \Rightarrow \sin Q = \dfrac{7.3 \sin 37.5°}{7.72} = 0.5756$

Hence $\quad Q = 35.1°$. (Q cannot be obtuse as q is not the longest side of the triangle.)

Finally P can be found using $180° - Q - R$.

EXERCISE 6e

Solve the following triangles.

1) $a = 4$, $b = 7$, $c = 5$.

2) $A = 79°$, $b = 27 \, \text{cm}$, $c = 19.8 \, \text{cm}$.

3) $a = 15.73 \, \text{m}$, $B = 121°$, $c = 23.15 \, \text{m}$.

4) $a = 12.84 \, \text{m}$, $b = 6.58 \, \text{m}$, $c = 8.13 \, \text{m}$.

5) Using the cosine formula $\quad \cos A = \dfrac{b^2 + c^2 - a^2}{2bc}$:

(a) show that A is acute if $\quad a^2 < b^2 + c^2$,

(b) show that A is obtuse if $\quad a^2 > b^2 + c^2$,

(c) verify Pythagoras' Theorem by taking $\quad A = 90°$.

Illustrate each of these cases by a diagram.

6) Find the angles of a triangle whose sides are in the ratio $2:3:4$.

THE GRAPHS OF THE CIRCULAR FUNCTIONS

We have seen (p. 145) that there is a unique value for each trig ratio of any angle. Thus the mappings $\quad \theta \to \sin \theta$, $\theta \to \cos \theta$,   etc., are functions and we are now able to plot a graph showing the behaviour of each trigonometric function as θ varies. The graphs of the three major circular functions are particularly important and these will now be examined.

1) $f: \theta \rightarrow \sin \theta, \quad \theta \in \mathbb{R}$.

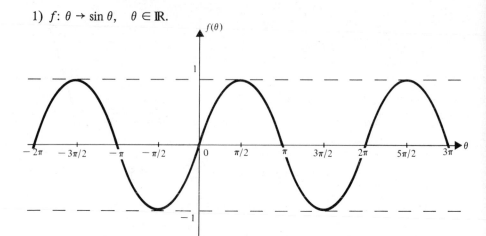

The above graph shows that the sine function has the following characteristics:

(a) It is continuous (i.e. its graph has no breaks).

(b) Its range is $-1 \leqslant \sin \theta \leqslant +1$.

(c) The shape of the graph from $\theta = 0$ to $\theta = 2\pi$ is repeated for each further complete revolution. Such a function is said to be *periodic* or *cyclic*. The width of the repeating pattern, as measured on the horizontal axis, is called the *period*.

Thus $f: \theta \rightarrow \sin \theta$ is a periodic function with a period of 2π, a maximum value of 1 and a minimum value of -1. A graph of this shape, for obvious reasons, is known as a *sine wave*.

The greatest value of $|\sin \theta|$ is called the *amplitude* of the sine wave and its value is 1.

2) $f: \theta \rightarrow \cos \theta, \quad \theta \in \mathbb{R}$.

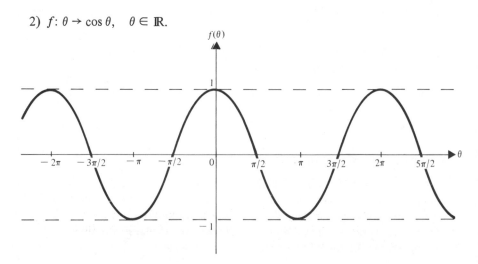

The characteristics of the graph of the cosine function are as follows:

(a) It is continuous.

(b) It lies entirely within the range $-1 \leqslant \cos \theta \leqslant 1$.

(c) It is periodic with a period of 2π.

(d) It has the same shape as the sine graph but is displaced a distance $\dfrac{\pi}{2}$ to the left on the horizontal axis. Such a displacement is known as a phase difference or phase shift.

Thus $f: \theta \to \cos \theta$ is a cyclic function with period 2π, and range $[-1, 1]$.

3) $f: \theta \to \tan \theta, \quad \theta \in \mathbb{R}$.

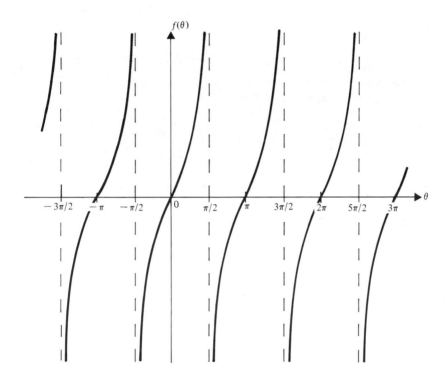

The behaviour of the tangent function is different from that of the sine and cosine functions in several respects.

(a) It is not continuous, being *undefined* when $\theta = -\dfrac{\pi}{2}, \dfrac{\pi}{2}, \dfrac{3\pi}{2},$ etc.

(b) The range of possible values of $\tan \theta$ is unlimited.

(c) The tangent function is periodic but the period in this case is π (not 2π as in the previous cases).

Special Values

It is useful to note the angles whose trig ratios have the values $0, \pm 1$, and (for $\tan \theta$ only) $\pm \infty$.

Reference to the graph of $f(\theta) \equiv \sin \theta$ shows that

$$\sin \theta = 0 \qquad \text{when} \quad \theta = \ldots -2\pi, -\pi, 0, \pi, 2\pi, 3\pi \ldots$$

$$\text{i.e. when} \quad \theta = \text{any whole multiple of } \pi$$

Thus $\sin \theta = 0$ when $\theta = n\pi$ where $n \in \mathbb{Z}$

$$\sin \theta = 1 \qquad \text{when} \quad \theta = \ldots -\frac{3\pi}{2}, \frac{\pi}{2}, \frac{5\pi}{2} \ldots$$

$$= \ldots \left(-2\pi + \frac{\pi}{2}\right), \frac{\pi}{2}, \left(2\pi + \frac{\pi}{2}\right) \ldots$$

$$= (\text{an even multiple of } \pi) + \frac{\pi}{2}$$

Thus $\sin \theta = 1$ when $\theta = 2n\pi + \dfrac{\pi}{2}$

$$\sin \theta = -1 \qquad \text{when} \quad \theta = \ldots -\frac{\pi}{2}, \frac{3\pi}{2}, \frac{7\pi}{2} \ldots$$

$$= (\text{an even multiple of } \pi) - \frac{\pi}{2}$$

Thus $\sin \theta = -1$ when $\theta = 2n\pi - \dfrac{\pi}{2}$

Similar examination of the graph $f(\theta) \equiv \cos \theta$ shows that

$$\cos \theta = 0 \qquad \text{when} \quad \theta = \ldots -\frac{3\pi}{2}, -\frac{\pi}{2}, \frac{\pi}{2}, \frac{3\pi}{2}, \frac{5\pi}{2} \ldots$$

$$\text{i.e. when} \quad \theta = \text{an odd multiple of } \frac{\pi}{2}$$

Thus $\cos \theta = 0$ when $\theta = (2n + 1)\dfrac{\pi}{2}$

$$\cos \theta = 1 \qquad \text{when} \quad \theta = \ldots -2\pi, 0, 2\pi, 4\pi \ldots$$

$$= \text{an even multiple of } \pi$$

Thus $\cos \theta = 1$ when $\theta = 2n\pi$

$$\cos \theta = -1 \qquad \text{when} \quad \theta = \ldots -\pi, \pi, 3\pi, 5\pi \ldots$$

$$= \text{an odd multiple of } \pi$$

Thus $\cos \theta = -1$ when $\theta = (2n + 1)\pi$

Now considering the graph of $f(\theta) \equiv \tan \theta$

$$\tan \theta = 0 \qquad \text{when} \quad \theta = \ldots - \pi, 0, \pi, 2\pi \ldots$$

Thus $\tan \theta = 0$ when $\theta = n\pi$

$$\tan \theta = \pm \infty \quad \text{when} \quad \theta = \ldots \frac{\pi}{2}, \frac{3\pi}{2}, \frac{5\pi}{2} \ldots$$

Thus $\tan \theta = \pm \infty$ when $\theta = (2n + 1)\dfrac{\pi}{2}$

Note: Frequent use will be made, throughout any study of mathematics, of the expressions used above for odd and even numbers. In general we use $2n$ to represent all even numbers and $(2n + 1)$ to represent all odd numbers, provided that (as in the previous analysis) n is an integer, i.e. $n \in \mathbb{Z}$.

THE RECIPROCAL (MINOR) TRIGONOMETRIC RATIOS

The three major trig ratios, $\sin \theta$, $\cos \theta$ and $\tan \theta$, are used much more frequently than the three minor trig ratios, cosec θ, sec θ and cot θ. Consequently our analysis will tend to concentrate on the major ratios. But the reciprocal ratios must not be overlooked completely so at this stage we will examine the graphs and properties of these functions.

The graph of the function $f(\theta) \equiv \text{cosec } \theta$ can be drawn without any reference to tables of values, simply by observing the graph of $f(\theta) \equiv \sin \theta$ and using the following properties of *any* reciprocal functions.

1) (a) The reciprocal of zero is $\pm \infty$.
   (b) The reciprocal of $\pm \infty$ is zero.

2) (a) The reciprocal of 1 is 1.
   (b) The reciprocal of -1 is -1.

3) If one expression has a maximum value its reciprocal has a minimum value (and conversely).

4) If one expression is increasing, its reciprocal is decreasing (and conversely).

5) An expression and its reciprocal have the same sign.

(**Note:** The above properties are logical enough to accept without detailed analysis at this stage. A fuller treatment of reciprocal curves is covered in Chapter 11).

Considering the range of values $-\pi \leqslant \theta \leqslant 2\pi$ for $f(\theta) \equiv \sin\theta$ and $f(\theta) \equiv \operatorname{cosec}\theta$ we have:

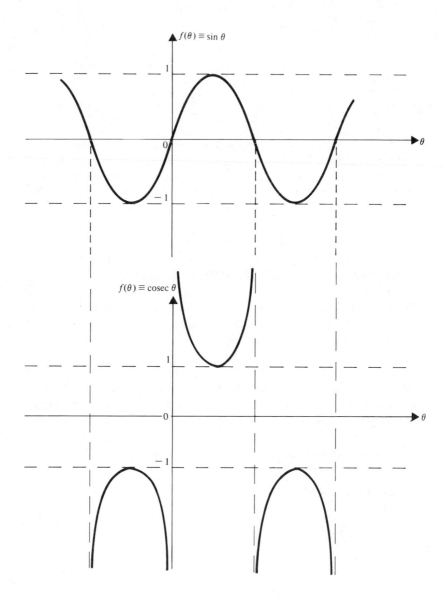

Similarly we can deduce the graphs of $f(\theta) \equiv \sec\theta$ and $f(\theta) \equiv \cot\theta$. These curves are shown opposite.

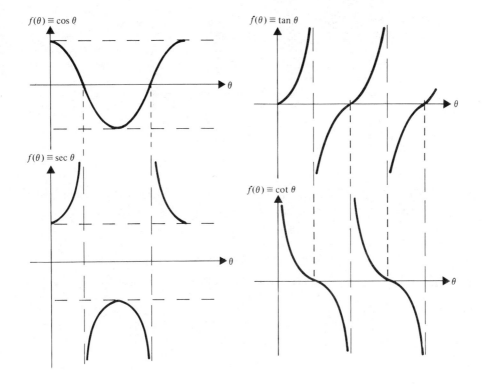

INVERSE CIRCULAR FUNCTIONS

The function $f: x \to \sin x$ is a many-one mapping for the domain $x \in \mathbb{R}$
and so does not have an inverse function.

However, by redefining the domain as $\left[-\dfrac{\pi}{2}, \dfrac{\pi}{2}\right]$ the function $f: x \to \sin x$
is a one-one mapping and now does have an inverse. This inverse sine function
is denoted by arcsin   or $\sin^{-1}$.

Now, for $f: x \to \sin x$,   the input is an angle and the output is a number.
So, for the inverse function $f^{-1}: x \to \arcsin x$,   the input is a number and the
output is an angle. Thus arcsin x is an angle function and means
'the angle whose sine is x'.

Thus   if $f: x \to \sin x$, $-\dfrac{\pi}{2} \leqslant x \leqslant \dfrac{\pi}{2}$

then $f^{-1}: x \to \arcsin x$, $-1 \leqslant x \leqslant 1$

The graphs of $\sin x$ and its inverse, arcsin x,   are shown overleaf.

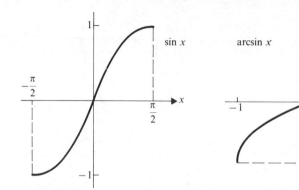

Similarly, if $f: x \to \cos x$, $0 \leqslant x \leqslant \pi$, then f^{-1} exists and is denoted by arccos or $\cos^{-1}$ where $\arccos x$ means 'the angle whose cosine is x',

Thus   if $f: x \to \cos x$, $0 \leqslant x \leqslant \pi$

                then $f^{-1}: x \to \arccos x$, $-1 \leqslant x \leqslant 1$

The graphs of $\cos x$ and its inverse, $\arccos x$, are shown below.

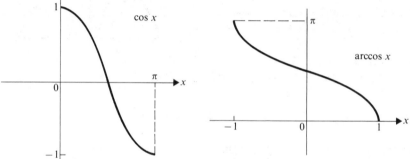

Further, if $f: x \to \tan x$ for $-\dfrac{\pi}{2} < x < \dfrac{\pi}{2}$, the inverse function, f^{-1},

exists and is written arctan or $\tan^{-1}$ where $\arctan x$ means 'the angle whose tangent is x'.

In this case the graphs of f and f^{-1} are

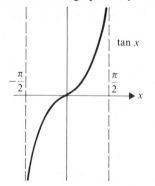
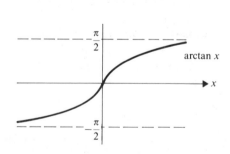

Note that the domain of $f^{-1}: x \to \arctan x$ is $x \in \mathbb{R}$.

It is most important, if the notation $\sin^{-1}$, etc. is adopted, to appreciate that $\sin^{-1}x$ is *not* the same as $\dfrac{1}{\sin x}$.

Common Trigonometric Ratios

The angles $30°$, $45°$ and $60°$ (and other angles for which these are the associated acute angles, e.g. $150°, 225°, 300° \ldots$) are used frequently in problems so it is well worth noting the values of these trig ratios.
Consider first an equilateral triangle ABC which is bisected by the line AD.

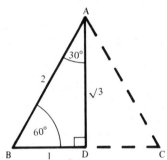

Then,

 if AB $=$ 2 units

 BD $=$ 1 unit

and AD $= \sqrt{3}$ units (Pythagoras)

In $\triangle$ BAD

$$\angle B = 60° \left(\frac{\pi}{3}\right), \quad \text{since } \triangle ABC \text{ is equilateral}$$

and $\quad \angle A = 30° \left(\frac{\pi}{6}\right), \quad$ since angle BAC is bisected

Therefore

$$\sin 30° = \sin\frac{\pi}{6} = \frac{1}{2} \qquad \sin 60° = \sin\frac{\pi}{3} = \frac{\sqrt{3}}{2}$$

$$\cos 30° = \cos\frac{\pi}{6} = \frac{\sqrt{3}}{2} \qquad \cos 60° = \cos\frac{\pi}{3} = \frac{1}{2}$$

$$\tan 30° = \tan\frac{\pi}{6} = \frac{1}{\sqrt{3}} \qquad \tan 60° = \tan\frac{\pi}{3} = \sqrt{3}$$

Alternatively we can say

$$\arcsin\frac{1}{2} = \frac{\pi}{6} \qquad \arcsin\frac{\sqrt{3}}{2} = \frac{\pi}{3}$$

$$\arccos\frac{\sqrt{3}}{2} = \frac{\pi}{6} \qquad \arccos\frac{1}{2} = \frac{\pi}{3}$$

$$\arctan\frac{1}{\sqrt{3}} = \frac{\pi}{6} \qquad \arctan\sqrt{3} = \frac{\pi}{3}$$

Now consider a triangle ABC in which AB = BC and ∠B is a right angle. If AB = BC = 1 unit, then AC = √2 units (Pythagoras).

Also since AB = BC,

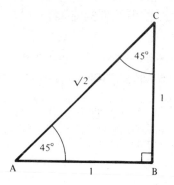

$$\angle A = \angle C = 45° \left(\frac{\pi}{4}\right)$$

Therefore

$$\sin 45° = \sin \frac{\pi}{4} = \frac{1}{\sqrt{2}} = \frac{\sqrt{2}}{2}$$

$$\cos 45° = \cos \frac{\pi}{4} = \frac{1}{\sqrt{2}} = \frac{\sqrt{2}}{2}$$

$$\tan 45° = \tan \frac{\pi}{4} = 1$$

Also

$$\arcsin \frac{1}{\sqrt{2}} = \frac{\pi}{4} = \arccos \frac{1}{\sqrt{2}}$$

and

$$\arctan 1 = \frac{\pi}{4}$$

Complementary Angles

If the sum of two acute angles is 90° $\left(\frac{\pi}{2}\right)$ they are said to be complementary and each is the complement of the other.

Consider a right angled triangle ABC containing the angles α and β as shown.

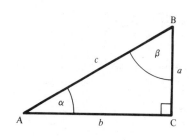

$$\sin \alpha = \frac{a}{c} = \cos \beta$$

$$\cos \alpha = \frac{b}{c} = \sin \beta$$

$$\tan \alpha = \frac{a}{b} = \cot \beta$$

$$\cot \alpha = \frac{b}{a} = \tan \beta$$

But α and β are complementary. Therefore we have shown that

$\left\{\begin{array}{l}\text{the sine of an angle is the cosine of its complement,} \\ \text{the tangent of an angle is the cotangent of its complement.}\end{array}\right.$

Because of this property, the sine and cosine of an angle are called complementary ratios.

Similarly the tangent and cotangent ratios are also said to be complementary.

Thorough familiarity with these special values and relationships and with the shapes and properties of the graphs of the trigonometric functions is essential in all further trigonometric analysis.

THE SOLUTION OF TRIGONOMETRIC EQUATIONS

If at least one term in an equation contains a trig ratio it is called a trig equation. Its solution involves finding the angle or angles for which it is valid. Consider the simple equation $\sin \theta = 0$

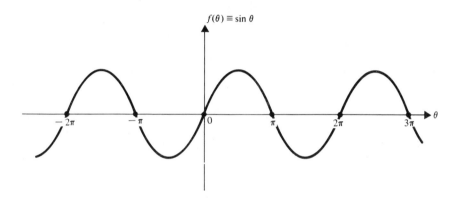

Referring to the graph of the sine function we see that

$$\sin \theta = 0 \quad \text{when} \quad \theta \text{ is any multiple of } \pi,$$

i.e.   when $\theta = n\pi$ where $n \in \mathbb{Z}$.

The full, or general, solution to this simple equation is therefore the infinite set of angles $\theta = n\pi$ (or $\theta = 180n°$).

Sometimes it is necessary to extract certain specific values of θ from the infinite set. Consider, for instance, a modified form of the above problem, viz:

Solve the equation $\sin \theta = 0$ for $-\pi \leqslant \theta \leqslant \pi$.

This time the finite solution set is $\theta = -\pi, 0, \pi$.

There are two basic approaches to finding the solution of a trig equation. One of them was used above and involves reference to the graph of the appropriate circular function. This approach is usually best for dealing with trig ratios of value ± 1 (for sine and cosine), $\pm \infty$ (for tangent) and zero.

Alternatively, the position of a rotating line OP in the appropriate quadrants can lead to a clear solution.

But in all cases, the first step in solving a trig equation is to find the *principal solution* which is the *principal value*, PV, of θ.

Principal Values

1) $f(\theta) \equiv \sin \theta$.

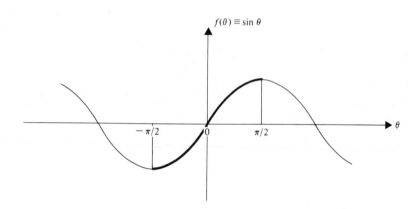

From the graph of the sine function it can be seen that, for the domain $\left[-\dfrac{\pi}{2}, \dfrac{\pi}{2}\right]$, every possible value of $\sin \theta$ occurs once and only once.

Thus any equation

$$\sin \theta = s \qquad (-1 \leqslant s \leqslant 1)$$

has one and only one solution in this interval and this is the *principal value* of θ. The position of this principal solution is therefore in either the first or the fourth quadrant.

e.g.   if $\sin \theta = \frac{1}{2}$ the principal solution is $\theta = \dfrac{\pi}{6}$

         if $\sin \theta = -\frac{1}{2}$ the principal solution is $\theta = -\dfrac{\pi}{6}$

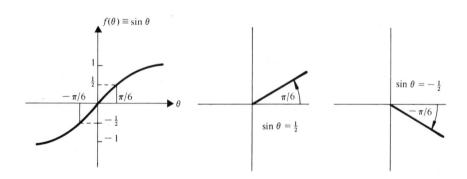

2) $f(\theta) \equiv \cos \theta$.

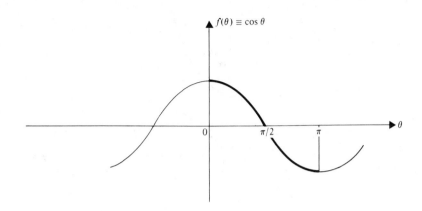

Every possible value of $\cos \theta$ occurs once and only once in the domain $[0, \pi]$ so there is one and only one solution of the equation

$$\cos \theta = c \qquad (-1 \leqslant c \leqslant 1)$$

within this interval. This is the principal value of θ and in this case the principal solution is in either the first or the second quadrant,

e.g.　　　if $\cos \theta = \frac{1}{2}$　　the principal solution is $\theta = \dfrac{\pi}{3}$

　　　　　if $\cos \theta = -\frac{1}{2}$　the principal solution is $\theta = \dfrac{2\pi}{3}$

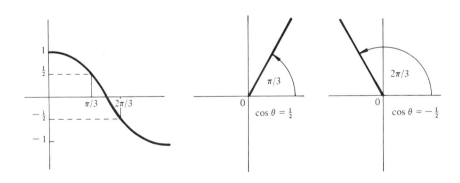

3) $f(\theta) \equiv \tan \theta$.

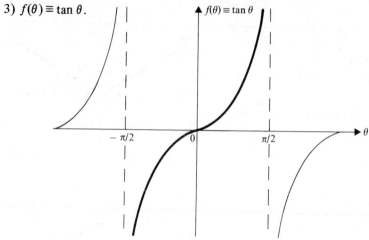

Every possible value of $\tan \theta$ occurs once and only once for angles in the interval $-\dfrac{\pi}{2} \leqslant \theta \leqslant \dfrac{\pi}{2}$. One and only one solution of the equation

$$\tan \theta = t$$

is in this interval and it is the principal value of θ. The principal solution is therefore in the first or the fourth quadrant,

e.g.   if $\tan \theta = 1$ the principal solution is $\theta = \dfrac{\pi}{4}$

              if $\tan \theta = -1$ the principal solution is $\theta = -\dfrac{\pi}{4}$

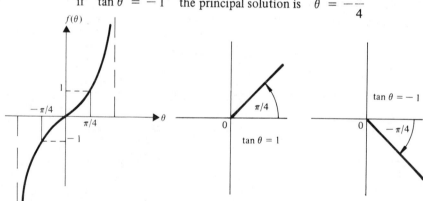

Secondary Values

Having determined the quadrant in which the principal value of a trig equation lies, it is usually found that a second angle with the same trig ratio occurs in the interval $-\pi \leqslant \theta \leqslant \pi$.

This solution lies in a different quadrant and is called the *secondary value*, SV, of θ, or the *secondary solution* of the equation,

e.g.

if $\sin \theta = \frac{1}{2}$ the secondary
solution is in the second
quadrant where the sine ratio is also
positive.

The secondary value is $\theta = \dfrac{5\pi}{6}$.

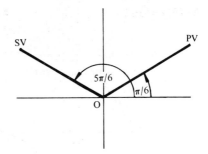

If $\sin \theta = -\frac{1}{2}$ the secondary
solution is in the third quadrant
where the sine ratio is also
negative.

The secondary value is $\theta = -\dfrac{5\pi}{6}$.

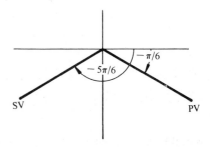

If $\cos \theta = \frac{1}{2}$ the secondary
value is in the fourth quadrant

and is $\theta = -\dfrac{\pi}{3}$.

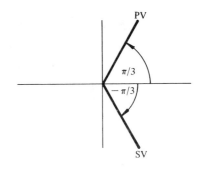

If $\cos \theta = -\frac{1}{2}$ the secondary
value is in the third quadrant

and is $\theta = -\dfrac{2\pi}{3}$.

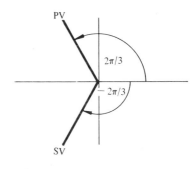

Note that, for an equation of the form $\cos \theta = c$, $\mathrm{SV} = -\mathrm{PV}$.

If $\tan \theta = 1$ the secondary
solution is in the third quadrant,
and is $\theta = \dfrac{-3\pi}{4}$.

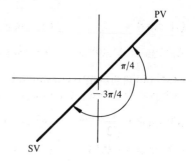

If $\tan \theta = -1$ the secondary
value is in the second quadrant
and is $\theta = \dfrac{3\pi}{4}$.

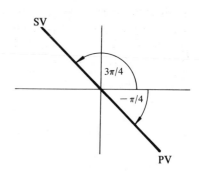

EXERCISE 6f

1) Write down the values of the following angles:

(a) $\arcsin \dfrac{1}{2}$

(b) $\arctan 1$

(c) $\arccos(-1)$

(d) $\arctan \dfrac{1}{\sqrt{3}}$

(e) $\arcsin 0$

(f) $\arcsin 1$

(g) $\arccos\left(-\dfrac{1}{2}\right)$

(h) $\arctan \infty$

(i) $\arccos \dfrac{1}{\sqrt{2}}$

(j) $\arctan 0.8$

(k) $\arcsin\left(-\dfrac{\sqrt{3}}{2}\right)$

(l) $\arccos 0.45$

(m) $\arctan(-2.6)$

(n) $\arctan(-1)$

(o) $\arccos\left(-\dfrac{\sqrt{3}}{2}\right)$

2) Determine the principal solutions of the following equations. In each case indicate your solution on the graph of the appropriate circular function.

(a) $\sin \theta = \dfrac{\sqrt{3}}{2}$

(b) $\cos \theta = -\dfrac{\sqrt{2}}{2}$

(c) $\tan \theta = \sqrt{3}$

(d) $\cos \theta = \dfrac{\sqrt{3}}{2}$

(e) $\sin \theta = -1$

(f) $\tan \theta = 0$

(g) $\sin \theta = -\dfrac{\sqrt{2}}{2}$

(h) $\cos \theta = 0$

(i) $\sin \theta = 0$

(j) $\cos \theta = -1$ (k) $\tan \theta = -\dfrac{\sqrt{3}}{3}$ (l) $\sin \theta = 1$

(m) $\sin \theta = 0.63$ (n) $\cos \theta = 0.44$ (p) $\tan \theta = 2.84$

3) Find the principal and secondary solutions of the following equations. In each case draw a quadrant diagram showing your solutions:

(a) $\sin \theta = \dfrac{1}{\sqrt{2}}$ (b) $\cos \theta = -\dfrac{\sqrt{3}}{2}$ (c) $\tan \theta = -\sqrt{3}$

(d) $\sin \theta = -0.54$ (e) $\cos \theta = 0.63$ (f) $\tan \theta = 1.5$

4) By referring to the graph of the appropriate circular function, explain why the following equations have no secondary solution:
(a) $\sin \theta = -1$ (b) $\cos \theta = 1$
Write down two more equations which have no secondary solution.

SOLUTION OF TRIGONOMETRIC EQUATIONS IN A SPECIFIED RANGE

In attempting to solve a trig equation we find first the principal angle and the secondary angle (except in those cases where there is no secondary angle).
At this stage a quadrant diagram can be drawn showing the principal and secondary solution positions. Then any angle measured from the positive x-axis to either of the solution positions is a solution to the given equation.

EXAMPLES 6g

1) Solve the equation $\sin \theta = 0.4$ within the interval $[-360°, 360°]$.

The principal solution is $\theta = 23.58°$.
The secondary solution is in the second quadrant (sine ratio is positive in first and second quadrants) and is $\theta = 156.42°$.

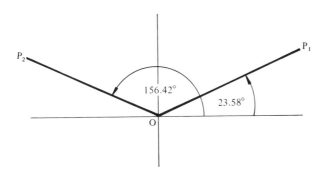

Therefore the following angles are the solutions within the specified interval.

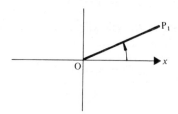
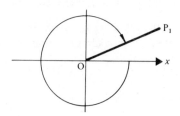
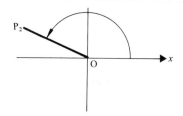
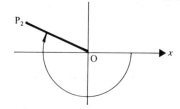

i.e. $\theta = 23.58°,\quad -336.42°,\quad 156.42°,\quad -203.58°$

2) Solve the equation $\tan\theta = -\dfrac{1}{\sqrt{3}}$ in the interval $[0, 2\pi]$.

If $\tan\theta = -\dfrac{1}{\sqrt{3}}$,   the principal solution is in the fourth quadrant and the secondary solution is in the second quadrant.

i.e.   PV is $-\dfrac{\pi}{6}$

and   SV is $\dfrac{5\pi}{6}$

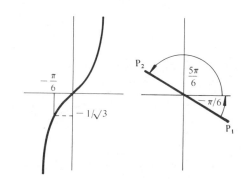

Within the specified interval, the solution set is

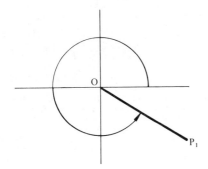
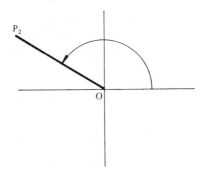

i.e. $\theta = \dfrac{5\pi}{6}, \dfrac{11\pi}{6}$.

(**Note**. The principal value is not always included in the solution set.)

3) Find the angles in the interval $[-360°, 0]$ which satisfy the equation

$$\cos \theta = 0.7$$

Since $\cos \theta$ is positive we have the principal solution in the first quadrant and the secondary solution in the fourth quadrant.

i.e.   PV is $45.57°$

SV is $-45.57°$

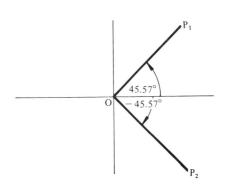

In the interval $-360° \leqslant \theta \leqslant 0$,   the solution set is

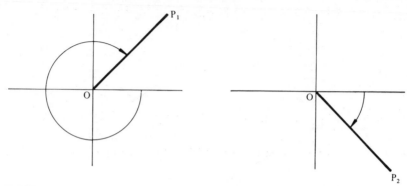

i.e. $\theta = -314.43°, -45.57°$.

4) Solve, within the interval $0 \leqslant \theta \leqslant 360°$,   the equation
$$\sin \theta + 3 \sin \theta \cos \theta = 0.$$

First the equation must be factorised

$\Rightarrow$ $\sin \theta \, (1 + 3 \cos \theta) = 0$

Therefore either $\sin \theta = 0$ [1]

or $1 + 3 \cos \theta = 0$ $\Rightarrow$ $\cos \theta = -\frac{1}{3}$ [2]

Now considering [1] by referring to the sine graph

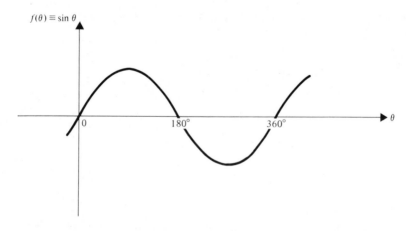

we see that $\theta = 0, 180°, 360°$.

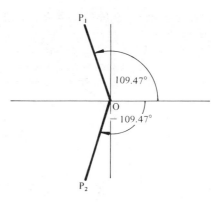

For [2] the principal solution is in the
second quadrant
and the secondary solution is in the
third quadrant

i.e.   PV is $109.47°$

SV is $-109.47°$

Within the specified range the solutions are

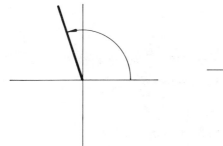 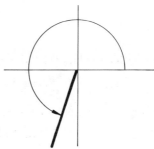

i.e. $\theta = 109.47°, 250.53°$.
So the complete solution set from 0 to $360°$ is

$$\theta = 0, 109.47°, 180°, 250.53°, 360°.$$

(**Note** that, although the solution to [1] can conveniently be expressed in
radians, degrees are used because the required range is specified in degrees. In
any case, units must not be mixed in any one example.)

GENERAL SOLUTION OF TRIGONOMETRIC EQUATONS

The general solution is an expression which represents all angles which satisfy
the given equation, i.e. the general solution is an infinite set of angles.

In looking for a general solution to a given trig equation, use can be made of:

1) the circular function graphs,

2) the period of each circular function,

3) the principal solution and, except when the tangent ratio is involved, the
   secondary solution.

Consider the equation $\sin \theta = s$ where $|s| \leqslant 1$.

The period, 2π, of the sine function is covered by the interval $[-\pi, \pi]$ which includes both the PV and the SV of θ.

So, by adding (or subtracting) any multiple of 2π to either the PV or the SV, we get another angle with the same sine ratio.

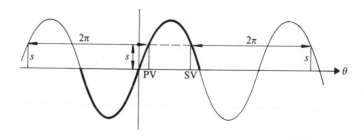

Thus the complete solution of the equation $\sin \theta = s$ where $|s| \leqslant 1$ is

$$\theta = \begin{cases} \text{PV} + 2n\pi \\ \text{SV} + 2n\pi \end{cases} \text{or} \quad \begin{cases} \text{PV} + 360n° \\ \text{SV} + 360n° \end{cases} \text{where} \quad n \in \mathbb{Z}$$

A similar situation arises when we consider the equation $\cos \theta = c$ because both the PV and the SV of the cosine function occur within one period, which is again 2π.

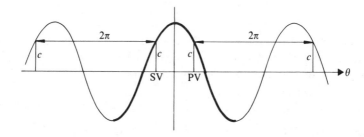

So the complete solution of the equation $\cos \theta = c$ where $|c| \leqslant 1$ is also given by adding multiples of 2π to either the PV or the SV.

But remembering that, for cosines, the PV and SV are equal in value but opposite in sign, (i.e. SV $= -$ PV),

the general solution of the equation $\cos \theta = c$ can be given in the form

$$\theta = \pm \text{PV} + 2n\pi \quad \text{or} \quad \pm \text{PV} + 360n° \quad \text{where} \quad n \in \mathbb{Z}$$

When we consider the equation $\tan \theta = t$,   we have already seen that only the principal value is included in the complete period $\left[-\dfrac{\pi}{2}, \dfrac{\pi}{2}\right]$. All further angles with the same tangent are given by combining multiples of π, i.e. the period, with the PV.

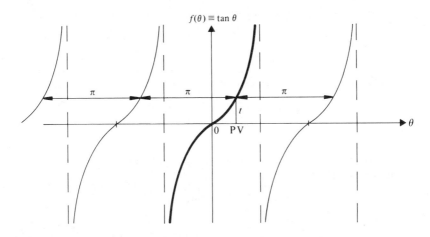

$f(\theta) \equiv \tan \theta$

So the general solution of the equation $\tan \theta = t$ is

$$\theta = \text{PV} + n\pi \quad \text{or} \quad \text{PV} + 180n° \quad \text{where} \quad n \in \mathbb{Z}$$

EXAMPLES 6g (continued)

5) Find the general solution of the equation
(a) $\tan \theta = 1$ (b) $\tan \theta = -\sqrt{3}$.

(a) The principal solution of the equation $\tan \theta = 1$ is $\theta = \dfrac{\pi}{4}$.

So the general solution is $\dfrac{\pi}{4} + n\pi$ where $n \in \mathbb{Z}$.

(b) If $\tan \theta = -\sqrt{3}$,   the principal solution is $\theta = -\dfrac{\pi}{3}$ and the general solution is therefore $\theta = -\dfrac{\pi}{3} + n\pi$.

6) Find the general solution of the equation
(a) $\cos \theta = \dfrac{1}{\sqrt{2}}$ (b) $\cos \theta = -\frac{1}{2}$.

(a) The principal solution, when $\cos \theta = \dfrac{1}{\sqrt{2}}$,   is $\theta = \dfrac{\pi}{4}$.

So the general solution is $\theta = \pm\dfrac{\pi}{4} + 2n\pi$.

(b) When $\cos \theta = -\frac{1}{2}$, the principal value of θ is $\frac{2\pi}{3}$ and the general solution is therefore $\theta = \pm\frac{2\pi}{3} + 2n\pi$.

7) Find the general solution set of the equation $\sin \theta = \frac{1}{2}$.

The principal value of θ for which $\sin \theta = \frac{1}{2}$ is $\frac{\pi}{6}$ and the secondary value, in the second quadrant, is $\frac{5\pi}{6}$.

So the general solution set includes $\theta = \dfrac{\pi}{6} + 2n\pi$

and $\theta = \dfrac{5\pi}{6} + 2n\pi$

Note. Although not obvious, there is a way of combining the two parts of this general solution giving,

when $\sin \theta = \frac{1}{2}$, $\theta = (-1)^n \dfrac{\pi}{6} + n\pi$

In general

when $\sin \theta = s$ $\theta = (-1)^n \text{PV} + n\pi$

Students who prefer this quotable form may use it instead of the two-part formula.

8) Find the general solution of the equation

$$4 \sin \theta \, (2 \tan \theta + 3) + 6 \tan \theta + 9 = 0.$$

First we must simplify and factorise the equation.

$$4 \sin \theta (2 \tan \theta + 3) + 3(2 \tan \theta + 3) = 0$$

$\Rightarrow$ $(4 \sin \theta + 3)(2 \tan \theta + 3) = 0$

So   either $4 \sin \theta + 3 = 0$ or $2 \tan \theta + 3 = 0$

Therefore either $\sin \theta = -\frac{3}{4}$ [1]

or $\tan \theta = -\frac{3}{2}$ [2]

For [1] the principal solution is $\theta = -48.59°$

and the secondary solution, in the third quadrant, is $\theta = -131.41°$.

Hence the general solution is $\begin{cases} \theta = -48.59° + 360n° \\ \theta = -131.41° + 360n° \end{cases}$

For [2]   the principal solution is the only one we need and it is $\theta = -56.31°$.
The general solution then is $\theta = -56.31° + 180n°$

Combining the results of [1] and [2], the general solution set of the given equation is

$$\theta = \begin{cases} -48.59° + 360n° \\ -131.41° + 360n° \\ -56.31° + 180n° \end{cases}$$

EXERCISE 6g

1) Solve the following equations for angles in the interval $[0, 2\pi]$, or $[0, 360°]$.

(a) $\cos\theta = \dfrac{1}{2}$ (b) $\tan\theta = -\dfrac{1}{\sqrt{3}}$ (c) $\sin\theta = \dfrac{\sqrt{2}}{2}$

(d) $\sec\theta = -3$ (e) $\cot\theta = 2$ (f) $\operatorname{cosec}\theta = 4$

(g) $\cos\theta = 0.84$ (h) $\operatorname{cosec}\theta = -2.5$ (i) $\tan\theta = 0.75$

(j) $\sqrt{3}\tan\theta = 2\sin\theta$ $\left(\text{Hint: } \tan\theta \equiv \dfrac{\sin\theta}{\cos\theta}\right)$

(k) $2\sin\theta\cos\theta + \sin\theta = 0$ (l) $4\cos\theta = \cos\theta\operatorname{cosec}\theta$

2) Find the general solution of the following equations, illustrating your results by reference to the graphs of the circular functions and/or quadrant diagrams.

(a) $\sin\theta = \dfrac{\sqrt{3}}{2}$ (b) $\cos\theta = 0$ (c) $\tan\theta = -\sqrt{3}$

(d) $\sin\theta = -\frac{1}{4}$ (e) $\cos\theta = 0.371$ (f) $\cot\theta = \frac{1}{2}$

(g) $\operatorname{cosec}\theta = 3$ (h) $\sec\theta = 1$ (i) $(\sin\theta)^2 = \frac{1}{4}$

MULTIPLE ANGLES

Equations are frequently met in which the angle involved is a multiple of θ,

e.g. $\cos 2\theta = \frac{1}{2}$, $\tan 3\theta = -2$

Such equations are solved by determining first the necessary values of the multiple angle and then, by division, the corresponding values of θ.

EXAMPLES 6h

1) Find the general solution of the equation $\cos 2\theta = \frac{1}{2}$.

Let $2\theta \equiv \phi$ so that $\cos\phi = \frac{1}{2}$.

The principal value of ϕ is $\dfrac{\pi}{3}$ so the general solution for ϕ is

$$\phi = \pm\frac{\pi}{3} + 2n\pi$$

i.e.
$$2\theta = \pm\frac{\pi}{3} + 2n\pi$$

Hence
$$\theta = \pm\frac{\pi}{6} + n\pi$$

2) Find the angles within the range $-180° \leqslant \theta \leqslant 180°$ which satisfy the equation $\tan 3\theta = -2$.

Let $3\theta \equiv \phi$ so that $\tan \phi = -2$.
The principal value is in the fourth quadrant and the secondary value is in the second quadrant. All angles which satisfy the equation therefore correspond to the solution positions shown

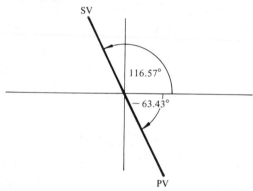

Values of θ are required in the range $-180° \leqslant \theta \leqslant 180°$. But $\phi \equiv 3\theta$ so we need to consider values of ϕ in the range $3(-180°) \leqslant \phi \leqslant 3(180°)$.
Hence, within the range $-540° \leqslant \phi \leqslant 540°$, we have

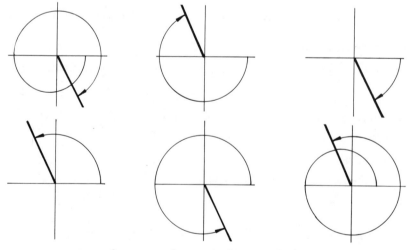

i.e. $\phi = -423.43°, -243.43°, -63.43°, 116.57°, 296.57°, 476.57°$

Therefore $\theta = -141.14°, -81.14°, -21.14°, 38.86°, 98.86°, 158.86°$.

Alternatively, quoting the general solution for ϕ gives

$$\phi = -63.43° + 180n° \Rightarrow \theta = -21.14° + 60n°$$

Giving n the values which cover the required range for θ
i.e. $n = -2, -1, 0, 1, 2, 3$, the correct values of θ can be found.

3) Find the solution of the equation $\sin\dfrac{\theta}{2} = 0.6$ giving values of θ between 0 and 360°.

Let $\dfrac{\theta}{2} \equiv \phi$ so that $\sin\phi = 0.6$

The principal value of ϕ is $36.87°$ and the secondary value, in the second quadrant, is $143.13°$.
So the solution positions for ϕ are

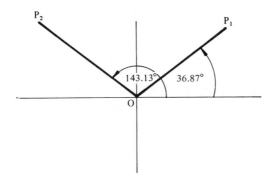

Now the required range of values of θ is from 0 to 360°. Hence the range of values of ϕ $\left(\equiv\dfrac{\theta}{2}\right)$ is from 0 to 180°,

i.e.

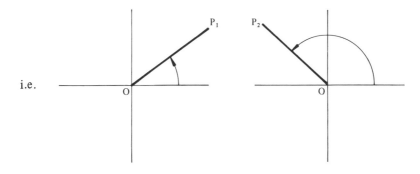

Hence $\phi = 36.87°, 143.13°$.
Therefore $\theta = 73.74°, 286.26°$.

EXERCISE 6h

1) Within the interval $[0, 360°]$, solve the following equations:
(a) $\tan 2\theta = 1$ (b) $\sin 3\theta = 0.7$ (c) $\cos \frac{1}{2}\theta = 0.85$ (d) $\sec 5\theta = 2$

2) Find the general solution of the following equations:

(a) $\operatorname{cosec} \dfrac{\theta}{3} = 1.5$ (b) $\cot 4\theta = 3$ (c) $\cos 2\theta = 0.63$

(d) $\sin \dfrac{\theta}{3} = 0.5$

3) Sketch the graph of $f: \theta \to \cos \theta$ for $-3\pi \leqslant \theta \leqslant 3\pi$. *Use your graph* to derive a general solution of the equation $\cos \theta = -1$.

4) Sketch the graph of $f: \theta \to \tan \theta$ for the domain $[-2\pi, 2\pi]$. Mark on the graph the points where $\tan \theta = \sqrt{3}$.   Explain why these points are equally spaced and derive the general solution of the equation

$$\tan \theta = \sqrt{3}.$$

THE EQUATION   cos A = cos B

This type of equation can be solved using a factor formula but we will now look at an alternative, very neat, method of solution.
Let $\cos A = \cos B = c$ say $(-1 \leqslant c \leqslant 1)$.
In general, for $\cos B = c$ there are two solution positions, OP_1 and OP_2.

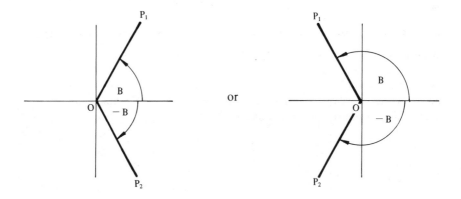

The set of angles represented by OP_1 and OP_2 in either case is $2n\pi \pm B$. But we also know that $\cos A = c$ so OP_1 and OP_2 together represent all possible values of A.

Thus $A = 2n\pi \pm B$

i.e.   A = general solution set for B

The same conclusion is reached when considering the equations

$$\tan A = \tan B \implies A = n\pi + B$$

and

$$\sin A = \sin B \implies A = \begin{cases} 2n\pi + B \\ (2n+1)\pi - B \end{cases}$$

EXAMPLES 6i

1) Solve the equation $\cos 4\theta = \cos \theta$.

Using only the conclusion to the argument above, we can write

$$4\theta = 2n\pi \pm \theta$$

Hence $\qquad\qquad 5\theta = 2n\pi \quad \text{or} \quad 3\theta = 2n\pi$

$$\implies \qquad\qquad \theta = \frac{2n\pi}{5}, \frac{2n\pi}{3}$$

2) Find the values in the range $0 \leqslant \theta \leqslant 360°$ which satisfy the equation $\tan(3\theta - 40°) = \tan \theta$.

The general solution is

$$3\theta - 40° = 180n° + \theta$$

$$\implies \qquad\qquad 2\theta = 180n° + 40°$$

$$\implies \qquad\qquad \theta = 90n° + 20°$$

For $\quad 0 \leqslant \theta \leqslant 360°, \quad$ let $\quad n = 0, 1, 2, 3$

giving $\qquad\qquad \theta = 20°, 110°, 200°, 290°$

This very neat method can be used only for equations containing two terms involving the same trig ratio. This situation can sometimes be arranged in an apparently unsuitable case as in Example 3.

3) Find the general solution of $\cos 3\theta = \sin \theta$.

We know that $\quad \sin \theta \equiv \cos\left(\dfrac{\pi}{2} - \theta\right) \quad$ (complementary angles)

So the equation can be written

$$\cos 3\theta = \cos\left(\frac{\pi}{2} - \theta\right)$$

giving $\qquad\qquad 3\theta = 2n\pi \pm \left(\dfrac{\pi}{2} - \theta\right)$

Therefore either $\quad 4\theta = 2n\pi + \dfrac{\pi}{2} \quad$ or $\quad 2\theta = 2n\pi - \dfrac{\pi}{2}$

$\Rightarrow \qquad\qquad \theta = \dfrac{n\pi}{2} + \dfrac{\pi}{8}, \; n\pi - \dfrac{\pi}{4}$

EXERCISE 6i

Find the general solution of the following equations:

1) $\cos 4\theta = \cos 3\theta$ 2) $\tan 7\theta = \tan 2\theta$ 3) $\cos 3\theta = \sin 2\theta$

4) $\cot 4\theta = \tan 5\theta$ 5) $\sin 4\theta = \sin 3\theta$ 6) $\cos 5\theta \sec \theta = 1$

7) $\cos\left(\theta - \dfrac{\pi}{4}\right) = \cos\left(4\theta + \dfrac{\pi}{4}\right)$ 8) $\tan 2\theta = \cot \theta$

Find the solutions, from 0 to π inclusive, of the following equations:

9) $\cos 3\theta = \cos 7\theta$ 10) $\tan 3\theta = \cot 2\theta$ 11) $\sin 7\theta = \sin 2\theta$

12) $\sec 6\theta = \sec 5\theta$

Solve the following equations giving values from $-180°$ to $+180°$:

13) $\cos (2\theta + 60°) = \cos \theta$ 14) $\tan 3\theta = \tan (\theta - 50°)$

15) $\sin \theta = \cos (2\theta + 60°)$

THE GRAPHS OF CIRCULAR FUNCTIONS OF COMPOUND ANGLES

Consider $\quad f: \theta \rightarrow \sin 2\theta$.

The following table gives pairs of corresponding values of θ and $f(\theta)$.

θ	0	$\dfrac{\pi}{4}$	$\dfrac{\pi}{2}$	$\dfrac{3\pi}{4}$	π	$\dfrac{5\pi}{4}$	$\dfrac{3\pi}{2}$	$\dfrac{7\pi}{4}$	2π
2θ	0	$\dfrac{\pi}{2}$	π	$\dfrac{3\pi}{2}$	2π	$\dfrac{5\pi}{2}$	3π	$\dfrac{7\pi}{2}$	4π
$f(\theta)$	0	1	0	-1	0	1	0	-1	0

Thus

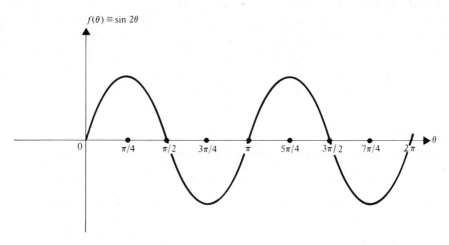

$f(\theta) \equiv \sin 2\theta$

The following characteristics can be observed.

(a) The function $\sin 2\theta$ is cyclic and its period is π $\left(\text{i.e. } \dfrac{2\pi}{2}\right)$.

(b) The range is $[-1, 1]$.

(c) The shape is a sine wave.

(d) Within the domain $0 \leqslant \theta \leqslant 2\pi$, there are two complete curve patterns compared with only one for the basic function $f: \theta \to \sin \theta$, i.e. the complete *cycle* appears with twice the frequency.

When the same investigation is carried out on $f: \theta \to \sin 3\theta$ we find that the function is cyclic with a period $\dfrac{2\pi}{3}$, so that three complete cycles occur between 0 and 2π.

It seems likely then (although it has not been generally proved) that the graph of the function $f: \theta \to \sin k\theta$ is a sine wave with a period of $\dfrac{2\pi}{k}$ and a frequency k times that of $f: \theta \to \sin \theta$.

These properties are, in fact, valid for all values of k and similar properties can be deduced for $f: \theta \to \cos k\theta$ and $f: \to \tan k\theta$,

e.g.

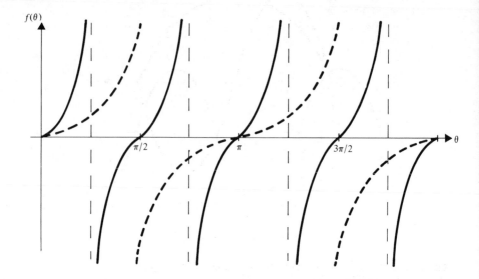

$$f(\theta) \equiv \tan 2\theta \quad \text{———}$$

$$f(\theta) \equiv \tan \theta \quad \text{— — — —}$$

The period of $\tan 2\theta$ is $\dfrac{\pi}{2}$ and the frequency is 2.

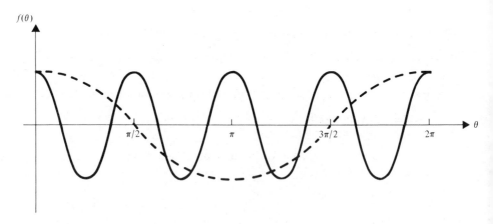

$$f(\theta) \equiv \cos 4\theta \quad \text{———}$$

$$f(\theta) \equiv \cos \theta \quad \text{— — — —}$$

The period of $\cos 4\theta$ is $\dfrac{\pi}{2}$ and the frequency is 4.

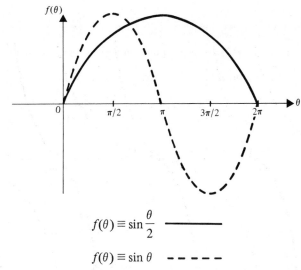

$$f(\theta) \equiv \sin \frac{\theta}{2} \quad \underline{\qquad}$$

$$f(\theta) \equiv \sin \theta \quad \text{-- -- --}$$

The period of $\sin \dfrac{\theta}{2}$ is 4π and the frequency is $\frac{1}{2}$.

Now consider the function $\quad f(\theta) \equiv \cos(\theta - \alpha)$.
This is clearly a cosine function but

(a) $\quad f(\theta) = 0 \qquad$ when $\quad (\theta - \alpha) = \dfrac{\pi}{2}, \dfrac{3\pi}{2}$, etc.

$\qquad\qquad\qquad$ i.e. when $\qquad \theta = \dfrac{\pi}{2} + \alpha, \dfrac{3\pi}{2} + \alpha$ etc.

(b) $\quad f(\theta) = 1 \qquad$ when $\quad (\theta - \alpha) = 0, 2\pi, 4\pi$, etc.

$\qquad\qquad\qquad$ i.e. when $\qquad \theta = \alpha, 2\pi + \alpha, 4\pi + \alpha$, etc.

(c) $\quad f(\theta) = -1 \qquad$ when $\quad (\theta - \alpha) = \pi, 3\pi, 5\pi$, etc.

$\qquad\qquad\qquad$ i.e. when $\qquad \theta = \pi + \alpha, 3\pi + \alpha, 5\pi + \alpha$, etc.

e.g. if $\quad \alpha = \dfrac{\pi}{4}$, we have

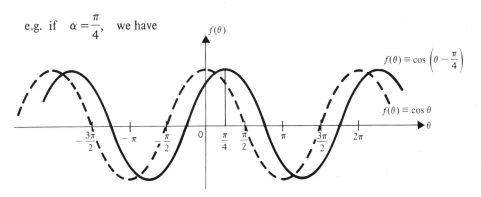

$$f(\theta) \equiv \cos\left(\theta - \frac{\pi}{4}\right)$$

$$f(\theta) \equiv \cos \theta$$

From the sketch we see that the graph of $f(\theta) \equiv \cos\left(\theta - \dfrac{\pi}{4}\right)$ is identical in shape to the graph of $f(\theta) \equiv \cos\theta$ but is in a position given by moving the standard cosine curve a horizontal distance $\dfrac{\pi}{4}$ to the *right*.

Similarly it can be shown that the graph of the function $f(\theta) \equiv \cos(\theta + \alpha)$ is given by moving the standard cosine curve a horizontal distance α to the *left*.

e.g. the graph of $f(\theta) \equiv \cos\left(\theta + \dfrac{\pi}{3}\right)$ is

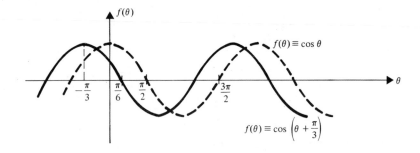

Now consider the function $f(\theta) \equiv \sin(2\theta + \alpha)$.
We know that *adding* a constant angle causes a curve to move *leftward*.

In this case $\sin(2\theta + \alpha) = 0$ when $\theta = -\dfrac{\alpha}{2}$ so the graph of this function is obtained by moving the graph of $\sin 2\theta$ a distance $\dfrac{\alpha}{2}$ to the left

e.g. if $f(\theta) \equiv \sin\left(2\theta + \dfrac{\pi}{2}\right)$, the graph is

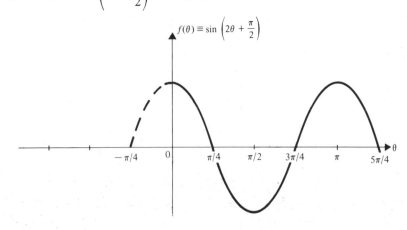

Similarly if $f(\theta) \equiv \cos\left(3\theta - \dfrac{\pi}{2}\right)$, by putting $3\theta - \dfrac{\pi}{2} = 0$, we see that the required graph is obtained by moving the graph of $f(\theta) \equiv \cos 3\theta$ a distance $\dfrac{\pi}{6}$ to the *right*

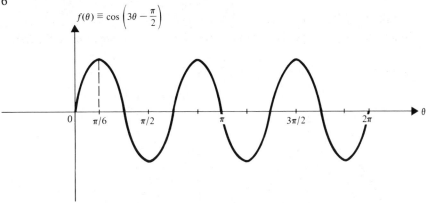

$$f(\theta) \equiv \cos\left(3\theta - \frac{\pi}{2}\right)$$

and if $f(\theta) \equiv \tan\left(\dfrac{\theta}{2} + \dfrac{\pi}{6}\right)$ we have a tangent curve with a frequency of $\dfrac{1}{2}$ which is moved a distance $\dfrac{\pi}{3}$ to the left $\left(\dfrac{\theta}{2} + \dfrac{\pi}{6} = 0 \Rightarrow \theta = -\dfrac{\pi}{3}\right)$

i.e.

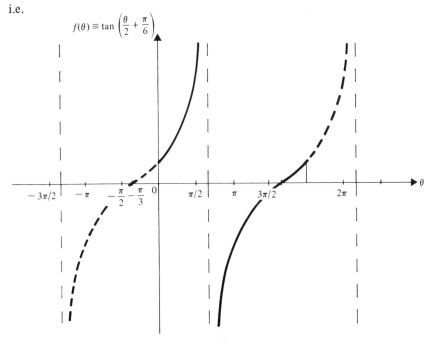

$$f(\theta) \equiv \tan\left(\frac{\theta}{2} + \frac{\pi}{6}\right)$$

Note. An equation containing a compound angle can be solved in the following way.

If
$$\cos\left(2\theta - \frac{\pi}{6}\right) = \tfrac{1}{2}$$

the principal value of $\left(2\theta - \frac{\pi}{6}\right)$ is $\frac{\pi}{3}$

i.e.
$$2\theta - \frac{\pi}{6} = \pm\frac{\pi}{3} + 2n\pi$$

$$\Rightarrow \qquad 2\theta = \pm\frac{\pi}{3} + 2n\pi + \frac{\pi}{6}$$

$$\Rightarrow \qquad \theta = \pm\frac{\pi}{6} + n\pi + \frac{\pi}{12}$$

i.e.
$$\theta = \begin{cases} n\pi + \dfrac{\pi}{4} \\[2mm] n\pi - \dfrac{\pi}{12} \end{cases}$$

EXERCISE 6j

Sketch the graphs of the following functions in the domain $[0, 2\pi]$, in each case state the period of the function and its frequency.

1) $\sin 4\theta$ 2) $\sec 2\theta$ 3) $\tan\dfrac{\theta}{4}$ 4) $\sin 2\theta$ 5) $\cot 2\theta$

6) $\sin\left(\theta + \dfrac{\pi}{3}\right)$ 7) $\cos\left(3\theta - \dfrac{\pi}{4}\right)$ 8) $\csc 3\theta$ 9) $-\sin\dfrac{\theta}{4}$

10) $\cos\left(2\theta + \dfrac{\pi}{2}\right)$ 11) $\tan\left(\theta - \dfrac{\pi}{6}\right)$

Find the general solutions of the following equations:

12) $\cos\left(\theta + \dfrac{\pi}{4}\right) = \dfrac{1}{2}$ 13) $\tan\left(2\theta - \dfrac{\pi}{3}\right) = -1$

14) $\cos\left(3\theta - \dfrac{\pi}{3}\right) = -\dfrac{\sqrt{3}}{2}$ 15) $\sin\left(2\theta + \dfrac{\pi}{6}\right) = \dfrac{1}{2}$

SUMMARY

An angle of 1 radian is subtended at the centre of a circle of radius r by an arc of length r.

π radians $= 180°$.

For a circle of radius r
the length of an arc subtending θ radians at the centre is $r\theta$.

For a circle of radius r
the area of a sector containing θ radians at the centre is $\frac{1}{2}r^2\theta$.

For any angle θ, the associated acute angle α is the difference between θ and the nearest multiple of $180°$ (π).

If the principal solution of $\cos\theta = c$ is $\theta = \theta_1$

the general solution is $\qquad\qquad \theta = 2n\pi \pm \theta_1$

If the principal solution of $\tan\theta = t$ is $\theta = \theta_1$

the general solution is $\qquad\qquad \theta = n\pi + \theta_1$

If the principal solution of $\sin\theta = s$ is $\theta = \theta_1$

the general solution is
$$\theta = \begin{cases} 2n\pi + \theta_1 \\ (2n+1)\pi - \theta_1 \end{cases}$$

MULTIPLE CHOICE EXERCISE 6

(The instructions for answering these questions are on p. xii)

TYPE I

1) An angle of 1 radian is equivalent to:
(a) $90°$ (b) $60°$ (c) $67°\,18'$ (d) $57°\,18'$ (e) $45°$.

2) An arc PQ subtends an angle of $60°$ at the centre of a circle of radius 1 cm. The length of PQ is:

(a) 60 cm   (b) 30 cm   (c) $\dfrac{\pi}{6}$ cm   (d) $\dfrac{\pi}{3}$ cm   (e) $\dfrac{\pi^2}{18}$ cm.

3) If $\theta = \dfrac{13\pi}{6}$, $\cos\theta$ is:

(a) $\dfrac{1}{2}$ (b) $-\dfrac{1}{2}$ (c) $\dfrac{\sqrt{3}}{2}$ (d) $-\dfrac{\sqrt{3}}{2}$ (e) $\dfrac{\sqrt{2}}{2}$.

4) The associated acute angle for $280°$ is:
(a) $100°$ (b) $10°$ (c) $80°$ (d) $-80°$ (e) $190°$.

5) If $\cos\theta = \frac{1}{2}$, the general solution is:

(a) $\theta = 2n\pi \pm \dfrac{\pi}{6}$ (b) $\theta = n\pi + \dfrac{\pi}{3}$ (c) $\theta = 2n\pi + \dfrac{\pi}{3}$

(d) $\theta = 2n\pi \pm \dfrac{\pi}{3}$ (e) $\theta = n\pi \pm \dfrac{\pi}{6}$.

6) There is a solution of the equation $4 \sin \theta + 1 = 0$ in quadrants:
(a) 1 and 2   (b) 1 and 3   (c) 3 and 4   (d) 2 and 3.

7) The graph of the function $f(\theta) \equiv \cos \left(2\theta - \dfrac{\pi}{2}\right)$ has a period

(a) 2π (b) π (c) $\dfrac{\pi}{2}$ (d) $-\dfrac{\pi}{2}$ (e) none of these.

TYPE II

8) $\theta = 60°$.

(a) $\sin \theta = \dfrac{1}{2}$.   (b) $\tan \theta = \cot 30°$.   (c) $\theta = \dfrac{\pi}{3}$.   (d) $\sec \theta = 2$.

9)   An angle θ is such that $\tan \theta = 1$ and $\cos \theta$ is negative.

(a) $\sin \theta$ is positive.   (b) $\cos \theta = -\dfrac{\sqrt{2}}{2}$.   (c) $\cot \theta = -1$.
(d) $\sec \theta$ is negative.

10) $f(\theta) \equiv \cos \theta$.

(a) For $-\dfrac{\pi}{2} < \theta < \dfrac{\pi}{2},\ f(\theta) > 0$.

(b) $f(\theta)$ is undefined when $\theta = (2n + 1) \dfrac{\pi}{2}$.

(c) $-1 \leqslant f(\theta) \leqslant 1$.

(d) $f(\theta)$ is periodic with a period of π.

11) π can represent:
(a) the ratio of the circumference to the radius of a circle,
(b) half a revolution,
(c) the period of the function $f(\theta) \equiv \tan \theta$,
(d) an angle whose cosine is zero.

12) Given that $\tan \theta = 1$
(a) θ lies in quadrants 1 and 2,

(b) the principal solution is $\theta = \dfrac{\pi}{4}$,

(c) $\cos \theta = \sqrt{2}$,

(d) the general solution is $\theta = n\pi \pm \dfrac{\pi}{4}$.

13)

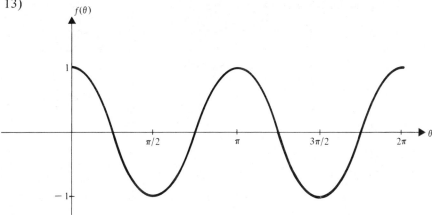

In the sketch above $f(\theta)$ could be:

(a) $\cos 2\theta$ (b) $\cos \dfrac{\theta}{2}$ (c) $\sin \left(2\theta - \dfrac{\pi}{4}\right)$ (d) $\sin \left(2\theta + \dfrac{\pi}{2}\right)$

(e) $\sin \left(2\theta + \dfrac{\pi}{4}\right)$.

14) The graph of $f(\theta) \equiv \cos\theta$ compared with the graph of $f(\theta) \equiv \sin\theta$ is:

(a) inverted   (b) $90°$ to the left
(c) $90°$ to the right   (d) of equal period.

TYPE III

15) (a) $\sin\theta = 1$.

    (b) $\theta = \dfrac{\pi}{2}$.

16) (a) The graph of a trigonometric function $f(\theta)$ is continuous.
    (b) $f(\theta) \equiv \cot\theta$.

17) In a circle of radius r an arc PQ subtends an angle θ at the centre 0.
    (a) The length of PQ is $4r$.
    (b) The area of the sector POQ is $2r^2$.

18) $f(\theta)$ is a circular function.
    (a) $f(\theta) \equiv \sec\theta$.

    (b) $f(\theta)$ is undefined at $\theta = \dfrac{\pi}{2}$.

19) (a) $\cos\theta = a$.
    (b) $\sin \left(\dfrac{\pi}{2} - \theta\right) = a$.

20) (a) $\arcsin \frac{1}{2} = \theta$.

   (b) $\theta = \arccos \dfrac{\sqrt{3}}{2}$.

21) (a) A circular function $f(\theta)$ has a period of π.
    (b) $f(\theta) \equiv \cos 2\theta$.

TYPE IV

22) Find the length of an arc PQ of a circle if:

(a) the angle POQ is $\dfrac{\pi}{2}$,

(b) the area of the circle is $9\pi \, \text{cm}^2$,

(c) O is the centre of the circle,

(d) the radius of the circle is measured in cm.

23) Evaluate the angle θ,
(a) $\tan \theta$ is given,
(b) $\cos \theta$ is positive,
(c) $0 \leqslant \theta \leqslant 2\pi$,
(d) there is only one value of θ.

24) Sketch the graph of $f(\theta)$ given that:
(a) $f(\theta)$ is a major circular function,
(b) $f(\theta)$ is continuous,
(c) $f(\theta)$ is zero when $\theta = n\pi$,
(d) $f(\theta)$ is periodic.

TYPE V

25) $\sin^{-1} \theta \geqslant 1$.

26) π radians $= 360°$.

27) The function $f(\theta) \equiv \cos \theta$ is such that $|\theta| \leqslant 1$.

28) $\sin \theta = 0$ when $\theta = n\pi$.

29) The function $f(\theta) \equiv \sec \theta$ is undefined when $\theta = (2n + 1)\dfrac{\pi}{2}$.

30) If a function $f(x)$ is positive and decreasing for $1 < x < 2$ then the function $\dfrac{1}{f(x)}$ is negative and increasing for $1 < x < 2$.

MISCELLANEOUS EXERCISE 6

1) A chord of a circle subtends an angle of θ radians at the centre of the circle. If the area of the minor segment cut off by the chord is one sixth of the area of the circle prove that $\sin \theta = \theta - \dfrac{\pi}{3}$.

2) Three circular discs each of radius a lie on a table touching each other. Find, in terms of a, the area enclosed between them.

3) A chord AB of a circle of radius $5a$ is of length $3a$. The tangents to the circle at A and B meet at T. Find the area enclosed by TA, TB and the *major* arc AB.

4) Three cylinders are placed in contact with each other with their axes parallel. The radii of the cylinders are 3 cm, 4 cm, 5 cm. An elastic band is stretched round the three cylinders so that the plane of the elastic band is perpendicular to the axes of the cylinders. Calculate the length of the part of the band in contact with the largest cylinder.   (U of L)p

5) P and Q are points on a circle of radius r, and the chord PQ subtends an angle 2θ radians at its centre O. If A is the area enclosed by the minor arc PQ and the chord PQ, and if B is the area enclosed by the arc PQ and the tangents to the circle at P and Q, prove that

$$A - B \equiv r^2(2\theta - \tan \theta - \sin \theta \cos \theta)$$

6) Find the general solutions of the equations:

(a) $\sin \left(\theta + \dfrac{\pi}{4} \right) = \dfrac{1}{2}$,

(b) $\cos \left(\theta - \dfrac{\pi}{4} \right) = \dfrac{1}{2}$.

By drawing sketches of $f(\theta) \equiv \sin \left(\theta + \dfrac{\pi}{4} \right)$ and $g(\theta) \equiv \cos \left(\theta - \dfrac{\pi}{4} \right)$, explain your answers to (a) and (b).

7) Sketch the function $f(\theta) \equiv 2 \sin \left(2\theta + \dfrac{\pi}{2} \right)$ for values of θ in the range $-\pi \leqslant \theta \leqslant \pi$.

Solve the equation $f(\theta) = 1$, indicating your solutions on the sketch.

Find the general solutions of the equations:

8) $\tan \left(3\theta - \dfrac{\pi}{2} \right) = \sqrt{3}$

9) $\sec \left(2\theta + \dfrac{\pi}{2} \right) = 2$.

10) $\tan 5\theta = \cot 2\theta$.

11) Explain briefly what is meant by:
(a) a periodic function,
(b) the amplitude of a sine wave,
(c) the frequency of a circular function,
(d) the general solution of a trigonometric equation,
(e) a radian.

12) What is the period of the function $f(\theta) \equiv \cos k\theta$? Draw sketches to illustrate your answer when $k = 2$ and $k = \frac{1}{2}$. In each of these cases, write down the general solution of the equations $f(\theta) = 0$, $f(\theta) = 1$, $f(\theta) = -1$.

13) Without the use of tables or calculator find, for each of the following equations, all the solutions in the interval $0° \leqslant x \leqslant 180°$.
(a) $\cos (x + 30°) = \cos (60° - 3x)$
(b) $\sin (x + 20°) = \cos 3x$.   (JMB)

14) Without using tables or calculator, show that $x = \dfrac{\pi}{14}$ is a solution of the equation
$$\sin 3\theta = \cos 4\theta$$
Find the general solution.

15) (a) Find the solutions of the equation $\cos 3\theta = \cos 2\theta$ for which $0 < \theta \leqslant 2\pi$.

(b) Write down the general solution of the equation
$$\sin k\theta = \sin \theta$$
Hence find the value of k for which this equation and the equation $\cos 3\theta = \cos 2\theta$ have, within the range $0 < \theta \leqslant 2\pi$,
(i) five common solutions,
(ii) one and only one common solution.

CHAPTER 7

TRIGONOMETRIC IDENTITIES

In the previous chapter we saw that any one angle is associated with six different trigonometric ratios and that a particular value of one trigonometric ratio applies to an infinite set of angles. It is not surprising, therefore, to find that relationships exist between the various circular functions. Some of these are very useful in the development of trigonometry. This chapter deals with their derivation and applications.

Consider, for instance, the simple relationship between the sine, cosine and tangent of any angle.

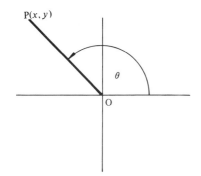

For any angle θ we have:

$$\sin \theta = \frac{y}{OP}, \quad \cos \theta = \frac{x}{OP},$$

$$\tan \theta = \frac{y}{x}$$

But $\quad \dfrac{\sin \theta}{\cos \theta} = \dfrac{y}{OP} \Big/ \dfrac{x}{OP} = \dfrac{y}{x}$

Thus for all angles $\qquad \tan \theta \equiv \dfrac{\sin \theta}{\cos \theta}.$

THE PYTHAGOREAN OR 'SQUARED RATIO' GROUP

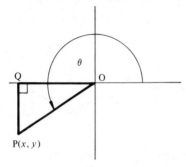

Q

O

θ

P(x, y)

For any position of OP a right angled triangle OPQ can be drawn for which

$$x^2 + y^2 = OP^2$$

Dividing throughout, in turn, by OP^2, x^2 and y^2 gives

$$\left(\frac{x}{OP}\right)^2 + \left(\frac{y}{OP}\right)^2 = 1 \qquad [1]$$

$$1 + \left(\frac{y}{x}\right)^2 = \left(\frac{OP}{x}\right)^2 \qquad [2]$$

$$\left(\frac{x}{y}\right)^2 + 1 = \left(\frac{OP}{y}\right)^2 \qquad [3]$$

[1] becomes $\qquad (\cos \theta)^2 + (\sin \theta)^2 = 1$

[2] becomes $\qquad 1 + (\tan \theta)^2 = (\sec \theta)^2$

[3] becomes $\qquad (\cot \theta)^2 + 1 = (\csc \theta)^2$

The use of brackets when raising a trigonometric ratio to a power can be avoided by writing $\cos^2\theta$ for $(\cos \theta)^2$ etc.

The above relationships are valid for any position of OP, i.e. for any angle θ, so for all angles:

$$\cos^2\theta + \sin^2\theta \equiv 1$$

$$1 + \tan^2\theta \equiv \sec^2\theta$$

$$\cot^2\theta + 1 \equiv \csc^2\theta$$

These identities are very useful in the solution of certain trig equations.

EXAMPLES 7a

1) Solve the equation $\quad 2\cos^2\theta - \sin \theta = 1 \quad$ for values of θ between 0 and 2π.

Using $\quad \cos^2\theta + \sin^2\theta \equiv 1 \quad$ gives

$$2(1 - \sin^2\theta) - \sin \theta = 1$$

or $\qquad\qquad\qquad 2\sin^2\theta + \sin \theta - 1 = 0$

This is now a quadratic equation (of the form $2x^2 + x - 1 = 0$)

Hence $\qquad$ $(2 \sin \theta - 1)(\sin \theta + 1) = 0$

$\Rightarrow$ $\qquad$ $\sin \theta = \frac{1}{2}$ or -1

If $\sin \theta = \frac{1}{2}$, $\quad \theta = \dfrac{\pi}{6}, \dfrac{5\pi}{6}$

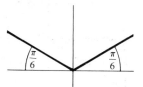

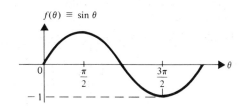

$f(\theta) \equiv \sin \theta$

If $\sin \theta = -1$, $\quad \theta = \dfrac{3\pi}{2}$

Therefore the solution of the equation is

$$\theta = \frac{\pi}{6}, \frac{3\pi}{2}, \frac{5\pi}{6}$$

2) If $\quad 3 \sec^2\theta - 5 \tan \theta - 4 = 0 \quad$ find the general solution.

Using $\quad 1 + \tan^2\theta \equiv \sec^2\theta \quad$ we have

$$3(1 + \tan^2\theta) - 5 \tan \theta - 4 = 0$$

i.e. $\qquad 3 \tan^2\theta - 5 \tan \theta - 1 = 0$

Again we have a quadratic equation but because it has no simple factors we solve it by formula

$\Rightarrow \qquad\qquad \tan \theta = \dfrac{5 \pm \sqrt{25 + 12}}{6}$

$\Rightarrow \qquad\qquad \tan \theta = 1.8471 \quad$ or $\quad -0.1805$

If $\quad \tan \theta = 1.8471$, $\quad$ the principal solution is $\quad \theta = 61.57°$.

If $\quad \tan \theta = -0.1805 \quad$ the principal solution is $\quad \theta = -10.23°$.

The complete general solution is therefore

$$\theta = \begin{cases} 180n° + 61.57° \\ 180n° - 10.23° \end{cases}$$

Other applications of the standard identities include:
derivation of a variety of further trigonometric relationships,
elimination of trigonometric terms from pairs of equations,
calculation of the remaining trigonometric ratios of any angle for which only
one trigonometric ratio is known.

3) Prove that $(1 - \cos A)(1 + \sec A) \equiv \sin A \, \tan A.$

Because this relationship has yet to be proved, we must not assume its truth by
using the complete identity in our working. The left and right hand sides must
be isolated throughout the proof.
Consider the L.H.S.

$$(1 - \cos A)(1 + \sec A) \equiv 1 + \sec A - \cos A - \cos A \sec A$$

$$\equiv 1 + \sec A - \cos A - \cos A\left(\frac{1}{\cos A}\right)$$

$$\equiv \sec A - \cos A$$

$$\equiv \frac{1 - \cos^2 A}{\cos A}$$

$$\equiv \frac{\sin^2 A}{\cos A} \qquad\qquad (\text{since } \cos^2 A + \sin^2 A \equiv 1)$$

$$\equiv \sin A\left(\frac{\sin A}{\cos A}\right)$$

$$\equiv \sin A \, \tan A \quad \text{which is identical to the R.H.S.}$$

4) Prove that $(\csc A - \sin A)(\sec A - \cos A) \equiv \dfrac{1}{\tan A + \cot A}$

Considering the L.H.S:

$$(\csc A - \sin A)(\sec A - \cos A) \equiv \left(\frac{1}{\sin A} - \sin A\right)\left(\frac{1}{\cos A} - \cos A\right)$$

$$\equiv \left(\frac{1 - \sin^2 A}{\sin A}\right)\left(\frac{1 - \cos^2 A}{\cos A}\right)$$

$$\equiv \left(\frac{\cos^2 A}{\sin A}\right)\left(\frac{\sin^2 A}{\cos A}\right)$$

$$\equiv \cos A \, \sin A$$

Now this is already a very simple form but is not obviously identical to the
given R.H.S. So this time we begin working independently on the R.H.S.

$$\frac{1}{\tan A + \cot A} \equiv 1 \div \left(\frac{\sin A}{\cos A} + \frac{\cos A}{\sin A}\right)$$

$$\equiv 1 \div \left(\frac{\sin^2 A + \cos^2 A}{\cos A \sin A}\right)$$

$$\equiv \cos A \sin A$$

Since both L.H.S and R.H.S reduce to cos A sin A they are identical.

5) Eliminate θ from the equations $x = 2\cos\theta$ and $y = 3\sin\theta$.

Now $\cos\theta = \dfrac{x}{2}$ and $\sin\theta = \dfrac{y}{3}$

Using $\qquad\qquad \cos^2\theta + \sin^2\theta \equiv 1$

gives $\qquad\qquad \left(\dfrac{x}{2}\right)^2 + \left(\dfrac{y}{3}\right)^2 = 1$

$\Rightarrow \qquad\qquad 9x^2 + 4y^2 = 36$

Note. In this problem we see that both x and y initially depend on θ, a variable angle. Used in this way, θ is called a *parameter*, a type of variable which plays an important part in the analysis of curves and functions.

6) If $\sin A = \frac{1}{3}$ and A is obtuse, find cos A and cot A without using tables or calculator.

Using $\qquad\qquad \cos^2 A + \sin^2 A \equiv 1$

$\Rightarrow \qquad\qquad \cos^2 A + \frac{1}{9} = 1$

Therefore $\qquad\qquad \cos A = \pm\sqrt{\dfrac{8}{9}} = \pm\dfrac{2\sqrt{2}}{3}$

But A is obtuse so cos A is negative,

i.e. $\qquad\qquad \cos A = -\dfrac{2\sqrt{2}}{3}$

Also $\qquad\qquad \cot A \equiv \dfrac{\cos A}{\sin A}$

Hence $\qquad\qquad \cot A = -\dfrac{2\sqrt{2}}{3}\bigg/\dfrac{1}{3} = -2\sqrt{2}$

Note: This type of problem can often be done more directly by drawing the appropriate right angled triangle and using Pythagoras' Theorem.

e.g.

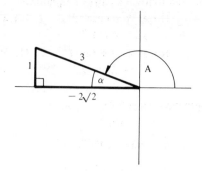

EXERCISE 7a

Solve the following equations for angles in the range $-180° \leqslant \theta \leqslant 180°$.

1) $\sec^2\theta + \tan^2\theta = 6$

2) $4\cos^2\theta + 5\sin\theta = 3$

3) $\cot^2\theta = \csc\theta$

4) $\tan\theta + \cot\theta = 2$

5) $\tan\theta + 3\cot\theta = 5\sec\theta$

6) $\sec\theta = 1 - 2\tan^2\theta$

Find the general solution of the following equations.

7) $5\cos\theta - 4\sin^2\theta = 2$

8) $4\cot^2\theta + 12\csc\theta + 1 = 0$

9) $4\sec^2\theta - 3\tan\theta = 5$

10) $2\cos\theta - 4\sin^2\theta + 2 = 0$

Prove the following identities:

11) $\cot\theta + \tan\theta \equiv \sec\theta\csc\theta$

12) $\dfrac{\cos A}{1 - \tan A} + \dfrac{\sin A}{1 - \cot A} \equiv \sin A + \cos A$

13) $\tan^2\theta + \cot^2\theta \equiv \sec^2\theta + \csc^2\theta - 2$

14) $\dfrac{\sin A}{1 + \cos A} \equiv \dfrac{1 - \cos A}{\sin A}$ (*Hint*: Multiply L.H.S by $(1 - \cos A)$ in both numerator and denominator)

15) $(\sec^2\theta + \tan^2\theta)(\csc^2\theta + \cot^2\theta) \equiv 1 + 2\sec^2\theta\csc^2\theta$

16) $\dfrac{\sin A}{1 + \cos A} + \dfrac{1 + \cos A}{\sin A} \equiv \dfrac{2}{\sin A}$

17) $\sec^2 A \equiv \dfrac{\csc A}{\csc A - \sin A}$

18) $(1 + \sin\theta + \cos\theta)^2 \equiv 2(1 + \sin\theta)(1 + \cos\theta)$

19) $\dfrac{\tan^2 A + \cos^2 A}{\sin A + \sec A} \equiv \sec A - \sin A$

20) Eliminate θ from the following pairs of equations:

(a) $x = 4 \sec \theta$
$y = 5 \tan \theta$

(b) $x = a \operatorname{cosec} \theta$
$y = b \cot \theta$

(c) $x = 2 \tan \theta$
$y = 3 \cos \theta$

(d) $x = 1 - \sin \theta$
$y = 1 + \cos \theta$

(e) $x = 2 + \tan \theta$
$y = 2 \cos \theta$

(f) $x = a \sec \theta$
$y = b \sin \theta$

21) Simplify the following expressions:

(a) $\dfrac{1 - \sec^2 A}{1 - \operatorname{cosec}^2 A}$

(b) $\dfrac{\sin \theta}{\sqrt{(1 - \cos^2 \theta)}}$

(c) $\dfrac{\sin \theta}{\cos \theta} + \dfrac{\cos \theta}{\sin \theta}$

(d) $\dfrac{\sqrt{(1 + \tan^2 \theta)}}{\sqrt{(1 - \sin^2 \theta)}}$

(e) $\dfrac{1}{\cos \theta \sqrt{(1 + \cot^2 \theta)}}$

22) Without reference to tables or calculator, complete the following table:

	$\sin \theta$	$\cos \theta$	$\tan \theta$	type of angle
a)		$-\dfrac{5}{13}$		reflex
b)	$\dfrac{3}{5}$			obtuse
c)			$\dfrac{7}{24}$	acute
d)		-1		

COMPOUND ANGLE IDENTITIES

It is often useful to be able to express the trig ratios of angles such as $A + B$ or $A - B$ in terms of the trig ratios of A and of B.

At first sight it is dangerously easy to think, for instance, that $\sin (A + B)$ is $\sin A + \sin B$.

That this is false can be seen by considering $\sin (45° + 45°) = \sin 90° = 1$

whereas $\sin 45° + \sin 45° = \dfrac{\sqrt{2}}{2} + \dfrac{\sqrt{2}}{2} = \sqrt{2} \neq 1$.

So we see that the sine function is *not distributive* (and similarly for the other trig ratios).

The correct expression is $\sin (A + B) \equiv \sin A \cos B + \cos A \sin B$.

This formula can be proved geometrically when A and B are both acute, by using the diagram overleaf.

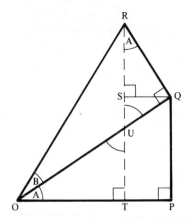

The right angled triangles OPQ and OQR contain angles A and B as shown. The dotted lines are construction lines and the angle URQ is equal to A.

$$\sin (A + B) \equiv \frac{TR}{OR} \equiv \frac{TS + SR}{OR} \equiv \frac{PQ + SR}{OR}$$

$$\equiv \frac{PQ}{OQ} \times \frac{OQ}{OR} + \frac{SR}{QR} \times \frac{QR}{OR}$$

$$\equiv \sin A \cos B + \cos A \sin B$$

Accepting at this stage the validity of this formula for all angles, it can be adapted to give the full set of compound angle identities. The reader is given the opportunity in the following exercise to derive these for himself.

EXERCISE 7b

1) In the identity $\sin (A + B) \equiv \sin A \cos B + \cos A \sin B$,   replace B by $-B$ to show that $\sin (A - B) \equiv \sin A \cos B - \cos A \sin B$.

2) In the identity derived in (1), replace A by $\left(\frac{\pi}{2} - A\right)$ to show that $\cos (A + B) \equiv \cos A \cos B - \sin A \sin B$.

3) In the identity derived in (2), replace B by $-B$ to show that $\cos (A - B) \equiv \cos A \cos B + \sin A \sin B$.

4) Use $\dfrac{\sin (A + B)}{\cos (A + B)}$ to show that $\tan (A + B) \equiv \dfrac{\tan A + \tan B}{1 - \tan A \tan B}$.

5) In the identity derived in (4), replace B by $-B$ to show that

$\tan (A - B) \equiv \dfrac{\tan A - \tan B}{1 + \tan A \tan B}$

Collating these results we have:

$$\left.\begin{array}{l} \sin (A + B) \equiv \sin A \cos B + \cos A \sin B \\ \sin (A - B) \equiv \sin A \cos B - \cos A \sin B \end{array}\right\}$$

$$\left.\begin{array}{l} \cos (A + B) \equiv \cos A \cos B - \sin A \sin B \\ \cos (A - B) \equiv \cos A \cos B + \sin A \sin B \end{array}\right\}$$

$$\left.\begin{array}{l} \tan (A + B) \equiv \dfrac{\tan A + \tan B}{1 - \tan A \tan B} \\[4mm] \tan (A - B) \equiv \dfrac{\tan A - \tan B}{1 + \tan A \tan B} \end{array}\right\}$$

The similarity between the pairs of identities for $A + B$ and $A - B$ makes it clear that care must be taken with signs when using these formulae.

EXAMPLES 7c

1) Without using tables or calculator, evaluate:
(a) $\sin 75°$ (b) $\cos 105°$ (c) $\tan (- 15°)$

(a) $\sin 75° = \sin (45° + 30°) = \sin 45° \cos 30° + \cos 45° \sin 30°$

$$= \left(\frac{\sqrt{2}}{2}\right)\left(\frac{\sqrt{3}}{2}\right) + \left(\frac{\sqrt{2}}{2}\right)\left(\frac{1}{2}\right)$$

$$= (\sqrt{3} + 1)\frac{\sqrt{2}}{4}$$

(b) $\cos 105° = \cos (60° + 45°) = \cos 60° \cos 45° - \sin 60° \sin 45°$

$$= \left(\frac{1}{2}\right)\left(\frac{\sqrt{2}}{2}\right) - \left(\frac{\sqrt{3}}{2}\right)\left(\frac{\sqrt{2}}{2}\right)$$

$$= \frac{(1 - \sqrt{3})\sqrt{2}}{4}$$

Note that this result is negative which is consistent with the cosine of an angle in the second quadrant.

(c) $\tan (- 15°) = \tan (45° - 60°) = \dfrac{\tan 45° - \tan 60°}{1 + \tan 45° \tan 60°}$

$$= \frac{1 - \sqrt{3}}{1 + (1)(\sqrt{3})}$$

$$= \sqrt{3} - 2$$
(rationalising the denominator)

Note: In each part of this example there are alternative compound angles which could be used,

e.g.　$75° = 120° - 45°$;　$105° = 150° - 45°$;　$-15° = 30° - 45°$.

2) A is obtuse and　$\sin A = \frac{3}{5}$,　B is acute and　$\sin B = \frac{12}{13}$.　Without finding the values of A and B, evaluate:

(a)　$\cos(A + B)$,　　　(b)　$\tan(A - B)$

In order to use the relevant compound angle formulae we need values for $\cos A$, $\cos B$, $\tan A$ and $\tan B$. These are most simply obtained by using Pythagoras in the appropriate right angled triangles.

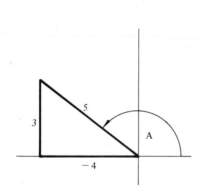
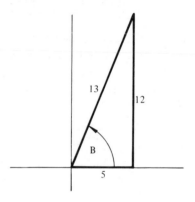

(a)　$\cos(A + B)$

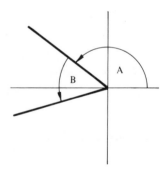

$$\equiv \cos A \cos B - \sin A \sin B$$

$$= \left(-\frac{4}{5}\right)\left(\frac{5}{13}\right) - \left(\frac{3}{5}\right)\left(\frac{12}{13}\right)$$

$$= -\frac{56}{65}$$

(b)　$\tan(A - B)$

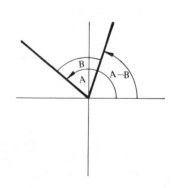

$$\equiv \frac{\tan A - \tan B}{1 + \tan A \tan B}$$

$$= \frac{\left(-\frac{3}{4}\right) - \left(\frac{12}{5}\right)}{1 + \left(-\frac{3}{4}\right)\left(\frac{12}{5}\right)}$$

$$= \frac{63}{16}$$

3) Prove that $\dfrac{\sin(A-B)}{\cos A \cos B} + \dfrac{\sin(B-C)}{\cos B \cos C} + \dfrac{\sin(C-A)}{\cos C \cos A} \equiv 0.$

L.H.S. becomes:

$$\frac{\sin A \cos B - \cos A \sin B}{\cos A \cos B} + \frac{\sin B \cos C - \cos B \sin C}{\cos B \cos C} + \frac{\sin C \cos A - \cos C \sin A}{\cos C \cos A}$$

$$\equiv \frac{\sin A \cos B}{\cos A \cos B} - \frac{\cos A \sin B}{\cos A \cos B} + \frac{\sin B \cos C}{\cos B \cos C} - \frac{\cos B \sin C}{\cos B \cos C} + \frac{\sin C \cos A}{\cos C \cos A} - \frac{\cos C \sin A}{\cos C \cos A}$$

$$\equiv \tan A - \tan B + \tan B - \tan C + \tan C - \tan A$$

$$\equiv 0$$

4) Solve the equation $2\cos\theta = \sin(\theta + 30°)$ giving the general values of θ.

It is very important to appreciate that this is *not an identity*. Only certain distinct values of θ satisfy this *equation*.

$$2\cos\theta = \sin(\theta + 30°)$$

$$= \sin\theta \cos 30° + \cos\theta \sin 30°$$

$$= \frac{\sqrt{3}}{2}\sin\theta + \frac{1}{2}\cos\theta$$

Therefore $\dfrac{3}{2}\cos\theta = \dfrac{\sqrt{3}}{2}\sin\theta$

$\Rightarrow$ $\dfrac{3}{\sqrt{3}} = \dfrac{\sin\theta}{\cos\theta}$

$\Rightarrow$ $\tan\theta = \sqrt{3}$

The principal solution is $\theta = \dfrac{\pi}{3}$

So the general solution is $\theta = n\pi + \dfrac{\pi}{3}$

EXERCISE 7c

(Do not use tables or a calculator in Questions 1–3.)

1) Evaluate:
(a) $\cos 80° \cos 20° + \sin 80° \sin 20°$
(b) $\sin 37° \cos 7° - \cos 37° \sin 7°$
(c) $\cos 15°$ (d) $\sin 165°$ (e) $\cos 75°$ (f) $\tan 75°$

2) Complete the following table:

	(a)	(b)	(c)	(d)	(e)	(f)
A	60°	$\frac{3\pi}{4}$	Acute	Obtuse	Acute	
sin A			$\frac{7}{25}$	$\frac{3}{5}$		$-\frac{5}{13}$
cos A						$\frac{12}{13}$
tan A					$\frac{7}{24}$	
B	150°	$\frac{7\pi}{6}$	Obtuse	Obtuse	Acute	
sin B			$\frac{4}{5}$			
cos B					$\frac{24}{25}$	$\frac{4}{5}$
tan B				$-\frac{12}{5}$		
sin (A + B)						$\frac{16}{65}$
cos (A + B)						
tan (A − B)						

Prove the following identities:

3) $\cot (A + B) \equiv \dfrac{\cot A \cot B - 1}{\cot A + \cot B}$

4) $\sin \left(\dfrac{\pi}{4} + A\right) + \sin \left(\dfrac{\pi}{4} - A\right) \equiv \sqrt{2} \cos A$

5) $(\sin A + \cos A)(\sin B + \cos B) \equiv \sin (A + B) + \cos (A - B)$

6) $\dfrac{\sin (A + B)}{\cos A \cos B} \equiv \tan A + \tan B$

7) $\sin (\theta + 60°) \equiv \sin (120° - \theta)$

8) $\tan (x + y) - \tan x \equiv \dfrac{\sin y}{\cos x \cos (x + y)}$

Solve the following equations, giving angles from 0° to 360°:

9) $\cos (45° - \theta) = \sin (30° + \theta)$

10) $3 \sin x = \cos (x + 60°)$

11) $\tan (A - \theta) = \frac{2}{3}$ and $\tan A = 3$

12) $\sin (x + 60°) = \cos x$

THE DOUBLE ANGLE IDENTITIES

The compound angle formulae deal with any two angles A and B and can therefore be used for two equal angles $(B = A)$.

Replacing B by A in the compound angle formulae for $(A + B)$ gives

$$\sin 2A \equiv 2 \sin A \cos A$$

$$\cos 2A \equiv \cos^2 A - \sin^2 A$$

$$\tan 2A \equiv \frac{2 \tan A}{1 - \tan^2 A}$$

The second of this group can be expressed in several forms because

$$\cos^2 A - \sin^2 A \equiv (1 - \sin^2 A) - \sin^2 A \equiv 1 - 2 \sin^2 A$$

$$\cos^2 A - \sin^2 A \equiv \cos^2 A - (1 - \cos^2 A) \equiv 2 \cos^2 A - 1$$

Thus

$$\cos 2A \equiv \begin{cases} \cos^2 A - \sin^2 A \\ 1 - 2 \sin^2 A \\ 2 \cos^2 A - 1 \end{cases}$$

These alternative expressions for cos 2A can themselves be rearranged to give

$$2 \sin^2 A \equiv 1 - \cos 2A$$

$$2 \cos^2 A \equiv 1 + \cos 2A$$

Complete familiarity with all the double angle formulae, including *all* the alternative forms involving cos 2A, is essential. These are probably the most useful of all the trig identities, being a powerful tool for simplifying trig functions.

EXAMPLES 7d

1) Find the general solution of the equation $\cos 2x + 3 \sin x = 2$.

Using $\cos 2x \equiv 1 - 2 \sin^2 x$ gives

$$1 - 2 \sin^2 x + 3 \sin x = 2$$

i.e.

$$2 \sin^2 x - 3 \sin x + 1 = 0$$

$$(2 \sin x - 1)(\sin x - 1) = 0$$

$$\Rightarrow \qquad \qquad \sin x = \tfrac{1}{2} \text{ or } 1$$

For $\sin x = \frac{1}{2}$ the principal solution is $\dfrac{\pi}{6}$

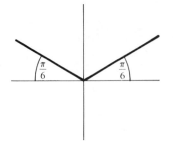

General solution is $x = \begin{cases} 2n\pi + \dfrac{\pi}{6} \\ \\ (2n+1)\pi - \dfrac{\pi}{6} \end{cases}$

For $\sin x = 1$ the principal solution is $x = \dfrac{\pi}{2}$

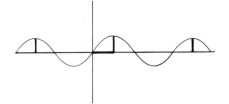

General solution is $x = 2n\pi + \dfrac{\pi}{2}$

The full solution is therefore

$$x = 2n\pi + \frac{\pi}{6}, \quad (2n+1)\pi - \frac{\pi}{6}, \quad 2n\pi + \frac{\pi}{2}$$

2) If $\tan \theta = \frac{3}{4}$ and θ is acute, find the values of $\tan 2\theta$, $\tan 4\theta$ and $\tan \dfrac{\theta}{2}$.

In this problem we use $\tan 2A \equiv \dfrac{2 \tan A}{1 - \tan^2 A}$,

for three cases: $A = \theta$, $A = 2\theta$, $A = \dfrac{\theta}{2}$

If $A = \theta$, $\qquad \tan 2\theta \equiv \dfrac{2 \tan \theta}{1 - \tan^2 \theta} = \dfrac{\frac{3}{2}}{1 - \frac{9}{16}} = \dfrac{24}{7}$

If $A = 2\theta$, $\qquad \tan 4\theta \equiv \dfrac{2 \tan 2\theta}{1 - \tan^2 2\theta} = \dfrac{2(\frac{24}{7})}{1 - (\frac{24}{7})^2} = -\dfrac{336}{527}$

If $A = \dfrac{\theta}{2}$, $\qquad \tan \theta \equiv \dfrac{2 \tan \dfrac{\theta}{2}}{1 - \tan^2 \dfrac{\theta}{2}} = \dfrac{3}{4}$

or $$\frac{2t}{1-t^2} = \frac{3}{4} \quad \text{where} \quad t \equiv \tan\frac{\theta}{2}$$

$\Rightarrow$ $$8t = 3 - 3t^2$$

$\Rightarrow$ $$3t^2 + 8t - 3 = 0$$

$\Rightarrow$ $$(3t - 1)(t + 3) = 0$$

$\Rightarrow$ $$t = \frac{1}{3} \quad \text{or} \quad -3$$

i.e. $$\tan\frac{\theta}{2} = \frac{1}{3} \quad \text{or} \quad -3$$

But θ is acute, so $\dfrac{\theta}{2}$ is acute and $\tan\dfrac{\theta}{2} \neq -3$.

Therefore $\tan\dfrac{\theta}{2} = \dfrac{1}{3}$

3) Prove that $\sin 3A \equiv 3 \sin A - 4 \sin^3 A$.

$$\sin 3A \equiv \sin(2A + A)$$
$$\equiv \sin 2A \cos A + \cos 2A \sin A$$
$$\equiv (2 \sin A \cos A) \cos A + (1 - 2 \sin^2 A) \sin A$$
$$\equiv 2 \sin A \cos^2 A + \sin A - 2 \sin^3 A$$
$$\equiv 2 \sin A(1 - \sin^2 A) + \sin A - 2 \sin^3 A$$
$$\equiv 3 \sin A - 4 \sin^3 A$$

Note: This is quite a useful identity and is worth remembering.

4) Eliminate θ from the equations $x = \cos 2\theta$, $y = \sec\theta$.

Using $\cos 2\theta \equiv 2\cos^2\theta - 1$

$$x = 2\cos^2\theta - 1 \quad \text{and} \quad y = \frac{1}{\cos\theta}$$

Hence $$x = 2\left(\frac{1}{y}\right)^2 - 1$$

$\Rightarrow$ $$(x + 1)y^2 = 2$$

Note: This is a *Cartesian equation* which we have obtained by eliminating the *parameter* θ from a *pair of parametric equations*.

EXERCISE 7d

(Do not use tables or calculator in Questions 1–3.)

1) Express as a single trig ratio:

(a) $2 \sin 14° \cos 14°$

(b) $\dfrac{2 \tan 35°}{1 - \tan^2 35°}$

(c) $1 - 2 \sin^2 4\theta$

(d) $\dfrac{2 \tan 3\theta}{1 - \tan^2 3\theta}$

(e) $\sqrt{1 + \cos 6\theta}$

(f) $\cos^2 26° - \sin^2 26°$

(g) $\sin \theta \cos \theta$

(h) $2 \cos^2 34° - 1$

(i) $\dfrac{1 + \tan x}{1 - \tan x}$ (*Hint*: $\tan 45° = 1$)

2) Find the values of $\sin 2\theta$ and $\cos 2\theta$ given:

(a) $\cos \theta = \frac{3}{5}$ (b) $\sin \theta = \frac{7}{25}$ (c) $\tan \theta = \frac{12}{5}$

assuming that   (i) in all cases θ is acute,
                (ii) in all cases θ is not acute.

3) If $\tan \theta = -\frac{7}{24}$ and θ is obtuse, find the value of $\tan \dfrac{\theta}{2}$ and hence find $\sin \dfrac{\theta}{2}$ and $\cos \dfrac{\theta}{2}$. Use these results to evaluate $\sin \theta$ and $\cos \theta$ and check that they are consistent with the given value of $\tan \theta$.

4) By eliminating θ from the following pairs of parametric equations, find the corresponding Cartesian equation:

(a) $x = \tan 2\theta, \ y = \tan \theta$ (b) $x = \cos 2\theta, \ y = \cos \theta$
(c) $x = \cos 2\theta, \ y = \operatorname{cosec} \theta$ (d) $x = \sin 2\theta, \ y = \sec 4\theta$

Prove the following identities:

5) $\dfrac{1 - \cos 2A}{\sin 2A} \equiv \tan A$

6) $\tan \theta + \cot \theta \equiv 2 \operatorname{cosec} 2\theta$

7) $\sec 2A + \tan 2A \equiv \dfrac{\cos A + \sin A}{\cos A - \sin A}$

8) $\dfrac{1 - \cos 2A + \sin 2A}{1 + \cos 2A + \sin 2A} \equiv \tan A$

9) $\cos 4A \equiv 8 \cos^4 A - 8 \cos^2 A + 1$

10) $\sin 2\theta \equiv \dfrac{2 \tan \theta}{1 + \tan^2 \theta}$

11) $\cos 2\theta \equiv \dfrac{1 - \tan^2\theta}{1 + \tan^2\theta}$

12) $\cos 3\theta \equiv 4 \cos^3\theta - 3 \cos \theta$

Solve the following equations giving angles within the range $0°$ to $360°$. Also in each case state the general solution.

13) $\cos 2x = \sin x$

14) $\sin 2x + \cos x = 0$

15) $4 - 5 \cos \theta = 2 \sin^2\theta$

16) $\tan \theta \tan 2\theta = 2$

17) $\sin 2\theta - 1 = \cos 2\theta$

18) $5 \cos x \sin 2x + 4 \sin^2 x = 4$

IDENTITIES APPLIED TO INVERSE TRIGONOMETRIC FUNCTIONS

The double angle and compound angle identities are often useful in simplifying expressions or solving equations containing inverse trig functions. The following examples illustrate the appropriate techniques.

EXAMPLES 7e

1) Find x if $\arcsin x + \arccos \dfrac{x}{2} = \dfrac{5\pi}{6}$

Let $\arcsin x \equiv \theta \;\Rightarrow\; \sin \theta \equiv x \;\Rightarrow\;$

and $\arccos \dfrac{x}{2} \equiv \phi \;\Rightarrow\; \cos \phi \equiv \dfrac{x}{2} \;\Rightarrow\;$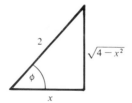

The given equation then becomes

$$\theta + \phi = \frac{5\pi}{6}$$

Hence $\qquad\qquad \sin (\theta + \phi) = \tfrac{1}{2}$

$\Rightarrow \qquad\qquad \sin \theta \cos \phi + \cos \theta \sin \phi = \tfrac{1}{2}$

$\Rightarrow \qquad (x)\left(\dfrac{x}{2}\right) + \sqrt{(1 - x^2)} \left(\dfrac{\sqrt{4 - x^2}}{2}\right) = \tfrac{1}{2}$

$\Rightarrow \qquad\qquad \sqrt{(1 - x^2)}\sqrt{(4 - x^2)} = 1 - x^2$

Squaring both sides (*not* cancelling $\sqrt{1 - x^2}$)

gives $\qquad\qquad 4 - 5x^2 + x^4 = 1 - 2x^2 + x^4$

Therefore, either $\quad x^4 = \infty \quad$ (impossible since x is a sine ratio)

or $\qquad\qquad\qquad\qquad 3x^2 = 3$

$\Rightarrow \qquad\qquad\qquad\qquad x = \pm 1$

But the value of $\quad \arcsin(-1)$ is $-\dfrac{\pi}{2}$ and the value of $\arccos(-\tfrac{1}{2})$ is $\dfrac{2\pi}{3}$.

So $\quad x = -1 \quad$ does not satisfy the given equation and the only solution is

$$x = 1.$$

2) Prove that $\quad \arctan 3 + 2 \arctan 2 = \pi + \text{arccot } 3$

Let $\qquad\qquad \arctan 3 = \theta \quad \Rightarrow \quad \tan \theta = 3 \quad \Rightarrow \quad \tfrac{1}{4}\pi < \theta < \tfrac{1}{2}\pi$

and $\qquad\qquad \arctan 2 = \phi \quad \Rightarrow \quad \tan \phi = 2 \quad \Rightarrow \quad \tfrac{1}{4}\pi < \phi < \tfrac{1}{2}\pi$

The L.H.S then becomes $\qquad\qquad \theta + 2\phi$

Now $\qquad\qquad \tan(\theta + 2\phi) \equiv \dfrac{\tan \theta + \tan 2\phi}{1 - \tan \theta \tan 2\phi}$

Also $\qquad\qquad \tan 2\phi \equiv \dfrac{2 \tan \phi}{1 - \tan^2 \phi} = \dfrac{2 \times 2}{1 - 4} = -\dfrac{4}{3}$

Hence $\qquad\qquad \tan(\theta + 2\phi) = \dfrac{3 - \tfrac{4}{3}}{1 - 3(-\tfrac{4}{3})} = \dfrac{\tfrac{5}{3}}{5} = \dfrac{1}{3}$

But $\qquad \tfrac{1}{4}\pi < \theta < \tfrac{1}{2}\pi \quad$ and $\quad \tfrac{1}{4}\pi < \phi < \tfrac{1}{2}\pi \quad \Rightarrow \quad \tfrac{3}{4}\pi < \theta + 2\phi < \tfrac{3}{2}\pi$

$\Rightarrow \qquad\qquad \theta + 2\phi = \pi + \arctan \tfrac{1}{3} = \pi + \text{arccot } 3$

i.e. $\qquad\qquad \arctan 3 + 2 \arctan 2 = \pi + \text{arccot } 3$

3) Simplify $\quad \arctan x + \arctan\left(\dfrac{1-x}{1+x}\right)$

Let $\qquad\qquad \alpha \equiv \arctan x \quad \Rightarrow \quad \tan \alpha \equiv x$

and $\qquad\qquad \beta \equiv \arctan\left(\dfrac{1-x}{1+x}\right) \quad \Rightarrow \quad \tan \beta \equiv \dfrac{1-x}{1+x}$

So we have to simplify $\alpha + \beta$.

Using $\qquad\qquad \tan(\alpha + \beta) \equiv \dfrac{\tan \alpha + \tan \beta}{1 - \tan \alpha \tan \beta}$

gives
$$\tan(\alpha + \beta) \equiv \dfrac{x + \dfrac{1-x}{1+x}}{1 - x\left(\dfrac{1-x}{1+x}\right)}$$

$$\equiv \dfrac{x^2 + 1}{1 + x^2} \equiv 1$$

Hence
$$\alpha + \beta \equiv \arctan 1 \equiv \dfrac{\pi}{4}$$

Thus
$$\arctan x + \arctan \dfrac{1-x}{1+x} \equiv \dfrac{\pi}{4}$$

EXERCISE 7e

Prove the following relationships:

1) $\arcsin x + \arccos x \equiv \dfrac{\pi}{2}$

2) $\arctan \dfrac{1}{3} + \arctan \dfrac{1}{2} = \dfrac{\pi}{4}$

3) $2 \arctan \dfrac{1}{2} = \arccos \dfrac{3}{5}$

4) $2 \arctan \dfrac{3}{4} = \arccos \dfrac{7}{25}$

5) $4 \operatorname{arccot} 2 + \arctan \dfrac{24}{7} = \pi$

Simplify:

6) $\sin(2 \arctan x)$

7) $\tan^{-1} x + \tan^{-1} \dfrac{1}{x}$

8) $\tan\left(\arctan \tfrac{1}{3} + \arctan \tfrac{1}{4}\right)$

Solve the following equations:

9) $\tan^{-1} 2 = \tan^{-1} 4 - \tan^{-1} x$

10) $\arctan(1+x) + \arctan(1-x) = \arctan 2$

11) $\sin^{-1}\left(\dfrac{x}{x-1}\right) + 2\tan^{-1}\left(\dfrac{1}{x+1}\right) = \dfrac{\pi}{2}$

THE HALF ANGLE IDENTITIES

We already know that $\tan 2A \equiv \dfrac{2 \tan A}{1 - \tan^2 A}$ and, from Exercise 7d

(Nos. 10 and 11), that $\sin 2A \equiv \dfrac{2 \tan A}{1 + \tan^2 A}$ and $\cos 2A \equiv \dfrac{1 - \tan^2 A}{1 + \tan^2 A}$.

If we replace $2A$ by θ and use t to denote $\tan \dfrac{\theta}{2}$ we have

$$\tan \theta \equiv \frac{2t}{1 - t^2}$$

$$\sin \theta \equiv \frac{2t}{1 + t^2}$$

$$\cos \theta \equiv \frac{1 - t^2}{1 + t^2}$$

These three identities allow *all* the trig ratios of any one angle to be expressed in terms of a common variable t. In problems where it is not possible to apply any of the identities used previously, this group can be helpful.

EXAMPLE

Solve the equation $\sin \theta + 2 \cos \theta = 1$ for angles between $0°$ and $360°$.

$$\sin \theta + 2 \cos \theta = 1$$

Therefore $\dfrac{2t}{1 + t^2} + 2\left(\dfrac{1 - t^2}{1 + t^2}\right) = 1$ where $t = \tan \dfrac{\theta}{2}$

$\Rightarrow$ $\qquad 2t + 2 - 2t^2 = 1 + t^2$

$\Rightarrow$ $\qquad 3t^2 - 2t - 1 = 0$

$\Rightarrow$ $\qquad (3t + 1)(t - 1) = 0$

Therefore either $3t + 1 = 0$ or $t - 1 = 0$

i.e. $\qquad \tan \dfrac{\theta}{2} = -\dfrac{1}{3}$ or 1

The range of values specified for θ is $0°$ to $360°$,

so the range of values required for $\dfrac{\theta}{2}$ is $0°$ to $180°$.

Within this range

$\qquad \tan \dfrac{\theta}{2} = -\dfrac{1}{3}$ gives $\dfrac{\theta}{2} = 161.57°$

$\qquad \tan \dfrac{\theta}{2} = 1$ gives $\dfrac{\theta}{2} = 45°$

Thus $\qquad\qquad\qquad\qquad \theta = 323.14°, 90°$

If we now look at a very similar equation we will see that extra care is sometimes needed in using this method. Consider the equation

$$\sin \theta - \cos \theta = 1$$

Using $\quad t = \tan\dfrac{\theta}{2} \qquad\qquad \dfrac{2t}{1 + t^2} - \dfrac{1 - t^2}{1 + t^2} = 1$

$\Rightarrow \qquad\qquad\qquad\qquad 2t - 1 + t^2 = 1 + t^2$

Now it was seen on page 11 that if we simply cancel out the two t^2 terms, we lose the solution $t = \pm\infty$, so the solution proceeds

$\Rightarrow \qquad\qquad\qquad\qquad 2t = 2 \quad \text{or} \quad t = \pm\infty$

Hence $\qquad\qquad\qquad\qquad \tan\dfrac{\theta}{2} = 1 \quad \text{or} \quad \pm\infty$

Giving $\qquad\qquad\qquad\qquad \dfrac{\theta}{2} = n\pi + \dfrac{\pi}{4}, \quad n\pi + \dfrac{\pi}{2}$

Note: When using the half angle identities, t does not always represent $\tan\dfrac{\theta}{2}$.

For instance, in solving the equation $\quad \sin 4\theta + \tan 2\theta = 0 \quad$ we would use $t \equiv \tan 2\theta$.

THE EXPRESSION $\quad a \cos \theta + b \sin \theta$

It is often useful to reduce $\quad a \cos \theta + b \sin \theta$
to a single term such as $\quad r \cos (\theta - \alpha)$.
This is possible, provided that we can find values of r and α for which

$$r[\cos \theta \cos \alpha + \sin \theta \sin \alpha] \equiv a \cos \theta + b \sin \theta$$

Comparing the coefficients of $\cos \theta$ and $\sin \theta$

we have $\qquad\qquad\qquad\qquad r \cos \alpha = a \qquad\qquad\qquad\qquad$ [1]

and $\qquad\qquad\qquad\qquad r \sin \alpha = b \qquad\qquad\qquad\qquad$ [2]

[2] $\div$ [1] gives $\qquad\qquad \tan \alpha = \dfrac{b}{a} \quad \Rightarrow$

From the triangle $\qquad\qquad \cos \alpha = \dfrac{a}{\sqrt{(a^2 + b^2)}}$

Hence from [1] $\qquad\qquad\qquad r = \sqrt{a^2 + b^2}$

i.e. r is equal to the length of the hypotenuse of the triangle containing α.

Thus
$$a \cos \theta + b \sin \theta \equiv r \cos (\theta - \alpha)$$

where
$$r = \sqrt{a^2 + b^2} \text{ and } \alpha = \arctan \frac{b}{a}$$

EXAMPLE

Express $3 \cos \theta + 4 \sin \theta$ in the form $r \cos (\theta - \alpha)$ giving values for r and α.

Let
$$r[\cos \theta \cos \alpha + \sin \theta \sin \alpha] \equiv 3 \cos \theta + 4 \sin \theta$$

So that
$$\begin{cases} r \cos \alpha = 3 \\ r \sin \alpha = 4 \end{cases}$$

Thus
$$r = 5$$

and
$$\tan \alpha = \tfrac{4}{3} \quad \Rightarrow$$

i.e.
$$3 \cos \theta + 4 \sin \theta \equiv 5 \cos (\theta - \alpha)$$

where
$$\alpha = \arctan \tfrac{4}{3}$$

It is sometimes more convenient to begin by comparing $a \cos \theta + b \sin \theta$ with $r \sin (\theta + \alpha)$ so that

$$r[\sin \theta \cos \alpha + \cos \theta \sin \alpha] \equiv a \cos \theta + b \sin \theta$$

Then, since
$$\begin{cases} r \sin \alpha = a \\ r \cos \alpha = b \end{cases}$$

we get
$$\tan \alpha = \frac{a}{b}$$

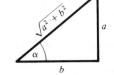

and
$$r = \sqrt{a^2 + b^2}$$

Note that the value of α is not the same as it was when we used $r \cos (\theta - \alpha)$. Further variations that could be used are $r \sin (\theta - \alpha)$ and $r \cos (\theta + \alpha)$. When using this method, it is better to work from the basic comparison, as we did in the examples above, rather than to quote values of r and α. Before going any further, the reader is recommended to carry out the following transformations:

$$5 \cos \theta + 12 \sin \theta \quad \text{to the forms} \quad r \cos (\theta - \alpha), \; r \sin (\theta + \alpha)$$
$$7 \cos \theta - 24 \sin \theta \quad \text{to the forms} \quad r \cos (\theta + \alpha), \; r \sin (\theta - \alpha)$$
$$3 \sin \theta + 4 \cos \theta \quad \text{to the forms} \quad r \sin (\theta + \alpha), \; r \cos (\theta - \alpha)$$

THE GRAPH OF THE FUNCTION $a \cos \theta + b \sin \theta$

First consider the function $f(\theta) \equiv k \cos \theta \; (k > 0)$ which has the following characteristics:

(a) $f(\theta) = \quad 0$ when $\theta = \dfrac{\pi}{2}, \dfrac{3\pi}{2}, \dfrac{5\pi}{2}$, etc.

(b) $f(\theta) = \quad k$ when $\theta = 0, 2\pi, 4\pi$, etc.

(c) $f(\theta) = -k$ when $\theta = \pi, 3\pi, 5\pi$, etc.

So we see that the graph of this function is very similar to a standard cosine curve but has maximum and minimum values $\pm k$. (We say that the curve has an amplitude k),

e.g. if $k = 5$,   we have

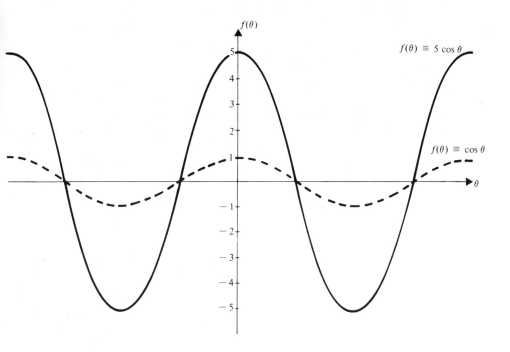

Also we saw in Chapter Six that the graph of $\cos(\theta - \alpha)$ is given by moving the graph of $\cos \theta$ a distance α to the right.

Now combining these two modifications to a standard cosine curve, the graph of the function $f(\theta) \equiv k \cos(\theta - \alpha)$ can be sketched.

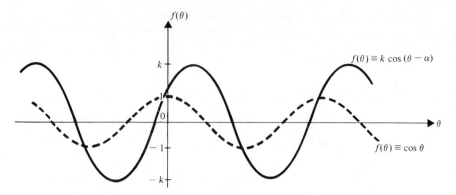

Consider now the function $a \cos \theta + b \sin \theta$. At first sight its graph is not easy to visualise, but using $a \cos \theta + b \sin \theta \equiv r \cos (\theta - \alpha)$, we see that the graph of $f(\theta) \equiv a \cos \theta + b \sin \theta$ is a cosine curve modified as follows:

(a) its maximum and minimum values are $\pm r$,
   i.e. its *amplitude* is r;

(b) its position is a distance α to the right of the standard curve.

EXAMPLE

Sketch the graph of the function $3 \cos \theta + 4 \sin \theta$ from $-180°$ to $+180°$.

Let $\qquad\qquad 3 \cos \theta + 4 \sin \theta \equiv r \cos (\theta - \alpha)$

so that $\qquad\qquad r = 5 \quad$ and $\quad \tan \alpha = \dfrac{4}{3} \Rightarrow \alpha = 53.13°$

Hence the graph is a cosine curve with an amplitude of 5 and a rightward phase shift of 53.13°,

i.e.

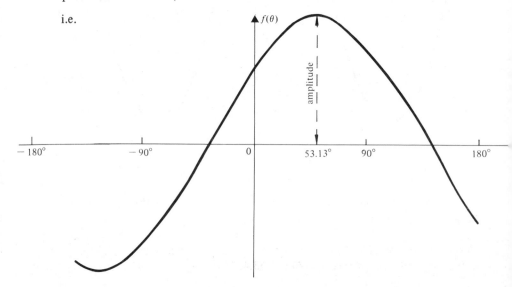

It is interesting to see how the same graph is produced if the alternative transformation $3 \cos \theta + 4 \sin \theta = r \sin (\theta + \alpha')$ is used. With this approach we have

$$r = 5 \quad \text{and} \quad \tan \alpha' = \frac{3}{4}$$

and $3 \cos \theta + 4 \sin \theta \equiv 5 \sin (\theta + \alpha')$

The R.H.S gives a sine curve with an amplitude of 5 and a displacement of α' to the left. But since $\tan \alpha' = \cot \alpha,$ α' and α are complementary angles,

i.e. $\alpha' + \alpha = \dfrac{\pi}{2}$

We also know that a cosine curve is the same as a sine curve displaced $\dfrac{\pi}{2}$ to the left.

Thus a cosine curve moved a distance α to the right, coincides with a sine curve moved a distance α' to the left.

So any correct transformation of $a \cos \theta + b \sin \theta$ into a compound angle form provides a quick method of sketching the graph of that function and, in particular, of evaluating its maximum and minimum values (which, for sine and cosine functions, are also the greatest and least values).

THE EQUATION $a \cos \theta + b \sin \theta = c$

One way of solving an equation of this type has already been used. It depends on using the half angle formulae. An alternative method is now available using a compound angle form such as $r \cos (\theta - \alpha).$

This approach, applied to one of the examples already solved using the 'little t' formulae, (see page 218) gives

$$\sin \theta + 2 \cos \theta = 1$$

But $2 \cos \theta + \sin \theta \equiv r[\cos \theta \cos \alpha + \sin \theta \sin \alpha] \equiv r \cos (\theta - \alpha)$

where $\begin{cases} r \cos \alpha = 2 \\ r \sin \alpha = 1 \end{cases}$

i.e. where $\tan \alpha = \frac{1}{2}$ and $r = \sqrt{5}$

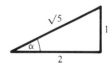

Hence $\sqrt{5} \cos (\theta - \alpha) = 1$

$$\cos (\theta - \alpha) = \frac{1}{\sqrt{5}}$$

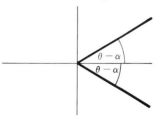

$\Rightarrow$ $\theta - \alpha = 360n° \pm 63.43°$

from which $$\theta = 360n° \pm 63.43° + \alpha$$

But $$\alpha = \arctan \tfrac{1}{2} = 26.57°$$

Hence $$\theta = 360n° + 90° \text{ or } 360n° - 36.86°$$

Using values of n which give values of θ between $0°$ and $360°$ $(n = 0, 1)$

we have $$\theta = 90°, 323.14°$$

EXAMPLES 7f

1) Express $\sqrt{\dfrac{1 - \sin 2\theta}{1 + \sin 2\theta}}$ in terms of $\tan \theta$.

Using $\sin 2\theta \equiv \dfrac{2t}{1 + t^2}$ where $t \equiv \tan \theta$ gives

$$1 - \sin 2\theta \equiv 1 - \frac{2t}{1 + t^2} \equiv \frac{1 + t^2 - 2t}{1 + t^2} \equiv \frac{(1 - t)^2}{1 + t^2}$$

and $$1 + \sin 2\theta \equiv 1 + \frac{2t}{1 + t^2} \equiv \frac{1 + t^2 + 2t}{1 + t^2} \equiv \frac{(1 + t)^2}{1 + t^2}$$

Hence $$\frac{1 - \sin 2\theta}{1 + \sin 2\theta} \equiv \frac{(1 - t)^2}{1 + t^2} \bigg/ \frac{(1 + t)^2}{1 + t^2} \equiv \left(\frac{1 - t}{1 + t}\right)^2$$

$$\Rightarrow \qquad \sqrt{\frac{1 - \sin 2\theta}{1 + \sin 2\theta}} \equiv \frac{1 - \tan \theta}{1 + \tan \theta}$$

2) Find the general solution of the equation $\cos \theta - \sqrt{3} \sin \theta = 1$,
(a) by using half angle formulae,
(b) by using a compound angle transformation.

(a) $$\cos \theta - \sqrt{3} \sin \theta = 1$$

Therefore $$\frac{1 - t^2}{1 + t^2} - \frac{2\sqrt{3}t}{1 + t^2} = 1 \quad \text{where} \quad t \equiv \tan \frac{\theta}{2}.$$

$$\Rightarrow \qquad 1 - t^2 - 2\sqrt{3}t = 1 + t^2$$

$$\Rightarrow \qquad 2t(t + \sqrt{3}) = 0$$

Hence, either $$t = 0 \quad \text{or} \quad t + \sqrt{3} = 0$$

$$\Rightarrow \qquad \tan \frac{\theta}{2} = 0 \quad \text{or} \quad -\sqrt{3}$$

Principal values of $\dfrac{\theta}{2}$ are 0 and $-\dfrac{\pi}{3}$

So the general solution is

$$\dfrac{\theta}{2} = n\pi, \quad n\pi - \dfrac{\pi}{3}$$

$\Rightarrow$

$$\theta = 2n\pi, \quad 2n\pi - \dfrac{2\pi}{3}$$

(b) Let $\cos\theta - \sqrt{3}\sin\theta \equiv r(\cos\theta\,\cos\alpha - \sin\theta\,\sin\alpha) \equiv r\cos(\theta + \alpha)$

where

$$\left. \begin{array}{l} r\cos\alpha = 1 \\[2mm] r\sin\alpha = \sqrt{3} \end{array} \right\} \;\Rightarrow\; \tan\alpha = \sqrt{3}$$

Hence

$$\alpha = \dfrac{\pi}{3} \text{ and } r = 2$$

The equation can now be written as

$$2\cos\left(\theta + \dfrac{\pi}{3}\right) = 1$$

$\Rightarrow$

$$\cos\left(\theta + \dfrac{\pi}{3}\right) = \dfrac{1}{2}$$

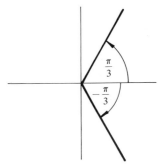

The principal value of $\theta + \dfrac{\pi}{3}$ is $\dfrac{\pi}{3}$

Therefore

$$\theta + \dfrac{\pi}{3} = 2n\pi \pm \dfrac{\pi}{3}$$

$\Rightarrow$

$$\theta = 2n\pi \pm \dfrac{\pi}{3} - \dfrac{\pi}{3}$$

$\Rightarrow$

$$\theta = \begin{cases} 2n\pi \\[2mm] 2n\pi - \dfrac{2\pi}{3} \end{cases}$$

3) Express $5\sin\theta + 12\cos\theta$ in the form $r\sin(\theta + \alpha)$ giving the values
of r and α.
Show that $5\sin\theta + 12\cos\theta + 7 \leqslant 20$ and find the minimum value of

$5 \sin \theta + 12 \cos \theta + 7$. Sketch the graph of the function $\dfrac{1}{5 \sin \theta + 12 \cos \theta}$

for $0 \leqslant \theta \leqslant 2\pi$.

Let $\quad 5 \sin \theta + 12 \cos \theta \equiv r[\sin \theta \cos \alpha + \cos \theta \sin \alpha] \equiv r \sin (\theta + \alpha)$

so that $\qquad\qquad \left.\begin{array}{l} r \cos \alpha = 5 \\[2mm] r \sin \alpha = 12 \end{array}\right\} \Rightarrow$

giving $\qquad\qquad r = 13 \quad$ and $\quad \tan \alpha = \frac{12}{5} \Rightarrow \alpha = 67.38°$

Hence $\qquad\qquad 5 \sin \theta + 12 \cos \theta \equiv 13 \sin (\theta + \alpha)$

But $\qquad\qquad -1 \leqslant \sin (\theta + \alpha) \leqslant 1$

so $\qquad\qquad -13 \leqslant 13 \sin (\theta + \alpha) \leqslant 13$

i.e. $\qquad\qquad -13 \leqslant 5 \sin \theta + 12 \cos \theta \leqslant 13$

Adding 7 throughout gives

$$-6 \leqslant 5 \sin \theta + 12 \cos \theta + 7 \leqslant 20$$

This shows that $\quad 5 \sin \theta + 12 \cos \theta + 7 \leqslant 20 \quad$ and that the minimum value of $\quad 5 \sin \theta + 12 \cos \theta + 7 \quad$ is -6.

Now $\quad 5 \sin \theta + 12 \cos \theta \equiv 13 \sin (\theta + \alpha)$

Therefore the graph of the function $\dfrac{1}{5 \sin \theta + 12 \cos \theta}$ is also the graph of the

function $\dfrac{1}{13 \sin (\theta + \alpha)} \quad$ or $\quad \dfrac{1}{13} \csc (\theta + \alpha)$

The general shape of the graph is a typical cosec curve (see page 162) except that the maximum and minimum values are $-\frac{1}{13}$ and $+\frac{1}{13}$ and its position is a distance α to the left of the standard curve.

Thus the graph of the function $\dfrac{1}{5 \sin \theta + 12 \cos \theta}$ is deduced to be

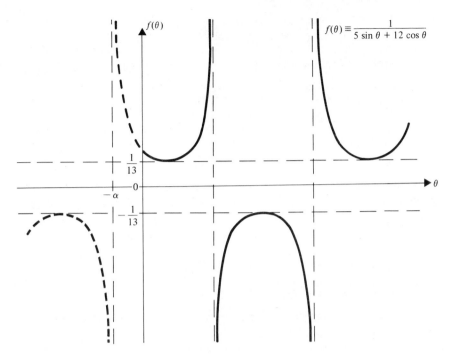

$$f(\theta) \equiv \frac{1}{5 \sin \theta + 12 \cos \theta}$$

EXERCISE 7f

1) If $\tan \theta = \dfrac{4}{3}$ and θ is acute, find the value of:

(a) $\sin 2\theta$ (b) $\tan \dfrac{\theta}{2}$ (c) $\cot 2\theta$

2) If $t \equiv \tan \dfrac{\theta}{2}$ express in terms of t:

(a) $\dfrac{1 - \cos \theta}{1 + \cos \theta}$ (b) $\dfrac{\sin \theta}{1 - \cos \theta}$ (c) $\cot \theta \cot \dfrac{\theta}{2}$

(d) $\dfrac{\cos^2 \dfrac{\theta}{2}}{3 \sin \theta + 4 \cos \theta - 1}$ (e) $\dfrac{1 - 2 \sin \theta}{2 \cos \theta + 1}$

3) Prove that $\operatorname{cosec} A + \cot A \equiv \cot \dfrac{A}{2}$

4) If $\sec \theta - \tan \theta = x$ prove that $\tan \dfrac{\theta}{2} = \dfrac{1 - x}{1 + x}$

5) Using $t \equiv \tan \dfrac{\theta}{2},$ solve the following equations giving values of θ from $-180°$ to $180°$:

(a) $3 \cos \theta + 2 \sin \theta = 3$ (b) $5 \cos \theta - \sin \theta + 4 = 0$
(c) $\cos \theta + 7 \sin \theta = 5$ (d) $2 \cos \theta - \sin \theta = 1$

6) Transform each of the following expressions into the compound angle form suggested.

(a) $\sqrt{3} \cos \theta - \sin \theta$ $r \cos (\theta + \alpha)$
(b) $\cos \theta + 3 \sin \theta$ $r \cos (\theta - \alpha)$
(c) $4 \sin \theta - 3 \cos \theta$ $r \sin (\theta - \alpha)$
(d) $\cos 2\theta - \sin 2\theta$ $r \cos (2\theta + \alpha)$
(e) $2 \cos 3\theta + 5 \sin 3\theta$ $r \sin (3\theta + \alpha)$

7) Find the maximum and minimum values of the following functions, stating in each case the values (from $0°$ to $360°$) of θ at which the turning points occur:

(a) $\cos \theta - \sqrt{3} \sin \theta$ (b) $7 \cos \theta - 24 \sin \theta + 3$

(c) $\dfrac{1}{\cos 2\theta + \sin 2\theta}$ (d) $\dfrac{\sqrt{2}}{\cos \theta - \sqrt{2} \sin \theta}$

(e) $(3 \cos \theta + 4 \sin \theta)^2$

8) Use a compound angle transformation to find the general solution of the following equations:

(a) $\cos x + \sin x = \sqrt{2}$ (b) $7 \cos x + 6 \sin x = 2$
(c) $\cos x - 3 \sin x = 1$ (d) $2 \cos x - \sin x = 2$

9) Sketch the graphs of the functions in Question 6.

THE FACTOR FORMULAE

To factorise is to express in the form of a product. The set of identities called the factor formulae (which, the reader may be relieved to learn, is the last set in this work) converts expressions such as $\sin A + \sin B$ into a product, so *factorising* the expression.

To derive these identities, we use the compound angle group.

$$\sin A \cos B + \cos A \sin B \equiv \sin (A + B)$$

$$\sin A \cos B - \cos A \sin B \equiv \sin (A - B)$$

Adding: $2 \sin A \cos B \equiv \sin (A + B) + \sin (A - B)$ [1]

Subtracting: $2 \cos A \sin B \equiv \sin (A + B) - \sin (A - B)$ [2]

Similar treatment of $\cos (A + B)$ and $\cos (A - B)$ gives:

$$2 \cos A \cos B \equiv \cos (A + B) + \cos (A - B)$$ [3]

$$-2 \sin A \sin B \equiv \cos (A + B) - \cos (A - B)$$ [4]

The R.H.S of each of these formulae can be simplified by putting

$$\begin{cases} A + B = P \\ A - B = Q \end{cases} \Rightarrow \begin{cases} A = \frac{1}{2}(P + Q) \\ B = \frac{1}{2}(P - Q) \end{cases}$$

Then

$$\sin P + \sin Q \equiv 2 \sin \frac{P + Q}{2} \cos \frac{P - Q}{2} \qquad [5]$$

$$\sin P - \sin Q \equiv 2 \cos \frac{P + Q}{2} \sin \frac{P - Q}{2} \qquad [6]$$

$$\cos P + \cos Q \equiv 2 \cos \frac{P + Q}{2} \cos \frac{P - Q}{2} \qquad [7]$$

$$\cos P - \cos Q \equiv -2 \sin \frac{P + Q}{2} \sin \frac{P - Q}{2} \qquad [8]$$

Identities [5]–[8] are best used when a sum or difference is to be expressed as a product, while identities [1]–[4] should be used when a given product is to be changed to a sum or difference,

e.g. to express $\sin 6\theta - \sin 4\theta$ as a product we would use [6] to give

$$2 \cos \frac{6\theta + 4\theta}{2} \sin \frac{6\theta - 4\theta}{2} \equiv 2 \cos 5\theta \sin \theta$$

But to express $2 \cos 7\theta \cos 2\theta$ as a sum we would use [3] to give

$$\cos (7\theta + 2\theta) + \cos (7\theta - 2\theta) \equiv \cos 9\theta + \cos 5\theta$$

When these identities are being used regularly, it is not too difficult to remember them. Most people find it best to memorise them in words rather than as symbols,
e.g. [5] would be remembered in the form:

sum of sines $\equiv$ twice sin(semi-sum) cos(semi difference)

And [1] would be:

twice sin cos $\equiv$ sin (sum) + sin (difference)

Note: Numbers [4] and [8] require special care because of the minus sign. This group of identities will prove to be particularly useful when integrating certain trig functions later on.

EXAMPLES 7g

1) Prove that $\dfrac{\sin A + \sin B}{\cos A + \cos B} \equiv \tan \dfrac{A + B}{2}$

If A, B and C are the angles of a triangle, deduce that

$$\frac{\sin A + \sin B}{\cos A + \cos B} = \cot \frac{C}{2}$$

Considering the L.H.S

$$\frac{\sin A + \sin B}{\cos A + \cos B} \equiv \frac{2 \sin \dfrac{A + B}{2} \cos \dfrac{A - B}{2}}{2 \cos \dfrac{A + B}{2} \cos \dfrac{A - B}{2}}$$

$$\begin{bmatrix} [5] \\ [7] \end{bmatrix}$$

$$\equiv \tan \frac{A + B}{2}$$

If A, B and C are angles in a triangle

$$A + B + C = 180°$$

$\Rightarrow$

$$\frac{A + B}{2} + \frac{C}{2} = 90°$$

$\Rightarrow$

$$\left(\frac{A + B}{2}\right) \text{ and } \frac{C}{2} \text{ are complementary}$$

therefore

$$\tan \frac{A + B}{2} = \cot \frac{C}{2}$$

$\Rightarrow$

$$\frac{\sin A + \sin B}{\cos A + \cos B} = \cot \frac{C}{2}$$

2) Solve the equation $\sin 5x - \sin 3x = 0$ giving the general solution.

$$\sin 5x - \sin 3x = 0$$

Therefore

$$2 \cos \frac{5x + 3x}{2} \sin \frac{5x - 3x}{2} = 0$$

Therefore   either $\cos 4x = 0$ or $\sin x = 0$

If $\cos 4x = 0$, let $\theta = 4x$ so that $\cos \theta = 0$

then

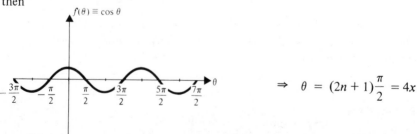

$\Rightarrow \quad \theta = (2n + 1)\dfrac{\pi}{2} = 4x$

If $\sin x = 0$

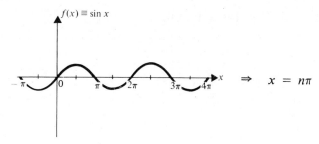

$\Rightarrow \quad x = n\pi$

The general solution is therefore

$$x = (2n + 1)\frac{\pi}{8}, \; n\pi$$

3) Factorise $\cos\theta - \cos 3\theta - \cos 5\theta + \cos 7\theta$

Grouping in pairs we have

$$(\cos 7\theta + \cos\theta) - (\cos 5\theta + \cos 3\theta) \equiv f(\theta)$$

But $\qquad \cos 7\theta + \cos\theta \equiv 2\cos\dfrac{7\theta + \theta}{2}\cos\dfrac{7\theta - \theta}{2}$

$$\equiv 2\cos 4\theta \cos 3\theta$$

and $\qquad \cos 5\theta + \cos 3\theta \equiv 2\cos\dfrac{5\theta + 3\theta}{2}\cos\dfrac{5\theta - 3\theta}{2}$

$$\equiv 2\cos 4\theta \cos\theta$$

So $\qquad f(\theta) \equiv 2\cos 4\theta(\cos 3\theta - \cos\theta)$

$$\equiv -2\cos 4\theta\left(2\sin\frac{3\theta + \theta}{2}\sin\frac{3\theta - \theta}{2}\right)$$

$\Rightarrow \qquad \cos\theta - \cos 3\theta - \cos 5\theta + \cos 7\theta \equiv -4\cos 4\theta \sin 2\theta \sin\theta$

Note. Other groupings of the four given terms can be used, but arranging them so that pairs of cosines are *added* makes the factorising simplest.

4) If A, B, C are the angles of a triangle show that

$$\sin A + \sin B + \sin C = 4\cos\frac{A}{2}\cos\frac{B}{2}\cos\frac{C}{2}$$

Considering the L.H.S

$$\sin A + \sin B + \sin C = (\sin A + \sin B) + \sin C$$

$$= 2 \sin \frac{A+B}{2} \cos \frac{A-B}{2} + 2 \sin \frac{C}{2} \cos \frac{C}{2}$$

Now $A + B + C = 180° \Rightarrow \dfrac{A+B}{2} + \dfrac{C}{2} = 90°$

Therefore $\sin \dfrac{A+B}{2} = \cos \dfrac{C}{2}$ and $\sin \dfrac{C}{2} = \cos \dfrac{A+B}{2}$

So $\sin A + \sin B + \sin C = 2 \cos \dfrac{C}{2} \cos \dfrac{A-B}{2} + 2 \cos \dfrac{A+B}{2} \cos \dfrac{C}{2}$

$$= 2 \cos \frac{C}{2} \left[\cos \frac{A-B}{2} + \cos \frac{A+B}{2} \right]$$

But $\cos \dfrac{A-B}{2} + \cos \dfrac{A+B}{2} = 2 \cos \dfrac{1}{2} \left(\dfrac{A-B}{2} + \dfrac{A+B}{2} \right) \cos \dfrac{1}{2} \left(\dfrac{A-B}{2} - \dfrac{A+B}{2} \right)$

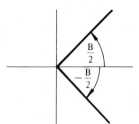

$$= 2 \cos \frac{A}{2} \cos \left(-\frac{B}{2} \right)$$

$$= 2 \cos \frac{A}{2} \cos \frac{B}{2}$$

So $\qquad \sin A + \sin B + \sin C = 4 \cos \dfrac{A}{2} \cos \dfrac{B}{2} \cos \dfrac{C}{2}$

EXERCISE 7g

1) Factorise:
(a) $\sin 3A + \sin A$
(b) $\cos 5A + \cos 3A$
(c) $\sin 4A - \sin 2A$
(d) $\cos 7A - \cos A$
(e) $\sin 3A - \sin 5A$
(f) $\cos A - \cos 5A$
(g) $\sin 30° + \sin 60°$
(h) $\cos 70° + \cos 50°$
(i) $\sin 2A + 1$ (*Hint*: $\sin 90° = 1$)
(j) $1 + \cos 4A$

2) Express as a sum or difference:
(a) $2 \sin 2\theta \cos \theta$
(b) $2 \cos 3\theta \cos 2\theta$
(c) $2 \cos \theta \sin 4\theta$
(d) $-2 \sin 3\theta \sin \theta$
(e) $2 \sin 4\theta \sin 2\theta$
(f) $\cos \theta \cos 4\theta$
(g) $2 \sin 30° \cos 60°$
(h) $2 \cos 20° \cos 40°$

3) Prove the following identities:

(a) $\dfrac{\sin 2A + \sin 2B}{\sin 2A - \sin 2B} \equiv \dfrac{\tan (A + B)}{\tan (A - B)}$

(b) $\dfrac{\cos 2A + \cos 2B}{\cos 2B - \cos 2A} \equiv \cot (A + B) \cot (A - B)$

(c) $\dfrac{\sin A \sin 2A + \sin 3A \sin 6A}{\sin A \cos 2A + \sin 3A \cos 6A} \equiv \tan 5A$

(d) $\dfrac{\sin 3x + \sin 5x}{\sin 4x + \sin 6x} \equiv \dfrac{\sin 4x}{\sin 5x}$

(e) $\dfrac{\sin A + \sin 3A + \sin 5A}{\cos A + \cos 3A + \cos 5A} \equiv \tan 3A$

(f) $\dfrac{\sin A - \sin B}{\sin A + \sin B} \equiv \cot \dfrac{A + B}{2} \tan \dfrac{A - B}{2}$

(g) $\sin \theta + \sin 2\theta + \sin 3\theta \equiv \sin 2\theta \, (1 + 2 \cos \theta)$

(h) $1 + 2 \cos 2A + \cos 4A \equiv 4 \cos^2 A \cos 2A$

(i) $\cos 2\theta + \cos 4\theta + \cos 6\theta + \cos 12\theta \equiv 4 \cos 3\theta \cos 4\theta \cos 5\theta$

(j) $\dfrac{\cos A - \cos B}{\sin A + \sin B} \equiv \tan \dfrac{B - A}{2}$

4) Simplify:

(a) $\cos (\theta - 60°) + \cos (\theta + 60°)$

(b) $\sqrt{3} \cos x - \sin (x + 60°) - \sin (x + 120°)$

5) If A, B and C are the angles of a triangle, prove that:

(a) $\cos (B + C) = -\cos A$ (b) $\sin C = \sin (A + B)$

(c) $\sin \dfrac{A + B}{2} = \cos \dfrac{C}{2}$ (d) $\sin \dfrac{B}{2} = \cos \dfrac{A + C}{2}$

(e) $\sin B + \sin (A - C) = 2 \sin A \cos C$

(f) $\cos (A - B) - \cos C = 2 \cos A \cos B$

(g) $\sin (A + B) + \sin (B + C) = 2 \cos \dfrac{B}{2} \cos \dfrac{A - C}{2}$

(h) $\cos A + \cos B + \cos C = 1 + 4 \sin \dfrac{A}{2} \sin \dfrac{B}{2} \sin \dfrac{C}{2}$

(i) $\sin 2A + \sin 2B + \sin 2C = 4 \sin A \sin B \sin C$

(j) $\sin \dfrac{A}{2} - \cos \dfrac{B - C}{2} = -2 \sin \dfrac{B}{2} \sin \dfrac{C}{2}$

(k) $1 + \cos 2C - \cos 2A - \cos 2B = 4 \sin A \sin B \cos C$

Solve the following equations, giving values from $0°$ to $360°$:

6) $\cos 2x + \cos 4x = 0$ 7) $\sin 3x - \sin x = 0$

8) $\sin 4\theta + \sin 2\theta = 0$ 9) $\cos x = \cos 2x + \cos 4x$

10) $\cos x + \cos 3x = \sin x + \sin 3x$ 11) $\sin 3\theta + \sin 6\theta + \sin 9\theta = 0$

12) $\sin 3\theta - \sin \theta = \cos 2\theta$ 13) $\cos 5\theta - \cos \theta = \sin 3\theta$

14) $\cos 2x = \cos (30° - x)$

SMALL ANGLES

A glance at the values of $\sin \theta$ and $\tan \theta$ when θ is a very small positive angle shows that these two trig ratios are almost equal. Further, if the small angle is measured in radians it is found that $\sin \theta \simeq \tan \theta \simeq \theta$.
These relationships are important in the development of analytical trigonometry and can be demonstrated as follows.

Consider a small angle θ measured in radians subtended by an arc AB at the centre, O, of a circle of radius r. The area of the sector OAB is $\frac{1}{2}r^2\theta$.

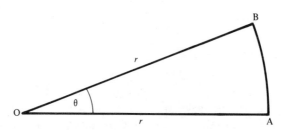

Now if AC is drawn perpendicular to OA, to cut OB produced at C, OAC is a right angled triangle with base r and height $r \tan \theta$.
Its area is therefore $\frac{1}{2}r^2 \tan \theta$.
Further, when the chord AB is drawn, an isosceles triangle OAB is formed with area $\frac{1}{2}r^2 \sin \theta$.

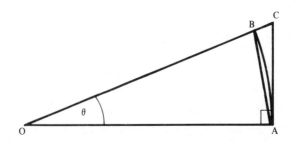

Now $\qquad$ area $\triangle OAB <$ area sector $OAB <$ area $\triangle OAC$

i.e. $\qquad \frac{1}{2}r^2 \sin\theta < \frac{1}{2}r^2\theta < \frac{1}{2}r^2 \tan\theta$

Dividing throughout by $\frac{1}{2}r^2$ (which is positive) we have

$$\sin\theta < \theta < \tan\theta$$

But $\sin\theta$, θ and $\tan\theta$ are all positive since θ is a small positive angle. Therefore we can divide throughout by any of these terms.

Thus $\qquad\qquad \dfrac{\sin\theta}{\sin\theta} < \dfrac{\theta}{\sin\theta} < \dfrac{\tan\theta}{\sin\theta}$

$\Rightarrow \qquad\qquad\qquad 1 < \dfrac{\theta}{\sin\theta} < \sec\theta$

For small positive values of θ, $\sec\theta \to 1$ as $\theta \to 0$.

Hence, as $\theta \to 0$, $\dfrac{\theta}{\sin\theta}$ lies between 1 and a number which approaches 1 and we can say

$$\text{as } \theta \to 0, \quad \frac{\theta}{\sin\theta} \to 1$$

Similarly, by dividing the first inequalities by $\tan\theta$ we can show that

$$\text{as } \theta \to 0, \quad \frac{\theta}{\tan\theta} \to 1$$

These limiting values verify that, for small positive values of θ,

$$\sin\theta \simeq \theta$$

and $\qquad\qquad\qquad\qquad \tan\theta \simeq \theta$

So far we have not found an approximate value for $\cos\theta$ when θ is small. To do this we use the double angle identity

$$\cos\theta \equiv 1 - 2\sin^2\frac{\theta}{2}$$

But, if $\dfrac{\theta}{2}$ is small, $\quad \sin\dfrac{\theta}{2} \simeq \dfrac{\theta}{2}$

Hence $\qquad\qquad\qquad\qquad \cos\theta \simeq 1 - 2\left(\dfrac{\theta}{2}\right)^2$

i.e. if θ is small $\quad \cos\theta \simeq 1 - \dfrac{\theta^2}{2}$

It can be shown, in a similar way, that the same results are obtained if θ is a small negative angle.

Thus, when θ is measured in radians,

$$\lim_{\theta \to 0} \left(\frac{\theta}{\sin \theta} \right) = 1$$

and

$$\lim_{\theta \to 0} \left(\frac{\theta}{\tan \theta} \right) = 1$$

Also, for any small angle, θ, measured in radians,

$$\sin \theta \simeq \theta$$

$$\tan \theta \simeq \theta$$

$$\cos \theta \simeq 1 - \frac{\theta^2}{2}$$

These approximations are correct to 3 s.f. for angles in the range

$$-0.105^c < \theta < 0.105^c$$

i.e.
$$-6° < \theta < 6°$$

Note that these and other, more accurate, approximations for the sine, cosine and tangent of a small angle are derived from series expansions at a later stage.

EXAMPLES 7h

1) Find an approximation for the expression $\dfrac{\sin 3\theta}{1 + \cos 2\theta}$ when θ is small.

When θ is small, 3θ is small, so $\quad \sin 3\theta \simeq 3\theta$

also 2θ is small, so $\qquad \cos 2\theta \simeq 1 - \dfrac{(2\theta)^2}{2}$

so when θ is small, $\qquad \dfrac{\sin 3\theta}{1 + \cos 2\theta} \simeq \dfrac{3\theta}{2(1 - \theta^2)}$

2) Without using tables find an approximate value for $\tan 61°$ given $\sqrt{3} = 1.732$ and $1° = 0.017^c$ giving your answer to 3 decimal places.

$$\tan(60° + 1°) = \frac{\tan 60° + \tan 1°}{1 - \tan 60° \tan 1°}$$

$$= \frac{\sqrt{3} + \tan 1°}{1 - \sqrt{3} \tan 1°}$$

Now $\tan \theta \simeq \theta$ when θ is small and measured in radians

Therefore $\tan 1° = \tan 0.017^c \simeq 0.017$

Hence $$\tan 61° \simeq \frac{1.732 + 0.017}{1 - (1.732)(0.017)}$$

$\Rightarrow$ $$\tan 61° \simeq 1.802$$

EXERCISE 7h

1) If θ is small enough to neglect θ^3, find approximations for the following expressions:

(a) $\dfrac{2\theta}{\sin 4\theta}$
(b) $\dfrac{\theta \sin \theta}{\cos 2\theta}$
(c) $\sin \dfrac{\theta}{2} \sec \theta$

(d) $\dfrac{\theta \tan \theta}{1 - \cos \theta}$
(e) $\sin \left(\alpha + \dfrac{\theta}{2} \right) \sin \dfrac{\theta}{2}$
(f) $\dfrac{2 \sin \dfrac{\theta}{2}}{\theta}$

(g) $\dfrac{\sin \theta \tan \theta}{\theta^2}$
(h) $\dfrac{\cos \left(\alpha + \dfrac{\theta}{2} \right) - \cos \alpha}{\theta}$

2) If θ is small enough to neglect θ^2 show that:

(a) $2 \cos \left(\dfrac{\pi}{3} + \theta \right) \simeq 1 - \sqrt{3}\theta$
(b) $\tan \left(\dfrac{\pi}{4} + \theta \right) \simeq \dfrac{1 + \theta}{1 - \theta}$

(c) $4 \sin \left(\dfrac{\pi}{4} - \theta \right) \simeq 2\sqrt{2}(1 - \theta)$

3) Using only the values $\sqrt{3} = 1.732\,05$, $5° = 0.087\,27^c$,
$\sin 2^c = 0.909\,30$, $\cos 2^c = -0.416\,15$, $\tan 2^c = -2.185\,04$,
$\sin 3^c = 0.141\,12$, $\cos 3^c = -0.989\,99$, $\tan 3^c = -0.142\,55$,   find,
to three decimal places, approximate values for:

(a) $\tan 2.01^c$ (b) $\sin 55°$ (c) $\cos 115°$ (d) $\sin 3.005^c$

(e) $\tan 175°$ (f) $\cos 1.98^c$

FURTHER PROPERTIES OF TRIANGLES

There is a great variety of relationships between the sides and angles of a triangle. Some of these were given in Chapter 6. Others can be derived from the identities proved in this chapter and some of these are given overleaf.

1) If D divides the side AB of triangle ABC in the ratio $m:n$ then, with the notation shown in the diagram below,

$$(m + n) \cot \theta = m \cot \alpha - n \cot \beta$$

This relationship is known as the *cotangent formula* and is proved as follows.

In triangle ACD, $\dfrac{CD}{\sin A} = \dfrac{AD}{\sin \alpha}$

$\Rightarrow$ $CD = \dfrac{AD \sin (\theta - \alpha)}{\sin \alpha}$

In triangle BCD, $\dfrac{CD}{\sin B} = \dfrac{BD}{\sin \beta}$

$\Rightarrow$ $CD = \dfrac{BD \sin (\theta + \beta)}{\sin \beta}$

Hence $\dfrac{AD \sin (\theta - \alpha)}{\sin \alpha} = \dfrac{BD \sin (\theta + \beta)}{\sin \beta}$

But $AD:DB = m:n$

Hence $\dfrac{m \sin (\theta - \alpha)}{\sin \alpha} = \dfrac{n \sin (\theta + \beta)}{\sin \beta}$

$\Rightarrow$ $m \left\{ \dfrac{\sin \theta \cos \alpha - \cos \theta \sin \alpha}{\sin \alpha} \right\} = n \left\{ \dfrac{\sin \theta \cos \beta + \cos \theta \sin \beta}{\sin \beta} \right\}$

$m \{ \cot \alpha - \cot \theta \} = n \{ \cot \beta + \cot \theta \}$

i.e. $(m + n) \cot \theta = m \cot \alpha - n \cot \beta$

Note. If D is the midpoint of AB we have

$$2 \cot \theta = \cot \alpha - \cot \beta$$

The cotangent formula, particularly in this special case, is valuable in the solution of certain equilibrium problems in mechanics.

Although the sine and cosine formulae are the two formulae most frequently used there are a number of other relationships between the sides and angles of a triangle.
A selection of some of the most useful of these alternative formulae is given below.

2) $a \cos B = c - b \cos A$

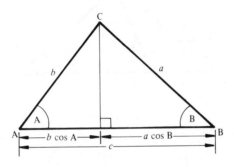

3) $\dfrac{a-b}{a+b} = \tan \dfrac{A-B}{2} \tan \dfrac{C}{2}.$

The proof of this formula is derived by using the sine rule in the form

$$\frac{a}{\sin A} = \frac{b}{\sin B} = k, \text{ say,} \qquad \Rightarrow \qquad \begin{cases} a = k \sin A \\ b = k \sin B \end{cases}$$

Hence

$$\frac{a-b}{a+b} = \frac{k(\sin A - \sin B)}{k(\sin A + \sin B)}$$

$$= \frac{2 \cos \dfrac{A+B}{2} \sin \dfrac{A-B}{2}}{2 \sin \dfrac{A+B}{2} \cos \dfrac{A-B}{2}}$$

$$= \tan \frac{A-B}{2} \cot \frac{A+B}{2}$$

But $A + B + C = \pi \qquad \Rightarrow \qquad \dfrac{A+B}{2} = \dfrac{\pi}{2} - \dfrac{C}{2}$

i.e. $\qquad \cot \dfrac{A+B}{2} = \tan \dfrac{C}{2}$

So $\qquad \dfrac{a-b}{a+b} = \tan \dfrac{A-B}{2} \tan \dfrac{C}{2}$

4) In triangle ABC the bisector of the angle A divides BC in the ratio $c:b$.

Proof

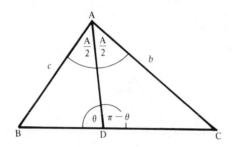

In triangle ABD,
$$\frac{BD}{\sin \frac{1}{2}A} = \frac{c}{\sin \theta}$$

and in triangle ACD
$$\frac{DC}{\sin \frac{1}{2}A} = \frac{b}{\sin (\pi - \theta)} = \frac{b}{\sin \theta}$$

Hence
$$\frac{\sin \frac{1}{2}A}{\sin \theta} = \frac{BD}{c} = \frac{DC}{b}$$

$\Rightarrow$
$$BD : DC = c : b$$

The reader should also be familiar with the following geometric properties of a triangle.

The *perpendicular bisectors of the sides* of a triangle ABC are concurrent at a point called the *circumcentre*, which is the centre of the circle through A, B and C (the circumcircle).

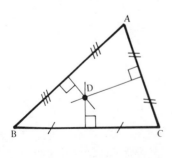

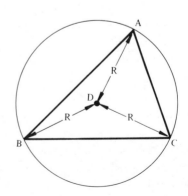

The *bisectors of angles* A, B and C are concurrent at a point called the *incentre* which is the centre of the circle that touches all three sides (the inscribed circle).

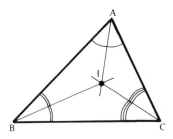 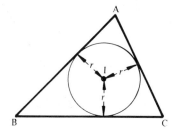

The *altitudes* are concurrent, at a point H called the *orthocentre* and the *medians* are concurrent, at a point G called the *centroid*.

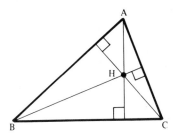 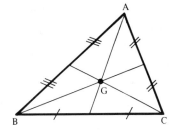

Area of a Triangle

The area of a triangle can be found using:
(a) half base × perpendicular height,
(b) $\frac{1}{2}ab \sin C$ etc.,
(c) $\sqrt{s(s-a)(s-b)(s-c)}$ where $s = \frac{1}{2}(a+b+c)$.

EXAMPLES 7i

1) In a surveying exercise, P and Q are two points on land which is inaccessible. To find the distance PQ, a line AB of length 200 metres is drawn so that P and Q are on opposite sides of AB. The following angles are measured:

$\angle\,ABP = 60°$, $\angle\,ABQ = 46°$, $\angle\,BAP = 30°$ and $\angle\,BAQ = 67°$. Find the distance PQ.

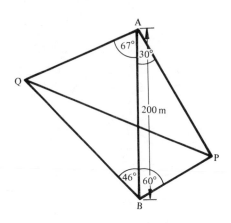

In triangle APB,

$$\angle\,APB \;=\; 90°$$

Hence $\quad AP \;=\; 200\cos 30°$

$$=\; 173.2\,\text{m}$$

In triangle ABQ,

$$\angle\,AQB \;=\; 67°$$

Hence

$$\frac{AQ}{\sin 46°} \;=\; \frac{200}{\sin 67°}$$

$$\Rightarrow \qquad AQ \;=\; 156.3\,\text{m}$$

Then in triangle APQ, the cosine rule gives

$$PQ^2 \;=\; AP^2 + AQ^2 - 2\cdot AP\cdot AQ \cos 97°$$

$$=\; (173.2)^2 + (156.3)^2 - 2(173.2)(156.3)(-0.1219)$$

Hence $\quad PQ \;=\; 247\,\text{m}$

2) In a triangle ABC, $BC = 7.4\,\text{cm}$, $AC = 4.1\,\text{cm}$ and $\angle ACB = 66°$. Calculate:
(a) the other angles of the triangle,
(b) the area of the triangle,
(c) the distance of I, the incentre of the triangle, from BC.

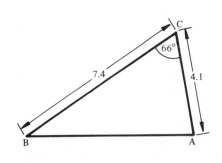

(a) Using the cosine formula,

$$c^2 \;=\; (7.4)^2 + (4.1)^2$$

$$- 2(7.4)(4.1)\cos 66°$$

gives $\quad c = 6.85\,\text{cm}$

Then the sine rule gives

$$\frac{\sin A}{7.4} \;=\; \frac{\sin B}{4.1} \;=\; \frac{\sin 66°}{6.85}$$

Hence

$$A \;=\; 80.8°$$

and $\quad B \;=\; 33.2°$

(b) The area of the triangle, Δ, is given by

$$\Delta = \tfrac{1}{2}ab \sin C = \tfrac{1}{2}(7.4)(4.1) \sin 66°$$
$$= 13.86 \text{ cm}^2$$

(c) The incentre, I, is the centre of the inscribed circle and is therefore equidistant (r) from all three sides.

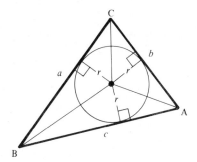

Joining AI, BI and CI we see that

Area of $\triangle$ AIB $= \tfrac{1}{2}cr$

Area of $\triangle$ BIC $= \tfrac{1}{2}ar$

Area of $\triangle$CIA $= \tfrac{1}{2}br$

Thus the total area of $\triangle$ ABC $= \tfrac{1}{2}r(a + b + c) = 9.18r$
But this area is 13.86
Hence the distance of I from BC is $r = 1.5$ cm.

EXERCISE 7i

1) Calculate the area of triangle ABC if:
   (a) $B = 57°$, $a = 3.4$, $c = 14.1$,
   (b) $a = 11$, $b = 10$, $c = 15$,
   (c) $a = 8$, $c = 9$, $A = 52°$.

2) In a triangle PQR, $p = 2.8$ m and $q = 4.5$ m. If the area of the triangle is 5.84 m^2 find the two possible values of r.

3) Prove that the area of a triangle ABC can be found from

$$\frac{1}{2} c^2 \frac{\sin A \sin B}{\sin (A + B)}$$

4) Given a triangle ABC prove that:
   (a) $c \sin \dfrac{A - B}{2} = (a - b) \cos \dfrac{C}{2}$,

   (b) $\cot C = \dfrac{a}{c} \operatorname{cosec} B - \cot B$,

   (c) $abc = 4\Delta R$ where Δ is the area of the triangle and R is the radius of the circumcircle of the triangle.

5) ABC is a triangle and D is a point on BC such that $BD = \frac{1}{3}DC$. Angle BAD is $20°$ and angle CAD is $30°$. Find angle ACB.

6) A quadrilateral ABCD is right-angled at B and D whilst the angle $DAB = 132°$. If $DA = 4$ and $AB = 7$, find the lengths of the diagonals of the quadrilateral and the radius of the inscribed circle of the triangle ABC.

(U of L)

7) ABC is a triangle in which none of the angles is obtuse. The perpendicular AD from A to BC is produced to meet the circumcircle of the triangle at E. If D is equidistant from A and E prove that the triangle must be right-angled. If, alternatively, the incentre of the triangle is equidistant from A and E, prove that $\cos B + \cos C = 1$.

(U of L)

SUMMARY

$$\tan \theta \equiv \frac{\sin \theta}{\cos \theta}$$

$$\begin{cases} \cos^2\theta + \sin^2\theta \equiv 1 \\ \tan^2\theta + 1 \equiv \sec^2\theta \\ \cot^2\theta + 1 \equiv \text{cosec}^2\theta \end{cases}$$

$$\begin{cases} \sin(A \pm B) \equiv \sin A \cos B \pm \cos A \sin B \\ \cos(A \pm B) \equiv \cos A \cos B \mp \sin A \sin B \\ \tan(A \pm B) \equiv \dfrac{\tan A \pm \tan B}{1 \mp \tan A \tan B} \end{cases}$$

$$\begin{cases} \sin 2A \equiv 2 \sin A \cos A \\ \cos 2A \equiv \cos^2 A - \sin^2 A \equiv 2\cos^2 A - 1 \equiv 1 - 2\sin^2 A \\ \tan 2A \equiv \dfrac{2 \tan A}{1 - \tan^2 A} \end{cases}$$

$$\begin{cases} \sin^2\theta \equiv \frac{1}{2}(1 - \cos 2\theta) \\ \cos^2\theta \equiv \frac{1}{2}(1 + \cos 2\theta) \end{cases}$$

$$\begin{cases} \sin P + \sin Q \equiv 2 \sin \dfrac{P+Q}{2} \cos \dfrac{P-Q}{2} \\[2mm] \sin P - \sin Q \equiv 2 \cos \dfrac{P+Q}{2} \sin \dfrac{P-Q}{2} \\[2mm] \cos P + \cos Q \equiv 2 \cos \dfrac{P+Q}{2} \cos \dfrac{P-Q}{2} \\[2mm] \cos P - \cos Q \equiv -2 \sin \dfrac{P+Q}{2} \sin \dfrac{P-Q}{2} \end{cases}$$

$$\begin{cases} 2 \sin A \cos B \equiv \sin (A + B) + \sin (A - B) \\ 2 \cos A \sin B \equiv \sin (A + B) - \sin (A - B) \\ 2 \cos A \cos B \equiv \cos (A + B) + \cos (A - B) \\ -2 \sin A \sin B \equiv \cos (A + B) - \cos (A - B) \end{cases}$$

$$\begin{cases} \sin \theta \equiv \dfrac{2t}{1 + t^2} \\[2mm] \cos \theta \equiv \dfrac{1 - t^2}{1 + t^2} \quad \text{where} \quad t \equiv \tan \dfrac{\theta}{2} \\[2mm] \tan \theta \equiv \dfrac{2t}{1 - t^2} \end{cases}$$

$$\sin 3\theta \equiv 3 \sin \theta - 4 \sin^3 \theta$$
$$\cos 3\theta \equiv 4 \cos^3 \theta - 3 \cos \theta$$

For any small angle, θ, measured in radians,

$$\sin \theta \simeq \theta$$
$$\tan \theta \simeq \theta$$
$$\cos \theta \simeq 1 - \frac{\theta^2}{2}$$

and $\qquad \lim\limits_{\theta \to 0} \dfrac{\sin \theta}{\theta} = 1$

MULTIPLE CHOICE EXERCISE 7

(Instructions for answering these questions are given on p. xii.)

TYPE I

1) $\cos (A + B) + \cos (A - B) \equiv$
(a) $2 \cos A \sin B$ (b) $-2 \sin A \cos B$ (c) $2 \cos A \cos B$ (d) $-2 \sin A \sin B$.

2) Using $t \equiv \tan \dfrac{\theta}{2}$ converts the equation $2 \cos \theta + 3 \sin \theta + 4 = 0$
into:

(a) $2t^2 + 3t + 6 = 0$ (b) $3 + 3t + t^2 = 0$
(c) $6 + 6t - 2t^2 = 0$ (d) $t^2 + 6t + 5 = 0$.

3) One of the following expressions is not identical to any of the others. Which one is it?

(a) $\dfrac{2\tan\theta}{1+\tan^2\theta}$ (b) $2\cos^2\dfrac{\theta}{2}$ (c) $1-\sin^2\theta$

(d) $\dfrac{1}{1+\tan^2\theta}$ (e) $\sin 2\theta$.

4) The approximate value, when θ is small, of the expression $\dfrac{2\theta-\sin\theta}{\sin 2\theta-\theta}$ is:

(a) 1   (b) 2   (c) -1 (d) -2 (e) none of these.

5)

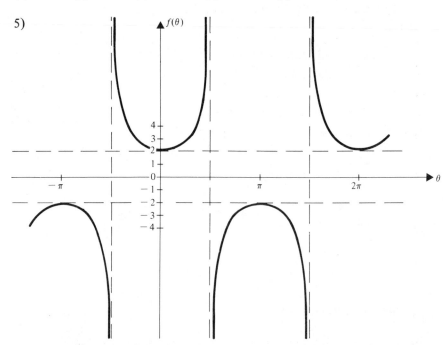

$f(\theta)$ could be:
(a) cosec 2θ (b) $2\sec\theta$ (c) $\sec 2\theta$ (d) 2 cosec θ (e) $2\cot\theta$.

6) The greatest value of $5\cos\theta-4\sin\theta$ is:
(a) 3   (b) 1   (c) $\sqrt{41}$ (d) ± 5 (e) $\pm\sqrt{41}$.

TYPE II

7) $3\cos\theta-4\sin\theta\equiv$
(a) $5\cos(\theta+\alpha)$ where $\tan\alpha=\frac{3}{4}$ (b) $5\sin(\alpha-\theta)$ where $\tan\alpha=\frac{3}{4}$
(c) $5\cos(\theta+\alpha)$ where $\tan\alpha=\frac{4}{3}$ (d) $-5\cos(\theta-\alpha)$ where $\tan\alpha=\frac{4}{3}$.

8) The general solution of the equation $\cos 2\theta = \frac{1}{2}$ is the same as the solution of the equation

(a) $\sin 2\theta = \dfrac{\sqrt{3}}{2}$ (b) $\tan 2\theta = \sqrt{3}$

(c) $\cos \theta = \frac{1}{4}$ (d) $\cos(-2\theta) = \frac{1}{2}$ (e) $4\cos^2 \theta = 3$.

9) If $x = 1 - \tan \theta$ and $y = \sec \theta$ the Cartesian equation given by eliminating θ is:
(a) $x^2 + y^2 = 2x$ (b) $x^2 - y^2 = 2x$ (c) $x^2 - y^2 + 2 = 2x$
(d) $(1 - x)^2 = (y - 1)(y + 1)$ (e) $(x - 1)^2 = (1 - y)(1 + y)$.

10) Given that θ is a small positive angle measured in radians:

(a) $\dfrac{\sin \theta}{\theta}$ is very slightly greater than 1.

(b) $\cos \theta$ is very slightly less than 1.

(c) $\dfrac{\tan \theta}{\theta}$ is very slightly greater than 1.

(d) $\dfrac{\sin \theta}{\cos \theta}$ is very slightly greater than 1.

(e) $\dfrac{\sin \theta}{\tan \theta}$ is very slightly greater than 1.

11) Given that α is a very small angle measured in radians:
(a) $\sin(2\pi + \alpha) = \sin \alpha$ (b) $\sin \alpha \simeq \alpha$
(c) $\sin(2\pi + \alpha) \simeq 2\pi + \alpha$ (d) $\cos \alpha \simeq \alpha$
(e) $\tan \alpha \simeq \alpha$.

12) $f(x) \equiv \arccos x$:

(a) $-1 \leqslant f(x) \leqslant 1$ (b) $f(x) \equiv \dfrac{1}{\cos x}$

(c) $f(0) = 1$ (d) $f(x) \equiv \operatorname{arcsec} \dfrac{1}{x}$.

13) $\sin\left(\theta - \dfrac{\pi}{2}\right) \equiv$

(a) $\cos\left(\theta + \dfrac{\pi}{2}\right)$ (b) $\sin\left(\dfrac{\pi}{2} - \theta\right)$ (c) $\cos \theta$ (d) $\sin\left(\theta + \dfrac{3\pi}{2}\right)$

14) If $f(\theta) \equiv \cos \theta - \sin \theta$
(a) $-1 \leqslant f(\theta) \leqslant 1$ (b) $f(\theta)$ is cyclic
(c) $[f(\theta)]^2 \equiv 1 - \sin 2\theta$ (d) The amplitude of $f(\theta)$ is $\sqrt{2}$.

15)

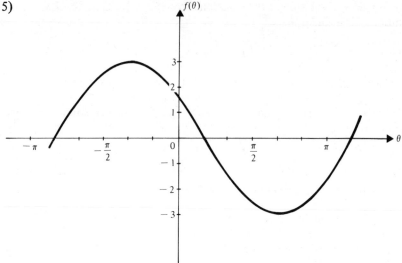

In the sketch $f(\theta)$ could be:

(a) $3 \sin\left(\theta - \dfrac{\pi}{6}\right)$

(b) $3 \cos\left(\theta - \dfrac{\pi}{6}\right)$

(c) $3 \sin\left(\theta - \dfrac{5\pi}{6}\right)$

(d) $3 \cos\left(\theta + \dfrac{\pi}{6}\right)$

(e) $3 \cos\left(\theta + \dfrac{\pi}{3}\right)$.

TYPE III

16) (a) $\theta = n\pi$.

   (b) $\cos\theta = 0$.

17) (a) $\cos A = \sin B$.

   (b) $A + B = 90°$.

18) (a) $\sin\theta \simeq \tan\theta$.

   (b) θ is very small.

19) (a) $\theta = 0.5°$.

   (b) $\sin\theta \simeq 0.5$.

20) (a) $\tan\theta \to 0$.

   (b) $\theta \to 0$.

21) (a) $\theta = \arctan 1$.

   (b) $\theta = \dfrac{\pi}{4} + n\pi$.

22) (a) $f(\theta) \equiv \cos^{-1}x$.

    (b) $f(\theta) \equiv \sin^{-1}\left(\dfrac{\pi}{2}-x\right)$.

23) (a) $f(\theta) \equiv \cos\theta$.

    (b) $-1 \leqslant f(\theta) \leqslant 1$.

24) (a) $f(\theta) \equiv \sqrt{\dfrac{1-\cos\theta}{1+\cos\theta}}$.

    (b) $f(\theta) \equiv \tan\dfrac{\theta}{2}$.

25) (a) $\sin 3\theta = x$.

    (b) $3\sin\theta = x$.

TYPE IV

26) Identify the function $f(\theta)$ given that:

(a) $f(\theta)$ is a trig function   (b) $f(\theta)$ is cyclic

(c) $f(\theta)$ is continuous   (d) $f(\theta) = 1$ when $\theta = 0$.

27) Find the value of θ if:

(a) $\cos 2\theta$ is given   (b) $0 < \theta < 180°$

(c) $\cos\theta$ is positive   (d) $\cos 2\theta \equiv 2\cos^2\theta - 1$.

TYPE V

28) The graph of $\tan\theta$ is continuous because it is cyclic.

29) The general solution of the equation $\cos\theta = -1$ is $\theta = (2n+1)\pi$.

30) $\sin 2\theta \equiv \dfrac{2t}{1+t^2}$ where $t \equiv \tan\dfrac{\theta}{2}$.

31) $\cos 4\theta = \sin 3\theta \;\Rightarrow\; \cos 4\theta = \sin\left(3\theta - \dfrac{\pi}{2}\right)$.

MISCELLANEOUS EXERCISE 7

1) (a) If $\sin\alpha = \tfrac{2}{3}$ and $\cos\beta = -\tfrac{2}{7}$,   find the possible values of $\cos(\alpha+\beta)$.

    (b) Find the values of θ between $-180°$ and $180°$ which satisfy the equation $3\cos\theta - 5\sin\theta = 2$.   (C)

2) (a) Solve the equations:

    (i) $\cos\dfrac{3x}{4} = \tan 163°$ (ii) $7\cos x - 24\sin x = 12.5$

    giving in each case all the solutions between $0°$ and $360°$.

(b) Without using tables find the numerical value of

$$\sin^2 \frac{\pi}{8} - \cos^4 \frac{3\pi}{8}$$ (U of L)

3) (a) If A is the acute angle such that $\sin A = \frac{3}{5}$ and B is the obtuse angle such that $\sin B = \frac{5}{13}$, find without using tables the values of $\cos(A + B)$ and $\tan(A - B)$.

(b) Find the solutions of the equation $\tan \theta + 3 \cot \theta = 5 \sec \theta$ for which $0 < \theta < 2\pi$.   (U of L)

4) Calculate the values of θ in the range $0 \leqslant \theta \leqslant 180°$ which satisfy the equations:

(a) $2 \sin \theta + \cos \theta = 1$ (b) $2 \sin \theta + \cos 2\theta = 1$.   (AEB)'72p

5) (a) Find the values of θ between $0°$ and $180°$ for which
$\tan^2 \theta = 5 - \sec \theta$.

(b) Find the values of θ between $0°$ and $360°$ which satisfy the equation

$$6 \cos \theta + 7 \sin \theta = 4$$ (C)

6) (a) Find the values of x between $0°$ and $360°$ which satisfy the equation
$$\sin 2x + 2 \cos 2x = 1$$

(b) Find the general solution of the equation $\cos 3x + \cos x = \sin 2x$.
     (U of L)p

7) (a) If $\sin(\theta - \alpha) = k \sin(\theta + \alpha)$ find $\tan \theta$ in terms of $\tan \alpha$ and k and so determine the possible values of θ between 0 and $360°$ when $k = \frac{1}{2}$ and $\alpha = 150°$.

(b) Show without the use of tables or calculator, that $x = \frac{\pi}{10}$ satisfies the equation $\cos 3x = \sin 2x$. By expressing this equation in terms of $\sin x$ and $\cos x$ show that $\sin \frac{\pi}{10}$ is a root of the equation

$$4s^2 + 2s - 1 = 0$$ (C)

8) (a) Find the values of θ between 0 and 2π for which $\sin 2\theta = \sin \frac{\pi}{6}$.

(b) Show that $(2 \cos \phi + 3 \sin \phi)^2 \leqslant 13$ for all values of ϕ.   (U of L)p

9) Find all the values of θ in the range $0 \leqslant \theta \leqslant 2\pi$ for which
$$\sin \theta + \sin 3\theta = \cos \theta + \cos 3\theta$$ (JMB)

10) Find, to the nearest minute, the acute angle α for which
$$4 \cos \theta - 3 \sin \theta \equiv 5 \cos(\theta + \alpha)$$

Calculate the values of θ in the interval $-180° \leqslant \theta \leqslant 180°$ for which the function $f(\theta) \equiv 4 \cos \theta - 3 \sin \theta - 4$ attains its greatest value, its least value and the value zero. (JMB)

11) (a) Prove that $(\sin 2\theta - \sin \theta)(1 + 2 \cos \theta) \equiv \sin 3\theta.$

(b) Find the values of x between $0°$ and $360°$ which satisfy the equation
$$3 \cos x + 1 = 2 \sin x \qquad \text{(C)}$$

12) Find all the solutions of the following equations for which $-180° < \theta \leqslant 180°.$

(a) $3 \sin \theta + 4 \cos \theta = 2$ (b) $7 \tan 2\theta + 4 \sin \theta = 0$ (JMB)p

13) Prove that $\sec x + \tan x = \tan \left(\dfrac{\pi}{4} + \dfrac{x}{2} \right)$ and deduce a similar expression for $\sec x - \tan x.$

Hence find in surd form the values of $\tan \dfrac{7\pi}{12}$ and $\tan \dfrac{\pi}{12}.$ (AEB)'75p

14) (a) Prove that $\cos 3\theta - \sin 3\theta \equiv (\cos \theta + \sin \theta)(1 - 4 \cos \theta \sin \theta).$

(b) Prove that if $\sec A = \cos B + \sin B$

   (i) $\tan^2 A = \sin 2B$ (ii) $\cos 2A = \tan^2 \left(\dfrac{\pi}{4} - B \right)$ (C)

15) (a) Find, in radians, the general solution of the equation
$$\sin x + \sin 2x = \sin 3x$$

(b) By expressing $\sec 2x$ and $\tan 2x$ in terms of $\tan x$, or otherwise, solve the equation $2 \tan x + \sec 2x = 2 \tan 2x$, giving all solutions between $-180°$ and $+180°$. (U of L)

16) Express $\sqrt{3} \sin \theta - \cos \theta$ in the form $R \sin (\theta - \alpha)$ where R is positive. Find all values of θ in the range $0° \leqslant \theta \leqslant 360°$ which satisfy the equation
$$4 \sin \theta \cos \theta = \sqrt{3} \sin \theta - \cos \theta \qquad \text{(JMB)}$$

17) (a) Prove that $(\cot \theta + \operatorname{cosec} \theta)^2 \equiv \dfrac{1 + \cos \theta}{1 - \cos \theta}$ and hence, or otherwise, solve the equation $(\cot 2\theta + \operatorname{cosec} 2\theta)^2 = \sec 2\theta$ for values of θ between $0°$ and $180°.$

(b) Find the general solution of the equation $\sin 2x + \sin 3x + \sin 5x = 0.$ (AEB)'73

18) By using the formulae expressing $\sin \theta$ and $\cos \theta$ in terms of $t \left(\equiv \tan \dfrac{\theta}{2} \right)$ or otherwise, show that $\dfrac{1 + \sin \theta}{5 + 4 \cos \theta} \equiv \dfrac{(1 + t)^2}{9 + t^2}.$

Deduce that $0 \leqslant \dfrac{1 + \sin \theta}{5 + 4 \cos \theta} \leqslant \dfrac{10}{9}$ for all values of $\theta.$ (C)

19) (a) Show that $\cos^6 x + \sin^6 x \equiv 1 - \frac{3}{4}\sin^2 2x$.

(b) Solve, for $0° \leqslant x \leqslant 180°$, the equation $\sin x + \sin 5x = \sin 3x$.

(c) Find the general solution of the equation $3\cos x + 4\sin x = 2$

(U of L)

20) Prove that $\operatorname{cosec} \theta + \cot \theta \equiv \cot \dfrac{\theta}{2}$.

Hence (a) deduce the values, in surd form, of $\cot \dfrac{\pi}{8}$ and $\cot \dfrac{\pi}{12}$

(b) express $\operatorname{cosec} \theta + \operatorname{cosec} 2\theta + \operatorname{cosec} 4\theta$ as the difference of two cotangents.

(c) prove, without using tables (or calculator), that

$$\operatorname{cosec} \frac{4\pi}{15} + \operatorname{cosec} \frac{8\pi}{15} + \operatorname{cosec} \frac{16\pi}{15} + \operatorname{cosec} \frac{32\pi}{15} = 0$$ (C)

21) (Tables should not be used for this question.)

Prove that $\tan 3\theta \equiv \dfrac{3t - t^3}{1 - 3t^2}$, where $t \equiv \tan \theta$.

Hence, or otherwise, show that $\tan \frac{1}{12}\pi = 2 - \sqrt{3}$.

Give the angle θ, between 0 and $\frac{1}{2}\pi$, for which $\tan \theta = 2 + \sqrt{3}$.   (O)

22) (a) Express $7\sin x - 24\cos x$ in the form $R\sin(x - \alpha)$, where R is positive and α is an acute angle.

Hence or otherwise solve the equation

$$7\sin x - 24\cos x = 15, \quad \text{for} \quad 0° < x < 360°$$

(b) Solve the simultaneous equations

$$\cos x + \cos y = 1, \quad \sec x + \sec y = 4$$

for $0° < x < 180°, \quad 0° < y < 180°$.   (AEB)'76

23) Express $\cos 2x - \sin 2x$ in the form $R\cos(2x + \alpha)$, giving values of R and α.

Hence find the general solution of each of the following equations:

(a) $\cos 2x - \sin 2x = 1$ (b) $\cos 2x - \sin 2x = \sqrt{2}\cos 4x$.   (U of L)p

24) (a) Find in the range $-180° < x < 180°$ the solutions of the equation $\cos 5x = \cos x$.

(b) Prove that $\dfrac{1 + \cos \theta + \sin \theta}{1 - \cos \theta + \sin \theta} \equiv \dfrac{1 + \cos \theta}{\sin \theta}$ (JMB)p

25) (a) Find, in radians, the general solution of the equation $4\sin \theta = \sec \theta$.

(b) If $\sin \theta + \sin 2\theta + \sin 3\theta + \sin 4\theta = 0$, show that θ is either a multiple of $\frac{1}{2}\pi$ or a multiple of $\frac{2}{5}\pi$.   (U of L)

26) (a) Find the values of x, for angles between $0°$ and $360°$ inclusive, for which $3\sin 2x = 2\tan x$.

(b) Find the values of x, for angles between $0°$ and $360°$ inclusive, for
which $4 \cos x - 6 \sin x = 5$.   (C)

27) (a) Find the general solution of the equation

$$10 \sin \frac{\pi x}{3} + 24 \cos \frac{\pi x}{3} = 13$$

(b) Solve the equation $2 \cos \theta \cos 2\theta + \sin 2\theta = 2(3 \cos^3\theta - \cos \theta)$
for values of θ within the range $0 < \theta < 2\pi$.   (AEB)'67

28) (a) Find, in radians, the general solution of the equation
$2 \sin \theta = \sqrt{3} \tan \theta$.
(b) Express $4 \sin \theta - 3 \cos \theta$ in the form $R \sin (\theta - \alpha)$,   where α is
an acute angle.
(i) Solve the equation $4 \sin \theta - 3 \cos \theta = 3$,   giving all solutions
between $0°$ and $360°$.
(ii) Find the greatest and least values of $\dfrac{1}{4 \sin \theta - 3 \cos \theta + 6}$ (U of L)

29) (a) Given $x = 2 \sin \left(nt + \dfrac{\pi}{3} \right)$ and $y = 4 \sin \left(nt + \dfrac{\pi}{6} \right)$,   express x
and y in terms of $\sin nt$ and $\cos nt$. Find the Cartesian equation of
the locus of the point (x, y) as t varies.
(b) Solve the equation $3 \cos^2\theta + 5 \sin \theta - 1 = 0$ for
$0° < \theta < 360°$.   (AEB)'75

30) (a) Prove that $\sin^2 2\theta \, (\cot^2\theta - \tan^2\theta) = 4 \cos 2\theta$.
(b) Solve the equation $\sec \theta \tan \theta = 2$,   giving solutions for
$0° \leqslant \theta < 360°$.   (C)

31) Write down the expansions of $\cos (A + B)$ and $\cos (A - B)$ in terms
of cosines and sines of A and B.
(a) Find angles x and y, each between $0°$ and $90°$, which satisfy the
simultaneous equations $\cos x \cos y = 0.6$, $\sin x \sin y = 0.2$.
(b) Prove that $\cos 3x \equiv 4 \cos^3 x - 3 \cos x$.   Hence find all the solutions,
in the range $-180° < x \leqslant +180°$,   of the equation
$2 \cos 3x + \cos 2x + 1 = 0$.   (JMB)

32) (a) Solve the equations for values of x between $0°$ and $360°$:
(i) $\cos 2x° + \sin x° = 0$
(ii) $\sin x° - \sin 2x° + \sin 3x° = 0$.
(b) If $A = 36°$,   show that $\sin 3A = \sin 2A$,   and deduce that

$$\cos 36° = \frac{\sqrt{5} + 1}{4}$$

(c) If $A + B = \dfrac{\pi}{4}$ and $\tan A = \dfrac{n}{n + 1}$,   find $\tan B$ and
$\tan (A - B)$.

33) Find approximations, when θ is very small, for the following expressions

(a) $\cos\theta + \sin\theta$

(b) $\dfrac{2\tan\theta - \theta}{\sin 2\theta}$

(c) $\cot\theta\,(1-\cos\theta)$

(d) $\dfrac{\sqrt{2}-\sin\theta}{\cos\theta}$

34) A circle has centre O and radius r. Two parallel chords AB and CD are on the same side of O: the angle AOB is $\frac{1}{3}\pi$ and the angle COD is $(\frac{1}{3}\pi + 2\theta)$. Show that the area of the part of the circle between AB and CD is

$$\tfrac{1}{4}r^2\left[4\theta + \sqrt{3} - 2\sin\left(\tfrac{1}{3}\pi + 2\theta\right)\right]$$

If θ is small deduce an approximation for this area in the form $a + b\theta + c\theta^2$ and state the values of the constants a, b, c.   (JMB)

35) By means of the substitution $\tan\theta \equiv t$, or otherwise, find the values of θ in the range $0 \leqslant \theta \leqslant \dfrac{\pi}{2}$ such that

$$(2-\tan\theta)(1+\sin 2\theta) - 2 = 0$$

Show that, when θ is small

$$(2-\tan\theta)(1+\sin 2\theta) - 2 \simeq 3\theta \qquad \text{(JMB)}$$

36) Prove that $\arctan x + \arctan y \equiv \arctan\dfrac{x+y}{1-xy}$

Use this relationship to show that, if $\arctan x + \arctan y + \arctan z = \dfrac{\pi}{2}$

then $xy + yz + zx = 1$.

37) Find, without using tables, the value of x when

$$\tan^{-1}\tfrac{1}{2} - \tan^{-1}\tfrac{1}{3} = \sin^{-1}x. \qquad \text{(AEB)'75p}$$

38) Solve the equation $\arctan\left(\dfrac{1-x}{1+x}\right) = \dfrac{1}{2}\arctan x.$ (U of L)p

39) Show that for any plane triangle ABC

$$\tan\frac{A}{2}\tan\frac{B-C}{2} = \frac{b-c}{b+c}$$

Three towns A, B and C are all at sea level. The bearings of towns B and C from A are $36°$ and $247°$ respectively. If B is $120\,\text{km}$ from A and C is $234\,\text{km}$ from A calculate the distance and bearing of town B from C.

(AEB)'72

CHAPTER 8

DIFFERENTIATION II

DIFFERENTIATION OF TRIGONOMETRIC FUNCTIONS

For any function $f(x)$, $\dfrac{d}{dx} f(x) = \lim_{\delta x \to 0} \left[\dfrac{f(x + \delta x) - f(x)}{\delta x} \right]$

If $f(x) \equiv \sin x$, $\dfrac{d}{dx} \sin x = \lim_{\delta x \to 0} \left[\dfrac{\sin (x + \delta x) - \sin x}{\delta x} \right]$

Now $\sin (x + \delta x) - \sin x \equiv 2 \cos \left(x + \dfrac{\delta x}{2} \right) \sin \dfrac{\delta x}{2}$ (factor formula)

Therefore $\dfrac{d}{dx} \sin x = \lim_{\delta x \to 0} \left[\cos \left(x + \dfrac{\delta x}{2} \right) \dfrac{\sin \delta x/2}{\delta x/2} \right]$

Provided that x is measured in radians,

$$\lim_{\delta x \to 0} \left[\cos \left(x + \dfrac{\delta x}{2} \right) \right] = \cos x \quad \text{and} \quad \lim_{\delta x \to 0} \left[\dfrac{\sin \delta x/2}{\delta x/2} \right] = 1$$

Hence $\dfrac{d}{dx} \sin x = \cos x$.

Similarly it is found that $\dfrac{d}{dx} (\cos x) = - \sin x$.

Hence, provided that x is measured in radians,

$$\frac{d}{dx} (\sin x) = \cos x \quad \text{and} \quad \frac{d}{dx} (\cos x) = - \sin x$$

The derivatives of the remaining trig functions are dealt with later in this chapter.

EXAMPLE 8a

Find the smallest positive value of θ for which the curve $y = 2\theta - 3 \sin \theta$ has a gradient of $\frac{1}{2}$.

$$y = 2\theta - 3 \sin \theta \quad \text{gives} \quad \frac{dy}{d\theta} = 2 - 3 \cos \theta$$

when
$$\frac{dy}{d\theta} = \frac{1}{2} \qquad 2 - 3 \cos \theta = \frac{1}{2}$$
$$3 \cos \theta = \frac{3}{2}$$
$$\cos \theta = \frac{1}{2}$$

The smallest positive value of θ for which $\cos \theta = \frac{1}{2}$, is $\dfrac{\pi}{3}$.

Note that the answer *must* be given in radians as the differentiation is valid only if θ is measured in radians.

EXERCISE 8a

1) Find $\dfrac{d}{dx} (\cos x)$ from first principles.

2) Write down the derivatives of the following functions:
(a) $\cos x + \sin x$ (b) $3 - \cos x$ (c) $2 \sin \theta$
(d) $4 \cos \theta$ (e) $2 \sin \theta - 3 \cos \theta$ (f) $4 \sin t - 6$

3) Expand and simplify the following functions and write down their derivatives:

(a) $\sin \left(x + \dfrac{\pi}{2} \right)$ (b) $\cos \left(x - \dfrac{\pi}{4} \right)$ (c) $\sin \left(\dfrac{\pi}{3} - x \right)$

4) Find the gradient of the following curves at the point whose x coordinate is given:

(a) $y = \sin x, \quad \pi$ (b) $y = \cos x + \sin x, \quad \dfrac{\pi}{2}$

(c) $y = 3 \sin x - 2 \cos x, \quad \dfrac{\pi}{4}$ (d) $y = x^2 - \sin x, \quad \dfrac{\pi}{3}$

5) Find the smallest positive value of x for which the gradient of each of the following curves has the value indicated:
(a) $y = x + \sin x, \quad \frac{1}{2}$ (b) $y = \cos x - 2 \sin x, \quad 1$
(c) $y = \sin x - 3 \cos x, \quad 0$

6) Find the values of θ for which the following functions have
(a) maximum values,   (b) minimum values:
$\cos \theta - 3, \qquad 2 \sin \theta + \cos \theta, \qquad 3 \cos \theta - 4 \sin \theta$

7) Find the equation of the tangent to the following curves at the point indicated.

(a) $y = 3 \cos x - 2 \sin x$ where $x = \dfrac{\pi}{4}$

(b) $y = \theta^2 - 3 \sin \theta$ where $\theta = \dfrac{\pi}{3}$

INTRODUCING THE EXPONENTIAL FUNCTION

We have already met exponential functions such as 2^x (Chapter 3) and the diagram below shows a few members of the family of curves $y = a^x$ $(a \geqslant 0)$

viz: $y = 1^x (= 1)$, $y = 2^x$, $y = 3^x$, $y = 4^x$.

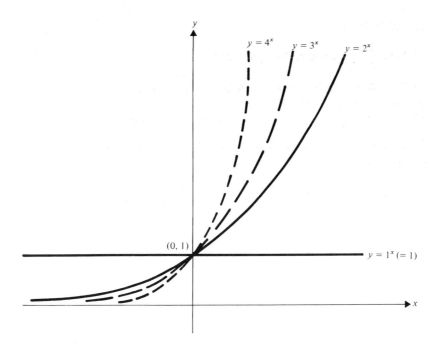

Note that all the curves pass through (0, 1).
In fact any member of the family of curves $y = a^x$ passes through this point (when $x = 0$, $y = a^0 = 1$).
Note also that the higher the value of the base (a), the greater is the gradient of the curve at the point where it cuts the y-axis.

By plotting these curves accurately we find the following approximate values for the gradient of each curve as it cuts the y-axis:

	gradient at $(0, 1)$
$y = 1^x$	0
$y = 2^x$	0.7
$y = 3^x$	1.1
$y = 4^x$	1.4

From this table we deduce that there is a number between 2 and 3, which we call e, for which the gradient of $y = e^x$ is 1 at the point $C(0, 1)$.
Now consider the general form $y = a^x$:

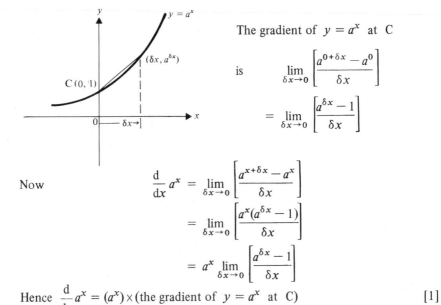

The gradient of $y = a^x$ at C

is $\displaystyle \lim_{\delta x \to 0} \left[\frac{a^{0+\delta x} - a^0}{\delta x} \right]$

$\displaystyle = \lim_{\delta x \to 0} \left[\frac{a^{\delta x} - 1}{\delta x} \right]$

Now $\displaystyle \frac{d}{dx} a^x = \lim_{\delta x \to 0} \left[\frac{a^{x+\delta x} - a^x}{\delta x} \right]$

$\displaystyle = \lim_{\delta x \to 0} \left[\frac{a^x(a^{\delta x} - 1)}{\delta x} \right]$

$\displaystyle = a^x \lim_{\delta x \to 0} \left[\frac{a^{\delta x} - 1}{\delta x} \right]$

Hence $\dfrac{d}{dx} a^x = (a^x) \times$ (the gradient of $y = a^x$ at C)   [1]

But we saw earlier that there is a number e whose value lies between 2 and 3, for which $y = e^x$ has a gradient of 1 at $(0, 1)$.
In this special case, [1] becomes

$$\frac{d}{dx} e^x = (e^x) \times (1) = e^x$$

The function e^x is clearly important as it is the only function that remains unaltered when differentiated.
The number e came to the notice of many mathematicians at about the same time in the early 18th century. It appeared as a result of several different lines of investigation and became known as a natural number.
The number e is irrational, i.e. it cannot be given an exact numerical value

(like $\pi, \sqrt{2}$, etc.) but, to 4 significant figures, $e = 2.718$.

Working with the function e^x involves evaluating powers of e, such as e^3, e^{-4}, $e^{2.72}$ ... etc. These values are obtainable from books of 4 figure tables or from a suitable calculator.

To summarise

$$a^x \ (a > 0) \text{ is an exponential function}$$
$$e^x \ (e \simeq 2.718) \text{ is } \textit{the} \text{ exponential function}$$

and if $y = e^x$ then

$$\frac{d}{dx} e^x = e^x \quad \text{and} \quad \frac{dy}{dx} = y$$

The diagrams below show sketches of the graphs of e^x and some simple variations.

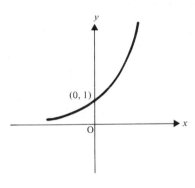

$$f(x) \equiv e^x$$

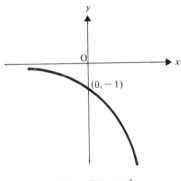

$$g(x) \equiv -f(x) \equiv -e^x$$

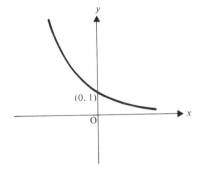

$$h(x) \equiv f(-x) \equiv e^{-x}$$

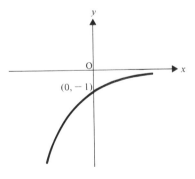

$$j(x) \equiv -h(x) \equiv -e^{-x}$$

EXAMPLE 8b

Find the coordinates of the point on the curve

$$y = x - e^x$$

at which y has a stationary value and sketch the curve.

For $y = x - e^x$, $\dfrac{dy}{dx} = 1 - e^x$

y is stationary when $1 - e^x = 0$

$\Rightarrow$ $e^x = 1$

$\Rightarrow$ $x = 0$ and $y = 0 - e^0 = -1$

Therefore y has a stationary value at $(0, -1)$.

The diagram shows the line $y = x$ and the curve $y = -e^x$.

We may obtain the curve $y = x - e^x$ by adding the ordinates of $y = x$ and $y = -e^x$.

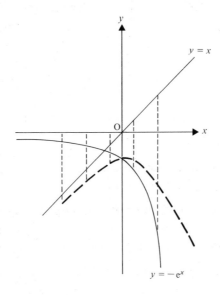

Alternatively we may observe that

for large $+$ve values of x, $x - e^x \simeq -e^x$

for large $-$ve values of x, $x - e^x \simeq -x$

We also know that $(0, -1)$ is a stationary point and that $\dfrac{d^2 y}{dx^2} = -e^x$,

$\Rightarrow$ $\qquad\qquad$ when $\quad x = 0, \quad \dfrac{d^2 y}{dx^2} = -1$

so $(0, -1)$ is maximum turning point.

From this information the sketch of $\quad y = x - e^x \quad$ can be drawn.

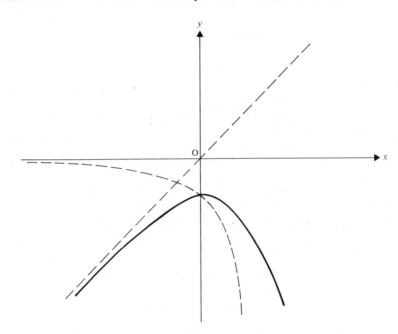

EXERCISE 8b

1) Evaluate (to 3 s.f.) e^3, e^{-2}, $e^{1.7}$, $e^{-0.2}$.

2) Write down the differential coefficients of:

(a) $2e^x$ $\qquad$ (b) $x^2 - e^x$ $\qquad$ (c) $e^x + \cos x$.

3) Find the values of x for which the following functions have stationary values:

(a) $e^x - x$ $\qquad$ (b) $2e^x - x - 1$ $\qquad$ (c) $e^x + 1$.

4) Draw a sketch of the curves whose equations are:

(a) $y = e^x - x$ $\qquad$ (b) $y = e^x + 1$ $\qquad\quad$ (c) $y = 1 - e^x$

(d) $y = x^2 + e^x$ $\qquad$ (e) $y = 1 + x - e^x$ $\qquad$ (f) $y = 1 - e^{-x}$

Naperian Logarithms

Consider the equation $\quad e^x = 0.78$.

We can solve this equation by taking logs of both sides,

i.e. $$x \log e = \log 0.78$$

giving $$x = \frac{\log 0.78}{\log e}$$

Equations, such as this one, arise frequently when working with exponential functions.

Introducing logarithms with base e makes the calculation simpler. We can then write

$$x \log_e e = \log_e 0.78$$

$\Rightarrow$ $$x = \log_e 0.78 \qquad (\log_e e = 1)$$

Such logarithms are called natural, or Naperian, logarithms. To avoid confusion with common logarithms (base 10) we denote

$$\log_e a \quad \text{by} \quad \ln a$$

where $$\ln a = b \iff a = e^b$$

The values of Naperian logarithms can be found from a suitable calculator and they are also given in most books of four figure tables.

The three laws of logarithms (which apply to any base) are useful when working with Naperian logarithms and, as a reminder, they are listed below.

$$\log_c a + \log_c b \equiv \log_c ab$$

$$\log_c a - \log_c b \equiv \log_c \frac{a}{b}$$

$$\log_c a^n \equiv n \log_c a$$

EXAMPLE 8c

Express $\ln (\tan x)$ as a sum or difference of logarithms.

$$\ln (\tan x) \equiv \ln \left(\frac{\sin x}{\cos x} \right) \equiv \ln \sin x - \ln \cos x$$

EXERCISE 8c

1) Evaluate: $\ln 7.821$, $\ln 0.483$, $\ln 21.5$.

2) Express as a sum or difference of logarithms:

(a) $\ln \left(\dfrac{x}{x + 1} \right)$ (b) $\ln (3x^2)$ (c) $\ln (x^2 - 4)$

(d) $\ln \sqrt{\left(\dfrac{x - 1}{x + 1} \right)}$ (e) $\ln \cot x$ (f) $\ln (4 \cos^2 x)$

3) Express as a single logarithm:
(a) $\ln x - 2 \ln (1 - x)$ (b) $1 - \ln x$
(c) $\ln \sin x + \ln \cos x$ (d) $3 \ln x - \frac{1}{2} \ln (x - 1)$

4) Solve the following equations for x:

(a) $e^x = 8.2$ (b) $e^{2x} + e^x - 2 = 0$ (*Hint*: see Ch. 2, Examples 2d)

(c) $e^{2x-1} = 3$ (d) $e^{4x} - e^x = 0$

THE LOGARITHMIC FUNCTION

Consider the function $f(x) \equiv \ln x$ and the curve whose equation is $y = \ln x$.

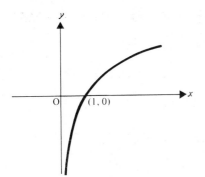

If $y = \ln x$ then $x = e^y$,
i.e. the logarithmic function is the
inverse exponential function.
So the curve $y = \ln x$ has the
same shape as $y = e^x$ with the
x- and y-axes interchanged
(this is achieved by reflecting
in the line $y = x$).

Note that there is no part of the curve in the 2nd and 3rd quadrants.
If $\log_a b = c$,   i.e. $b = a^c$,   there is no real value of c for which b is
negative.

So $\ln x$ does not exist for negative values of x

If $y = f(x)$ where $f(x)$ is any function of x, then

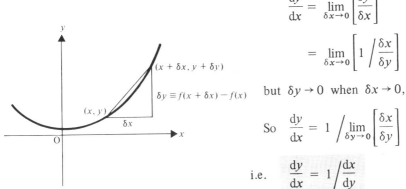

$$\frac{dy}{dx} = \lim_{\delta x \to 0} \left[\frac{\delta y}{\delta x} \right]$$

$$= \lim_{\delta x \to 0} \left[1 \Big/ \frac{\delta x}{\delta y} \right]$$

but $\delta y \to 0$ when $\delta x \to 0$,

So $\quad \dfrac{dy}{dx} = 1 \Big/ \lim_{\delta y \to 0} \left[\dfrac{\delta x}{\delta y} \right]$

i.e. $\quad \dfrac{dy}{dx} = 1 \Big/ \dfrac{dx}{dy}$

When $y = \ln x$ we may write $x = e^y$.

Differentiating e^y w.r.t. y we have

$$\frac{dx}{dy} = e^y = x$$

so
$$\frac{dy}{dx} = 1 \Big/ \frac{dx}{dy} = \frac{1}{x}$$

i.e.
$$\frac{d}{dx} \ln x = \frac{1}{x}$$

This result can be used to differentiate compound logarithmic functions if they are first simplified using the identities on page 262.

EXAMPLE 8d

Write down the derivative of:

(a) $\ln (2x^3)$ (b) $\ln (1/\sqrt{x})$

(a) $\ln (2x^3) = \ln 2 + \ln x^3 = \ln 2 + 3 \ln x$

Therefore $\dfrac{d}{dx} \ln 2x^3 = \dfrac{d}{dx} [\ln 2] + \dfrac{d}{dx} [3 \ln x]$

$$= 0 + 3\left(\frac{1}{x}\right) = \frac{3}{x}$$

(b) $\ln (1/\sqrt{x}) = \ln 1 - \ln x^{\frac{1}{2}} = 0 - \tfrac{1}{2} \ln x$

Therefore $\dfrac{d}{dx} [\ln (1/\sqrt{x})] = \dfrac{d}{dx} [-\tfrac{1}{2} \ln x] = -\dfrac{1}{2x}$

EXERCISE 8d

1) Write down the derivatives of:

(a) $2 \ln x$ (b) $\ln x^5$ (c) $\ln (3x^2)$ (d) $\ln (x^{-\frac{3}{2}})$

(e) $\ln (3/\sqrt{x})$ (f) $\ln (3x/\sqrt{x})$ (g) $\ln \sqrt{(2/x^5)}$ (h) $\ln \sqrt{(\tfrac{1}{3}x^{-\frac{1}{2}})}$

(i) $\lg x$ (*Hint*: change the base to e.)

2) Find the coordinates of points at which the following curves have zero gradient.

(a) $y = x - \ln x$ (b) $y = \ln (x^2) - x^2$

3) Sketch the curves whose equations are:

(a) $y = \ln (-x)$ (b) $y = -\ln x$ (c) $y = x - \ln x$

(d) $y = 1 + \ln x$ (e) $y = \log x$

COMPOUND FUNCTIONS

We have shown that differentiation is distributive across addition and subtraction,

i.e. $$\frac{d}{dx}[f(x)+g(x)] = \frac{d}{dx}[f(x)] + \frac{d}{dx}[g(x)]$$

but that differentiation is *not* distributive across multiplication or division of functions, and so we have not attempted to differentiate functions such as xe^x. Nor have we attempted to differentiate compound functions of the form e^{x^2-1}, $\sin(2x-4)\ldots$ etc.
Before we derive rules to differentiate such compound functions it is important to recognise into which category a particular function can be placed.

1) *Products*
Functions such as $e^x \sin x$, $x \ln x \ldots$ are of the form $[f(x)] \times [g(x)]$, i.e. each is a product of two functions, and by the substitutions $u \equiv f(x)$ and $v \equiv g(x)$ such a function may be written as uv,

e.g. $e^x \sin x \equiv uv$ where $u \equiv e^x$ and $v \equiv \sin x$.

2) *Quotients*
Functions such as $\dfrac{e^x}{\sin x}$, $\dfrac{\sin x}{\cos x}$, ... are of the form $\dfrac{f(x)}{g(x)}$, i.e. each is a quotient of two functions, and such a function may be written as $\dfrac{u}{v}$,

e.g. $\dfrac{e^x}{\sin x} \equiv \dfrac{u}{v}$ where $u \equiv e^x$ and $v \equiv \sin x$.

3) *Function of a Function*
Consider the two functions $f: x \to x^2 - 1$ and $g: x \to e^x$. If the output of f is made the input of g we get the compound relationship $x \to x^2 - 1 \to e^{x^2-1}$ or more briefly $x \to e^{x^2-1}$. This function is obtained by taking the function g of the function f and is known as a function of a function. We denote it by

$$g[f(x)] \equiv e^{x^2-1} \quad \text{or} \quad gf: x \to e^{x^2-1}$$

Thus if f is the function $f: x \to \sin x$ and g is the function $g: x \to x^2$ then the function $f[g(x)]$ is the function in which the output of g becomes the input of f,

i.e. $$f[g(x)] \equiv \sin(x^2)$$

Similarly for $g[f(x)]$ the output of f becomes the input of g giving:

$$g[f(x)] \equiv (\sin x)^2 \equiv \sin^2 x$$

Functions such as $\sin(3x-2)$, $(x^2+1)^{\frac{1}{2}}$, $\ln(x+1)$ are all of the form $g[f(x)]$, sometimes written $gf(x)$, i.e. each is a 'function of (a function of x)' and, by substituting u for $f(x)$, such functions may be expressed as $g(u)$,

e.g. $$(x^2 + 1)^{\frac{1}{2}} \equiv u^{\frac{1}{2}} \quad \text{where} \quad u \equiv x^2 + 1$$

Some functions may fall into more than one of these categories, for example

$$e^x \sqrt{1-x} \text{ is a product of } e^x \text{ and } \sqrt{1-x}$$

where $\sqrt{1-x}$ is a function of a function.
Substituting u for e^x and v for $1-x$

$$e^x \sqrt{1-x} \text{ may be written as } u\sqrt{v}$$

EXERCISE 8e

Write the following functions in terms of u and or v, stating clearly the substitutions that you have made.

1) $e^x(x^2 - 1)$ 2) e^{3x^2} 3) $\sin(x^2 - 2)$ 4) $x^3 \cos x$

5) $\sqrt{\dfrac{x-1}{x+1}}$ 6) $(x+1)^4$ 7) $\ln(x^2 - 1)$ 8) $(x^2 - 1)(x - 2)^5$

9) $\tan x^2$ 10) $\ln\left(\dfrac{x+1}{x-1}\right)$ 11) $\cos^2 x$ 12) $\sqrt{\ln x}$

13) If f and g are the functions defined by $f: x \to x^2$ and $g: x \to x + 1$ write down the functions $f[g(x)]$ and $g[f(x)]$.

14) The functions f, g and h are defined as follows:

$$f: x \to \cos x, \qquad g: x \to \frac{1}{x}, \qquad h: x \to \ln x$$

Write down the functions equivalent to

(a) $g[f(x)]$ (b) $h[g(x)]$ (c) $f[h(x)]$ (d) $fgh(x)$

DIFFERENTIATION OF A FUNCTION OF A FUNCTION

Consider $y = g(u)$ where $u = f(x)$.
If δx is a small increase in x and δu, δy are the corresponding increases in u, y, then as $\delta x \to 0$, δu and δy also tend to zero.

Hence $$\frac{dy}{dx} = \lim_{\delta x \to 0}\left[\frac{\delta y}{\delta x}\right] = \lim_{\delta x \to 0}\left[\frac{\delta y}{\delta u} \times \frac{\delta u}{\delta x}\right]$$

$$= \lim_{\delta u \to 0}\left[\frac{\delta y}{\delta u}\right] \times \lim_{\delta x \to 0}\left[\frac{\delta u}{\delta x}\right]$$

i.e. $$\frac{dy}{dx} = \frac{dy}{du} \times \frac{du}{dx}$$

This rule may now be used to differentiate a function of a function.

EXAMPLES 8f

1) Differentiate $\sqrt{(x^2-1)}$ w.r.t. x.

$$y = \sqrt{(x^2-1)} = (x^2-1)^{\frac{1}{2}}$$

Let $\qquad u \equiv x^2 - 1 \qquad \Rightarrow \qquad y = u^{\frac{1}{2}}$

Therefore $\qquad \dfrac{du}{dx} = 2x \qquad$ and $\qquad \dfrac{dy}{du} = \tfrac{1}{2}u^{-\frac{1}{2}}$

Now $\qquad \dfrac{dy}{dx} = \dfrac{dy}{du} \times \dfrac{du}{dx}$

$$= (\tfrac{1}{2}u^{-\frac{1}{2}})(2x)$$

$$= \frac{x}{u^{\frac{1}{2}}}$$

$$= \frac{x}{\sqrt{(x^2-1)}}$$

Therefore $\qquad \dfrac{d}{dx}\sqrt{(x^2-1)} = \dfrac{x}{\sqrt{(x^2-1)}}$

2) Differentiate $\sin\left(2\theta - \dfrac{\pi}{4}\right)$ w.r.t. θ.

$$y = \sin\left(2\theta - \frac{\pi}{4}\right)$$

Let $\qquad u \equiv 2\theta - \dfrac{\pi}{4} \qquad \Rightarrow \qquad y = \sin u$

So that $\qquad \dfrac{du}{d\theta} = 2 \qquad$ and $\qquad \dfrac{dy}{du} = \cos u$

Then $\qquad \dfrac{dy}{d\theta} = \dfrac{dy}{du} \times \dfrac{du}{d\theta}$

$$= 2 \times \cos u$$

$$= 2\cos\left(2\theta - \frac{\pi}{4}\right)$$

Therefore $\qquad \dfrac{d}{d\theta}\left[\sin\left(2\theta - \dfrac{\pi}{4}\right)\right] = 2\cos\left(2\theta - \dfrac{\pi}{4}\right)$

The reader will find that, with a little practice, the substitution $u \equiv f(x)$ can often be carried out mentally so that $\dfrac{d}{dx}gf(x)$ can be written down directly as $\left(\dfrac{d}{du}g(u)\right)\left(\dfrac{d}{dx}f(x)\right)$ and it is important that this facility should be acquired.

For example, if $y = \ln(x^3 + 1)$ then, by mentally substituting u for $x^2 + 1$, we get $\dfrac{dy}{dx} = \left(\dfrac{1}{x^3 + 1}\right)(3x^2)$.

3) Find the derivative of $2\ln(x\sqrt{x^2 - 1})$.

$$2\ln(x\sqrt{x^2 - 1}) \equiv 2\ln x + 2\ln(x^2 - 1)^{\frac{1}{2}} \equiv 2\ln x + \ln(x^2 - 1)$$

Therefore $\quad \dfrac{d}{dx}[2\ln(x\sqrt{x^2 - 1})] = \dfrac{d}{dx}[2\ln x] + \dfrac{d}{dx}[\ln(x^2 - 1)]$

$$= \dfrac{2}{x} + \left(\dfrac{1}{x^2 - 1}\right)(2x)$$

$$= \dfrac{2(2x^2 - 1)}{x(x^2 - 1)}$$

In the example above, transforming the given log function into a sum at the start made differentiating it simpler. *Any* function should be simplified, whenever possible, *before* differentiation.

4) Differentiate $\cos^3 x$ with respect to x.
Note that $\cos^3 x$ means $(\cos x)^3$

Let $\quad\quad\quad\quad\quad y = (\cos x)^3 \quad$ and $\quad u \equiv \cos x \quad$ so that $\quad y = u^3$

then $\quad\quad\quad\quad \dfrac{dy}{dx} = 3u^2(-\sin x)$

$$= -3\cos^2 x \sin x$$

In general, $\quad$ if $\quad y = \cos^n x \quad$ then $\quad \dfrac{dy}{dx} = -n\cos^{n-1}x \sin x$

Similarly, $\quad$ if $\quad y = \sin^n x \quad$ then $\quad \dfrac{dy}{dx} = n\sin^{n-1}x \cos x$

EXERCISE 8f

Differentiate the following functions:

1) e^{3x}

2) $\ln(x - 1)^2$

3) $(x + 1)^5$

4) $\cos\left(3\theta - \dfrac{\pi}{4}\right)$

5) $(x^2 + 1)^5$

6) e^{x^2}

7) $4\ln(x^2 + 1)$

8) $(3x^2 + 4)^{\frac{1}{2}}$

9) $5e^{2x}$

10) $\sin(\theta^2)$

11) $\sin^2\theta$

12) $(e^x)^3$

13) $(x + 1)^{-1}$

14) $\dfrac{1}{x^2 + 1}$

15) $\sqrt{\dfrac{1}{x - 1}}$

Write down directly the derivatives of the following functions:

16) $(x + 1)^6$ 17) $(2x - 4)^3$ 18) $e^{(x^2+2)}$ 19) $\sin\left(3\theta + \dfrac{\pi}{4}\right)$

20) $\cos^2\theta$ 21) $\ln(x^2 + 2)$ 22) $(2x + 3)^{-2}$ 23) $\sqrt{\dfrac{1}{2x - 1}}$

24) $e^x - e^{-x}$ 25) $\dfrac{e^x - 1}{e^{2x}}$ 26) $3\sin\left(2\theta - \dfrac{\pi}{4}\right)$ 27) $\dfrac{1}{\sin\theta}$

28) $\dfrac{1}{\cos\theta}$ 29) $2e^{-3x} + e^{4x}$ 30) $\ln(\sin x)$ 31) e^{-kt}

32) $\ln(x - 1)^3$ 33) $e^{\sin x}$ 34) $\dfrac{1}{\sin^2 x}$ 35) $\sec^2 t$

36) $2e^{\sin 3\theta}$ 37) $4\ln\sqrt{x - 1}$ 38) $2\sin(\theta^2 + 4)^{\frac{1}{2}}$ 39) e^{e^x}

40) $2\cos^2(x^2 + 1)$ 41) $3\sin(e^x)$ 42) $\sqrt{(3 - \cos^2\theta)}$ 43) $\ln(\sin x \cos x)$

44) $e^{\sqrt{1-x}}$ 45) $\sin^2(x^2 + 1)$ 46) $2e^{\frac{1}{x}}$

47) $\ln\dfrac{\sin x}{1 - \cos x}$ 48) $(e^x - e^{-x})^{-1}$

DIFFERENTIATION OF PRODUCTS

Consider $y = uv$ where $u \equiv f(x)$ and $v \equiv g(x)$.
If δx is a small increase in x, and $\delta y, \delta u, \delta v$ are the corresponding increases in y, u, v

then $y + \delta y = (u + \delta u)(v + \delta v) = uv + u\delta v + v\delta u + \delta u\delta v$

As $y = uv$, $\delta y = u\delta v + v\delta u + \delta u\delta v$

Therefore $\dfrac{\delta y}{\delta x} = u\dfrac{\delta v}{\delta x} + v\dfrac{\delta u}{\delta x} + \delta u\dfrac{\delta v}{\delta x}$

When $\delta x \to 0$: $\dfrac{\delta y}{\delta x} \to \dfrac{dy}{dx}$, $\dfrac{\delta u}{\delta x} \to \dfrac{du}{dx}$, $\dfrac{\delta v}{\delta x} \to \dfrac{dv}{dx}$, $\delta u \to 0$.

Therefore $\dfrac{dy}{dx} = \lim_{\delta x \to 0}\left[\dfrac{\delta y}{\delta x}\right]$

$= u\dfrac{dv}{dx} + v\dfrac{du}{dx} + 0$

i.e.

$$\frac{d}{dx}[uv] = v\frac{du}{dx} + u\frac{dv}{dx}$$

Thus if

$$y = e^x \sin 2x$$

and

$$u \equiv e^x, \qquad v \equiv \sin 2x$$

$$\frac{du}{dx} = e^x, \qquad \frac{dv}{dx} = 2 \cos 2x$$

$\Rightarrow$
$$\frac{dy}{dx} = 2e^x \cos 2x + (\sin 2x) e^x = e^x(2 \cos 2x + \sin 2x)$$

As with functions of a function, after some practice in the use of the rule for differentiating a product, such derivatives can be written down directly.

e.g.
$$\frac{d}{dx}[x \sin 3x] = x(3 \cos 3x) + (\sin 3x)(1)$$
$$= 3x \cos 3x + \sin 3x$$

DIFFERENTIATION OF QUOTIENTS

Consider $\quad y = \dfrac{u}{v} \quad$ where $\quad u \equiv f(x) \quad$ and $\quad v \equiv g(x) \quad$ as above

then
$$y + \delta y = \frac{u + \delta u}{v + \delta v}$$

as $\quad y = \dfrac{u}{v}$,
$$\delta y = \frac{u + \delta u}{v + \delta v} - \frac{u}{v}$$
$$= \frac{v\delta u - u\delta v}{v^2 + v\delta v}$$

Therefore
$$\frac{\delta y}{\delta x} = \frac{v\dfrac{\delta u}{\delta x} - u\dfrac{\delta v}{\delta x}}{v^2 + v\delta v}$$

and
$$\frac{dy}{dx} = \lim_{\delta x \to 0}\left[\frac{\delta y}{\delta x}\right] = \frac{v\dfrac{du}{dx} - u\dfrac{dv}{dx}}{v^2}$$

or
$$\frac{d}{dx}\left[\frac{u}{v}\right] = \frac{v\dfrac{du}{dx} - u\dfrac{dv}{dx}}{v^2}$$

Thus if $\quad y = \dfrac{e^x}{\sin x} \quad$ where $\quad u \equiv e^x \quad$ and $\quad v \equiv \sin x$

$$\frac{du}{dx} = e^x \qquad \frac{dv}{dx} = \cos x$$

$$\frac{dy}{dx} = \frac{(\sin x)(e^x) - (e^x)(\cos x)}{\sin^2 x}$$

$$= \frac{e^x(\sin x - \cos x)}{\sin^2 x}$$

Use of Partial Fractions

Rational functions with two, or more, factors in the denominator may be differentiated by first expressing the function as partial fractions (see Chapter 1).

For example, $f(x) \equiv \dfrac{x}{(x-2)(x-3)}$

$$\equiv \frac{-2}{x-2} + \frac{3}{x-3}$$

Hence $\quad \dfrac{d}{dx} f(x) = \dfrac{d}{dx} [-2(x-2)^{-1}] + \dfrac{d}{dx} [3(x-3)^{-1}]$

$\Rightarrow \qquad \dfrac{d}{dx} f(x) = 2(x-2)^{-2} - 3(x-3)^{-2} = \dfrac{2}{(x-2)^2} - \dfrac{3}{(x-3)^2}$

The 'Cover Up' Method for Expressing a Function in Partial Fractions

Consider again the function

$$f(x) \equiv \frac{x}{(x-2)(x-3)} \equiv \frac{A}{x-2} + \frac{B}{x-3} \equiv \frac{A(x-3) + B(x-2)}{(x-2)(x-3)}$$

so that $\qquad x \equiv A(x-3) + B(x-2)$

when $\quad x = 2,$ we have $\quad A = \dfrac{2}{(2-3)} = -2$

i.e. A is the value of $\dfrac{x}{(x-3)}$ when $\quad x = 2$

i.e. $\qquad A = f(2)$ with the factor $(x-2)$ omitted or 'covered up'

similarly $\quad B = f(3)$ with the factor $(x-3)$ covered up

$$= \frac{3}{(3-2)} = 3$$

This method gives a quick way of expressing a function in partial fractions when linear factors only are in the denominator, e.g.

$$f(x) \equiv \frac{1}{(2x-1)(x+3)}$$

$$\equiv \frac{f(\frac{1}{2}) \text{ with } (2x-1) \text{ covered up}}{2x-1} + \frac{f(-3) \text{ with } (x+3) \text{ covered up}}{x+3}$$

$$\equiv \frac{\frac{2}{7}}{(2x-1)} - \frac{\frac{1}{7}}{(x+3)}$$

The cover up method can be used to find the numerator of a partial fraction with a linear denominator, but for quadratic denominators, the methods described in Chapter 1 must be used,

e.g. $$f(x) \equiv \frac{x}{(x-1)(x^2+1)} \equiv \frac{f(1) \text{ with } (x-1) \text{ covered up}}{x-1} + \frac{Ax+B}{x^2+1}$$

$$\equiv \frac{\frac{1}{2}}{(x-1)} + \frac{Ax+B}{x^2+1}$$

thus $$x \equiv \frac{1}{2}(x^2+1) + (Ax+B)(x-1)$$

⇒ $$A = -\frac{1}{2}, \quad B = \frac{1}{2} \quad \text{by comparing coefficients.}$$

EXERCISE 8g

Differentiate the following functions w.r.t. x.

1) $x \ln x$

2) $x^2(x-1)^{\frac{1}{2}}$

3) $\dfrac{x}{\ln x}$

4) $3 \sin x \cos 2x$

5) $\tan x$

6) $\dfrac{e^x}{x-1}$

7) $(x-1) \ln (x-1)$

8) $\sin x \ln x$

9) $x \sec x$

10) $\operatorname{cosec} x \cos x$

11) $\dfrac{x-1}{x+1}$

12) $e^x \sin x$

13) $\dfrac{e^x}{\sin x}$

14) $x \log x$

15) $\dfrac{e^x}{e^x - e^{-x}}$

16) $\dfrac{x}{x^2+1}$

17) $\dfrac{x^2+1}{x}$

18) $\dfrac{\sqrt{x-1}}{x}$

19) $\dfrac{\sin x}{\sqrt{x}}$

20) $\dfrac{\ln x}{\ln (x-1)}$

21) $\cot x$

Use the cover up method to express the following functions in partial fractions and hence differentiate them.

22) $\dfrac{1}{(x-1)(x+1)}$
23) $\dfrac{x}{(x-2)(x+1)}$
24) $\dfrac{2x}{(2x-1)(x-3)}$

25) $\dfrac{3x-1}{(x+2)(2x-1)}$
26) $\dfrac{5x}{(x-1)(x-2)(x-3)}$

27) $\dfrac{x^2+1}{(2x-1)(x-1)(x+1)}$
28) $\dfrac{x^2}{(x+3)(x^2+1)}$
29) $\dfrac{1}{(x-1)^2(x+1)}$

30) $\dfrac{3x}{(x+3)(x^2+1)}$

By taking results from the last two exercises we may complete the list of derivatives of trig functions:

$$\frac{d}{dx}(\tan x) = \sec^2 x$$

$$\frac{d}{dx}(\cot x) = -\text{cosec}^2 x$$

$$\frac{d}{dx}(\sec x) = \sec x \tan x$$

$$\frac{d}{dx}(\text{cosec } x) = -\text{cosec } x \cot x$$

Also from the 'function of a function' rule we obtain the following general results

$$\frac{d}{dx}\sin f(x) = f'(x)\cos f(x) \qquad \text{where} \quad f'(x) \equiv \frac{d}{dx}[f(x)]$$

$$\frac{d}{dx}\cos f(x) = -f'(x)\sin f(x)$$

$$\frac{d}{dx}e^{f(x)} = f'(x)e^{f(x)}$$

$$\frac{d}{dx}\ln f(x) = \frac{f'(x)}{f(x)}$$

Any of these results may now be used to differentiate a particular function.

EXERCISE 8h

Find the derivatives of the following functions:

1) $3e^{\sin x}$
2) $e^{-x}\sin x$
3) $\dfrac{1}{1+x+x^2}$

4) $\cos x \sin^3 x$ 5) $\ln (\sec x)$ 6) $2 \sin 3t \cos 4t$

7) $e^{x^2} \ln 2x^2$ 8) $\tan^2 2x$ 9) $\ln \left(\dfrac{1 + \cos x}{1 + \sin x} \right)$

10) Find the value of $\dfrac{d^2 y}{dx^2}$ when $x = 0$ if $y = e^x \sin 2x$.

11) If $y = e^x \sin x$ show that $\dfrac{d^2 y}{dx^2} - 2 \dfrac{dy}{dx} + 2y = 0$.

12) If $y = \dfrac{1 + x}{1 - x}$ find $\dfrac{d^2 y}{dx^2}$

13) Find the gradient of the curve $y = \ln \sqrt{(1 + \sin 2x)}$ at the point where $x = \dfrac{\pi}{2}$.

14) If $y = \sqrt{(x - 1)} e^x \ln x$, find $\dfrac{dy}{dx}$

15) Using the identity $\dfrac{dy}{dx} \equiv 1 \Big/ \dfrac{dx}{dy}$ find $\dfrac{dy}{dx}$ when $\sin y = x$,

giving your answer as a function of x.

16) Find the gradient of the curve $y = \dfrac{1}{\sqrt{(x - 1)} \sin x}$ when $x = \dfrac{\pi}{2}$.

17) If $y = \sin^2 ax$ find a if $\dfrac{dy}{dx} = 1$ when $x = \dfrac{\pi}{4}$, given that a is an integer.

18) Find the values of k for which $\dfrac{x}{(x + 1)^2 (x - k)}$ has one stationary value.

19) Find the value of x in the range $0 \leqslant x \leqslant 2\pi$ for which $e^x \sin x$ has a minimum value. Sketch the curve $y = e^x \sin x$ for $0 \leqslant x \leqslant 2\pi$.

20) Show that $x \ln x$ has only one stationary value and find it.

IMPLICIT FUNCTIONS

All the differentiation carried out so far has involved equations of the form $y = f(x)$.

Now consider the curve whose equation is $y + xy + y^2 = 2$.

This equation is not easily transposed to the form $y = f(x)$ and we say that $y = f(x)$ is implied by the equation $y + xy + y^2 = 2$, i.e. $f(x)$ is an implicit function.

Differentiation of Implicit Functions

Consider again the equation $y + xy + y^2 = 2$.
The equation may be rewritten as

$$f(x) + xf(x) + [f(x)]^2 = 2$$

where $y = f(x)$ and $\dfrac{dy}{dx} = f'(x)$.

Differentiating term by term we have

(i) $\dfrac{d}{dx} f(x) = f'(x) = \dfrac{dy}{dx}$

(ii) $xf(x)$ is a product

therefore $\dfrac{d}{dx} [xf(x)] = (1)f(x) + (x)f'(x) = y + x\dfrac{dy}{dx}$

i.e. $\dfrac{d}{dx}(xy) = (1)y + (x)\dfrac{dy}{dx}$

(iii) $[f(x)]^2$ is a function of a function.

Therefore $\dfrac{d}{dx} [f(x)]^2 = [2f(x)] f'(x) = 2y\dfrac{dy}{dx}$

i.e. $\dfrac{d}{dx}(y^2) = 2y\dfrac{dy}{dx}$

(iv)   and finally $\dfrac{d}{dx}(2) = 0$

Therefore differentiating $y + xy + y^2 = 2$ w.r.t. x

we get $\dfrac{dy}{dx} + \left(y + x\dfrac{dy}{dx}\right) + 2y\dfrac{dy}{dx} = 0$

or $(1 + x + 2y)\dfrac{dy}{dx} + y = 0$

Note that *every* term in the equation is differentiated w.r.t. x.
Note that if $g(y)$ is any function of y where $y = f(x)$ then $g(y)$ is a function of a function of x.
Thus the derivative of $g(y)$ w.r.t. x is

$$\dfrac{d}{dx} g(y) = g'(y)\dfrac{dy}{dx}$$

where $g'(y)$ is the derivative of $g(y)$ w.r.t. y

For example
$$\frac{d}{dx}\sin y = (\cos y)\frac{dy}{dx}$$

EXAMPLES 8i

1) Differentiate the following w.r.t. x:

(a) $x^2 + xy^2 + y^3 = 2$ (b) $x = ye^x$

(a) If $x^2 + xy^2 + y^3 = 2$ then

$$\frac{d}{dx}(x^2) + \frac{d}{dx}(xy^2) + \frac{d}{dx}(y^3) = \frac{d}{dx}(2)$$

i.e.
$$2x + \left(x(2y)\frac{dy}{dx} + y^2\right) + 3y^2\frac{dy}{dx} = 0$$

$\Rightarrow$
$$\frac{dy}{dx}(2xy + 3y^2) + 2x + y^2 = 0$$

(b) $\dfrac{dy}{dx}$ can be found as a function of x by rewriting $x = ye^x$ as $y = xe^{-x}$

giving
$$\frac{dy}{dx} = -xe^{-x} + e^{-x}$$

or alternatively, using the method for implicit functions, we have

$$\frac{d}{dx}(x) = \frac{d}{dx}(ye^x) \Rightarrow 1 = ye^x + \frac{dy}{dx}e^x$$

2) If $e^x y = \sin x$ show that $\dfrac{d^2 y}{dx^2} + 2\dfrac{dy}{dx} + 2y = 0.$

The temptation with a problem of this type is to use $e^x y = \sin x$ in the form $y = e^{-x}\sin x$, find $\dfrac{dy}{dx}$ and $\dfrac{d^2 y}{dx^2}$ as functions of x and then show that they satisfy the given equation (called a differential equation).
However a more direct method is to differentiate the given implicit equation.

Thus if
$$e^x y = \sin x$$

then
$$e^x y + e^x\frac{dy}{dx} = \cos x$$

and differentiating again w.r.t. x gives

$$e^x y + e^x\frac{dy}{dx} + e^x\frac{dy}{dx} + e^x\frac{d^2 y}{dx^2} = -\sin x = -e^x y$$

hence
$$e^x\frac{d^2 y}{dx^2} + 2e^x\frac{dy}{dx} + 2e^x y = 0$$

There is no finite value of x for which $e^x = 0$ so we may divide the equation by e^x, giving

$$\frac{d^2y}{dx^2} + 2\frac{dy}{dx} + 2y = 0$$

The Equation of a Tangent

Consider the curve whose equation is $3x^2 - 7y^2 + 4xy - 8x = 0$.
For brevity we may write this as $f(x, y) = 0$ where
$f(x, y) \equiv 3x^2 - 7y^2 + 4xy - 8x$.
The equation of a tangent, or normal, to $f(x, y) = 0$ can be found, *without* transposing the equation to the form $y = g(x)$, as follows.
Differentiating w.r.t. x gives

$$6x - 14y\frac{dy}{dx} + 4\left(y + x\frac{dy}{dx}\right) - 8 = 0$$

$$\Rightarrow \qquad \frac{dy}{dx} = \frac{4 - 2y - 3x}{2x - 7y}$$

i.e. $\dfrac{4 - 2y - 3x}{2x - 7y}$ is the gradient function of $f(x, y) = 0$.

At the point $(-1, 1)$ on the curve

$$\frac{dy}{dx} = \frac{4 - 2 + 3}{-2 - 7} = -\frac{5}{9}$$

So the equation of the tangent to the curve at this point is

$$y - 1 = -\tfrac{5}{9}(x + 1)$$

$$\Rightarrow \qquad 5x + 9y - 4 = 0$$

EXAMPLES 8i (continued)

3) Find the equation of the tangent to $3x^2 - 4y^2 = 9$ at the point (x_1, y_1) on the curve.

Differentiating the equation w.r.t. x gives

$$6x - 8y\frac{dy}{dx} = 0$$

$$\frac{dy}{dx} = \frac{3x}{4y}$$

At the point (x_1, y_1), the gradient is $\dfrac{3x_1}{4y_1}$ and the equation of the tangent is

$$(y - y_1) = \frac{3x_1}{4y_1}(x - x_1)$$

$\Rightarrow$

$$3xx_1 - 4yy_1 = 3x_1{}^2 - 4y_1{}^2$$

but as (x_1, y_1) is on the curve, $3x_1{}^2 - 4y_1{}^2 = 9$, so the equation of the tangent can be written

$$3xx_1 - 4yy_1 = 9$$

Note that the equation of the tangent is the same as the equation of the curve except that x^2 is replaced by xx_1 and y^2 is replaced by yy_1.
In general, if $f(x, y)$ is of degree two, the equation of the tangent at (x_1, y_1) to the curve $f(x, y) = 0$ can be written down directly by replacing, in the equation of the curve

$$x^2, y^2 \text{ by } xx_1, yy_1 \text{ respectively}$$
$$xy \text{ by } \tfrac{1}{2}(xy_1 + yx_1)$$
$$x, y \text{ by } \tfrac{1}{2}(x + x_1), \tfrac{1}{2}(y + y_1) \text{ respectively}$$

It is left to the reader to prove this in Exercise 8i.
Using this result, the equation of the tangent at (x_1, y_1) to

$$3x^2 - 7y^2 + 4xy - 8x = 0$$

can be written down as

$$3xx_1 - 7yy_1 + 2(xy_1 + yx_1) - 4(x + x_1) = 0$$
$\Rightarrow$
$$(3x_1 + 2y_1 - 4)x + (2x_1 - 7y_1)y - 4x_1 = 0$$

and at the point $(-1, 1)$ this equation becomes

$$-5x - 9y + 4 = 0 \quad \text{or} \quad 5x + 9y - 4 = 0$$

which is consistent with the result on page 277.
Note that this result gives a quick method for *writing down* the equation of a tangent. It must not be used when the derivation of the equation is required.

The Inverse Trigonometric Functions

The inverse trigonometric functions, viz. arcsin, arccos, arctan, are introduced in Chapter 6 where 'arcsin' means 'the principal angle whose sine is' and is called *the inverse sine function*.

i.e. $y = \arcsin x \iff \sin y = x$

similarly $y = \arccos x \iff \cos y = x$

and $y = \arctan x \iff \tan y = x$

Differentiation of   arcsin x

We have already shown that $\dfrac{dy}{dx} \equiv 1 \left/ \dfrac{dx}{dy} \right.$

and used it to find $\dfrac{d}{dx} \ln x$ (page 263).

This relationship is also used to find $\dfrac{d}{dx} \arcsin x$.

Let $y = \arcsin x$
so that $x = \sin y$
differentiating w.r.t. y gives

$$\frac{dx}{dy} = \cos y$$

therefore

$$\frac{dy}{dx} = \frac{1}{\cos y}$$

but $\cos^2 y \equiv 1 - \sin^2 y = 1 - x^2$

Hence

$$\frac{dy}{dx} = \frac{1}{\sqrt{(1 - x^2)}}$$

i.e.

$$\frac{d}{dx}(\arcsin x) = \frac{1}{\sqrt{(1 - x^2)}} \qquad [1]$$

Note that $y = \arcsin x \;\Rightarrow\; -\dfrac{\pi}{2} \leqslant y \leqslant \dfrac{\pi}{2}$,   because y is a principal value angle.
For this range of values of y, $\cos x \geqslant 0$.
i.e. $\cos y = \sqrt{(1 - x^2)}$ [not $-\sqrt{(1 - x^2)}$].

Note also that, when asked to find $\dfrac{d}{dx}[f(x)]$,   the answer should be given as a function of x.

Similarly it can be shown that

$$\frac{d}{dx}(\arccos x) = -\frac{1}{\sqrt{(1 - x^2)}} \qquad [2]$$

and

$$\frac{d}{dx}(\arctan x) = \frac{1}{1 + x^2} \qquad [3]$$

Note. If $y = f(x)$,   the inverse function of $x, f^{-1}(x)$,   provided it exists, is implied by $f(y) = x$

or $g(y) = x \;\Rightarrow\; y = g^{-1}(x)$

Differentiation of a^x

It was shown earlier in this chapter that

$$\frac{d}{dx} a^x = (a^x) \times [\text{the gradient of } y = a^x \text{ at } (0, 1)]$$

The number e was introduced such that $y = e^x$ has a gradient of unity at $(0, 1)$, but we did not at that stage find $\frac{d}{dx} a^x$.

If $y = a^x$ then $\ln y = x \ln a$

differentiating w.r.t. y gives $\quad \dfrac{1}{y} = \dfrac{dx}{dy} \ln a$

$\Rightarrow \qquad \qquad \qquad \dfrac{dx}{dy} = \dfrac{1}{y \ln a}$

Therefore $\qquad \qquad \dfrac{dy}{dx} = y \ln a = a^x \ln a$

i.e. $\qquad \qquad \dfrac{d}{dx} a^x = a^x \ln a$ [4]

The results [1], [2], [3] and [4] are standard results and may be quoted *unless* more explanation is required.

Logarithmic differentiation

We differentiated the function a^x by first taking logs of the equation $y = a^x$. This method is known as logarithmic differentiation and is necessary whenever an exponent contains a variable (unless the base is e). Logarithmic differentiation is also a useful way of simplifying the differentiation of some of the more complicated compound functions.

EXAMPLES 8i (continued)

4) Differentiate x^x w.r.t. x.

Let $\qquad \qquad \qquad y = x^x$

so $\qquad \qquad \qquad \ln y = x \ln x$

thus $\qquad \qquad \dfrac{1}{y} \dfrac{dy}{dx} = x \dfrac{1}{x} + \ln x$

$\Rightarrow \qquad \qquad \qquad \dfrac{dy}{dx} = y(1 + \ln x)$

Therefore $\qquad \dfrac{d}{dx}(x^x) = x^x(1 + \ln x)$

5) Differentiate $\dfrac{x-1}{x\sqrt{(x^2+1)}}$

Let $y = \dfrac{x-1}{x\sqrt{(x^2+1)}}$ so that $\ln y = \ln(x-1) - \ln x - \tfrac{1}{2}\ln(x^2+1)$.

Therefore $\dfrac{1}{y}\dfrac{dy}{dx} = \dfrac{1}{x-1} - \dfrac{1}{x} - \dfrac{x}{x^2+1}$

$$\dfrac{dy}{dx} = y\left[\dfrac{-x^3+2x^2+1}{x(x-1)(x^2+1)}\right]$$

$$= \dfrac{(x-1)(-x^3+2x^2+1)}{x(x^2+1)^{\frac{1}{2}}x(x-1)(x^2+1)}$$

$$= \dfrac{1+2x^2-x^3}{x^2(x^2+1)^{3/2}}$$

6) Find $\dfrac{d}{dx}[\arcsin(ax+b)]$.

Let $y = \arcsin(ax+b)$ and $u \equiv ax+b$

so that $\qquad\qquad y = \arcsin u$

Therefore $\qquad \dfrac{dy}{du} = \dfrac{1}{\sqrt{(1-u^2)}}$ and $\dfrac{du}{dx} = a$

so that $\qquad \dfrac{dy}{dx} = \dfrac{1}{\sqrt{(1-u^2)}} \times a = \dfrac{a}{\sqrt{[1-(ax+b)^2]}}$

i.e. $\qquad \dfrac{d}{dx}\arcsin(ax+b) = \dfrac{a}{\sqrt{[1-(ax+b)^2]}}$

EXERCISE 8i

Differentiate the following equations w.r.t. x.

1) $x^2 + y^2 = 4$ 2) $x^2 + xy + y^2 = 0$

3) $x(x+y) = y^2$ 4) $\dfrac{1}{x} + \dfrac{1}{y} = e^y$

5) $\dfrac{1}{x^2} + \dfrac{1}{y^2} = \dfrac{1}{4}$ 6) $\dfrac{x^2}{4} - \dfrac{y^2}{9} = 1$

7) $\sin x + \sin y = 1$ 8) $\sin x \cos y = 2$

9) $xe^y = x + 1$ 10) $\sqrt{(1 + y)(1 + x)} = x$

11) Find $\dfrac{dy}{dx}$ as a function of x if $y^2 = 2x + 1$.

12) Show that $\dfrac{d}{dx}(\arctan x) = \dfrac{1}{1 + x^2}$

13) Show that $\dfrac{d}{dx}(\arccos x) = -\dfrac{1}{\sqrt{(1 - x^2)}}$

14) Find $\dfrac{d^2y}{dx^2}$ as a function of x if $\sin y + \cos y = x$.

15) Find the gradient of $x^2 + y^2 = 9$ at the point where $x = 1$.

16) If $y \cos x = e^x$ show that $\dfrac{d^2y}{dx^2} - 2\tan x\,\dfrac{dy}{dx} - 2y = 0$.

17) Find the differential equation given by differentiating $y2^x = 1$ w.r.t. x and find the gradient at the point where $x = 2$.

18) If $y = 3^x$ find $\dfrac{d^2y}{dx^2}$ when $x = -1$.

19) Find the gradient of $y = \arctan x$ when $x = 1$.

20) Sketch the curves:
(a) $y = \arccos x$ (b) $y = \arctan x$.

21) If $\sin y = 2 \sin x$ show that:

(a) $\left(\dfrac{dy}{dx}\right)^2 = 1 + 3 \sec^2 y$

(b) by differentiating (a) w.r.t. x show that $\dfrac{d^2y}{dx^2} = 3 \sec^2 y \tan y$

and hence that $\cot y\,\dfrac{d^2y}{dx^2} - \left(\dfrac{dy}{dx}\right)^2 + 1 = 0$.

22) Differentiate the following functions w.r.t. x:
(a) $a^x + b^x$ (b) $x^{\sin x}$ (c) $(\sin x)^x$

(d) $(x + x^2)^x$ (e) $\dfrac{x}{(x - 1)(x + 3)^2(x^2 - 1)}$ (f) $\sqrt{\dfrac{1}{(x^2 - 1)(3x - 4)^4}}$

23) Differentiate the following equations w.r.t. x.

(a) $y^x = x$ (b) $x^y = \sin x$

(c) $(1 + x)y = \sin^{\frac{1}{2}}x$ (d) $y^{\sin x} = \sqrt{x}$

24) Write down the equation of the tangent to:

(a) $x^2 - 3y^2 = 4y$ (b) $x^2 + xy + y^2 = 3$

at the point (x_1, y_1).

25) Show that the equation of the tangent to $x^2 + xy + y = 0$ at the point (x_1, y_1) is

$$x(2x_1 + y_1) + y(x_1 + 1) + y_1 = 0$$

26) Write down the equation of the tangent at $(1, \frac{1}{3})$ to the curve whose equation is $2x^2 + 3y^2 - 3x + 2y = 0$.

27) Show that the equation of the tangent at (x_1, y_1) to the curve $ax^2 + by^2 + cxy + dx = 0$ is

$$axx_1 + byy_1 + \tfrac{1}{2}c(xy_1 + yx_1) + \tfrac{1}{2}d(x + x_1) = 0$$

PARAMETRIC EQUATIONS

Some relationships between x and y are so complicated that it is easier to express x and y each in terms of a third variable, called a *parameter*.

For example:
$$\begin{cases} x = t^3 & [1] \\ y = t^2 - t & [2] \end{cases}$$

The equations [1] and [2] are called the parametric equations of the curve. By eliminating t from [1] and [2] it is possible to get a direct relationship between x and y, which is the Cartesian equation of the curve. In this case we get $y = x^{\frac{2}{3}} - x^{\frac{1}{3}}$ which is clearly awkward to analyse and it is much simpler to use the parametric equations.

To get an idea of the shape of the curve whose parametric equations are
$$\begin{cases} x = t^3 \\ y = t^2 - t \end{cases}$$

we can find some points on the curve by assigning various values to t, viz:

t	-2	-1	0	1	2
x	-8	-1	0	1	8
y	6	2	0	0	2

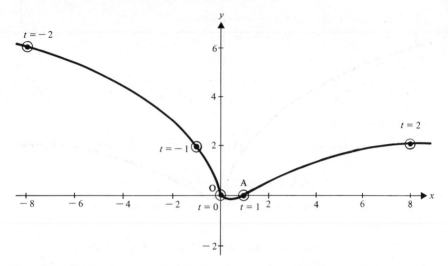

There is no finite value of t for which either x or y is infinite, so it is reasonable to assume that the curve is continuous.

We have, however, assumed that there is one turning point (between O and A). We can use differentiation to confirm this.

Differentiation of a Curve whose Equation is given Parametrically

By using $\quad \dfrac{dy}{dx} = \dfrac{dy}{dt}\dfrac{dt}{dx}\quad$ and $\quad \dfrac{dt}{dx} = 1\bigg/\dfrac{dx}{dt}$

we can find $\dfrac{dy}{dx}$ in terms of the parameter t, as follows,

$$y = t^2 - t \quad \text{and} \quad x = t^3$$

$$\frac{dy}{dt} = 2t - 1 \quad \text{and} \quad \frac{dx}{dt} = 3t^2 \quad \Rightarrow \quad \frac{dt}{dx} = \frac{1}{3t^2}$$

$$\frac{dy}{dx} = \frac{dy}{dt}\frac{dt}{dx} = \frac{2t - 1}{3t^2}$$

From this we note that:

1) $\dfrac{dy}{dx} = 0 \quad$ when $\quad t = \tfrac{1}{2}$

i.e. there *is* a turning point between O and A.

2) $\dfrac{dy}{dx}$ is infinite when $\quad t = 0$.

i.e. the tangent to the curve at O is parallel to the y-axis.

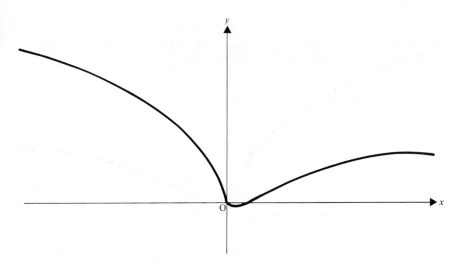

Tangents and Normals

If a curve has an equation expressed in the form $y = f(x)$, we may use $[x, f(x)]$ as the coordinates of a general point on that curve. Similarly, for a curve expressed parametrically

i.e.
$$\begin{cases} x = f(t) \\ y = g(t) \end{cases}$$

we may use $[f(t), g(t)]$ as the coordinates of a general point on the curve.

Thus for
$$\begin{cases} x = t^3 \\ y = t^2 - t \end{cases}$$

$(t^3, t^2 - t)$ are the parametric coordinates of any point on the given curve. Using the coordinates of a general point enables us to find an equation for a tangent to the curve at any point.

For example, for the curve
$$\begin{cases} x = t^2 - 1 \\ y = 3t \end{cases}$$

we find that
$$\frac{dy}{dx} = \frac{3}{2t}$$

This is the gradient function of the curve, i.e. the gradient at a general point on the curve.

Therefore, using $y - y_1 = m(x - x_1)$, the general equation of the tangent at $(t^2 - 1, 3t)$ is

$$y - 3t = \frac{3}{2t}[x - (t^2 - 1)]$$

$$\Rightarrow \qquad\qquad 3x - 2ty + (3t^2 + 3) = 0$$

The equation of the tangent at a particular point can be found by substituting the value of t at this point into the general equation.

The Second Derivative of Parametric Equations

Consider the equations $\quad\begin{cases} x = \sin\theta \\ y = \cos 2\theta \end{cases}$

$$\frac{dx}{d\theta} = \cos\theta \quad \text{and} \quad \frac{dy}{d\theta} = -2\sin 2\theta$$

Therefore $\quad \dfrac{dy}{dx} = \dfrac{dy}{d\theta}\Big/\dfrac{dx}{d\theta} = \dfrac{-2\sin 2\theta}{\cos\theta} = -4\sin\theta$

now $\quad \dfrac{d^2y}{dx^2} = \dfrac{d}{dx}\left(\dfrac{dy}{dx}\right) = \dfrac{d}{dx}(-4\sin\theta) = \dfrac{d}{d\theta}(-4\sin\theta)\dfrac{d\theta}{dx}$

$$= (-4\cos\theta)\left(\frac{1}{\cos\theta}\right) = -4$$

Note that if $\quad x = f(t) \quad$ and $\quad y = g(t)$

then $\quad \dfrac{d^2y}{dx^2} = \dfrac{d}{dx}\left[\dfrac{dy}{dx}\right] = \dfrac{d}{dx}\left[\dfrac{dy}{dt}\dfrac{dt}{dx}\right]$

$$= \frac{d}{dx}\,[\text{a product}]$$

so that $\dfrac{d^2y}{dx^2}$ is *not* equal to $\dfrac{d^2y}{dt^2}\times\dfrac{d^2t}{dx^2}$, *nor* to $1\Big/\dfrac{d^2x}{dy^2}$.

EXAMPLES 8j

1) Find the turning points on the curve $\quad\begin{cases} x = t \\ y = t^3 - 3t \end{cases}$

and distinguish between them. Sketch the curve.

$$\left.\begin{array}{l} y = t^3 - 3t \quad \Rightarrow \quad \dfrac{dy}{dt} = 3t^2 - 3 \\[3mm] x = t \qquad\quad \Rightarrow \quad \dfrac{dx}{dt} = 1 \end{array}\right\} \quad \Rightarrow \quad \dfrac{dy}{dx} = \dfrac{3t^2 - 3}{1}$$

At turning points $\quad \dfrac{dy}{dx} = 0, \quad$ i.e. $\quad 3t^2 - 3 = 0 \Rightarrow (t+1)(t-1) = 0$

$$\Rightarrow t = \pm 1$$

When $\quad t = 1, \quad x = 1 \quad$ and $\quad y = -2.$
When $\quad t = -1, \quad x = -1 \quad$ and $\quad y = 2.$

To determine the nature of these stationary points we can either

(a) find the sign of $\dfrac{d^2y}{dx^2}$ at these values of t, or

(b) find the signs of $\dfrac{dy}{dx}$ in the neighbourhood of $x = 1$ and of $x = -1$.

(a) $\dfrac{d^2y}{dx^2} = \dfrac{d}{dx}\left(\dfrac{dy}{dx}\right) = \dfrac{d}{dx}(3t^2-3) = \dfrac{d}{dt}(3t^2-3)\dfrac{dt}{dx} = (6t)(1) = 6t.$

When $t = -1$, $\dfrac{d^2y}{dx^2} < 0$ so $(-1,2)$ is a maximum turning point.

When $t = 1$, $\dfrac{d^2y}{dx^2} > 0$ so $(1,-2)$ is a minimum turning point.

(b)

Values of x	$-1\frac{1}{2}$	-1	$-\frac{1}{2}$	$\frac{1}{2}$	1	$1\frac{1}{2}$
Values of t	$-1\frac{1}{2}$	-1	$-\frac{1}{2}$	$\frac{1}{2}$	1	$1\frac{1}{2}$
Sign of $\dfrac{dy}{dx}$	$+$	0	$-$	$-$	0	$+$
	$/$	$-$	$\backslash$	$\backslash$	$-$	$/$

Hence $(-1,2)$ is a maximum turning point
and $(1,-2)$ is a minimum turning point.

From $\left.\begin{array}{l} x = t \\ y = t^3 - 3t \end{array}\right\}$ there are no finite values of t for which either x or y

is undefined, so the curve is continuous.

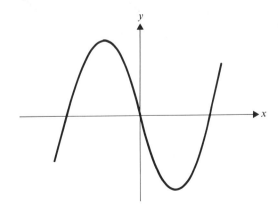

2) The parametric equations of a curve are: $x = \sin^2\theta$, $y = 2\sin\theta$.
Show that the tangent to the curve at the point P $(\sin^2\theta, 2\sin\theta)$ has an
equation $x - y\sin\theta + \sin^2\theta = 0$ and find the points on the curve at which
the tangent is parallel to the y-axis.

$$\frac{dy}{d\theta} = 2\cos\theta \quad \text{and} \quad \frac{dx}{d\theta} = 2\sin\theta\cos\theta$$

Therefore $\quad \dfrac{dy}{dx} = \dfrac{dy}{d\theta} \Big/ \dfrac{dx}{d\theta} = \dfrac{2\cos\theta}{2\sin\theta\cos\theta} = \dfrac{1}{\sin\theta}$

Using $\;y - y_1 = m(x - x_1),\;$ the general tangent to the curve has equation

$$y - 2\sin\theta = \frac{1}{\sin\theta}(x - \sin^2\theta)$$

$$y\sin\theta - 2\sin^2\theta = x - \sin^2\theta$$

$$x - y\sin\theta + \sin^2\theta = 0$$

The tangents which are parallel to the y-axis have infinite gradient.

i.e. $\qquad \dfrac{1}{\sin\theta} = \infty \;\Rightarrow\; \sin\theta = 0 \;\Rightarrow\; x = 0 \;\text{ and }\; y = 0$

So the point where the tangent is parallel to the y-axis is the origin.

3) Find the equation of the normal to the curve $\;x = t^2, \;\; y = t + \dfrac{1}{t}\;$ at the point on the curve where $\;t = 2.$

$$\left. \begin{array}{l} y = t + \dfrac{1}{t} \;\Rightarrow\; \dfrac{dy}{dt} = 1 - \dfrac{1}{t^2} \\[2mm] x = t^2 \;\Rightarrow\; \dfrac{dx}{dt} = 2t \end{array} \right\} \;\Rightarrow\; \dfrac{dy}{dx} = \dfrac{dy}{dt} \Big/ \dfrac{dx}{dt} = \dfrac{t^2 - 1}{2t^3}$$

When $\;t = 2: \;\; x = 4, \;\; y = \dfrac{5}{2}, \;\; \dfrac{dy}{dx} = \dfrac{3}{16}$

so the required normal has gradient $-\frac{16}{3}$ and cuts the curve at $(4, \frac{5}{2})$. Hence its equation is

$$y - \tfrac{5}{2} = -\tfrac{16}{3}(x - 4)$$

$\Rightarrow \qquad\qquad 6y + 32x + 143 = 0$

EXERCISE 8j

1) Find the Cartesian equation of the curves whose parametric equations are:

(a) $x = t^2$ (b) $x = \cos\theta$ (c) $x = 2t$

$\quad\;\; y = 2t$ $y = \sin\theta$ $y = \dfrac{1}{t}$

2) By taking values of t from -2 to $+2$ at intervals of one unit for (a) and (c), and values of θ from $-\pi$ to $+\pi$ at intervals of $\dfrac{\pi}{4}$ for (b), draw sketches of the curves in (1).

3) Find the gradient function of each curve in (1) as a function of the parameter.

4) Find $\dfrac{d^2 y}{dx^2}$ for each curve in (1) as a function of the parameter.

5) Find the turning points of the curve whose parametric equations are $x = t, \quad y = t^3 - t,$ and distinguish between them.

6) A curve has parametric equations $x = \theta - \cos\theta, \quad y = \sin\theta.$ Find the coordinates of the points at which the gradient of this curve is zero.

7) Find the equation of the tangent to the curve $x = t^2, \quad y = 4t$ at the point where $t = -1$.

8) Find the equation of the general normal to the curve $x = t, \quad y = \dfrac{1}{t}$.

9) Find the equation of the general tangent to the curve $x = t^2, \quad y = 4t$.

10) Find the equation of the normal to the curve $x = \cos\theta, \quad y = \sin\theta$ at the point where $\theta = \dfrac{\pi}{4}$. Find the coordinates of the point where this normal cuts the curve again.

11) The parametric equations of a curve are $x = e^t, \quad y = \sin t$.

Find $\dfrac{dy}{dx}$ and $\dfrac{d^2 y}{dx^2}$ as functions of t.

Hence show that $x^2 \dfrac{d^2 y}{dx^2} + x \dfrac{dy}{dx} + y = 0$.

12) The parametric equations of a curve are $x = t, \quad y = \dfrac{1}{t}$. Find the equation of the general tangent to this curve [i.e. the tangent at the point $(t, 1/t)$].
Find in terms of t the coordinates of the points at which the tangent cuts the coordinate axes. Hence show that the area enclosed by this tangent and the coordinate axes is constant.

13) A curve has parametric equations $x = t^2, \quad y = 4t$.
Find the equation of the normal to this curve at $(t^2, 4t)$.
Find the coordinates of the points where the normal cuts the coordinate axes.
Hence find, in terms of t, the area of the triangle enclosed by the normal and the axes.

SUMMARY

Standard Results

$f(x)$	$\dfrac{\mathrm{d}}{\mathrm{d}x} f(x)$
$\sin x$	$\cos x$
$\cos x$	$-\sin x$
$\tan x$	$\sec^2 x$
$\sec x$	$\sec x \tan x$
$\operatorname{cosec} x$	$-\operatorname{cosec} x \cot x$
$\cot x$	$-\operatorname{cosec}^2 x$
e^x	e^x
$\ln x$	$\dfrac{1}{x}$
a^x	$a^x \ln a$
$\arcsin x$	$1/\sqrt{1 - x^2}$
$\arccos x$	$-1/\sqrt{1 - x^2}$
$\arctan x$	$1/(1 + x^2)$

General Results

$$\frac{\mathrm{d}y}{\mathrm{d}x} = 1 \bigg/ \frac{\mathrm{d}x}{\mathrm{d}y}$$

$$\frac{\mathrm{d}y}{\mathrm{d}x} = \frac{\mathrm{d}y}{\mathrm{d}u}\frac{\mathrm{d}u}{\mathrm{d}x} = \frac{\mathrm{d}y}{\mathrm{d}u} \bigg/ \frac{\mathrm{d}x}{\mathrm{d}u}$$

$$\frac{\mathrm{d}}{\mathrm{d}x}(uv) = v\frac{\mathrm{d}u}{\mathrm{d}x} + u\frac{\mathrm{d}v}{\mathrm{d}x}$$

$$\frac{\mathrm{d}}{\mathrm{d}x}\left(\frac{u}{v}\right) = \frac{v\dfrac{\mathrm{d}u}{\mathrm{d}x} - u\dfrac{\mathrm{d}v}{\mathrm{d}x}}{v^2}$$

Any of these results may be quoted, unless their derivation is asked for.

MULTIPLE CHOICE EXERCISE 8

(Instructions for answering these questions are given on page xii.)

TYPE I

1) $\dfrac{\mathrm{d}}{\mathrm{d}x}(e^{x^2+1})$ is:

(a) $2x$ (b) $2x\,e^{x^2+1}$ (c) $2x\,e^{2x}$ (d) $(x^2 + 1)\,e^{x^2}$

(e) $\ln(x^2 + 1)\,e^{x^2+1}$.

2) If $x^2 + y^2 = 4$ then $\dfrac{dy}{dx}$ is:

(a) $2x + 2y$ (b) $4 - x^2$ (c) $-\dfrac{x}{y}$ (d) $\dfrac{y}{x}$ (e) $\dfrac{4 - x}{y}$.

3) If $y = \cos x + \sin x,$ $\dfrac{d^2y}{dx^2}$ is:

(a) $\cos x - \sin x$ (b) $-y$ (c) $\cos 2x$ (d) y^2 (e) $\cos x + \sin x$.

4) $\dfrac{d}{dx}\left(\dfrac{1}{1 + x}\right)$ is:

(a) $\dfrac{-1}{(1 + x)^2}$ (b) $\dfrac{1}{1 - x}$ (c) $\ln(1 + x)$ (d) $\dfrac{-1}{1 + x^2}$ (e) 1.

5) $\dfrac{d}{dx}\ln\left(\dfrac{x + 1}{2x}\right)$ is:

(a) $\dfrac{1}{2}$ (b) $\dfrac{1}{x + 1} - \dfrac{1}{2x}$ (c) $\dfrac{2x}{x + 1}$ (d) $\dfrac{1}{x + 1} + \dfrac{1}{x}$ (e) $\dfrac{1}{x + 1} - \dfrac{1}{x}$.

6) $\dfrac{d}{dx}a^x$ is:

(a) xa^{x-1} (b) a^x (c) $x \ln a$ (d) $a^x \ln a$ (e) none of these.

7) If $x = \cos \theta$ and $y = \cos \theta + \sin \theta,$ $\dfrac{dy}{dx}$ is:

(a) $1 - \cot \theta$ (b) $1 - \tan \theta$ (c) $\cot \theta - 1$ (d) $\cot \theta + 1$

(e) $\dfrac{1}{1 - \cot \theta}$.

8) If $e^x = 7.3,$ the value of x, to 2 s.f. is:

(a) 2.0 (b) 1.0 (c) -2.0 (d) 0 (e) 0.2.

9) $\dfrac{d}{dx}(\ln 2x^{\frac{1}{3}})$ is:

(a) $2x^{-\frac{1}{3}}$ (b) $\dfrac{1}{3x}$ (c) $\dfrac{2}{3x}$ (d) $\dfrac{x}{3}$ (e) $\dfrac{2}{3x^{2/3}}$.

10) If $f: x \to x^2 - x$ and $g: x \to \sqrt{x},$ fg is the function defined by:

(a) $fg: x \to \sqrt{x^2 - x}$ (b) $fg: x \to x - \sqrt{x}$ (c) $fg: x \to x - x^2$

(d) $fg: x \to \sqrt{x} - x^2$

11) If $y = x^2 - 4$, the graph of the curve $y^2 - 4 = x$ is:

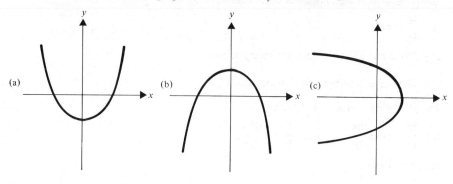

(a)   (b)   (c)

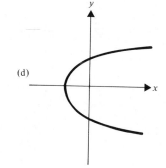
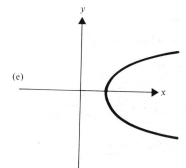

(d)   (e)

TYPE II

12) $f(x) \equiv \dfrac{u}{v}$ where u and v are both functions of x.

(a) $\ln [f(x)] \equiv \ln u - \ln v$.

(b) $\dfrac{d}{dx}[f(x)] = \dfrac{1}{v}\dfrac{du}{dx} - \dfrac{u}{v^2}\dfrac{dv}{dx}$ (c) $e^{f(x)} \equiv e^u - e^v$.

13) $x = f(y)$.

(a) $\dfrac{dy}{dx} = f'(y)$ (b) $\dfrac{d}{dx}[f(y)] = f'(y)\dfrac{dy}{dx}$ (c) $\dfrac{d^2y}{dx^2} = \left(\dfrac{d^2}{dy^2}f(y)\right)\left(\dfrac{dy}{dx}\right)$

14) $xy = e^x$.

(a) $\dfrac{d^2y}{dx^2} + 2\dfrac{dy}{dx} - y = 0$.

(b) $\dfrac{dy}{dx} = \dfrac{1}{x}(e^x - y)$.   (c) $\ln y = x - \ln x$.

15) The parametric equations of a curve are $x = t^2$, $y = t^3$.

(a) $\dfrac{dy}{dx} = 6t^3$.

(b) The curve has only one stationary value.

(c) The curve is symmetrical about the x-axis.

16) $y = g(u)$ and $u = f(x)$.

(a) $\dfrac{dy}{dx} = g'(u)\dfrac{du}{dx}$ (b) $\dfrac{d^2y}{dx^2} = \dfrac{d^2u}{dx^2}$.

(c) $y = fg(x)$.

TYPE III

17) (a) $f(x) \equiv e^x \cos x$

   (b) $f(-x) \equiv \dfrac{\cos x}{e^x}$

18) (a) $f(x) \equiv \dfrac{3x}{(x-2)(x+1)}$

   (b) $f(x) \equiv \dfrac{2}{x-2} + \dfrac{1}{x+1}$

19) (a) $y = \ln 3x$

   (b) $\dfrac{dy}{dx} = \dfrac{1}{x}$

20) (a) $y = \sin\theta$, $x = \sin 2\theta$.

   (b) $x^2 + y^2 = 1$.

21) (a) y has a maximum value when $t = 2$.

   (b) $\dfrac{dy}{dx} = 4 - t^2$.

TYPE IV

22) If $x = a\cos\theta$ and $y = b\sin\theta$, find a and b.

(a) $x = 2$ when $y = 3$ (b) $\dfrac{dy}{dx} = 1$ when $\theta = \dfrac{\pi}{4}$

(c) $-5 \leqslant x \leqslant 5$ for all values of θ.

23) Does $y = ae^x$ satisfy the differential equation $\dfrac{d^2y}{dx^2} + k\dfrac{dy}{dx} + cy = 0$?

(a) $k = -2$ (b) $c = 1$ (c) $a = 2$.

24) Find the maximum value of $y = f(x)$.

(a) $\dfrac{d^2y}{dx^2} < 0$ for all values of x (b) $\dfrac{dy}{dx} = ax^3$

(c) the curve $y = f(x)$ passes through the origin.

25) Sketch the curve whose parametric equations are $x = f(t), \ y = g(t)$.
(a) $f(t)$ is a quadratic function of t.
(b) $g(t)$ is a linear function of t.
(c) The curve passes through the points $(1, 0), (0, 1), (2, 4), (3, 5)$.

TYPE V

26) $\dfrac{d}{dx}(s \times t) = \dfrac{ds}{dx} \times \dfrac{dt}{dx}$.

27) $\dfrac{d}{dy}[f(x)] = f'(x) \Big/ \dfrac{dy}{dx}$.

28) $\dfrac{d}{dx}\left[\ln\dfrac{x}{1+x}\right] = \dfrac{d}{dx}[\ln x] - \dfrac{d}{dx}[\ln(1+x)]$.

29) $y = e^{x^2+1}$ has intercept 1 on the y-axis.

30) $f(\theta) \equiv \sin 2\theta$ has a maximum value of 2.

31) If $f(x) \equiv 1 + x$, $f^{-1}(x) \equiv \dfrac{1}{1+x}$.

32) If $f(x) \equiv \cos x$ and $g(x) \equiv x^2$, then $fg(x) \equiv \cos^2 x$.

33) If f is the function defined as $f: x \to \ln x$, $x \in \mathbb{R}^+$ the range of f is $f \in \mathbb{R}$.

MISCELLANEOUS EXERCISE 8

1) (a) If $y = e^{ax} \sin bx$, express $\dfrac{d^2y}{dx^2}$ in the form

$$e^{ax} (A \sin bx + B \cos bx)$$

giving A and B in terms of a and b.

(b) If $y = \dfrac{(1 + 2x)}{(1 - 2x)}$, find $\dfrac{d^2y}{dx^2}$ in its simplest form.

(c) If $x^2 - y^2 = a^2$, find $\dfrac{dy}{dx}$ and $\dfrac{d^2y}{dx^2}$ in terms of x and y.   (U of L)

2) Show that $\dfrac{d}{dx}\left(\dfrac{x}{1+x}\right) = \dfrac{1}{(1+x)^2}$.

A curve is described by the equation

$$\frac{y}{1+y} + \frac{x}{1+x} - x^2 y^3 = 0$$

Find the equation of the tangent to the curve at the point $(1, 1)$.   (JMB)

3) (a) Differentiate the following functions with respect to x simplifying your answers where possible:

(i) $\dfrac{1}{x^2}\sqrt{(1+x^3)}$ (ii) $\ln\left(\dfrac{2+\cos x}{3-\sin x}\right)$

(b) If $y = e^{3x}\sin 4x$ show that $\dfrac{d^2 y}{dx^2} - 6\dfrac{dy}{dx} + 25y = 0$.   (U of L)

4) (a) Find $\dfrac{dy}{dx}$ when:

(i) $y = \ln(\sec 2x + \tan 2x)$ (ii) $y = \dfrac{(1+2x^2)}{(1+x^2)}$

and simplify your answers.

(b) If $y = \cos\left(e^x + \dfrac{\pi}{4}\right)$, show that $\dfrac{d^2 y}{dx^2} = \dfrac{dy}{dx} - e^{2x}y$.

Find the least positive value of x for which y is a minimum.   (AEB)'76

5) Differentiate with respect to x:

(a) x^x (b) $\arctan\left(\dfrac{1-x}{1+x}\right)$

simplifying the results where possible.   (U of L)p

6) (a) For values of $x > 0$, the equation of a curve is $y = x\ln x$.
Find the coordinates of the turning point on this curve, and determine whether it is a maximum or a minimum.
Sketch the graph. (You may assume that $y \to 0$ as $x \to 0$.)

(b) A curve is given parametrically by

$$x = t - \frac{1}{t}, \quad y = t + \frac{1}{t}, \quad \text{where} \quad t \neq 0$$

Find the coordinates of the points on the curve where the gradient is zero, and find the equation of the tangent at the point where $t = 2$.   (C)

7) Show that the graph of $y = \dfrac{ax+b}{cx+d}$ has, in general, no turning points and that

$$2\left(\frac{dy}{dx}\right)\left(\frac{d^3 y}{dx^3}\right) = 3\left(\frac{d^2 y}{dx^2}\right)^2$$ (U of L)

8) Given that $x = \sec \theta + \tan \theta$ and $y = \operatorname{cosec} \theta + \cot \theta$, show that
$x + \dfrac{1}{x} = 2 \sec \theta$ and $y + \dfrac{1}{y} = 2 \operatorname{cosec} \theta$. Find $\dfrac{dx}{d\theta}$ and $\dfrac{dy}{d\theta}$ in terms of θ,

and hence show that $\dfrac{dy}{dx} = -\dfrac{1 + y^2}{1 + x^2}$.   (JMB)

9) (a) Find $\dfrac{dy}{dx}$ when $2y\, e^{3x} + \dfrac{1}{x^2} \sin 2x = 0$.

(b) If $x = \dfrac{1 + t}{1 - 2t}$ and $y = \dfrac{1 + 2t}{1 - t}$, where t is variable find the value

of $\dfrac{dy}{dx}$ when $t = 0$.

(c) Differentiate with respect to x, $f(x) \equiv 3x + \sin x - 8 \sin \tfrac{1}{2}x$ and
deduce that $f(x)$ is positive for $x > 0$.   (AEB)'72

10) Prove that $\dfrac{d}{dx} \arcsin x = \dfrac{1}{(1 - x^2)^{\frac{1}{2}}}$.

Given that the variables x and y satisfy the equation

$$\arcsin 2x + \arcsin y + \arcsin (xy) = 0$$

find $\dfrac{dy}{dx}$ when $x = y = 0$.   (JMB)

11) Sketch the curves with the equations $y = e^x$ and $y = x^2 - 1$ on the
same diagram.
Using the information gained from your sketches, redraw part of the curves
more accurately to find the negative real root of the equation $e^x = x^2 - 1$
to two decimal places.   (U of L)

12) (a) If $y = \sqrt{(5x^2 + 3)}$, show that $y \dfrac{d^2y}{dx^2} + \left(\dfrac{dy}{dx}\right)^2 = 5$.

(b) The parametric equations of a curve are

$$x = 3(2\theta - \sin 2\theta)$$
$$y = 3(1 - \cos 2\theta)$$

The tangent and the normal to the curve at the point P where $\theta = \dfrac{\pi}{4}$
meet the y-axis at L and M respectively. Show that the area of triangle
PLM is $\dfrac{9}{4}(\pi - 2)^2$.   (AEB)'74

13) Express the function $y = \dfrac{2x^2}{(2x - 1)(x + 1)}$ in partial fractions. Hence, or

otherwise, find the value of $\dfrac{d^2y}{dx^2}$ when $x = 1$.   (JMB)

14) (a) Differentiate $\ln(k \sec x) + a^x$ with respect to x, a and k being
constants.

(b) If $x = \sin t$ and $y = \cos 2t$, prove that $\dfrac{d^2y}{dx^2} + 4 = 0$.

(c) Find the maximum and minimum values of $\dfrac{x-3}{x^2-x-2}$ and distinguish between them.   (U of L)

15) (a) If $x^2 y = a \cos nx$, show that
$$x^2 \frac{d^2y}{dx^2} + 4x \frac{dy}{dx} + (n^2 x^2 + 2)y = 0.$$

(b) A curve is given by the parametric equations
$$x = t^2 + 3, \quad y = t(t^2 + 3).$$
   (i) Show that the curve is symmetrical about the x-axis.
   (ii) Show that there is no part of the curve for which $x < 3$.

   (iii) Find $\dfrac{dy}{dx}$ in terms of t, and show that $\left(\dfrac{dy}{dx}\right)^2 \geqslant 9$.

Using the results (i), (ii), and (iii), sketch the curve.   (C)

16) (a) Differentiate with respect to x:
   (i) $\ln [x + \sqrt{(x^2 + 1)}]$
   (ii) $\sec^2 2x$
   (iii) 10^{3x}
simplifying your answer to (a).

(b) If $x^2 + y^2 = 2y$, find $\dfrac{dy}{dx}$ in terms of x and y without first

finding y in terms of x. Prove that $\dfrac{d^2y}{dx^2} = \dfrac{1}{(1-y)^3}$.   (U of L)

17) The point P moves in such a way that at time t its Cartesian coordinates with respect to an origin O are $x = e^{-t}$, $y = 2t\,e^{-t}$.
The distance OP is denoted by r and the angle between OP and the x-axis by θ.
Find in terms of t:
(a) the rate of change of r^2 with respect to t,
(b) the rate of change of θ with respect to t.   (JMB)

18) If $y = x \arctan x$, show that:

(a) $x(1 + x^2) \dfrac{dy}{dx} = x^2 + (1 + x^2)y$

(b) $(1 + x^2) \dfrac{d^2y}{dx^2} + 2x \dfrac{dy}{dx} - 2y = 2.$

Draw a rough sketch of the curve.   (U of L)p

19) If $\tan y = x$ find the value of $\dfrac{d^2y}{dx^2}$ when $y = \dfrac{\pi}{4}$.   (U of L)p

20) (a) Differentiate with respect to x:

(i) $x^2 \sin 3x$

(ii) $e^{-2/x}$

(iii) $\left\{ \dfrac{x-1}{2-x} \right\}^2$

(b) Given that $y = \ln (1 + \sin x)$, prove that $\dfrac{d^2y}{dx^2} + e^{-y} = 0$. What

can be deduced from the equation about all the stationary values of y?

(AEB)'74

21) Given that u and v are functions of x, prove from first principles that

$$\frac{d}{dx} \left(\frac{u}{v} \right) = \frac{v \, du/dx - u \, dv/dx}{v^2}$$

Find, in a simplified form, the derivative of the function

$$\frac{2 + \ln (1 + x)^2}{2 - \ln (1 - x)^2} \qquad \text{(JMB)}$$

22) If $x = a(\theta - \sin \theta)$, $y = a(1 - \cos \theta)$, show that $\dfrac{dy}{dx} = \cot \tfrac{1}{2}\theta$. As θ
varies, the point $P(x, y)$ traces out a curve. When $\theta = \tfrac{1}{2}\pi$, P is at the point A
and when $\theta = \tfrac{3}{2}\pi$, P is at the point B. Find the coordinates of the points A
and B and the equations of the tangents to the curve at these two points.

(U of L)

23) Given that $-1 < x < 1$, that $0 < \arccos x < \pi$ and that $(1 - x^2)^{\frac{1}{2}}$
denotes the positive square root of $1 - x^2$, find the derivative of the function

$$f(x) \equiv \arccos x - x(1 - x^2)^{\frac{1}{2}},$$

expressing your answer as simply as possible.
Prove that, as x increases in the interval $-1 < x < 1$, $f(x)$ decreases, and
sketch the graph of $f(x)$ in this interval. (JMB)

24) Two functions are defined on the domain $0 \leqslant x \leqslant \pi$ by

$$f : x \to \sin x \quad \text{and} \quad g : x \to \cos x.$$

Explain why one of these functions has an inverse while the other does not.
When the domain is restricted to $0 \leqslant x \leqslant \tfrac{1}{2}\pi$, calculate $f^{-1}g(\tfrac{1}{3}\pi)$. (C)

CHAPTER 9

INTEGRATION

INDEFINITE INTEGRATION

When x^2 is differentiated with respect to x the derived function is $2x$. Conversely, given that an unknown function has a derived function of $2x$, it is clear that the unknown function could be x^2. This process of finding a function from its derived function is called *integration* and it reverses the operation of differentiation.

The Constant of Integration

Now consider the functions $x^2 + 1$ and $x^2 - 7$.

We have already noted that $\dfrac{d}{dx}(x^2) = 2x$

But we also see that

$$\frac{d}{dx}(x^2 + 1) = 2x \quad \text{and} \quad \frac{d}{dx}(x^2 - 7) = 2x$$

Clearly $2x$ is the derivative, not only of x^2, but also of x^2 *plus any constant.* Thus the result of integrating $2x$, which is called the *integral* of $2x$, is not a unique function but is of the form $x^2 + K$ where K is an arbitrary constant called *the constant of integration.*

This is written

$$\int 2x \, dx = x^2 + K$$

where $\int \ldots dx$ means 'the integral of $\ldots$ w.r.t. x'.

As integration reverses the process of differentiation, for a function $f(x)$ we have

$$\int \frac{d}{dx} f(x)\, dx = f(x) + K$$

Similarly, since differentiating x^3 w.r.t. x gives $3x^2$, we can say

$$\int 3x^2\, dx = x^3 + K$$

or

$$\int x^2\, dx = \tfrac{1}{3}x^3 + K$$

(It is not necessary to write $K/3$ in the second form as K represents *any* constant in either expression.)

In general, since differentiating x^{n+1} w.r.t. x gives $(n+1)x^n$, we have

$$\int x^n\, dx = \frac{1}{(n+1)} x^{n+1} + K$$

i.e. to integrate a power of x, *increase* the power by 1 and divide by the *new* power.

This rule can be used to integrate x^n for any value of n *except* -1 (which will be considered later),

e.g.

$$\int x^7\, dx = \tfrac{1}{8}x^8 + K$$

$$\int x^{-4}\, dx = -\tfrac{1}{3}x^{-3} + K$$

$$\int x^{\frac{1}{2}}\, dx = \tfrac{2}{3}x^{\frac{3}{2}} + K$$

Integrating a Constant, c

The result of differentiating cx is c.

Thus

$$\int c\, dx = cx + K$$

Integrating cx^n

Differentiating cx^{n+1} gives $(n+1)(c)x^n$.

Hence

$$\int (n+1)(c)x^n\, dx = cx^{n+1} + K$$

or

$$\int cx^n\, dx = \frac{c}{(n+1)} x^{n+1} + K$$

Integrating a Sum or Difference of Functions

It was shown in Chapter 5 that differentiation is a distributive process across addition or subtraction. So, since integration reverses the differential operation, integration also is distributive in this respect,

e.g. $$\int \left(x^3 + \frac{1}{x^2} + \sqrt{x} \right) dx = \int (x^3 + x^{-2} + x^{\frac{1}{2}}) \, dx$$

$$= \int x^3 \, dx + \int x^{-2} \, dx + \int x^{\frac{1}{2}} \, dx$$

$$= \tfrac{1}{4} x^4 - x^{-1} + \tfrac{2}{3} x^{\frac{3}{2}} + K$$

EXERCISE 9a

Integrate the following functions with respect to x.

1) x^6; $x^{\frac{1}{3}}$; x^{-4}; $x^{-\frac{3}{2}}$; $\sqrt[4]{x}$; $\dfrac{1}{x^7}$

2) $2x^2 - \dfrac{1}{x^2} + x$ 3) $\sqrt{x} + \dfrac{1}{\sqrt[3]{x}}$ 4) $3x^3 + x^{-3} + 3$

5) $\dfrac{4}{x^3} - \dfrac{1}{x^2} - x^2$ 6) $2x^{\frac{5}{2}} - x^{-\frac{2}{5}}$ 7) $5x^4 - 3x^2 + 7$

8) $4x^{-3} + x^{-4} + 1$ 9) $3x^{-\frac{1}{2}} - x^{-\frac{3}{2}}$ 10) $\tfrac{1}{2} x - \dfrac{2}{\sqrt{x}} - 1$

11) $\dfrac{1}{x^4} + \dfrac{1}{\sqrt[4]{x}} - 4$ 12) $6\sqrt{x} - 3x^3 + x^{-2} + 2$

Integrating $(ax + b)^n$

First consider the function $f(x) \equiv (2x + 3)^4$. The derivative of $f(x)$ is found by using the substitution $u \equiv 2x + 3$ so that $f(x) \equiv u^4$, giving

$$\frac{d}{dx} (2x + 3)^4 = (4)(2)(2x + 3)^3$$

Conversely $$\int (4)(2)(2x + 3)^3 \, dx = (2x + 3)^4 + K$$

or $$\int (2x + 3)^3 \, dx = \frac{1}{(4)(2)} (2x + 3)^4 + K$$

Similarly considering the general case where $f(x) \equiv (ax + b)^{n+1}$ we find that

$$\frac{d}{dx} (ax + b)^{n+1} = (n + 1)(a)(ax + b)^n$$

Hence
$$\int (ax + b)^n \, dx = \frac{1}{(n+1)(a)} (ax + b)^{n+1} + K$$

e.g.
$$\int (3x - 8)^6 \, dx = \frac{1}{(7)(3)} (3x - 8)^7 + K$$

$$\int (2 - 5x)^9 \, dx = \frac{1}{(10)(-5)} (2 - 5x)^{10} + K$$

$$\int (\tfrac{1}{2} - \tfrac{1}{3}x)^7 \, dx = \frac{1}{(8)(-\tfrac{1}{3})} (\tfrac{1}{2} - \tfrac{1}{3}x)^8 + K$$

EXERCISE 9b

Integrate the following functions with respect to x.

1) $3;\ 3x^5;\ (3x)^5;\ 3(x+1)^5$

2) $4x^3 - 5x + 6$ 3) $(2x - 1)^2$ 4) $(2 + 7x)^3$

5) $4(1 - x)^{\frac{1}{2}}$ 6) $3\sqrt{2 - 5x}$ 7) $\dfrac{1}{(4x + 5)^3}$

8) $\dfrac{1}{\sqrt{(1 - 2x)}}$ 9) $\dfrac{1}{\sqrt[3]{(3 - 7x)}}$ 10) $\dfrac{3}{(2x + 1)^3} + \sqrt{1 - 2x}$

11) $4\sqrt{x} + \sqrt{4x + 1} - 4(1 - 3x)^3$ 12) $(px + q)^r$

13) $\dfrac{1}{\sqrt[4]{(3 + 5x)}}$ 14) $\dfrac{3}{\sqrt[3]{(4 - 5x)}}$ 15) $\sqrt{(1 - x)} + \dfrac{1}{\sqrt{(1 - x)}} - \dfrac{1}{(1 - x)^2}$

INTEGRATION OF TRIGONOMETRIC FUNCTIONS

Whenever a function $f'(x)$ is *recognised* as the derivative of a function $f(x)$, then

$$\frac{d}{dx} f(x) = f'(x) \Leftrightarrow \int f'(x) \, dx = f(x) + K$$

Thus, referring to the derivatives of trig functions derived in Chapter 8, we see that

$$\frac{d}{dx} (\sin x) = \cos x \Leftrightarrow \int \cos x \, dx = \sin x + K$$

$$\frac{d}{dx} (\cos x) = -\sin x \Leftrightarrow \int \sin x \, dx = -\cos x + K$$

$$\frac{d}{dx} (\tan x) = \sec^2 x \Leftrightarrow \int \sec^2 x \, dx = \tan x + K$$

$$\frac{d}{dx} (\sec x) = \sec x \tan x \iff \int \sec x \tan x \, dx = \sec x + K$$

$$\frac{d}{dx} (\operatorname{cosec} x) = -\operatorname{cosec} x \cot x \iff \int \operatorname{cosec} x \cot x \, dx = -\operatorname{cosec} x + K$$

$$\frac{d}{dx} (\cot x) = -\operatorname{cosec}^2 x \iff \int \operatorname{cosec}^2 x \, dx = -\cot x + K$$

(The reader should not regard these integrals as six more facts to be memorised. Knowledge of the standard derivatives is clearly sufficient.)
It can be shown, by considering the corresponding differentiation, that

$$\int c \cos x \, dx = c \sin x + K$$

and
$$\int \cos (ax + b) \, dx = \frac{1}{a} \sin (ax + b) + K$$

with similar results for the remaining trig integrals.

e.g.
$$\int 3 \sec^2 x \, dx = 3 \tan x + K$$

$$\int \sin 4\theta \, d\theta = -\tfrac{1}{4} \cos 4\theta + K$$

$$\int \operatorname{cosec}^2 (2x + 3\pi/4) \, dx = -\tfrac{1}{2} \cot (2x + 3\pi/4) + K$$

$$\int \operatorname{cosec} 5\theta \cot 5\theta \, d\theta = -\tfrac{1}{5} \operatorname{cosec} 5\theta + K$$

$$\int \sec (\pi/2 - 6x) \tan (\pi/2 - 6x) \, dx = -\tfrac{1}{6} \sec (\pi/2 - 6x) + K$$

EXERCISE 9c

Integrate the following expressions with respect to x.

1) $\sin 2x$ 2) $3 \cos (4x - \pi/2)$ 3) $\sec^2 (\pi/3 - 2x)$

4) $\operatorname{cosec}^2 (3x + \pi/6)$ 5) $5 \sec (\pi/4 - x) \tan (\pi/4 - x)$

6) $2 \operatorname{cosec} 3x \cot 3x$ 7) $2 \sin (3x + \alpha)$

8) $5 \cos (\alpha - x/2)$ 9) $\cos 3x + 3 \sin x$

10) $\sec^2 2x - \operatorname{cosec}^2 4x$

INTEGRATION OF EXPONENTIAL FUNCTIONS

It is already known that $\dfrac{d}{dx} e^x = e^x$,

hence

$$\int e^x \, dx = e^x + K$$

Further, we have $\dfrac{d}{dx} (ce^x) = ce^x$

and $\dfrac{d}{dx} e^{(ax+b)} = ae^{(ax+b)}$

Hence

$$\int ce^x \, dx = ce^x + K$$

and

$$\int e^{ax+b} \, dx = \frac{1}{a} e^{(ax+b)} + K$$

e.g.

$$\int e^{3x} \, dx = \tfrac{1}{3} e^{3x} + K$$

$$\int 2e^{-5x} \, dx = (2)(-\tfrac{1}{5}) e^{-5x} + K$$

To integrate an exponential function in which the given base is not e but is a, say, the base must first be changed to e, as follows.

Let $a^x \equiv e^z$

Taking logs to the base e gives

$$x \ln a \equiv z$$

Hence $a^x \equiv e^{x \ln a}$

Then $\displaystyle\int a^x \, dx \equiv \int e^{x \ln a} \, dx$

$$= \frac{1}{\ln a} e^{x \ln a} + K$$

i.e.

$$\int a^x \, dx = \frac{1}{\ln a} a^x + K$$

Alternatively, remembering that

$$\frac{d}{dx} (a^x) = (\ln a) a^x$$

it follows that

$$\frac{1}{\ln a} \frac{d}{dx} (a^x) = a^x \Leftrightarrow \int a^x \, dx = \frac{a^x}{\ln a} + K$$

EXERCISE 9d

Integrate the following expressions with respect to x.

1) e^{2x}

2) $3e^{-x}$

3) e^{4x+1}

4) $4e^{5-3x}$

5) 2^x

6) 3^{2x}

7) 3^{1-x}

8) $\sqrt{e^x}$

9) $e^{2x} + \dfrac{1}{e^{2x}}$

10) $3e^{-3x} - \tfrac{1}{2}e^{2x}$

Integration of $\dfrac{1}{x}$

(a) $x > 0$

At first sight it may appear that, since $1/x \equiv x^{-1}$, integrating $1/x$ requires use of the rule $\displaystyle\int x^n \, dx = \dfrac{1}{n+1} x^{n+1} + K$. But when $n = -1$, this method fails.

Taking a second look at $1/x$ it can be *recognised* as the derived function of $\ln x$. But $\ln x$ is defined only when $x > 0$. Hence, provided $x > 0$ we have

$$\frac{d}{dx} \ln x = \frac{1}{x} \; \Leftrightarrow \; \int \frac{1}{x} \, dx = \ln x + K$$

(b) $x < 0$

If x is negative the statement that $\displaystyle\int \frac{1}{x} \, dx = \ln x + K$ is not valid because the logarithm of a negative number does not exist. This problem is dealt with as follows

$$\int \frac{1}{x} \, dx = \int \frac{-1}{(-x)} \, dx \qquad \{-x > 0\}$$

$$= \ln(-x) + K$$

Thus

when $x > 0$ $\qquad \displaystyle\int \frac{1}{x} \, dx = \ln x + K$

when $x < 0$ $\qquad \displaystyle\int \frac{1}{x} \, dx = \ln(-x) + K$

Combining these results, for all values of x we have

$$\int \frac{1}{x} \, dx = \ln |x| + K$$

where $|x|$ denotes the positive magnitude of x,

e.g.
$$|-3| = 3; \qquad |-1| = 1$$

This result, being fully comprehensive, will be used in all future work of this nature in this book.

Note. If the constant K is replaced by $\ln A$, $(A > 0)$, which also represents any real number, we can say

$$\int \frac{1}{x} dx = \ln |x| + \ln A = \ln A |x|$$

Further,
$$\frac{d}{dx} [\ln x^c] = \frac{d}{dx} [c \ln x] = \frac{c}{x}$$

$\Leftrightarrow$
$$\int \frac{c}{x} dx = c \ln |x| + K$$

Now consider
$$\frac{d}{dx} \ln (ax + b) = \frac{a}{ax + b}$$

Hence
$$\int \frac{1}{ax + b} dx = \frac{1}{a} \ln |ax + b| + K$$

e.g.
$$\int \frac{4}{x} dx = 4 \ln |x| + K$$

$$\int \frac{1}{2x + 5} dx = \tfrac{1}{2} \ln |2x + 5| + K$$

$$\int \frac{3}{4 - 2x} dx = -\tfrac{3}{2} \ln |4 - 2x| + K$$

EXERCISE 9e

Integrate the following expressions with respect to x.

1) $\dfrac{1}{2x}$ 2) $\dfrac{2}{x}$ 3) $\dfrac{1}{3x + 1}$ 4) $\dfrac{1}{1 - 3x}$

5) $\dfrac{4}{1 + 2x}$ 6) $\dfrac{3}{4 - 2x}$ 7) $\dfrac{3x}{(x - 1)(x - 2)}$

(Hint. Use partial fractions.*)*

Questions 8–20 are miscellaneous examples of the types considered so far in this chapter.

8) $\dfrac{1}{(2 - 3x)^3}$ 9) $\dfrac{1}{\sqrt{2 - 3x}}$ 10) $\dfrac{1}{2 - 3x}$

11) $\sin (\pi/2 - 3x)$ 12) $(4x + 1)^2$ 13) x^2 14) 4^x

15) e^{4-5x} 16) $\sec^2 4x$ 17) $\dfrac{4}{1-x}$ 18) $\sqrt{2x+3}$

19) e^{6x} 20) $\dfrac{5}{6-7x}$

Integration of $\dfrac{1}{\sqrt{(1-x^2)}}$ **and** $\dfrac{1}{1+x^2}$

It was shown in Chapter 8 that

$$\frac{d}{dx}(\arcsin x) = \frac{1}{\sqrt{(1-x^2)}}$$

and

$$\frac{d}{dx}(\arctan x) = \frac{1}{1+x^2}$$

Thus

$$\int \frac{1}{\sqrt{(1-x^2)}}\, dx = \arcsin x + K$$

and

$$\int \frac{1}{1+x^2}\, dx = \arctan x + K$$

Further, as

$$\frac{d}{dx}(\arcsin ax) = \frac{a}{\sqrt{(1-a^2 x^2)}}$$

and

$$\frac{d}{dx}(\arctan ax) = \frac{a}{1+a^2 x^2}$$

we have

$$\int \frac{1}{\sqrt{(1-a^2 x^2)}}\, dx = \frac{1}{a}\arcsin ax + K$$

and

$$\int \frac{1}{1+a^2 x^2}\, dx = \frac{1}{a}\arctan ax + K$$

A method for dealing with further examples of this type is given later in this chapter. For the moment it is sufficient to be able to integrate only straightforward examples.

EXERCISE 9f

Integrate the following expressions with respect to x.

1) $\dfrac{-1}{\sqrt{(1-x^2)}}$ 2) $\dfrac{2}{1+4x^2}$ 3) $\dfrac{3}{\sqrt{(1-9x^2)}}$ 4) $\dfrac{3}{1+x^2}$

5) $\dfrac{1}{1+16x^2}$ 6) $\dfrac{6}{\sqrt{(1-4x^2)}}$ 7) $\dfrac{1}{4+x^2}$ 8) $\dfrac{1}{\sqrt{(4-x^2)}}$

(*Hint.* In Questions 7 and 8, take 4 as a common factor of the denominator.)

THE RECOGNITION ASPECT OF INTEGRATION

The ability to integrate a given function often depends primarily on *recognising it as a derived function*. Such identification applied to certain groups of functions leads to simple integration rules, as we have seen in the preceding paragraphs.

It is equally important to use the recognition process to avoid serious errors in integration. Consider, for instance, the derived function of the product $\dfrac{x^2}{2} \cos x$.

Using

$$\frac{d}{dx}(uv) = v\frac{du}{dx} + u\frac{dv}{dx}$$

gives

$$\frac{d}{dx}\left(\frac{x^2}{2}\cos x\right) = x\cos x - \frac{x^2}{2}\sin x$$

Clearly the derived function is not a simple product.

Conversely the integral of a simple product is not itself a product,

i.e. integration is not distributive when applied to a product.

INTEGRATING PRODUCTS

Consider the function e^u where u is a function of x. Differentiating this function of a function gives

$$\frac{d}{dx}(e^u) = \frac{du}{dx}e^u$$

In this case the derived function *is* a product.

Thus any product of the form $\left(\dfrac{du}{dx}\right)e^u$ can be integrated by recognition, since

$$\int\left(\frac{du}{dx}\right)e^u\, dx = e^u + K$$

e.g.

$$\int 2x\, e^{x^2}\, dx = e^{x^2} + K \qquad (u \equiv x^2)$$

$$\int \cos x\, e^{\sin x}\, dx = e^{\sin x} + K \qquad (u \equiv \sin x)$$

$$\int x^2\, e^{x^3}\, dx = \tfrac{1}{3}\int 3x^2\, e^{x^3}\, dx = \tfrac{1}{3}e^{x^3} + K \qquad (u \equiv x^3)$$

In these simple examples, the method of integration uses a *mental* change of variable (from x to u). A similar approach can be made to the integration of similar, but slightly less simple, functions.

Changing the Variable

Let us consider a more general function $f(u)$ where u is a function of x, so that

$$\frac{d}{dx} f(u) = \frac{du}{dx} f'(u) \qquad\qquad \text{where } f'(u) \text{ denotes } \frac{d}{du} f(u).$$

From this we see that any product of the form $\left(\dfrac{du}{dx}\right) f'(u)$ can be integrated using

$$\int \left(\frac{du}{dx}\right) f'(u)\, dx = f(u) + K \qquad\qquad [1]$$

But

$$f'(u) \equiv \frac{d}{du} f(u)$$

so

$$\int f'(u)\, du = f(u) + K \qquad\qquad [2]$$

Comparing [1] and [2] gives

$$\int \frac{du}{dx} f'(u)\, dx = \int f'(u)\, du$$

or

$$\int f'(u) \frac{du}{dx}\, dx = \int f'(u)\, du$$

i.e.

$$\int \ldots \frac{du}{dx}\, dx = \int \ldots du$$

Thus $\dfrac{du}{dx}\, dx$ and du are equivalent operators (i.e. give the same results when applied to the same function).

In practice such an equivalence can be found more simply.
Suppose we have to integrate $2x(x^2 + 3)^7$ w.r.t. x.
Making the substitution $u \equiv x^2 + 3$ we have

$$\frac{du}{dx} = 2x$$

But

$$\ldots \frac{du}{dx}\, dx \equiv \ldots du$$

So

$$\ldots 2x\, dx \equiv \ldots du$$

This pair of equivalent operators can be obtained directly from

$$\frac{du}{dx} = 2x$$

if we *separate the variables*, i.e. treat $\dfrac{du}{dx}$ as though it were a fraction, giving

$$\ldots du \equiv \ldots 2x\, dx$$

Note that this is not an equation or an identity; it is an *equivalence of operators*. Certain functions can be integrated in this way by choosing a suitable substitution (i.e. change of variable). Products in which one factor is basically the derivative of the function in the other factor, respond to this approach.

EXAMPLES 9g

1) Integrate $x^2\sqrt{x^3 + 5}$ w.r.t. x.

Let $\qquad\qquad u \equiv x^3 + 5 \ \Rightarrow\ \dfrac{du}{dx} = 3x^2$

or $\qquad\qquad\qquad\qquad \ldots du \equiv \ldots 3x^2\, dx$

Thus $\qquad\qquad \int x^2\sqrt{x^3 + 5}\ dx \equiv \tfrac{1}{3}\int (x^3 + 5)^{\frac{1}{2}} 3x^2\, dx$

$$\equiv \tfrac{1}{3}\int u^{\frac{1}{2}}\, du$$

$$= (\tfrac{1}{3})(\tfrac{2}{3})u^{\frac{3}{2}} + K$$

i.e. $\qquad\qquad \int x^2\sqrt{x^3 + 5}\ dx = \tfrac{2}{9}(x^3 + 5)^{\frac{3}{2}} + K$

2) Find $\displaystyle\int \cos x \sin^3 x\, dx$.

Writing the given integral in the form

$$\int \cos x\, (\sin x)^3\, dx,$$

we see that a suitable substitution is

$$u \equiv \sin x \ \Rightarrow\ \ldots du \equiv \ldots \cos x\, dx$$

Thus $\qquad\qquad \int \cos x \sin^3 x\, dx \equiv \int (\sin x)^3 \cos x\, dx$

$$\equiv \int u^3\, du$$

$$= \frac{u^4}{4} + K$$

i.e. $\qquad\qquad \int \cos x \sin^3 x\, dx = \tfrac{1}{4}\sin^4 x + K$

Note. The method used above also shows that, in general

$$\int \cos \theta \, \sin^n\theta \, d\theta \; = \; \frac{1}{n+1} \, \sin^{n+1} \theta + K$$

Similarly we can show that

$$\int \sin \theta \, \cos^n\theta \, d\theta \; = \; \frac{-1}{n+1} \, \cos^{n+1} \theta + K$$

3) Find $\int \dfrac{\ln x}{x} \, dx$

As

$$\int \frac{\ln x}{x} \, dx \equiv \int \left(\frac{1}{x}\right) \ln x \, dx$$

we see that a suitable change of variable is

$$u \equiv \ln x \quad \Rightarrow \quad \dots du \equiv \dots \frac{1}{x} \, dx$$

Thus

$$\int \frac{1}{x} \ln x \, dx \equiv \int u \, du$$

$$= \frac{u^2}{2} + K$$

i.e.

$$\int \frac{\ln x}{x} \, dx \; = \; \tfrac{1}{2}(\ln x)^2 + K$$

Note. $(\ln x)^2$ is *not* the same as $\ln x^2$.

The reader may well find that, after some practice, certain problems where a change of variable is appropriate can be integrated at sight.

EXERCISE 9g

Integrate the following expressions with respect to x.

1) $4x^3 \, e^{x^4}$ 2) $\sin x \, e^{\cos x}$ 3) $\sec^2 x \, e^{\tan x}$

4) $(2x + 1) \, e^{(x^2 + x)}$ 5) $\csc^2 x \, e^{(1 - \cot x)}$

Find the following integrals by making the substitution suggested.

6) $\displaystyle\int x(x^2 - 3)^4 \, dx$ $u \equiv x^2 - 3$

7) $\displaystyle\int x\sqrt{1 - x^2} \, dx$ $u \equiv 1 - x^2$

8) $\displaystyle\int \cos 2x(\sin 2x + 3)^2 \, dx$ $u \equiv \sin 2x + 3$

9) $\int x^2(1-x^3)\,dx$ $u \equiv 1 - x^3$

10) $\int e^x\sqrt{1+e^x}\,dx$ $u \equiv 1 + e^x$

11) $\int \cos x \sin^4 x\,dx$ $u \equiv \sin x$

12) $\int \sec^2 x \tan^3 x\,dx$ $u \equiv \tan x$

13) $\int x^n(1+x^{n+1})^2\,dx$ $u \equiv 1 + x^{n+1}$

14) $\int \csc^2 x \cot^2 x\,dx$ $u \equiv \cot x$

15) $\int \sqrt{x}\sqrt{(1+x^{\frac{3}{2}})}\,dx$ $u \equiv 1 + x^{\frac{3}{2}}$

By using a suitable substitution, or by integrating at sight, find

16) $\int x^3(x^4+4)^2\,dx$ 17) $\int e^x(1-e^x)^3\,dx$

18) $\int \sin\theta\sqrt{1-\cos\theta}\,d\theta$ 19) $\int (x+1)\sqrt{x^2+2x+3}\,dx$

20) $\int x\,e^{x^2+1}\,dx$

FURTHER INTEGRATION OF PRODUCTS

Integration by Parts

Many products cannot be expressed in the form $\dfrac{du}{dx}f'(u)$ and so cannot be integrated by the previous method.

A different approach being needed, we look again at the differentiation of a product uv where u and v are both functions of x,

i.e. $\dfrac{d}{dx}(uv) = v\dfrac{du}{dx} + u\dfrac{dv}{dx}$

Isolating one of the products on the R.H.S. gives

$$v\frac{du}{dx} = \frac{d}{dx}(uv) - u\frac{dv}{dx}$$

Now $v\dfrac{du}{dx}$ can be taken to represent a product which is to be integrated w.r.t. x.

Thus
$$\int v \frac{du}{dx} \, dx = \int \frac{d}{dx}(uv) \, dx - \int u \frac{dv}{dx} \, dx$$

i.e.
$$\int v \frac{du}{dx} \, dx = uv - \int u \frac{dv}{dx} \, dx$$

Integrating a product by using this formula is called integrating by parts. Care must be exercised in the choice of the factor to be replaced by v. The aim must be to ensure that $u \dfrac{dv}{dx}$ is simpler to integrate than $v \dfrac{du}{dx}$.

EXAMPLES 9h

1) Integrate $x \, e^x$ w.r.t. x.

Let
$$\begin{cases} v = x \\ \dfrac{du}{dx} = e^x \end{cases} \Rightarrow \begin{cases} \dfrac{dv}{dx} = 1 \\ u = e^x \end{cases}$$

Using
$$\int v \frac{du}{dx} \, dx = uv - \int u \frac{dv}{dx} \, dx$$

gives
$$\int x \, e^x \, dx = (e^x)(x) - \int (e^x)(1) \, dx$$

$$= x \, e^x - e^x + K$$

Hence
$$\int x \, e^x \, dx = e^x(x - 1) + K$$

2) Find $\int x^2 \sin x \, dx$.

Let
$$\begin{cases} v = x^2 \\ \dfrac{du}{dx} = \sin x \end{cases} \Rightarrow \begin{cases} \dfrac{dv}{dx} = 2x \\ u = -\cos x \end{cases}$$

Then
$$\int v \frac{du}{dx} \, dx = uv - \int u \frac{dv}{dx} \, dx$$

gives
$$\int x^2 \sin x \, dx = (-\cos x)(x^2) - \int (-\cos x)(2x) \, dx$$

$$= -x^2 \cos x + 2 \int x \cos x \, dx$$

At this stage the integral on the R.H.S. cannot be found without *repeating* the process of integrating by parts.

Thus, for $\int x \cos x \, dx$,

let
$$\begin{cases} v = x \\ \dfrac{du}{dx} = \cos x \end{cases} \Rightarrow \begin{cases} \dfrac{dv}{dx} = 1 \\ u = \sin x \end{cases}$$

giving
$$\int x \cos x \, dx = (\sin x)(x) - \int (\sin x)(1) \, dx$$
$$= x \sin x + \cos x + K$$

Hence
$$\int x^2 \sin x \, dx = -x^2 \cos x + 2x \sin x + 2 \cos x + K$$

3) Find $\int x^4 \ln x \, dx$.

Because $\ln x$ can be differentiated but *not* integrated, we are obliged to take $v = \ln x$.

Thus, let
$$\begin{cases} v = \ln x \\ \dfrac{du}{dx} = x^4 \end{cases} \Rightarrow \begin{cases} \dfrac{dv}{dx} = \dfrac{1}{x} \\ u = \tfrac{1}{5}x^5 \end{cases}$$

The formula for integrating by parts then gives

$$\int x^4 \ln x \, dx = (\tfrac{1}{5}x^5)(\ln x) - \int (\tfrac{1}{5}x^5)(1/x) \, dx$$
$$= \tfrac{1}{5}x^5 \ln x - \tfrac{1}{5} \int x^4 \, dx$$

$\Rightarrow$
$$\int x^4 \ln x \, dx = \tfrac{1}{5}x^5 \ln x - \tfrac{1}{25}x^5 + K$$

Special Cases of Integration by Parts

An interesting situation arises when an attempt is made to integrate $e^x \cos x$.

Integrating by parts, let

$$\begin{cases} v = e^x \\ \dfrac{du}{dx} = \cos x \end{cases} \Rightarrow \begin{cases} \dfrac{dv}{dx} = e^x \\ u = \sin x \end{cases}$$

Hence
$$\int e^x \cos x \, dx = e^x \sin x - \int e^x \sin x \, dx \qquad [1]$$

But $\int e^x \sin x \, dx$ is very similar to $\int e^x \cos x \, dx$, so apparently we have made no progress.

However if we now apply integration by parts to $\int e^x \sin x \, dx$

we have

$$\begin{cases} v = e^x \\ \dfrac{du}{dx} = \sin x \end{cases} \Rightarrow \begin{cases} \dfrac{dv}{dx} = e^x \\ u = -\cos x \end{cases}$$

so that $\int e^x \sin x = -e^x \cos x + \int e^x \cos x \, dx$

or $\int e^x \cos x \, dx = e^x \cos x + \int e^x \sin x \, dx$ [2]

Adding [1] and [2] gives

$$2 \int e^x \cos x \, dx = e^x (\sin x + \cos x) + K$$

Clearly the same two equations also give

$$2 \int e^x \sin x \, dx = e^x (\sin x - \cos x) + K$$

(**Note.** Neither of the equations [1] and [2] contains a completed integration process, so the constant of integration is introduced when these two equations are combined.)

Integration of ln x

So far we have found no means of integrating ln x. But now, by regarding ln x as a product $(1)(\ln x)$, the method of integration by parts can be applied as follows:

Let

$$\begin{cases} v = \ln x \\ \dfrac{du}{dx} = 1 \end{cases} \Rightarrow \begin{cases} \dfrac{dv}{dx} = \dfrac{1}{x} \\ u = x \end{cases}$$

Then $\int v \dfrac{du}{dx} \, dx = uv - \int u \dfrac{dv}{dx} \, dx$

becomes $\int \ln x \, dx = x \ln x - \int x \left(\dfrac{1}{x} \right) dx$

$$= x \ln x - x + K$$

i.e. $\int \ln x \, dx = x (\ln x - 1) + K$

This 'trick' of multiplying a function by 1 to convert it into a product can also be used to integrate arcsin x and other inverse trig functions. (See Exercise 9i, Questions 25 to 27.)

EXERCISE 9h

Integrate the following functions w.r.t. x.

1) $x \cos x$ 2) $x^2 e^x$ 3) $x^3 \ln 3x$

4) $x e^{-x}$ 5) $3x \sin x$ 6) $e^x \sin 2x$

7) $e^{2x} \cos x$ 8) $x^2 e^{4x}$ 9) $e^{-x} \sin x$

10) $\ln 2x$ 11) $e^x(x + 1)$ 12) $x(1 + x)^7$

13) $x \sin\left(x + \dfrac{\pi}{6}\right)$ 14) $x \cos nx$ 15) $x^n \ln x$

16) $3x \cos 2x$ 17) $2e^x \sin x \cos x$ 18) $x^2 \sin x$

19) $e^{ax} \sin bx$

20) By writing $\cos^3\theta$ as $(\cos^2\theta)(\cos\theta)$ use integration by parts to find $\displaystyle\int \cos^3\theta \, d\theta$.

Each of the following products can be integrated either:

(a) by immediate recognition, or

(b) by a suitable change of variable, or

(c) by parts.

Choose the best method in each case and hence integrate each function.

21) $(x - 1) e^{x^2 - 2x + 4}$ 22) $(x + 1)^2 e^x$

23) $\sin x(4 + \cos x)^3$ 24) $\cos x \, e^{\sin x}$

25) $x^4\sqrt{1 + x^5}$ 26) $e^x(e^x + 2)^4$

27) $x \cos\left(\dfrac{\pi}{4} - x\right)$ 28) $x \, e^{2x - 1}$

29) $x(1 - x^2)^9$ 30) $\cos x \sin^5 x$

INTEGRATING FRACTIONS

Type I

Consider first a function $\ln u$ where u is a function of x. Differentiating with respect to x gives

$$\frac{d}{dx} \ln u = \left(\frac{1}{u}\right)\left(\frac{du}{dx}\right) \quad \text{or} \quad \frac{du/dx}{u}$$

So

$$\int \frac{du/dx}{u} \, dx = \ln |u| + K$$

or, writing $f(x)$ and $f'(x)$ for u and $\dfrac{du}{dx}$ respectively

$$\int \frac{f'(x)}{f(x)}\, dx = \ln|f(x)| + K$$

Thus all fractions of the form $f'(x)/f(x)$ can be integrated immediately by recognition

e.g. $\displaystyle\int \frac{\cos x}{1 + \sin x}\, dx = \ln|1 + \sin x| + K \qquad \left(\frac{d}{dx}(1 + \sin x) = \cos x\right)$

$$\int \frac{x^2}{1 + x^3}\, dx \equiv \frac{1}{3}\int \frac{3x^2}{1 + x^3}\, dx = \frac{1}{3}\ln|1 + x^3| + K$$

$$\left(\frac{d}{dx}(1 + x^3) = 3x^2\right)$$

$$\int \frac{e^x}{e^x + 4}\, dx = \ln|e^x + 4| + K$$

But $\displaystyle\int \frac{x}{\sqrt{1 + x}}\, dx \neq \ln|\sqrt{1 + x}| + K \qquad \left(\frac{d}{dx}\sqrt{1 + x} \neq x\right)$

i.e. integrals are of Type I *only* when the numerator is basically the derivative of the *complete denominator*.

An integral whose numerator is the derivative, not of the complete denominator, but of a function *within* the denominator, belongs to the following group.

Type II (Changing the Variable)

An example of this category is $\displaystyle\int \frac{2x}{\sqrt{(x^2 + 1)}}\, dx$.

Noting that $2x$ is the derivative of $x^2 + 1$ we use a change of variable based on $x^2 + 1$.

Let $\qquad\qquad u \equiv x^2 + 1 \Rightarrow \ldots du \equiv \ldots 2x\, dx$

Thus $\qquad\qquad \displaystyle\int \frac{2x}{\sqrt{(x^2 + 1)}}\, dx \equiv \int \frac{du}{\sqrt{u}}$

But $\qquad\qquad \displaystyle\int u^{-\frac{1}{2}}\, du = 2u^{\frac{1}{2}} + K$

Hence $\qquad\qquad \displaystyle\int \frac{2x}{\sqrt{(x^2 + 1)}}\, dx = 2\sqrt{x^2 + 1} + K$

Similarly for $\displaystyle\int \frac{\cos x}{\sin^4 x}\, dx$ we note that $\cos x$ is the derivative of $\sin x$ (which is a function within the denominator) and proceed as follows:

Let $\qquad\qquad u \equiv \sin x \Rightarrow \ldots du \equiv \ldots \cos x\, dx$

Then
$$\int \frac{\cos x}{\sin^4 x} \, dx \equiv \int \frac{1}{u^4} \, du$$
$$= -\tfrac{1}{3} u^{-3} + K$$

$\Rightarrow$
$$\int \frac{\cos x}{\sin^4 x} \, dx = K - \frac{1}{3 \sin^3 x}$$

Type III

A fraction which has not already fallen into one of the earlier categories, may be suitable for conversion to partial fractions before integration is attempted,

e.g.
$$\frac{1}{(x+1)(x+2)} \equiv \frac{1}{x+1} - \frac{1}{x+2}$$

so
$$\int \frac{1}{(x+1)(x+2)} \, dx \equiv \int \frac{1}{x+1} \, dx - \int \frac{1}{x+2} \, dx$$
$$= \ln |x+1| - \ln |x+2| + K$$
$$= \ln \left| \frac{x+1}{x+2} \right| + K$$

Note. Only proper fractions can be converted directly into partial fractions. *An improper fraction must first be divided out until the remaining fraction is proper.*

e.g. to find $\int \frac{x^2 + 1}{(x+1)(x-1)} \, dx$, we first convert $\frac{x^2 + 1}{(x+1)(x-1)}$ as follows:

$$\frac{x^2 + 1}{(x+1)(x-1)} \equiv \frac{x^2 + 1}{x^2 - 1} \equiv 1 + \frac{2}{x^2 - 1} \equiv 1 + \frac{2}{(x-1)(x+1)}$$
$$\equiv 1 + \frac{1}{x-1} - \frac{1}{x+1}$$

Hence
$$\int \frac{x^2 + 1}{(x+1)(x-1)} \, dx \equiv \int \left(1 + \frac{1}{x-1} - \frac{1}{x+1} \right) dx$$
$$= x + \ln |x-1| - \ln |x+1| + K$$

Some very simple improper fractions do not require transformation into partial fractions, but it is still essential to reduce them to proper fraction form.

e.g.
$$\frac{x+2}{x+1} \equiv \frac{x+1+1}{x+1} \equiv 1 + \frac{1}{x+1}$$

Hence
$$\int \frac{x+2}{x+1} \, dx \equiv \int \left(1 + \frac{1}{x+1} \right) dx$$
$$= x + \ln |x+1| + K$$

It is very important to identify the correct category when integrating a given fraction. Otherwise lengthy or fruitless attempts are likely to be made. Correct identification requires really careful scrutiny as fractions requiring different integration techniques often *look* very similar. This is demonstrated by the following example

EXAMPLES 9i

1) Integrate the following expressions with respect to x:

(a) $\dfrac{x+1}{x^2+2x-8}$ (b) $\dfrac{x+1}{(x^2+2x-8)^4}$ (c) $\dfrac{x+2}{x^2+2x-8}$

When integrating any fraction, the check-points used to identify the type of integral are, in order:

(i) is the numerator the basic derivative of the complete denominator

i.e. is the integral of the form $\displaystyle\int \dfrac{f'(x)}{f(x)}\,dx$?

(ii) is the numerator the derivative of a function within the denominator? (use a change of variable)

(iii) are partial fractions possible?

So we apply these checks to the given problem.

(a) As the derivative of x^2+2x-8 is $2x+2$ or $2(x+1)$ we identify this integral as belonging to the group $\displaystyle\int \dfrac{f'(x)}{f(x)}\,dx$.

Thus $\displaystyle\int \dfrac{\frac{1}{2}(2x+2)}{x^2+2x-8}\,dx = \frac{1}{2}\ln|x^2+2x-8| + K$

(b) This time the numerator is basically the derivative of the function x^2+2x-8 *within* the denominator so we use

$$u \equiv x^2+2x-8 \Rightarrow \ldots du \equiv \ldots (2x+2)\,dx$$

Thus $\displaystyle\frac{1}{2}\int \dfrac{2x+2}{(x^2+2x-8)^4}\,dx \equiv \frac{1}{2}\int \frac{1}{u^4}\,du$

$$= \frac{1}{2}\left(\frac{u^{-3}}{-3}\right) + K$$

i.e. $\displaystyle\int \dfrac{x+1}{(x^2+2x-8)^4}\,dx = K - \dfrac{1}{6(x^2+2x-8)^3}$

(c) In $\displaystyle\int \dfrac{x+2}{x^2+2x-8}\,dx$ the numerator is not related to the derivative of the denominator so we try partial fractions (using the cover-up method).

$$\dfrac{x+2}{x^2+2x-8} \equiv \dfrac{x+2}{(x+4)(x-2)} \equiv \dfrac{1/3}{x+4} + \dfrac{2/3}{x-2}$$

Thus $\displaystyle\int \frac{x+2}{x^2+2x-8}\,dx \equiv \frac{1}{3}\int \frac{1}{x+4}\,dx + \frac{2}{3}\int \frac{1}{x-2}\,dx$

$\displaystyle = \tfrac{1}{3}\ln|x+4| + \tfrac{2}{3}\ln|x-2| + K$

or $\ln A\,|(x+4)^{\frac{1}{3}}(x-2)^{\frac{2}{3}}|$

2) Find $\displaystyle\int \tan x\,dx$ by writing $\tan x$ as $\dfrac{\sin x}{\cos x}$

$$\int \frac{\sin x}{\cos x}\,dx \equiv -\int \frac{-\sin x}{\cos x}\,dx$$

$$\equiv -\int \frac{f'(x)}{f(x)}\,dx \quad \text{where} \quad f(x) \equiv \cos x$$

Thus $\displaystyle\int \tan x\,dx = -\ln|\cos x| + K$

$= K - \ln|\cos x|$

or $K + \ln|\sec x|$

Note. This result is important and quotable.

EXERCISE 9i

Integrate the following functions w.r.t. x.

1) $\dfrac{\cos x}{3+\sin x}$ 2) $\dfrac{e^x}{1-e^x}$ 3) $\dfrac{2x}{1+x^2}$ 4) $\dfrac{\sec^2 x}{1-3\tan x}$

5) $\dfrac{1}{1+e^x}$ (*Hint*: multiply throughout by e^{-x})   6) $\dfrac{x^3}{1+x^4}$

7) $\dfrac{\cos 2x}{4\sin x\cos x+1}$ 8) $\dfrac{\tan x}{1+\cos x}$ (*Hint*: multiply throughout by $\sec x$)

9) $\dfrac{1}{x\ln x}$ $\left(\text{i.e. } \dfrac{1/x}{\ln x}\right)$ 10) $\dfrac{2x+3}{x^2+3x+4}$

Integrate the following trig functions by writing each as a fraction as indicated.

11) $\cot x$ $\left(\dfrac{\cos x}{\sin x}\right)$ 12) $\sec x$ $\left(\dfrac{\sec x\,(\sec x+\tan x)}{(\tan x+\sec x)}\right)$

13) $\operatorname{cosec} x$ $\left(\dfrac{\operatorname{cosec} x\,(\operatorname{cosec} x+\cot x)}{(\cot x+\operatorname{cosec} x)}\right)$

Integrate the following w.r.t. x.

14) $\dfrac{x}{\sqrt{(x^2 + 1)}}$

15) $\dfrac{\cos x}{\sin^6 x}$

16) $\dfrac{\cos x}{\sqrt{(1 + \sin x)}}$

17) $\dfrac{e^x}{(e^x + 4)^2}$

18) $\dfrac{\sec^2 x}{\tan^3 x}$

19) $\dfrac{\sin x}{\cos^n x}$

20) $\dfrac{\cos x}{\sin^n x}$

21) $\dfrac{e^x}{\sqrt{(1 + e^x)}}$

22) $\dfrac{\sec x \tan x}{3 - \sec x}$

23) $\dfrac{\csc^2 x}{(2 + \cot x)^4}$

24) $\dfrac{x - 1}{3x^2 - 6x + 1}$

25) $\arcsin x$

26) $\arctan x$

27) $\arccos x$

Use partial fractions to find the following integrals.

28) $\displaystyle\int \dfrac{x}{x + 1}\, dx$

29) $\displaystyle\int \dfrac{x^2 - 2}{x^2 - 1}\, dx$

30) $\displaystyle\int \dfrac{x^2}{(x + 1)(x + 2)}\, dx$

31) $\displaystyle\int \dfrac{x + 4}{x}\, dx$

32) $\displaystyle\int \dfrac{x + 4}{x + 1}\, dx$

33) $\displaystyle\int \dfrac{2x}{(x - 2)(x + 2)}\, dx$

34) $\displaystyle\int \dfrac{3u + 4}{u(u + 1)}\, du$

35) $\displaystyle\int \dfrac{x^2 + x + 5}{x(x + 1)}\, dx$

36) $\displaystyle\int \dfrac{3 - y}{(y - 1)(y - 2)}\, dy$

37) $\displaystyle\int \dfrac{2z - 5}{z^2 - 5z + 6}\, dz$

38) $\displaystyle\int \dfrac{12x}{(2 - x)(3 - x)(4 - x)}\, dx$

39) $\displaystyle\int \dfrac{x^2 + 2x + 4}{(2x - 1)(x^2 - 1)}\, dx$

40) $\displaystyle\int \dfrac{4u^2 + 3u - 2}{(u + 1)(2u + 3)}\, du$

Standard Results

Some of the results obtained in this exercise, together with the integral of $\tan x$ [Example 9i (2)] are useful and can be quoted.

$$\int \tan x \, dx = \ln |\sec x| + K$$

$$\int \cot x \, dx = \ln |\sin x| + K$$

$$\int \sec x \, dx = \ln |\sec x + \tan x| + K$$

$$\int \csc x \, dx = -\ln |\csc x + \cot x| + K$$

INTEGRATION OF SOME TRIGONOMETRIC EXPRESSIONS

Even Powers of $\sin \theta$ or $\cos \theta$

The double angle trig identities are useful here,

e.g. to find $\int \cos^4\theta \, d\theta$ we use

$$\cos^2\theta \equiv \tfrac{1}{2}(1 + \cos 2\theta)$$

$\Rightarrow$
$$\cos^4\theta \equiv \tfrac{1}{4}(1 + 2\cos 2\theta + \cos^2 2\theta)$$

$$\equiv \tfrac{1}{4}[1 + 2\cos 2\theta + \tfrac{1}{2}(1 + \cos 4\theta)]$$

Thus
$$\int \cos^4\theta \, d\theta \equiv \int (\tfrac{3}{8} + \tfrac{1}{2}\cos 2\theta + \tfrac{1}{8}\cos 4\theta) \, d\theta$$

$$= \tfrac{3}{8}\theta + \tfrac{1}{4}\sin 2\theta + \tfrac{1}{32}\sin 4\theta + K$$

Odd Powers of $\sin \theta$ or $\cos \theta$

In this case we can use the identity $\cos^2\theta + \sin^2\theta \equiv 1$

e.g. to find $\int \sin^5\theta \, d\theta$ we use

$$\sin^5\theta \equiv (\sin \theta)(\sin^2\theta)^2$$

$$\equiv (\sin \theta)(1 - \cos^2\theta)^2$$

Hence
$$\int \sin^5\theta \, d\theta \equiv \int (\sin \theta - 2\sin \theta \cos^2\theta + \sin \theta \cos^4\theta) \, d\theta$$

Now we saw in Example 9g (2) that

$$\int \sin \theta \cos^n\theta \, d\theta = -\frac{1}{n + 1}\cos^{n+1}\theta + K$$

so
$$\int \sin^5\theta \, d\theta = -\cos \theta + \tfrac{2}{3}\cos^3\theta - \tfrac{1}{5}\cos^5\theta + K$$

Powers of tan θ

The identity $\tan^2\theta + 1 \equiv \sec^2\theta$ is helpful in integrating any power of $\tan\theta$,

e.g.
$$\int \tan^3\theta \, d\theta \equiv \int (\tan\theta)(\tan^2\theta) \, d\theta$$

$$\equiv \int \tan\theta \, (\sec^2\theta - 1) \, d\theta$$

$$\equiv \int \tan\theta \sec^2\theta \, d\theta - \int \tan\theta \, d\theta$$

Now consider $\int \tan\theta \sec^2\theta \, d\theta$ in which the substitution $u \equiv \tan\theta$ is appropriate $\left(\text{since}\quad \sec^2\theta \equiv \dfrac{du}{d\theta}\right),$ thus

$$u \equiv \tan\theta \Rightarrow \dots du \equiv \dots \sec^2\theta \, d\theta$$

Hence
$$\int \tan\theta \sec^2\theta \, d\theta \equiv \int u \, du$$

$$= \frac{u^2}{2} + K$$

i.e.
$$\int \tan\theta \sec^2\theta \, d\theta = \tfrac{1}{2}\tan^2\theta + K$$

Further we know that $\int \tan\theta \, d\theta = \ln|\sec\theta| + K$ so, finally

$$\int \tan^3\theta \, d\theta \equiv \int \tan\theta \, (\sec^2\theta - 1) \, d\theta$$

$$= \tfrac{1}{2}\tan^2\theta - \ln|\sec\theta| + K$$

Multiple Angles

When integrating products such as $\sin 5\theta \cos 3\theta$, one of the factor formulae should be used

e.g.
$$\int \sin 5\theta \cos 3\theta \, d\theta \equiv \tfrac{1}{2} \int (\sin 8\theta + \sin 2\theta) \, d\theta$$

$$= \tfrac{1}{2}[-\tfrac{1}{8}\cos 8\theta - \tfrac{1}{2}\cos 2\theta] + K$$

$$= K - \tfrac{1}{16}(\cos 8\theta + 4\cos 2\theta)$$

The basic ideas used in the examples above can be applied to the integration of a variety of trig functions. The aims when dealing with trig integrals are usually:

(a) to convert the integral to the form $\int \dfrac{du}{dx} f'(u)\, dx$,

(b) to reduce the trig expression to a number of single trig ratios.

For instance, to find $\int \sin^2\theta \cos^3\theta\; d\theta$

we use $\cos^3\theta \equiv \cos\theta \cos^2\theta \equiv \cos\theta(1 - \sin^2\theta)$

Hence $\int \sin^2\theta \cos^3\theta\; d\theta \equiv \int (\sin^2\theta \cos\theta - \sin^4\theta \cos\theta)\, d\theta$

$$= \tfrac{1}{3}\sin^3\theta - \tfrac{1}{5}\sin^5\theta + K$$

But to find $\int \sin 3\theta \cos^2\theta\; d\theta$ it is better to transform
$\cos^2\theta$ into $\tfrac{1}{2}(1 + \cos 2\theta)$ so that

$$\int \sin 3\theta \cos^2\theta\; d\theta \equiv \tfrac{1}{2}\int (\sin 3\theta + \sin 3\theta \cos 2\theta)\, d\theta$$

$$\equiv \tfrac{1}{2}\int [\sin 3\theta + \tfrac{1}{2}(\sin 5\theta + \sin\theta)]\; d\theta$$

$$= -\tfrac{1}{6}\cos 3\theta - \tfrac{1}{20}\cos 5\theta - \tfrac{1}{4}\cos\theta + K$$

EXERCISE 9j

1) Find:

(a) $\int \sin^4\theta\; d\theta$ (b) $\int \cos^3\theta\; d\theta$ (c) $\int \tan^4\theta\; d\theta$

(d) $\int \sin^3\theta\; d\theta$ (e) $\int \cos^4\theta\; d\theta$ (f) $\int \tan^5\theta\; d\theta$

2) Find:

(a) $\int 2\sin 4\theta \cos 3\theta\; d\theta$ (b) $\int 2\cos 2\theta \cos 5\theta\; d\theta$ (c) $\int \sin 2\theta \cos 6\theta\; d\theta$

(d) $\int \sin\theta \sin 3\theta\; d\theta$ (e) $\int 2\sin nx \cos mx\; dx$ (f) $\int 2\cos\dfrac{u}{2}\cos\dfrac{u}{3}\; du$

(g) $\int \cos nx \cos mx\; dx$

3) Integrate w.r.t. x:

(a) $\sin^2 x \cos^3 x$ 　　　(b) $\sin^{10} x \cos^3 x$ 　　　(c) $\sin^2 x \cos^2 x$ (use double angle formulae)

(d) $\tan^2 x \sec^4 x$ 　　　(e) $\sin^3 2x \cos^2 2x$ 　　　(f) $\sin^n x \cos^3 x$

(g) $\dfrac{\cos^2 x}{\operatorname{cosec}^3 x}$ 　　　(h) $\dfrac{\tan^3 x}{\cos^2 x}$

SYSTEMATIC INTEGRATION

At this stage it is possible to classify most of the integrals which are likely to arise.

Once correctly classified, the given expression can be integrated using the method best suited to its category.

The simplest category comprises the standard integrals listed below.

Function	Integral
x^n	$\dfrac{1}{n+1} x^{n+1} \quad (n \neq -1)$
e^x	e^x
$\dfrac{1}{x}$	$\ln |x|$
$\cos x$	$\sin x$
$\sin x$	$-\cos x$
$\sec^2 x$	$\tan x$
$-\operatorname{cosec}^2 x$	$\cot x$
$\tan x$	$\ln |\sec x|$
$\sec x$	$\ln |\sec x + \tan x|$
$\dfrac{1}{1+x^2}$	$\arctan x$
$\dfrac{1}{\sqrt{(1-x^2)}}$	$\arcsin x$

Each of these results should also be recognised when x is replaced by $ax + b$, as in the following shortened list.

Function	Integral
$(ax + b)^n$	$\dfrac{1}{a(n+1)} (ax + b)^{n+1}$
e^{ax+b}	$\dfrac{1}{a} e^{ax+b}$

$$\frac{1}{ax + b} \qquad\qquad \frac{1}{a} \ln |ax + b|$$

$$\cos (ax + b) \qquad\qquad \frac{1}{a} \sin (ax + b)$$

When attempting to classify a given integral, and so determine the best method of integration, the following points should be considered.

(a) Is the integral a standard form?

(b) If it is a product, is it of the form $\dfrac{du}{dx} f'(u)$? If so, integrate at sight or change the variable. If it is not of this form, try integrating by parts.

(c) If it is a quotient, is it of the form:

(i) $\dfrac{f'(x)}{f(x)}$ (integrate at sight giving $\ln |f(x)|$) or,

(ii) $\dfrac{du/dx}{f(u)}$ (change the variable) or,

(iii) will partial fractions help?

(iv) be prepared for fractions whose *numerators* can be separated, thus producing two distinct integrals which may be of completely different types,

e.g. $\displaystyle \int \frac{x + 1}{\sqrt{(1 - x^2)}}\,dx \equiv \int \frac{x}{\sqrt{(1 - x^2)}}\,dx + \int \frac{1}{\sqrt{(1 - x^2)}}\,dx$

(d) If it is a trig function, and it has not already been classified, try to use, or adapt, one of the methods suggested on pages 322 to 324.

Although this systematic approach deals successfully with many integrals, inevitably the reader will encounter some integrals for which no method is obvious. Many expressions other than products and quotients can be integrated if an appropriate substitution is made. Because at this stage the reader cannot always be expected to 'spot' a suitable change of variable, a substitution will be suggested in most cases.

The following examples demonstrate a variety of integrals in which this technique can be applied.

EXAMPLES 9k

1) Find $\displaystyle \int x(2 - 3x)^{11}\,dx$.

Let $\qquad\qquad\qquad u \equiv 2 - 3x \Rightarrow \ldots du \equiv \ldots - 3\,dx$

Hence $\qquad\qquad \displaystyle \int x(2 - 3x)^{11}\,dx \equiv \int \left(\frac{2 - u}{3}\right)(u^{11})\left(-\frac{du}{3}\right)$

$$\equiv \frac{1}{9} \int (u^{12} - 2u^{11})\, du$$

$$= \frac{1}{9} \left[\frac{u^{13}}{13} - \frac{u^{12}}{6} \right] + K$$

$$= \left(\frac{u^{12}}{9} \right) \frac{(6u - 13)}{78} + K$$

i.e. $$\int x(2 - 3x)^{11}\, dx = -\frac{(2 - 3x)^{12}(1 + 18x)}{702} + K$$

2) Integrate $\dfrac{3x}{\sqrt{(4 - x)}}$ w.r.t. x.

Let $u \equiv \sqrt{4 - x} \Rightarrow u^2 \equiv 4 - x \Rightarrow \ldots 2u\, du \equiv \ldots - dx$

Hence $$\int \frac{3x}{\sqrt{(4 - x)}}\, dx \equiv \int \frac{3(4 - u^2)(- 2u\, du)}{u}$$

$$\equiv -6 \int (4 - u^2)\, du$$

$$= -6 \left(4u - \frac{u^3}{3} \right) + K$$

$$= -2u(12 - u^2) + K$$

i.e. $$\int \frac{3x}{\sqrt{(4 - x)}}\, dx = K - 2(8 + x)\sqrt{(4 - x)}$$

3) Integrate $\dfrac{1}{\sqrt{(9 - 16x^2)}}$ w.r.t. x by using $x \equiv \frac{3}{4}\sin\theta$.

Let $x \equiv \left| \frac{3}{4} \right| \sin\theta \quad \Rightarrow \quad \ldots dx \equiv \ldots \frac{3}{4}\cos\theta\, d\theta$

Hence $$\int \frac{1}{\sqrt{(9 - 16x^2)}}\, dx \equiv \int \frac{1}{\sqrt{(9 - 9\sin^2\theta)}}\, \frac{3}{4}\cos\theta\, d\theta$$

$$\equiv \int \frac{3\cos\theta}{12\cos\theta}\, d\theta$$

$$= \frac{1}{4}\theta + K$$

$\Rightarrow$ $$\int \frac{1}{\sqrt{(9 - 16x^2)}}\, dx = \frac{1}{4}\arcsin\frac{4x}{3} + K$$

4) By completing the square in the denominator and using the substitution $u \equiv x+1$, find $\int \dfrac{1}{x^2+2x+2}\,dx$.

First we deal with the denominator,

i.e. $\qquad\qquad\qquad x^2+2x+2 \equiv (x+1)^2+1$

Then let $\qquad\qquad u \equiv x+1 \quad \Rightarrow \quad \dots du \equiv \dots dx$

Hence $\qquad \displaystyle\int \dfrac{1}{x^2+2x+2}\,dx \equiv \int \dfrac{1}{(x+1)^2+1}\,dx$

$$\equiv \int \dfrac{1}{u^2+1}\,du$$

$$= \arctan u + K$$

So $\qquad \displaystyle\int \dfrac{1}{x^2+2x+2}\,dx = \arctan(x+1)+K$

5) Integrate $\sqrt{1-x^2}$ w.r.t. x by using the substitution $x \equiv \sin\theta$.

Let $\qquad\qquad x \equiv \sin\theta \Rightarrow \dots dx \equiv \dots \cos\theta\,d\theta$

Hence $\qquad \displaystyle\int \sqrt{1-x^2}\,dx \equiv \int \sqrt{(1-\sin^2\theta)}\cos\theta\,d\theta$

$$\equiv \int \cos^2\theta\,d\theta$$

$$\equiv \tfrac{1}{2}\int (1+\cos 2\theta)\,d\theta$$

$$= \dfrac{\theta}{2} + \dfrac{1}{4}\sin 2\theta + K$$

$$= \dfrac{\theta}{2} + \dfrac{1}{2}\sin\theta\,\cos\theta + K$$

$$= \dfrac{\theta}{2} + \dfrac{1}{2}\sin\theta\,\sqrt{1-\sin^2\theta} + K$$

$\Rightarrow \qquad \displaystyle\int \sqrt{1-x^2}\,dx = \tfrac{1}{2}(\arcsin x + x\sqrt{1-x^2}) + K$

EXERCISE 9k

Find the following integrals using the suggested substitution.

1) $\displaystyle\int (x+1)(x+3)^5\,dx; \qquad x+3 \equiv u$

2) $\int \dfrac{1}{4+x^2}\,dx;$ $x \equiv 2\tan\theta$

3) $\int \dfrac{x}{\sqrt{(3-x)}}\,dx;$ $3-x \equiv u^2$

4) $\int x\sqrt{x+1}\,dx;$ $x+1 \equiv u^2$

5) $\int \dfrac{2x+1}{(x-3)^6}\,dx;$ $x-3 \equiv u$

6) $\int \dfrac{1}{\sqrt{(1+x^2)}}\,dx;$ $x \equiv \tan\theta$

7) $\int 2x\sqrt{3x-4}\,dx;$ $3x-4 \equiv u^2$

8) $\int \dfrac{3}{25+4x^2}\,dx;$ $2x \equiv 5\tan\theta$

9) $\int \dfrac{1}{\sqrt{(x^2+4x+3)}}\,dx;$ $x+2 \equiv \sec\theta,$ after 'completing the square' in the denominator.

Devise a suitable substitution and hence find:

10) $\int 2x(1-x)^7\,dx$ 11) $\int \dfrac{1}{\sqrt{(9-x^2)}}\,dx$ 12) $\int \dfrac{x+3}{(4-x)^5}\,dx$

Classify each of the following integrals. Hence perform each integration using an appropriate method.

13) $\int e^{2x+3}\,dx$ 14) $\int x\sqrt{2x^2-5}\,dx$ 15) $\int \sin^2 3x\,dx$

16) $\int xe^{-x^2}\,dx$ 17) $\int \sin 3\theta\,\cos\theta\,d\theta$ 18) $\int u(u+7)^9\,du$

19) $\int \dfrac{x^2}{(x^3+9)^5}\,dx$ 20) $\int \dfrac{\sin 2y}{1-\cos 2y}\,dy$ 21) $\int \dfrac{1}{2x+7}\,dx$

22) $\int \dfrac{1}{\sqrt{(1-u^2)}}\,du$ 23) $\int \sin 3x\sqrt{1+\cos 3x}\,dx$

24) $\int x\sin 4x\,dx$ 25) $\int \dfrac{x+2}{x^2+4x-5}\,dx$ 26) $\int \dfrac{x+1}{x^2+4x-5}\,dx$

27) $\int \dfrac{x+2}{(x^2+4x-5)^3}\,dx$ 28) $\int 3y\sqrt{9-y^2}\,dy$ 29) $\int e^{2x}\cos 3x\,dx$

30) $\int \ln 5x \, dx$ 31) $\int \cos^3 2x \, dx$ 32) $\int \csc^2 x \, e^{\cot x} \, dx$

33) $\int \dfrac{\sin y}{\sqrt{(7 + \cos y)}} \, dy$ 34) $\int x^2 \, e^x \, dx$ 35) $\int \dfrac{x}{x^2 - 4} \, dx$

36) $\int \dfrac{x^2}{x^2 - 4} \, dx$ 37) $\int \dfrac{1}{x^2 - 4} \, dx$ 38) $\int \cos 4x \cos x \, dx$

39) $\int \sin^5 2\theta \, d\theta$ 40) $\int \cos^2 u \, \sin^3 u \, du$ 41) $\int \tan^4 \theta \, d\theta$

42) $\int \tan^5 \theta \, d\theta$ 43) $\int \dfrac{1 - 2x}{\sqrt{(1 - x^2)}} \, dx$ 44) $\int \dfrac{1}{u \ln u} \, du$

45) $\int y^2 \cos 3y \, dy$ 46) $\int \dfrac{\sec^2 x}{1 - \tan x} \, dx$ 47) $\int x\sqrt{(7 + x^2)} \, dx$

48) $\int \sin (5\theta - \pi/4) \, d\theta$ 49) $\int \cos \theta \, \ln \sin \theta \, d\theta$ 50) $\int \sec^2 u \, e^{\tan u} \, du$

51) $\int \dfrac{x}{(3 - x)^7} \, dx$ 52) $\int \tan^2 x \, \sec^2 x \, dx$

CALCULATION OF THE CONSTANT OF INTEGRATION

When an expression is integrated the result includes a constant of unknown value.

In order to determine its value, further information is needed. For example, if a curve has a gradient function $(x + 4)$ and also passes through the point $(2, 5)$, the equation of the curve can be found as follows:

If the equation of the curve is $y = f(x)$ the gradient function is $\dfrac{dy}{dx}$ where

$$\frac{dy}{dx} = x + 4 \implies y = \int (x + 4) \, dx$$

$$\implies \qquad y = \frac{x^2}{2} + 4x + K$$

But we also know that when $x = 2, \quad y = 5$

$$\implies \qquad 5 = \frac{2^2}{2} + 8 + K \implies K = -5$$

Thus the equation of the curve is

$$y = \frac{x^2}{2} + 4x - 5$$

EXERCISE 9I

In the following problems the gradient function of a curve is given, together with the coordinates of a point on the curve. Find the equation of the curve in each case.

1) $3x - 4$; $(1, 2)$　　　2) $3x^2 - 5x + 1$; $(0, 3)$

3) $6e^{2x}$; $(0, 2)$　　　4) $(7 - 5x)^2$; $(1, \frac{7}{15})$

5) $\cos 3x$; $(\pi/2, 1)$

6) A curve that passes through the origin has a gradient function $2x - 1$. Find its equation and sketch the curve.

7) If $\dfrac{dy}{dx} = e^{3x}$　and　$y = 2$　when　$x = 1$　find the coordinates of the point where the curve crosses the y axis.

DIFFERENTIAL EQUATIONS

An equation in which at least one term contains $\dfrac{dy}{dx}$, $\dfrac{d^2y}{dx^2}$ etc, is called a *differential equation*. For instance

$$x + 3\frac{dy}{dx} = 4y$$ is a *first order* differential equation because it

contains only *a first differential* coefficient $\left(\dfrac{dy}{dx}\right)$,

$$\frac{d^2y}{dx^2} + 3\frac{dy}{dx} + 4y = x^2$$ is a *second order* differential equation

because it contains a *second differential* coefficient $\left(\dfrac{d^2y}{dx^2}\right)$.

Each of these examples is a *linear* differential equation because none of the differential coefficients is raised to a power other than 1.

Conversely $x^2 + 2\left(\dfrac{dy}{dx}\right)^2 + 3y = 0$　is *not* a linear differential equation.

Any differential equation represents a relationship between two variables, x and y, say.

The same relationship can often be expressed in a form that does not contain a differential coefficient,

e.g. $y = x^2 + K$　and　$\dfrac{dy}{dx} = 2x$　express the same relationship between x

and y but $\dfrac{dy}{dx} = 2x$　is a differential equation while $y = x^2 + K$　is not.

Converting a differential relationship into a direct one is called *solving a differential equation* and this clearly involves some form of integration. There are many different types of differential equation, each requiring a particular approach and technique for its solution. At this stage however, we are going to solve only one simple type belonging to the first order, linear group.

Differential Equations with Variables Separable

Consider the differential equation

$$3y \frac{dy}{dx} = 5x^2 \qquad [1]$$

If we integrate both sides of this equation w.r.t. x we have

$$\int 3y \frac{dy}{dx} dx = \int 5x^2 dx$$

But we saw earlier in this chapter (p. 309) that

$$\int \ldots \frac{dy}{dx} dx \equiv \int \ldots dy$$

so

$$\int 3y \, dy = \int 5x^2 \, dx \qquad [2]$$

Temporarily removing the integral signs from equation [2] leaves

$$3y \, dy = 5x^2 \, dx \qquad [3]$$

This form can be obtained direct from equation [1] by *separating the variables*, i.e. by separating dy from dx and collecting on one side all terms involving y together with dy, while all the x terms together with dx are collected separately.

It is important to appreciate that what we have written in [3] above does not, in itself, have any meaning. It nevertheless provides a quick method of converting the given equation [1] into the integral form [2].

Thus, the solution of the given equation is carried out as follows

$$3y \frac{dy}{dx} = 5x^2$$

Separating the variables gives　　$3y \, dy = 5x^2 \, dx.$

Hence　　$\int 3y \, dy = \int 5x^2 \, dx$

$\Rightarrow$　　$\dfrac{3y^2}{2} = \dfrac{5x^3}{3} + A$

The constant of integration is usually represented by A, B, etc. rather than K when solving differential equations and is called an *arbitrary constant*.
The solution of a differential equation including the arbitrary constant is called the *general solution* (or sometimes the *complete primitive*).

The equation $\dfrac{3y^2}{2} = \dfrac{5x^2}{3} + A$ represents a family of curves with similar

characteristics. Each value of A gives one particular curve of the family.
Further information is required if A is to be evaluated, providing a *particular solution*.

EXAMPLES 9m

1) Solve the differential equation $y = x\dfrac{dy}{dx}$ and describe the family it

represents. Sketch any three members of this family.
Find the particular solution for which $y = 2$ when $x = 1$ and sketch
this member of the family on the same axes as before.

$$y = x\frac{dy}{dx}$$

$$\left(\text{Separating the variables gives} \quad \frac{dy}{y} = \frac{dx}{x}\right)$$

Thus $$\int \frac{1}{y}\,dy = \int \frac{1}{x}\,dx$$

$\Rightarrow$ $$\ln y = \ln x + \ln A \quad \text{(using } \ln A \text{ for the arbitrary constant)}$$

$\Rightarrow$ $$y = Ax$$

This equation represents a family of straight lines through the origin, each line
having a gradient A.

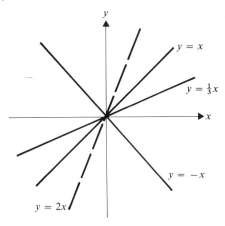

If $y = 2$ when $x = 1$ then $A = 2$ and in this case the particular member of the family is $y = 2x$.

2) A curve is such that at any point the gradient multiplied by the x coordinate is equal to three times the y coordinate of that point. If the curve passes through the point $(1, 4)$ find its equation.

The given condition can be written

$$x \frac{dy}{dx} = 3y$$

$$\left[\text{Separating the variables gives} \quad \frac{dy}{y} = 3 \frac{dx}{x} \right]$$

Thus
$$\int \frac{1}{y} \, dy = \int \frac{3}{x} \, dx$$

$\Rightarrow$
$$\ln |y| = 3 \ln |x| + A$$

But $y = 4$ when $x = 1$, so $A = \ln 4$

Thus
$$\ln |y| = 3 \ln |x| + \ln 4$$

$\Rightarrow$
$$\ln |y| = \ln |4x^3|$$

$\Rightarrow$
$$y = 4x^3$$

3) Find the general solution of the differential equation

$$2(x^2 + 1) \frac{dy}{dx} = x(4 - y^2)$$

If $y = 1$ when $x = 0$, express y as a function of x.

$$2(x^2 + 1) \frac{dy}{dx} = x(4 - y^2)$$

$\Rightarrow$
$$\frac{2}{4 - y^2} \frac{dy}{dx} = \frac{x}{x^2 + 1}$$

So, after separating the variables, we have,

$$\int \frac{2}{4 - y^2} \, dy = \int \frac{x}{x^2 + 1} \, dx$$

i.e.
$$\int \frac{\frac{1}{2}}{2 - y} + \frac{\frac{1}{2}}{2 + y} \, dy = \int \frac{x}{x^2 + 1} \, dx$$

Hence $-\frac{1}{2} \ln (2 - y) + \frac{1}{2} \ln (2 + y) = \frac{1}{2} \ln (x^2 + 1) + \ln A$

$$\Rightarrow \qquad \ln \sqrt{\left(\frac{2+y}{2-y}\right)} = \ln A \sqrt{(x^2+1)}$$

$$\Rightarrow \qquad 2+y = A^2(2-y)(x^2+1)$$

Now if $y = 1$ when $x = 0$, $A^2 = 3$, so

$$2+y = 3(2-y)(x^2+1)$$

$$\Rightarrow \qquad y + 3y(x^2+1) = 6(x^2+1) - 2$$

$$\Rightarrow \qquad y = \frac{6x^2+4}{3x^2+4}$$

Natural Occurrence of Differential Equations

Differential equations often arise when a physical situation is interpreted mathematically (i.e. when a mathematical model is made of the physical situation).
For example:

(a) Suppose that a body falls from rest in a medium which causes the velocity to decrease at a rate proportional to the velocity.
Using v for velocity and t for time, the rate of *decrease* of velocity can be written as $-\dfrac{dv}{dt}$.
Thus the motion of the body satisfies the differential equation

$$-\frac{dv}{dt} = kv$$

(b) During the initial stages of the growth of yeast cells in a culture, the number of cells present increases in proportion to the number already formed.
Thus n, the number of cells at a particular time t, can be found from the differential equation

$$\frac{dn}{dt} = kn$$

(c) Suppose that a chemical mixture contains two substances A and B whose weights are W_A and W_B and whose combined weight remains constant. B is converted into A at a rate which is inversely proportional to the weight of B and proportional to the square of the weight of A in the mixture at any time t. The weight of B present at time t can be found using

$$\frac{d}{dt}(W_B) = \frac{k}{W_B} \times (W_A)^2$$

But $W_A + W_B$ is constant, W say

Hence
$$\frac{d}{dt}(W_B) = \frac{k(W - W_B)^2}{W_B}$$

This differential equation now relates W_B and t.

Note. In forming (and subsequently solving) differential equations from naturally occurring data, it is not necessary to understand the background of the situation or experiment.

EXERCISE 9m

Find the general solutions of the following differential equations.

1) $y\dfrac{dy}{dx} = \sin x$

2) $\dfrac{dy}{dx} = y^2$

3) $\dfrac{1}{x}\dfrac{dy}{dx} = \dfrac{1}{x^2 + 1}$

4) $(x - 3)\dfrac{dy}{dx} = y$

5) $\tan y \dfrac{dx}{dy} = 4$

6) $u\dfrac{du}{dv} = v + 2$

7) $\dfrac{y^2}{x^3}\dfrac{dy}{dx} = \ln x$

8) $e^x \dfrac{dy}{dx} = \dfrac{x}{y}$

9) $\sec x \dfrac{dy}{dx} = e^y$

10) $r\dfrac{dr}{d\theta} = \sin^2\theta$

11) $\dfrac{dv}{du} = \dfrac{v + 1}{u + 2}$

12) $xy\dfrac{dy}{dx} = \ln x$

13) $y(x + 1) = (x^2 + 2x)\dfrac{dy}{dx}$

14) $v^2 \dfrac{dv}{dt} = (2 + t)^3$

15) $x\dfrac{dy}{dx} = \dfrac{1}{y} + y$

16) $e^x \dfrac{dy}{dx} = e^{y-1}$

17) $\tan x \dfrac{dy}{dx} = 2y^2 \sec^2 x$

18) $r\dfrac{d\theta}{dr} = \cos^2\theta$

19) $y \sin^3 x \dfrac{dy}{dx} = \cos x$

20) $\dfrac{uv}{u - 1} = \dfrac{du}{dv}$

Find the particular solutions of the following differential equations:

21) $\dfrac{y}{x}\dfrac{dy}{dx} = \dfrac{y^2 + 1}{x^2 + 1}$; $y = 0$ when $x = 1$.

22) $e^t \dfrac{ds}{dt} = \sqrt{s}$; $s = 4$ when $t = 0$.

23) $y^2 \dfrac{dy}{dx} = x^2 + 1$; $y = 1$ when $x = 2$.

Find the equation of each of the following curves:

24) A curve passes through the points $(1, 2)$ and $(\frac{1}{4}, -10)$ and has a gradient which is inversely proportional to x^2.

25) The gradient function of a curve is $\dfrac{y+1}{x^2-1}$ and the curve passes through $(-3, 1)$.

26) A curve passes through the point $(0, -1)$ and $e^{-x}\dfrac{dy}{dx} = 1$.

Form, *but do not solve*, the differential equations representing the following data:

27) A body moves with a velocity v which is inversely proportional to its displacement s from a fixed point.

28) The rate at which the height h of a certain plant increases is proportional to the natural logarithm of the difference between its present height and its final height H.

29) The manufacturers of a certain brand of soap powder are concerned that the number, n, of people buying their product at any time t has remained constant for some months. They launch a major advertising programme which results in the number of customers increasing at a rate proportional to the square root of n.
Express as differential equations the progress of sales:
(a) before advertising
(b) after advertising.

30) In an isolated community, the number, n, of people suffering from an infectious disease is N_1 at a particular time. The disease then becomes epidemic and spreads so that the number of sick people increases at a rate proportional to n, until the total number of sufferers is N_2. The rate of increase then becomes inversely proportional to n until N_3 people have the disease. After this, the total number of sick people decreases at a constant rate. Write down the differential equation governing the incidence of the disease:
(a) for $N_1 \leqslant n < N_2$,
(b) for $N_2 \leqslant n < N_3$,
(c) for $n \geqslant N_3$.

INTEGRATION AS A PROCESS OF SUMMATION

Consider the area bounded by the x-axis, the lines $x = a$ and $x = b$ and the curve $y = f(x)$ that is continuous for $a \leqslant x \leqslant b$.

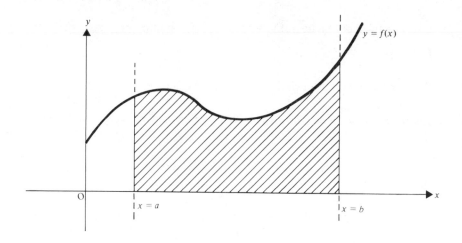

There are several ways in which this area can be estimated (e.g. counting squares on graph paper). The method we are going to use this time is to split the area into thin vertical strips and treat each strip as being approximately rectangular.

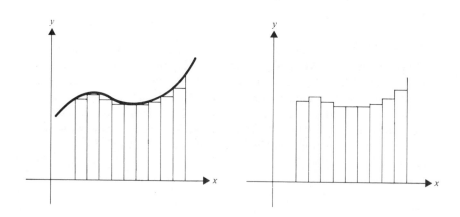

The sum of the areas of the rectangular strips then gives an approximate value for the required area. The thinner the strips are, the better is the approximation.

Note that every strip has one end on the x-axis, one end on the curve and two vertical sides, i.e., they all have the same type of boundaries.

Now consider one typical strip
(or element) PP′Q′Q where P is
the point (x, y). The width of the
strip, P′Q′, is a small increment in
x and can be called δx.
Also, if A represents the area of all
the strips up to PP′ then adding
the area of the strip PP′Q′Q
causes a small increase in A.
Thus the area of the strip can be
called δA.

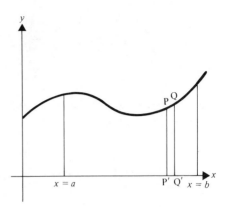

The area of the strip is approximately that of a rectangle of width δx and
length y.

Thus for every strip $\qquad\qquad \delta A \simeq y\,\delta x$ [1]

Now if we sum the areas of all the strips from $\quad x = a \quad$ to $\quad x = b$,

$\left(\text{which we denote by } \displaystyle\sum_{x=a}^{b} \delta A\right)$, the total area is obtained

i.e. $\qquad\qquad\qquad$ total area $= \displaystyle\sum_{x=a}^{x=b} \delta A$

$\Rightarrow \qquad\qquad\qquad$ total area $\simeq \displaystyle\sum_{x=a}^{x=b} y\,\delta x$

As δx gets smaller the accuracy of the results increases until, in the limit,

$$\text{total area} = \lim_{\delta x \to 0} \sum_{x=a}^{x=b} y\,\delta x$$

Alternatively we can return to equation [1] above and consider it in the form

$$\frac{\delta A}{\delta x} \simeq y$$

This form too becomes more accurate as δx gets smaller giving, in the limiting
case,

$$\lim_{\delta x \to 0} \frac{\delta A}{\delta x} = y$$

But $\qquad\qquad\qquad \lim_{\delta x \to 0} \dfrac{\delta A}{\delta x} \quad$ is $\quad \dfrac{dA}{dx}$

so
$$\frac{dA}{dx} = y$$

Hence
$$A = \int y \, dx$$

The boundary values of x defining the total area are $x = a$ and $x = b$ and we indicate this by writing

$$\text{total area} = \int_a^b y \, dx$$

The total area can therefore be found in two ways, either by a process of summation or by integration

i.e.
$$\lim_{\delta x \to 0} \sum_{x=a}^{x=b} y \, \delta x = \int_a^b y \, dx$$

and we conclude that integration is a process of summation.

The application of integration to problems involving summation continues in Chapter 17. In this chapter we will use integration only to find areas bounded by straight lines and a curve, but first we must investigate the meaning of

$$\int_a^b y \, dx$$

DEFINITE INTEGRATION

Suppose that we wish to find the area bounded by the x-axis, the lines $x = a$ and $x = b$ and the curve $y = 3x^2$.

Using the method above we find that $A = \int 3x^2 \, dx$

$$\text{i.e. } A = x^3 + K$$

From this area function we can find the value of A corresponding to a particular value of x.

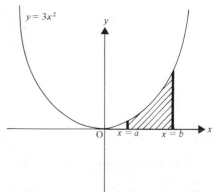

Hence, using $x = a$ gives

$$A_a = a^3 + K$$

Similarly using $x = b$ gives

$$A_b = b^3 + K$$

Then the area between $x = a$ and $x = b$ is given by $A_b - A_a$ where

$$A_b - A_a = (b^3 + K) - (a^3 + K)$$
$$= b^3 - a^3$$

Now $A_b - A_a$ is referred to as *the definite integral from* a *to* b of $3x^2$ and is denoted by $\displaystyle\int_a^b 3x^2 \, dx$

i.e. $$\int_a^b 3x^2 \, dx = (x^3)_{x=b} - (x^3)_{x=a}$$

The R.H.S. of this equation is usually written in the form $\left[x^3\right]_a^b$ where a and b are called the boundary values or limits of the integration. (b is the upper limit and a the lower limit.)

Whenever a definite integral is calculated, the constant of integration disappears.

Note. A definite integral can be found in this way only if the function to be integrated is defined for every value of x from a to b.

e.g. $\displaystyle\int_{-1}^1 \frac{1}{x} \, dx$ cannot, as yet, be found because $\dfrac{1}{x}$ is undefined when $x = 0$.

EXAMPLES 9n

1) Evaluate $\displaystyle\int_1^4 \frac{1}{(x+3)^2} \, dx$.

$$\int_1^4 \frac{1}{(x+3)^2} \, dx \equiv \int_1^4 (x+3)^{-2} \, dx$$

$$= \left[-(x+3)^{-1}\right]_1^4$$

$$= \{-(4+3)^{-1}\} - \{-(1+3)^{-1}\}$$

$$= -\tfrac{1}{7} + \tfrac{1}{4} = \tfrac{3}{28}$$

2) Evaluate $\displaystyle\int_0^{\frac{\pi}{2}} \cos x \, dx$.

$$\int_0^{\frac{\pi}{2}} \cos x \, dx = \left[\sin x\right]_0^{\frac{\pi}{2}}$$

$$= \sin\frac{\pi}{2} - \sin 0$$

$$= 1 - 0 = 1$$

Definite Integration with Change of Variable

A definite integral can be evaluated only after the appropriate integration has been performed. In some problems, as we have already seen, this may require a change of variable, e.g. from x to u. In such cases the limits of integration, originally values of x, can be transformed into values of u allowing direct calculation of the required definite integral.

EXAMPLES 9n (continued)

3) By using the substitution $u \equiv x^3 + 1$, evaluate $\int_0^1 x^2 \sqrt{x^3 + 1} \, dx$.

If $u \equiv x^3 + 1$ then $\ldots du \equiv \ldots 3x^2 \, dx$

and $\begin{cases} x = 0 \Rightarrow u = 1 \\ x = 1 \Rightarrow u = 2 \end{cases}$

Hence
$$\int_0^1 x^2 \sqrt{x^3 + 1} \, dx \equiv \int_1^2 \sqrt{u} \, \frac{du}{3}$$

$$= \left[\frac{2}{3} \frac{u^{\frac{3}{2}}}{3} \right]_1^2$$

$$= \tfrac{2}{9} \{ 2\sqrt{2} - 1 \}$$

4) Evaluate $\int_0^1 \dfrac{1}{(4 - x^2)^{\frac{3}{2}}} \, dx$ by using the substitution $x \equiv 2 \sin \theta$.

If $x \equiv 2 \sin \theta$ then $\ldots dx \equiv \ldots 2 \cos \theta \, d\theta$

and $\begin{cases} x = 0 \Rightarrow \sin \theta = 0 \Rightarrow \theta = 0 \\ x = 1 \Rightarrow \sin \theta = \frac{1}{2} \Rightarrow \theta = \dfrac{\pi}{6} \end{cases}$

Hence
$$\int_0^1 \frac{1}{(4 - x^2)^{\frac{3}{2}}} \, dx \equiv \int_0^{\frac{\pi}{6}} \frac{1}{8 \cos^3 \theta} 2 \cos \theta \, d\theta$$

$$\equiv \frac{1}{4} \int_0^{\frac{\pi}{6}} \sec^2 \theta \, d\theta$$

$$= \frac{1}{4} \left[\tan \theta \right]_0^{\frac{\pi}{6}}$$

$$= \frac{1}{4} \left\{ \frac{1}{\sqrt{3}} - 0 \right\} = \frac{\sqrt{3}}{12}$$

Definite Integration by Parts

When using the formula

$$\int v \frac{du}{dx} \, dx = uv - \int u \frac{dv}{dx} \, dx$$

it must be appreciated that the term uv on the R.H.S. is fully integrated. Consequently in a definite integration, uv must be *evaluated between the appropriate boundaries*

i.e. $$\int_a^b v \frac{du}{dx} dx = \left[uv\right]_a^b - \int_a^b u \frac{dv}{dx} dx$$

EXAMPLES 9n (continued)

5) Evaluate $\int_0^1 x e^x dx$

$$\int x e^x dx \equiv \int v \frac{du}{dx} dx$$

where $$v = x \quad \text{and} \quad \frac{du}{dx} = e^x$$

Hence $$\int_0^1 x e^x dx = \left[x e^x\right]_0^1 - \int_0^1 e^x dx$$

$$= \left[x e^x\right]_0^1 - \left[e^x\right]_0^1$$

$$= (e^1 - 0) - (e^1 - e^0)$$

$$= e - e + 1$$

i.e. $$\int_0^1 x e^x dx = 1$$

EXERCISE 9n

Evaluate the following definite integrals

1) $\int_2^3 (x^2 + 2x - 1) \, dx$

2) $\int_{\frac{\pi}{6}}^{\frac{\pi}{3}} \cos 2\theta \, d\theta$

3) $\int_0^1 e^{4x} \, dx$

4) $\int_3^4 \frac{1}{x+1} \, dx$

5) $\int_2^7 \sqrt{(x+2)} \, dx$

6) $\int_0^{\frac{\pi}{4}} \sec^2 3\theta \, d\theta$

7) $\int_0^{\frac{\pi}{6}} \sin^2 \theta \, d\theta$

8) $\int_1^2 x e^{x^2} \, dx$

9) $\int_{\frac{\pi}{6}}^{\frac{\pi}{2}} \sin^4 x \cos x \, dx$

10) $\int_{\frac{\pi}{3}}^{\frac{\pi}{2}} \frac{\sin \theta}{1 - \cos \theta} \, d\theta$

11) $\int_0^{\frac{\pi}{2}} x \sin x \, dx$

12) $\int_0^1 \frac{x+1}{(x+2)(x+3)} \, dx$

13) $\int_1^2 \frac{x}{x^2+1} \, dx$

14) $\int_0^{\frac{\pi}{4}} \cos 3\theta \cos \theta \, d\theta$

15) $\int_0^1 x^2 (x^3 + 1)^4 \, dx$

16) $\int_1^2 x^3 \ln x \, dx$

17) $\int_0^{\frac{\pi}{2}} \cos x \sqrt{\sin x} \, dx$

18) $\int_1^3 \left(x^2 - \frac{1}{x^2} \right) dx$ 19) $\int_1^2 \frac{(e^u + e^{-u})^2}{e^u} du$ 20) $\int_0^{\frac{\pi}{2}} e^x \sin x \, dx$

Evaluate the following, using the suggested change of variable, or otherwise.

21) $\int_0^{\sqrt{3}} \frac{x}{\sqrt{(x^2 + 1)}} dx$; $u^2 \equiv x^2 + 1$ 22) $\int_3^4 x(x - 3)^7 dx$; $u \equiv x - 3$

23) $\int_0^{\frac{\pi}{6}} \cos \theta \sqrt{1 - 2 \sin \theta} \, d\theta$; $x \equiv 1 - 2 \sin \theta$

24) $\int_0^{\frac{\pi}{6}} \sec^3 \theta \tan \theta \, d\theta$; $u \equiv \sec \theta$ 25) $\int_{1.5}^3 \frac{1}{\sqrt{(9 - x^2)}} dx$; $x \equiv 3 \sin \theta$

26) $\int_{\frac{1}{4}}^{\frac{\sqrt{3}}{4}} \frac{1}{1 + 16u^2} du$; $4u \equiv \tan x$ 27) $\int_0^\pi \frac{\sin x}{\cos^4 x} dx$; $\cos x \equiv u$

28) $\int_0^2 (x + 1)\sqrt{(x^2 + 2x)} \, dx$; $x^2 + 2x \equiv 0$

29) $\int_0^2 \frac{x + 1}{\sqrt{(x^2 + 2x + 8)}} dx$; $u \equiv x^2 + 2x + 8$

30) $\int_{\pi/4}^{\arctan 2} \frac{1}{3 \sin 2\theta - 4 \cos 2\theta} d\theta$; $\tan \theta \equiv t$

$\left(\text{use also}\quad \sin 2\theta \equiv \frac{2t}{1 + t^2}, \quad \cos 2\theta \equiv \frac{1 - t^2}{1 + t^2} \quad \text{and} \quad \tan^2 \theta + 1 \equiv \sec^2 \theta \right)$

AREA BOUNDED PARTLY BY A CURVE

If the area, A, between two specified values of x, ($x = a$ and $x = b$)   is bounded by the x-axis and a curve $y = f(x)$,   we have seen that A can be found from $\int_a^b f(x) \, dx$

The reader is now in a position to evaluate this definite integral and so can use this method for determining certain areas.

It is recommended that the required area is first treated as a summation of areas of elements, a typical element being indicated on a diagram.

EXAMPLES 9p

1) Find the area bounded by the x-axis, the y-axis, the curve $y = e^x$ and the line $x = 2$.

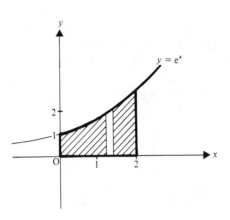

This area can be divided into vertical strips of approximate area area $y\,\delta x$.

Thus the required area

$$= \lim_{\delta x \to 0} \sum_{x=0}^{x=2} y\,\delta x$$

$$= \int_0^2 y\,dx$$

But $\quad \int_0^2 y\,dx = \int_0^2 e^x\,dx$

$$= \left[e^x \right]_0^2 = e^2 - e^0$$

So the required area is $(e^2 - 1)$.

2) Find the area between the curve $\quad x = 4 - y^2\quad$ and the y-axis.

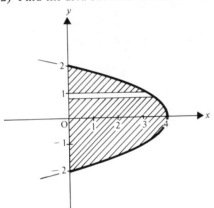

A vertical strip in this case would have *both ends on the same curve* Its length therefore cannot be defined easily so we use a horizontal strip instead. Its approximate area is $x\,\delta y$.

The curve crosses the y-axis at $\quad y = -2\quad$ and $\quad y = 2,\quad$ so our summation is bounded by these values,

i.e. $\qquad$ required area $= \lim_{\delta y \to 0} \sum_{y=-2}^{y=2} x\,\delta y$

$$= \int_{-2}^{2} x\,dy$$

But $\qquad \int_{-2}^{2} x\,dy = \int_{-2}^{2} (4 - y^2)\,dy = \left[4y - \frac{y^3}{3} \right]_{-2}^{2}$

$$= (8 - \tfrac{8}{3}) - (-8 + \tfrac{8}{3})$$

Thus the required area is $10\tfrac{2}{3}$ square units.

The Meaning of a Negative Result

Consider the area bounded by $y = 4x^3$ and the x-axis if the other boundaries are the lines

(a) $x = -2$ and $x = -1$ (b) $x = 1$ and $x = 2$

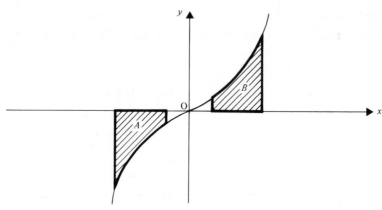

This curve is symmetrical about the origin so the two shaded areas are equal.

(a) Considering A

$$\lim_{\delta x \to 0} \sum_{x=-2}^{-1} y\, \delta x = \int_{-2}^{-1} y\, dx$$

$$= \int_{-2}^{-1} 4x^3\, dx$$

$$= \left[x^4 \right]_{-2}^{-1}$$

$$= 1 - 16 = -15$$

(b) Considering B

$$\lim_{\delta x \to 0} \sum_{x=1}^{2} y\, \delta x = \int_{1}^{2} y\, dx$$

$$= \int_{1}^{2} 4x^3\, dx$$

$$= \left[x^4 \right]_{1}^{2}$$

$$= 16 - 1 = 15$$

So we see that, while the magnitudes of the two areas are equal, the result for the area of A which is below the x-axis, is negative. This is explained by the fact that the length of a strip in A was taken as y, which is negative for the part of the curve bounding A.

Note. Care must be taken with problems involving a curve that crosses the x-axis between the boundary values.

EXAMPLES 9p (continued)

3) Find the area enclosed between the curve $y = x(x-1)(x-2)$ and the x-axis.

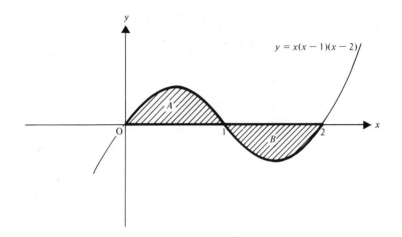

The area enclosed between the curve and the x-axis is the sum of the areas A and B.

For A we use

$$\int_0^1 y \, dx = \int_0^1 (x^3 - 3x^2 + 2x) \, dx$$

$$= \left[\frac{x^4}{4} - x^3 + x^2 \right]_0^1$$

$$= \tfrac{1}{4}$$

For B we use

$$\int_1^2 (x^3 - 3x^2 + 2x) \, dx = \left[\frac{x^4}{4} - x^3 + x^2 \right]_1^2$$

$$= (4 - 8 + 4) - (\tfrac{1}{4} - 1 + 1)$$

$$= -\tfrac{1}{4}$$

The minus sign refers only to the *position* of area B relative to the x-axis. The actual area is $\tfrac{1}{4}$ square unit.
So the total shaded area is $\tfrac{1}{4} + \tfrac{1}{4} = \tfrac{1}{2}$ square unit.

EXERCISE 9p

Find the areas bounded by the specified lines and curves in Questions 1–10.

1) The x- and y-axes, the line $x = 3$ and the curve $y = x^2 + 1$.

2) The x-axis, the lines $x = 1$ and $x = 4$ and the curve $xy = 2$.

3) The y-axis, the lines $y = 2$ and $y = 4$ and the curve $y = e^{-x}$.

4) The x-axis and the curve $y = 1 - x^2$.

5) The y-axis and the curve $x = 9 - y^2$.

6) The curve $y = \sin x$, the x-axis and the line $x = \dfrac{\pi}{2}$.

7) The x-axis, the line $x = \dfrac{\pi}{4}$ and the curve $y = \tan x$.

8) The curve $y = x^2 - 1$, the positive x-axis and the negative y-axis.

9) The curve $y = \ln x$, the y-axis, the x-axis and the line $y = 1$.

10) The curve $y = 9 - x^2$ and the x-axis.

11) Evaluate:

(a) $\displaystyle\int_0^2 (x - 2)\,dx$ (b) $\displaystyle\int_2^4 (x - 2)\,dx$ (c) $\displaystyle\int_0^4 (x - 2)\,dx$

Interpret your results with the help of a graph.

12) Find the area enclosed between the curve $y = \cos x$ and the x-axis bounded by the lines $x = 0$ and $x = \pi$.

13) If $y = x^2$, show by means of sketch graphs and *not* by evaluating the integrals, that

$$\int_0^1 y\,dx = 1 - \int_0^1 x\,dy$$

SUMMARY

1)

Function	Integral	
x^n	$\dfrac{1}{n+1}x^{n+1}$	$(n \neq -1)$
e^x	e^x	
$\dfrac{1}{x}$	$\ln|x|$	
$\cos x$	$\sin x$	
$\sin x$	$-\cos x$	
$\sec^2 x$	$\tan x$	

Function	Integral	
$\cosec^2 x$	$-\cot x$	
$\tan x$	$\ln|\sec x|$	
$\cot x$	$\ln|\sin x|$	
$\dfrac{1}{\sqrt{(1-x^2)}}$	$\arcsin x$	
$\dfrac{1}{1+x^2}$	$\arctan x$	
$\sin^p x \cos x$	$\dfrac{1}{p+1}\sin^{p+1}x$	$(p \neq -1)$
$\cos^p x \sin x$	$-\dfrac{1}{p+1}\cos^{p+1}x$	$(p \neq -1)$
$\tan^p x \sec^2 x$	$\dfrac{1}{p+1}\tan^{p+1}x$	$(p \neq -1)$

2) $\int f'(x)\, e^{f(x)}\, dx = e^{f(x)} + K$

3) $\int \dfrac{f'(x)}{f(x)}\, dx = \ln|f(x)| + K$

4) $\int v\, \dfrac{du}{dx}\, dx = uv - \int u\, \dfrac{dv}{dx}\, dx$

5) $\displaystyle\lim_{\delta x \to 0} \sum_{x=a}^{x=b} f(x)\, \delta x = \int_a^b f(x)\, dx$

MULTIPLE CHOICE EXERCISE 9

(*Instructions for answering these questions are given on p. xii.*)

TYPE I

1) $\int \left(x^2 - \dfrac{1}{x^2} + \sin x \right) dx$ is:

(a) $\dfrac{x^3}{3} + \dfrac{1}{x} + \cos x + K$ (b) $\dfrac{x^3}{3} - \dfrac{2}{x} - \cos x + K$

(c) $2x - \dfrac{3}{x^2} + \cos x + K$ (d) $\dfrac{x^3}{3} + \dfrac{1}{x} - \cos x + K$

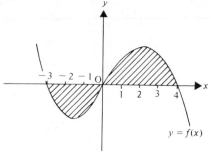

2) The shaded area in the diagram is given by:

(a) $\int_{-3}^{4} f(x)\,dx$

(b) $\int_{-3}^{0} f(x)\,dx + \int_{0}^{4} f(x)\,dx$

(c) $\int_{-3}^{1} f(x)\,dx + \int_{1}^{4} f(x)\,dx$

(d) none of these.

3) e^{x^2} could be the integral w.r.t. x of:

(a) e^{2x} (b) $2x\,e^{x^2}$ (c) $\dfrac{e^{x^2}}{2x}$ (d) $x^2\,e^{x^2} - 1$ (e) none of these.

4) $x - \ln x^2 + K$ is the result of integrating w.r.t. x:

(a) $\dfrac{1}{1-x^2}$ (b) $\dfrac{1-2x}{x^2}$ (c) $\dfrac{x-2}{x}$

(d) $1 - \dfrac{2}{x^2}$ (e) none of these.

5) If $\int_{1}^{5} \dfrac{dx}{2x-1} = \ln K$, the value of K is:

(a) 9   (b) 3   (c) undefined   (d) 81   (e) 8.

6) $I = \int_{1}^{2} x\sqrt{x^2 - 1}\,dx$ is found as follows. Where does an error first occur?

(a) Let $u \equiv x^2 - 1$ (b) $\ldots du \equiv \ldots 2x\,dx$

(c) $I = \dfrac{1}{2} \int_{1}^{2} u^{\frac{1}{2}}\,du$ (d) $I = \dfrac{3}{4}\left[u^{\frac{3}{2}}\right]_{1}^{2}$

7) $\int_{0}^{\frac{\pi}{6}} \sin^n x \cos x \, dx = \frac{1}{64}$; n is:

(a) 6   (b) 5   (c) 4   (d) 3   (e) none of these.

8) The differential equation of a curve is $\dfrac{x}{y}\dfrac{dy}{dx} = 1$.   The sketch of the curve could be:

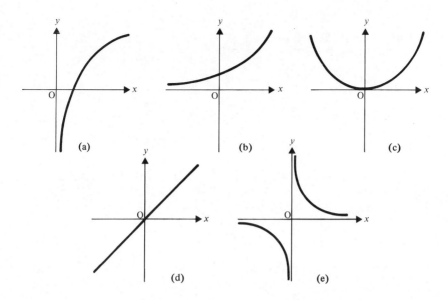

(a)   (b)   (c)

(d)   (e)

9) The value of $\displaystyle\int_0^2 2e^{2x}\,dx$ is:

(a) e^4 (b) $e^4 - 1$ (c) ∞ (d) $4e^4$ (e) $\tfrac{1}{2}e^4$.

TYPE II

10) $\displaystyle\int_0^n \tan x\,dx$ can be evaluated if:

(a) $n = \dfrac{\pi}{4}$ (b) $n = -\dfrac{\pi}{3}$ (c) $n = \dfrac{\pi}{2}$ (d) $n = -\dfrac{\pi}{2}$.

11) Which of the following differential equations can be solved by separating the variables:

(a) $x\dfrac{dy}{dx} = y + x$ (b) $xy\dfrac{dy}{dx} = x + 1$

(c) $e^{x+y} = y\dfrac{dy}{dx}$ (d) $x + \dfrac{dy}{dx} = \ln y$

12) $\displaystyle\int_1^2 xe^x\,dx$:

(a) is a definite integral   (b) is equal to $xe^x - e^x$

(c) is equal to $\left[\tfrac{1}{2}e^{x^2}\right]_1^2$ (d) can be integrated by parts.

13) Using $x \equiv \sin \theta$ transforms $\displaystyle \int \frac{x^2}{\sqrt{(1-x^2)}}\,dx$ into:

(a) $\displaystyle \int \frac{\sin^2 \theta}{\cos \theta}\,d\theta$ (b) $\displaystyle \tfrac{1}{2} \int (1 - \cos 2\theta)\,d\theta$

(c) $\displaystyle - \int \sin^2 \theta\,d\theta$ (d) $\displaystyle \tfrac{1}{2} \int (1 + \cos 2\theta)\,d\theta$.

14) Integration by parts can be used to find:

(a) $\displaystyle \int x^2\, e^x\,dx$ (b) $\displaystyle \int e^x \ln x\,dx$ (c) $\displaystyle \int \ln x\,dx$ (d) $\displaystyle \int (\ln x)(\sin x)\,dx$.

15) Which of the following definite integrals can be evaluated:

(a) $\displaystyle \int_0^1 \frac{1}{x-1}\,dx$ (b) $\displaystyle \int_0^{\frac{\pi}{2}} \sin x\,dx$

(c) $\displaystyle \int_1^2 \sqrt{1-x^2}\,dx$ (d) $\displaystyle \int_{-2}^1 \ln x\,dx$

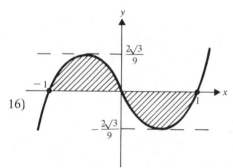

16)   The shaded area is equal to:

(a) $\displaystyle \int_{-1}^1 y\,dx$ (b) zero   (c) $\displaystyle 2 \int_0^{(2\sqrt{3})/9} x\,dy$

(d) $\displaystyle \int_{-1}^0 y\,dx + \int_0^1 y\,dx$ (e) $\displaystyle \int_{-1}^0 y\,dx - \int_0^1 y\,dx$.

TYPE III

17) (a) $\displaystyle \frac{dy}{dx} = \cos x\, e^{\sin x}$

(b) $y = e^{\sin x}$

18) (a) An area can be found by evaluating $\displaystyle \int_0^p f(x)\,dx$.

(b) An area is bounded by the curve $y = f(x)$, the x- and y-axes and the line $y = p$.

19) (a) $y = \int x \cos x \, dx$.

(b) $y = K + \cos x + x \sin x$.

20) (a) x and y are related by a linear differential equation.

(b) y is a quadratic function of x.

TYPE IV

21) Find the shaded area.

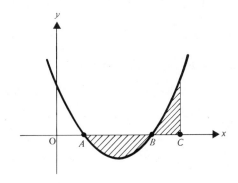

(a) The equation of the curve is $y = x^2 + ax + b$.
(b) At A, $x = 1$.
(c) The curve crosses the y-axis at the point $(0, 3)$.
(d) At C, $x = 4$.

22) Find the equation of a curve.
(a) Its gradient is proportional to x.
(b) It passes through the origin.
(c) y is a quadratic function of x.
(d) It passes through the point $(1, 2)$.

23) Evaluate $\int_p^q f(x) \, dx$.

(a) The constant of integration is zero.
(b) $f(x)$ is fully defined for $p \leqslant x \leqslant q$.
(c) p is given.
(d) q is given.

TYPE V

24) A differential equation must contain $\dfrac{dy}{dx}$.

25) $\int \tan x \, dx = \sec^2 x + K.$

26) $\int_0^a f(y) \, dy = \lim_{\delta y \to 0} \sum_{y=0}^a f(y) \, \delta y.$

27) $\left[f(x) \right]_0^a \equiv f(a) - 0.$

MISCELLANEOUS EXERCISE 9

1) Evaluate:

(a) $\int_1^3 \frac{x+2}{x+1} \, dx$ (b) $\int_0^{\frac{\pi}{2}} \sin^4 x \cos x \, dx.$ (U of L)p

2) (a) Evaluate $\int_0^1 \frac{x^2 \, dx}{(1+x^3)^2}$ (b) Evaluate $\int_5^6 \frac{dx}{(x-2)(x-4)}$

giving the answer to three decimal places.   (JMB)p

3) Evaluate the integrals:

(a) $\int x(1+x^2)^5 \, dx$ (b) $\int x \, e^{-2x} \, dx$

(c) $\int_2^3 \frac{dx}{x(x^2-1)}$ (d) $\int_0^{\frac{\pi}{2}} \sin^2 x \cos^3 x \, dx$ (O)

4) (a) Find: (i) $\int (e^{2x} - 1)^2 \, dx$ (ii) $\int \frac{\cos \theta \, d\theta}{\sqrt{\sin \theta}}$.

    (b) Evaluate: (i) $\int_0^{\frac{\pi}{2}} 2 \cos 2\theta \cos \theta \, d\theta$ (ii) $\int_3^4 \frac{5 \, dx}{x^2 + x - 6}$.   (AEB)'74

5) Evaluate:

(a) $\int_0^{\frac{\pi}{6}} \tan 2\theta \, d\theta$ (b) $\int_0^1 x(1-x)^{\frac{1}{2}} \, dx.$ (JMB)

6) Evaluate the integrals:

(a) $\int_0^2 x(x^2+1)^3 \, dx$ (b) $\int_2^3 \frac{dx}{(4-x)(x-1)}$ (c) $\int_0^{\frac{\pi}{2}} \tan \tfrac{1}{2} x \, dx.$ (O)

7) (a) Evaluate:

$$\int_{\frac{\pi}{6}}^{\frac{\pi}{4}} 2 \sin 3x \cos 2x \, dx$$

     giving your answer correct to two significant figures.

    (b) Using the substitution $t = \tan x,$ or otherwise, find:

$$\int \frac{dx}{4 \cos^2 x - 9 \sin^2 x}$$ (C)

8) Evaluate the definite integrals:

(a) $\int_1^2 \frac{(x^2-1)^2}{x^3}\,dx$ (b) $\int_{\frac{\pi}{2}}^{\pi} (\sin 3x + \cos \tfrac{1}{2}x)\,dx$ (c) $\int_1^2 \frac{x+2}{x(x+4)}\,dx$.

(U of L)

9) (a) By use of partial fractions, or otherwise, find:

$$\int \frac{x}{x^2-2x-3}\,dx.$$

(b) Using the substitution $z = 1-x$, or otherwise, evaluate:

$$\int_0^1 x^2(1-x)^{\frac{1}{2}}\,dx.$$

(c) Solve the differential equation:

$$\frac{dy}{dx} = 1-y$$

given that $y < 1$ and that $y = 0$ when $x = 0$.   (C)

10) Evaluate, correct to three decimal places:

(a) $\int_0^1 \frac{2x^2\,dx}{2x+1}$ (b) $\int_0^{\frac{\pi}{3}} \sin^2 x \cos^2 x\,dx$ (c) $\int_e^{e^2} \frac{dx}{x\ln x}$.   (U of L)

11) (a) Evaluate $\int_4^5 \frac{2x\,dx}{x^2-4x+3}$.

(b) By using the substitution $x \equiv \sec^2 y$, or otherwise, evaluate

$$\int_2^5 \frac{dx}{x^2\sqrt{(x-1)}}.$$ (U of L)p

12) (a) Find $\int \frac{2x-1}{(x+1)^2}\,dx$ and $\int \left(e^x - \frac{1}{e^x}\right)^2\,dx$.

(b) Evaluate $\int_0^{\frac{\pi}{4}} \tan^2 x\,dx$.   (C)

13) Evaluate the following integrals:

(a) $\int_0^1 \frac{8}{3+4x}\,dx$ (b) $\int_0^1 \frac{8}{\sqrt{(3+4x)}}\,dx$ (c) $\int_0^1 \frac{8x}{3+4x}\,dx$.

(U of L)p

14) (a) If $e^x = \tan 2y$ prove that $\dfrac{d^2y}{dx^2} = \dfrac{e^x - e^{3x}}{2(1+e^{2x})^2}$.

(b) Find (i) $\int \frac{dx}{\sqrt{x}-x}$ (ii) $\int (1-3\cos^2 x)^{\frac{1}{2}} \sin 2x\,dx$.   (AEB)'69

15) (a) Evaluate $\int_1^2 \frac{(x-1)^2}{x^3}\,dx$ giving the answer correct to three decimal places.

(b) Find $\int_{\alpha}^{7\alpha} \sin 2(x + \alpha)\, dx$ where $\alpha = \dfrac{\pi}{24}$.

(c) $\int_{0}^{3} (x + 1)^{\frac{3}{2}}\, dx$, and hence find $\int_{0}^{3} x\sqrt{(x + 1)}\, dx$.   (U of L)

16) (a) Find $\int 2\cos 3x \sin x\, dx$ and $\int \dfrac{x - 2}{\sqrt{(x - 1)}}\, dx$.

   (b) Using the substitution $t = \tan x$, or otherwise, evaluate:
   $$\int_{0}^{\frac{\pi}{3}} \frac{dx}{9 - 8\sin^2 x}.$$ (C)

17) Find:

(a) $\int \dfrac{x + 1}{x(2x + 1)}\, dx$ and   (b) $\int \dfrac{x(2x + 1)}{x + 1}\, dx$.   (JMB)

18) (a) If $a > 1$ and $\int_{1}^{a} \dfrac{x^4 - 1}{x^3}\, dx = \dfrac{9}{8}$, find a.

   (b) If n is a positive integer, find in terms of n the three possible values
   of $\int_{\frac{\pi}{2}}^{\pi} \cos nx\, dx$.

   (c) Evaluate $\int_{0}^{\frac{\pi}{4}} \dfrac{2\cos x - \sin x}{2\sin x + \cos x}\, dx$, correct to three decimal places.

   (U of L)

19) Express $\dfrac{x - 2}{2x^2 - x - 3}$ in partial fractions and hence evaluate:
   $$\int_{2}^{3} \frac{(x - 2)\, dx}{2x^2 - x - 3}.$$ (AEB)'73

20) (a) Find $\int \dfrac{x - 7}{2x^2 - 3x - 2}\, dx$.

   (b) Evaluate $\int_{0}^{\frac{\pi}{3}} \sin^3 x\, dx$.

   (c) Using the substitution $x = 4\sin^2\theta$, or otherwise, show that:
   $$\int_{0}^{2} \sqrt{x(4 - x)}\, dx = \pi.$$ (C)

21) Evaluate:

(a) $\int_{0}^{\frac{\pi}{4}} \sin 5x \cos 3x\, dx$ (b) $\int_{0}^{1} xe^{-3x}\, dx$.   (JMB)p

22) Evaluate the integrals:

(a) $\int_0^{\frac{\pi}{2}} \sin 2x \cos 3x \, dx$ (b) $\int_0^{\frac{\pi}{2}} \sin^2 x \cos^2 x \, dx$

(c) $\int_0^3 \frac{x \, dx}{\sqrt{(25 - x^2)}}$ (d) $\int_0^1 \frac{x \, dx}{6 - 5x + x^2}$.   (O)

23) (a) Evaluate $\int_0^{\frac{\pi}{4}} \sin 3\theta \sin \theta \, d\theta$.

    (b) Find $\int \sin^2 \theta \cos^3 \theta \, d\theta$.

    (c) Find $\int \frac{x}{\sqrt{(3 + x)}} \, dx$.   (C)

24) Evaluate:

(a) $\int_1^{15} \frac{x + 2}{(x + 1)(x + 3)} \, dx$ (b) $\int_0^{\frac{\pi}{3}} x \sin 3x \, dx$ (c) $\int_{-1}^2 x^2 \sqrt{(x^3 + 1)} \, dx$.

                                                                (U of L)

25) Find y in terms of x given that:

$$x \frac{dy}{dx} = (1 - 2x^2)y, \quad (x > 0)$$

and that $y = 1$ when $x = 1$.   (U of L)p

26) (a) Evaluate $\int_0^1 \frac{1 - 4x}{3 + x - 2x^2} \, dx$.

    (b) Find $\int xe^{2x} \, dx$.

    (c) Solve the differential equation $\frac{dy}{dx} = xy$ given that $y = 2$ when

    $x = 0$.   (C)

27) Solve the differential equation $(1 + \cos 2x) \frac{dy}{dx} - (1 + e^y) \sin 2x = 0$

given that $y = 0$ when $x = \frac{\pi}{4}$.   (AEB)'73

28) Solve the differential equation $(3x + 5)^2 \frac{dy}{dx} = \frac{1 + 4y^2}{1 + y}$, given that

$y = 0$ when $x = 0$.   (AEB)'76

29) (a) Evaluate:

    (i) $\int_0^1 \frac{1 + x}{1 + 2x} \, dx$ (ii) $\int_0^{\frac{\pi}{2}} \sin x \cos^2 x \, dx$.

(b) Sketch the arc of the curve $y = 2x - x^2$ for which y is positive. Find the area of the region which lies between this arc and the x-axis.

(U of L)p

30) Solve the differential equation $(1 + x^2)\dfrac{dy}{dx} - y(y + 1)x = 0$, given that $y = 1$ when $x = 0$.

(AEB)'76

31) P is a point on a curve. The normal at P meets the x-axis at a point Q, and the curve is such that PQ is always equal to OP where O is the origin. Show that

$$y\frac{dy}{dx} = x$$

Hence find the equation of the curve.

32) Find the solution of the differential equation

$$\frac{dy}{dx} = \frac{x(y^2 - 1)}{y(x^2 + 1)}$$

for which $y = 3$ when $x = 1$.

(O)p

33) (a) Find y in terms of x given that

$$(1 + x)\frac{dy}{dx} = (1 - x)y$$

and that $y = 1$ when $x = 0$.

(b) A curve passes through $(2, 2)$ and has gradient at (x, y) given by the differential equation

$$ye^{y^2}\frac{dy}{dx} = e^{2x}$$

Find the equation of the curve. Show that the curve also passes through the point $(1, \sqrt{2})$ and sketch the curve.

(U of L)

34) If $\dfrac{dy}{dx} = \dfrac{y(y + 1)}{x(x + 1)}$, and $y = 2$ when $x = 1$, find y in terms of x.

(O)p

35) (a) Find:

(i) $\displaystyle\int \frac{x^2}{(x + 2)}\,dx$ (ii) $\displaystyle\int \sin 3x \cos 2x \,dx$.

(b) Find the coordinates of P, the point of intersection of the curves

$$y = e^x, \quad y = 2 + 3e^{-x}$$

If these curves cut the y-axis at the points A and B, calculate the area bounded by AB and the arcs AP and BP.

(U of L)

36) The curve $y = 2 \cos x - 1$ cuts the axis of x at two points P and Q whose abscissae lie in the range $-\frac{1}{2}\pi < x < \frac{1}{2}\pi$.   Find the area enclosed by the straight line PQ and the arc of the curve between P and Q.   (C)p

37) Calculate the area bounded by the lines $x = 0$, $x = 1$, $y = 1$,   and the part of the graph of

$$y = \frac{x^2}{x^2 + 1}$$

between $x = 0$ and $x = 1$.   (JMB)p

38) The equations of the tangents to the curve

$$y = x(x - a)(x - b),$$

where a and b are constants, at the points $(0, 0)$ and $(a, 0)$ are $y = 4x$ and $y = 3 - 3x$ respectively. Calculate the values of a and b and sketch the curve.

Find the area in the right half plane $x > 0$ which is bounded by the x-axis and the curve, and which lies below the x-axis.   (U of L)p

39) The curves $y = 3 \sin x$, $y = 4 \cos x$ $(0 \leqslant x \leqslant \frac{1}{2}\pi)$ intersect at the point A, and meet the axis of x at the origin O and the point $B(\frac{1}{2}\pi, 0)$ respectively. Prove that the area enclosed by the arcs OA, AB and the line OB is 2 square units.   (C)p

40) (a) Find y as a function of x when

$$\frac{1}{y}\frac{dy}{dx} - x = xy$$

and $y = 1$ when $x = 0$.

(b) If $x - y = z$,   express $\dfrac{dy}{dx}$ in terms of $\dfrac{dz}{dx}$. Using the substitution

$x - y = z$,   or otherwise, solve the differential equation

$$\frac{dy}{dx} = x - y$$

given that $y = 0$ when $x = 0$.   (U of L)

41) In a chemical reaction in which a compound X is formed from a compound Y and other substances, the masses of X and Y present at time t are x and y respectively. The sum of the two masses is constant and at any time the rate at which x is increasing is proportional to the product of the two masses at that time. Show that the equation governing the reaction is of the form

$$\frac{dx}{dt} = kx(a - x)$$

and interpret the constant a. If $x = \dfrac{a}{10}$ at time $t = 0$,   find in terms of k and a the time at which $y = \dfrac{a}{10}$.   (U of L)

CHAPTER 10

COORDINATE GEOMETRY II

ANALYTICAL GEOMETRY IN TWO DIMENSIONS

Any line, straight or curved, is made up of an infinite set of points. Any point, P, can be located by a pair of coordinates in a suitable frame of reference. If there is an equation whose solution set is made up of the coordinates of every point P on the line, but excludes the coordinates of any point not on the line, we say that this is the equation of the line. (The form of the equation depends upon the shape of the line.)
Conversely, an equation which relates the coordinates of a point P can be displayed graphically by plotting corresponding coordinates on a suitable graph. (The shape of the line depends upon the form of the equation.)
Hence:

(a) the geometric properties of a line can be expressed in the form of an equation,

(b) the geometric properties of a line can be found by analysing its equation.

There have been references, in earlier chapters, to coordinates of three types; Cartesian, parametric and polar. Thus lines can have Cartesian, parametric or polar equations.
The techniques used in the analysis of straight lines and curves depend, to some extent, on the frame of reference being used. In this chapter we will examine a variety of the methods adopted when using Cartesian coordinates.

THE STRAIGHT LINE

Any geometric work makes frequent use of straight lines. Their analysis, which began in Chapter 4, will now be developed. Before proceeding, the reader should remind himself of the results already obtained.

Division of a Straight Line in a Given Ratio

Suppose that a point $P(X, Y)$ divides the line joining $A(x_1, y_1)$ to $B(x_2, y_2)$ in the ratio $\lambda : \mu$

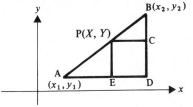

In the diagram

triangles $\begin{cases} APE \\ PBC \end{cases}$ are similar.

Therefore $\dfrac{AP}{PB} = \dfrac{AE}{PC}$.

But $AE = X - x_1$, $PC = x_2 - X$ and $\dfrac{AP}{PB} = \dfrac{\lambda}{\mu}$

Therefore $\dfrac{X - x_1}{x_2 - X} = \dfrac{\lambda}{\mu}$

$\Rightarrow$ $$X = \frac{\lambda x_2 + \mu x_1}{\lambda + \mu}$$

Similarly $$Y = \frac{\lambda y_2 + \mu y_1}{\lambda + \mu}$$

These formulae, which are quotable, apply to both internal and external division. Their use is not always necessary however. When the coordinates of A and B are known numbers, a diagram, together with simple mental arithmetic, is often adequate,

e.g. to divide the line joining $A(-2, 5)$ to $B(4, 2)$ internally and externally in the ratio $2:1$ the following diagrams are effective.

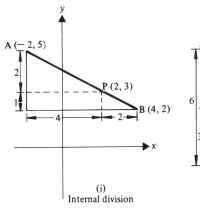

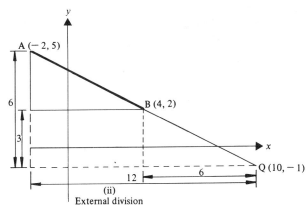

(i) Internal division

(ii) External division

Alternatively the division formulae can be used as follows:

(a) For the internal ratio $2:1$, $\lambda = 2$ and $\mu = 1$.

Thus, at P, $x = \dfrac{2(4) + 1(-2)}{2 + 1} = 2$

and
$$y = \frac{2(2) + 1(5)}{2 + 1} = 3$$

(b) For the external ratio $2:1$, $\lambda = 2$ and $\mu = -1$

Thus, at Q, $\quad x = \dfrac{2(4) - 1(-2)}{2 - 1} = 10$

and
$$y = \frac{2(2) - 1(5)}{2 - 1} = -1$$

(**Note:** In the external division the sign of μ is opposite to the sign of λ because the direction of the line segment QB is opposite to the direction of the line segment AQ.

For the same reason, external division is sometimes denoted by a negative ratio or fraction, e.g. $2:-1$ or $-\frac{2}{1}$.)

The Centroid of a Triangle

The centroid, G, of a triangle is the point of intersection of the medians and G divides each median in the ratio $2:1$.

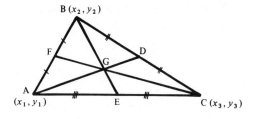

i.e. $\dfrac{AG}{GD} = \dfrac{BG}{GE} = \dfrac{CG}{GF} = \dfrac{2}{1}$

If the coordinates of A, B and C are (x_1, y_1), (x_2, y_2) and (x_3, y_3) respectively then

the coordinates of D, the midpoint of BC, are $\left(\dfrac{x_2 + x_3}{2}, \dfrac{y_2 + y_3}{2} \right)$

But $\quad \dfrac{AG}{GD} = \dfrac{2}{1}$

Hence, at G, $\qquad x = \left[2\left(\dfrac{x_2 + x_3}{2} \right) + x_1 \right] \div 3$

and $\qquad y = \left[2\left(\dfrac{y_2 + y_3}{2} \right) + y_1 \right] \div 3$

This result shows that the coordinates of the centroid of a triangle are the average of the coordinates of the vertices,

i.e.

$$\left(\tfrac{1}{3}(x_1 + x_2 + x_3), \tfrac{1}{3}(y_1 + y_2 + y_3)\right)$$

The Angle between Two Straight Lines

Consider two lines with gradients m_1 ($= \tan\theta_1$) and m_2 ($= \tan\theta_2$). The angle, α, between the lines is given by

$$\alpha = \theta_1 - \theta_2$$

Therefore

$$\tan\alpha = \tan(\theta_1 - \theta_2)$$

$$= \frac{\tan\theta_1 - \tan\theta_2}{1 + \tan\theta_1 \tan\theta_2}$$

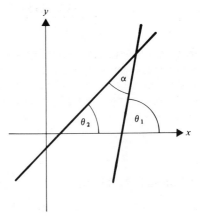

$\Rightarrow$

$$\tan\alpha = \frac{m_1 - m_2}{1 + m_1 m_2}$$

This result is quotable:
e.g. if two straight lines l_1 and l_2 have equations $3x - 2y = 5$ and $4x + 5y = 1$ then

$$\text{gradient of } l_1 = m_1 = \tfrac{3}{2}$$
$$\text{gradient of } l_2 = m_2 = -\tfrac{4}{5}$$

Thus if α is the angle between l_1 and l_2,

$$\tan\alpha = \frac{\tfrac{3}{2} - (-\tfrac{4}{5})}{1 + (\tfrac{3}{2})(-\tfrac{4}{5})} = \frac{23}{10}\bigg/\left(\frac{-2}{10}\right) = -\frac{23}{2}$$

Note: If, as in this example, $\tan\alpha$ is negative, α is the obtuse angle between the lines. The acute angle between them is $\arctan\tfrac{23}{2}$.

The Distance of a Point from a Straight Line

It is understood that the distance from a point to a line is the perpendicular distance.
Consider the line with equation $y = mx + c$ i.e. $mx - y + c = 0$

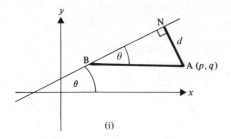

A is distant d from the line where

$$d = \text{AN} = \text{AB} \sin \theta$$

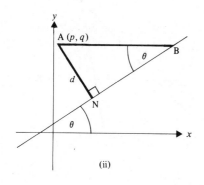

AB is horizontal therefore

at **B** $y = q$

and $x = \dfrac{q - c}{m}$

Hence, in diagram (i) $\text{AB} = p - \left(\dfrac{q - c}{m}\right)$

and in diagram (ii) $\text{AB} = \left(\dfrac{q - c}{m}\right) - p$

i.e. for any point **A** $\text{AB} = \pm\left[\dfrac{mp - q + c}{m}\right]$

The gradient of the line is $m = \tan \theta$

Hence $\sin \theta = \dfrac{m}{\sqrt{(m^2 + 1)}}$

Therefore $\text{AN} = \pm\left[\dfrac{mp - q + c}{m}\right]\left[\dfrac{m}{\sqrt{(m^2 + 1)}}\right]$

i.e. $\text{AN} = \pm\dfrac{mp - q + c}{\sqrt{(m^2 + 1)}}$

Note that this result is equally valid for lines with negative gradient since
$\text{AN} = \text{AB} \sin (\pi - \theta) = \text{AB} \sin \theta$.

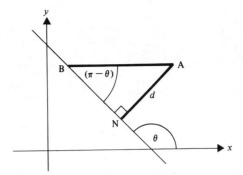

The two signs ($\pm$) in the formula arise from points which are on opposite sides of the line. If we are interested only in the length d of AN, we take the positive value of the formula. This is written

$$d = \left| \frac{mp - q + c}{\sqrt{(m^2 + 1)}} \right|$$

When the equation of the line is given in the form $ax + by + c = 0$, the argument used above leads to the result that

$$d = \left| \frac{ap + bq + c}{\sqrt{(a^2 + b^2)}} \right|$$

where d is the length of AN.

Note: If the formula is used in the form $AN = \dfrac{ap + bq + c}{\sqrt{(a^2 + b^2)}}$ then for two points on opposite sides of the line $ax + by + c = 0$, the two values given for AN are opposite in sign.

EXAMPLES 10a

1) The lines with equations $y = 2x$, $2y + x = 5$ and $4y = 3x - 5$ form a triangle. Find the coordinates of each vertex. By finding the length of one altitude and one side calculate the area of the triangle.

Let the lines $y = 2x$ and $2y + x = 5$ meet at A,

then at A, $\qquad\qquad 2(2x) + x = 5 \implies x = 1, y = 2$

If the lines $2y + x = 5$ and $4y = 3x - 5$ meet at B

then at B $\qquad\qquad 3x - 5 = 10 - 2x \implies x = 3, y = 1$

If the lines $y = 2x$ and $4y = 3x - 5$ meet at C,

then at C, $\qquad\qquad 4(2x) = 3x - 5 \implies x = -1, y = -2$

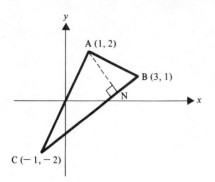

AN is an altitude whose length is the distance from A(1, 2) to the line CB with equation $4y = 3x - 5$.

Re-arranging this equation in the form $3x - 4y - 5 = 0$ and using the formula

$$d = \left| \frac{ap + bq + c}{\sqrt{[a^2 + b^2]}} \right|$$

we have length of AN $= \left| \dfrac{3(1) - 4(2) - 5}{\sqrt{[3^2 + (-4)^2]}} \right| = \left| -\dfrac{10}{5} \right| = 2$

The area of $\triangle ABC = \frac{1}{2}(CB)(AN)$

and $CB = \sqrt{[1 - (-2)]^2 + [3 - (-1)]^2} = 5$

Therefore the area is $\frac{1}{2}(5)(2) = 5$ square units.

2) Show that the point $(-1, 5)$ is the reflection (image) of the point $(3, -3)$ in the line $2y = x + 1$.

A point A is the reflection of a point B in a given line if, and only if,

  (a) the given line bisects AB, *and*

  (b) AB is perpendicular to the given line.

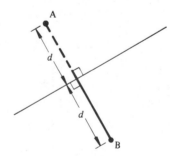

(a) The midpoint of the line joining A(−1, 5) and B(3, −3) is

$$\left(\frac{-1 + 3}{2}, \frac{5 - 3}{2} \right) \quad \text{i.e.} \quad (1, 1)$$

$x = 1$, $y = 1$ is a solution of the equation $2y = x + 1$.

Therefore the midpoint of AB is on the given line.

(b) The gradient of $AB = \dfrac{5 - (-3)}{-1 - 3} = -2 = m_1$.

The gradient of the given line, $y = \frac{1}{2}(x + 1)$, is $\frac{1}{2} = m_2$,

i.e. $\qquad\qquad\qquad\qquad m_1 m_2 = -1$,

showing that AB is perpendicular to the given line.
Therefore $(-1, 5)$ *is* the reflection of $(3, -3)$ in the line $2y = x + 1$.

3) Determine, without drawing a diagram, whether the points $A(-3, 4)$ and $B(-2, 3)$ are on the same side of the line $y + 3x + 4 = 0$. Find the acute angle between this line and the line AB.

Using the formula $\dfrac{ap + bq + c}{\sqrt{[a^2 + b^2]}}$ to find the directed distances of A and B from the line $3x + y + 4 = 0$ gives

for A, $\qquad\qquad\qquad\qquad \dfrac{3(-3) + 4 + 4}{\sqrt{[3^2 + 1^2]}} < 0$

for B, $\qquad\qquad\qquad\qquad \dfrac{3(-2) + 3 + 4}{\sqrt{[3^2 + 1^2]}} > 0$

The signs are opposite so A and B are on opposite sides of the line.
The gradient m_1 of the given line is -3,

the gradient m_2 of AB is $\qquad \dfrac{3 - 4}{-2 - (-3)} = -1$.

If an angle between the given line and AB is α

then $\qquad\qquad tan\, \alpha = \dfrac{m_1 - m_2}{1 + m_1 m_2} = \dfrac{-3 + 1}{1 + 3} = -\dfrac{1}{2}$

Thus the acute angle between the lines is $\arctan \frac{1}{2}$.

EXERCISE 10a

1) Find the coordinates of the point that divides AB in the given ratio in each of the following cases:
(a) $A(2, 4)$, $B(-3, 9)$ $1 : 4$ internally
(b) $A(-3, -4)$, $B(3, 5)$ $3 : 1$ externally
(c) $A(1, 5)$, $B(8, -2)$ $4 : 3$
(d) $A(-1, 6)$, $B(3, -2)$ $3 : -2$

2) A is the midpoint of BC. If A is (X, Y) and B is (x_1, y_1) show that C has coordinates $(2X - x_1, 2Y - y_1)$.

3) Find the distance from A to the given line in each of the following cases:

(a) A(3, 4); $2x - y = 3$ (b) A($-1, -2$); $3y = 4x - 1$
(c) A(a, b); $y = mx + c$ (d) A(4, -1); $x + y = 6$
(e) A(x, y); $ax + by + c = 0$ (f) A(0, 0); $ax + by + c = 0$

4) Determine whether A and B are on the same or opposite sides of the given line in each of the following cases:
(a) A(1, 2), B(4, -3); $3x + y = 7$
(b) A(0, 3), B(7, 6); $x - 4y + 1 = 0$
(c) A($-5, 1$), B($-2, 3$); $7x + y - 6 = 0$

5) Find the tangents of the acute angles between the following pairs of lines:
(a) $2x + 3y = 7$, $x - 6y = 5$
(b) $x + 4y - 1 = 0$, $3x + 7y = 2$
(c) $a_1x + b_1y + c_1 = 0$, $a_2x + b_2y + c_2 = 0$

6) A(0, 1), B(3, 7) and C($-4, -4$) are the vertices of a triangle. Find the tangent of each of the three vertex angles and the length of each altitude.

7) Find the image of the point (5, 6) in the line:
(a) $3x - y + 1 = 0$ (b) $y = 4x + 20$ (c) $2x + 5y + 18 = 0$

8) Find the equations of the two lines through the origin which are inclined at $45°$ to the line $2x + 3y - 4 = 0$.

9) A point P(X, Y) is equidistant from the line $x + 2y = 3$ and from the point (2, 0). Find an equation relating X and Y.

10) Show that A(4, 1) and B(2, -3) are equidistant from the line $2x + 5y = 1$. Is A the reflection of B in this line?

11) Write down the distance of the point P(X, Y) from each of the lines $5x - 12y + 3 = 0$ and $3x + 4y - 6 = 0$. By equating these distances find the equations of two lines that bisect the angles between the two given lines. [i.e. the equations of the set of points P(X, Y).]

12) A(4, 4) and B(7, 0) are two vertices of a triangle OAB. Find the equation of the line that bisects the angle OBA. If this line meets OA at C show that C divides OA in the ratio OB : BA.

LOCI

When the possible positions of a point P are restricted to a straight line or curve, the set of such points is called the *locus of P*.
Further, if P is the point (x, y), the relationship between x and y which applies only to the set of points P is called the Cartesian equation of the locus of P.

EXAMPLES 10b

1) A point P moves so that it is equidistant from the points A(1, 2) and B($-2, -1$). Find the Cartesian equation of the locus of P.

Consider one of the possible positions of $P(x, y)$.
The given condition can be written

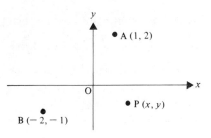

$$PA = PB$$

or $$PA^2 = PB^2$$

i.e. $$(x - 1)^2 + (y - 2)^2 = (x + 2)^2 + (y + 1)^2$$

$\Rightarrow$ $$0 = 6x + 6y$$

This equation is satisfied by every point on the locus and by no other point.
Thus $x + y = 0$ is the equation of the locus of the specified set of points.
Note The line $x + y = 0$ is the perpendicular bisector of AB.

2) A point $P(x, y)$ is twice as far from the point $(3, 0)$ as it is from the line $x = 5$. Find the Cartesian equation of the set of points P.

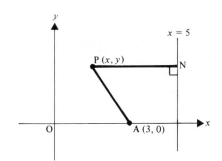

In the diagram, A is the point $(3, 0)$.
PN is the distance of P from the line $x = 5$.
For the set of point P,

$$PA = 2PN$$

or $$PA^2 = 4PN^2$$

Thus P satisfies the given condition $\iff y^2 + (x - 3)^2 = 4(5 - x)^2$

i.e. the equation of the set of points P is

$$y^2 - 3x^2 + 34x = 91$$

EXERCISE 10b

Find the Cartesian equation of the locus of the set of points P in each of the following cases.

1) P is equidistant from the point $(4, 1)$ and the line $x = -2$.

2) P is equidistant from $(3, 5)$ and $(-1, 1)$.

3) P is three times as far from the line $x = 8$ as from the point $(2, 0)$.

4) P is equidistant from the lines $3x + 4y + 5 = 0$ and $12x - 5y + 13 = 0$.

5) P is at a constant distance of two units from the point $(3, 5)$.

6) P is at a constant distance of five units from the line $4x - 3y = 1$.

7) A is the point $(-1, 0)$, B is the point $(1, 0)$ and angle APB is a right angle.

THE CIRCLE

By definition, a circle is the locus of the set of points P which are at a constant distance r from a fixed point C. This geometric property can be used to derive the Cartesian equation of a circle.

If C is the point (p, q) and P(x, y) is any point satisfying the definition,

then

$$CP = r$$

or $\quad CP^2 = r^2$

But

$$CP^2 = (x - p)^2 + (y - q)^2$$

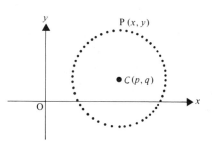

Therefore the coordinates of P must satisfy the equation

$$(x - p)^2 + (y - q)^2 = r^2 \qquad [1]$$

and this is the equation of the circle which is the locus of P.

e.g. the circle with centre $(-2, 3)$ and radius 1 has an equation

$$[x - (-2)]^2 + [y - 3]^2 = 1$$

$\Rightarrow \qquad\qquad x^2 + y^2 + 4x - 6y + 12 = 0$

Conversely, an equation of the form

$$x^2 + y^2 + 2gx + 2fy + c = 0 \qquad [2]$$

where g, f and c are constants, can be rearranged as follows

$$x^2 + 2gx + g^2 + y^2 + 2fy + f^2 + c = g^2 + f^2$$

i.e. $\qquad\qquad (x + g)^2 + (y + f)^2 = g^2 + f^2 - c \qquad [3]$

Comparing [1] and [3] we see that equation [3], and hence equation [2], is the equation of a circle with centre $(-g, -f)$ and radius $\sqrt{g^2 + f^2 - c}$ (provided that $g^2 + f^2 > c$).

Thus $\quad x^2 + y^2 + 2gx + 2fy + c = 0 \quad$ is called the *general equation of a circle*.

(**Note** that the coefficients of x^2 and y^2 are equal and that no xy term is present.)

e.g. $x^2 + y^2 + 8x - 2y + 13 = 0$ is the equation of a circle whose centre and radius can be found in one of the following ways:

(a) by forming perfect squares as illustrated in the general case

i.e. $x^2 + 8x + 16 + y^2 - 2y + 1 = 16 + 1 - 13$

$\Rightarrow$ $(x + 4)^2 + (y - 1)^2 = 4$

comparing with $(x - p)^2 + (y - q)^2 = r^2$ we see that the centre is $(-4, 1)$ and the radius is 2.

(b) by comparing the given equation with the general equation and noting that $2g = 8$, $2f = -2$ and $c = 13$,
from which the centre $(-g, -f)$, is $(-4, 1)$
and the radius, $\sqrt{g^2 + f^2 - c}$, is 2.

Note. A similar equation, e.g. $2x^2 + 2y^2 - 6x + 10y = 1$ also represents a circle but the centre and radius can be found *only after dividing the whole equation by 2* as follows

$$2x^2 + 2y^2 - 6x + 10y - 1 = 0$$

$\Rightarrow$ $$x^2 + y^2 - 3x + 5y - \tfrac{1}{2} = 0$$

Comparing this form with the general equation we see that

$$2g = -3, \quad 2f = 5, \quad c = -\tfrac{1}{2} \Rightarrow g = -\tfrac{3}{2}, \quad f = \tfrac{5}{2}, \quad \sqrt{g^2 + f^2 - c} = 3$$

So the centre of the circle is the point $(\tfrac{3}{2}, -\tfrac{5}{2})$ and its radius is 3.

The analytical geometry of circles makes use of some of the simple geometric properties of the circle, such as:

(a) the tangent at a point P on the circle is perpendicular to the radius PC,

(b) the distance of the centre, C, from any tangent is equal to the radius,

(c) the angle subtended at the circumference by a diameter is $90°$, i.e. an angle in a semi-circle is a right angle.

The Equation of a Tangent at a Given Point

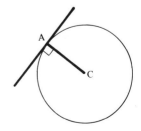

Suppose that $A(x_1, y_1)$ is a point on a circle with centre $C(p, q)$.
The gradient of AC is
$$m = \frac{y_1 - q}{x_1 - p}$$

The tangent at A is perpendicular to AC, so its gradient is $-\dfrac{1}{m} = -\dfrac{x_1 - p}{y_1 - q}$

It is therefore unnecessary to use $\dfrac{dy}{dx}$ to find the gradient of the tangent at a point on a *circle*.

EXAMPLES 10c

1) Find the equation of the tangent at the point $(3, 1)$ on the circle
$$x^2 + y^2 - 4x + 10y - 8 = 0.$$

The centre of the circle is
$$C(2, -5)$$
Hence the gradient of AC is
$$\frac{1 - (-5)}{3 - 2} = 6$$

Therefore the tangent at A has a gradient of $-\frac{1}{6}$, so its equation is
$$y - 1 = -\tfrac{1}{6}(x - 3)$$
i.e. $6y + x = 9$

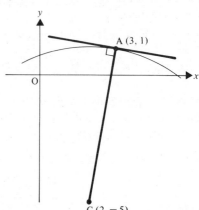

Condition for a Line to be Tangential to a Circle

A line is a tangent to a circle if and only if the distance of that line from the centre of the circle is equal to the radius.

EXAMPLES 10c (continued)

2) Determine whether the lines $5y = 12x - 33$ and $3x + 4y = 9$,   are tangents to the circle $x^2 + y^2 + 2x - 8y = 8$.

The equation of the circle can be written
$$x^2 + 2x + 1 + y^2 - 8y + 16 = 1 + 16 + 8$$
$\Rightarrow$ $(x + 1)^2 + (y - 4)^2 = 25$

Therefore $C(-1, 4)$ is the centre of the circle and the radius is 5 units.
For the line $5y = 12x - 33$,   i.e. $12x - 5y - 33 = 0$,   the distance d_1
from the centre $(-1, 4)$ of the circle is given by
$$d_1 = \left| \frac{12(-1) - 5(4) - 33}{\sqrt{[12^2 + (-5)^2]}} \right| = \left| -\frac{65}{13} \right|$$
i.e. $d_1 = 5 = r$

Thus $12x - 5y - 33 = 0$ *is* a tangent.

For the line $3x + 4y = 9$,   i.e. $3x + 4y - 9 = 0$,

the distance d_2 from $(-1, 4)$ is given by

$$d_2 = \left| \frac{3(-1) + 4(4) - 9}{\sqrt{[3^2 + 4^2]}} \right| = \frac{4}{5}$$

i.e. $d_2 = \frac{4}{5} \neq 5$

Thus $3x + 4y - 9 = 0$ *is not* a tangent.

3) Find the equations of the tangents from the origin to the circle
$x^2 + y^2 - 5x - 5y + 10 = 0$.

The given circle has centre $(\frac{5}{2}, \frac{5}{2})$

and radius $\sqrt{(\frac{5}{2})^2 + (\frac{5}{2})^2 - 10} = \frac{\sqrt{10}}{2}$

Any line through the origin has equation $y = mx$

i.e. $mx - y = 0$

If the line is a tangent to the circle, the distance from the centre, $(\frac{5}{2}, \frac{5}{2})$, to the
line is equal to the radius

i.e. $\left| \dfrac{m(\frac{5}{2}) - (\frac{5}{2})}{\sqrt{[(m)^2 + (-1)^2]}} \right| = \dfrac{\sqrt{10}}{2}$

$\Rightarrow$ $(\frac{5}{2}m - \frac{5}{2})^2 = \frac{10}{4}(m^2 + 1)$

$\Rightarrow$ $3m^2 - 10m + 3 = 0$

$\Rightarrow$ $(3m - 1)(m - 3) = 0$

$\Rightarrow$ $m = \frac{1}{3}$ or 3

So the two tangents from the origin to the given circle are

$$y = 3x \quad \text{and} \quad 3y = x.$$

Orthogonal Circles

Two circles which intersect at right angles cut *orthogonally* and are called
orthogonal circles.
Consider two circles, with centres A and B, which cut orthogonally at P
and Q.

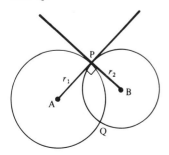

The tangent at P to the circle with
centre B is perpendicular to PB.
Similarly the tangent at P to the
circle with centre A is perpendicular to PA.

But the circles are orthogonal so the tangents at P are perpendicular, hence PA is perpendicular to PB. Thus each tangent at P passes through the centre of the other circle.

Also $$AP^2 + BP^2 = AB^2$$

i.e. $$r_1{}^2 + r_2{}^2 = AB^2$$

Thus, if two circles are orthogonal, the square of the distance between their centres is equal to the sum of the squares of the two radii.

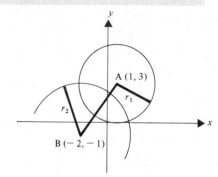

e.g. $(x-1)^2 + (y-3)^2 = 9$

and $(x+2)^2 + (y+1)^2 = 16$

are orthogonal because

$$AB^2 = 3^2 + 4^2$$

and $r_1{}^2 + r_2{}^2 = 9 + 16$

i.e. $$AB^2 = r_1{}^2 + r_2{}^2$$

Circles that Touch

Consider two circles with centres A and B, and radii r_1 and r_2.

If the circles touch externally, then

$$AB = r_1 + r_2$$

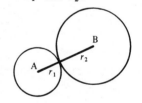

and if they touch internally, then

$$AB = r_1 \sim r_2$$

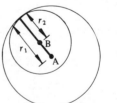
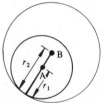

Thus two circles touch if the distance between their centres is the sum or the difference of their radii.

EXAMPLES 10c (continued)

4) Show that the circles $x^2 + y^2 - 16x - 12y + 75 = 0$ and $5x^2 + 5y^2 - 32x - 24y + 75 = 0$ touch each other and find the equation of the common tangent at their point of contact.

The circle $x^2 + y^2 - 16x - 12y + 75 = 0$ has centre A(8, 6) and radius

$$r_1 = \sqrt{8^2 + 6^2 - 75} = 5$$

The circle $5x^2 + 5y^2 - 32x - 24y + 75 = 0$ can be written as

$$x^2 + y^2 - \tfrac{32}{5}x - \tfrac{24}{5}y + 15 = 0$$

and has centre B$(\tfrac{16}{5}, \tfrac{12}{5})$ and radius

$$r_2 = \sqrt{(\tfrac{16}{5})^2 + (\tfrac{12}{5})^2 - 15} = 1$$

The distance between the centres is given by

$$AB^2 = (8 - \tfrac{16}{5})^2 + (6 - \tfrac{12}{5})^2 = (\tfrac{24}{5})^2 + (\tfrac{18}{5})^2 = 36$$

Thus   AB $= 6$

and $r_1 + r_2 = 5 + 1 = 6$

Therefore the circles touch externally.

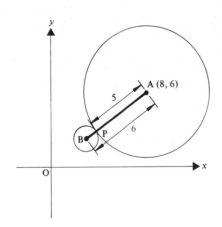

The point of contact, P, divides BA in the ratio 1:5. Therefore the coordinates of P are

$$\begin{cases} x = \dfrac{1(8) + 5(\tfrac{16}{5})}{1 + 5} \\[2mm] y = \dfrac{1(6) + 5(\tfrac{12}{5})}{1 + 5} \end{cases}$$

i.e. $x = 4,\ \ y = 3$

The common tangent at P is perpendicular to BA.

The gradient of BA is $\dfrac{6 - 12/5}{8 - 16/5} = \dfrac{3}{4}$.

Thus the gradient of the common tangent is $-\tfrac{4}{3}$ and its equation is

$$y - 3 = -\tfrac{4}{3}(x - 4)$$

$\Rightarrow$ $3y + 4x = 25$

5) A circle, whose centre is in the first quadrant, touches the x and y axes
and the line $3x - 4y = 12$.   Find its equation.

In this problem there is no information which leads directly to the
centre and radius of the circle. So
we represent the radius by r. Then,
since the centre of the circle is in
the first quadrant, its coordinates
are both positive. The circle also
touches both axes so its centre is

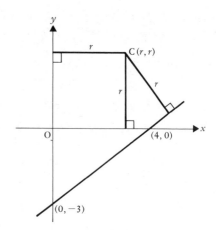

$$(r, r).$$

But the circle also touches the line
$3x - 4y - 12 = 0$,   so the
distance CN is equal to r.

i.e.
$$r = \pm \frac{3(r) - 4(r) - 12}{\sqrt{(3^2 + 4^2)}}$$

$\Rightarrow$
$$\pm(-r - 12) = 5r$$

Thus either
$$6r = -12 \quad \Rightarrow \quad r = -2$$

or
$$4r = 12 \quad \Rightarrow \quad r = 3$$

But   C   is in the first quadrant therefore its coordinates are positive, i.e. r is
positive.

Thus $r = 3$,   the centre of the circle is $(3, 3)$ and its equation is

$$(x - 3)^2 + (y - 3)^2 = 3^2$$

$\Rightarrow$
$$x^2 + y^2 - 6x - 6y + 9 = 0$$

6) Find the equation of a circle with centre on the y-axis, which cuts
orthogonally each of the circles $x^2 + y^2 + 6x + 2y - 9 = 0$ and
$x^2 + y^2 - 2x - 2y + 1 = 0$.

The required circle has its centre on the y-axis, so let its centre be the point
$(0, q)$ and let its radius be r.
If it is orthogonal to the first circle

$$r^2 + 19 = [0 - (-3)]^2 + [q - (-1)]^2$$

i.e.
$$r^2 = q^2 + 2q - 9 \tag{1}$$

and if it is orthogonal to the second circle

$$r^2 + 1^2 = [0 - 1]^2 + [q - 1]^2$$

i.e.
$$r^2 = q^2 - 2q + 1 \tag{2}$$

Solving [1] and [2] gives $q = \frac{5}{2}$, $r^2 = \frac{9}{4}$.

Thus the equation of the circle which is orthogonal to both the given circles is

$$[x - 0]^2 + [y - \tfrac{5}{2}]^2 = \tfrac{9}{4}$$

i.e. $$x^2 + y^2 - 5y + 4 = 0$$

There are many different ways in which a circle can be specified. The examples that follow illustrate some of varied approaches which can lead to the equation of a circle.

7) Find the equation of a circle given that it passes through the points $(0, 1)$, $(4, -1)$, $(4, 7)$.

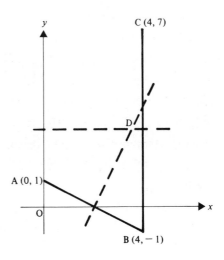

The centre of a circle lies on the perpendicular bisector of any chord.

Using chords AB and BC we have

AB has midpoint $(2, 0)$ and gradient $-\frac{1}{2}$, so its perpendicular bisector is

$$(y - 0) = 2(x - 2)$$
$$\Rightarrow \quad y = 2x - 4 \qquad [1]$$

BC has midpoint $(4, 3)$ and it is vertical, so its perpendicular bisector is horizontal and its equation is

$$y = 3 \qquad\qquad\qquad [2]$$

Solving [1] and [2] gives $y = 3$, $x = \frac{7}{2}$ and these are the coordinates of the centre, D, of the circle.

The radius, r, is the length of DA (or DC, or DB)

i.e. $$r^2 = (3 - 1)^2 + (\tfrac{7}{2} - 0)^2 = \tfrac{65}{4}$$

Therefore the equation of the circle is

$$(x - \tfrac{7}{2})^2 + (y - 3)^2 = \tfrac{65}{4}$$

$\Rightarrow$ $$x^2 + y^2 - 7x - 6y + 5 = 0$$

8) The points $A(x_1, y_1)$ and $B(x_2, y_2)$ are the ends of a diameter of a circle. Find the equation of the circle.

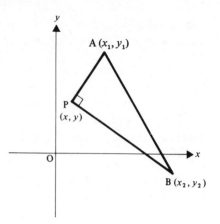

Let $P(x, y)$ be any point on the required circle. Since AB is a diameter, angle APB is $90°$.

Gradient of PA is $\dfrac{y - y_1}{x - x_1} = m_1$

Gradient of PB is $\dfrac{y - y_2}{x - x_2} = m_2$

But PA and PB are perpendicular, so $m_1 m_2 = -1$

i.e.
$$\left(\frac{y - y_1}{x - x_1}\right)\left(\frac{y - y_2}{x - x_2}\right) = -1$$

$\Rightarrow$
$$(y - y_1)(y - y_2) + (x - x_1)(x - x_2) = 0$$

This is the equation of the circle on AB as diameter.
(**Note** that this method is neater than finding the midpoint and half the length of AB to determine the centre and radius of the circle.)

EXERCISE 10c

1) Write down the equation of the circle with:
(a) centre $(1, 2)$, radius 3   (b) centre $(0, 4)$, radius 1
(c) centre $(-3, -7)$, radius 2   (d) centre $(4, 5)$, radius 3

2) Find the centre and radius of the circle whose equation is:
(a) $x^2 + y^2 + 8x - 2y - 8 = 0$ (b) $x^2 + y^2 + x + 3y - 2 = 0$
(c) $x^2 + y^2 + 6x - 5 = 0$ (d) $2x^2 + 2y^2 - 3x + 2y + 1 = 0$
(e) $x^2 + y^2 = 4$ (f) $(x - 2)^2 + (y + 3)^2 = 9$
(g) $2x + 6y - x^2 - y^2 = 1$ (h) $3x^2 + 3y^2 + 6x - 3y - 2 = 0$

3) Determine whether the given line is a tangent to the given circle in each of the following cases:
(a) $3x - 4y + 14 = 0$; $x^2 + y^2 + 4x + 6y - 3 = 0$
(b) $5x + 12y = 4$; $x^2 + y^2 - 2x - 2y + 1 = 0$
(c) $x + 2y + 6 = 0$; $x^2 + y^2 - 6x - 4y + 8 = 0$
(d) $x + 2y + 6 = 0$; $x^2 + y^2 - 6x + 4y + 8 = 0$

4) Write down the equation of the tangent to the given circle at the given point:
(a) $x^2 + y^2 - 2x + 4y - 20 = 0$; $(5, 1)$

(b) $x^2 + y^2 - 10x - 22y + 129 = 0$; $(6, 7)$

(c) $x^2 + y^2 - 8y + 3 = 0$; $(-2, 7)$

5) Some of the following pairs of circles touch and some pairs are orthogonal. Determine whether each of the pairs: (i) touch,   (ii) cut orthogonally,   (iii) do neither of these.

(a) $x^2 + y^2 + 2x - 4y + 1 = 0$; $x^2 + y^2 - 6x - 10y + 25 = 0$

(b) $x^2 + y^2 + 8x + 2y - 8 = 0$; $x^2 + y^2 - 16x - 8y = 64$

(c) $x^2 + y^2 + 6x = 0$; $x^2 + y^2 + 6x - 4y + 12 = 0$

(d) $x^2 + y^2 + 2x - 8y + 1 = 0$; $x^2 + y^2 - 6y = 0$

(e) $x^2 + y^2 + 2x = 3$; $x^2 + y^2 - 6x - 3 = 0$

6) If $y = 2x + c$ is a tangent to the circle $x^2 + y^2 + 4x - 10y - 7 = 0$ find the value(s) of c.

7) Find the condition that m and c satisfy if the line $y = mx + c$ touches the circle $x^2 + y^2 - 2ax = 0$.

Find the equations of the following circles (in some cases more than one circle is possible).

8) A circle touches the negative x- and y-axes and also the line $7x + 24y + 12 = 0$.

9) A circle passes through the points $(1, 4)$, $(7, 5)$ and $(1, 8)$.

10) A circle has its centre on the line $x + y = 1$ and passes through the origin and the point $(4, 2)$.

11) The line joining (a, b) to $(3a, 5b)$ is a diameter of a circle.

12) A circle whose centre is in the first quadrant touches the y-axis at the point $(0, 3)$ and is orthogonal to the circle $x^2 + y^2 - 8x + 4y - 5 = 0$.

13) A circle with centre $(2, 7)$ passes through the point $(-3, -5)$.

14) A circle intersects the y-axis at the origin and at the point $(0, 6)$ and also touches the x-axis.

Find the equations of the tangents specified in Questions 15–18.

15) Tangents *from* the origin to the circle $x^2 + y^2 - 10x - 6y + 25 = 0$.

16) The tangent *at* the origin to the circle $x^2 + y^2 + 2x + 4y = 0$.

17) Tangents to the circle $x^2 + y^2 - 4x + 6y - 7 = 0$ which are parallel to the line $2x + y = 3$.

18) Show that if the point P is twice as far from the point $(4, -2)$ as it is from the origin then P lies on a circle. Find the centre and radius of this circle.

THE PARABOLA

If a point P is always equidistant from a fixed point and a fixed straight line, the locus of the set of points P is called a parabola. The general shape of a parabola can be seen by plotting some of the possible positions of P,

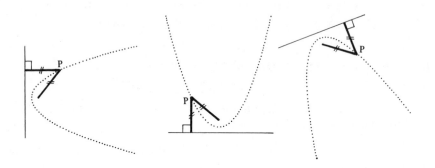

Although the shape varies slightly according to the relative positions of the fixed point (called the *focus*) and the fixed line (called the *directrix*), all parabolas have similar properties.
For instance, every parabola is symmetrical about the line through the focus which is perpendicular to the directrix. This line is called the axis of the parabola.

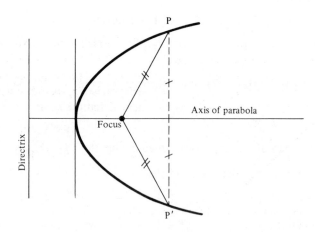

The point where a parabola crosses its axis is the *vertex*.
The distance between the vertex and the focus is the *focal length*.

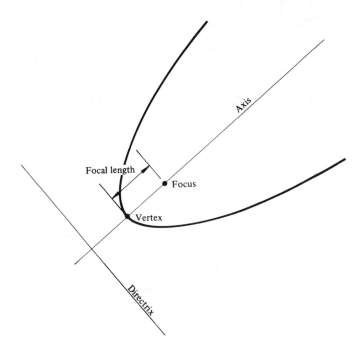

Equation of a Parabola

Consider a parabola whose focus is the point $(3, 4)$ and whose directrix is the line $x + y = 1$.

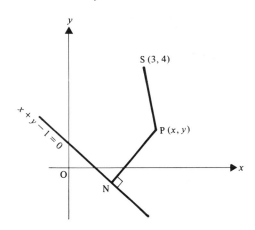

If S is the point $(3, 4)$, P is any point (x, y) on the parabola and PN is the perpendicular from P to the directrix, then P must satisfy the definition

$$PS = PN$$

or $\qquad PS^2 = PN^2$

Now PN is the distance from (x, y) to $x + y - 1 = 0$

i.e. $\qquad PN = \left| \dfrac{x + y - 1}{\sqrt{(1^2 + 1^2)}} \right|$

and $\qquad PS^2 = (x - 3)^2 + (y - 4)^2$

Thus, for all points on the parabola

$$(x-3)^2 + (y-4)^2 = \left(\frac{x+y-1}{\sqrt{2}}\right)^2$$

$\Rightarrow \qquad\qquad x^2 + y^2 - 10x - 14y - 2xy + 49 = 0$

and this is the equation of the parabola.

(**Note.** This cannot be the equation of a circle as it contains the term $2xy$.)

Rather than analyse a parabola with such an awkward equation, it is usual to study one with a simpler equation. This can be achieved by choosing the point $S(a, 0)$ for the focus and the line $x = -a$ for the directrix.

Thus, for any point P on this parabola

$$PS = PN$$

or $\qquad PS^2 = PN^2$

i.e.

$$(x-a)^2 + y^2 = [x - (-a)]^2$$

$\Rightarrow \qquad\qquad y^2 = 4ax$

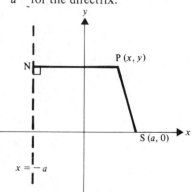

So, for a parabola with focus $(a, 0)$ and directrix $x = -a$

$$P(x, y) \text{ is on the parabola} \iff y^2 = 4ax$$

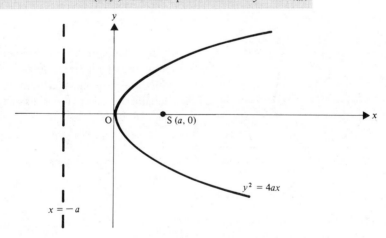

$y^2 = 4ax$

$x = -a$

Tangent and Normal

The equation $y^2 = 4ax$ can be differentiated implicitly,

i.e. $$2y\frac{dy}{dx} = 4a$$

Thus, at any point on the parabola

$$\frac{dy}{dx} = \frac{2a}{y}$$

Hence the gradient at a particular point can be determined.
For instance, to find the equations of the tangent and the normal to the parabola
$y^2 = 8x$ at the point $(2, 4)$ we say

$$y^2 = 8x \quad \Rightarrow \quad 2y\frac{dy}{dx} = 8$$

$$\Rightarrow \qquad\qquad \frac{dy}{dx} = \frac{4}{y}$$

At the point $(2, 4)$ $$\frac{dy}{dx} = \frac{4}{4} = 1$$

Thus the equation of the tangent is $\quad y - 4 = (1)(x - 2) \quad$ i.e. $\quad y = x + 2$
and the equation of the normal is $\quad y - 4 = (-1)(x - 2) \quad$ i.e. $\quad y + x = 6$.

EXERCISE 10d

1) Find the Cartesian equation of the locus of the set of points P if P is equidistant from the given point and the given line. In each case sketch the locus of P.

(a) $(1, 1);$ $\quad x + y + 3 = 0$
(b) $(3, 4);$ $\quad y = 0$
(c) $(-a, 0);$ $\quad x = a$

2) Find the equations of the tangent and the normal to the given parabola at the given point in each of the following cases:

(a) $y^2 = 4x;$ $\quad (1, 2)$ $\qquad$ (b) $(y - 1)^2 = 6x;$ $\quad (0, 1)$
(c) $y = 8x^2;$ $\quad (\frac{1}{2}, 2)$ $\qquad$ (d) $y^2 + 4x = 0;$ $\quad (-4, -4)$

3) Sketch each of the following parabolas, marking the focus, the directrix, the vertex, the axis of the parabola and the focal length.

(a) $y^2 = 12x$ $\qquad$ (b) $y^2 = 4x$ $\qquad\qquad$ (c) $y^2 = -4x$
(d) $4y = x^2$

CARTESIAN ANALYSIS OF A GENERAL CURVE

The Cartesian equation of any two dimensional curve expresses a relationship between x and y and can be given in the form

$$f(xy) = 0$$

The gradient function is given by the equation

$$\frac{d}{dx} f(xy) = 0$$

INTERSECTION

The points of intersection of any two lines or curves whose equations are $f(xy) = 0$ and $g(xy) = 0$ are found by solving simultaneously the two equations.
Each real solution gives a point of intersection.

Coincident Points of Intersection

The equation obtained by eliminating x (or y) from the equations $f(xy) = 0$ and $g(xy) = 0$ in order to solve them simultaneously may have two or more equal roots.

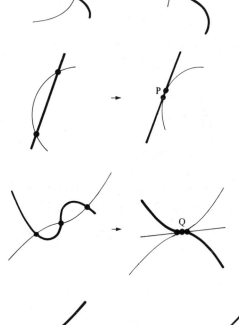

If there are two equal roots the curves meet twice at the same point P, i.e. they touch at P and have a common tangent at P.
In particular, when a line and a curve meet twice at the same point P, the line is a tangent to the curve at P.
If there are three equal roots the curves meet three times at the same point Q. The curves have a common tangent at Q but this time each curve crosses, at Q, to the opposite side of the common tangent. In particular, when a line and a curve meet at three coincident points Q, the line is a tangent to the curve at Q and the curve has a point of inflexion at Q.

Taking this argument further it becomes clear that,

(a) when the number of coincident points of intersection of a line and a curve is even, the curve touches the line and remains on the same side of the line;

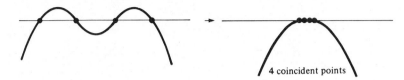

4 coincident points

(b) when the number of coincident points of intersection is odd, the curve touches the line and crosses it. Thus the curve has a point of inflexion.

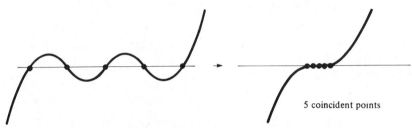

5 coincident points

so if the solution of $\begin{cases} y = mx + c \\ f(x, y) = 0 \end{cases}$

has $\begin{cases} \text{distinct roots, the curve crosses the line at distinct points.} \\ \text{repeated roots, the line touches the curve.} \end{cases}$

The Curves $y = kx^n$ where n is a Positive Integer

If $y = kx^n$,
the points where the curve meets the x-axis are given by $x^n = 0$
i.e. by n equal zero values of x.
Thus if n is even, the curve touches the x-axis at the origin at a minimum point $(k > 0)$ or a maximum point $(k < 0)$

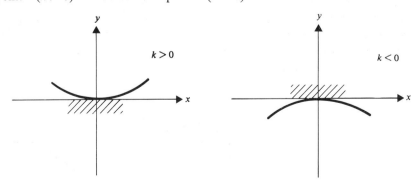

But if n is odd the curve has a point of inflexion at the origin.

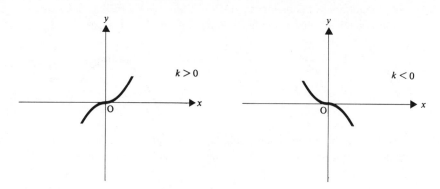

Further, since $\dfrac{\mathrm{d}y}{\mathrm{d}x} = knx^{n-1}$, the only stationary point is where $x = 0$.

Thus the graph of $y = kx^n$ is

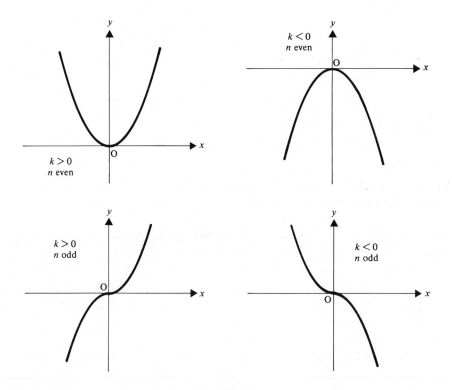

The graph of $y = k(x-a)^n$ is very similar to the graphs shown above. But the intersection with the x-axis occurs when $x = a$ in this case.

Thus, for $y = k(x-a)^n$ we have

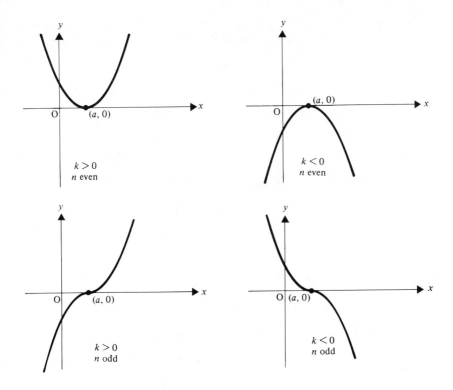

EXAMPLES 10e

1) Show that the line $y = mx + c$ touches the parabola $y^2 = 4ax$ if and only if $mc = a$.

The line $y = mx + c$ intersects the parabola $y^2 = 4ax$ at points whose y-coordinates are the roots of the equation

$$y^2 = 4a\left(\frac{y - c}{m}\right)$$

i.e. $my^2 - 4ay + 4ac = 0$ [1]

This equation is quadratic so it cannot have more than two roots. Therefore the line and the parabola cannot have more than two points of intersection. The line touches the parabola if the two points of intersection coincide, i.e. if and only if equation [1] has equal roots.

$\Rightarrow$ $(-4a)^2 = 4(m)(4ac)$

$\Rightarrow$ $a = mc$

2) Find the equations of the tangents with gradient 1 to the curve with equation $x^2 + 2y^2 = 6$ and find their points of contact.

Any line with gradient 1 has equation $y = x + c$. This line meets the curve $x^2 + 2y^2 = 6$ at points whose x-coordinates are the roots of the equation

$$x^2 + 2(x + c)^2 = 6$$

i.e.
$$3x^2 + 4cx + (2c^2 - 6) = 0 \qquad [1]$$

If the line is a tangent, then equation [1] has equal roots.

So
$$(4c)^2 - 4(3)(2c^2 - 6) = 0$$

$\Rightarrow$
$$c = \pm 3$$

So the equations of the tangents are $y = x \pm 3$.

Now when c has either of these values, equation [1] has equal roots whose value is given by $\dfrac{`-b\text{'}}{2a}$ i.e. $\dfrac{-4c}{6}$, and these are the x-coordinates of the points of contact.

When $c = 3$, $x = -2$ and $y = 1$.
When $c = -3$, $x = 2$ and $y = -1$.
So the points of contact of the two tangents are $(2, -1)$ and $(-2, 1)$.

Note that the curve $x^2 + 2y^2 = 6$ is an ellipse. In general any curve whose equation is of the form $\dfrac{x^2}{a^2} + \dfrac{y^2}{b^2} = 1$ is an ellipse and its shape is shown below.

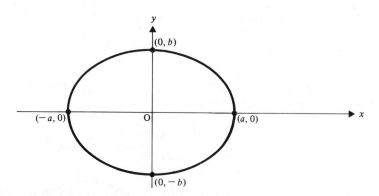

3) Find the point(s) of intersection of the curves $xy = 1$ and $x^2 + y^2 + 2x + 2y - 6 = 0$.
Use sketches to illustrate your solution.

Points of intersection satisfy both equations simultaneously.
Eliminating y from the two equations we have

$$x^2 + \left(\frac{1}{x}\right)^2 + 2x + 2\left(\frac{1}{x}\right) - 6 = 0$$

i.e. $$x^4 + 2x^3 - 6x^2 + 2x + 1 = 0 \qquad [1]$$

The roots of this equation are the x coordinates of the points of intersection of the two curves.

Using the factor theorem to factorise the L.H.S. we find that [1] becomes

$$(x - 1)(x - 1)(x^2 + 4x + 1) = 0$$

Hence $x = 1, 1, -2 \pm \sqrt{3}$,

i.e. there are two real equal roots and two other real distinct roots.

Therefore, calculating the corresponding y-coordinates from the equation $xy = 1$, we see that

the two curves *touch* at the point $A(1, 1)$ and *cut* at the points

$$B\left(-2 + \sqrt{3}, \ \frac{1}{-2 + \sqrt{3}}\right) \quad \text{and} \quad C\left(-2 - \sqrt{3}, \ \frac{1}{-2 - \sqrt{3}}\right).$$

The curve with equation

$$xy = 1, \quad \text{or} \quad y = \frac{1}{x},$$

is called a rectangular hyperbola.
(Any curve with equation $xy = K$ is a rectangular hyperbola.)

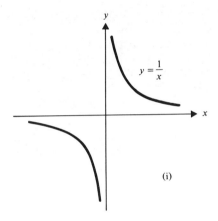

(i)

The equation
$x^2 + y^2 + 2x + 2y - 6 = 0$,
or $(x + 1)^2 + (y + 1)^2 = 8$, can
be recognised as a circle.
The points common to the two
curves are shown in diagram (iii)

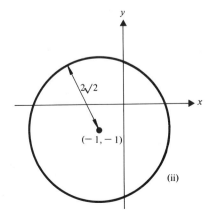

(ii)

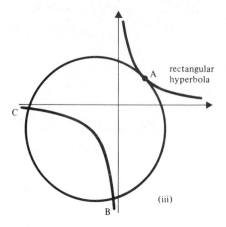

rectangular
hyperbola

(iii)

4) Find the midpoint of the chord cut off by the curve $y^2 - 3y = 5x$ on the line $3x + y = 1$.

Points common to the line and the curve have y-coordinates given by the equation

$$y^2 - 3y = 5\left(\frac{1-y}{3}\right)$$

i.e. $3y^2 - 4y - 5 = 0$

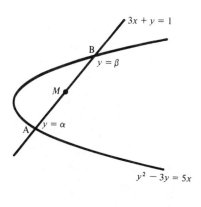

If the roots are α and β
the line cuts the curve at A and B
where $y = \alpha$ and $y = \beta$.
Thus at M, the midpoint of AB

$$y = \tfrac{1}{2}(\alpha + \beta)$$

But $\alpha + \beta = -\dfrac{b}{a} = \dfrac{4}{3}$

Therefore, at M,
$y = \tfrac{2}{3}$ and $x = \tfrac{1}{9}$ (from $3x + y = 1$). Thus M is the point $(\tfrac{1}{9}, \tfrac{2}{3})$.

Note: In this problem there is the alternative method of actually solving the equation $3y^2 - 4y - 5 = 0$ to determine α and β. This is practical only when the coefficients are numbers. The method shown above however is equally suitable when the line and curve have general equations.

5) Find the points in which the line $y = x + 2$ meets the curve $y = x^4 - 2x^3 + 3x + 1$ showing that one of them is an inflexion on the curve.

The line and curve meet at points whose x-coordinates are given by

$$x^4 - 2x^3 + 3x + 1 = x + 2$$

i.e. $$x^4 - 2x^3 + 2x - 1 = 0$$

Using the factor theorem to factorise we have

$$(x - 1)^3(x + 1) = 0$$

Thus the curve and the line meet three times at the point where $x = 1$ and once at the point where $x = -1$.
Thus there is a point of inflexion at $P(1, 3)$ and a single crossing at $Q(-1, 1)$.

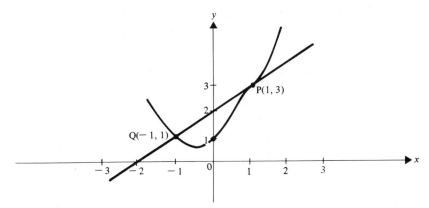

EXERCISE 10e

Investigate the possible intersection of the following lines and curves giving the coordinates of all common points. State clearly those cases where the line touches the curve.

1) $y = x + 1;$ $y^2 = 4x$

2) $2y + x = 3;$ $x^2 - y^2 - 3y + 3 = 0$

3) $y = x - 5;$ $x^2 + 2y^2 = 7$

4) $2y - x = 4;$ $x^2 + y^2 - 4x = 4$

5) $y = 0;$ $y = x^2 - 3x + 2$

6) $y = 0;$ $y = x^3 + 5x^2 + 6x$

7) $y = 0;$ $y = (x - 1)^2(x - 2)^2$

8) $y = 0;$ $y = (x + 3)^3(x + 2)$

9) $x = 0;$ $x = y^4$

Find the value of k such that the given line shall touch the given curve.

10) $y = x + 2$; $y^2 = kx$

11) $y = kx + 3$; $xy + 9 = 0$

12) $y = 3x - k$; $x^2 + 2y^2 = 8$

Without finding the coordinates of the points of intersection of the following lines and curves, find the coordinates of the midpoint of the chord cut off by the curve on the line.

13) $y = 2x + 3$; $y^2 = 4x + 8$

14) $y = x$; $2x^2 + 5y^2 = 10$

15) $3y + x + 7 = 0$; $xy = 1$

Find the points of intersection or points of contact (if any) of the following pairs of curves. Illustrate your results by drawing diagrams.

16) $y^2 = 8x$; $xy = 1$

17) $x^2 + y^2 + 2x - 7 = 0$; $y^2 = 4x$

18) $xy = 2$; $2x^2 + 2y^2 - 6x + 3y - 10 = 0$

19) $9x^2 = 2y$; $y^2 = 6x$

20) Find the value(s) or ranges of values of λ for which the line $y = 2x + \lambda$:
(a) touches   (b) cuts in real points
(c) does not meet, the curve $y^2 + 2x^2 = 4$.

21) Sketch the curves $y = 3x^4$, $y = 4(2 - x)^5$, $y = 2(x + 3)^7$,
$y = -5x^6$.

22) Find the equation(s) of the tangent(s):
  (i) from the point $(1, 0)$,
  (ii) with gradient $-\frac{1}{2}$,
  to each of the following curves,
  (a) $y^2 + 4x = 0$ (b) $xy = 9$ (c) $x^2 = 6y$

THE CUBIC CURVE

Observation of the equation of the general cubic curve

$$y = ax^3 + bx^2 + cx + d$$

shows that:

(i) if x is very large, $y \simeq ax^3$,   thus

as	$x \to \infty$	$y \to \infty$	$\left.\phantom{\begin{matrix}a\\a\end{matrix}}\right\}$ if $a > 0$
and as	$x \to -\infty$	$y \to -\infty$	

while as	$x \to \infty$	$y \to -\infty$	$\left.\phantom{\begin{matrix}a\\a\end{matrix}}\right\}$ if $a < 0$
and as	$x \to -\infty$	$y \to \infty$	

(ii)   there are no finite values of x for which y is undefined so the curve is continuous.

(iii) $\dfrac{dy}{dx} = 3ax^2 + 2bx + c$,   therefore solutions of the equation

$$3ax^2 + 2bx + c = 0$$

are the values of x for which y is stationary.
This quadratic equation may have:

(a) two real distinct roots, in which case the curve has two distinct turning points which must be a maximum and a minimum (as the curve is continuous),

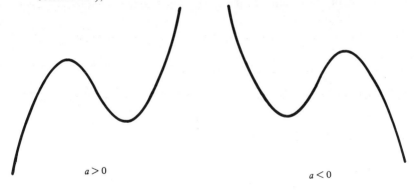

$a > 0$ $a < 0$

(b) two equal real roots, in which case the maximum and minimum coincide and the curve therefore has a point of inflexion,

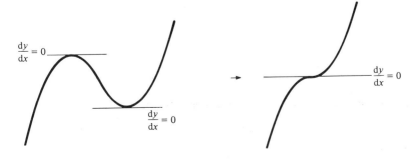

(c) no real roots, indicating that the curve has no real stationary points.

Combining these observations we deduce that a cubic curve has one of the following shapes.

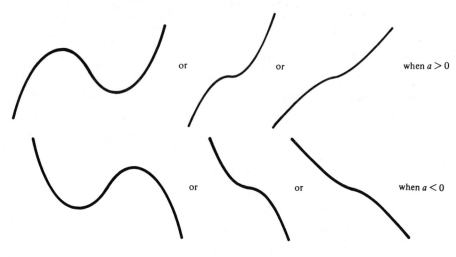

Depending upon the position of the curve relative to the x-axis we also see that a cubic curve can

cross the x-axis three times

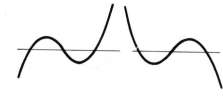

or cross and touch the x-axis

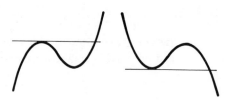

or touch the x-axis at a point of inflexion

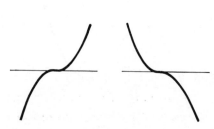

or cross the x-axis once only

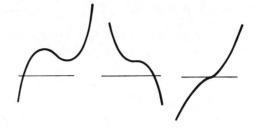

Note: There is always *at least one point of intersection* with the x-axis. Therefore, since a value of x where the curve meets the x-axis is a solution of the equation $ax^3 + bx^2 + cx + d = 0$, we see that *a cubic equation has at least one real root.*

EXAMPLES 10f

Sketch the curves   (1) $y = x(x-1)(x-2)$
                     (2) $y = x(x-1)^2$
                     (3) $y = (1-x)^3$

1) If $y = x(x-1)(x-2) = x^3 - 3x^2 + 2x$

      $y = 0$ when $x = 0, 1, 2$

Also $a = 1$ i.e. $a > 0$

Thus

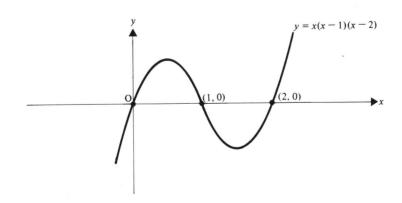

2) If $y = x(x-1)^2$

    $y = 0$ when $x = 0, 1, 1$

Hence the curve *touches* the x-axis when $x = 1$.

Again $a > 0$

Thus

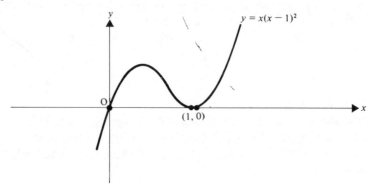

3) If $y = (1-x)^3 = 1 - 3x + 3x^2 - x^3$

   $y = 0$ when $x = 1, 1, 1$

The triple solution $x = 1$ indicates a point of inflexion.

In this case $a = -1$ i.e. $a < 0$

so the graph is

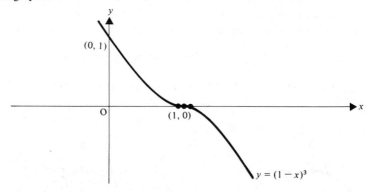

EXERCISE 10f

Sketch the graphs of the following cubic curves showing clearly the behaviour of the curve at the points where it meets the x-axis and at stationary points.

1) $y = (x-1)(x-2)(x-3)$ 2) $y = 1 - x - x^2 + x^3$

3) $y = x^2(1-x)$ 4) $y = (x-2)^3$

5) $y = (x+1)(x-3)^2$ 6) $y = x^3 - 1$

7) $y = a + x^2 - x - x^3$ 8) $y = (1+x)(1-x)(2-x)$

9) $y = x^3 + 3x^2 + 3x + 1$ 10) $y + x^3 = 0$

PARAMETRIC COORDINATES

The relationship between the x- and y-coordinates of a point can often be expressed more simply in the form of two equations

$$\begin{cases} x = f(p) \\ y = g(p) \end{cases}$$

where p is a parameter.

The use of parametric equations to find the gradient function of a curve, and hence the equations of the tangent and normal at a particular point, was demonstrated in Chapter 8. There are many other ways in which the parametric approach helps with problems on curve analysis, and we will now explore some of them.

Relation between Cartesian and Parametric Equations

When the parametric equations of a curve are given, the Cartesian equation can be found by eliminating the parameter.

It is not always so easy to convert a Cartesian equation into a suitable pair of parametric equations. In some cases, however, the form of a Cartesian equation suggests clearly what parameter should be used,

e.g. if $x^2 + y^2 = 4$, the similarity to the trig identity $\cos^2\theta + \sin^2\theta \equiv 1$

indicates that $\begin{cases} x = 2\cos\theta \\ y = 2\sin\theta \end{cases}$ are suitable parametric equations.

In this case the parameter, θ, has graphical significance, as can be seen in the diagram below.

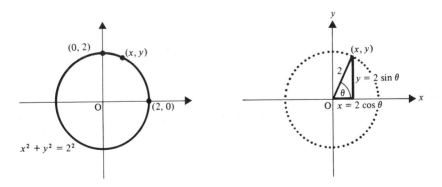

Frequently, however, a parameter cannot be represented graphically,

e.g. the Cartesian equation of the parabola $y^2 = 4ax$ can be replaced by the parametric equations $x = at^2$, $y = 2at$, where t is simply a real number.

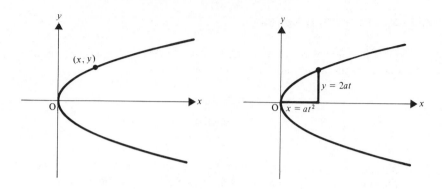

In those cases where the parameter has a geometric meaning, its *geometric* properties can be used in the solution of problems.

EXAMPLE

Show that the curve whose parametric equations are $x = a(1 + \cos \theta)$ and $y = a \sin \theta$, represents a circle. Indicate on a diagram the significance of the parameter θ.

Hence find, in terms of θ, the intercepts on the x- and y-axes made by the tangent to the circle at the point $P[a(1 + \cos \theta), a \sin \theta]$.

To convert the parametric equations to Cartesian form we eliminate θ as follows

$$x = a(1 + \cos \theta) \quad \Rightarrow \quad \cos \theta = \left(\frac{x}{a} - 1\right) \qquad [1]$$

$$y = a \sin \theta \qquad \Rightarrow \quad \sin \theta = \frac{y}{a} \qquad [2]$$

But $\cos^2 \theta + \sin^2 \theta \equiv 1$

Hence squaring and adding equations [1] and [2] gives

$$\left(\frac{x}{a} - 1\right)^2 + \left(\frac{y}{a}\right)^2 = 1$$

i.e. $(x - a)^2 + y^2 = a^2$

This represents a circle with centre $C(a, 0)$ and radius a.

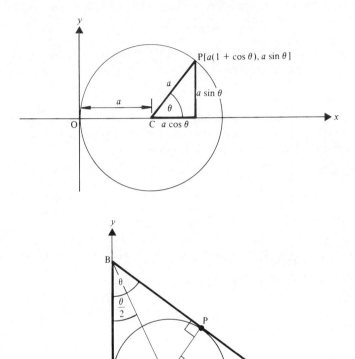

The tangent at P is perpendicular to CP and so is inclined at an angle θ to the y-axis.

Also BC bisects the angle OBP so angle $OBC = \dfrac{\theta}{2}$

Then

$$OB = OC \cot \frac{\theta}{2} = a \cot \frac{\theta}{2}$$

and

$$OA = OB \tan \theta = a \cot \frac{\theta}{2} \tan \theta$$

Thus (without finding the equation of the tangent at P) the intercepts made by the tangent on the x- and y-axes are

$$a \cot \frac{\theta}{2} \tan \theta \quad \text{and} \quad a \cot \frac{\theta}{2}$$

respectively.

INTERSECTION

Consider the line $y = mx + c$ and the parabola with parametric equations

$$x = at^2, \quad y = 2at$$

Any point which is common to the line and the parabola must have coordinates $(at^2, 2at)$ and these coordinates must also satisfy the equation $y = mx + c$.

Hence $\qquad\qquad\qquad\qquad 2at = mat^2 + c$

The roots of this quadratic equation in t give the values of the parameter at points of intersection. If the roots are equal, the two points of intersection coincide and the line is a tangent to the parabola.

EXAMPLES 10g

1) Find the value of k if the line $y = k - 2x$ touches the curve with parametric equations $x = t, \quad y = \dfrac{1}{t}$.

Points on the curve have coordinates $\left(t, \dfrac{1}{t}\right)$ so, where the line meets the curve, these coordinates also satisfy the equation of the line

$$y = k - 2x$$

i.e. $\qquad\qquad\qquad\qquad \dfrac{1}{t} = k - 2t$

$\Rightarrow \qquad\qquad\qquad\qquad 2t^2 - kt + 1 = 0 \qquad\qquad\qquad\qquad [1]$

If the line *touches* the curve there is *only one* common point, so equation [1] has equal roots

i.e. $\qquad\qquad\qquad (-k)^2 - 4(2)(1) = 0$

$\Rightarrow \qquad\qquad\qquad\qquad k = \pm 2\sqrt{2}$

Alternatively, since the gradient of the curve at any point $\left(t, \dfrac{1}{t}\right)$ is $-\dfrac{1}{t^2}$, we can deduce that a tangent with gradient -2 touches the curve where $-\dfrac{1}{t^2} = -2$, i.e. where $t = \pm \dfrac{1}{\sqrt{2}}$.

The point where $t = \dfrac{1}{\sqrt{2}}$ is $\left(\dfrac{1}{\sqrt{2}}, \sqrt{2}\right)$ and the tangent at this point has equation

$$y - \sqrt{2} = -2\left(x - \dfrac{1}{\sqrt{2}}\right) \quad \Rightarrow \quad y = 2\sqrt{2} - 2x$$

So $\quad t = \dfrac{1}{\sqrt{2}} \quad \Rightarrow \quad k = 2\sqrt{2}$.

Similarly $t = -\dfrac{1}{\sqrt{2}} \Rightarrow k = -2\sqrt{2}.$

2) Find the coordinates of the points of intersection of two curves, C_1 and C_2, where

C_1 has parametric equations
$$\begin{cases} x = t^2 + 1 \\ y = 2t \end{cases}$$

and C_2 has parametric equations
$$\begin{cases} x = 2s \\ y = \dfrac{2}{s} \end{cases}$$

If C_1 and C_2 meet at a point then, at that point,

$$x = t^2 + 1 \quad and \quad x = 2s \Rightarrow t^2 + 1 = 2s \qquad [1]$$

also $$y = 2t \quad and \quad y = \dfrac{2}{s} \Rightarrow 2t = \dfrac{2}{s} \qquad [2]$$

Solving [1] and [2] by eliminating s gives

$$t^3 + t - 2 = 0$$

$$\Rightarrow \qquad (t - 1)(t^2 + t + 2) = 0$$

So either $t = 1$ or $t^2 + t + 2 = 0$, which gives no real values of t. Hence the given curves have only one point of intersection.

At this point $t = 1$ so $x = 2$ and $y = 2$.

$$\left[\text{As a check we can use the corresponding value of } s \; \left(= \dfrac{1}{t} \right) \text{ in } C_2. \right]$$

A sketch to illustrate this result can be drawn by recognising that:

(a) Since 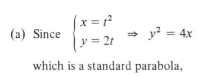 $\Rightarrow y^2 = 4x$

which is a standard parabola,

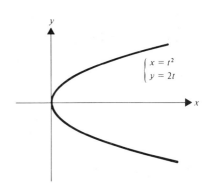

then $\begin{cases} x = t^2 + 1 \\ y = 2t \end{cases}$ represent

a similar parabola where each x
coordinate is increased by
1 unit.

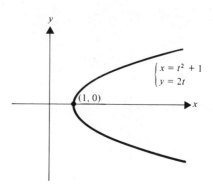

(b) $\begin{cases} x = 2s \\ y = \dfrac{2}{s} \end{cases} \Rightarrow xy = 4$

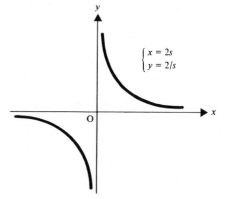

which is a rectangular hyperbola.
(See page 389.)

Thus these curves meet only
once, as shown.

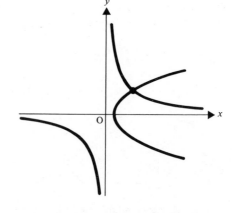

Note. This example shows that it is useful to recognise a standard parabola or
rectangular hyperbola in parametric as well as Cartesian form.

The standard ellipse $\dfrac{x^2}{a^2} + \dfrac{y^2}{b^2} = 1$ can also be recognised from its

parametric equations which are usually given as $\begin{cases} x = a\cos\theta \\ y = b\sin\theta \end{cases}$

EXERCISE 10g

1) Eliminate the parameter from the following pairs of parametric equations:

(a) $x = t^2 - 1;$ $y = 3 + t$

(b) $x = t^3;$ $y = t^2$

(c) $x = 3t;$ $y = \dfrac{3}{t}$

(d) $x = \dfrac{1+t}{t};$ $y = \dfrac{1-t}{t^2}$

(e) $x = 2\cos\theta;$ $y = 3\sin\theta$

(f) $x = \sec\theta;$ $y = 2\tan\theta$

(g) $x = a\cos 2\theta;$ $y = a\cos\theta$

2) Plot the curves represented by the equations given in Question 1, parts (a) to (d) taking values in the range $-3 \leqslant t \leqslant 3$ and parts (e) and (f) taking $0 \leqslant \theta \leqslant 2\pi$.

A curve has parametric equations $x = 2t^2$, $y = 4t$.   Find:

3) The Cartesian equation of the curve.

4) The equation of the tangent at the point where $y = 8$.

5) The equation of the chord joining the points on the curve where $t = p$ and $t = q$.

6) The coordinates of the points where the line $y = x - 6$ meets the curve.

7) The value(s) of k if $y = x + k$ is a tangent to the curve.

8) The coordinates of the point(s) of intersection of the curve and the circle $x^2 + y^2 - 2x = 16$.

9) The coordinates of the point(s) of intersection of the curve and the curve whose parametric equations are $x = 8s$, $y = \dfrac{8}{s}$.

Find the equation of the chord AB in Questions 10–12.

10) At A, $(t = t_1)$ and at B, $(t = t_2)$ on the curve with parametric equations $x = ct$, $y = \dfrac{c}{t}$

11) $A(ap^2, 2ap)$ and $B(aq^2, 2aq)$ are on the parabola $x = at^2$, $y = 2at$.

12) The curve has equations $x = t^3$, $y = t^2$. At A and B, $x = T^3$ and 8 respectively.

13) Find the length of OP where P is a point on the curve given by:

(a) $x = at^2$, $y = 2at$ (b) $x = ct$, $y = \dfrac{c}{t}$

(c) $x = t - 1$, $y = t^2 + 1$ (d) $x = a \cos \theta$, $y = b \sin \theta$.

Find the equation of the normal to the given curve at the given point and find the coordinates of the point where this normal meets the curve again.

14) $x = at^2$, $y = 2at$; $(at_1^2, 2at_1)$.

15) $x = ct$, $y = \dfrac{c}{t}$; $\left(cp, \dfrac{c}{p}\right)$

16) $x = b(1 - t^2)$, $y = 2b(t - 2)$; $[b(1 - p^2), 2b(p - 2)]$.

Find the coordinates of the midpoint of the chord cut off on the line $2x + y = 7$ by the following curves:

17) $x = 4t^2$, $y = 8t$ 18) $x = 2t$, $y = \dfrac{2}{t}$

19) $x = t^2 - 2$, $y = 2t + 1$

FURTHER LOCI

When a point Q is constrained to move in a particular way, its possible positions may be defined in terms of another point P which itself satisfies a specified condition.

The examples which follow show how the Cartesian equation of Q can be found in such cases.

EXAMPLES 10h

1) A circle has centre $(2, 0)$ and radius 2 and P is any point on this circle. OP is produced to Q so that $OQ = 3OP$. Find the equation of the locus of Q.

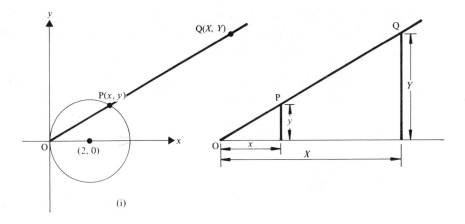

(i)

The equation of the given circle is

$$(x-2)^2 + y^2 = 2^2$$

i.e. $$x^2 + y^2 - 4x = 0 \qquad [1]$$

$P(x, y)$ is any point on the circle.
$Q(X, Y)$ is any point on the required locus.

The given condition is $OQ = 3OP$

Hence
$$\begin{cases} X = 3x \\ Y = 3y \end{cases} \Rightarrow \begin{cases} x = \tfrac{1}{3}X \\ y = \tfrac{1}{3}Y \end{cases}$$

But P is on the given circle so its coordinates satisfy equation [1].

Thus
$$\left(\frac{X}{3}\right)^2 + \left(\frac{Y}{3}\right)^2 - 4\left(\frac{X}{3}\right) = 0$$

$$\Rightarrow \qquad X^2 + Y^2 - 12X = 0 \qquad [2]$$

Equation [1] contains the coordinates of every point P on the circle, so equation [2] contains the coordinates of every possible point Q and is therefore the equation of the locus of Q.

Note: Because $Q(X, Y)$ is defined in terms of $P(x, y)$ it is necessary to find separate relationships between x and X, y and Y. In this way the given equation linking x and y leads to another equation linking X and Y.

2) $P(ap^2, 2ap)$ is any point on the parabola whose parametric equations are $x = at^2$, $y = 2at$ and S is the point $(a, 0)$. PS is produced to cut the parabola again at Q. Find the Cartesian equation of the locus of M, the midpoint of PQ.

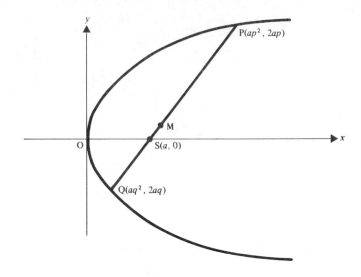

As Q is a point on the parabola, its parametric coordinates are $(aq^2, 2aq)$. Also, since Q is on PS produced, the gradients of SQ and SP are equal,

i.e. $$\frac{2aq}{aq^2 - a} = \frac{2ap}{ap^2 - a}$$

$\Rightarrow$ $$q(p^2 - 1) = p(q^2 - 1)$$

$\Rightarrow$ $$(pq + 1)(p - q) = 0$$

But $p \neq q$ therefore $pq = -1$.   [1]

(**Note.** When the chord joining $P(ap^2, 2ap)$ and $Q(aq^2, 2aq)$ passes through the focus $S(a, 0)$, PQ is a focal chord and the property $pq = -1$, derived above, can be quoted in other problems.)

At M
$$\begin{cases} x = \tfrac{1}{2}(ap^2 + aq^2) \quad \Rightarrow \quad \dfrac{2x}{a} = p^2 + q^2 & [2] \\[4mm] y = \tfrac{1}{2}(2ap + 2aq) \quad \Rightarrow \quad \dfrac{y}{a} = p + q & [3] \end{cases}$$

Equations [2] and [3] are parametric equations for the locus of M. To find the Cartesian equation of this locus we must eliminate p and q.

[3] $\Rightarrow$ $$\left(\frac{y}{a}\right)^2 = p^2 + 2pq + q^2$$

But $p^2 + q^2 = \dfrac{2x}{a}$ and $pq = -1$ (from [2] and [1])

Hence $\quad \dfrac{y^2}{a^2} = \dfrac{2x}{a} - 2$

Thus the Cartesian equation of the locus of M is $\quad y^2 = 2ax - 2a^2$

(**Note.** In this problem the position of M depends upon the positions of *two* moving points P and Q. *Three* equations are needed in these circumstances to derive the Cartesian equation of M.)

3) Show that the equation of the normal at $P(a\cos\theta, b\sin\theta)$, to the curve $b^2x^2 + a^2y^2 = a^2b^2$ is

$$ax \sin\theta - by \cos\theta = (a^2 - b^2)\sin\theta \cos\theta$$

The normal at P meets the x-axis at Q and the y-axis at R. Find:

(a) the greatest value of the area of the triangle OQR, where O is the origin,

(b) the equation of the locus of the centroid of triangle OQR.

(**Note.** Although the reader may recognise, either from the Cartesian equation or from the parametric coordinates, that this curve is an ellipse, it is not necessary to know any geometry of the ellipse, or even its shape, in order to solve this problem, since the parametric analysis of all curves utilises the same methods.)
Using the parametric equations (based on the given parametric coordinates of P)

we have $\quad \begin{cases} x = a\cos\theta \\ y = b\sin\theta \end{cases}$

Hence $\qquad \dfrac{dy}{dx} = \dfrac{dy}{d\theta} \Big/ \dfrac{dx}{d\theta} = -\dfrac{b\cos\theta}{a\sin\theta}$

The gradient of the normal at P is therefore $\dfrac{a\sin\theta}{b\cos\theta}$

and the equation of this normal is

$$y - b\sin\theta = \dfrac{a\sin\theta}{b\cos\theta}(x - a\cos\theta)$$

$\Rightarrow \qquad by\cos\theta - b^2\sin\theta\cos\theta = ax\sin\theta - a^2\sin\theta\cos\theta$

$\Rightarrow \qquad ax\sin\theta - by\cos\theta = (a^2 - b^2)\sin\theta\cos\theta$

At Q, where the normal meets Ox, $\quad y = 0$

Hence $\qquad\qquad\qquad x = \left(\dfrac{a^2 - b^2}{a}\right)\cos\theta$

Similarly at R, $\quad x = 0$

Hence $\qquad\qquad\qquad y = \left(\dfrac{a^2 - b^2}{-b}\right)\sin\theta$

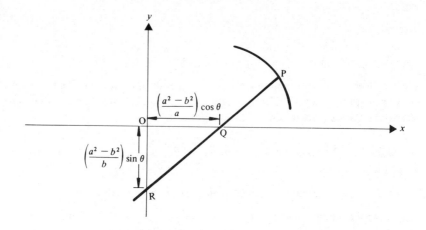

(i) The area, Δ, of triangle OQR is

$$\frac{1}{2}(OQ)(OR) = \frac{1}{2}\left(\frac{a^2 - b^2}{a}\right)\cos\theta\left(\frac{a^2 - b^2}{b}\right)\sin\theta$$

$$= \frac{(a^2 - b^2)^2}{2ab}\sin\theta\cos\theta$$

$$= \frac{(a^2 - b^2)^2}{4ab}\sin 2\theta$$

But the greatest value of $\sin 2\theta$ is 1.
Hence the greatest value of Δ is $(a^2 - b^2)^2/4ab$.

(ii) The coordinates of O, Q and R are:

$$(0, 0), \left(\frac{a^2 - b^2}{a}\cos\theta, 0\right), \left(0, \frac{a^2 - b^2}{-b}\sin\theta\right)$$

Hence at G, the centroid of triangle OQR,

$$\begin{cases} x = \frac{1}{3}\left(0 + \frac{a^2 - b^2}{a}\cos\theta + 0\right) = \left(\frac{a^2 - b^2}{3a}\right)\cos\theta \\ y = \frac{1}{3}\left(0 + 0 + \frac{a^2 - b^2}{-b}\sin\theta\right) = \left(\frac{a^2 - b^2}{-3b}\right)\sin\theta \end{cases}$$

These equations represent the coordinates of G for *any* point P on the given curve, so they are the parametric equations of the locus of G. The corresponding Cartesian equation is given by eliminating θ from these two equations as follows:

$$\cos\theta = \frac{3ax}{a^2 - b^2} \quad \text{and} \quad \sin\theta = \frac{-3by}{a^2 - b^2}$$

But $$\cos^2\theta + \sin^2\theta \equiv 1$$

Hence $$\left(\frac{3ax}{a^2-b^2}\right)^2 + \left(\frac{-3by}{a^2-b^2}\right)^2 = 1$$

for all possible positions of G.
Thus the Cartesian equation of the locus of G is

$$9a^2x^2 + 9b^2y^2 = (a^2-b^2)^2$$

(which may be recognised as the equation of another ellipse).

EXERCISE 10h

1) P is any point on the curve $xy = 4$ and O is the origin. Find the equation of the locus of the midpoint of OP.

2) P is any point on the parabola $y^2 = 4x$ and A is the point $(4, 0)$. Q divides AP in the ratio $1:2$. Find the equation of the locus of Q.

3) A parabola has parametric equations $x = 3t^2$, $y = 6t$ and P is any point on the parabola. Find the Cartesian equation of the locus of the midpoint of OP.

4) $P\left(ct, \dfrac{c}{t}\right)$ is any point on the rectangular hyperbola $xy = c^2$. The tangent at P cuts the y-axis at T and the normal at P cuts the x-axis at N. Find the Cartesian equation of the locus of the midpoint of TN.

5) A circle touches the x-axis and also touches the circle $x^2 + y^2 = 4$. Find the equation of the locus of its centre.

6) P is any point on the curve whose parametric equations are $x = 3\cos\theta$, $y = 2\sin\theta$. The line joining the origin, O, to P is produced to Q where $OQ = 2OP$. Find the Cartesian equation of the locus of Q.

7) A line parallel to the x-axis cuts the curve $y^2 = 4x$ at P and cuts the line $x = -2$ at Q. Find the equation of the locus of the midpoint of PQ.

8) $P(ap^2, 2ap)$ and $Q(aq^2, 2aq)$ are two points on the parabola $y^2 = 4ax$ and PQ passes through the focus $(a, 0)$. Show that, if the tangents at P and Q intersect at T, the locus of T is a straight line and identify this line.

9) A point $P(x, y)$ moves so that its distance from the point $(2, 2)$ is $\sqrt{2}$ times its distance from the line $x + y = 2$.
Find and sketch the locus of P.

SUMMARY

1) If $P(X, Y)$ divides the line joining $A(x_1, y_1)$ and $B(x_2, y_2)$ in the ratio $\lambda:\mu$ then

$$X = \frac{\mu x_1 + \lambda x_2}{\mu + \lambda}, \quad Y = \frac{\mu y_1 + \lambda y_2}{\mu + \lambda}$$

2) The centroid of a triangle with vertices (x_1, y_1), (x_2, y_2), (x_3, y_3) has coordinates

$$[\tfrac{1}{3}(x_1 + x_2 + x_3), \tfrac{1}{3}(y_1 + y_2 + y_3)]$$

3) If θ is the acute angle between $y = m_1 x + c_1$ and $y = m_2 x + c_2$ then

$$\tan \theta = \left| \frac{m_1 - m_2}{1 + m_1 m_2} \right|$$

4) If d is the distance from (p, q) to the line $ax + by + c = 0$, then

$$d = \left| \frac{ap + bq + c}{\sqrt{(a^2 + b^2)}} \right|$$

5) $x^2 + y^2 + 2gx + 2fy + c = 0$ is the equation of a circle with centre $(-g, -f)$ and radius $\sqrt{g^2 + f^2 - c}$.

6) Two circles with centres A, B and radii r_1, r_2 respectively:
(a) are orthogonal if $r_1{}^2 + r_2{}^2 = (AB)^2$,
(b) touch if $r_1 + r_2 = AB$ or $|r_1 - r_2| = AB$.

7) $y^2 = 4ax$ is the equation of a parabola with focus $(a, 0)$ and directrix $x = -a$.

8) Two curves or lines $y = f(x)$, $y = g(x)$ meet where $f(x) = g(x)$. If this equation has equal roots, the curves touch.

MULTIPLE CHOICE EXERCISE 10

(*Instructions for answering these questions are given on p. xii.*)

(All questions relate to figures in one plane.)

TYPE I

1) A point P moves so that it is equidistant from A and B. The locus of the set of points P is:
(a) a circle on AB as diameter,

(b) a line parallel to **AB**,

(c) the perpendicular bisector of **AB**,

(d) a parabola with focus **A** and directrix **B**,

(e) none of these.

2) The point dividing $A(1, 2)$ and $B(7, -4)$ in the ratio $1:2$ has coordinates:

(a) $(3, -2)$　　(b) $(5, -2)$　　(c) $(\frac{8}{3}, -\frac{4}{3})$　　(d) $(3, 0)$

(e) $(13, -10)$.

3) The gradient of the normal to the parabola $x = 3t^2$, $y = 6t$ at the point where $t = -2$ is:

(a) $-\frac{1}{2}$　　(b) -2　　(c) 2　　(d) $\frac{1}{2}$　　(e) 1.

4) The radius of the circle $2x^2 + 2y^2 - 4x + 12y + 11 = 0$ is:

(a) $\sqrt{29}$　　(b) $\sqrt{51}$　　(c) $\sqrt{4.5}$　　(d) 29　　(e) $\sqrt{15.5}$.

5) The parametric equations of a curve are $x = \sec\theta + 1$, $y = \tan\theta - 1$. Its Cartesian equation is:

(a) $y^2 + 3 = x^2$　　　　　　　(b) $x^2 - y^2 - 2x - 2y = 1$

(c) $x^2 - y^2 + 2x + 2y + 1 = 0$　　(d) $x^2 - y^2 = 1$　　(e) $(x - 1)^2 = y^2$.

6) S is the focus of a parabola $\begin{cases} x = at^2 \\ y = 2at \end{cases}$ and P is any point on the parabola.

The equation of the locus of the midpoint of SP is:

(a) $y^2 = a(2x - a)$　　(b) $2y^2 = ax$　　(c) $y^2 = 2ax - a$

(d) $(x - a)^2 = 4y^2$.

7)

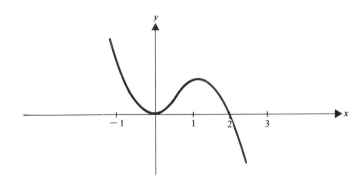

The equation of this curve could be:

(a) $y = x(x - 2)^2$　　(b) $y = 2 - x + 2x^2 - x^3$　　(c) $y = 2x - x^3$

(d) $y = x^2(x + 2)$　　(e) none of these.

8) In an attempt to find the condition that the line $y = mx + c$ should touch the curve $y^2 + 4x = 0$, the following solution was given. State the line where an error first appears.

(a) The line meets the curve where $y^2 + 4\left(\dfrac{y-c}{m}\right) = 0.$

(b) i.e. where $my^2 + 4y - 4c = 0$.

(c) If the line is a tangent this equation has real roots.

(d) The required condition is therefore $mc = 1$.

9) $P(ap^2, ap^3)$ and $Q(aq^2, aq^3)$ are two points on a curve.

(a) The curve is a parabola.

(b) The equation of PQ is $y(p + q) = x(p^2 + pq + q^2) - ap^2q^2$.

(c) The Cartesian equation of the curve is $ax^3 = y^2$.

(d) The curve cuts the x-axis three times.

10) A line with gradient 2 which passes through the point $(k, 0)$ will touch the parabola $y^2 = 4ax$ if and only if:

(a) $k > 0$ (b) $k < 0$ (c) $k = -a$

(d) $k < -a$ (e) $4k = -a$.

TYPE II

11) If the equation $ax^2 + by^2 + 2gx + 2fy + c = 0$ represents a circle through the origin,

(a) $g = 0$ and $f = 0$.   (b) $c = 0$.

(c) $a = b$.   (d) $a = -b$.

12) The equations of two curves, C_1 and C_2, are $xy = 1$ and $x^2 + y^2 = 2$.

(a) C_1 is a rectangular hyperbola.

(b) C_1 and C_2 intersect at four distinct points.

(c) C_1 and C_2 have two common tangents.

(d) C_1 is a continuous curve.

13) The parabola $y^2 + 8x = 0$:

(a) has a focal length of two units,

(b) can be represented by $\begin{cases} x = 2t^2 \\ y = 4t, \end{cases}$

(c) is the locus of a point equidistant from the point $(2, 0)$ and the line $x = -2$,

(d) is symmetrical about the x-axis.

TYPE III

14) (a) A point P is equidistant from a fixed straight line and a fixed point.
 (b) The locus of a point P is a parabola.

15) (a) A certain cubic equation is satisfied by $x = 1$ and $x = 2$.
 (b) A certain cubic equation has three real roots.

16) (a) The point $(1, 2)$ is on a certain parabola.
 (b) The equation of a certain parabola is $y^2 = 4x$.

17) (a) The circles $x^2 + y^2 = r^2$ and $(x - 3)^2 + (y + 4)^2 = 16$ touch
 each other.
 (b) $r = 1$.

18) A, B and P are points on a circle.
 (a) Angle APB is $90°$.
 (b) AB is a diameter of the circle.

19) (a) $f(x) = 0$ has two equal roots.
 (b) $y = f(x)$ touches the x-axis.

20) (a) $f(x, y) = 0$ is the equation of a circle.
 (b) $f(x, y) \equiv x^2 + 2y^2 - 3x - y + 2$.

TYPE IV

21) Find the equation of the tangent at a point A on the curve
$ax^2 + by^2 + 2gx + 2fy + c = 0$ if:
(a) the coordinates of A are given,
(b) the values of g and f are given,
(c) the values of a and b are equal,
(d) the curve passes through the origin.

22) Find the equation of a circle whose centre is in the first quadrant given that:
(a) it touches the x-axis,
(b) its centre is on the line $y = x$,
(c) it touches the y-axis,
(d) it touches the line $x + y = 6$.

23) Determine whether the line $y = mx + c$ touches the parabola
$y^2 = 4x$ if:
(a) the line is parallel to $3x + 2y = 7$,
(b) the line passes through the focus of the parabola,
(c) the focus of the parabola is on the x-axis,
(d) the line has a negative gradient.

24) $y = x^3 + ax^2 + bx + c$.   Determine the values of a, b and c if:
(a) the curve touches the x-axis at the point $(3, 0)$,
(b) the curve crosses the y-axis at the point $(0, 9)$,
(c) the gradient of the curve when $x = 0$ is 3.

25) Find the equation of the bisector of the angle A in triangle ABC, given:
(a) the equation of BC,
(b) the coordinates of A,
(c) the distance of A from BC,
(d) the coordinates of the centroid of triangle ABC.

TYPE V

26) $x^2 + y^2 - 2x - 4y + 6 = 0$ is the equation of a circle.

27) If the line $y = mx + 2$ is a tangent to the parabola $y^2 = 4x$ there are two possible values of m.

28) No cubic equation has exactly two real roots.

29) The gradient of every tangent to the rectangular hyperbola $xy = c^2$ is negative.

30) The parametric equations $x = 2 \cos \theta$, $y = 3 \sin \theta$,   represent a circle.

MISCELLANEOUS EXERCISE 10

1) A triangle has its vertices at $A(4, 4)$, $B(-4, 0)$, $C(6, 0)$.
(a) Find the equation of the circle through the points A, B, C.
(b) Find the coordinates of the point where the internal bisector of the angle BAC meets the x-axis.
(c) Find the equation of the circle which passes through B and touches AC at C.   (U of L)

2) The vertex A of a square ABCD, lettered in the anticlockwise sense, has coordinates $(-1, -3)$. The diagonal BD lies along the line $x - 2y + 5 = 0$.
(a) Prove, by calculation, that the coordinates of C are $(-5, 5)$, and find the coordinates of B and D.
(b) Find the equation of the circle which touches all four sides of the square, confirming that this circle passes through the origin.
(c) Calculate the area of that portion of the square which lies in the first quadrant $(x > 0, y > 0)$.   (C)

3) Show that the points $(-1, 0)$ and $(1, 0)$ are on the same side of the line $y = x - 3$.
Find the equations of the two circles each passing through the points $(-1, 0)$, $(1, 0)$ and touching the line $y = x - 3$.   (U of L)

4) If P, Q are the points (x_1, y_1), (x_2, y_2) respectively, show that the equation of the circle on PQ as diameter is

$$(x - x_1)(x - x_2) + (y - y_1)(y - y_2) = 0.$$

A triangle has vertices at the origin O and the points $A(4, 3)$, $B(\frac{6}{5}, \frac{17}{5})$. The line through O perpendicular to AB and the line through A perpendicular to OB meet at H. Show that the coordinates of H are $(\frac{3}{5}, \frac{21}{5})$. If BH meets OA at D, find the coordinates of D.
Write down the equation of the circle on OB as diameter and verify that this circle passes through D. (U of L)

5) Prove that the point $B(1, 0)$ is the mirror-image of the point $A(5, 6)$ in the line $2x + 3y = 15$.
Find the equation of:
(a) the circle on AB as diameter,
(b) the circle which passes through A and B and touches the x-axis. (U of L)

6) A straight line of gradient m passes through the point $(1, 1)$ and cuts the x- and y-axes at A and B respectively. The point P lies on AB and is such that $AP : PB = 1 : 2$. Show that, as m varies, P moves on the curve whose equation is $3xy - x - 2y = 0$.
Find the perpendicular distance of the point $(1, 1)$ from the tangent to this curve at the origin. (JMB)p

7) Find the equation of the circle S which passes through $A(0, 4)$ and $B(8, 0)$ and has its centre on the x-axis. If the point C lies on the circumference of S, find the greatest possible area of the triangle ABC. (U of L)

8) A rhombus ABCD is such that the coordinates of A and C are $(-3, -4)$ and $(5, 4)$ respectively. Find the equation of the diagonal BD of the rhombus. If the side BC has gradient 2, obtain the coordinates of B and of D.
Prove that the rhombus has area $21\frac{1}{3}$. (C)

9) A circle with centre P and radius r touches externally both the circles $x^2 + y^2 = 4$ and $x^2 + y^2 - 6x + 8 = 0$. Prove that the x-coordinate of P is $\frac{1}{3}r + 2$, and that P lies on the curve $y^2 = 8(x - 1)(x - 2)$. (U of L)

10) The circle S_1 with centre $C_1(a_1, b_1)$ and radius r_1 touches externally the circle S_2 with centre $C_2(a_2, b_2)$ and radius r_2. The tangent at their common point passes through the origin. Show that

$$(a_1{}^2 - a_2{}^2) + (b_1{}^2 - b_2{}^2) = (r_1{}^2 - r_2{}^2).$$

If, also, the other two tangents from the origin to S_1 and S_2 are perpendicular, prove that $|a_2 b_1 - a_1 b_2| = |a_1 a_2 + b_1 b_2|$.

Hence show that, if C_1 remains fixed but S_1 and S_2 vary, then C_2 lies on the curve

$$(a_1{}^2 - b_1{}^2)(x^2 - y^2) + 4a_1 b_1 xy = 0. \qquad \text{(U of L)}$$

11) Given the points A(2, 14), B(− 6, 2), C(12, − 10) verify that the triangle ABC is right angled. Calculate the coordinates of:
(a) the point D on AB produced such that   AC = CD,
(b) the point of intersection of the perpendiculars from A, C, D to the
   opposite sides of the triangle ACD. (JMB)

12) Find the equation of the two circles which each satisfy the following conditions:
(a) the axis of x is a tangent to the circle,
(b) the centre of the circle lies on the line $2y = x$,
(c) the point (14, 2) lies on the circle.
Prove that the line $3y = 4x$ is a common tangent to these circles. (C)

13) On the same diagram sketch the circles $x^2 + y^2 = 4$ and
$x^2 + y^2 - 10x = 0$.   The line $ax + by + 1 = 0$ is a tangent to both
these circles. State the distances of the centres of the circles from this tangent.
Hence, or otherwise, find the possible values of a and b and show that, if 2ϕ
is the angle between the common tangents, then $\tan \phi = \frac{3}{4}$. (U of L)

14) Find the equation of the circle which touches the line $3y - 4x - 24 = 0$
at the point (0, 8) and also passes through the point (7, 9). Prove that this
circle also touches the axis of x. Find the equations of the tangents to this
circle which are perpendicular to the line $3y - 4x - 24 = 0$. (C)

15) (a) A circle with centre (3, 2) touches the line $4x - 3y + 4 = 0$.   Find
   the equation of the circle and show that it touches the x-axis.
   (b) In each of the following pairs of equations, t is a parameter. Sketch
   the locus given by each pair of equations:
   (i) $x = 3 + 5 \cos t$, $y = 4 + 5 \sin t$ $(0 \leqslant t \leqslant \pi)$
   (ii) $x = 3 \cos t$, $y = 4 \cos t$ $(0 \leqslant t \leqslant \pi)$
   (iii) $x = 3 + t \cos \dfrac{\pi}{3}$, $y = 4 + t \sin \dfrac{\pi}{3}$ $(-\infty < t < \infty)$. (JMB)

16) Find the coordinates of the foot of the perpendicular from the point
(2, − 6) to the line $3y - x + 2 = 0$. (U of L)

17) Find the points of intersection of the circle

$$x^2 + y^2 - 6x + 2y - 17 = 0$$

and the line $x - y + 2 = 0$.   Show that an equation of the circle which has
these points as the ends of a diameter is

$$x^2 + y^2 - 4y - 5 = 0$$

Show also that this circle and the circle
$$x^2 + y^2 - 8x + 2y + 13 = 0$$
touch externally.   (U of L)

18) If the normal at $P(ap^2, 2ap)$ to the parabola $y^2 = 4ax$ meets the curve again at $Q(aq^2, 2aq)$ prove that $p^2 + pq + 2 = 0$. Prove that the equation of the locus of the point of intersection of the tangents to the parabola at P and Q is
$$y^2(x + 2a) + 4a^3 = 0$$ (U of L)

19) Obtain the equation of the tangent to the parabola $y^2 = 4x$ at the point $(t^2, 2t)$. The tangents to the parabola at the points $P(p^2, 2p)$ and $Q(q^2, 2q)$ meet on the line $y = 3$. Find the equation of the locus of the midpoint of PQ.
If PQ intersects the x-axis and the y-axis at R and S respectively, find also the equation of the locus of the midpoint of RS.   (AEB)'74

20) Show that the equation of the normal to the parabola $y^2 = 4ax$ at the point $(at^2, 2at)$ is $y + tx = 2at + at^3$.
If this normal meets the parabola again at the point $(aT^2, 2aT)$ show that
$$t^2 + tT + 2 = 0$$
and deduce that T^2 cannot be less that 8.
The line $3y = 2x + 4a$ meets the parabola at the points P and Q. Show that the normals at P and Q meet on the parabola.   (U of L)

21) The points $P(ap^2, 2ap)$ and $Q(aq^2, 2aq)$ move on the parabola $y^2 = 4ax$ and $p + q = 2$. Show that the chord PQ makes a constant angle with the x-axis and that the locus of the midpoint M of PQ is part of a line which is parallel to the x-axis.
If also the point $R(ar^2, 2ar)$ moves so that $p - r = 2$, find in its simplest form the Cartesian equation of the locus of the midpoint N of PR.   (JMB)

22) The parabolas $x^2 = 4ay$ and $y^2 = 4ax$ meet at the origin and at the point P. The tangent to $x^2 = 4ay$ at P meets $y^2 = 4ax$ again at A, and the tangent to $y^2 = 4ax$ at P meets $x^2 = 4ay$ again at B. Prove that the angle APB is $\arctan\left(\frac{3}{4}\right)$ and that AB is a common tangent to the two parabolas.   (U of L)

23) Show that the equation of the normal to the parabola $y^2 = 4ax$ at the point $P(at^2, 2at)$ is $y + tx = 2at + at^3$.
The normal at P meets the x-axis at G and the midpoint of PG is N.
(a) Find the equation of the locus of N as P moves on the parabola.
(b) The focus of the parabola is S. Prove that SN is perpendicular to PG.
(c) If the triangle SPG is equilateral, find the coordinates of P.   (AEB)'75

24) Prove that the line $y = mx + \dfrac{15}{4m}$ is a tangent to the parabola $y^2 = 15x$

for all non-zero values of m. Using this result, or otherwise, find the equations of the common tangents to this parabola and the circle $x^2 + y^2 = 16$. (U of L)

25) A variable line $y = mx + c$ cuts the fixed parabola $y^2 = 4ax$ in two points P and Q. Show that the coordinates of M, the midpoint of PQ are

$$\left(\frac{2a - mc}{m^2}, \frac{2a}{m} \right)$$

Find one equation satisfied by the coordinates of M in each of the following cases:
(a) if the line has fixed gradient m,
(b) if the line passes through the fixed point $(0, -a)$.   (U of L)

26) Sketch the curve given by the parametric equations $x = \sin \theta$, $y = \sin 2\theta$.

27) Show that the tangent at the point P, with parameter t, on the curve $x = 3t^2$, $y = 2t^3$ has equation $y = tx - t^3$. Prove that this tangent will cut the curve again at the point Q with coordinates $(3t^2/4, -t^3/4)$.
Find the coordinates of the possible positions of P if the tangent to the curve at P is the normal to the curve at Q.   (U of L)

28) Show that the point with coordinates $(2 + 2 \cos \theta, 2 \sin \theta)$ lies on the circle $x^2 + y^2 = 4x$, and obtain the equation of the tangent to the circle at this point.
The tangents at the points P and Q on this circle touch the circle $x^2 + y^2 = 1$ at the points R and S. Find the coordinates of the point of intersection of these tangents, and obtain the equation of the circle through the points P, Q, R and S.   (U of L)

29) The parametric equations of a curve are $x = \cos 2t$, $y = 4 \sin t$. Sketch the curve for $0 \leqslant t \leqslant \frac{1}{2}\pi$.

Show that $\dfrac{dy}{dx} = -\operatorname{cosec} t$ and find the equation of the tangent to the curve

at the point $A(\cos 2T, 4 \sin T)$.
The tangent at A crosses the x-axis at the point M and the normal at A crosses the x-axis at the point N. If the area of the triangle AMN is $12 \sin T$, find the value of T between 0 and $\frac{1}{2}\pi$.   (AEB)'76

30) The foot of the perpendicular from a point P to the straight line $x + y = \sqrt{2}$ is the point R and Q is the point with coordinates $(\sqrt{2}, \sqrt{2})$. If P varies in such a way that $PQ^2 = 2PR^2$, show that its locus is the rectangular hyperbola $xy = 1$.

Find the equation of the tangent to this hyperbola at the point $\left(t, \dfrac{1}{t} \right)$.

This tangent cuts the x-axis at A and the y-axis at B, and C is the point on AB such that $AC:CB = a:b$. Show that the locus of C as t varies is the

rectangular hyperbola $xy = \dfrac{4ab}{(a+b)^2}$.

Determine the two possible values of the ratio $a:b$ such that the straight line $x+y = \sqrt{2}$ is a tangent to the locus of C. (JMB)

31) A curve is given parametrically by $x = a(5\cos\theta + \cos 5\theta)$,
$$y = a(5\sin\theta - \sin 5\theta).$$
Find the equation of the normal to the curve at the point with parameter θ. Find those points of the curve in the first quadrant at which the normal is also normal to the curve at another point. (O)

32) Find the equation of the tangent to the curve $y = \frac{5}{12}x^3 - \frac{13}{9}x$ at the point P at which $x = x_0$.
Show that the x-coordinate of the point Q where this tangent meets the curve again is $-2x_0$, and find the values of x_0 for which the tangent at P is the normal at Q. (O)

33) Find the equation of the tangent to the circle $x^2 + y^2 = a^2$ at the point $T(a\cos\theta, a\sin\theta)$. This tangent meets the line $x + a = 0$ at R. If RT is produced to P so that $RT = TP$, find the coordinates of P in terms of θ and find the coordinates of the points in which the locus of P meets the y-axis. (U of L)

34) Find the equation of the tangent to the curve $ay^2 = x^3$ at the point (at^2, at^3), where $a > 0$ and t is a parameter. (U of L)

35) A curve is given by the parametric equations $x = t^2 - 3$, $y = t(t^2 - 3)$.
(a) Find its Cartesian equation, in a form clear of surds and fractions.
(b) Prove that it is symmetrical about the x-axis.
(c) Show that there are no points on the curve for which $x < -3$. (C)

36) The tangent and the normal at a point $P(t, e^{-t^2/2})$ on the curve $y = e^{-x^2/2}$ meet the x-axis in T and G respectively, and N is the foot of the ordinate from P. Show that $G = [t(1 - e^{-t^2}), 0]$ and if $NT.GN = e^{-1}$, find the length of PN. (AEB)'75

CHAPTER 11

CURVE SKETCHING

In this chapter we are going to investigate a variety of ways in which the graph of a function can be determined.

First, consider a group of functions with some specially interesting features.

EVEN FUNCTIONS

A function $f(x)$ is said to be *even* if $f(x) = f(-x)$ for all values of x. The graphs of all even functions are therefore symmetrical about the vertical axis. Some common even functions are:

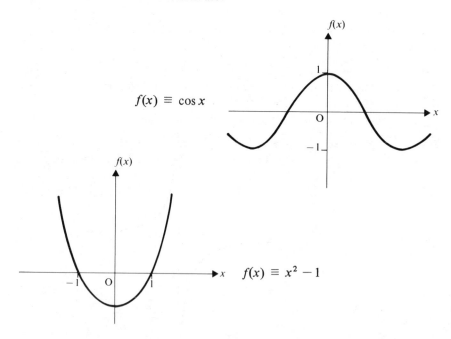

$f(x) \equiv \cos x$

$f(x) \equiv x^2 - 1$

ODD FUNCTIONS

A function $f(x)$ is said to be *odd* if $f(x) = -f(-x)$ for all values of x. The graphs of odd functions are 'symmetrical about the origin', i.e. the origin, O, divides the graph into two sections each of which can be rotated about O, through an angle π, to give the other section.

Some common odd functions are:

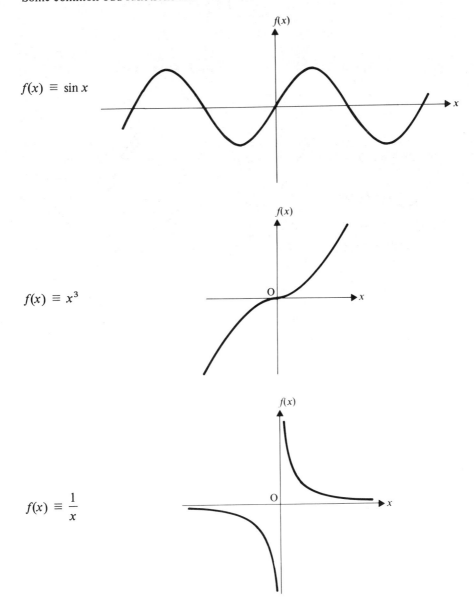

$f(x) \equiv \sin x$

$f(x) \equiv x^3$

$f(x) \equiv \dfrac{1}{x}$

CONTINUOUS FUNCTIONS

A function $f(x)$ is *continuous* when $x = a$ if $f(a)$ is defined and

$\lim\limits_{x \to a,} f(x)$ is defined and is equal to $f(a)$

i.e. $f(x) \to f(a)$ as $x \to a$ from above and from below.

A continuous function satisfies this condition for all values of a in its domain, so the graph of a continuous function is unbroken.

The graphs of some common continuous functions are shown below.

(a) $f : x \to \sin x, x \in \mathbb{R}$ (b) $f : x \to x(x-1)(x-2), x \in \mathbb{R}$

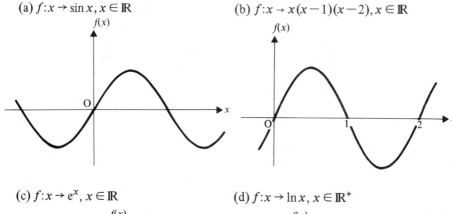

(c) $f : x \to e^x, x \in \mathbb{R}$ (d) $f : x \to \ln x, x \in \mathbb{R}^+$

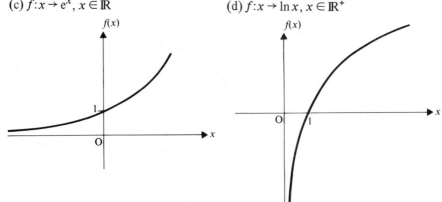

PERIODIC FUNCTIONS

A function whose graph consists of a basic pattern which repeats at regular intervals is said to be *periodic*. The width of the basic pattern is the *period* of the function, e.g. $\sin x$ is periodic and its period is 2π

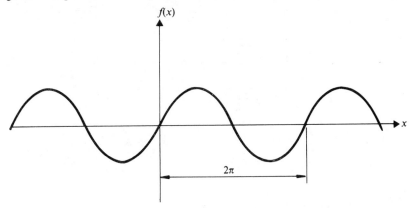

and $\tan x$ is periodic with a period π

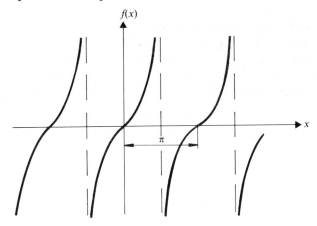

For a periodic function with period a,

$$f(x) = f(x + a) = f(x + 2a) = \ldots$$

Similarly $\qquad f(x) = f(x - a) = f(x - 2a) \ldots .$

i.e. $\qquad\qquad f(x) = f(x + ka) \quad$ where $k \in \mathbb{Z}$

So a periodic function is defined by the condition $\quad f(x + ka) = f(x), \quad$ or $f(x \pm a) = f(x) \quad$ for all values of x, where a is the period of the function. The definition of the function within one period, e.g. the interval $\quad 0 < x \leqslant a$, then defines the whole function.

e.g. if $\quad f(x) \equiv 2x - 1 \quad$ for $\quad 0 < x \leqslant 1$
and $\quad f(x + 1) = f(x) \quad$ for all values of x
then we know that the function is periodic with a period of 1.
So if we draw $\quad f(x) \equiv 2x - 1 \quad$ for the range $\quad 0 < x \leqslant 1 \quad$ this pattern then repeats at unit intervals,

i.e.

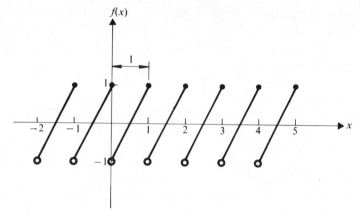

Note that this function is not continuous.

COMPOUND PERIODIC FUNCTIONS

The basic pattern in the graph of a periodic function may be made up of two or more different definitions.

For example, a function $f(x)$ is defined by:

$$f(x) \equiv x \qquad \text{for} \quad 0 \leqslant x < 1$$

$$f(x) \equiv 4 - x^2 \quad \text{for} \quad 1 \leqslant x < 2$$

$$f(x+2) = f(x) \quad \text{for all values of } x.$$

The graph representing this function can be derived as follows

for $0 \leqslant x < 1$ we have

$$f(x) \equiv x \qquad \Rightarrow$$

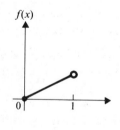

for $1 \leqslant x < 2$ we have

$$f(x) \equiv 4 - x^2 \qquad \Rightarrow$$

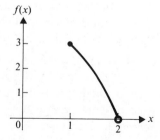

Combining these two sections,
for　$0 \leqslant x < 2$　　the graph is:

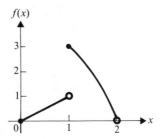

But the statement that　　$f(x+2) = f(x)$　　for all values of x

tells us that the function is periodic with a period of 2, so we repeat the above
pattern at regular intervals, giving

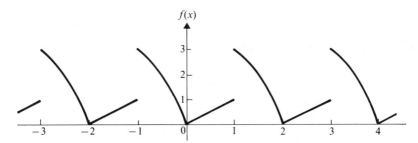

EXERCISE 11a

1) Sketch each of the following functions and state which are continuous.
(a) $\cos x$　　　(b) $\tan x$　　　(c) e^{-x}　　(d) $\ln(1+x)$
(e) $\cot x$　　　(f) $(x-1)^2$　　(g) $x-1$　　(h) $1/x$

2) State which, if any, of the functions in Question 1 is
(a) even　　　(b) odd　　　(c) periodic

3) Which of the following graphs represent functions that are
(a) continuous　　　(b) even　　　(c) odd　　　(d) periodic

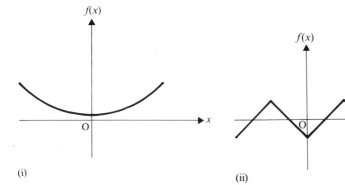

(i)

(ii)

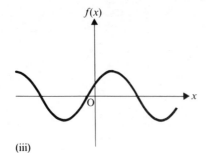

(iii)

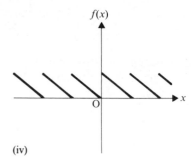

(iv)

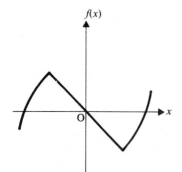

(v)

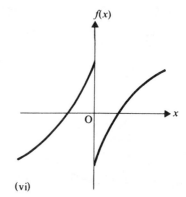

(vi)

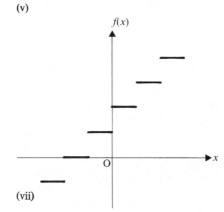

(vii)

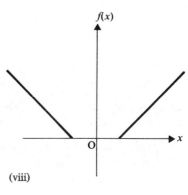

(viii)

4) Sketch the graph of $f(x)$ within the interval $-4 < x \leqslant 6$ if

$$f(x) \equiv 4-x^2 \quad \text{for} \quad 0 < x \leqslant 2$$

and $f(x) = f(x-2)$ for all values of x.

5) If $f(\theta) \equiv \sin \theta \quad \text{for} \quad 0 < \theta \leqslant \dfrac{\pi}{2}$

$$f(\theta) \equiv \cos \theta \quad \text{for} \quad \dfrac{\pi}{2} < \theta \leqslant \pi$$

and $\qquad\qquad f(\theta + \pi) = f(\theta) \quad$ for all values of θ,

sketch the function $f(\theta)$ for the range $\quad -2 < \theta \leqslant 2\pi$.

6) A function $f(x)$ is periodic with a period of 4. Sketch the graph of the function for $-6 \leqslant x \leqslant 6$, given that

$$f(x) \equiv -x \quad \text{for} \quad 0 < x \leqslant 3$$
$$f(x) \equiv 3x \quad \text{for} \quad 3 < x \leqslant 4.$$

GENERAL CURVE SKETCHING

The frequent use made of curve sketching throughout this text serves to emphasise the importance of this aspect of mathematical understanding. When the equation of a curve is unfamiliar, determining the shape of the curve can require varied and sometimes extensive consideration. It is impossible to study every type of curve but if we discuss some of those which occur quite frequently the techniques used in these cases can be applied to other curves.

Certain features of a curve can be observed, or easily calculated, from the equation of the curve, for example:

(a) the points where the curve crosses the coordinate axes,
(b) the stationary points,
(c) the linear asymptotes,
(d) regions of the xy plane where the graph does not exist,
(e) the behaviour of y as $x \to \pm\infty$.

It is not always necessary to consider *all* of these features when examining a particular curve, different equations requiring different approaches. But as each of the above observations will be used in some problems, let us consider the application of each in turn.

(a) *Axes intercepts*

A curve meets (i.e. cuts or touches) the x-axis if the equation $\quad y = 0 \quad$ has real solutions,

e.g. for $\quad y = 2 - \dfrac{1}{x}$,

$$y = 0 \Rightarrow 2 - \frac{1}{x} = 0 \Rightarrow x = \frac{1}{2}$$

Thus the curve *cuts* the x-axis at $(\frac{1}{2}, 0)$.

Similarly the intercepts on the y-axis are given by the real solution(s) of the equation $\quad x = 0$,

e.g. for $\quad y(y - 2) = x - 1, \quad x = 0 \quad \Rightarrow \quad y^2 - 2y + 1 = 0 \quad \Rightarrow \quad (y - 1)^2 = 0$

i.e. when $\quad x = 0, \quad y = 1, 1.$

Thus the curve *touches* the y-axis at $(0, 1)$.

(b) *Stationary points*

The analysis of stationary points is dealt with in Chapter 5.

But in curve sketching the *absence* of stationary points is just as important as their presence,

e.g. for $y = 2 - \dfrac{1}{x}$, we note that $\dfrac{dy}{dx} = \dfrac{1}{x^2}$. Hence there is no finite value

of x for which $\dfrac{dy}{dx} = 0$, and we deduce that there are no finite stationary

points.

(c) *Linear asymptotes*

If, as either x or y becomes very large the equation of a curve approximates to the equation of a straight line, that straight line is an asymptote. For instance

(i) if $y = 2 - \dfrac{1}{x}$ then, as

$x \to \pm \infty$ the equation
approximates to $y = 2$.
So the line $y = 2$ is an
asymptote,
i.e. the curve approaches this
line as $x \to \pm \infty$.

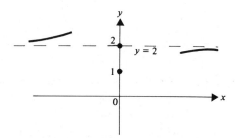

(ii) if $y = \dfrac{1}{x - 2}$,

$y \to \pm \infty$ when $x \to 2$.
So the line $x = 2$ is an
asymptote.

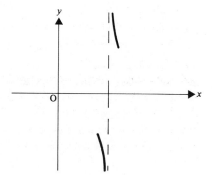

(d) *Empty sections*, i.e. regions where no part of the graph lies, can be located using the method adopted in the following example.

If $y = \dfrac{x - 2}{x + 1}$ we see that $y = 0$ when $x = 2$ and that y is

undefined when $x = -1$.

So the values $x = 2$ and $x = -1$ are significant and we investigate the sign of y in the regions of the xy plane separated by the lines $x = -1$ and $x = 2$ by completing the table below.

	$x < -1$	$-1 < x < 2$	$x > 2$
$x + 1$	$-$	$+$	$+$
$x - 2$	$-$	$-$	$+$
y	$+$	$-$	$+$

From these results we can shade the regions of the xy plane where no part of the curve lies.

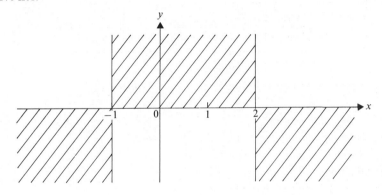

Note. The series of signs of y found by using the above table can be determined mentally and the appropriate empty sections shaded immediately.

(e) *The behaviour of y as $x \to \pm\infty$.* If a curve has no obvious asymptotes, the behaviour of y as $x \to \pm\infty$ is often useful,

e.g. if $y = xe^{x^2}$, e^{x^2} is always positive
thus $y \to \infty$ when $x \to \infty$
and $y \to -\infty$ when $x \to -\infty$.

The information gained from the above investigations is usually sufficient for a *sketch* of the corresponding curve to be drawn. So many variations are possible, however, that no standard procedure can be formulated to cover all cases. The examples that follow indicate some of the approaches that can be used.

EXAMPLES 11b

1) Sketch the curve whose equation is $y = \dfrac{x-2}{x-1}$.

(a) The curve crosses the axes where $\begin{cases} x = 0 & y = 2 \\ x = 2 & y = 0 \end{cases}$

(b) There is no real value of x for which $\dfrac{dy}{dx} = 0$ so there are no stationary

points.

(c) Rewriting the equation as $y = 1 - \dfrac{1}{x-1}$

we see that
$$\begin{cases} y \to 1 & \text{when} \quad x \to \pm\infty \\ y \to \pm\infty & \text{when} \quad x \to 1. \end{cases}$$

So $y = 1$ and $x = 1$ are asymptotes.

(d)

	$x < 1$	$1 < x < 2$	$2 < x$
$x - 2$	$-$	$-$	$+$
$x - 1$	$-$	$+$	$+$
y	$+$	$-$	$+$

We now have sufficient information to sketch the curve.

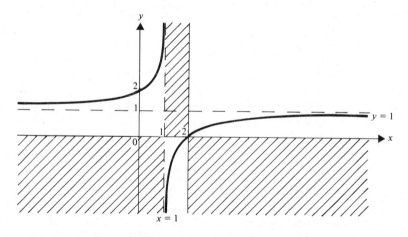

Note that y is undefined when $x = 1$ and x is undefined when $y = 1$.

2) Sketch the curve $y = \dfrac{x}{(x+1)^2}$.

The following observations can be made.

(a) The curve crosses the axes only at the origin.

(b) To find any stationary points we use $\dfrac{dy}{dx} = 0$.

$$\frac{dy}{dx} = \frac{(x+1)^2 - 2x(x+1)}{(x+1)^4} = \frac{(x+1)(1-x)}{(x+1)^4} = \frac{1-x}{(x+1)^3}$$

So there is only one stationary point at $(1, \frac{1}{4})$.

Now
$$\frac{d^2y}{dx^2} = \frac{-(x+1)-3(1-x)(x+1)^2}{(x+1)^6} = \frac{2x-4}{(x+1)^4}$$

This is negative when $x = 1$ so $(1, \frac{1}{4})$ is a maximum point.

(c) When
$$\begin{cases} x \to -1, & y \to -\infty \\ x \to \pm\infty, & y \to 0 \end{cases}$$

So $x = -1$ and $y = 0$ are asymptotes.

(d) As $(x+1)^2$ cannot be negative, we see that
$$\begin{cases} x > 0 & \Rightarrow & y > 0 \\ x < 0 & \Rightarrow & y < 0 \end{cases}$$

From the information above the graph can now be drawn.

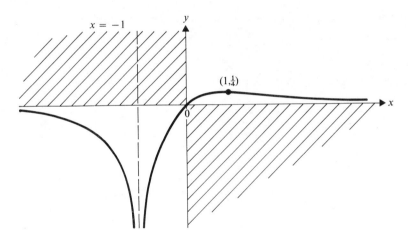

Note that the curve approaches the *lower* end of the asymptote $x = -1$ from *both* sides. This is because it is a *double* asymptote, obtained from $(x+1)^2$.

3) Find the turning point(s) on the curve with equation $y = xe^{-x}$ and hence sketch the curve.

If $y = xe^{-x}$, $\dfrac{dy}{dx} = e^{-x} + x(-e^{-x})$

At turning points
$$\frac{dy}{dx} = e^{-x}(1-x) = 0 \quad \Rightarrow \quad x = 1$$

Further,
$$\frac{d^2y}{dx^2} = -e^{-x}(1-x) + e^{-x}(-1)$$

$$= -\frac{1}{e} \quad \text{when} \quad x = 1$$

Since $\dfrac{dy}{dx} = 0$ and $\dfrac{d^2y}{dx^2} < 0$ when $x = 1$, there is a maximum turning point at $\left(1, \dfrac{1}{e}\right)$.

As y is defined for all real values of x, there are no vertical asymptotes and the curve is continuous.

Now consider the behaviour of y when $x \to \pm\infty$.

When $x \to \infty$, $y = \dfrac{x}{e^x} \to 0$ (since e^n is very much greater than n when n is large and positive).

When $x \to -\infty$, $y \to -\infty$.

Note also, since e^{-x} is positive for all values of x, that when $x < 0$, $y < 0$, when $x > 0$, $y > 0$ and when $x = 0$, $y = 0$.

Thus the sketch of the curve with equation $y = xe^{-x}$ is as shown.

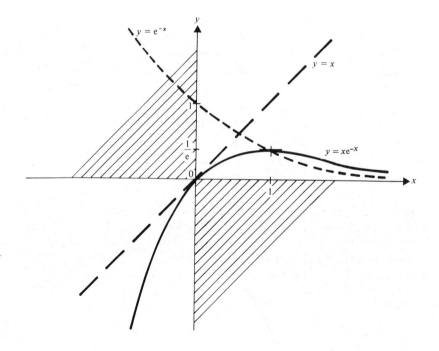

4) Sketch the graph of $y = \dfrac{1}{x} \ln x$, showing clearly any turning points.

There are no values of $\ln x$ for negative values of x, so the curve $y = \dfrac{1}{x} \ln x$ does not exist when $x < 0$.

To determine any turning points we use

$$\frac{dy}{dx} = -\frac{1}{x^2}\ln x + \left(\frac{1}{x}\right)\left(\frac{1}{x}\right) = 0 \quad \Rightarrow \quad \frac{1}{x^2}(1 - \ln x) = 0$$

i.e. $\ln x = 1 \quad \Rightarrow \quad x = e$.

$$\left(\text{**Note** also that } \frac{dy}{dx} \to 0 \quad \text{when} \quad x \to \infty.\right)$$

$$\frac{d^2y}{dx^2} = -\frac{2}{x^3}(1 - \ln x) + \frac{1}{x^2}\left(-\frac{1}{x}\right) < 0 \quad \text{when} \quad x = e$$

Thus $\left(e, \dfrac{1}{e}\right)$ is a maximum turning point.

When $x \to \infty$, $y \to 0$ (since n is much greater than $\ln n$ when n is large).
When $x = 1$, $\ln x = 0 \quad \Rightarrow \quad y = 0$.
When $x < 1$, $\ln x < 0 \quad \Rightarrow \quad y < 0$.
When $x > 1$, $\ln x > 0 \quad \Rightarrow \quad y > 0$.

When $x \to 0$, $\dfrac{1}{x} \to \infty$, $\ln x \to -\infty \quad \Rightarrow \quad y \to -\infty$.

The curve can now be sketched.

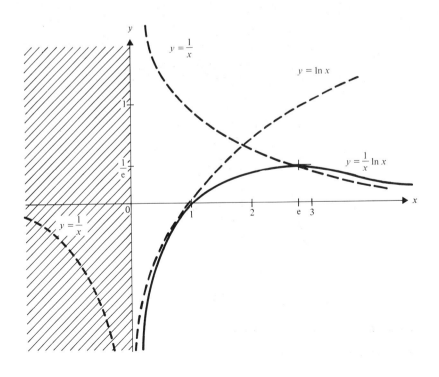

EXERCISE 11b

1) Write down the equations of the linear asymptotes of the curves whose equations are:

(a) $y = \dfrac{1}{1-x}$

(b) $y = \dfrac{x}{1-x}$

(c) $y = \dfrac{1-x}{x}$

(d) $y = \dfrac{x+1}{x^2}$

(e) $y = \dfrac{x-2}{x-4}$

(f) $y = \dfrac{x+3}{2x+1}$

(g) $y = \dfrac{3x}{x-4}$

(h) $y = \dfrac{4x-3}{2x+5}$

(i) $y = \dfrac{x}{2x+3}$

2) Write down the coordinates of the point(s) where each of the curves in Question (1) crosses the coordinate axes (i.e. the x- and y-axes).

3) Determine the behaviour of y as $x \to \infty$ and $x \to -\infty$ if:

(a) $y = \dfrac{x}{1+x^2}$

(b) $y^2 = 4ax$

(c) $y = x^2 e^{-x}$

(d) $y = x \ln x$

(e) $y = x^3 + x - 1$

(f) $y = x^3 + 4x^2 + 2x - 3$

4) For what range(s) of values of x is y positive, when:

(a) $y = x^2 + 5x - 6$

(b) $y = x^3 - 6x^2 + 11x - 6$

(c) $y = (x-1) e^x$

(d) $y = \dfrac{x}{1-x}$

(e) $y = \dfrac{x}{(1-x)^2}$

(f) $y = \dfrac{x-2}{(x+1)(x-1)}$

(g) $y = x^2 + 4x + 5$

(h) $\dfrac{x-2}{x-4}$

Sketch the graphs of the following functions showing clearly any asymptotes, turning points, points of intersection with the coordinate axes and the behaviour of the curve when x and/or y are very large.

5) $y = \dfrac{1}{1-x}$ [Use (1a) and (2a)]

6) $y = \dfrac{x}{1-x}$ [Use (1b) and (4d)]

7) $y = \dfrac{x-2}{(x-4)}$ [Use (2e) and (4h)]

8) $y = x^2 e^x$

9) $y = \dfrac{\ln x}{x-1}$

10) $y = \dfrac{x+3}{2x+1}$

11) $y = x^3 + 2x^2 - x - 2$

12) $y = \dfrac{x}{(1-x)^2}$

13) $y = \dfrac{x^2}{(1-x)}$

14) $y = \dfrac{3x}{x-4}$

15) $y = \dfrac{e^x}{x}$

16) $y = \dfrac{2x}{x^2 + 1}$

RECIPROCAL GRAPHS

Consider the curve whose equation is

$$y = \frac{1}{f(x)}$$

when the graph of the function $f(x)$ is familiar.
The following simple properties provide the means to adapt the known graph of $f(x)$ to provide the reciprocal graph of $1/f(x)$.

(a) For a given value of x, $f(x)$ and $\dfrac{1}{f(x)}$ have the same sign.

(b) If the numerical value of $f(x)$ increases, the numerical value of $\dfrac{1}{f(x)}$ decreases (and conversely).

(c) If $f(x) = 1$ then $\dfrac{1}{f(x)} = 1$ also, and

   if $f(x) = -1$ then $\dfrac{1}{f(x)} = -1$ also.

(d) If $f(x) \to \pm\infty$, $\dfrac{1}{f(x)} \to 0$

   and if $f(x) \to 0$, $\dfrac{1}{f(x)} \to \pm\infty$.

(e) If $f(x)$ has a maximum turning point, i.e. $f(x)$ stops increasing and begins to decrease then, at the same value of x, $\dfrac{1}{f(x)}$ stops decreasing and begins to increase, i.e. $\dfrac{1}{f(x)}$ has a minimum turning point (and conversely).

EXAMPLES 11c

1) Use the graph of the function $x^2 + 2$ to sketch the graph of the function
$$\frac{1}{x^2 + 2}.$$

If $f(x) \equiv x^2 + 2$ then,

when $x \in \mathbb{R}$, $f(x) > 0$ so $\dfrac{1}{f(x)} > 0$

for $x < 0$, $f'(x)$ is negative so the gradient of $\dfrac{1}{f(x)}$ is positive,

for $x > 0$, $f'(x)$ is positive so the gradient of $\dfrac{1}{f(x)}$ is negative.

$f(x)$ is minimum when $x = 0$ $\Rightarrow$ $\dfrac{1}{f(x)}$ is maximum when $x = 0$,

when $x \to \pm\infty$, $f(x) \to \infty$, so $\dfrac{1}{f(x)} \to 0$.

From this information the graph of $\dfrac{1}{f(x)}$ can be drawn as shown below.

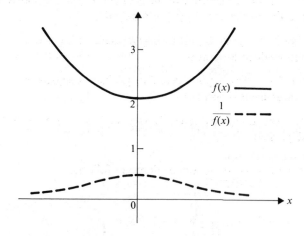

Note that the minimum value of $f(x)$ and the maximum value of $\dfrac{1}{f(x)}$ are not equal although they occur at the same value of x.

2) Sketch the graph of the function $f(x) \equiv 4 + 3x - x^2$.

Hence deduce the graph of the function $\dfrac{1}{f(x)}$.

Noting that the coefficient of x^2 is negative and that the factors of $4 + 3x - x^2$ are $(4 - x)$ and $(1 + x)$, the graph of this quadratic function can be drawn.

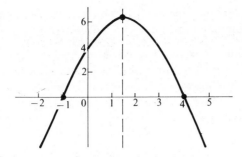

In order to deduce the graph of $\dfrac{1}{f(x)}$ the following points are noted.

When $x \to -1$ or 4 the value of $f(x) \to 0 \Rightarrow$ the value of $1/f(x) \to \pm\infty$ so $x = -1$ and $x = 4$ are asymptotes to the graph of $1/f(x)$.

When $x \to \pm\infty$ the value of $f(x) \to -\infty$ so the value of $1/f(x) \to 0$.

When $x = 1.5$ $f(x)$ has a maximum value so $1/f(x)$ has a minimum value.

For $x < -1$ $f(x)$ is negative and the gradient of $f(x)$ is positive so $1/f(x)$ is negative and the gradient of $1/f(x)$ is negative.

For $-1 < x < 1.5$ $f(x)$ is positive and its gradient is positive so $1/f(x)$ is positive and its gradient is negative.

For $1.5 < x < 4$ $f(x)$ is positive and its gradient is negative so $1/f(x)$ is positive and its gradient is positive.

For $x > 4$ $f(x)$ is negative and so is its gradient so $1/f(x)$ is negative and has a positive gradient.

Thus the graph of $\dfrac{1}{4 + 3x - x^2}$ can be drawn as follows.

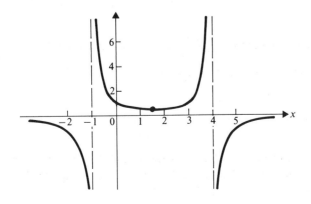

3) Sketch the graph of the function $(x+1)(x-1)(x-2)$ and hence sketch the graph of the reciprocal function $1/(x+1)(x-1)(x-2)$.

The cubic function can be sketched at once as we have its factors and we see that the coefficient of x^3 is positive. Then the graph of the reciprocal function can be deduced in the same way as in the previous examples.

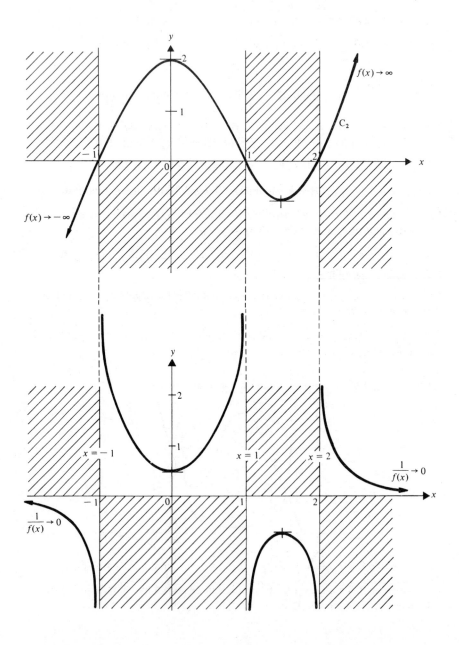

EXERCISE 11c

Sketch the graph of $f(x)$, and hence the graph of $\dfrac{1}{f(x)}$, when $f(x)$ is:

1) $x^2 - 3x + 2$ 2) $x^3 - 3x^2 + 2x$ 3) $\dfrac{x}{x-1}$

4) $x^2 - 4$ 5) $\sin x$ 6) $1 - x^2$

7) $x^3 - x^2 - 4x + 4$ 8) $x^2 - x$ 9) $\ln x$

10) $3 + 5x - 2x^2$ 11) $\tan x$ 12) $-1 - x^2$

INEQUALITIES

Many inequalities can be solved directly with the aid of a sketch graph. The solution of an inequality that involves a function of x is the set of values of x for which the graph of the function lies on the appropriate side of the x-axis,

i.e.   above the x-axis if $f(x) > 0$

below the x-axis if $f(x) < 0$.

For example, to find the set of values of x for which

$$x^3 - 9x^2 + 23x - 15 \leqslant 0$$

the function $x^3 - 9x^2 + 23x - 15$ is first factorised in order to identify the points where its graph crosses the x-axis.

$$x^3 - 9x^2 + 23x - 15 \equiv (x - 1)(x - 3)(x - 5)$$

so the intercepts made by the graph on the x-axis are $1, 3$ and 5, and a sketch can be made as follows,

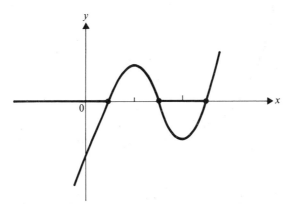

Hence $x^3 - 9x^2 + 23x - 15 \leqslant 0$ for $x \leqslant 1$ and $3 \leqslant x \leqslant 5$.

Inequalities which are not quite so simple can be solved by a combination of algebraic and graphical methods.

EXAMPLES 11d

1) Find the set of values of x for which

$$\frac{(x-1)^2}{(x+5)} < 1$$

$(x+5)$ may be positive or negative, so we avoid multiplying by $x+5$ and proceed as follows

$$\frac{(x-1)^2}{(x+5)} - 1 < 0$$

$\Rightarrow \qquad \dfrac{x^2 - 2x + 1 - (x+5)}{x+5} < 0$

$\Rightarrow \qquad \dfrac{x^2 - 3x - 4}{x+5} < 0$

$\Rightarrow \qquad \dfrac{(x-4)(x+1)}{x+5} < 0 \qquad \text{or} \qquad \dfrac{f(x)}{g(x)} < 0$

This fraction is negative (< 0) if the numerator, $f(x)$
                          and the denominator, $g(x)$, have opposite signs.

Consider $\quad f(x) \equiv (x-4)(x+1)$.

From the graph we see that

$f(x) > 0 \quad$ for $\quad x < -1 \quad$ and $\quad x > 4$

Considering $\quad g(x) \equiv x + 5$

$\qquad g(x) > 0 \quad$ for $\quad x > -5$

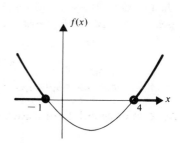

Illustrating these sets on a number line

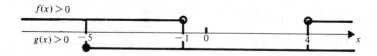

we see that $f(x)$ and $g(x)$ have opposite signs (where the positive ranges do *not* coincide) for the set of values

$$x < -5 \qquad \text{and} \qquad -1 < x < 4$$

Alternatively, the set of values of x that satisfy $\quad \dfrac{(x-4)(x+1)}{(x+5)} < 0 \quad$ can be found from a sign table for $h(x)$, where $h(x) \equiv \dfrac{(x-4)(x+1)}{(x+5)}$, in which

the significant values of x are -5, -1 and 4.

	$x < -5$	$-5 < x < -1$	$-1 < x < 4$	$x < 4$
$x+5$	$-$	$+$	$+$	$+$
$x+1$	$-$	$-$	$+$	$+$
$x-4$	$-$	$-$	$-$	$+$
$h(x)$	$-$	$+$	$-$	$+$

Thus $\qquad h(x) < 0 \quad$ for $\quad x < -5 \quad$ and $\quad -1 < x < 4$.

2) Find the range of the function f defined by

$$f: x \to \frac{x+1}{2x^2+x+1}, \quad x \in \mathbb{R}.$$

Let p be a possible value of the function. Then the values of x corresponding to $f(x) = p$ can be found from the equation

$$\frac{x+1}{2x^2+x+1} = p$$

$$p(2x^2+x+1) = x+1$$

$$2px^2 + (p-1)x + (p-1) = 0$$

As x must be real, this equation must have real roots

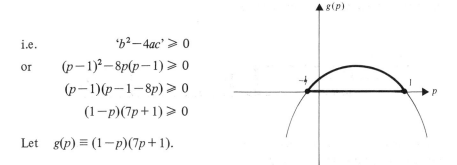

i.e. $\qquad$ '$b^2 - 4ac$' $\geqslant 0$

or $\qquad (p-1)^2 - 8p(p-1) \geqslant 0$

$\qquad\quad (p-1)(p-1-8p) \geqslant 0$

$\qquad\quad (1-p)(7p+1) \geqslant 0$

Let $\quad g(p) \equiv (1-p)(7p+1)$.

From the sketch, $\quad g(p) \geqslant 0 \quad$ when $\quad -\frac{1}{7} \leqslant p \leqslant 1$
Therefore the range of f is $\quad -\frac{1}{7} \leqslant f(x) \leqslant 1$.

3) Show that, if x is real, the function $\dfrac{2-x}{x^2-4x+1}$ can take any real value.

Let n be a possible value of the given function

so that
$$n = \frac{2-x}{x^2-4x+1}$$

$\Rightarrow$
$$nx^2 + (1-4n)x + n - 2 = 0$$

The values of x have to be real,

so
$$(1-4n)^2 - 4n(n-2) \geqslant 0$$

$\Rightarrow$
$$12n^2 + 1 \geqslant 0$$

But $12n^2 + 1$ is always positive, so there is no restriction on the value of n.

Hence $\dfrac{2-x}{x^2-4x-1}$ can take any real value.

EXERCISE 11d

Find the set of values of x for which,

1) $\dfrac{2+x}{x-7} > 3$

2) $\dfrac{x+5}{4x-1} < 2$

3) $\dfrac{x}{2x-8} > 3$

4) $\dfrac{5-2x}{1-7x} > 1$

5) $\dfrac{x-1}{2x-3} < 2$

6) $\dfrac{x}{x-2} < \dfrac{x}{x-1}$

7) $(x+1)(x+3)(x+5) > 0$

8) $(1-x)(1+x)^2 < 0$

9) $\dfrac{(x-2)}{(x-1)(x-3)} > 0$

10) $\dfrac{x^2-2}{2} < \dfrac{6x^2-8x-1}{x+5}$

11) If x is real, find the range of possible values of $\dfrac{4(x-2)}{4x^2+9}$.   (AEB) 76p

12) If x is real, find the set of possible values of $\dfrac{(2x+1)}{(x^2+2)}$.   (U of L)p

13) Find the solution set of the inequality
$$\frac{12}{x-3} < x+1$$ (C)p

In Questions 14 to 22 find the range of the function f defined by $f:x \rightarrow$

14) $\dfrac{1}{1-x}$

15) $\dfrac{x}{1+x^2}$

16) $\dfrac{1+x^2}{x}$

17) $\dfrac{x-2}{(x+1)(x-1)}$ 18) $\dfrac{x}{(1-x)^2}$ 19) $\dfrac{x^2}{x+1}$

20) $\dfrac{x+3}{x}$ 21) $\dfrac{1+x^2}{1-x^2}$ 22) $\dfrac{x-2}{(x+2)(x+3)}$

23) Show that, for real values of x, the function $\dfrac{2(3x+1)}{3(x^2-9)}$ can take all real values.

24) Find the value of p for which $\dfrac{x^2-5x+p}{x-3}$ can take all real values provided that x is real.

25) Given that $f(x) \equiv \dfrac{x-2}{x^2-5x+k}$ and that x is real, find:

(a) the range of values of k such that the value of $f(x)$ is unrestricted,

(b) the set of values of x for which $f(x) < -1$ when $k = 6$.

THE MODULUS OF A FUNCTION

If $y = x$, y is negative when x is negative. But if $y = |x|$, y takes the numerical value and *not the sign* of x, i.e. y is *always positive*. $|x|$ is called the *modulus of x*.

Taking values of x from -3 to $+3$ we have

x	-3	-2	-1	0	1	2	3
$|x|$	3	2	1	0	1	2	3

Thus the graphs of $y = x$ and $y = |x|$ are

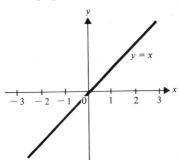
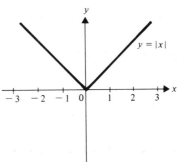

It can be seen that, for negative values of x, $y = |x|$ is the reflection of $y = x$ in the x-axis.

In general, the graph C_1 with equation $y = |f(x)|$ can be obtained from the graph C_2 of $y = f(x)$ by reflecting in the x-axis those points on C_2 for which $f(x) < 0$, the remaining sections of C_2 being retained without change.

For instance, to sketch $y = |x^2 - 3x + 2|$ the graph of $y = x^2 - 3x + 2$ is adapted as follows

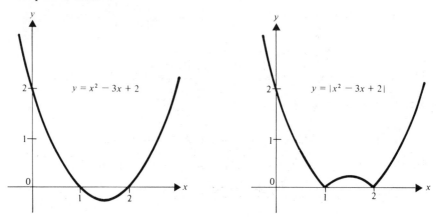

Note that, for any function $f(x)$, the mapping defined by $x \to |f(x)|$ is itself a function.

FURTHER INEQUALITIES

We saw in Chapter 4 that an equation $y = f(x)$, where x and y are the coordinates of a point P in the xy plane, describes a straight line or a curve. The inequalities $y < f(x)$ and $y > f(x)$ describe areas in the xy plane which are bounded by the above line or curve,

e.g. $y < x^2 + 1$ defines the area shaded in the diagram below, while the unshaded area is defined by $y > x^2 + 1$. The boundary between these areas is the curve $y = x^2 + 1$.

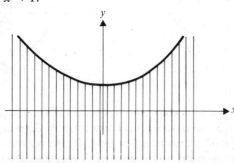

These interpretations also apply to equations and inequalities involving moduli.

An *equation* involving a modulus describes a line or curve in the xy plane. An *inequality* involving a modulus describes an area in this plane,

e.g. $y > |x^2 - 3x + 2|$ defines the area shaded in the following diagram while the unshaded area corresponds to $y < |x^2 - 3x + 2|$.

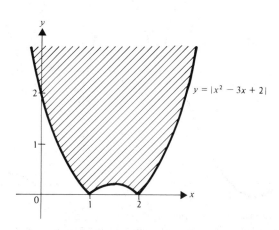

The Effect of a Modulus Sign on a Cartesian Equation

When a section of a graph $y = f(x)$ is reflected in the x-axis, the y coordinate of every point on that section of the graph changes sign.

Thus the equation of the *reflected section* becomes
$$y = -f(x)$$
e.g. when $y = |x|$
the Cartesian equations are

$$\begin{cases} y = x & \text{for} \quad x > 0 \\ y = -x & \text{for} \quad x < 0 \end{cases}$$

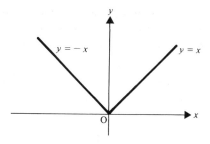

Intersection

If intersection occurs between two lines or curves whose equations involve a modulus, care must be taken to use the correct pair of Cartesian equations when locating the point(s) of intersection. For example, the points common to $y = x - 1$ and $y = |x^2 - 3|$ are seen by sketching these two graphs on the same axes as shown.

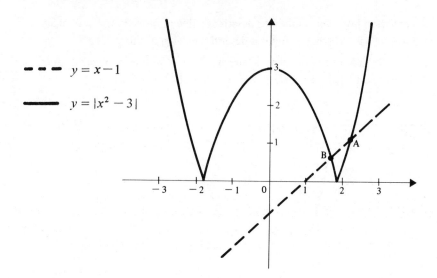

The Cartesian equations of the various sections of the line and curve show that

at A $\qquad$ $y = x - 1$ $\qquad$ cuts $\quad y = x^2 - 3$ $\qquad\qquad$ [1]

at B $\qquad$ $y = x - 1$ $\qquad$ cuts $\quad y = -(x^2 - 3)$ $\qquad$ [2]

Thus the coordinates of A are given by the equation

$$x^2 - x - 2 = 0 \quad \Rightarrow \quad x = -1, 2$$

But it is clear from the diagram that $\quad x \neq -1$.

So, at A, $\qquad\qquad$ $x = 2$ $\quad$ and $\quad y = x - 1 \Rightarrow y = 1$

Similarly the coordinates of B are given by

$$x^2 + x - 4 = 0 \quad \Rightarrow \quad x = -\tfrac{1}{2}(1 \pm \sqrt{17})$$

Again from the diagram we see that $\quad x \neq -\tfrac{1}{2}(1 + \sqrt{17})$,

so, at B, $\qquad$ $x = -\tfrac{1}{2}(1 - \sqrt{17}) \quad \Rightarrow \quad y = \tfrac{1}{2}(-3 + \sqrt{17})$

Alternatively, as both sections of the graph of $y = |x^2 - 3|$ are included in the equation $y^2 = (x^2 - 3)^2$, the coordinates of both points of intersection are given by solving

$$(x - 1)^2 = (x^2 - 3)^2$$

This method is attractively concise but is not always suitable. For instance this equation is quartic and can be solved only if it can be factorised.

EXAMPLES 11e

1) Sketch on the same axes the graphs $y = f(x)$ and $y = g(x)$ where $f(x) \equiv 2 - |x - 1|$ and $g(x) \equiv |x + 2| - 3$. Indicate on your graph the set(s) of values of x for which $g(x) < f(x)$.

In order to sketch $y = f(x)$ we first draw the graph of the function $|x - 1|$ diagram (i)

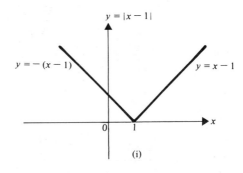

Then the function $-|x - 1|$ can be drawn as shown in diagram (ii)

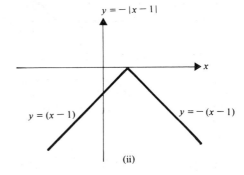

When this graph is raised vertically by two units we have the graph of $f(x) \equiv 2 - |x - 1|$, as shown in diagram (iii)

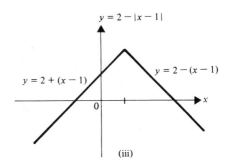

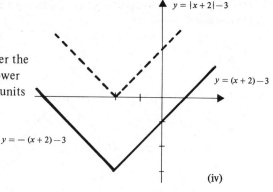

Similarly for $g(x)$ we consider the graph of $|x + 2|$ and then lower this graph vertically by three units as in diagram (iv).

(iv)

Sketching the graphs of $y = f(x)$ and $y = g(x)$ on the same axes we have

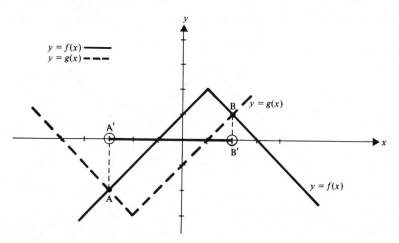

Between A and B, $f(x) > g(x)$.

Thus the set of values of x for which $g(x) < f(x)$ is from A' to B' as shown.

Note. The actual coordinates of A' and B' can be found, if required, by solving the equations

$$2 - (x - 1) = (x + 2) - 3 \qquad \text{for B'}$$
$$2 + (x - 1) = -(x + 2) - 3 \qquad \text{for A'}$$

2) Shade the area in which a point P can lie if the coordinates of P satisfy the following conditions

$$\begin{cases} y \geqslant |2 - x| \\ y \leqslant 4 - |x| \\ 0 \leqslant y < \sqrt{4x} \end{cases}$$

Considering the conditions separately and shading the area in which P can lie in each case, we have

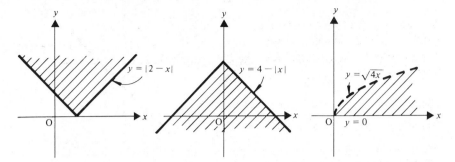

Combining all three conditions shows that P can lie only in the area shaded in the diagram below.

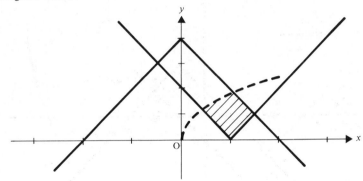

EXERCISE 11e

Sketch the following graphs:

1) $y = |2x - 1|$

2) $y = |x(x - 1)(x - 2)|$

3) $y = |x^2 - 1|$

4) $y = |x^2 + 1|$

5) $y = |\sin x|$

6) $y = |\ln x|$

7) $y = -|\cos x|$

8) $y = 3 + |x + 1|$

9) $y = |2x + 5| - 4$

10) $y = |x^2 - x - 20|$

Indicate on a diagram the set(s) of values of x that satisfy the following inequalities. (Do not calculate the boundary values of x.)

11) $|x + 1| > |3x - 5|$

12) $|x| < |x^2 - 7x + 6|$

13) $|(x - 1)(x - 2)(x - 3)| > e^{-x}$

14) $|\sin x| < |\cos x|$ [for $0 \leqslant x \leqslant 2\pi$]

Solve the following equations

15) $x = |x^2 - 2|$ 16) $2 - |x + 1| = |4x - 3|$

17) $|x^2 - 1| - 1 = 3x - 2$

18) $2|x| = 3 + 2x - x^2$

Shade the areas indicated by the following sets of conditions

19) $0 < x < 2\pi$, $0 < y < |\sin x|$

20) $y > |x|$, $y < 4 - |x|$

21) $y < 4 - x^2$, $y > |x - 1|$

22) Each of the four diagrams corresponds to one of the following seven equations.

Arrange the correct equations in the correct order.

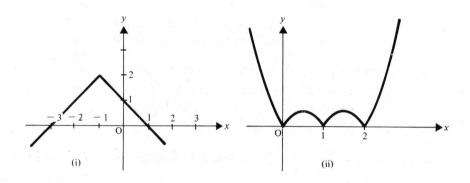

(i)   (ii)

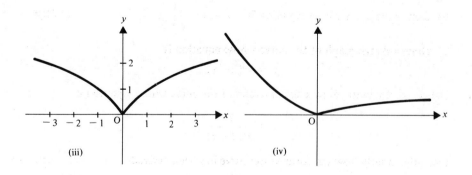

(iii)   (iv)

(a) $y = |x - 1|$ (b) $y = |2x - 3x^2 + x^3|$ (c) $y = |e^{-x}|$
(d) $y = 2 - |x + 1|$ (e) $y = \sqrt{|x|}$
(f) $y = |e^{-x} - 1|$ (g) $y = |x^3|$

MISCELLANEOUS EXERCISE 11

1) The function f is periodic with period π and

$$f(x) = \sin x \quad \text{for} \quad 0 \leqslant x \leqslant \pi/2,$$
$$f(x) = 4(\pi^2 - x^2)/(3\pi^2) \quad \text{for} \quad \pi/2 < x < \pi.$$

Sketch the graph of $f(x)$ in the range $-\pi \leqslant x \leqslant 2\pi$.   (U of L)

2) The function $f(x)$ is defined for $0 \leqslant x \leqslant 2$ by

$$f(x) = x \quad \text{for} \quad 0 \leqslant x \leqslant 1,$$
$$f(x) = (2-x)^2 \quad \text{for} \quad 1 < x \leqslant 2.$$

Sketch the graph of this function for $0 \leqslant x \leqslant 2$.   (U of L)p

3) The function f is defined by

$$f(x) = \sin x \quad \text{for} \quad x \leqslant 0,$$
$$f(x) = x \quad \text{for} \quad x > 0.$$

Sketch the graphs of $f(x)$ and its derivative $f'(x)$ for $-\pi/2 < x < \pi/2$ and decide whether the functions f and f' are continuous at $x = 0$ or not.   (U of L)

4) The function f is periodic with period 3 and

$$f(x) = \sqrt{(9 - 4x^2)} \quad \text{for} \quad 0 < x \leqslant 1.5,$$
$$f(x) = 2x - 3 \quad \text{for} \quad 1.5 < x \leqslant 3.$$

Sketch the graph of $f(x)$ in the range $-3 \leqslant x \leqslant 6$.   (U of L)

5) Given that $y = \dfrac{4x^2 + 2x + 1}{x^2 - x + 1}$, determine the range of values taken by y for real values of x.   (JMB)

6) Sketch the curve whose equation is $y = 1 - \dfrac{1}{x+2}$.   (JMB)p

7) Draw a sketch-graph of the curve whose equation is

$$y = x^2(2 - x)$$

Hence, or otherwise, draw a sketch-graph of the curve whose equation is

$$y = \frac{1}{x^2(2-x)}$$

indicating briefly how the form of the curve has been derived.

8) Find the stationary points on the graphs of

(a) $y = \dfrac{x-1}{x^2}$ (b) $y = \dfrac{1}{x} - 1 + \ln x$

Sketch the graphs of these functions.

9) R is the set of all real numbers. The mapping g is defined by

$$g: x \to \frac{2x+1}{x-1}, \quad (x \in R, \quad x \neq 1)$$

State the range of g and sketch the graph of $y = g(x)$.
Define the mapping g^{-1}.   (C)

10) Find the range of values of k for which the function

$$\frac{x^2-1}{(x-2)(x+k)}$$

where x is real, takes all real values.   (JMB)

11) Given that $f(x) \equiv 6x^2 + x - 12$, find the minimum value of $f(x)$ and the values of x for which $f(x) = 0$.

Using the same axes, sketch the curves $y = f(x)$ and $y = \dfrac{1}{f(x)}$, labelling each clearly.

Deduce that there are four values of x for which $[f(x)]^2 = 1$. Find these values, each to two decimal places.   (U of L)

12) For the curve with equation

$$y = \frac{x}{(x-2)},$$

find

(i) the equation of each of the asymptotes,
(ii) the equation of the tangent at the origin.

Sketch the curve, paying particular attention to its behaviour at the origin and as it approaches its asymptotes.

On a separate diagram, sketch the curve with equation

$$y = \left| \frac{x}{x-2} \right|. \quad\quad\quad \text{(U of L)}$$

13) The function f has the following properties:
(a) $f(x) = 4x$ for $0 \leqslant x < 2/3$,
(b) $f(x) = 8(1-x)$ for $2/3 \leqslant x \leqslant 1$,
(c) $f(x)$ is an odd function,
(d) $f(x)$ is periodic with period 2.

Sketch the graph of $f(x)$ for $-3 \leqslant x \leqslant 3$.   (U of L)

14) For each of the following expressions state whether or not it is periodic, and, if it is periodic, give the period.

(a) $\dfrac{\sin x}{x}$, $x \neq 0$,   (b) $|\sin x|$.   (U of L)

15) The function f is periodic with period π and

$$f(x) = \cos x \qquad \text{for } -\pi/2 < x \leqslant 0,$$
$$f(x) = 1 - x^2/2 \qquad \text{for } \quad 0 < x \leqslant \pi/2.$$

Sketch the graph of the function f for $-3\pi/2 < x < 3\pi/2$. (U of L)

16) The domain of a mapping m is the set of all real numbers except 1, and

the mapping is given by $m: x \rightarrow \left| \dfrac{x}{x-1} \right|$.

(a) State, with a reason, whether or not m is a function.

(b) State the range of m.

(c) Show that $\qquad (x > 1) \Rightarrow \{m(x) > 1\}$

and that $\qquad (x < 0) \Rightarrow \{m(x) < 1\}$

(d) Sketch the graph of $\quad y = m(x)$. (C)

17) Show that, if $\quad y = \dfrac{(x+1)(x-3)}{x(x-2)} \quad$ and x is real, y cannot lie between

1 and 4.

18) Find the set of values of x for which

$$|x-1| > 2|x-2|.$$ (U of L)

19) Solve the inequality

$$\frac{1}{x-4} > \frac{x}{x-6}.$$ (U of L)p

20) Solve the inequality

$$(x+1)(x-2)(x+3)(x-4) > 0.$$ (U of L)

21) Given that $px^2 + 2px - 3 < 0$ for all real values of x, determine the set of possible values of p. (U of L)

22) A function f is defined by

$$f: x \rightarrow \frac{x^2+a}{x+2}, \quad x \in \mathbb{R}$$

Find (i) the range of f when $a = 21$, (ii) the set of values of a for which the range of f is all real values.

23) Find the range of values of x for which $\dfrac{x^2+56}{x} > 15.$ (U of L)p

24) Given that $\quad y = \dfrac{x^2+\lambda}{x-1} \quad$ find the set of values of λ for which y can

take all real values when x is real. Find the set of values of y when $\lambda = 3$.

25) (a) Determine whether the functions f, g, h are odd, even or neither, given that

$$f(x) \equiv \sin(\cos x)$$
$$g(x) \equiv e^{\cos x} \qquad x \in \mathbb{R}$$
$$h(x) \equiv e^{\sin x}$$

(b) In the domain $-\dfrac{\pi}{2} < x < \dfrac{\pi}{2}$, p and q are defined by

$p: x \to (\sin x)/3$ and $q: x \to \cos(x/3)$.

Determine whether functions p^{-1} and q^{-1} exist and, if so, whether they are odd, even or neither.   (U of L)

26) For any real number x, $[x]$ denotes the greatest integer not exceeding x; e.g. $[3.6] = 3$, $[2] = 2$, $[-1.4] = -2$, etc.

Functions f and g are defined on the domain of all real numbers as follows:

$$f: x \to [x]; \qquad g: x \to x - [x]$$

Find the ranges of f and g and sketch the graph of g.

Determine the solution sets of the equations

(a) $f(x) = g(x)$ (b) $fg(x) = gf(x)$ (U of L)

CHAPTER 12

VECTORS AND THREE DIMENSIONAL COORDINATE GEOMETRY

VECTORS

Readers will by now appreciate that two and two do not always make four.

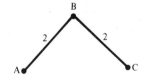

For example, consider three points A, B and C. If B is 2 cm from A and C is 2 cm from B, then C is not, in general, 4 cm from A.

AB, BC and AC are displacements. Each has a magnitude (e.g. the magnitude of AB is the distance from A to B) and is related to a definite direction in space. The displacement from A to B $(\overrightarrow{AB})$ followed by the displacement from B to C $(\overrightarrow{BC})$ is equivalent to the displacement from A to C $(\overrightarrow{AC})$, i.e. $\overrightarrow{AB}$ together with $\overrightarrow{BC}$ is equivalent to $\overrightarrow{AC}$, which is written

$$\overrightarrow{AB} + \overrightarrow{BC} \equiv \overrightarrow{AC}.$$

There are many other quantities which behave in the same way as these displacements and they can all be represented by vectors.

A vector quantity is one that has a magnitude and is also related to a definite direction in space.

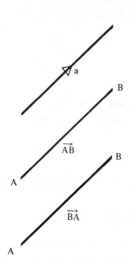

It follows that we may represent a vector by a section of a straight line where the length of the line represents the magnitude of the vector, and the direction of the line (together with the arrow) represents the direction of the vector.

Such vectors are denoted by bold type e.g. **a** or, when hand written, by a̲. Alternatively we may represent a vector by the straight line joining two points A and B, and denote the vector in the direction A to B by $\overrightarrow{AB}$ (or **AB**) and the vector in the opposite direction, B to A, by $\overrightarrow{BA}$ (or **BA**).

Scalars

A scalar quantity is one that is fully defined by magnitude alone and can therefore be represented completely by a real number. For example, length is a scalar quantity as a real number specifies the length of an object completely. Scalar quantities can be compounded using the familiar laws of algebra, and in this case two and two *do* make four.

MODULUS OF A VECTOR

The modulus of a vector **a** is the magnitude of **a**, i.e. the length of the line representing **a**. The modulus of **a** is written |**a**| or a.

EQUAL VECTORS

Two vectors are equal if they have the same magnitude and the same direction

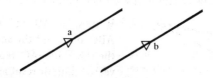

i.e. $\mathbf{a} = \mathbf{b} \iff \begin{cases} |\mathbf{a}| = |\mathbf{b}| \\ \text{the direction of } \mathbf{a} \text{ is the same as the direction of } \mathbf{b} \end{cases}$

It follows from this that a vector may be represented by any line of the right length and direction, i.e. the location of the line in space does not matter.

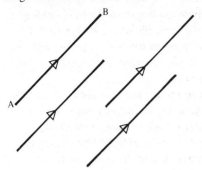

Thus any of the lines in the diagram may represent the vector $\overrightarrow{AB}$.

NEGATIVE VECTORS

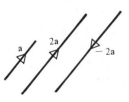

If two vectors **a** and **b** have the same magnitude but opposite directions then we say that

$$\mathbf{a} = -\mathbf{b}$$

i.e. $-\mathbf{a}$ is a vector of magnitude $|\mathbf{a}|$ and direction opposite to that of **a**.

MULTIPLICATION OF A VECTOR BY A SCALAR

If λ is a positive real number then

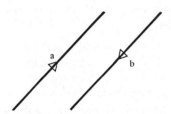

$\lambda\mathbf{a}$ is a vector of magnitude $\lambda|\mathbf{a}|$ and in the same direction as **a**. It also follows that $-\lambda\mathbf{a}$ is a vector of magnitude $\lambda|\mathbf{a}|$ and with direction opposite to that of **a**.

ADDITION OF VECTORS

All vectors are compounded in the same way as the displacements with which we began this chapter. The addition of vectors is defined as follows.

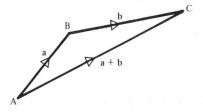

If the sides **AB** and **BC** of triangle ABC represent the vectors **a** and **b**, the third side **AC** represents the vector sum, or resultant, of **a** and **b** and is denoted by $\mathbf{a} + \mathbf{b}$.

Note that **a** and **b** have directions that go in one sense round the triangle (clockwise in the diagram) and that their resultant, **a** + **b**, has a direction in the opposite sense (anticlockwise in the diagram). This is known as the triangle law for addition of vectors. It can be extended to cover the addition of more than two vectors.

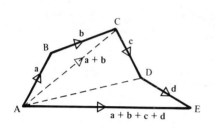

Consider a polygon whose sides AB, BC, CD, DE represent the vectors **a**, **b**, **c**, **d** respectively.

Using the triangle law gives

$$\overrightarrow{AC} = \mathbf{a} + \mathbf{b}$$

so

$$\overrightarrow{AD} = \overrightarrow{AC} + \overrightarrow{CD} = (\mathbf{a} + \mathbf{b}) + \mathbf{c}$$

and

$$\overrightarrow{AE} = \overrightarrow{AD} + \overrightarrow{DE} = (\mathbf{a} + \mathbf{b} + \mathbf{c}) + \mathbf{d}$$

i.e. if vectors **a, b, c, d**, ... are represented by the sides of an open polygon taken in order (i.e. the direction of the vectors all follow the same sense round the polygon) then the line that closes the polygon, in the opposite sense, represents the vector **a** + **b** + **c** + **d** + ...

Note that the vectors **a, b, c, d,** ... are not necessarily coplanar so the polygon is not necessarily a plane figure.

The diagram below shows that the operation of addition on vectors is commutative,

i.e. the order in which the addition is performed does not matter.

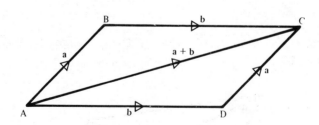

ABCD is a parallelogram, therefore $\overrightarrow{AB}$ and $\overrightarrow{DC}$ both represent **a**

similarly $\overrightarrow{BC}$ and $\overrightarrow{AD}$ both represent **b**

From △ABC, $\overrightarrow{AC} = \mathbf{a} + \mathbf{b}$

From △ADC, $\overrightarrow{AC} = \mathbf{b} + \mathbf{a}$

i.e. $\mathbf{a} + \mathbf{b} \equiv \mathbf{b} + \mathbf{a}$

EXAMPLES 12a

1) In the triangle ABC, D is the midpoint of AB. $\overrightarrow{AB}$ represents **a** and $\overrightarrow{BC}$ represents **b**.

Express in terms of **a** and **b** the vectors $\overrightarrow{CA}$ and $\overrightarrow{DC}$.

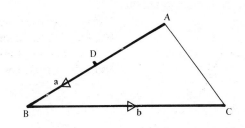

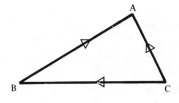

Now

$$\overrightarrow{CA} = \overrightarrow{CB} + \overrightarrow{BA}$$

$$\overrightarrow{BA} = -\overrightarrow{AB} = -\mathbf{a}$$

and

$$\overrightarrow{CB} = -\overrightarrow{BC} = -\mathbf{b}$$

Therefore

$$\overrightarrow{CA} = -\mathbf{a} - \mathbf{b} = -(\mathbf{a} + \mathbf{b})$$

$$\overrightarrow{DC} = \overrightarrow{DB} + \overrightarrow{BC}$$

$$= \tfrac{1}{2}\overrightarrow{AB} + \overrightarrow{BC}$$

$$= \tfrac{1}{2}\mathbf{a} + \mathbf{b}$$

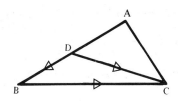

2) Show that $\overrightarrow{AB} + \overrightarrow{AC} = 2\overrightarrow{AD}$ where ABC is a triangle and D is the midpoint of BC.

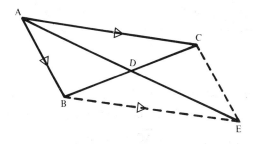

Completing the parallelogram ABEC we see that $\overrightarrow{BE} = \overrightarrow{AC}$.

Therefore $\overrightarrow{AB} + \overrightarrow{AC} = \overrightarrow{AB} + \overrightarrow{BE} = \overrightarrow{AE}$

The diagonals of a parallelogram bisect each other.

Therefore $\qquad\qquad\qquad \overrightarrow{AE} = 2\overrightarrow{AD}$

$\Rightarrow \qquad\qquad\qquad \overrightarrow{AB} + \overrightarrow{AC} = 2\overrightarrow{AD}$

3) If O, A, B, C are four points such that

$$\overrightarrow{OA} = 10a, \quad \overrightarrow{OB} = 5b, \quad \overrightarrow{OC} = 4a + 3b$$

show that A, B and C are collinear.

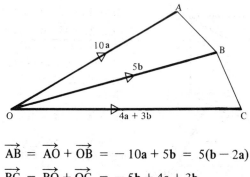

$$\overrightarrow{AB} = \overrightarrow{AO} + \overrightarrow{OB} = -10a + 5b = 5(b - 2a)$$
$$\overrightarrow{BC} = \overrightarrow{BO} + \overrightarrow{OC} = -5b + 4a + 3b$$
$$= 4a - 2b$$
$$= -2(b - 2a)$$

i.e. $\overrightarrow{BC} = -\frac{2}{5}\overrightarrow{AB}$, so $\overrightarrow{AB}$ and $\overrightarrow{BC}$ are in opposite directions, i.e. AB and CB are parallel. Hence since B is a common point A, B and C are collinear.

EXERCISE 12a

1) ABCD is a quadrilateral. Find the single vector which is equivalent to:

(a) $\overrightarrow{AB} + \overrightarrow{BC}$ (b) $\overrightarrow{BC} + \overrightarrow{CD}$ (c) $\overrightarrow{AB} + \overrightarrow{BC} + \overrightarrow{CD}$ (d) $\overrightarrow{AB} + \overrightarrow{DA}$.

2) ABCDEF is a regular hexagon in which $\overrightarrow{BC}$ represents b and $\overrightarrow{FC}$ represents 2a.
Express the vectors $\overrightarrow{AB}$, $\overrightarrow{CD}$ and $\overrightarrow{BE}$ in terms of a and b.

3) Draw diagrams representing the following vector equations.

(a) $\overrightarrow{AB} - \overrightarrow{CB} = \overrightarrow{AC}$ (b) $\overrightarrow{AB} = 2\overrightarrow{PQ}$ (c) $\overrightarrow{AB} + \overrightarrow{BC} = 3\overrightarrow{AD}$

(d) $a + b = -c$

4) If A, B, C, D are four points such that

$$\overrightarrow{AB} = \overrightarrow{DC} \quad \text{and} \quad \overrightarrow{BC} + \overrightarrow{DA} = 0$$

prove that ABCD is a parallelogram.

5) O, A, B, C, D are five points such that $\overrightarrow{OA} = a$, $\overrightarrow{OB} = b$, $\overrightarrow{OC} = a + 2b$, $\overrightarrow{OD} = 2a - b$.

Express $\overrightarrow{AB}, \overrightarrow{BC}, \overrightarrow{CD}, \overrightarrow{AC}, \overrightarrow{BD}$ in terms of a and b.

6) If a and b are represented by two adjacent sides of a regular hexagon, find, in terms of a and b, the vectors represented by the remaining sides taken in order.

7)

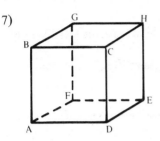

If a, b, c are represented by the edges $\overrightarrow{AB}, \overrightarrow{AD}, \overrightarrow{AF}$ of the cube in the diagram, find, in terms of a, b and c the vectors represented by the remaining edges.

8) If O, A, B, C are four points such that $\overrightarrow{OA} = a$, $\overrightarrow{OB} = 2a - b$, $\overrightarrow{OC} = b$. Show that A, B and C are collinear.

9)

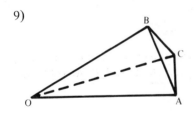

If OABC is a tetrahedron and $\overrightarrow{OA} = a$, $\overrightarrow{OB} = b$, $\overrightarrow{OC} = c$,   find $\overrightarrow{AC}, \overrightarrow{AB}, \overrightarrow{CB}$ in terms of a, b, c.

10) For the cube defined in Question 7, find the vectors $\overrightarrow{BE}, \overrightarrow{GD}, \overrightarrow{AH}, \overrightarrow{FC}$ in terms of a, b, and c.

11) If D is the midpoint of the edge OB of the tetrahedron in Question 9, find $\overrightarrow{CD}$ in terms of b and c. Also, if E is the midpoint of CB, find $\overrightarrow{AE}$ in terms of a, b and c.

POSITION VECTORS

In general a vector has no specific location in space. There are some quantities that compound in the same way as vectors but which are constrained to a specific position.

Consider, for example, the position of a point A relative to a fixed origin O.

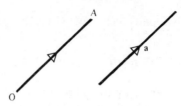

$\overrightarrow{OA}$ is called the position vector of A relative to O. This displacement is unique and *cannot* be represented by the line in the diagram denoting **a** even though $\mathbf{a} = \overrightarrow{OA}$.

Vectors such as $\overrightarrow{OA}$, representing quantities that have a specific location, are called *tied vectors* or *position vectors*.

Vectors such as **a**, representing quantities not related to a fixed position, are known as *free vectors*.

LOCATION OF A POINT IN A PLANE

The work has so far applied to vectors in either two or three dimensions. We will now, for the moment, restrict ourselves to two dimensions, i.e. to vectors in a plane.

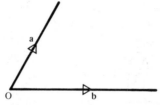

If **a** and **b** are *non-parallel* free vectors and O is a fixed point, then there is one and only one plane containing O and the vectors **a** and **b**. i.e. the vectors **a**, **b** and the fixed point O define a particular plane.

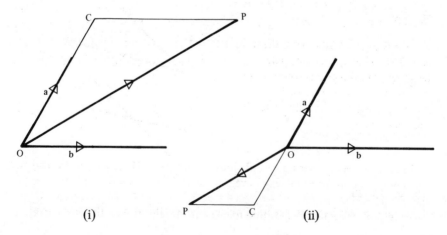

(i)   (ii)

If P is any other point in the plane,

$$\overrightarrow{OP} \text{ is the position vector of P.}$$

If C is the point such that $\overrightarrow{OC}$ is parallel to **a**,

i.e. $\overrightarrow{OC} = \lambda\mathbf{a}$

and $\overrightarrow{CP}$ is parallel to **b**,

i.e. $\overrightarrow{CP} = \mu\mathbf{b}$

where λ and μ are scalars (both having positive values in diagram (i) and both having negative values in diagram (ii)),

then $\overrightarrow{OP} = \overrightarrow{OC} + \overrightarrow{CP}$

$$= \lambda\mathbf{a} + \mu\mathbf{b}$$

i.e. the position vector of any point P can be expressed in terms of **a** and **b**. Thus **a**, **b** and O form a 'frame of reference' for the position vector of any point in the plane.

The vectors **a** and **b** are known as base or basis vectors for a plane.

It also follows that *any vector* equal to $\overrightarrow{OP}$ can be expressed in terms of **a** and **b**,

i.e. any vector (tied or free) that is parallel to the plane can be expressed in the form $\lambda\mathbf{a} + \mu\mathbf{b}$, where λ and μ are scalars.

ANGLE BETWEEN TWO VECTORS

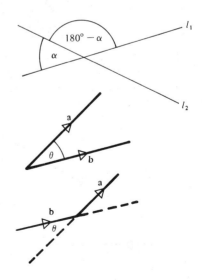

The angle between two lines l_1 and l_2 is ambiguous as it may be α or $180° - \alpha$.

But the angle between two vectors **a** and **b** is unique. It is the angle between their directions when those directions *both converge* or *both diverge* from a point, as illustrated in the diagrams.

EXAMPLES 12b

1) **a** and **b** are base vectors for a plane, where $|\mathbf{a}| = 3$, $|\mathbf{b}| = 5$ and

$\frac{\pi}{3}$ is the angle between **a** and **b**. The position vector of a point P is $3\mathbf{a} - 2\mathbf{b}$ relative to an origin O. Find $|\overrightarrow{OP}|$ and the angle between $\overrightarrow{OP}$ and **a**.

Drawing OQ parallel to **a** such that $\overrightarrow{OQ} = 3\mathbf{a}$

then QP parallel to **b** such that $\overrightarrow{QP} = -2\mathbf{b}$

gives $\overrightarrow{OP} = 3\mathbf{a} - 2\mathbf{b}$.

Now $|\overrightarrow{OQ}| = 3|\mathbf{a}| = 9$,

and $|\overrightarrow{QP}| = 2|\mathbf{b}| = 10$.

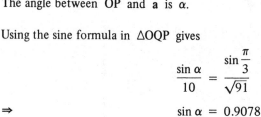

Using the cosine formula in $\triangle OPQ$ gives

$$OP^2 = 81 + 100 - 2(9)(10) \cos \frac{\pi}{3}$$

$$= 91$$

Therefore $OP = \sqrt{91}$

The angle between $\overrightarrow{OP}$ and **a** is α.

Using the sine formula in $\triangle OQP$ gives

$$\frac{\sin \alpha}{10} = \frac{\sin \frac{\pi}{3}}{\sqrt{91}}$$

$\Rightarrow \qquad\qquad\qquad \sin \alpha = 0.9078$

$\Rightarrow \qquad\qquad\qquad \alpha = 65.20°$

So $|\overrightarrow{OP}| = \sqrt{91}$ and $\overrightarrow{OP}$ is inclined at an angle of $65.20°$ to **a**.

CARTESIAN COMPONENTS

Calculations are greatly simplified when the base vectors for a plane are perpendicular and both have a magnitude of unity.

Note that *any* vector whose magnitude is one unit is called a *unit vector*.

A frame of reference consisting of an origin O and a pair of perpendicular vectors has an obvious similarity to the Cartesian *xy* plane. The two are related formally by taking as base vectors the unit vector **i** in the positive direction of the *x*-axis and the unit vector **j** in the positive direction of the *y*-axis.

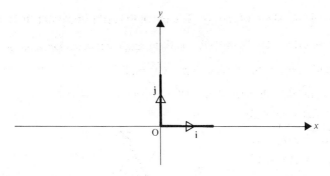

Thus if A is the point with coordinates $(3, 2)$

$$ON = 3, \quad so \quad \overrightarrow{ON} = 3i$$

$$NA = 2, \quad so \quad \overrightarrow{NA} = 2j$$

Therefore $\qquad \overrightarrow{OA} = \overrightarrow{ON} + \overrightarrow{NA} = 3i + 2j$

i.e. the position vector of $A(3, 2)$ is $\quad \mathbf{r} = 3i + 2j$.

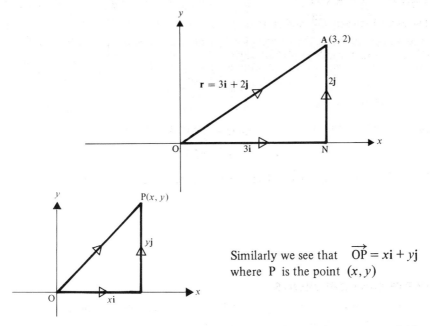

Similarly we see that $\quad \overrightarrow{OP} = xi + yj$
where P is the point (x, y)

so the position vector of any point $P(x, y)$ is $xi + yj$ and the converse of this statement is also true, i.e.

the position vector $\quad \overrightarrow{OP} = xi + yj \quad \Longleftrightarrow \quad$ P is the point (x, y).

Note that, unless we are told that a vector is a position vector or we are using vectors to represent a quantity that has a specific position, we may assume that a given vector is free,

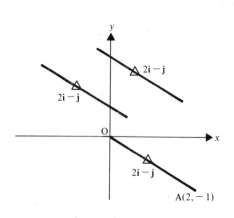

i.e. if a vector is given simply as
a = 2**i** − **j**   then any line section
in the xy plane of the same
magnitude and direction can represent
2**i** − **j**.

But if we are told that **a** is a position
vector 2**i** − **j** then only the line
$\overrightarrow{OA}$ can represent **a**.

Now $|\overrightarrow{OA}|$ = length of OA

$$= \sqrt{[(2)^2 + (-1)^2]}$$

i.e. $|2\mathbf{i} - \mathbf{j}| = \sqrt{[(2)^2 + (-1)^2]}$

$$= \sqrt{5}$$

and in general,

$$|a\mathbf{i} + b\mathbf{j}| = \sqrt{(a^2 + b^2)}$$

Alternative notation: the vector $a\mathbf{i} + b\mathbf{j}$ can be denoted by $\begin{pmatrix} a \\ b \end{pmatrix}$ which is
known as a column vector or column matrix.

EXAMPLES 12b (continued)

2) Find $\overrightarrow{AB}$ where A is the point $(2, -1)$ and B is the point $(-1, 3)$

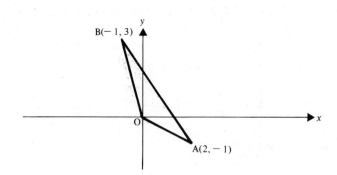

$$\overrightarrow{AB} = \overrightarrow{OB} + \overrightarrow{AO} = \overrightarrow{OB} - \overrightarrow{OA}$$
$$\overrightarrow{OA} = 2\mathbf{i} - \mathbf{j} \quad \text{and} \quad \overrightarrow{OB} = -\mathbf{i} + 3\mathbf{j}$$
Therefore $\overrightarrow{AB} = (-\mathbf{i} + 3\mathbf{j}) - (2\mathbf{i} - \mathbf{j})$
$$= -3\mathbf{i} + 4\mathbf{j}$$

Note that for any two points $A(x_1, y_1)$ and $B(x_2, y_2)$,

$$\overrightarrow{AB} = \overrightarrow{OB} - \overrightarrow{OA}$$

$$= (x_2 - x_1)i + (y_2 - y_1)j$$

and $|\overrightarrow{AB}| = $ length of $AB = \sqrt{[(x_2 - x_1)^2 + (y_2 - y_1)^2]}$

3) Find in terms of **a** and **b** the position vector of the point dividing the line AB in the ratio $m:n$, where **a** is the position vector of A and **b** is the position vector of B with respect to an origin O.

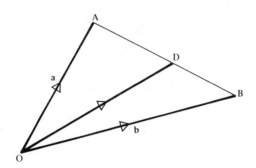

If D is the point on AB such that $\dfrac{AD}{DB} = \dfrac{m}{n}$

then $AD = \left(\dfrac{m}{m+n}\right) AB$ so $\overrightarrow{AD} = \left(\dfrac{m}{m+n}\right) \overrightarrow{AB}$

Now $\overrightarrow{AB} = b - a$,   so $\overrightarrow{AD} = \left(\dfrac{m}{m+n}\right) (b - a)$

$$\overrightarrow{OD} = \overrightarrow{OA} + \overrightarrow{AD}$$

$$= a + \left(\dfrac{m}{m+n}\right) (b - a)$$

$$= \dfrac{na + mb}{m + n}$$

This example gives an important result which may be quoted.
Now if **A** is the point (x_1, y_1) and **B** is the point (x_2, y_2) so that
$a = x_1 i + y_1 j$ and $b = x_2 i + y_2 j$.

then $\overrightarrow{OD} = \dfrac{n(x_1 i + y_1 j) + m(x_2 i + y_2 j)}{m + n}$

$$= \left(\dfrac{nx_1 + mx_2}{m + n}\right) i + \left(\dfrac{ny_1 + my_2}{m + n}\right) j$$

i.e.   D is the point $\left(\dfrac{nx_1 + mx_2}{m + n}, \dfrac{ny_1 + my_2}{m + n}\right)$,

confirming, by vector methods, a result obtained on p. 361.

The Position Vector of the Centroid of a Triangle

Consider a triangle ABC where a, b, c are the position vectors of A, B, C respectively.

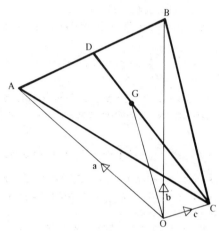

The centroid, G, of this triangle is the point of intersection of its medians, i.e. G is the point which divides CD in the ratio $2:1$, D being the midpoint of AB.

Hence $\overrightarrow{OG} = \dfrac{2\overrightarrow{OD} + \overrightarrow{OC}}{3}$

and $\overrightarrow{OD} = \dfrac{a + b}{2}$

Therefore $\overrightarrow{OG} = \frac{1}{3}(a + b + c)$

EXERCISE 12b

1) If $a = 2i + j$, $b = i - 2j$,   express, in terms of i and j:
(a) $a + b$ (b) $a - b$ (c) $2a + b$ (d) $-3a + 4b$

2) If the position vector of A is $i - 3j$ and the position vector of B is $2i + 5j$, find:

(a) $|\overrightarrow{AB}|$,
(b) the position vector of the midpoint of AB,
(c) the position vector of the point dividing AB in the ratio $1:3$.

3) If $a = i - 3j$, $b = -2i + 5j$, $c = 3i - j$ find:

(a) $a + b + c$ in terms of i and j,

(b) $|a - b|$,

(c) the position vector of the point dividing the line AC in the ratio $-2 : 3$, where a is position vector of A and c is the position vector of C.

4) Find a vector which is parallel to the line $y = 2x - 1$.

5) Show that $i + 2j$ is the position vector of a point on the line $x - 2y + 3 = 0$.

6) Find the angle that the vector $3i - 2j$ makes with the x-axis.

7) Show that $\begin{pmatrix} a \\ b \end{pmatrix}$ and $\begin{pmatrix} -b \\ a \end{pmatrix}$ are perpendicular.

8) Show that, for all values of t, the point whose position vector is $r = ti + (2t - 1)j$ lies on the line whose equation is $y = 2x - 1$.

9) Show that
the position vector of P is $t^2 i + 2tj \iff P$ is a point on $y^2 = 4x$.

10) a and b are base vectors where $|a| = 2$, $|b| = 5$ and the angle between a and b is $45°$. Draw diagrams showing the points whose position vectors, relative to an origin O, are:

(a) $a - b$ (b) $a + b$ (c) $2a - b$ (d) $-a + 2b$ (e) $4a + 3b$

THREE DIMENSIONAL SPACE

The analysis of certain lines and curves has been introduced at various stages throughout this book. In all cases so far this analysis was applied to problems involving only two variables. Any two variable problem can be represented visually in two dimensional space. In a similar way a three variable problem can be represented in three dimensional space and the methods we have used for working in a plane can be extended to analysis in three dimensions, i.e. we now develop both vector analysis and coordinate geometry to investigate three dimensional figures. The first step must be the unambiguous location of a particular point in space.

Base Vectors for Three Dimensions

Let a, b and c be any three *non-parallel* and *non-coplanar* vectors. If O is a fixed point in space and if P is *any* other point in space then we can draw the closed polygon OPQR such that

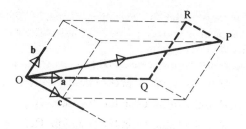

$\overrightarrow{OQ}$ is parallel to **a**, i.e. $\overrightarrow{OQ} = \lambda\mathbf{a}$,

$\overrightarrow{QR}$ is parallel to **b**, i.e. $\overrightarrow{QR} = \mu\mathbf{b}$,

$\overrightarrow{RP}$ is parallel to **c**, i.e. $\overrightarrow{RP} = \eta\mathbf{c}$.

Now
$$\overrightarrow{OP} = \overrightarrow{OQ} + \overrightarrow{QR} + \overrightarrow{RP}$$
$$= \lambda\mathbf{a} + \mu\mathbf{b} + \eta\mathbf{c}.$$

Thus the position vector of any point in space can be expressed in terms of **a, b** and **c**,

i.e. **a, b** and **c** together with O form a frame of reference for three dimensional space.

It also follows that any free vector equal to $\overrightarrow{OP}$ can be expressed in the form

$$\lambda\mathbf{a} + \mu\mathbf{b} + \eta\mathbf{c} \quad \text{where} \quad \lambda, \mu, \eta \text{ are scalar.}$$

Thus a fixed point O and any three vectors which are neither parallel to each other nor to one plane, form a frame of reference for three dimensional space. A set of three vectors satisfying these conditions is called a set of basis vectors for three dimensions.

Work within such a frame of reference is made easier if the basis vectors chosen are (a) mutually perpendicular and (b) of unit length.

The Cartesian Frame of Reference

Starting with three mutually perpendicular directions leads us to the Cartesian frame of reference. This consists of a fixed point O, the origin, and three mutually perpendicular axes, Ox, Oy and Oz.

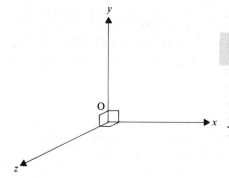

The axes are placed in such a way that they form a right-handed set.
This means that if a screw, placed at the origin, is turned in the sense from the positive x-axis to the positive y-axis, it moves in the direction of the positive z-axis.

A point is located within this frame by giving its directed distances from O in the directions of the positive x-axis, y-axis and z-axis. These coordinates are written as the ordered set (x, y, z).

Hence the point P whose coordinates are $(3, 2, 4)$ is

$\qquad$ 3 units from O in the direction Ox,

$\qquad$ 2 units from O in the direction Oy,

$\qquad$ 4 units from O in the direction Oz.

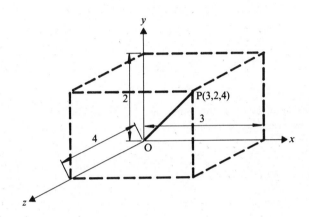

Cartesian Base Vectors

We now take unit vectors, **i**, **j** and **k** in the directions of the Cartesian axes as follows:

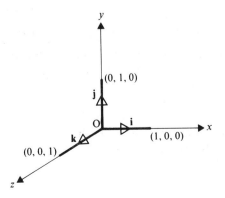

i is a unit vector in the direction Ox,

j is a unit vector in the direction Oy,

k is a unit vector in the direction Oz.

If $P(a, b, c)$ is any point and
r is the position vector of P
then **r** has components

$a\mathbf{i}$ in the direction Ox,

$b\mathbf{j}$ in the direction Oy,

$c\mathbf{k}$ in the direction Oz.

i.e. $\overrightarrow{OP} = \mathbf{r} = a\mathbf{i} + b\mathbf{j} + c\mathbf{k}$.

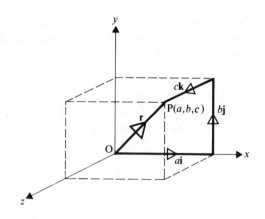

For a general point $P(x, y, z)$, the position vector, **r**, of P is given by

$$\mathbf{r} = x\mathbf{i} + y\mathbf{j} + z\mathbf{k}.$$

Hence the position vector of the point $(2, -1, 3)$ is $2\mathbf{i} - \mathbf{j} + 3\mathbf{k}$.
Conversely the point whose position vector is $-4\mathbf{i} + \mathbf{j} - 2\mathbf{k}$ has coordinates
$(-4, 1, -2)$.

However, given a vector such as $\mathbf{V} = 2\mathbf{i} + \mathbf{j} + 3\mathbf{k}$,   then without further
information we may assume that **V** is a free vector. Thus although we *can*
represent **V** by the position vector $\overrightarrow{OP}$ of the point $P(2, 1, 3)$, we can *also*
represent **V** by any other line of the same length and direction as $\overrightarrow{OP}$.

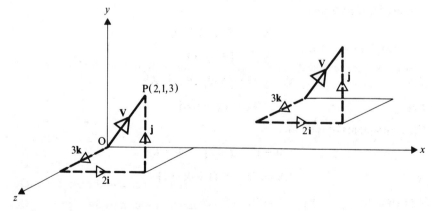

Note that wherever the line representing **V** is located the components of **V**
are $2\mathbf{i}, \mathbf{j}$ and $3\mathbf{k}$.

Alternative notation: The vector $\mathbf{V} = a\mathbf{i} + b\mathbf{j} + c\mathbf{k}$ may be written as

$\mathbf{V} = \begin{pmatrix} a \\ b \\ c \end{pmatrix}$ where $\begin{pmatrix} a \\ b \\ c \end{pmatrix}$ is called a column vector.

Modulus

The magnitude or *modulus* of $\mathbf{r}$ is denoted by $|\mathbf{r}|$ or r and is represented by the length d of OP.

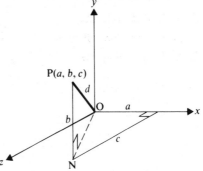

If P is the point (a, b, c) then OP can be found by Pythagoras' Theorem as follows:

$$ON^2 = a^2 + c^2$$
$$OP^2 = ON^2 + b^2 = a^2 + c^2 + b^2$$
$$\Rightarrow OP = \sqrt{(a^2 + b^2 + c^2)}.$$

Hence $$\boxed{|\mathbf{r}| = d = \sqrt{(a^2 + b^2 + c^2)}.}$$

Therefore, by its definition, $|\mathbf{r}|$ is always positive,

e.g. if $\mathbf{r} = 2\mathbf{i} - \mathbf{j} + 3\mathbf{k}$, $|\mathbf{r}| = \sqrt{[2^2 + (-1)^2 + 3^2]} = \sqrt{14}$.

The free vector $\mathbf{V} = a\mathbf{i} + b\mathbf{j} + c\mathbf{k}$ may also be represented by $\overrightarrow{OP}$, hence, for any vector $\mathbf{V}$,

$$|\mathbf{V}| = \sqrt{(a^2 + b^2 + c^2)}.$$

Resultant Vectors

Consider the vectors

$$\mathbf{V}_1 = 3\mathbf{i} + 2\mathbf{j} + 2\mathbf{k}$$
$$\mathbf{V}_2 = \mathbf{i} + 2\mathbf{j} + \mathbf{k}$$

If $$\mathbf{V}_1 = \overrightarrow{OA} \quad \text{and} \quad \mathbf{V}_2 = \overrightarrow{AB}$$
then $$\mathbf{V}_1 + \mathbf{V}_2 = \overrightarrow{OB}.$$

The components of $\overrightarrow{OB}$ are

$$3\mathbf{i} + \mathbf{i}, \quad 2\mathbf{j} + 2\mathbf{j}, \quad 2\mathbf{k} + \mathbf{k},$$

i.e. $$\mathbf{V}_1 + \mathbf{V}_2 = 4\mathbf{i} + 4\mathbf{j} + 3\mathbf{k}.$$

In general if $\mathbf{V}_1 = x_1\mathbf{i} + y_1\mathbf{j} + z_1\mathbf{k}$
$$\mathbf{V}_2 = x_2\mathbf{i} + y_2\mathbf{j} + z_2\mathbf{k}$$

then $$\mathbf{V}_1 + \mathbf{V}_2 = (x_1 + x_2)\mathbf{i} + (y_1 + y_2)\mathbf{j} + (z_1 + z_2)\mathbf{k}.$$

EXAMPLE 12c

A triangle ABC has its vertices at the points $A(2, -1, 4)$, $B(3, -2, 5)$, $C(-1, 6, 2)$.

Find, in the form $ai + bj + ck$, the vectors $\overrightarrow{AB}, \overrightarrow{BC}, \overrightarrow{CA}$ and hence find the lengths of the sides of the triangle.

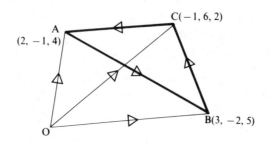

Note. The coordinate axes are not drawn in this diagram. When two, or more, points are illustrated, the presence of the axes tends to reduce the clarity of a diagram and so, in general, they are not introduced. However, the origin should be marked, as it provides a reference point.

Now
$$\overrightarrow{AB} = \overrightarrow{OB} - \overrightarrow{OA}$$
$$= (3i - 2j + 5k) - (2i - j + 4k)$$
$$= i - j + k$$
$$\overrightarrow{BC} = \overrightarrow{OC} - \overrightarrow{OB}$$
$$= (-i + 6j + 2k) - (3i - 2j + 5k)$$
$$= -4i + 8j - 3k$$
$$\overrightarrow{CA} = \overrightarrow{OA} - \overrightarrow{OC}$$
$$= (2i - j + 4k) - (-i + 6j + 2k)$$
$$= 3i - 7j + 2k.$$

Hence
$$AB = |\overrightarrow{AB}| = \sqrt{[(1)^2 + (-1)^2 + (1)^2]} = \sqrt{3}$$
$$BC = |\overrightarrow{BC}| = \sqrt{[(-4)^2 + (8)^2 + (-3)^2]} = \sqrt{89}$$
$$CA = |\overrightarrow{CA}| = \sqrt{[(3)^2 + (-7)^2 + (2)^2]} = \sqrt{62}$$

EXERCISE 12c

1) Write down, in the form $ai + bj + ck$, the vector represented by $\overrightarrow{OP}$ if P is a point with coordinates
(a) $(3, 6, 4)$ (b) $(1, -2, -7)$ (c) $(1, 0, -3)$.

2) $\overrightarrow{OP}$ represents a vector **r**. Write down the coordinates of P if
(a) $r = 5i - 7j + 2k$ (b) $r = i + 4j$ (c) $r = j - k$.

3) Find the length of the line OP if P is the point
(a) $(2, -1, 4)$ (b) $(3, 0, 4)$ (c) $(-2, -2, 1)$.

4) Find the modulus of the vector **V** if
(a) $\mathbf{V} = 2\mathbf{i} - 4\mathbf{j} + 4\mathbf{k}$ (b) $\mathbf{V} = 6\mathbf{i} + 2\mathbf{j} - 3\mathbf{k}$ (c) $\mathbf{V} = 11\mathbf{i} - 7\mathbf{j} - 6\mathbf{k}$.

5) If $\mathbf{a} = \mathbf{i} + \mathbf{j} + \mathbf{k}$, $\mathbf{b} = 2\mathbf{i} - \mathbf{j} + 3\mathbf{k}$, $\mathbf{c} = -\mathbf{i} + 3\mathbf{j} - \mathbf{k}$ find
(a) $\mathbf{a} + \mathbf{b}$ (b) $\mathbf{a} - \mathbf{c}$ (c) $\mathbf{a} + \mathbf{b} + \mathbf{c}$ (d) $\mathbf{a} - 2\mathbf{b} + 3\mathbf{c}$.

6) The triangle ABC has its vertices at the points $A(-1, 3, 0)$, $B(-3, 0, 7)$, $C(-1, 2, 3)$. Find in the form $a\mathbf{i} + b\mathbf{j} + c\mathbf{k}$ the vectors representing
(a) $\overrightarrow{AB}$ (b) $\overrightarrow{AC}$ (c) $\overrightarrow{CB}$.

7) Find the lengths of the sides of the triangle described in Question 6.

8) Find $|\mathbf{a} - \mathbf{b}|$ where $\mathbf{a} = \mathbf{i} - \mathbf{j} + 2\mathbf{k}$, $\mathbf{b} = 2\mathbf{i} - \mathbf{j}$.

9) A, B, C and D are the points $(0, 0, 2)$, $(-1, 3, 2)$, $(1, 0, 4)$ and $(-1, 2, -2)$ respectively. Find the vectors representing $\overrightarrow{AB}, \overrightarrow{BD}, \overrightarrow{CD}, \overrightarrow{AD}$.

DIRECTION RATIOS AND DIRECTION COSINES OF A VECTOR

Consider the vector $\overrightarrow{OP}$ where P is the point (a, b, c).
The coordinates of P determine the direction of OP relative to the axes and
the ratios $a : b : c$ are called the direction ratios of the vector $\overrightarrow{OP}$.

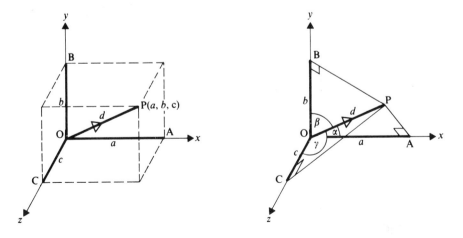

In the diagram, lines PA, PB and PC are drawn from P perpendicular to Ox,
Oy and Oz respectively, so that angles PAO, PBO and PCO are each $90°$.
But $OA = a$, $OB = b$, $OC = c$ and $OP = \sqrt{(a^2 + b^2 + c^2)} = d$.

So if OP makes angles α, β and γ with Ox, Oy and Oz respectively, then

$$\cos \alpha = \frac{a}{d} \qquad \cos \beta = \frac{b}{d} \qquad \cos \gamma = \frac{c}{d}$$

But the angles α, β and γ determine the direction of $\overrightarrow{OP}$, so their cosines are called the *direction cosines* of $\overrightarrow{OP}$ and are often given the symbols l, m and n, so that

$$\cos \alpha = \frac{a}{d} = l$$

$$\cos \beta = \frac{b}{d} = m$$

$$\cos \gamma = \frac{c}{d} = n.$$

So if the direction ratios of a vector are $a:b:c$
the direction cosines of that vector are

$$\frac{a}{\sqrt{(a^2+b^2+c^2)}}, \quad \frac{b}{\sqrt{(a^2+b^2+c^2)}}, \quad \frac{c}{\sqrt{(a^2+b^2+c^2)}}.$$

From these three relationships it follows that

1)
$$l^2 + m^2 + n^2 = \left(\frac{a}{d}\right)^2 + \left(\frac{b}{d}\right)^2 + \left(\frac{c}{d}\right)^2$$

$$= \frac{a^2 + b^2 + c^2}{d^2}$$

But
$$a^2 + b^2 + c^2 = d^2$$

Hence
$$l^2 + m^2 + n^2 = 1$$

Thus *the sum of the squares of the direction cosines of any vector is unity.*

2) $l:m:n = a:b:c.$

3) As $a = dl, \quad b = dm, \quad c = dn$

the coordinates of $P(a, b, c)$ may be written (dl, dm, dn).

Hence
$$\overrightarrow{OP} = dl\mathbf{i} + dm\mathbf{j} + dn\mathbf{k}$$

$$= d(l\mathbf{i} + m\mathbf{j} + n\mathbf{k})$$

Now if $\overrightarrow{OQ} = l\mathbf{i} + m\mathbf{j} + n\mathbf{k},$ Q is the point (l, m, n) and
$OQ = \sqrt{(l^2 + m^2 + n^2)} = 1,$

i.e. $\overrightarrow{OQ}$ is a unit vector in the direction $\overrightarrow{OP}$, where $\overrightarrow{OQ} = \dfrac{\overrightarrow{OP}}{|\overrightarrow{OP}|}.$

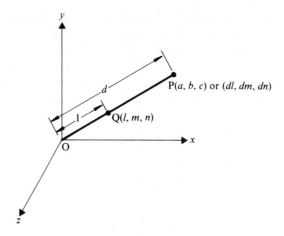

The free vector $\mathbf{v} = a\mathbf{i} + b\mathbf{j} + c\mathbf{k}$ has the same direction as the position vector $\mathbf{r} = a\mathbf{i} + b\mathbf{j} + c\mathbf{k}$ and may also be represented by $\overrightarrow{OP}$.

Hence for *any* vector $\mathbf{v} = a\mathbf{i} + b\mathbf{j} + c\mathbf{k}$
the direction ratios are $a:b:c$

the direction cosines are $l = \dfrac{a}{|\mathbf{v}|}$, $m = \dfrac{b}{|\mathbf{v}|}$, $n = \dfrac{c}{|\mathbf{v}|}$

and $l^2 + m^2 + n^2 = 1$.

Direction Ratios and Direction Cosines of Parallel Vectors

Consider two parallel vectors $\mathbf{v}_1$ and $\mathbf{v}_2$,

i.e. $\qquad\qquad\qquad \mathbf{v}_2 = \lambda\mathbf{v}_1$

So if $\qquad\qquad\qquad \mathbf{v}_1 = a\mathbf{i} + b\mathbf{j} + c\mathbf{k}$

then $\qquad\qquad\qquad \mathbf{v}_2 = \lambda a\mathbf{i} + \lambda b\mathbf{j} + \lambda c\mathbf{k}$

Now the direction ratios of $\mathbf{v}_1$ are $a:b:c$
and the direction ratios of $\mathbf{v}_2$ are $\lambda a:\lambda b:\lambda c$.
But $\lambda a:\lambda b:\lambda c = a:b:c$ whatever the value of λ.
Hence parallel vectors have equal direction ratios.
Now considering the direction cosines of $\mathbf{v}_1$ and $\mathbf{v}_2$ we see that:

$\mathbf{v}_1$ has direction cosines $\dfrac{a}{|\mathbf{v}_1|}$, $\dfrac{b}{|\mathbf{v}_1|}$, $\dfrac{c}{|\mathbf{v}_1|}$,

$\mathbf{v}_2$ has direction cosines $\dfrac{\lambda a}{|\lambda\mathbf{v}_1|}$, $\dfrac{\lambda b}{|\lambda\mathbf{v}_1|}$, $\dfrac{\lambda c}{|\lambda\mathbf{v}_1|}$.

As $|\lambda\mathbf{v}_1|$ is always positive, $\mathbf{v}_1$ and $\mathbf{v}_2$ have the same direction cosines when λ is positive. But if λ is negative the direction cosines of $\mathbf{v}_2$ are opposite in sign to those of $\mathbf{v}_1$.

Hence all parallel vectors have equal direction ratios, whereas like parallel vectors have equal direction cosines but unlike parallel vectors have direction cosines equal in magnitude but opposite in sign.

It follows that the direction ratios of a vector are not unique but its direction cosines are unique.

Unit Vectors

Any vector of magnitude 1 unit is a unit vector $(\mathbf{i}, \mathbf{j}$ and $\mathbf{k}$ are all unit vectors).

Consider a line OPQ where the position vector of P is $\mathbf{r}$ and OQ is of length 1 unit.

Then $\overrightarrow{OQ}$ represents a unit vector in the direction of $\mathbf{r}$.

Such a unit vector is written $\hat{\mathbf{r}}$.

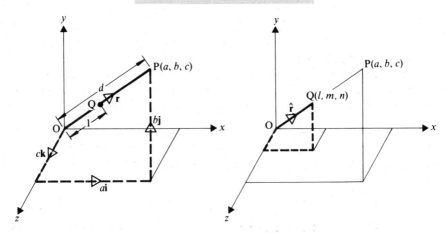

Now $OQ = 1$ and, if $OP = d$,

then $$\overrightarrow{OP} = d\,\overrightarrow{OQ}.$$

But $d = |\mathbf{r}|$.

Hence $$\mathbf{r} = |\mathbf{r}|\,\hat{\mathbf{r}}.$$

This is an important property of all vectors, i.e.
any vector, $\mathbf{v}$, can be expressed as the product of its magnitude and the unit vector in the same direction,

i.e. $$\mathbf{v} = |\mathbf{v}|\,\hat{\mathbf{v}}.$$

We have seen that Q is the point (l, m, n) where l, m, n are the direction cosines of $\overrightarrow{OP}$.

i.e. $$\hat{\mathbf{r}} = l\mathbf{i} + m\mathbf{j} + n\mathbf{k}.$$

Direction Vectors

A vector which is used to specify the direction of another vector can be called a *direction vector*.

For example if we are told that a vector $\mathbf{V}$ of magnitude 14 units is in the direction of the vector $3\mathbf{i} + 6\mathbf{j} + 2\mathbf{k}$ then $3\mathbf{i} + 6\mathbf{j} + 2\mathbf{k}$ is the direction vector for $\mathbf{V}$.

The unit direction vector $\hat{\mathbf{V}} = \frac{1}{7}(3\mathbf{i} + 6\mathbf{j} + 2\mathbf{k})$,

so $\qquad\qquad \mathbf{V} = |\mathbf{V}| \times \text{(unit direction vector)}$

$\Rightarrow \qquad\qquad \mathbf{V} = 14\left(\dfrac{3\mathbf{i} + 6\mathbf{j} + 2\mathbf{k}}{7}\right) = 6\mathbf{i} + 12\mathbf{j} + 4\mathbf{k}.$

Summarising we have found that

if $\qquad\qquad\qquad \mathbf{V} = a\mathbf{i} + b\mathbf{j} + c\mathbf{k},$

$$|\mathbf{V}| = \sqrt{(a^2 + b^2 + c^2)} = d$$

$\mathbf{V}$ has direction ratios $a:b:c$

$$\mathbf{V} \text{ has direction cosines } \frac{a}{d}, \frac{b}{d}, \frac{c}{d}$$

$$\text{or } l, m, n$$

$$\hat{\mathbf{V}} = \mathbf{V}/d = l\mathbf{i} + m\mathbf{j} + n\mathbf{k}$$

$$l^2 + m^2 + n^2 = 1$$

EXAMPLES 12d

1) Find the direction cosines of the vector $\overrightarrow{OP}$ where P is the point $(2, 3, -6)$.

$$\overrightarrow{OP} = 2\mathbf{i} + 3\mathbf{j} - 6\mathbf{k}$$

$$OP = \sqrt{[2^2 + 3^2 + (-6)^2]} = 7$$

$\overrightarrow{OP}$ has direction ratios $2:3:-6$

Therefore $\overrightarrow{OP}$ has direction cosines $\frac{2}{7}, \frac{3}{7}, -\frac{6}{7}$.

2) A vector $\mathbf{V}$ is inclined to Ox at $45°$ and to Oy at $60°$. Find its inclination to Oz.

If the magnitude of $\mathbf{V}$ is 12 units, express $\mathbf{V}$ in the form $a\mathbf{i} + b\mathbf{j} + c\mathbf{k}$.

The direction cosines of $\mathbf{V}$ are $\cos 45°, \cos 60°, \cos \gamma$.

So
$$l = \frac{1}{\sqrt{2}}, \quad m = \frac{1}{2}, \quad n = \cos\gamma.$$

But
$$l^2 + m^2 + n^2 = 1.$$

Hence
$$n^2 = 1 - \tfrac{1}{2} - \tfrac{1}{4} = \tfrac{1}{4}$$

⇒
$$n = \pm\tfrac{1}{2} = \cos\gamma.$$

Therefore V is inclined to Oz either at $60°$ or at $120°$.
Now $l\mathbf{i}, m\mathbf{j}$ and $n\mathbf{k}$ are the components of $\hat{\mathbf{V}}$, so

$$\hat{\mathbf{V}} = \frac{1}{\sqrt{2}}\mathbf{i} + \frac{1}{2}\mathbf{j} \pm \frac{1}{2}\mathbf{k}.$$

But
$$\mathbf{V} = |\mathbf{V}|\hat{\mathbf{V}}.$$

Hence
$$\mathbf{V} = 12\left(\frac{1}{\sqrt{2}}\mathbf{i} + \frac{1}{2}\mathbf{j} \pm \frac{1}{2}\mathbf{k}\right)$$

⇒
$$\mathbf{V} = 6\sqrt{2}\mathbf{i} + 6\mathbf{j} \pm 6\mathbf{k}.$$

3) Find the coordinates of P if OP is of length 5 units and is in the direction $\overrightarrow{OR}$ where R is the point $(2, -1, 4)$.

$|\overrightarrow{OR}| = \sqrt{21}$ so, the direction cosines of $\overrightarrow{OR}$ are

$$l = \frac{2}{\sqrt{21}}, \quad m = -\frac{1}{\sqrt{21}}, \quad n = \frac{4}{\sqrt{21}}.$$

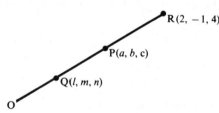

Hence the coordinates of P are (dl, dm, dn) where $d = |\overrightarrow{OP}| = 5$.

So P is the point $\left(\dfrac{10}{\sqrt{21}}, \dfrac{-5}{\sqrt{21}}, \dfrac{20}{\sqrt{21}}\right)$.

EXERCISE 12d

1) Find the direction cosines of the vectors
(a) $2i + 2j - k$,   (b) $6i - 2j - 3k$,   (c) $3i + 4k$,   (d) $i + 8j + 4k$.

2) Find the unit vectors in the direction of the vectors in Question 1.

3) Find the inclination to Oy of $\overrightarrow{OP}$, where P is the point
(a) $(1, 1, 1)$,   (b) $(1, 0, 1)$,   (c) $(0, 1, 1)$,   (d) $(2, -2, 1)$.

4) Find the coordinates of Q if $|\overrightarrow{OQ}| = 1$ and $\overrightarrow{OQ}$ is in the direction of
(a) $i + 2j - 2k$,   (b) $3i + 2j + 6k$.

5) Find the angle at which the following vectors are inclined to each of the coordinate axes:
(a) $i - j + k$,   (b) $4i + 8j + k$,   (c) $j - k$.

6) Find the coordinates of P, where
(a) $|\overrightarrow{OP}| = 6$ and $\overrightarrow{OP}$ is in the direction of $2i - 3j + 6k$,
(b) $|\overrightarrow{OP}| = 2$ and $\overrightarrow{OP}$ is in the direction of $8i + j - 4k$,
(c) $\overrightarrow{OP}$ is inclined at equal acute angles to Ox, Oy and Oz and $|\overrightarrow{OP}| = 4$.

7) Find the vector V if

(a) $V = \overrightarrow{OP}$ and P is the point $(0, 4, 5)$,
(b) $|V| = 24$ units and $\hat{V} = \frac{2}{3}i - \frac{2}{3}j - \frac{1}{3}k$,
(c) V is inclined at $60°$ to Oy and at $60°$ to Oz and is of magnitude 8 units,
(d) V is parallel to the vector $8i + j + 4k$ and is equal in magnitude to the vector $i - 2j + 2k$.

8) Find the magnitude and the inclination to each of the coordinate axes of a vector V if
(a) $V = 3i + 4j + 5k$,
(b) $V = -i + j - k$,

(c) V is represented by $\overrightarrow{OP}$ where P is the point $(5, 1, 4)$.

9) Find $\hat{r}$ in the form $aj + bj + ck$ if
(a) $r = i - j + k$,

(b) $r = \overrightarrow{OP}$ and P is the point $(3, 2, -6)$.

10) A vector V is inclined at equal acute angles to Ox, Oy and Oz. If the magnitude of V is 6 units, find V.

11) If $r_1 = 2i - j + k$ and $r_2 = i + 3j + 2k$ find the modulus and direction cosines of $r_1 + r_2$, $r_1 - r_2$.

PROPERTIES OF A LINE JOINING TWO POINTS

Consider the line joining the two points $A(x_1, y_1, z_1)$ and $B(x_2, y_2, z_2)$ whose position vectors are **a** and **b** respectively.

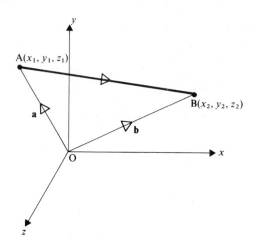

Length of AB

Now $\quad \mathbf{a} = \overrightarrow{OA} = x_1\mathbf{i} + y_1\mathbf{j} + z_1\mathbf{k} \quad$ and $\quad \mathbf{b} = \overrightarrow{OB} = x_2\mathbf{i} + y_2\mathbf{j} + z_2\mathbf{k}.$

Hence $\qquad \overrightarrow{AB} = \overrightarrow{OB} - \overrightarrow{OA}$

$$= (x_2 - x_1)\mathbf{i} + (y_2 - y_1)\mathbf{j} + (z_2 - z_1)\mathbf{k}.$$

Now the length of $AB = |\overrightarrow{AB}| = |\mathbf{a} - \mathbf{b}|$

$$= \sqrt{[(x_2 - x_1)^2 + (y_2 - y_1)^2 + (z_2 - z_1)^2]},$$

i.e. the length of the line joining $A(x_1, y_1, z_1)$ to $B(x_2, y_2, z_2)$ is

$$\sqrt{[(x_2 - x_1)^2 + (y_2 - y_1)^2 + (z_2 - z_1)^2]}$$

Direction Ratios and Direction Cosines of AB

$$\overrightarrow{AB} = (x_2 - x_1)\mathbf{i} + (y_2 - y_1)\mathbf{j} + (z_2 - z_1)\mathbf{k},$$

so the direction ratios of $\overrightarrow{AB}$ are $\quad (x_2 - x_1):(y_2 - y_1):(z_2 - z_1)$

and the direction cosines of $\overrightarrow{AB}$ are

$$l = \frac{x_2 - x_1}{AB}, \quad m = \frac{y_2 - y_1}{AB}, \quad n = \frac{z_2 - z_1}{AB}$$

Midpoint of AB

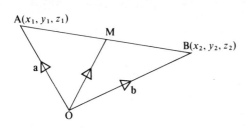

If M is the midpoint of AB

then $\overrightarrow{OM} = \overrightarrow{OA} + \overrightarrow{AM}$

$\qquad = \overrightarrow{OA} + \tfrac{1}{2}(\overrightarrow{AB})$

$\qquad = \overrightarrow{OA} + \tfrac{1}{2}(\overrightarrow{OB} - \overrightarrow{OA})$

$\qquad = \tfrac{1}{2}(\overrightarrow{OA} + \overrightarrow{OB})$

$\qquad = \tfrac{1}{2}(\mathbf{a} + \mathbf{b})$

i.e.

$$\overrightarrow{OM} = \tfrac{1}{2}[(x_1 + x_2)\mathbf{i} + (y_1 + y_2)\mathbf{j} + (z_1 + z_2)\mathbf{k}]$$

and

M has coordinates $\left(\dfrac{x_1 + x_2}{2},\ \dfrac{y_1 + y_2}{2},\ \dfrac{z_1 + z_2}{2} \right)$

Point Dividing AB in a Given Ratio

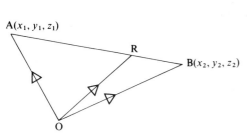

If R is the point dividing
AB internally in the ratio $\lambda : \mu$

then $\dfrac{AR}{RB} = \dfrac{\lambda}{\mu}$ and

$\overrightarrow{OR} = \overrightarrow{OA} + \overrightarrow{AR}$

$\qquad = \overrightarrow{OA} + \dfrac{\lambda}{\lambda + \mu}\ \overrightarrow{AB}$

$\qquad = \overrightarrow{OA} + \dfrac{\lambda}{\lambda + \mu}(\overrightarrow{OB} - \overrightarrow{OA})$

$\qquad = \dfrac{\lambda\overrightarrow{OB} + \mu\overrightarrow{OA}}{\lambda + \mu} = \dfrac{\lambda\mathbf{b} + \mu\mathbf{a}}{\lambda + \mu}$

i.e.

$$\overrightarrow{OR} = \dfrac{\lambda(x_2\mathbf{i} + y_2\mathbf{j} + z_2\mathbf{k}) + \mu(x_1\mathbf{i} + y_1\mathbf{j} + z_1\mathbf{k})}{\lambda + \mu}$$

$\Rightarrow$

R has coordinates $\left(\dfrac{\lambda x_2 + \mu x_1}{\lambda + \mu},\ \dfrac{\lambda y_2 + \mu y_1}{\lambda + \mu},\ \dfrac{\lambda z_2 + \mu z_1}{\lambda + \mu} \right).$

EXAMPLES 12e

1) Find the length of the median through O of the triangle OAB, where A is the point $(2, 7, -1)$ and B is the point $(4, 1, 2)$.

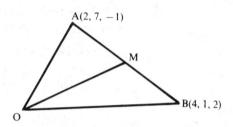

A(2, 7, −1)

M

B(4, 1, 2)

O

The coordinates of M, the midpoint of AB, are

$$\left(\frac{4+2}{2}, \frac{1+7}{2}, \frac{2-1}{2}\right) = (3, 4, \tfrac{1}{2})$$

So the length of OM is $\sqrt{[3^2 + 4^2 + (\tfrac{1}{2})^2]} = \dfrac{\sqrt{101}}{2}$

2) Find the length and direction cosines of the line LM where L is the midpoint of AB, M is the midpoint of BC, and A, B, C are the points $(3, -1, 5)$, $(7, 1, 3)$, $(-5, 9, -1)$ respectively.

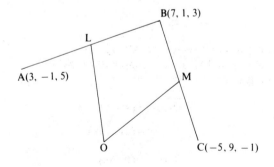

B(7, 1, 3)

L

A(3, −1, 5)

M

O

C(−5, 9, −1)

L is the point

$$\left(\frac{7+3}{2}, \frac{-1+1}{2}, \frac{5+3}{2}\right)$$

i.e. $(5, 0, 4)$

M is the point

$$\left(\frac{7-5}{2}, \frac{1+9}{2}, \frac{3-1}{2}\right)$$

i.e. $(1, 5, 1)$

Hence $LM = \sqrt{[(5-1)^2 + (0-5)^2 + (4-1)^2]} = 5\sqrt{2}$

Now $\overrightarrow{LM} = (1-5)\mathbf{i} + (5-0)\mathbf{j} + (1-4)\mathbf{k}$

$= -4\mathbf{i} + 5\mathbf{j} - 3\mathbf{k}$

Hence the direction cosines of $\overrightarrow{LM}$ are $-\dfrac{4}{5\sqrt{2}}, \dfrac{5}{5\sqrt{2}}, -\dfrac{3}{5\sqrt{2}}$

i.e. $-\dfrac{2\sqrt{2}}{5}, \dfrac{\sqrt{2}}{2}, -\dfrac{3\sqrt{2}}{10}$

3) A and B are two points whose position vectors are $3i + j - 2k$ and $i - 3j - k$ respectively. Find the position vectors of the points dividing AB
(a) internally in the ratio $1:3$,
(b) externally in the ratio $3:1$.

(a)

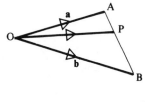

If P is the point dividing AB internally in the ratio $1:3$, then

$$AP:PB = 1:3.$$

So $\overrightarrow{OP} = \dfrac{3a + b}{3 + 1}$

$$= \dfrac{3(3i + j - 2k) + (i - 3j - k)}{4}$$

$$= \tfrac{5}{2}i - \tfrac{7}{4}k.$$

(b)

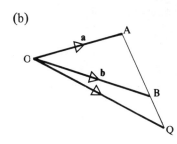

If Q is the point dividing AB externally in the ratio $3:1$, then

$$AQ:QB = 3:-1.$$

So $\overrightarrow{OQ} = \dfrac{-a + 3b}{-1 + 3}$

$$= \dfrac{-(3i + j - 2k) + 3(i - 3j - k)}{2}$$

$$= -5j - \tfrac{1}{2}k$$

4) A, B and C are the points whose position vectors are $2i - j + 5k$, $i - 2j + k$, $3i + j - 2k$ respectively. L and M are the midpoints of AC and CB. Show that LM is parallel to BA.

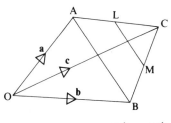

$\overrightarrow{OL} = \tfrac{1}{2}(a + c)$

$\qquad = \tfrac{1}{2}(5i + 3k)$

$\overrightarrow{OM} = \tfrac{1}{2}(b + c)$

$\qquad = \tfrac{1}{2}(4i - j - k).$

So $\overrightarrow{LM} = \overrightarrow{OM} - \overrightarrow{OL} = \tfrac{1}{2}(-i - j - 4k)$

Now $\overrightarrow{BA} = a - b \quad = i + j + 4k$

Therefore $\overrightarrow{BA} = -2\overrightarrow{LM}$

Hence BA and LM are parallel.

EXERCISE 12e

In Questions 1-7, A, B, C and D are the points with position vectors $i + j - k$, $i - j + 2k$, $j + k$ and $2i + j$ respectively.

1) Find $|\overrightarrow{AB}|$ and $|\overrightarrow{BD}|$.

2) Find the direction cosines of $\overrightarrow{CD}$ and $\overrightarrow{AC}$.

3) Find the position vector of the point which
(a) divides BC internally in the ratio $3:2$,
(b) divides AC externally in the ratio $3:2$.

4) Determine whether any of the following pairs of lines are parallel:
(a) AB and CD,   (b) AC and BD,   (c) AD and BC.

5) If L and M are the position vectors of the midpoints of AD and BD respectively, show that $\overrightarrow{LM}$ is parallel to $\overrightarrow{AB}$.

6) If H and K are the midpoints of AC and CD show that $\overrightarrow{HK} = \frac{1}{2}\overrightarrow{AD}$.

7) If L, M, N and P are the midpoints of AD, BD, BC and AC respectively, show that $\overrightarrow{LM}$ is parallel to $\overrightarrow{NP}$.

THE EQUATION OF A STRAIGHT LINE

A particular line is uniquely located in space if
(a) it has a known direction and passes through a known fixed point, or
(b) it passes through two known fixed points.

A Line with Known Gradient Passing through a Fixed Point

Consider a line which is parallel to a vector **m** and which passes through a fixed point A with position vector **a**.

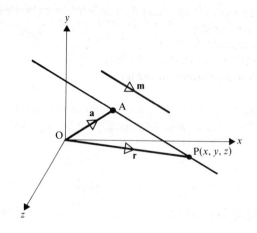

If $\mathbf{r}$ is the position vector, $\overrightarrow{OP}$, of a point P then

$$P \text{ is a point on this line } \iff \overrightarrow{AP} = \lambda\mathbf{m}$$

where λ is a variable scalar, i.e. a parameter.

Now $$\overrightarrow{AP} = \overrightarrow{OP} - \overrightarrow{OA}$$

i.e. $$\lambda\mathbf{m} = \mathbf{r} - \mathbf{a},$$

i.e. $$\boxed{P \text{ is on the line } \iff \mathbf{r} = \mathbf{a} + \lambda\mathbf{m}.}$$

For each value of the parameter λ this equation gives the position vector of one point on the line and it is called the vector equation of the line.

For example, the vector equation of the line which is parallel to the vector $2\mathbf{i} - \mathbf{j} + 3\mathbf{k}$ and which passes through the point $(5, -2, 4)$ is

$$\mathbf{r} = (5\mathbf{i} - 2\mathbf{j} + 4\mathbf{k}) + \lambda(2\mathbf{i} - \mathbf{j} + 3\mathbf{k}). \qquad [1]$$

Now $\mathbf{r}$ is the position vector of any point $P(x, y, z)$ on the line,

i.e. $$x\mathbf{i} + y\mathbf{j} + z\mathbf{k} = 5\mathbf{i} - 2\mathbf{j} + 4\mathbf{k} + \lambda(2\mathbf{i} - \mathbf{j} + 3\mathbf{k})$$

$$= (5 + 2\lambda)\mathbf{i} + (-2 - \lambda)\mathbf{j} + (4 + 3\lambda)\mathbf{k}$$

$$\Rightarrow \qquad \left. \begin{array}{l} x = 5 + 2\lambda \\ y = -2 - \lambda \\ z = 4 + 3\lambda. \end{array} \right\} \qquad [2]$$

Isolating λ in each of these equations gives

$$\frac{x-5}{2} = \frac{y+2}{-1} = \frac{z-4}{3} \quad (= \lambda) \qquad [3]$$

So there are three ways of expressing the relationships between the coordinates of any point P on this line:
[1] is the vector equation of the line,
[2] are the parametric equations of the line,
[3] are the Cartesian equations of the line.

Note that the line is parallel to $2\mathbf{i} - \mathbf{j} + 3\mathbf{k}$ so the direction ratios of the line are $2 : -1 : 3$. These direction ratios appear in all three forms of the equations of the line.

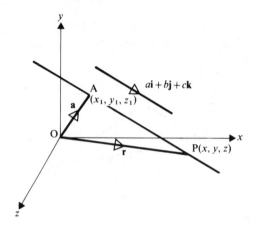

In general, if a line passes through $A(x_1, y_1, z_1)$ and is parallel to $a\mathbf{i} + b\mathbf{j} + c\mathbf{k}$ its equations may be written

$$\mathbf{r} = (x_1\mathbf{i} + y_1\mathbf{j} + z_1\mathbf{k}) + \lambda(a\mathbf{i} + b\mathbf{j} + c\mathbf{k})$$

or
$$\begin{cases} x = x_1 + \lambda a \\ y = y_1 + \lambda b \\ z = z_1 + \lambda c \end{cases}$$

or
$$\frac{x - x_1}{a} = \frac{y - y_1}{b} = \frac{z - z_1}{c} \quad (= \lambda).$$

Note that the direction ratios of the line, $a:b:c$, appear in all three forms.

Note also that although the direction ratios of a particular line are unique, the point (x_1, y_1, z_1) is only one of an infinite set of fixed points on the line. Hence the equations representing a particular line are not unique.

A Line Through Two Fixed Points

Now consider the definition of a particular line in space as the line passing through two fixed points $A(x_1, y_1, z_1)$ and $B(x_2, y_2, z_2)$ whose position vectors are $\mathbf{a}$ and $\mathbf{b}$ respectively.

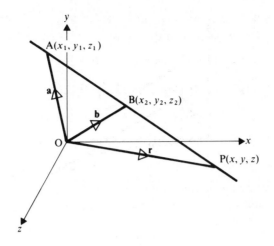

If **r** is the position vector, $\overrightarrow{OP}$, of the point P then

$$P \text{ is a point on the line} \iff \overrightarrow{AP} = \lambda \overrightarrow{AB}$$

Now $$\overrightarrow{AB} = \mathbf{b} - \mathbf{a} \quad \text{and} \quad \overrightarrow{AP} = \mathbf{r} - \mathbf{a}.$$

Therefore $$P \text{ is on the line} \iff \mathbf{r} - \mathbf{a} = \lambda(\mathbf{b} - \mathbf{a}),$$

i.e. $$\mathbf{r} = \mathbf{a} + \lambda(\mathbf{b} - \mathbf{a}). \tag{1}$$

Using the Cartesian components of **a** and **b**, [1] becomes

$$\mathbf{r} = x_1\mathbf{i} + y_1\mathbf{j} + z_1\mathbf{k} + \lambda[(x_2 - x_1)\mathbf{i} + (y_2 - y_1)\mathbf{j} + (z_2 - z_1)\mathbf{k}].$$

EXAMPLES 12f

1) (a) The Cartesian equations of a line are $\dfrac{x-5}{3} = \dfrac{y+4}{7} = \dfrac{z-6}{2}$

Find a vector equation for the line.

(b) If the vector equation of a line is $\mathbf{r} = \mathbf{i} - 3\mathbf{j} + 2\mathbf{k} + \lambda(5\mathbf{i} + 2\mathbf{j} - \mathbf{k})$, express the equation of the line in parametric form and hence find the coordinates of the point where the line crosses the xy plane.

(a) Comparing $$\frac{x-5}{3} = \frac{y+4}{7} = \frac{z-6}{2}$$

with $$\frac{x-x_1}{a} = \frac{y-y_1}{b} = \frac{z-z_1}{c}$$

we see that this line has direction ratios $3:7:2$. Hence its direction vector is $3\mathbf{i} + 7\mathbf{j} + 2\mathbf{k}$.
One point on the line has coordinates $(5, -4, 6)$.

The position vector of this point is $5i - 4j + 6k$.
Hence a vector equation of the line is

$$r = 5i - 4j + 6k + \lambda(3i + 7j + 2k).$$

(b) $r = i - 3j + 2k + \lambda(5i + 2j - k)$

$\Rightarrow$ $r = (1 + 5\lambda)i + (-3 + 2\lambda)j + (2 - \lambda)k.$

A general point on the line has coordinates

$$x = 1 + 5\lambda, \quad y = -3 + 2\lambda, \quad z = 2 - \lambda$$

and these are the parametric equations of the line.
At the point where the line crosses the xy plane, $z = 0$,
so $2 - \lambda = 0$.
When $\lambda = 2$, $x = 11$ and $y = 1$.
Therefore the line crosses the xy plane at the point $(11, 1, 0)$.

2) A line passes through the point with position vector $2i - j + 4k$ and is in the direction of $i + j - 2k$. Find equations for the line in vector and in Cartesian form.

The equation of the line in vector form is

$$r = 2i - j + 4k + \lambda(i + j - 2k).$$

This shows that the coordinates of any point P on the line are

$$[(2 + \lambda), (-1 + \lambda), (4 - 2\lambda)].$$

Hence the Cartesian equations are

$$x - 2 = y + 1 = \frac{z - 4}{-2} \quad (= \lambda).$$

3) Find a vector equation for the line through the points $A(3, 4, -7)$ and $B(1, -1, 6)$.

$$\overrightarrow{OA} = a = 3i + 4j - 7k,$$

$$\overrightarrow{OB} = b = i - j + 6k.$$

For any point P on the line, $\overrightarrow{OP} = r$,

so $r = a + \lambda(b - a)$

$$= 3i + 4j - 7k + \lambda(-2i - 5j + 13k).$$

4) Show that the line through the points $i + j - 3k$ and $4i + 7j + k$ is parallel to the line $r = i - k + \lambda(\frac{3}{2}i + 3j + 2k)$.

The line through the two given points has direction ratios

$$(1 - 4):(1 - 7):(-3 - 1) = 3:6:4.$$

The given line has direction ratios

$$\tfrac{3}{2}:3:2 \;=\; 3:6:4.$$

Hence the two lines are parallel.

5) Find the coordinates of the point where the line through $A(3, 4, 1)$ and $B(5, 1, 6)$ crosses the xy plane.

The vector equation of the line through A and B is

$$\mathbf{r} \;=\; 3\mathbf{i} + 4\mathbf{j} + \mathbf{k} + \lambda(2\mathbf{i} - 3\mathbf{j} + 5\mathbf{k}).$$

The line crosses the xy plane where $z = 0$.
Any point on the line has coordinates $[(3 + 2\lambda), (4 - 3\lambda), (1 + 5\lambda)]$.
If $z = 0$ then $1 + 5\lambda = 0 \;\Rightarrow\; \lambda = -\tfrac{1}{5}$

Hence $\qquad x = 3 - \tfrac{2}{5} = \tfrac{13}{5}; \quad y = 4 + \tfrac{3}{5} = \tfrac{23}{5}$

6) Write down the direction cosines of the line

$$\mathbf{r} \;=\; 3\mathbf{i} + 3\mathbf{j} - \mathbf{k} + \lambda(4\mathbf{i} + 3\mathbf{k})$$

and describe its position relative to the x, y, and z-axes.

From its equation, the line is seen to be parallel to $4\mathbf{i} + 0\mathbf{j} + 3\mathbf{k}$.
Hence the direction cosines of the line are $\tfrac{4}{5}, 0, \tfrac{3}{5}$

so $\qquad\qquad \cos\beta = 0 \;\Rightarrow\; \beta = \dfrac{\pi}{2}$

Therefore the line is perpendicular to Oy and hence parallel to the xz plane. As the point $(3, 3, -1)$ lies on the line, the line lies in the plane parallel to the xz plane cutting Oy at $y = 3$, and is inclined at $\arccos \tfrac{4}{5}$ to Ox and $\arccos \tfrac{3}{5}$ to Oz.

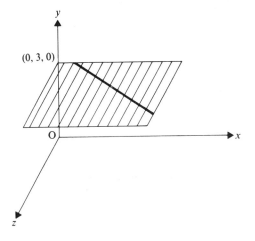

EXERCISE 12f

1) Convert the following vector equations to Cartesian form:
(a) $\mathbf{r} = 2\mathbf{i} + 3\mathbf{j} - \mathbf{k} + \lambda(\mathbf{i} + \mathbf{j} + \mathbf{k})$,
(b) $\mathbf{r} = 4\mathbf{j} + \lambda(3\mathbf{i} + 5\mathbf{k})$,
(c) $\mathbf{r} = \lambda(2\mathbf{i} + 3\mathbf{j} + 4\mathbf{k})$.

2) Convert to vector form, the following equations:

(a) $\dfrac{x-3}{4} = \dfrac{y-1}{2} = \dfrac{z-7}{6}$,

(b) $x = 3\lambda + 2, \quad y = \lambda - 5, \quad z = 4\lambda + 1$,

(c) $\dfrac{1-x}{3} = \dfrac{y}{5} = z$.

3) Write down equations, in vector and in Cartesian form, for the line through a point A with position vector $\mathbf{a}$ and with a direction vector $\mathbf{b}$ if:
(a) $\mathbf{a} = \mathbf{i} - 3\mathbf{j} + 2\mathbf{k}$ $\mathbf{b} = 5\mathbf{i} + 4\mathbf{j} - \mathbf{k}$,
(b) $\mathbf{a} = 2\mathbf{i} + \mathbf{j}$ $\mathbf{b} = 3\mathbf{j} - \mathbf{k}$,
(c) A is the origin $\mathbf{b} = \mathbf{i} - \mathbf{j} - \mathbf{k}$.

4) State whether or not the following pairs of lines are parallel:
(a) $\mathbf{r} = \mathbf{i} + \mathbf{j} - \mathbf{k} + \lambda(2\mathbf{i} - 3\mathbf{j} + \mathbf{k})$
   $\mathbf{r} = 2\mathbf{i} - 4\mathbf{j} + 5\mathbf{k} + \lambda(\mathbf{i} + \mathbf{j} - \mathbf{k})$,

(b) $\dfrac{x-1}{2} = \dfrac{y-4}{3} = \dfrac{z+1}{-4}$

   $\dfrac{x}{4} = \dfrac{y+5}{6} = \dfrac{3-z}{8}$,

(c) $\mathbf{r} = 2\mathbf{i} - \mathbf{j} + 4\mathbf{k} + \lambda(\mathbf{i} + \mathbf{j} + 3\mathbf{k})$

   $x - 4 = y + 7 = \dfrac{z}{3}$,

(d) $\mathbf{r} = \lambda(3\mathbf{i} - 3\mathbf{j} + 6\mathbf{k})$
   $\mathbf{r} = 4\mathbf{j} + \lambda(-\mathbf{i} + \mathbf{j} - 2\mathbf{k})$,

(e) $\mathbf{r} = 3\mathbf{i} + \mathbf{k} + \lambda(\mathbf{i} - \mathbf{j} - 2\mathbf{k})$

   $\dfrac{x-3}{1} = \dfrac{y}{1} = \dfrac{z-1}{2}$.

5) The points $A(4, 5, 10)$, $B(2, 3, 4)$ and $C(1, 2, -1)$ are three vertices of a parallelogram ABCD. Find vector and Cartesian equations for the sides AB and BC and find the coordinates of D.

6) Write down a vector equation for the line through A and B if

(a) $\overrightarrow{OA}$ is $3\mathbf{i} + \mathbf{j} - 4\mathbf{k}$ and $\overrightarrow{OB}$ is $\mathbf{i} + 7\mathbf{j} + 8\mathbf{k}$.

(b) A and B have coordinates $(1, 1, 7)$ and $(3, 4, 1)$.

Find, in each case, the coordinates of the points where the line crosses the
xy plane, the yz plane and the zx plane.

7) A line has Cartesian equations $\dfrac{x-1}{3} = \dfrac{y+2}{4} = \dfrac{z-4}{5}$.

Find a vector equation for a parallel line passing through the point with position
vector $5\mathbf{i} - 2\mathbf{j} - 4\mathbf{k}$ and find the coordinates of the point on this line
where $y = 0$.

8) The Cartesian equations of a line are $x - 2 = 2y + 1 = 3z - 2$.
Find the direction ratios of the line and write down the vector equation of the
line through $(2, -1, -1)$ which is parallel to the given line.

PAIRS OF LINES

The location of two lines in space may be such that:
(a) the lines are parallel,
(b) the lines are not parallel and intersect,
(c) the lines are not parallel and do not intersect. Such lines are called *skew*.

Parallel Lines
We have already seen that parallel lines have equal direction ratios. So if two
lines are parallel, this property can be observed from their equations.

Non-parallel Lines

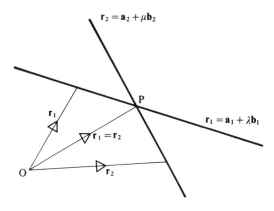

Consider two lines whose vector equations are $\mathbf{r}_1 = \mathbf{a}_1 + \lambda\mathbf{b}_1$
and $\mathbf{r}_2 = \mathbf{a}_2 + \mu\mathbf{b}_2$.

In order that these lines shall intersect there must be unique values of λ and μ such that

$$\mathbf{a}_1 + \lambda \mathbf{b}_1 \;=\; \mathbf{a}_2 + \mu \mathbf{b}_2.$$

If no such values can be found, the lines do not intersect.

EXAMPLE 12g

Find out whether the following pairs of lines are parallel, non-parallel and intersecting, or non-parallel and non-intersecting:
(a) $\mathbf{r}_1 = \mathbf{i} + \mathbf{j} + 2\mathbf{k} + \lambda(3\mathbf{i} - 2\mathbf{j} + 4\mathbf{k})$
   $\mathbf{r}_2 = 2\mathbf{i} - \mathbf{j} + 3\mathbf{k} + \mu(-6\mathbf{i} + 4\mathbf{j} - 8\mathbf{k})$,
(b) $\mathbf{r}_1 = \mathbf{i} - \mathbf{j} + 3\mathbf{k} + \lambda(\mathbf{i} - \mathbf{j} + \mathbf{k})$
   $\mathbf{r}_2 = 2\mathbf{i} + 4\mathbf{j} + 6\mathbf{k} + \mu(2\mathbf{i} + \mathbf{j} + 3\mathbf{k})$,
(c) $\mathbf{r}_1 = \mathbf{i} + \mathbf{k} + \lambda(\mathbf{i} + 3\mathbf{j} + 4\mathbf{k})$
   $\mathbf{r}_2 = 2\mathbf{i} + 3\mathbf{j} + \mu(4\mathbf{i} - \mathbf{j} + \mathbf{k})$.

(a) Checking first whether the lines are parallel we compare the direction ratios of the two lines.
   The first line has direction ratios $3 : -2 : 4$.
   The second line has direction ratios $-6 : 4 : -8 = 3 : -2 : 4$.
   Therefore these two lines are parallel.

(b) In this case the two sets of direction ratios are $1 : -1 : 1$ and $2 : 1 : 3$.
   These are not equal, so these two lines are not parallel.
   Now if the lines intersect it will be at a point where $\mathbf{r}_1 = \mathbf{r}_2$,
   i.e. where

$$(1 + \lambda)\mathbf{i} - (1 + \lambda)\mathbf{j} + (3 + \lambda)\mathbf{k} \;=\; 2(1 + \mu)\mathbf{i} + (4 + \mu)\mathbf{j} + (6 + 3\mu)\mathbf{k}.$$

   Equating the coefficients of $\mathbf{i}$ and $\mathbf{j}$, we have

$$1 + \lambda \;=\; 2(1 + \mu)$$

$$-(1 + \lambda) \;=\; 4 + \mu.$$

Hence $\mu = -2, \quad \lambda = -3$

With these values for λ and μ, the coefficients of $\mathbf{k}$ become

first line $3 + \lambda \;=\; 0$
                                         $\left.\right\}$ equal values.
second line $6 + 3\mu \;=\; 0$

So $\mathbf{r}_1 = \mathbf{r}_2$ when $\lambda = -3$ and $\mu = -2$.

Therefore the lines *do* intersect at the point with position vector

$$(1 - 3)\mathbf{i} - (1 - 3)\mathbf{j} + (3 - 3)\mathbf{k}, \hspace{3cm} (\lambda = -3 \text{ in } \mathbf{r}_1)$$

i.e. $-2\mathbf{i} + 2\mathbf{j}.$

(c) The direction ratios of these two lines are not equal so the lines are not parallel.
   If the lines intersect it will be where $\mathbf{r_1} = \mathbf{r_2}$,
   i.e. where

$$(1 + \lambda)\mathbf{i} + 3\lambda\mathbf{j} + (1 + 4\lambda)\mathbf{k} = (2 + 4\mu)\mathbf{i} + (3 - \mu)\mathbf{j} + \mu\mathbf{k}.$$

Equating the coefficients of $\mathbf{i}$ and $\mathbf{j}$ we have

$$1 + \lambda = 2 + 4\mu$$

$$3\lambda = 3 - \mu.$$

Hence $\mu = 0, \quad \lambda = 1.$

With these values of λ and μ, the coefficients of $\mathbf{k}$ become

first line $\qquad 1 + 4\lambda = 5$
$\left.\vphantom{\begin{matrix}a\\b\end{matrix}}\right\}$ unequal values.
second line $\qquad \mu = 0$

So there are no values of λ and μ for which $\mathbf{r_1} = \mathbf{r_2}$ and these lines do not intersect and are skew.

EXERCISE 12g

1) Find whether the following pairs of lines are parallel, intersecting or skew. In the case of intersection state the position vector of the common point.
(a) $\mathbf{r} = \mathbf{i} - \mathbf{j} + \mathbf{k} + \lambda(3\mathbf{i} - 4\mathbf{j} + \mathbf{k})$
   $\mathbf{r} = \mu(-9\mathbf{i} + 12\mathbf{j} - 3\mathbf{k}),$

(b) $\dfrac{x-4}{1} = \dfrac{y-8}{2} = \dfrac{z-3}{1}$

   $\dfrac{x-7}{6} = \dfrac{y-6}{4} = \dfrac{z-5}{5},$

(c) $\mathbf{r} = \mathbf{i} + 3\mathbf{k} + \lambda(2\mathbf{i} + \mathbf{j} + \mathbf{k})$
   $\mathbf{r} = 2\mathbf{i} - \mathbf{j} + \mathbf{k} + \mu(\mathbf{i} - 2\mathbf{j}).$

2) Two lines have equations

$$\mathbf{r} = 2\mathbf{i} + 9\mathbf{j} + 13\mathbf{k} + \lambda(\mathbf{i} + 2\mathbf{j} + 3\mathbf{k})$$

$$\mathbf{r} = a\mathbf{i} + 7\mathbf{j} - 2\mathbf{k} + \mu(-\mathbf{i} + 2\mathbf{j} - 3\mathbf{k}).$$

If they intersect, find the value of a and the position vector of the point of intersection.

3) Show that the lines

$$\mathbf{r} = 2\mathbf{i} - \mathbf{j} + \mathbf{k} + \lambda(\mathbf{i} - 2\mathbf{j} + 2\mathbf{k})$$

$$\mathbf{r} = \mathbf{i} - 3\mathbf{j} + 4\mathbf{k} + \mu(2\mathbf{i} + 3\mathbf{j} - 6\mathbf{k})$$

are skew.

$\overrightarrow{OQ}$ is the unit vector in the direction of the first line and $\overrightarrow{OR}$ is the unit vector in the direction of the second line. Write down the coordinates of Q and R. By using the cosine formula in triangle OQR find the angle between $\overrightarrow{OQ}$ and $\overrightarrow{OR}$ and hence the angle between the given lines.

4) If $\overrightarrow{OA}$ is the unit vector $l_1\mathbf{i} + m_1\mathbf{j} + n_1\mathbf{k}$ and $\overrightarrow{OB}$ is the unit vector $l_2\mathbf{i} + m_2\mathbf{j} + n_2\mathbf{k}$, by using the cosine formula in triangle OAB show that θ, the angle between $\overrightarrow{OA}$ and $\overrightarrow{OB}$, is given by $\cos\theta = l_1l_2 + m_1m_2 + n_1n_2$.

SCALAR PRODUCT

Geometric analysis often involves the use of expressions containing the sine or the cosine of an angle (for example, in the solution of triangles).
So we now look at a vector operation between two vectors **a** and **b** which, among other things, gives rise to $\cos\theta$ where θ is the angle between **a** and **b**. There is another vector operation which involves $\sin\theta$ and these two operations are both called products but, being vector operations, they are in no way related to the product of real numbers.
The first of these results in a scalar quantity and it is therefore known as the *scalar product* of two vectors.
The second operation, which is referred to as the *vector product* of two vectors, produces a vector quantity.
Distinction is drawn between the two processes by using the multiplication 'dot' symbol (**a.b**) exclusively for the scalar product and the multiplication 'cross' symbol (**a** × **b**) exclusively for the vector product.
(This can cause some difficulty in vector work since the 'dot' and 'cross' can cause confusion if used to represent multiplication of numbers in the context of vector analysis. This problem can be avoided by the use of brackets.)

Definition of the Scalar (or Dot) Product

The scalar product of two vectors **a** and **b** is denoted by **a.b** and defined as $|\mathbf{a}||\mathbf{b}|\cos\theta$ where θ is the angle between **a** and **b**.

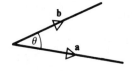

Thus $\mathbf{a.b} = ab\cos\theta$

Properties of the Scalar Product

1) Parallel Vectors.
If **a** and **b** are parallel

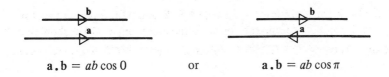

$$\mathbf{a}.\mathbf{b} = ab \cos 0 \qquad \text{or} \qquad \mathbf{a}.\mathbf{b} = ab \cos \pi$$

i.e. $\mathbf{a}.\mathbf{b} = ab$ for like parallel vectors

or $\mathbf{a}.\mathbf{b} = -ab$ for unlike parallel vectors.

In the special case when $\mathbf{a} = \mathbf{b}$, $\mathbf{a}.\mathbf{b} = \mathbf{a}.\mathbf{a} = a^2$ (sometimes $\mathbf{a}.\mathbf{a}$ is written as $\mathbf{a}^2$).

2) Perpendicular Vectors.
If **a** and **b** are perpendicular

$$\mathbf{a}.\mathbf{b} = ab \cos \frac{\pi}{2} = 0$$

$$\text{i.e.}\quad \mathbf{a}.\mathbf{b} = 0$$

In the special case of the Cartesian unit vectors these results give

$$\mathbf{i}.\mathbf{i} = \mathbf{j}.\mathbf{j} = \mathbf{k}.\mathbf{k} = 1$$
$$\mathbf{i}.\mathbf{j} = \mathbf{j}.\mathbf{k} = \mathbf{k}.\mathbf{i} = 0$$

3) The scalar product is commutative.
From the definition we have $\mathbf{a}.\mathbf{b} = ab \cos \theta$ and $\mathbf{b}.\mathbf{a} = ba \cos \theta$.

But $ab \cos \theta \equiv ba \cos \theta$

So $\mathbf{a}.\mathbf{b} = \mathbf{b}.\mathbf{a}$

4) The scalar product is distributive across addition,

i.e. $\mathbf{a}.(\mathbf{b}+\mathbf{c}) = \mathbf{a}.\mathbf{b} + \mathbf{a}.\mathbf{c}$

This property can be proved as follows.

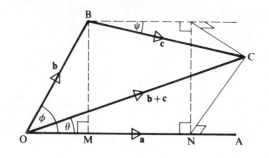

$$\mathbf{a}.(\mathbf{b} + \mathbf{c}) = |\mathbf{a}||\mathbf{b} + \mathbf{c}| \cos \theta$$

$$= (OA)(OC) \cos \theta$$

$$= (OA)(ON)$$

$$= (OA)(OM + MN)$$

$$= (OA)(OB \cos \phi) + (OA)(BC \cos \psi)$$

$$= \mathbf{a}.\mathbf{b} + \mathbf{a}.\mathbf{c}.$$

Calculation of a.b when a and b are in Cartesian Component Form

Using the properties of the scalar product

if $\qquad \mathbf{a} = x_1\mathbf{i} + y_1\mathbf{j} + z_1\mathbf{k}$

and $\qquad \mathbf{b} = x_2\mathbf{i} + y_2\mathbf{j} + z_2\mathbf{k}$

then $\qquad \mathbf{a}.\mathbf{b} = (x_1\mathbf{i} + y_1\mathbf{j} + z_1\mathbf{k}).(x_2\mathbf{i} + y_2\mathbf{j} + z_2\mathbf{k})$

$$= (x_1x_2\mathbf{i}.\mathbf{i} + y_1y_2\mathbf{j}.\mathbf{j} + z_1z_2\mathbf{k}.\mathbf{k})$$

$$+ (x_1y_2\mathbf{i}.\mathbf{j} + y_1z_2\mathbf{j}.\mathbf{k} + z_1x_2\mathbf{k}.\mathbf{i})$$

$$+ (y_1x_2\mathbf{j}.\mathbf{i} + z_1y_2\mathbf{k}.\mathbf{j} + x_1z_2\mathbf{i}.\mathbf{k})$$

$$= (x_1x_2 + y_1y_2 + z_1z_2) + (0) + (0),$$

i.e. $\qquad (x_1\mathbf{i} + y_1\mathbf{j} + z_1\mathbf{k}).(x_2\mathbf{i} + y_2\mathbf{j} + z_2\mathbf{k}) = x_1x_2 + y_1y_2 + z_1z_2.$

For example,

$$(2\mathbf{i} - 3\mathbf{j} + 4\mathbf{k}).(\mathbf{i} + 3\mathbf{j} - 2\mathbf{k}) = (2)(1) + (-3)(3) + (4)(-2) = -15$$

Applications of the Scalar Product

If θ is the angle between the two vectors $2\mathbf{i} - 3\mathbf{j} + 4\mathbf{k}$ and $\mathbf{i} + 3\mathbf{j} - 2\mathbf{k}$ then

$$(2\mathbf{i} - 3\mathbf{j} + 4\mathbf{k}).(\mathbf{i} + 3\mathbf{j} - 2\mathbf{k}) = |2\mathbf{i} - 3\mathbf{j} + 4\mathbf{k}||\mathbf{i} + 3\mathbf{j} - 2\mathbf{k}|\cos \theta$$

i.e. $$-15 = (\sqrt{29})(\sqrt{14}) \cos \theta$$

$\Rightarrow$ $$\theta = \arccos\left(-\frac{15}{\sqrt{406}}\right)$$

So we can use the scalar product to find the angle between a pair of vectors.

Angle Between a Pair of Lines

If two lines have equations

$$\begin{cases} \mathbf{r} = \mathbf{a}_1 + \lambda \mathbf{b}_1 \\ \mathbf{r} = \mathbf{a}_2 + \mu \mathbf{b}_2 \end{cases}$$

they are in the directions of $\mathbf{b}_1$ and $\mathbf{b}_2$ respectively.

The angle between any two lines is defined as the angle between their direction vectors (which are free vectors).

So the angle between a pair of lines depends only on their directions and not on their positions.

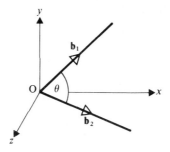

Drawing $\mathbf{b}_1$ and $\mathbf{b}_2$ from the origin, the angle θ between them is the angle between the two given lines.

As $\mathbf{b}_1 . \mathbf{b}_2 = b_1 b_2 \cos \theta$

$$\cos \theta = \frac{\mathbf{b}_1 . \mathbf{b}_2}{b_1 b_2}$$

This applies to a pair of skew lines as well as to a pair of intersecting lines. It follows that if the lines are perpendicular

$$\cos \theta = 0 \quad \Rightarrow \quad \mathbf{b}_1 . \mathbf{b}_2 = 0.$$

If two lines have equations

$$\mathbf{r} = x_1 \mathbf{i} + y_1 \mathbf{j} + z_1 \mathbf{k} + \lambda(a_1 \mathbf{i} + b_1 \mathbf{j} + c_1 \mathbf{k})$$

and $$\mathbf{r} = x_2\mathbf{i} + y_2\mathbf{j} + z_2\mathbf{k} + \mu(a_2\mathbf{i} + b_2\mathbf{j} + c_2\mathbf{k})$$

respectively, the angle θ between these lines is the angle between their direction vectors, $\mathbf{v}_1 = a_1\mathbf{i} + b_1\mathbf{j} + c_1\mathbf{k}$ and $\mathbf{v}_2 = a_2\mathbf{i} + b_2\mathbf{j} + c_2\mathbf{k}$.

Now

$$(a_1\mathbf{i} + b_1\mathbf{j} + c_1\mathbf{k}).(a_2\mathbf{i} + b_2\mathbf{j} + c_2\mathbf{k}) = a_1a_2 + b_1b_2 + c_1c_2 = |\mathbf{v}_1||\mathbf{v}_2|\cos\theta.$$

Hence $$\cos\theta = \frac{a_1a_2 + b_1b_2 + c_1c_2}{|\mathbf{v}_1||\mathbf{v}_2|}$$

But $a_1:b_1:c_1$ are the direction ratios of the first line

and $a_2:b_2:c_2$ are the direction ratios of the second line,

so $\dfrac{a_1}{|\mathbf{v}_1|}, \dfrac{b_1}{|\mathbf{v}_1|}, \dfrac{c_1}{|\mathbf{v}_1|}$ are the direction cosines l_1, m_1, n_1 of the first line,

and $\dfrac{a_2}{|\mathbf{v}_2|}, \dfrac{b_2}{|\mathbf{v}_2|}, \dfrac{c_2}{|\mathbf{v}_2|}$ are the direction cosines, l_2, m_2, n_2 of the second line.

Hence the angle θ between two lines is given by

$$\cos\theta = l_1l_2 + m_1m_2 + n_1n_2.$$

EXAMPLES 12h

1) Simplify (a) $(\mathbf{a} - \mathbf{b}).(\mathbf{a} + \mathbf{b})$

(b) $(\mathbf{a} + \mathbf{b}).\mathbf{c} - (\mathbf{a} + \mathbf{c}).\mathbf{b}$

(a) $$(\mathbf{a} - \mathbf{b}).(\mathbf{a} + \mathbf{b}) = \mathbf{a}.\mathbf{a} - \mathbf{b}.\mathbf{a} + \mathbf{a}.\mathbf{b} - \mathbf{b}.\mathbf{b}$$

but $$\mathbf{a}.\mathbf{b} = \mathbf{b}.\mathbf{a} \quad \text{hence} \quad \mathbf{a}.\mathbf{b} - \mathbf{b}.\mathbf{a} = 0$$

also $$\mathbf{a}.\mathbf{a} = a^2 \quad \text{and} \quad \mathbf{b}.\mathbf{b} = b^2$$

Therefore $$(\mathbf{a} - \mathbf{b}).(\mathbf{a} + \mathbf{b}) = a^2 - b^2$$

(b) $$(\mathbf{a} + \mathbf{b}).\mathbf{c} - (\mathbf{a} + \mathbf{c}).\mathbf{b}$$

$$= \mathbf{a}.\mathbf{c} + \mathbf{b}.\mathbf{c} - \mathbf{a}.\mathbf{b} - \mathbf{c}.\mathbf{b}$$

$$= \mathbf{a}.\mathbf{c} - \mathbf{a}.\mathbf{b}$$

$$= \mathbf{a}.(\mathbf{c} - \mathbf{b})$$

2) Find the scalar product of $\mathbf{a} = 2\mathbf{i} - 3\mathbf{j} + 5\mathbf{k}$ and $\mathbf{b} = \mathbf{i} - 3\mathbf{j} + \mathbf{k}$ and hence find the cosine of the angle between $\mathbf{a}$ and $\mathbf{b}$.

$$\mathbf{a}.\mathbf{b} = (2)(1) + (-3)(-3) + (5)(1) = 16$$

But $$\mathbf{a}.\mathbf{b} = |\mathbf{a}||\mathbf{b}|\cos\theta$$

$$|\mathbf{a}| = \sqrt{(4 + 9 + 25)} = \sqrt{38}$$

$$|\mathbf{b}| = \sqrt{(1 + 9 + 1)} = \sqrt{11}$$

Hence $$\cos \theta = \frac{\mathbf{a} \cdot \mathbf{b}}{|\mathbf{a}||\mathbf{b}|} = \frac{16}{\sqrt{11}\sqrt{38}} = \frac{16}{\sqrt{418}}$$

3) If $\mathbf{a} = 10\mathbf{i} - 3\mathbf{j} + 5\mathbf{k}$, $\mathbf{b} = 2\mathbf{i} + 6\mathbf{j} - 3\mathbf{k}$ and $\mathbf{c} = \mathbf{i} + 10\mathbf{j} - 2\mathbf{k}$, verify that $\mathbf{a} \cdot \mathbf{b} + \mathbf{a} \cdot \mathbf{c} = \mathbf{a} \cdot (\mathbf{b} + \mathbf{c})$.

$$\mathbf{a} \cdot \mathbf{b} = (10)(2) + (-3)(6) + (5)(-3) = -13$$
$$\mathbf{a} \cdot \mathbf{c} = (10)(1) + (-3)(10) + (5)(-2) = -30$$
$$\mathbf{b} + \mathbf{c} = 3\mathbf{i} + 16\mathbf{j} - 5\mathbf{k}$$

Hence $\mathbf{a} \cdot (\mathbf{b} + \mathbf{c}) = (10)(3) + (-3)(16) + (5)(-5) = -43$

But $\mathbf{a} \cdot \mathbf{b} + \mathbf{a} \cdot \mathbf{c} = -13 - 30 = -43$

Therefore $\mathbf{a} \cdot \mathbf{b} + \mathbf{a} \cdot \mathbf{c} = \mathbf{a} \cdot (\mathbf{b} + \mathbf{c})$

4) The resultant of two vectors $\mathbf{a}$ and $\mathbf{b}$ is perpendicular to $\mathbf{a}$. If $|\mathbf{b}| = \sqrt{2}|\mathbf{a}|$, show that the resultant of $2\mathbf{a}$ and $\mathbf{b}$ is perpendicular to $\mathbf{b}$.

The resultant of $\mathbf{a}$ and $\mathbf{b}$ is $\mathbf{a} + \mathbf{b}$.
Since $\mathbf{a} + \mathbf{b}$ is perpendicular to $\mathbf{a}$,

$$(\mathbf{a} + \mathbf{b}) \cdot \mathbf{a} = 0$$

Hence $\mathbf{a} \cdot \mathbf{a} + \mathbf{b} \cdot \mathbf{a} = 0$

But $\mathbf{a} \cdot \mathbf{a} = |\mathbf{a}|^2$

Hence $\mathbf{b} \cdot \mathbf{a} = -|\mathbf{a}|^2$

Now the resultant of $2\mathbf{a}$ and $\mathbf{b}$ is $2\mathbf{a} + \mathbf{b}$

and $\mathbf{b} \cdot (2\mathbf{a} + \mathbf{b}) = 2\mathbf{b} \cdot \mathbf{a} + \mathbf{b} \cdot \mathbf{b}$

But $\mathbf{b} \cdot \mathbf{b} = |\mathbf{b}|^2 = b^2$

and $2\mathbf{a} \cdot \mathbf{b} = -2|\mathbf{a}|^2 = -2a^2$

Hence $\mathbf{b} \cdot (2\mathbf{a} + \mathbf{b}) = -2a^2 + b^2$.

But $b = a\sqrt{2}$ (given)

so $\mathbf{b} \cdot (2\mathbf{a} + \mathbf{b}) = 0$.

Hence $2\mathbf{a} + \mathbf{b}$ is perpendicular to $\mathbf{b}$.

5) Find the angle between the lines
$$\mathbf{r}_1 = \mathbf{i} - 2\mathbf{j} + 3\mathbf{k} + \lambda(2\mathbf{i} - 3\mathbf{j} + 6\mathbf{k}) \qquad [1]$$
$$\mathbf{r}_2 = 2\mathbf{i} - 7\mathbf{j} + 10\mathbf{k} + \mu(\mathbf{i} + 2\mathbf{j} + 2\mathbf{k}). \qquad [2]$$

The angle between the lines depends only upon their directions.
Line [1] has direction cosines $\frac{2}{7}, -\frac{3}{7}, \frac{6}{7}$.

Line [2] has direction cosines $\frac{1}{3}, \frac{2}{3}, \frac{2}{3}$.

The angle θ between the lines is given by

$$\cos \theta = (\tfrac{2}{7})(\tfrac{1}{3}) + (-\tfrac{3}{7})(\tfrac{2}{3}) + (\tfrac{6}{7})(\tfrac{2}{3})$$

$\Rightarrow \qquad\qquad\qquad \theta = \arccos \tfrac{8}{21}.$

6) Find a unit vector which is perpendicular to AB and AC if $\overrightarrow{AB} = i + 2j + 3k$ and $\overrightarrow{AC} = 4i - j + 2k.$

Let $ai + bj + ck$ be a vector perpendicular to both AB and AC.

It is perpendicular to AB so $\quad (ai + bj + ck).(i + 2j + 3k) = 0.$

It is perpendicular to AC so $\quad (ai + bj + ck).(4i - j + 2k) = 0.$

Therefore
$$\begin{cases} a + 2b + 3c = 0 \\ 4a - b + 2c = 0. \end{cases}$$

Eliminating b gives $\qquad\qquad a = -\tfrac{7}{9}c.$

Eliminating a gives $\qquad\qquad b = -\tfrac{10}{9}c.$

Hence $\qquad\qquad ai + bj + ck = -\tfrac{7}{9}ci - \tfrac{10}{9}cj + ck$

$$= \tfrac{1}{9}c(-7i - 10j + 9k).$$

Thus $-7i - 10j + 9k$ is perpendicular to both AB and AC. A unit vector perpendicular to AB and AC is therefore

$$(-7i - 10j + 9k)/\sqrt{230}.$$

EXERCISE 12h

1) Calculate $\mathbf{a}.\mathbf{b}$ if
(a) $\mathbf{a} = 2i - 4j + 5k, \quad \mathbf{b} = i + 3j + 8k,$
(b) $\mathbf{a} = 3i - 7j + 2k, \quad \mathbf{b} = 5i + j - 4k,$
(c) $\mathbf{a} = 2i - 3j + 6k, \quad \mathbf{b} = i + j.$
What conclusion can you draw in (b)?

2) Find $\mathbf{p}.\mathbf{q}$ and the cosine of the angle between $\mathbf{p}$ and $\mathbf{q}$ if
(a) $\mathbf{p} = 2i + 4j + k, \quad \mathbf{q} = i + j + k, \qquad$ (b) $\mathbf{p} = -i + 3j - 2k, \quad \mathbf{q} = i + j - 6k,$
(c) $\mathbf{p} = -2i + 5j. \qquad\qquad \mathbf{q} = i + j.$

3) Simplify:
(a) $(\mathbf{a} - \mathbf{b}).\mathbf{a}$ $\qquad\qquad\qquad$ (b) $(\mathbf{a} - \mathbf{b}).(\mathbf{a} - \mathbf{b})$
(c) $(\mathbf{a} + \mathbf{b}).\mathbf{b} - (\mathbf{a} + \mathbf{b}).\mathbf{a}$ $\quad$ (d) $(\mathbf{a} + \mathbf{b}).\mathbf{c} - (\mathbf{a} - \mathbf{b}).\mathbf{c}$

4) Given that **a** and **b** are perpendicular, simplify:
(a) $\mathbf{a} \cdot \mathbf{b}$ (b) $(\mathbf{a} - \mathbf{b}) \cdot \mathbf{b}$ (c) $(\mathbf{a} + \mathbf{b}) \cdot \mathbf{a}$ (d) $(\mathbf{a} - \mathbf{b}) \cdot (2\mathbf{a} + \mathbf{b})$

5) The angle between two vectors $\mathbf{v}_1$ and $\mathbf{v}_2$ is $\arccos \frac{4}{21}$.
If $\mathbf{v}_1 = 6\mathbf{i} + 3\mathbf{j} - 2\mathbf{k}$ and $\mathbf{v}_2 = -2\mathbf{i} + \lambda\mathbf{j} - 4\mathbf{k}$, find the positive value of λ.

6) If $\mathbf{a} = 3\mathbf{i} + 4\mathbf{j} - \mathbf{k}$, $\mathbf{b} = \mathbf{i} - \mathbf{j} + 3\mathbf{k}$ and $\mathbf{c} = 2\mathbf{i} + \mathbf{j} - 5\mathbf{k}$, find
(a) $\mathbf{a} \cdot \mathbf{b}$ (b) $\mathbf{a} \cdot \mathbf{c}$ (c) $\mathbf{a} \cdot (\mathbf{b} + \mathbf{c})$ (d) $(2\mathbf{a} + 3\mathbf{b}) \cdot \mathbf{c}$ (e) $(\mathbf{a} - \mathbf{b}) \cdot \mathbf{c}$

7) In a triangle ABC, $\overrightarrow{AB} = \mathbf{i} + 2\mathbf{j} + 3\mathbf{k}$ and $\overrightarrow{BC} = -\mathbf{i} + 4\mathbf{j}$. Find the cosine of angle ABC.
Find the vector $\overrightarrow{AC}$ and use it to calculate the angle BAC.

8) A, B and C are points with position vectors **a**, **b** and **c** respectively, relative to the origin O. AB is perpendicular to OC and BC is perpendicular to OA.
Show that AC is perpendicular to OB.

9) Given two vectors **a** and **b** $(\mathbf{a} \neq 0, \ \mathbf{b} \neq 0)$, show that
(a) if $\mathbf{a} + \mathbf{b}$ and $\mathbf{a} - \mathbf{b}$ are perpendicular then $|\mathbf{a}| = |\mathbf{b}|$,
(b) if $|\mathbf{a} + \mathbf{b}| = |\mathbf{a} - \mathbf{b}|$ then **a** and **b** are perpendicular.

10) Three vectors **a**, **b** and **c** are such that $\mathbf{a} \neq \mathbf{b} \neq \mathbf{c} \neq 0$.
(a) If $\mathbf{a} \cdot (\mathbf{b} + \mathbf{c}) = \mathbf{b} \cdot (\mathbf{a} - \mathbf{c})$ prove that $\mathbf{c} \cdot (\mathbf{a} + \mathbf{b}) = 0$.
(b) If $(\mathbf{a} \cdot \mathbf{b})\mathbf{c} = (\mathbf{b} \cdot \mathbf{c})\mathbf{a}$ show that **c** and **a** are parallel.

11) Find the angle between each of the following pairs of lines:
(a) $\mathbf{r}_1 = 3\mathbf{i} + 2\mathbf{j} - 4\mathbf{k} + \lambda(\mathbf{i} + 2\mathbf{j} + 2\mathbf{k})$,
$\mathbf{r}_2 = 5\mathbf{j} - 2\mathbf{k} + \mu(3\mathbf{i} + 2\mathbf{j} + 6\mathbf{k})$.
(b) A line with direction ratios $2:2:1$,
A line joining $(3, 1, 4)$ to $(7, 2, 12)$.
(c) $\dfrac{x + 4}{3} = \dfrac{y - 1}{5} = \dfrac{z + 3}{4}$,

$\dfrac{x + 1}{1} = \dfrac{y - 4}{1} = \dfrac{z - 5}{2}$.

12) Find the angle between the following pairs of lines:
(a) $\dfrac{x - 2}{3} = \dfrac{y + 1}{-2}$, $z = 2$

$\dfrac{x - 1}{1} = \dfrac{2y + 3}{3} = \dfrac{z + 5}{2}$

(b) $\mathbf{r} = 4\mathbf{i} - \mathbf{j} + \lambda(\mathbf{i} + 2\mathbf{j} - 2\mathbf{k})$
$\mathbf{r} = \mathbf{i} - \mathbf{j} + 2\mathbf{k} - \mu(2\mathbf{i} + 4\mathbf{j} - 4\mathbf{k})$.

13) Show that $i + 7j + 3k$ is perpendicular to both $i - j + 2k$ and $2i + j - 3k$.

14) Show that $13i + 23j + 7k$ is perpendicular to both $2i + j - 7k$ and $3i - 2j + k$.

15) Find a unit vector which is perpendicular to a and b if
(a) $a = 6i + j + 3k$, $b = 5i + k$
(b) $a = i - j - k$, $b = 7i - 2j + 3k$.

EQUATION OF A PLANE

A particular plane can be specified in several ways, for example:

(a) one and only one plane can be drawn through three non-collinear points, therefore three given points specify a particular plane;

(b) one and only one plane can be drawn to contain two concurrent lines, therefore two given concurrent lines specify a particular plane;

(c) one and only one plane can be drawn perpendicular to a given direction at a given distance from the origin, therefore the normal to a plane and the distance of the plane from the origin specify a particular plane;

(d) one and only one plane can be drawn through a given point and perpendicular to a given direction, therefore a point on the plane and a normal to the plane specify a particular plane.

There are many other ways of specifying a particular plane but those described in (c) and (d) are particularly suitable for deriving the vector equation of a plane.

The Vector Equation of a Plane

Consider the plane which is at a distance d from the origin and which is perpendicular to the unit vector $\hat{n}$ ($\hat{n}$ being directed *away* from O).

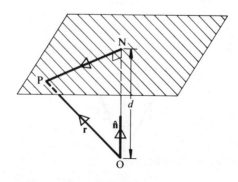

If ON is the perpendicular from the origin to the plane then $\overrightarrow{ON} = d\hat{n}$.
For any point, P, on the plane, NP is perpendicular to ON.
Conversely if P is not on the plane, NP is not perpendicular to ON. So

$$\text{P is on the plane} \iff \overrightarrow{NP}.\overrightarrow{ON} = 0. \qquad [1]$$

This equation is called the scalar product form of the vector equation of the plane.

If $\mathbf{r}$ is the position vector of P, $\overrightarrow{NP} = \mathbf{r} - d\hat{n}$.

Therefore [1] becomes $(\mathbf{r} - d\hat{n}).\, d\hat{n} = 0$

$\Rightarrow$ $\qquad\qquad\qquad \mathbf{r}.\hat{n} - d\hat{n}.\hat{n} = 0$

But $\hat{n}.\hat{n} = 1$

So $\qquad\qquad\qquad\qquad \mathbf{r}.\hat{n} = d \qquad\qquad [2]$

The equation $\mathbf{r}.\hat{n} = d$ is the standard form of the vector equation of the plane, where
$\mathbf{r}$ is the position vector of any point on the plane,
$\hat{n}$ is the unit vector perpendicular to the plane,
d is the distance of the plane from the origin.

This standard form of the vector equation of a plane can be multiplied by any scalar quantity, thus
any equation of the form $\mathbf{r}.\mathbf{n} = D$ represents a plane perpendicular to $\mathbf{n}$.

Converting back to standard form we have

$$\mathbf{r}.\hat{n} = \frac{D}{|\mathbf{n}|}$$

where $\dfrac{D}{|\mathbf{n}|}$ is the distance of the plane from the origin.

Now consider the plane which contains the point A whose position vector is $\mathbf{a}$ and which is perpendicular to the unit vector $\hat{n}$.

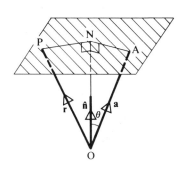

The distance of the plane from the origin is ON and ON $= |a| \cos \theta = \mathbf{a}.\hat{\mathbf{n}}$. So equation [2] becomes $\mathbf{r}.\hat{\mathbf{n}} = \mathbf{a}.\hat{\mathbf{n}}$.

Thus $\mathbf{r}.\hat{\mathbf{n}} = \mathbf{a}.\hat{\mathbf{n}}$ is the vector equation of a plane which is perpendicular to $\hat{\mathbf{n}}$ and which contains the point with position vector $\mathbf{a}$.

The Cartesian Equation of a Plane

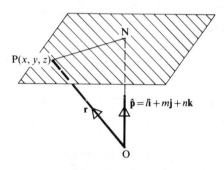

Consider a plane whose vector equation in standard form is

$$\mathbf{r}.\hat{\mathbf{p}} = d \qquad\qquad [1]$$

where $\hat{\mathbf{p}} = l\mathbf{i} + m\mathbf{j} + n\mathbf{k}$.
Now if a point $P(x, y, z)$ is on this plane its position vector, $\mathbf{r} = x\mathbf{i} + y\mathbf{j} + z\mathbf{k}$, satisfies equation [1], so

$$(x\mathbf{i} + y\mathbf{j} + z\mathbf{k}).(l\mathbf{i} + m\mathbf{j} + n\mathbf{k}) = d$$

$\Rightarrow$

$$lx + my + nz = d$$

which is the Cartesian equation of the plane.

Now if each term in this equation is multiplied by a constant we have

$$Ax + By + Cz = D$$

which is equivalent to the vector equation

$$\mathbf{r}.(A\mathbf{i} + B\mathbf{j} + C\mathbf{k}) = D$$

In either of these forms A, B and C are the direction ratios of the normal to the plane, but D has no geometrical significance. It is easy to convert back to standard form however, by dividing by $\sqrt{(A^2 + B^2 + C^2)}$, e.g. a plane whose equation is

$$2x + 3y + 6z = 21 \quad \text{or} \quad \mathbf{r}.(2\mathbf{i} + 3\mathbf{j} + 6\mathbf{k}) = 21$$

can be converted to the standard form

$$\tfrac{2}{7}x + \tfrac{3}{7}y + \tfrac{6}{7}z = 3 \quad \text{or} \quad \mathbf{r}.(\tfrac{2}{7}\mathbf{i} + \tfrac{3}{7}\mathbf{j} + \tfrac{6}{7}\mathbf{k}) = 3$$

Now we can see that the plane is 3 units from the origin and it is perpendicular to a vector with direction cosines $\frac{2}{7}, \frac{3}{7}, \frac{6}{7}$.

The Parametric Form for the Vector Equation of a Plane

Consider the plane which is parallel to the vectors **d** and **e** (**d** not parallel to **e**) and which also contains the point A whose position vector is **a**.

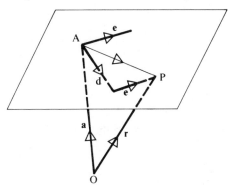

The vectors **d** and **e** determine the orientation of the plane and **a** fixes it in space, so these vectors specify a particular plane.
If P is any point on this plane
$\overrightarrow{AP} = \lambda\mathbf{d} + \mu\mathbf{e}$ where λ and μ are independent parameters.

If **r** is the position vector of P $\mathbf{r} = \mathbf{a} + \overrightarrow{AP}$

$$= \mathbf{a} + \lambda\mathbf{d} + \mu\mathbf{e}.$$

Thus any equation of the form $\mathbf{r} = \mathbf{a} + \lambda\mathbf{d} + \mu\mathbf{e}$,   where λ and μ are independent parameters, represents the plane parallel to the vectors **d** and **e** and containing the point **a**.

It should be noted that this parametric form is not a unique equation for a particular plane. This is because **a** is any one of an infinite number of points on the plane, and **d** and **e** are only one pair of the infinite set of vectors contained in the plane.

An interesting variation of the parametric form is found by considering the plane passing through three given points.

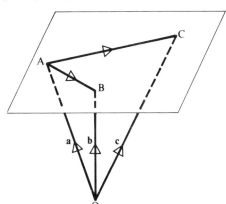

The vector equation of the plane in the diagram can be written

$$\mathbf{r} = \mathbf{a} + t\overrightarrow{AB} + u\overrightarrow{AC}$$

$$= \mathbf{a} + t(\mathbf{b} - \mathbf{a}) + u(\mathbf{c} - \mathbf{a})$$

$$= (1 - t - u)\mathbf{a} + t\mathbf{b} + u\mathbf{c}$$

$$= s\mathbf{a} + t\mathbf{b} + u\mathbf{c}$$

where $s = 1 - t - u$.

Therefore any equation of the form $r = sa + tb + uc$,   where $s + t + u = 1$, represents the plane passing through the points a, b, c.
(**Note:** In this form, s, t and u are *not* independent parameters.)

To summarise:

The *scalar product form* of the vector equation of a plane is

$$r.n = D$$

where n is a vector perpendicular to the plane and $\dfrac{D}{|n|}$ is the distance of the plane from the origin.

The *parametric form* of the vector equation of a plane is

$$r = a + \lambda b + \mu c$$

where a is a point on the plane and b and c are parallel to the plane.
The *Cartesian equation* of a plane is

$$Ax + By + Cz = D$$

where A, B, C are the direction ratios of n, the normal to the plane.

EXAMPLES 12i

1) Find the vector equation of the plane containing the points $A(0, 1, 1)$, $B(2, 1, 0), C(-2, 0, 3)$, (a) in parametric form, (b) in scalar product form.

(a) The parametric equation of this plane is

$$r = \lambda(j + k) + \mu(2i + j) + \eta(-2i + 3k)$$

where $\lambda + \mu + \eta = 1$.
Replacing η by $1 - (\lambda + \mu)$ and simplifying gives

$$r = (-2 + 2\lambda + 4\mu)i + (\lambda + \mu)j + (3 - 2\lambda - 3\mu)k. \qquad [1]$$

(b) If $P(x, y, z)$ is any point on this plane, then from equation [1] we have

$$x = -2 + 2\lambda + 4\mu$$
$$y = \lambda + \mu$$
$$z = 3 - 2\lambda - 3\mu.$$

Eliminating λ and μ gives $x + 2y + 2z = 4$ which is the Cartesian equation of the plane.
Therefore the scalar product form of the equation is

$$r.(i + 2j + 2k) = 4.$$

2) Show that the line L whose vector equation is

$$r = 2i - 2j + 3k + \lambda(i - j + 4k)$$

is parallel to the plane Π whose vector equation is

$$\mathbf{r}.(\mathbf{i} + 5\mathbf{j} + \mathbf{k}) = 5$$

and find the distance between them.

L is parallel to $\mathbf{i} - \mathbf{j} + 4\mathbf{k}$ and Π is normal to $\mathbf{i} + 5\mathbf{j} + \mathbf{k}$.
Now $(\mathbf{i} - \mathbf{j} + 4\mathbf{k}).(\mathbf{i} + 5\mathbf{j} + \mathbf{k}) = 0$.
So L is perpendicular to the normal to Π and hence parallel to Π.

The distance, d_1, of Π from the origin is $\dfrac{5}{\sqrt{27}}$.

As $2\mathbf{i} - 2\mathbf{j} + 3\mathbf{k}$ is a point on L we may write the equation of the plane which is parallel to Π and which contains L as

$$\mathbf{r}.(\mathbf{i} + 5\mathbf{j} + \mathbf{k}) = (2\mathbf{i} - 2\mathbf{j} + 3\mathbf{k}).(\mathbf{i} + 5\mathbf{j} + \mathbf{k})$$

$\Rightarrow$ $\mathbf{r}.(\mathbf{i} + 5\mathbf{j} + \mathbf{k}) = -5$ or $\mathbf{r}.(-\mathbf{i} - 5\mathbf{j} - \mathbf{k}) = 5$.

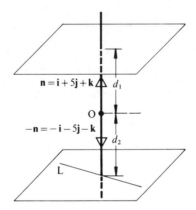

Since the normal vectors to these planes are in opposite directions we deduce that this plane and Π are on opposite sides of the origin.

The distance of this plane from the origin is given by $d_2 = \dfrac{5}{\sqrt{27}}$.

Hence the distance between the two planes, and hence between L and Π is

$$d_1 + d_2 = \frac{10}{\sqrt{27}}.$$

3) Show that the plane whose vector equation is $\mathbf{r}.(\mathbf{i} + 2\mathbf{j} - \mathbf{k}) = 3$ contains the line whose vector equation is $\mathbf{r} = \mathbf{i} + \mathbf{j} + \lambda(2\mathbf{i} + \mathbf{j} + 4\mathbf{k})$.

The line is contained in the plane if any two points on the line are on the plane. Taking $\lambda = 0$ and $\lambda = 1$ we find that $\mathbf{i} + \mathbf{j}$ and $3\mathbf{i} + 2\mathbf{j} + 4\mathbf{k}$ are two points on the line.

If $r = i + j$ then $r.(i + 2j - k) = (i + j).(i + 2j - k) = 3.$
Therefore $i + j$ is a point on the plane.
If $r = 3i + 2j + 4k$ then $r.(i + 2j - k) = (3i + 2j + 4k).(i + 2j - k) = 3.$
Therefore $3i + 2j + 4k$ is a point on the plane.
Therefore the line is contained in the plane.

4) Find the vector equation of line passing through the point $(3, 1, 2)$ and
perpendicular to the plane $r.(2i - j + k) = 4.$ Find also the point of
intersection of this line and the plane.

As $r.(2i - j + k) = 4$ is the equation of the plane, $2i - j + k$ is
perpendicular to the plane and therefore parallel to the required line.
As this line passes through the point $3i + j + 2k$ its equation is

$$r = (3i + j + 2k) + \lambda(2i - j + k).$$

Writing the equations of the line in parametric form

$$x = 3 + 2\lambda, \quad y = 1 - \lambda, \quad z = 2 + \lambda$$

we see that this line meets the plane, where

$$[(3 + 2\lambda)i + (1 - \lambda)j + (2 + \lambda)k].(2i - j + k) = 4$$

$\Rightarrow$ $\qquad\qquad 2(3 + 2\lambda) - (1 - \lambda) + (2 + \lambda) = 4$

$\Rightarrow$ $\qquad\qquad\qquad\qquad\qquad\qquad \lambda = -\frac{1}{2}$

$\Rightarrow$ $\qquad\qquad x = 2, \quad y = \frac{3}{2}, \quad z = \frac{3}{2}.$

So the point of intersection of the line and the plane is $(2, \frac{3}{2}, \frac{3}{2})$.

EXERCISE 12i

1) Find a vector equation of the plane containing the points A, B and C in
parametric form and in scalar product form where
(a) A, B and C are the points $(1, 2, -1), (1, 3, 2), (0, 2, 1)$ respectively,
(b) the position vectors of A, B and C are $i + j - 2k, \ i + k, \ -2i + j - 3k$.

2) Find the vector equation of the following planes in scalar product form:
(a) $r = i - j + \lambda(i + j + k) + \mu(i - 2j + 3k),$
(b) $r = 2i - k + \lambda(i) + \mu(i - 2j - k),$
(c) $r = (1 + s - t)i + (2 - s)j + (3 - 2s + 2t)k.$

3) Find the Cartesian equations of the following planes:
(a) $r.(i + j - k) = 2,$
(b) $r.(2i + 3j - 4k) = 1,$
(c) $r = (s - 2t)i + (3 - t)j + (2s + t)k.$

4) Find the vector equation in scalar product form of the plane that contains
the lines $r = (i + j) + s(i + 2j - k)$ and $r = (i + j) + t(-i + j - 2k).$

5) Find the vector equation in parametric form of the plane that contains the lines $r = -3i - 2j + t(i - 2j + k)$, $r = i - 11j + 4k + s(2i - j + 2k)$.

6) Find the vector equation in parametric form of the plane that goes through the point with position vector $i + j$ and which is parallel to the lines $r_1 = 2i - j + \lambda(i + k)$ and $r_2 = 2j - k + \mu(i - j + k)$.   Is either of these lines contained in the plane?

7) A plane goes through the three points whose position vectors are a, b and c

where$$a = i + j + 2k$$

$$b = 2i - j + 3k$$

$$c = -i + 2j - 2k.$$

Find the vector equation of this plane in scalar product form and hence find the distance of the plane from the origin.

8) A plane goes through the points whose position vectors are $i - 2j + k$ and $2i - j - k$ and is parallel to the line $r = i - j + \lambda(3i + j - 2k)$.   Find the distance of this plane from the origin.

9) Two planes Π_1 and Π_2 have vector equations $r.(2i + j - 2k) = 3$ and $r.(2i + j - 2k) = 9$.   Explain why Π_1 and Π_2 are parallel and hence find the distance between them.

10) Find the vector equation of the line through the origin which is perpendicular to the plane $r.(i - 2j + k) = 3$.

11) Find the vector equation of the line through the point $(2, 1, 1)$ which is perpendicular to the plane $r.(i + 2j - 3k) = 6$.

12) Find the vector equation of the plane which goes through the point $(0, 1, 6)$ and is parallel to the plane $r.(i - 2j) = 3$.

13) Find the vector equation of the plane which goes through the origin and which contains the line $r = 2i + \lambda(j + k)$.

14) Find the point of intersection of the line $r = (i + j - 2k) + \lambda(i - j + k)$ and the plane $r.(i + 2j - k) = 2$.

15) Find the point of intersection of the line $x - 2 = 2y + 1 = 3 - z$ and the plane $x + 2y + z = 3$.

16) Show that the line $x + 1 = y = \dfrac{z - 3}{2}$ is parallel to the plane

$r.(i + j - k) = 3$ and find the distance between them.

17) Determine whether the given lines are parallel to, contained in, or intersect the plane $r.(2i + j - 3k) = 5$:
(a) $r = 3i - j + k + \lambda(-2i + j - 3k)$,

(b) $\mathbf{r} = \mathbf{i} - \mathbf{j} + \mu(2\mathbf{i} + \mathbf{j} - 3\mathbf{k})$,
(c) $x = y = z$,
(d) $\mathbf{r} = (2\mathbf{i} + \mathbf{j}) + s(3\mathbf{i} + 2\mathbf{k})$.

THE ANGLE BETWEEN TWO PLANES

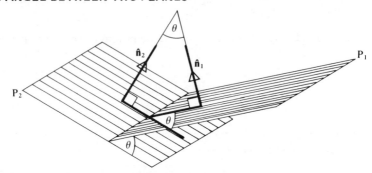

Consider two planes P_1 and P_2 whose vector equations are

$$\mathbf{r}.\hat{\mathbf{n}}_1 = d_1 \quad \text{and} \quad \mathbf{r}.\hat{\mathbf{n}}_2 = d_2.$$

The angle between P_1 and P_2 is equal to the angle between the normals to P_1 and P_2, i.e. the angle between $\hat{\mathbf{n}}_1$ and $\hat{\mathbf{n}}_2$.
Therefore if θ is the angle between P_1 and P_2,

$$\cos\theta = \hat{\mathbf{n}}_1.\hat{\mathbf{n}}_2 \qquad [1]$$

e.g. the angle between the planes whose vector equations are
$\mathbf{r}.(\mathbf{i} + \mathbf{j} - 2\mathbf{k}) = 3$ and $\mathbf{r}.(2\mathbf{i} - 2\mathbf{j} + \mathbf{k}) = 2$ is given by

$$\cos\theta = \frac{(\mathbf{i} + \mathbf{j} - 2\mathbf{k})}{\sqrt{6}}.\frac{(2\mathbf{i} - 2\mathbf{j} + \mathbf{k})}{3} = -\frac{\sqrt{6}}{9}$$

This is the cosine of the obtuse angle between the planes.
The acute angle is $\arccos \sqrt{6}/9$.

Two planes are perpendicular if $\hat{\mathbf{n}}_1.\hat{\mathbf{n}}_2 = 0$
and two planes are parallel if $\hat{\mathbf{n}}_1 = \pm\hat{\mathbf{n}}_2$.

THE ANGLE BETWEEN A LINE AND A PLANE

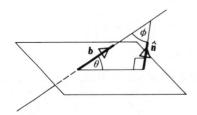

Consider the line $\mathbf{r} = \mathbf{a} + \lambda\mathbf{b}$ and the plane $\mathbf{r}.\hat{\mathbf{n}} = d$.
The angle ϕ between the line and the normal to the plane is given by

$$\cos\phi = \frac{\mathbf{b}.\hat{\mathbf{n}}}{|\mathbf{b}|}.$$

If θ is the angle between the line and the plane then $\quad \theta = \dfrac{\pi}{2} - \phi$

so $\qquad\qquad\qquad\qquad\qquad \sin \theta = \cos \phi.$

Therefore the angle between the line $\quad \mathbf{r} = \mathbf{a} + \lambda \mathbf{b} \quad$ and the plane $\quad \mathbf{r}.\hat{\mathbf{n}} = d$ is given by

$$\sin \theta = \frac{\mathbf{b}.\hat{\mathbf{n}}}{|\mathbf{b}|}$$

e.g. the angle θ between the line $\quad \mathbf{r} = (\mathbf{i} + 2\mathbf{j} - \mathbf{k}) + \lambda(\mathbf{i} - \mathbf{j} + \mathbf{k}) \quad$ and the plane $\quad \mathbf{r}.(2\mathbf{i} - \mathbf{j} + \mathbf{k}) = 4 \quad$ is given by

$$\sin \theta = \frac{(\mathbf{i} - \mathbf{j} + \mathbf{k})}{\sqrt{3}}.\frac{(2\mathbf{i} - \mathbf{j} + \mathbf{k})}{\sqrt{6}} = \frac{2\sqrt{2}}{3}$$

Therefore $\qquad\qquad\qquad \theta = \arcsin \dfrac{2\sqrt{2}}{3}$

THE DISTANCE OF A POINT FROM A PLANE

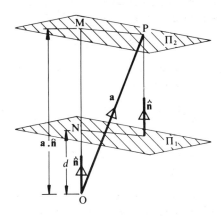

Consider a point P with position vector $\mathbf{a}$ and a plane Π_1 whose equation is

$$\mathbf{r}.\hat{\mathbf{n}} = d.$$

The equation of the plane Π_2 through P parallel to the plane Π_1 is

$$\mathbf{r}.\hat{\mathbf{n}} = \mathbf{a}.\hat{\mathbf{n}},$$

i.e. the distance OM of this plane from the origin is $\mathbf{a}.\hat{\mathbf{n}}$.
Therefore (assuming that P and O are on opposite sides of Π_1) the distance MN of P from the plane Π_1 is $\mathbf{a}.\hat{\mathbf{n}} - d$.

(If P and O are on the same side of the plane the use of this formula will give a negative result.)

Thus the distance of the point with position vector $i - 3j + 3k$ from the plane Π with equation $r.(2i + 3j - 6k) = 9$ is given by

$$\frac{(i - 3j + 3k).(2i + 3j - 6k)}{7} - \frac{9}{7} = -\frac{25}{7} - \frac{9}{7} = -\frac{34}{7}$$

The negative sign indicates that the point and the origin are on the same side of the plane.

THE INTERSECTION OF TWO PLANES

Unless two planes are parallel they will contain a common line which is the line of intersection of the two planes.

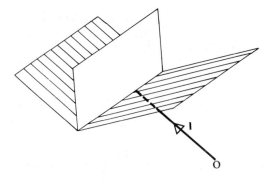

If Π_1 and Π_2 are planes with equations $r.\hat{n}_1 = d_1$ and $r.\hat{n}_2 = d_2$ respectively, the position vector of any point on the line of intersection must satisfy both equations.

If l is the position vector of a point on this line then

$l.\hat{n}_1 = d_1$ and $l.\hat{n}_2 = d_2$.

Therefore, for any value of k,

$$l.\hat{n}_1 - kl.\hat{n}_2 = d_1 - kd_2$$

or

$$l.(\hat{n}_1 - k\hat{n}_2) = d_1 - kd_2.$$

But the equation $r.(\hat{n}_1 - k\hat{n}_2) = d_1 - kd_2$ represents a plane Π_3.

It is also satisfied by the position vector, l, of any point which is on both of the planes Π_1 and Π_2. So

any plane passing through the intersection of the planes $r.\hat{n}_1 = d_1$ and $r.\hat{n}_2 = d_2$ has an equation

$$r.(\hat{n}_1 - k\hat{n}_2) = d_1 - kd_2.$$

Conversely, for all real values of k the equation $\mathbf{r}.(\hat{\mathbf{n}}_1 - k\hat{\mathbf{n}}_2) = d_1 - kd_2$
represents the family of planes passing through the line of intersection of the
planes $\mathbf{r}.\hat{\mathbf{n}}_1 = d_1$ and $\mathbf{r}.\hat{\mathbf{n}}_2 = d_2$.
This is a particular case of a more general fact which is,
if $E_1 = 0$ and $E_2 = 0$ are the equations of two members of a family of
curves (or surfaces) then the equation

$$E_1 = kE_2$$

represents, for all real values of k, those members of that family that contain
the point (or points) of intersection of E_1 and E_2.

The Line of Intersection of Two Planes

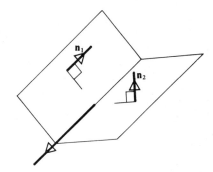

As the line of intersection of two
planes

$$\mathbf{r}.\mathbf{n}_1 = D_1$$

$$\mathbf{r}.\mathbf{n}_2 = D_2$$

is contained in both planes it is
perpendicular to both $\mathbf{n}_1$ and $\mathbf{n}_2$.

To find the equation of the line of intersection of two planes, consider, as an
example, the planes

$$\left.\begin{array}{l} \mathbf{r}.(\mathbf{i}+\mathbf{j}-3\mathbf{k}) = 6 \\ \mathbf{r}.(2\mathbf{i}-\mathbf{j}+\mathbf{k}) = 4 \end{array}\right\} \Rightarrow \left\{\begin{array}{ll} x+y-3z = 6 & \text{[1]} \\ 2x-y+z = 4. & \text{[2]} \end{array}\right.$$

These planes meet where $3x - 2z = 10$ [1] + [2]

and $7x - 2y = 18,$ [1] + 3[2]

i.e. where $x = \dfrac{2z + 10}{3} = \dfrac{2y + 18}{7} = \lambda.$

These are the Cartesian equations of the line.

So any point on the line has coordinates

$$x = \lambda, \quad y = \frac{7\lambda - 18}{2}, \quad z = \frac{3\lambda - 10}{2}$$

or $x = 2s, \quad y = 7s - 9 \qquad z = 3s - 5 \quad (\lambda = 2s).$

Hence the position vector of any point on the line is

$$\mathbf{r} = -9\mathbf{j} - 5\mathbf{k} + s(2\mathbf{i} + 7\mathbf{j} + 3\mathbf{k}).$$

This is the vector equation of the line.

If the equations of the planes are in parametric form it is not necessary to convert these equations into Cartesian form. The line of intersection may be found as follows.

Consider the planes

$$\mathbf{r} = \mathbf{i} + \mathbf{j} + \lambda(2\mathbf{i} - \mathbf{k}) + \mu(\mathbf{i} - \mathbf{j} + \mathbf{k}) \qquad\qquad [1]$$

$$\mathbf{r} = 3\mathbf{i} - \mathbf{k} + s(\mathbf{i} - \mathbf{j} + 2\mathbf{k}) + t(\mathbf{i} + 2\mathbf{j} - \mathbf{k}). \qquad\qquad [2]$$

Rearranging gives

$$\mathbf{r} = (1 + 2\lambda + \mu)\mathbf{i} + (1 - \mu)\mathbf{j} + (-\lambda + \mu)\mathbf{k}$$

$$\mathbf{r} = (3 + s + t)\mathbf{i} + (-s + 2t)\mathbf{j} + (-1 + 2s - t)\mathbf{k}.$$

These planes meet where

$$1 + 2\lambda + \mu = 3 + s + t$$

and

$$1 - \mu = -s + 2t$$

and

$$-\lambda + \mu = -1 + 2s - t.$$

Eliminating λ and μ from these equations gives $\quad 3 = 2s + 5t$,

i.e. the planes meet at all points where $\quad s = \dfrac{3 - 5t}{2}.$

Therefore substituting $\dfrac{3 - 5t}{2}$ for s in equation [2] gives the vector equation of the line of intersection of the planes,

i.e. $\qquad\qquad \mathbf{r} = \tfrac{1}{2}[(9\mathbf{i} - 3\mathbf{j} + 4\mathbf{k}) + t(-3\mathbf{i} + 9\mathbf{j} - 12\mathbf{k})].$

EXERCISE 12j

1) Find the cosine of the angle between the two planes whose equations are
(a) $\mathbf{r}.(\mathbf{i} - \mathbf{j} + 3\mathbf{k}) = 3, \quad \mathbf{r}.(2\mathbf{i} - \mathbf{j} + 2\mathbf{k}) = 5$;
(b) $\mathbf{r} = (\mathbf{i} + \mathbf{j}) + \lambda(\mathbf{i} + \mathbf{j} - \mathbf{k}) + \mu(2\mathbf{i} - \mathbf{j} + 3\mathbf{k})$,
$\quad \mathbf{r} = (\mathbf{i} - 2\mathbf{j} + \mathbf{k}) + s(2\mathbf{i} + \mathbf{k}) + t(\mathbf{i} - 2\mathbf{j} - \mathbf{k})$;
(c) $2x + 2y - 3z = 3, \quad x + 3y - 4z = 6.$

2) Find the sine of the angle between the line and plane whose equations are
(a) $\mathbf{r} = \mathbf{i} - \mathbf{j} + \lambda(\mathbf{i} + \mathbf{j} + \mathbf{k}), \quad \mathbf{r}.(\mathbf{i} - 2\mathbf{j} + 2\mathbf{k}) = 4$;
(b) $\mathbf{r} = \mathbf{i} - 2\mathbf{j} + \mathbf{k} + \lambda(2\mathbf{i} - \mathbf{j}), \quad \mathbf{r} = \mathbf{i} - \mathbf{j} + s(\mathbf{i} + \mathbf{k}) + t(\mathbf{j} - \mathbf{k})$;

(c) $\dfrac{x-2}{2} = \dfrac{y+1}{6} = \dfrac{z+3}{3},$ $2x - y - 2z = 4.$

3) Find the distance of the point $(1, 3, 2)$ from the following planes:
(a) $\mathbf{r}.(7\mathbf{i} + 4\mathbf{j} + 4\mathbf{k}) = 9$;
(b) $6x + 6y + 3z = 8$;
(c) $\mathbf{r} = \mathbf{i} - \mathbf{j} + \mathbf{k} + \lambda(\mathbf{i}) + \mu(\mathbf{j} - 2\mathbf{k}).$

4) Find the vector equation of the line of intersection of the following pairs of planes :
(a) $\mathbf{r}.(\mathbf{i} - 2\mathbf{j} + \mathbf{k}) = 3,$ $\mathbf{r}.(3\mathbf{i} + \mathbf{j} - 2\mathbf{k}) = 4$;
(b) $\mathbf{r} = (1 - \lambda + \mu)\mathbf{i} + (\lambda - \mu)\mathbf{j} + (2 - \mu)\mathbf{k},$
   $\mathbf{r} = (2 - s)\mathbf{i} + (1 - 3t)\mathbf{j} + (2s - 3t)\mathbf{k}.$

5) Prove that the line $\mathbf{r} = \mathbf{i} - 2\mathbf{j} + \lambda(\mathbf{i} - 3\mathbf{j} - \mathbf{k})$ is parallel to the intersection of the planes $\mathbf{r}.(\mathbf{i} + \mathbf{j} - 2\mathbf{k}) = 2$ and $\mathbf{r}.(2\mathbf{i} + \mathbf{j} - \mathbf{k}) = 0.$

The results given in this Chapter contain several formulae which must be used with caution. The use of a formula is not always the simplest method of solving a geometric problem. The particular information provided in that problem, illustrated on a diagram, should be considered first so that full use can be made of special properties.

EXAMPLES 12k

1) Show that the points $P(3, 0, 1), Q(2, 1, -2)$ lie on opposite sides of the plane Π whose equation is $\mathbf{r}.(2\mathbf{i} - \mathbf{j} + \mathbf{k}) = 3.$ Find the coordinates of the point of intersection of the plane Π and the line PQ.

Rewriting the equation of Π in the form $\mathbf{r}.\hat{\mathbf{n}} = d$ gives

$$\mathbf{r}.\frac{1}{\sqrt{6}}(2\mathbf{i} - \mathbf{j} + \mathbf{k}) = \frac{3}{\sqrt{6}}.$$

The distance of P from Π is

$$\overrightarrow{OP}.\hat{\mathbf{n}} - d = (3\mathbf{i} + \mathbf{k}).\frac{1}{\sqrt{6}}(2\mathbf{i} - \mathbf{j} + \mathbf{k}) - \frac{3}{\sqrt{6}} = \frac{4}{\sqrt{6}},$$

i.e. P and O are on opposite sides of Π.
The distance of Q from Π is

$$\overrightarrow{OQ}.\hat{\mathbf{n}} - d = (2\mathbf{i} + \mathbf{j} - 2\mathbf{k}).\frac{1}{\sqrt{6}}(2\mathbf{i} - \mathbf{j} + \mathbf{k}) - \frac{3}{\sqrt{6}} = -\frac{2}{\sqrt{6}},$$

i.e. Q and O are on the same side of Π.
Therefore P and Q are on opposite sides of Π.

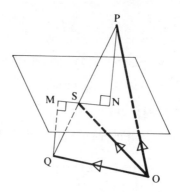

If S is the point of intersection of PQ and Π,   PS:SQ = PN:MQ = 2:1.
Therefore the position vector of S is given by $\frac{1}{3}(\overrightarrow{OP} + 2\overrightarrow{OQ})$.
Therefore the coordinates of S are $(\frac{7}{3}, \frac{2}{3}, -1)$.
(Alternatively $\overrightarrow{OS}$ can be found by solving simultaneously the equation of Π
and the equation of PQ, but this method is longer.)

2) A right circular cone has its vertex at the point $(2, 1, 3)$ and the centre of
its plane face at the point $(1, -1, 2)$. A generator of the cone has equation
$\mathbf{r} = (2\mathbf{i} + \mathbf{j} + 3\mathbf{k}) + \lambda(\mathbf{i} - \mathbf{j} - \mathbf{k})$.   Find the radius of the base of the cone and
hence its volume.

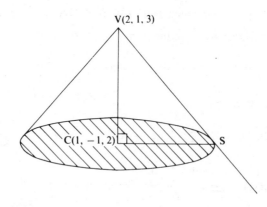

To find the radius of the base we need to find S, the point where the generator
VS meets the plane Π containing the base.
The equation of the plane Π is found as follows.

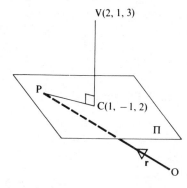

V(2, 1, 3)

C(1, −1, 2)

Π

r

O

P

If **r** is the position vector of any point P in the plane Π then $\overrightarrow{PC} \cdot \overrightarrow{VC} = 0$,

i.e. $[\mathbf{r} - (\mathbf{i} - \mathbf{j} + 2\mathbf{k})] \cdot (\mathbf{i} + 2\mathbf{j} + \mathbf{k}) = 0$

$\Rightarrow \qquad \mathbf{r} \cdot (\mathbf{i} + 2\mathbf{j} + \mathbf{k}) = 1.$

Any point on the given generator has coordinates $[(2 + \lambda), (1 - \lambda), (3 - \lambda)]$.
The coordinates of S also satisfy the equation of Π. So at S

$[(2 + \lambda)\mathbf{i} + (1 - \lambda)\mathbf{j} + (3 - \lambda)\mathbf{k}] \cdot (\mathbf{i} + 2\mathbf{j} + \mathbf{k}) = 1 \quad \Rightarrow \quad \lambda = 3.$

Therefore the coordinates of S are $(5, -2, 0)$.
The radius of the base is CS, where

$CS = \sqrt{[(1 - 5)^2 + (-1 + 2)^2 + (2 - 0)^2]} = \sqrt{21}.$

The volume of the cone is $\frac{1}{3}\pi r^2 h = \frac{1}{3}\pi(21)\sqrt{(1^2 + 2^2 + 1^2)}$
$$= 7\pi\sqrt{6}.$$

3) A tetrahedron has one vertex at O and the other vertices at the points A(2, 0, 0), B(0, 3, 0), C(0, 0, 1). Find the volume of the tetrahedron. This tetrahedron is divided into two parts by a plane Π which contains the line OB and which is inclined at 60° to the face OAB. Find the vector equation of Π in scalar product form and the ratio in which it divides the volume of the tetrahedron OABC.

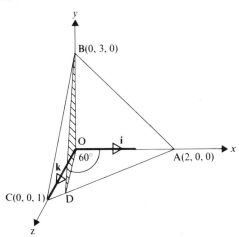

y

B(0, 3, 0)

O   i

60°

k

C(0, 0, 1)   D

A(2, 0, 0)   x

z

The volume of OABC $= (\frac{1}{3}$ area $\triangle$OAC$)($OB$)$

$$= \tfrac{1}{3}(1)(3) = 1.$$

The plane Π contains

the line OB (which is parallel to **j**)

the line OD (which is parallel to $\cos 60°\mathbf{i} + \cos 30°\mathbf{k}$, or $\mathbf{i} + \sqrt{3}\mathbf{k}$)

and the origin.

Therefore the equation of Π in parametric form is

$$\mathbf{r} = \lambda\mathbf{j} + \mu(\mathbf{i} + \sqrt{3}\mathbf{k}).$$

If (x, y, z) is any point on Π:

$$\left. \begin{array}{l} x = \mu \\ y = \lambda \\ z = \sqrt{3}\mu \end{array} \right\} \;\Rightarrow\; \sqrt{3}x - z = 0.$$

Therefore the equation of Π in scalar product form is

$$\mathbf{r}.(\sqrt{3}\mathbf{i} - \mathbf{k}) = 0.$$

If D is the point of intersection of Π and AC, OBCD and OABD are tetrahedrons with a common base OBD; therefore their volumes are in the ratio of the distances of C and A from the plane containing OBD (i.e. Π).

The distance of C from Π is $\;|\mathbf{k}.\frac{1}{2}(\sqrt{3}\mathbf{i} - \mathbf{k}) - 0| = \frac{1}{2}.$

The distance of A from Π is $\;|2\mathbf{i}.\frac{1}{2}(\sqrt{3}\mathbf{i} - \mathbf{k}) - 0| = \sqrt{3}.$

Therefore Π divides the volume of OABC in the ratio

$$\sqrt{3}:\tfrac{1}{2} = 2\sqrt{3}:1.$$

4) Find the reflection P$'$ of the point P whose position vector is $2\mathbf{i} + \mathbf{j} - 2\mathbf{k}$ in the plane $\;\mathbf{r}.(\mathbf{i} - \mathbf{j} + 4\mathbf{k}) = 0.$

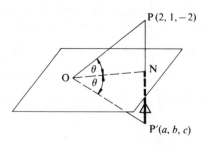

The given plane, Π, $\;\mathbf{r}.(\mathbf{i} - \mathbf{j} + 4\mathbf{k}) = 0,\;$ contains the origin.

If P$'$ is the reflection of P$(2, 1, -2)$ in the plane, then PP$'$ is perpendicular to the plane and hence parallel to $\mathbf{i} - \mathbf{j} + 4\mathbf{k}$.

So the equation of PP′ is

$$\mathbf{r} = 2\mathbf{i} + \mathbf{j} - 2\mathbf{k} + \lambda(\mathbf{i} - \mathbf{j} + 4\mathbf{k})$$
$$= (2 + \lambda)\mathbf{i} + (1 - \lambda)\mathbf{j} + (4\lambda - 2)\mathbf{k}.$$

PP′ cuts the plane at N, where

$$[(2 + \lambda)\mathbf{i} + (1 - \lambda)\mathbf{j} + (4\lambda - 2)\mathbf{k}] \cdot [\mathbf{i} - \mathbf{j} + 4\mathbf{k}] = 0$$

$$\Rightarrow \qquad\qquad\qquad\qquad \lambda = \tfrac{7}{18}$$

Hence N is the point $(\tfrac{43}{18}, \tfrac{11}{18}, -\tfrac{8}{18})$
As N is the midpoint of PP′, P′ is the point $(\tfrac{25}{9}, \tfrac{2}{9}, \tfrac{10}{9})$

5) Π is the plane whose equation is $\mathbf{r} \cdot (\mathbf{i} + \mathbf{j} - 3\mathbf{k}) = 5$.
Find the radius of the circle of intersection of the plane Π and the sphere
whose centre is $(1, 1, 1)$ and whose radius is 2.

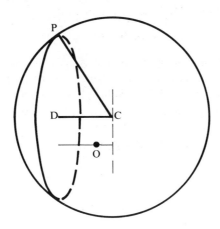

Let C be the centre of the sphere and D be the centre of the circle of
intersection. The equation of the plane through C parallel to Π is

$$\mathbf{r} \cdot (\mathbf{i} + \mathbf{j} - 3\mathbf{k}) = (\mathbf{i} + \mathbf{j} + \mathbf{k}) \cdot (\mathbf{i} + \mathbf{j} - 3\mathbf{k}) = -1$$

i.e. $\qquad\qquad\qquad\qquad \mathbf{r} \cdot (-\mathbf{i} - \mathbf{j} + 3\mathbf{k}) = 1$

Hence Π and this plane are on opposite sides of O and the distance between Π
and this plane is

$$\frac{5}{\sqrt{11}} + \frac{1}{\sqrt{11}} = \frac{6}{\sqrt{11}}$$

If P is any point on the circumference of the circle of intersection then
$$PC = 2 \quad \text{and} \quad \angle PDC = 90°.$$

From Pythagoras' Theorem

$$PD^2 = PC^2 - DC^2$$

$$= 4 - \frac{36}{11} = \frac{8}{11}$$

Hence the radius of the circle is given by $PD = 2\sqrt{\frac{2}{11}}$.

6) Show that the vectors $\mathbf{a} = \mathbf{i} - \mathbf{j} + 2\mathbf{k}$, $\mathbf{b} = \mathbf{i} + \mathbf{j} + \mathbf{k}$, $\mathbf{c} = 2\mathbf{i} - \mathbf{j} + \mathbf{k}$ form a set of base vectors for three dimensions. Express the vector $\mathbf{d} = 3\mathbf{i} - 2\mathbf{j} + 4\mathbf{k}$ in terms of $\mathbf{a}$, $\mathbf{b}$ and $\mathbf{c}$.

$\mathbf{a}$, $\mathbf{b}$ and $\mathbf{c}$ form a set of base vectors for three dimensions provided
(a) they are not parallel (which from their direction ratios they clearly are not),
(b) they are not coplanar.

We can show that $\mathbf{a}$, $\mathbf{b}$ and $\mathbf{c}$ are not coplanar by considering the angles between $\mathbf{a}$ and $\mathbf{b}$, $\mathbf{b}$ and $\mathbf{c}$, $\mathbf{a}$ and $\mathbf{c}$.

If θ is the angle between $\mathbf{a}$ and $\mathbf{b}$ then

$$\mathbf{a.b} = (\mathbf{i} - \mathbf{j} + 2\mathbf{k}).(\mathbf{i} + \mathbf{j} + \mathbf{k}) = 2$$

$\Rightarrow$ $\qquad \cos\theta = \dfrac{2}{\sqrt{18}} \quad \Rightarrow \quad \theta = 61.87°$

If ϕ is the angle between $\mathbf{b}$ and $\mathbf{c}$ then

$$\mathbf{b.c} = (\mathbf{i} + \mathbf{j} + \mathbf{k}) . (2\mathbf{i} - \mathbf{j} + \mathbf{k}) = 2$$

$\Rightarrow$ $\qquad \cos\phi = \dfrac{2}{\sqrt{18}} \quad \Rightarrow \quad \phi = 61.87°$

If ψ is the angle between $\mathbf{a}$ and $\mathbf{c}$ then

$$\mathbf{a.c} = (\mathbf{i} - \mathbf{j} + 2\mathbf{k}) . (2\mathbf{i} - \mathbf{j} + \mathbf{k}) = 5$$

$\Rightarrow$ $\qquad \cos\psi = \dfrac{5}{6} \quad \Rightarrow \quad \psi = 33.56°$

As $\theta + \phi \neq \psi$, it follows that $\mathbf{a}$, $\mathbf{b}$ and $\mathbf{c}$ are not coplanar.

So $\mathbf{a}$, $\mathbf{b}$ and $\mathbf{c}$ *do* form a set of base vectors for three dimensions. Now any other vector $\mathbf{d}$ can be expressed as $\lambda\mathbf{a} + \mu\mathbf{b} + \eta\mathbf{c}$.
Hence $3\mathbf{i} - 2\mathbf{j} + 4\mathbf{k} = \lambda(\mathbf{i} - \mathbf{j} + 2\mathbf{k}) + \mu(\mathbf{i} + \mathbf{j} + \mathbf{k}) + \eta(2\mathbf{i} - \mathbf{j} + \mathbf{k})$

$$= (\lambda + \mu + 2\eta)\mathbf{i} + (-\lambda + \mu - \eta)\mathbf{j} + (2\lambda + \mu + \eta)\mathbf{k}$$

$\Rightarrow \quad \left. \begin{array}{l} \lambda + \mu + 2\eta = 3 \\ -\lambda + \mu - \eta = -2 \\ 2\lambda + \mu + \eta = 4 \end{array} \right\} \Rightarrow \lambda = \frac{8}{5}, \ \mu = \frac{1}{5}, \ \eta = \frac{3}{5}$

Hence $\mathbf{d} = \frac{8}{5}\mathbf{a} + \frac{1}{5}\mathbf{b} + \frac{3}{5}\mathbf{c}$.

EXERCISE 12k

1) A tetrahedron has one vertex at O and the other vertices at the points A(1, 3, 2), B(1, − 1, 0), C(2, 3, 1). Find the distance of O from the face ABC.

2) Show that the points P(3, 2, − 2), Q(1, 2, 1) are on opposite sides of the plane $\mathbf{r} \cdot (\mathbf{i} - \mathbf{j} - \mathbf{k}) = 2$. Find the position vector of the point of intersection of the line PQ with the plane.

3) A tetrahedron has vertices at the points A(2, − 1, 0), B(3, 0, 1), C(1, − 1, 2), D(− 1, 3, 0). Find the cosine of the angle between the faces ABC and ABD.

4) OABC is one face of a cube, where A and C are the points (1, 4, − 1), (3, 0, 3) respectively. Find the coordinates of B. Find also the vector equation of the plane containing the other face of the cube of which AB is an edge.

5) A tetrahedron is bounded by the planes $\mathbf{r} \cdot \mathbf{i} = 0$, $\mathbf{r} \cdot \mathbf{j} = 0$, $\mathbf{r} \cdot \mathbf{k} = 0$, $\mathbf{r} \cdot (2\mathbf{i} - \mathbf{j} + \mathbf{k}) = 4$. Find the coordinates of the vertices and the volume of this tetrahedron.

6) A right circular cylinder has its plane faces contained in the planes $\mathbf{r} \cdot (2\mathbf{i} - \mathbf{j} + 2\mathbf{k}) = 10$, $\mathbf{r} \cdot (- 2\mathbf{i} + \mathbf{j} - 2\mathbf{k}) = 6$. Find the height of the cylinder. The lines $\mathbf{r} = (2\mathbf{i} + \mathbf{j}) + \lambda(2\mathbf{i} - \mathbf{j} + 2\mathbf{k})$, $\mathbf{r} = (\mathbf{i} - \mathbf{j} + \mathbf{k}) + \mu(2\mathbf{i} - \mathbf{j} + 2\mathbf{k})$ are generators of the curved surface of the cylinder, passing through opposite ends of a diameter of its plane face. Find the radius of the cylinder and hence its volume.

7) Find the radius of the circle in which the plane $\mathbf{r} \cdot (2\mathbf{i} + \mathbf{j} - 2\mathbf{k}) = 9$ cuts the sphere of radius 5 and centre the origin. Find the volume of the cone of which this circle is the base and whose vertex is the origin.

8) Show that the point C(2, 0, 1) lies in the plane Π whose equation is $\mathbf{r} \cdot (\mathbf{i} - 2\mathbf{j} + 2\mathbf{k}) = 4$.
A right circular cone has its plane face lying in the plane Π and its centre is at C. Find the vector equation of the axis of the cone.
The line $\mathbf{r} = 4\mathbf{i} - 3\mathbf{j} + 5\mathbf{k} + \lambda(\mathbf{i} + \mathbf{j} + 2\mathbf{k})$ is a generator of the cone. Find the coordinates of the vertex of the cone and the point where this generator meets the plane Π. Hence find the volume of the cone.

9) A tetrahedron has three of its vertices at the points A(3, 2, 0), B(1, 3, − 1), C(0, 2, 0). Find the unit vector perpendicular to the face ABC. The fourth vertex D is such that $\overrightarrow{DA} \cdot \overrightarrow{AB} = \overrightarrow{DA} \cdot \overrightarrow{AC} = 0$. Find the vector equation of AD.
If the volume of the tetrahedron is $3\sqrt{2}$ cubic units and if D is on the same side of the face ABC as the origin, find the coordinates of D.

10) Show that the lines

$$r = (i + j) + \lambda(3i - j + 5k)$$
$$r = (3i + 2j + 5k) + \mu(i - 2j)$$
$$r = (2i - j) + \eta(2i + j + 5k)$$

are coplanar and find in parametric form the vector equation of the plane containing them.

11) A right prism has a triangular cross-section. Two of its rectangular faces are contained in the planes

$$r.(i - 2j) = 0,$$
$$r.(3i - j + k) = 4.$$

The two edges of the prism which are parallel to the intersection of these two planes pass through the origin and the point $(1, 2, -1)$ respectively. Find the vector equations of these edges.

Find also the equation of the plane which contains the cross-section of this prism, one of whose vertices is the origin. Find the area of this cross-section.

THE USE OF VECTORS IN GENERAL GEOMETRIC PROBLEMS

Many of the theorems of Euclidean geometry can be proved easily and quickly using vector methods, as illustrated in the following examples.

EXAMPLES 12I

1) Prove that the diagonals of a parallelogram bisect each other.

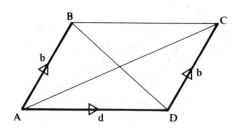

Taking one vertex A as origin and the position vectors of B and D as **b** and **d** respectively gives

$$\overrightarrow{AC} = \overrightarrow{AD} + \overrightarrow{DC} = d + b.$$

The position vector of the midpoint of BD is $\frac{1}{2}(b + d) = \frac{1}{2}\overrightarrow{AC}$.
Therefore the diagonals bisect each other.

2) Prove that the altitudes of any triangle are concurrent.

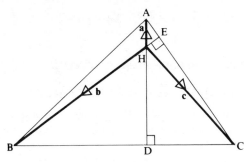

Let the altitudes AD and BE intersect at H.

Taking H as origin, let the position vectors of A, B, C be **a, b, c** respectively.

Now $\overrightarrow{HA}.\overrightarrow{BC} = 0$ i.e. $\mathbf{a}.(\mathbf{c}-\mathbf{b}) = 0$ [1]

and $\overrightarrow{HB}.\overrightarrow{AC} = 0$ i.e. $\mathbf{b}.(\mathbf{c}-\mathbf{a}) = 0$ [2]

$\Rightarrow$ $(\mathbf{a}.\mathbf{c}-\mathbf{a}.\mathbf{b})-(\mathbf{b}.\mathbf{c}-\mathbf{b}.\mathbf{a}) = 0$ ([1]−[2])

$\Rightarrow$ $\mathbf{a}.\mathbf{c}-\mathbf{b}.\mathbf{c} = 0$

$\Rightarrow$ $\mathbf{c}.(\mathbf{a}-\mathbf{b}) = 0.$

Therefore **c** is perpendicular to $\mathbf{a}-\mathbf{b}$, or $\overrightarrow{HC}$ is perpendicular to $\overrightarrow{BA}$ and so $\overrightarrow{HC}$ is the third altitude.

Therefore the three altitudes are concurrent.

3) Prove that in a skew quadrilateral the joins of the midpoints of opposite sides bisect each other.

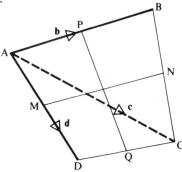

(A skew quadrilateral is one whose sides are not coplanar.)

Taking A as origin, the position vectors of B, C and D as **b, c, d**, and P, N, Q, M as the midpoints of AB, BC, CD, DA respectively we have

$$\overrightarrow{BC} = \mathbf{c}-\mathbf{b}, \qquad \overrightarrow{DC} = \mathbf{c}-\mathbf{d}.$$

Therefore $\overrightarrow{AM} = \tfrac{1}{2}\mathbf{d}$ and $\overrightarrow{AN} = \tfrac{1}{2}(\mathbf{b}+\mathbf{c})$.

Therefore the position vector of the midpoint of MN is

$$\tfrac{1}{2}(\overrightarrow{AM} + \overrightarrow{AN}) = \tfrac{1}{4}d + \tfrac{1}{4}b + \tfrac{1}{4}c.$$

Also $\overrightarrow{AP} = \tfrac{1}{2}b$, $\overrightarrow{AQ} = \tfrac{1}{2}(d + c)$.

Therefore the position vector of the midpoint of PQ is

$$\tfrac{1}{2}(\overrightarrow{AP} + \overrightarrow{AQ}) = \tfrac{1}{4}d + \tfrac{1}{4}b + \tfrac{1}{4}c.$$

As the midpoints of PQ and MN have the same position vector, these lines bisect each other.

MULTIPLE CHOICE EXERCISE 12

(The instructions for answering these questions are on p. xii.)

TYPE I

1) The modulus of the vector $6i - 2j - 3k$ is
(a) $\sqrt{23}$ (b) 7   (c) 1   (d) 49   (e) $\sqrt{11}$.

2) The direction cosines of the vector $i + j + k$ are
(a) $1, 1, 1$ (b) $\tfrac{1}{3}, \tfrac{1}{3}, \tfrac{1}{3}$ (c) $\sqrt{\tfrac{1}{3}}, \sqrt{\tfrac{1}{3}}, \sqrt{\tfrac{1}{3}}$ (d) $\sqrt{\tfrac{1}{2}}, \sqrt{\tfrac{1}{2}}, \sqrt{\tfrac{1}{2}}$
(e) $-\sqrt{\tfrac{1}{3}}, -\sqrt{\tfrac{1}{3}}, -\sqrt{\tfrac{1}{3}}$.

3) If $a = i - j + k$ and $b = i + j - k$ then $a.b$ is
(a) $2i$ (b) -1 (c) $-2j + 2k$ (d) 2   (e) 3.

4) The line whose equation is $r = \lambda(2i - j + k)$ has direction ratios
(a) $0, 0, 0$ (b) $-2 : 1 : -1$ (c) $2 : 1 : 1$ (d) $\tfrac{1}{3} : -\tfrac{1}{6} : \tfrac{1}{6}$.

5) The plane whose equation is $r.(i - j + k) = 2$ contains the point
(a) $(1, -1, 1)$ (b) $(-1, 1, 0)$ (c) $(0, 1, 1)$ (d) $(2, 0, 0)$
(e) $(0, 0, 0)$.

6) The angle between the lines whose equations are $r_1 = a_1 + \lambda b_1$ and
$r_2 = a_2 + \mu b_2$ is

(a) $\arccos \dfrac{b_1 . b_2}{b_1 b_2}$ (b) $b_1 . b_2$ (c) $\arccos \dfrac{a_1 . a_2}{a_1 a_2}$

(d) $r_1 . r_2$ (e) $\arccos \lambda . \mu$.

7) The equation of the plane normal to $4i + 3j$ and 5 units from 0 is
(a) $r.(4i + 3j) = 5$ (b) $r.5k = 5$ (c) $r.(4i + 3j) = 1$
(d) $r.(4i + 3j) = 25$ (e) $r.5k = 0$.

8) The points A, B and C are collinear and $\overrightarrow{OA} = i + j$, $\overrightarrow{OB} = 2i - j + k$,
$\overrightarrow{OC} = 3i + aj + bk$
(a) $a = -3$, $b = 2$ (b) $a = 3$, $b = -2$ (c) $a = 0$, $b = 1$
(d) $a = -1$, $b = 0$ (e) $a = 6$, $b = -1$.

TYPE II

9) A line has equation $r = i + 2j + 3k + \lambda(4i - j + 7k)$.
(a) The line passes through $(4, -1, 7)$.
(b) The length of the line is $\sqrt{14}$.
(c) The line passes through $(1, 2, 3)$.
(d) The line intersects the line $r = \lambda i$.

10) ABCD is a parallelogram and O is the origin.
(a) The area of ABCD is $\overrightarrow{AB}.\overrightarrow{AC}$.
(b) The equation of the line AC is $r = \lambda\overrightarrow{AC}$.
(c) The equation of the line AB is $r = \overrightarrow{OA} + \lambda\overrightarrow{AB}$.
(d) $\overrightarrow{AB} + \overrightarrow{BC} + \overrightarrow{CD} + \overrightarrow{DA} = 0$.

11) $V = 3i + 3j + 3k$.
(a) $\hat{V} = i + j + k$.
(b) V makes equal angles with i, j and k.
(c) $V.(i + j + k) = 0$.
(d) V is perpendicular to $2i + j - 3k$.

12) A, B and C have position vectors a, b and c respectively and $a = \lambda b + \mu c$.
(a) A, B, C are collinear.
(b) A, B, C are coplanar.
(c) A divides BC in the ratio $\mu : \lambda$.
(d) a is parallel to $b + c$.

TYPE III

13) (a) $a = \lambda b$.
    (b) $a.b = 0$.

14) A, B and C are collinear points with position vectors a, b and c respectively.
(a) $2c = a + b$.
(b) C is the midpoint of AB.

15) A, B and C are points with position vectors a, b and c respectively.
(a) The position vector of any point P is given by $r = \lambda a + \mu b + \eta c$.
(b) A, B and C are collinear.

16) (a) $a.(b + c) = 0$.
    (b) $b = -c$.

TYPE IV

17) Find the angle between the lines L_1 and L_2.
(a) The equation of L_1 is $r = a + \lambda b$.

(b) L_1 and L_2 intersect.

(c) L_2 is parallel to a vector **c**.

(d) L_2 passes through the point with position vector **d**.

18) ABC is a triangle. Find the direction cosines of a normal to the plane ABC.

(a) $\overrightarrow{AC}$ is parallel to $2\mathbf{i} + \mathbf{j}$.   (b) B is the point $(1, 0, 1)$.

(c) $\overrightarrow{OC}$ is parallel to $\mathbf{j} + 2\mathbf{k}$.   (d) A is the point $(3, 2, 1)$.

19) Determine whether two lines, L_1 and L_2, intersect.

(a) L_1 is parallel to **a**.   (b) L_2 is parallel to **b**.

(c) L_1 passes through O.   (d) L_2 is perpendicular to L_1.

20) Does the line $\dfrac{x - x_1}{a} = \dfrac{y - y_1}{b} = \dfrac{z - z_1}{c}$ lie in the plane

$Ax + By + Cz + D = 0$?

(a) $Ax_1 + By_1 + Cz_1 + D = 0$.

(b) $aA + bB + cC = 0$.

(c) The line and the plane are both 5 units from O.

(d) $a : b : c = 1 : 2 : 3$.

TYPE V

21) If two lines do not intersect they are parallel.

22) If three non-zero vectors are such that $\mathbf{a} . \mathbf{b} = \mathbf{a} . \mathbf{c}$ then either $\mathbf{b} = \mathbf{c}$
or   **a** and **b**−**c** are perpendicular.

23) $\mathbf{a} . \mathbf{b} = 0$ $\iff$ $\mathbf{a} = \mathbf{0}$ or $\mathbf{b} = \mathbf{0}$.

24) If $a : b : c$ are the direction ratios of a vector then $a^2 + b^2 + c^2 = 1$.

MISCELLANEOUS EXERCISE 12

In Questions 1–8 give proofs based on vector methods.

1) Prove that the line joining the midpoints of two sides of a triangle is parallel to the third side and equal to half of it.

2) Prove that the internal bisectors of the angles of a triangle are concurrent.

3) Prove that the joins of the midpoints of the opposite edges of a tetrahedron bisect each other.

4) Prove that the perpendicular bisectors of the sides of a triangle are concurrent.

5) Prove that the diagonals of a rhombus intersect at right angles.

6) Prove that the lines joining the midpoints of adjacent sides of a skew quadrilateral form a parallelogram.

7) ABCD is a parallelogram and M is the midpoint of AB. Prove that DM and AC cut each other at points of trisection.

8) Of the vectors

$$\mathbf{a} = \begin{pmatrix} 5 \\ 6 \\ -11 \end{pmatrix} \quad \mathbf{b} = \begin{pmatrix} 2 \\ -2 \\ 5 \end{pmatrix} \quad \mathbf{c} = \begin{pmatrix} 9 \\ 2 \\ -1 \end{pmatrix} \quad \mathbf{d} = \begin{pmatrix} -5 \\ -14 \\ 24 \end{pmatrix}$$

show that $\mathbf{a}, \mathbf{b}, \mathbf{d}$ form a set of basis vectors. Express $\mathbf{c}$ in terms of this basis. If $\mathbf{a}, \mathbf{b}, \mathbf{c}$ and $\mathbf{d}$ are the position vectors of points A, B, C and D respectively, show that the point $P(1, -2, 3)$ lies on AD, find the ratio AP:AD, and show that BP is perpendicular to PC.   (U of L)

9) Find a vector equation for the plane Π passing through the points A, B, C with position vectors $(\mathbf{i} - \mathbf{j} + 2\mathbf{k})$, $(2\mathbf{i} + \mathbf{j} + \mathbf{k})$, $(3\mathbf{i} - 2\mathbf{j} + 2\mathbf{k})$ respectively. Find the area of the triangle ABC and the distance from the plane Π of the point D with position vector $(3\mathbf{i} + \mathbf{j} + \mathbf{k})$.   (U of L)

10) Explain how the vector equations $\mathbf{r} = \mathbf{a} + t\mathbf{b}$ and $\mathbf{r} \cdot \mathbf{n} = p$ represent a straight line and a plane respectively, interpreting the symbols geometrically and using sketches if you wish. Prove that, provided $\mathbf{b} \cdot \mathbf{n} \neq 0$, the line and plane intersect in a point whose position vector is $\mathbf{a} + (p - \mathbf{a} \cdot \mathbf{n})\mathbf{b}/(\mathbf{b} \cdot \mathbf{n})$.   (JMB)

11) State a relation which exists between the vectors $\mathbf{p}$ and $\mathbf{q}$ when these vectors are (a) parallel, (b) perpendicular.
The position vectors of the vertices of a tetrahedron ABCD are

$$A: \quad -5\mathbf{i} + 22\mathbf{j} + 5\mathbf{k}, \qquad B: \quad \mathbf{i} + 2\mathbf{j} + 3\mathbf{k},$$
$$C: \quad 4\mathbf{i} + 3\mathbf{j} + 2\mathbf{k}, \qquad D: \quad -\mathbf{i} + 2\mathbf{j} - 3\mathbf{k}.$$

Find the angle CBD and show that AB is perpendicular to both BC and BD. Calculate the volume of the tetrahedron.
If ABDE is a parallelogram, find the position vector of E.   (U of L)

12) (a) Show that the plane which is at a distance d from the origin O and whose normal is in the direction of the unit vector $\mathbf{n}$, which points away from O, has equation $\mathbf{r} \cdot \mathbf{n} = d$.

   (b) Find the perpendicular distance between the planes
   $2x + 2y + z - 6 = 0$, $2x + 2y + z - 10 = 0$. Find also the area of the triangle whose vertices are $(2, 2, 2), (1, 1, 2)$ and $(1, -1, 6)$.   (U of L)

13) (a) In a triangle ABC the altitudes through A and B meet in a point O.
Let $\mathbf{a}, \mathbf{b}, \mathbf{c}$ be the position vectors of A, B, C relative to O as origin.
Show that $\mathbf{a} \cdot \mathbf{b} = \mathbf{a} \cdot \mathbf{c}$ and $\mathbf{b} \cdot \mathbf{a} = \mathbf{b} \cdot \mathbf{c}$.
Deduce that the altitudes of a triangle are concurrent.

(b) Find the point of intersection of the line through the points $(2, 0, 1)$
and $(-1, 3, 4)$ and the line through the points $(-1, 3, 0)$ and
$(4, -2, 5)$. Calculate the acute angle between the two lines.   (U of L)

14) Show that the equation of a plane can be expressed in the form $\mathbf{r} \cdot \mathbf{n} = p$.
Find the equation of the plane through the origin parallel to the lines

$$\mathbf{r} = 3\mathbf{i} + 3\mathbf{j} - \mathbf{k} + s(\mathbf{i} - \mathbf{j} - 2\mathbf{k}) \quad \text{and} \quad \mathbf{r} = 4\mathbf{i} - 5\mathbf{j} - 8\mathbf{k} + t(3\mathbf{i} + 7\mathbf{j} - 6\mathbf{k}).$$

Show that one of the lines lies in the plane, and find the distance of the other
line from the plane.   (U of L)

15) Points A, B, C have position vectors $\mathbf{a}, \mathbf{b}, \mathbf{c}$ and λ, μ, ν are variable
parameters subject to the condition $\lambda + \mu + \nu = 1$. If the points are not
collinear prove that the plane ABC is represented by the equation
$\mathbf{r} = \lambda\mathbf{a} + \mu\mathbf{b} + \nu\mathbf{c}$.
Prove that the equation of the line of intersection of the two planes:

$$\mathbf{r} = \lambda_1\mathbf{i} + 2\mu_1\mathbf{j} + 3\nu_1\mathbf{k}, \quad \lambda_1 + \mu_1 + \nu_1 = 1$$

and $\qquad\qquad \mathbf{r} = 2\lambda_2\mathbf{i} + \mu_2\mathbf{j} + 2\nu_2\mathbf{k}, \quad \lambda_2 + \mu_2 + \nu_2 = 1$

can be written in terms of a single parameter t as

$$6\mathbf{r} = (3 + t)\mathbf{i} + 4t\mathbf{j} + 9(1 - t)\mathbf{k}.$$ (U of L)

16) Obtain the equation of the straight line which passes through the point
$A(-1, 2, 3)$ and which is normal to the plane $2x - 3y + 4z + 8 = 0$.
Calculate the coordinates of P the point of intersection of this line and the
plane.
If the point $B(a, 2a, 3)$ lies on the plane, find the value of a and calculate the
angle between AP and AB.   (AEB '75)p

17) Prove that the points   A, B and C   whose coordinates are respectively
$(2, -1, 5)$, $(3, 1, -2)$ and $(1, -3, 12)$ are collinear. Find
(a) the sine of the angle between the line ABC and the plane
$7x + 2y + z = 9$,
(b) the coordinates of the point of intersection of ABC and this plane.
Find the mirror image in the plane of the point $(3, 1, -2)$ (AEB '75)

18) Obtain the equations of the straight line through the origin parallel to the
straight line through the points $P(3, 2, 3)$ and $Q(-1, -2, 1)$. Find the
equation of the plane which passes through both these two lines.
The line of intersection of a plane through P and a plane through Q is the
y-axis. Find the angle between these two planes.   (U of L)

19) The point $P(14 + 2\lambda, 5 + 2\lambda, 2 - \lambda)$ lies on a fixed straight line for all values of λ. Find Cartesian equations for this line and find the cosine of the acute angle between this line and the line $x = z = 0$.
Show that the line $2x = -y = -z$ is perpendicular to the locus of P.
Hence, or otherwise, find the equation of the plane containing the origin and all possible positions of P.   (U of L)

20) The position vectors of the points A, B, C with respect to the origin O are $\mathbf{a}, \mathbf{b}, \mathbf{c}$ respectively. If OA is perpendicular to BC, and OB is perpendicular to CA, show that OC is perpendicular to AB, and that

$$OA^2 + BC^2 = OB^2 + CA^2 = OC^2 + AB^2.$$

Show that the plane through BC perpendicular to OA meets the plane through AB perpendicular to OC in a line that lies in the plane through OB perpendicular to CA. If this line passes through the centroid of the triangle AOC, show that the angle AOC is $\pi/3$ radians.   (U of L)

21) Given that $\mathbf{i}, \mathbf{j}$ are perpendicular unit vectors and that $\mathbf{r}_1 = x_1\mathbf{i} + y_1\mathbf{j}$, $\mathbf{r}_2 = x_2\mathbf{i} + y_2\mathbf{j}$ are any two vectors, the scalar quantity $\mathbf{r}_1 \circ \mathbf{r}_2$ is defined for these two vectors by the equation

$$\mathbf{r}_1 \circ \mathbf{r}_2 = x_1 y_2 - x_2 y_1.$$

Deduce that $\mathbf{r}_2 \circ \mathbf{r}_1 = -\mathbf{r}_1 \circ \mathbf{r}_2$.
If $\mathbf{r}_3 = x_3\mathbf{i} + y_3\mathbf{j}$ is a further vector, prove that

$$(\mathbf{r}_1 + \mathbf{r}_2) \circ \mathbf{r}_3 = (\mathbf{r}_1 \circ \mathbf{r}_3) + (\mathbf{r}_2 \circ \mathbf{r}_3).   \text{(O)p}$$

22) In a triangle ABC the perpendicular from B to the side AC meets the perpendicular from C to the side AB at H. The position vectors of A, B and C relative to H are $\mathbf{a}, \mathbf{b}$ and $\mathbf{c}$ respectively.
Express $\overrightarrow{CA}$ in terms of $\mathbf{a}$ and $\mathbf{c}$ and deduce that $\mathbf{a} \cdot \mathbf{b} = \mathbf{b} \cdot \mathbf{c}$.
Prove that AH is perpendicular to BC.   (JMB)p

23) In a parallelogram ABCD X is the midpoint of AB and the line DX cuts the diagonal AC at P. Writing $\overrightarrow{AB} = \mathbf{a}$, $\overrightarrow{AD} = \mathbf{b}$, $\overrightarrow{AP} = \lambda\overrightarrow{AC}$ and $\overrightarrow{DP} = \mu\overrightarrow{DX}$,   express AP
(a) in terms of λ, $\mathbf{a}$ and $\mathbf{b}$,
(b) in terms of μ, $\mathbf{a}$ and $\mathbf{b}$.
Deduce that P is a point of trisection of both AC and DX.   (JMB)p

CHAPTER 13

COMPLEX NUMBERS

IMAGINARY NUMBERS

In all previous work it has been assumed that when any number, k say, is squared the result is either positive or zero, i.e. $k^2 \geqslant 0$. Such numbers are called *real* numbers.

We have already met equations such as $x^2 = -1$ whose roots are clearly not real since when squared they give -1 as the result. To work with equations of this type we need another category of numbers, namely the set of numbers whose squares are negative real numbers. Members of this set are called *imaginary* numbers, examples being $\sqrt{-1}$, $\sqrt{-7}$, $\sqrt{-20}$.

A general member of this set is $\sqrt{-n^2}$ where n is real.

But
$$\sqrt{-n^2} \equiv \sqrt{n^2 \times -1}$$
$$\equiv \sqrt{n^2} \times \sqrt{-1}$$
$$\equiv ni \quad \text{where} \quad i = \sqrt{-1}$$

So we see that every imaginary number can be written in the form ni where n is real and $i = \sqrt{-1}$ (or sometimes nj where $j = \sqrt{-1}$)

e.g.
$$\sqrt{-16} = 4i, \qquad \sqrt{-3} = i\sqrt{3}$$

Imaginary numbers can be added to and subtracted from other imaginary numbers,

e.g.
$$2i + 5i = 7i$$

$$\sqrt{7}i - i = (\sqrt{7} - 1)i$$

The product or quotient of two imaginary numbers is real,

e.g.
$$2i \times 5i = 10i^2$$

But $i^2 = -1$ since $i = \sqrt{-1}$

Hence
$$2i \times 5i = -10$$

Also $\qquad 6i \div 3i = 2$

Powers of i can be simplified,

e.g.

$$i^3 = (i^2)(i) = -i$$

$$i^4 = (i^2)^2 = (-1)^2 = 1$$

$$i^5 = (i^4)i = i$$

$$i^{-1} = \frac{i}{i^2} = \frac{i}{-1} = -i$$

COMPLEX NUMBERS

When a real number and an imaginary number are added or subtracted, the expression so formed, which cannot be simplified, is called a complex number, e.g. $2 + 3i, 4 - 7i, -1 + 4i$. A general complex number can be written in the form $a + bi$ where a and b can have any real value including zero.
If $a = 0$ we have numbers of the form bi, i.e. imaginary numbers.
If $b = 0$ we have numbers of the form a, i.e. real numbers.
Therefore *the field of complex numbers includes the real number set and the imaginary number set.*

OPERATIONS ON COMPLEX NUMBERS

Addition and Subtraction

Real terms and imaginary terms are compounded in two separate groups,

e.g.

$$(2 + 3i) + (4 - i) = (2 + 4) + (3i - i)$$
$$= 6 + 2i$$

and

$$(4 - 2i) - (3 + 5i) = (4 - 3) - (2i + 5i)$$
$$= 1 - 7i$$

Multiplication

The distributive law of multiplication applied to two complex numbers gives their product,

e.g.

$$(2 + 3i)(4 - i) = 8 - 2i + 12i - 3i^2$$
$$= 8 + 10i - 3(-1)$$
$$= 11 + 10i$$

and $$(2+3i)(2-3i) = 4-6i+6i-9i^2$$
$$= 4+9$$
$$= 13$$

Note that the product here is a real number. This is because of the special form of the given complex numbers, $2 \pm 3i$.

Any pair of complex numbers of the form $a \pm bi$ have a product which is real, since

$$(a+bi)(a-bi) \equiv a^2 - abi + abi - b^2 i^2$$
$$\equiv a^2 + b^2$$

Such complex numbers are said to be *conjugate* and each is the conjugate of the other.

Thus $4+5i$ and $4-5i$ are conjugate complex numbers and $4+5i$ is the conjugate of $4-5i$.

If $a+bi$ is denoted by z then its conjugate, $a-bi$, is denoted by $\bar{z}$ or z^*.

Division

Direct division by a complex number cannot be carried out, because the denominator is made up of two independent terms. This difficulty can be overcome by making the denominator real, a process called 'realising the denominator'.

This can be done by using the product property of conjugate complex numbers. Division can therefore be carried out as follows:

$$\frac{2+9i}{5-2i} = \frac{(2+9i)(5+2i)}{(5-2i)(5+2i)}$$

(multiplying numerator and denominator by the conjugate of the denominator)

$$= \frac{10+49i+18i^2}{25-4i^2}$$

$$= \frac{-8+49i}{29}$$

$$= -\frac{8}{29} + \frac{49}{29}i$$

(**Note.** The real term is given first even when it is negative.)

THE ZERO COMPLEX NUMBER

A complex number is zero if and only if the real term and the imaginary term are each zero,

i.e. $$X + Yi = 0 \iff X = 0 \quad and \quad Y = 0$$

EQUAL COMPLEX NUMBERS

Now consider the case

$$a + bi = c + di \qquad [1]$$

or $$(a + bi) - (c + di) = 0 \qquad [2]$$

[2] gives $$(a - c) + (b - d)i = 0$$

Hence $$a - c = 0 \quad and \quad b - d = 0$$

i.e. $$a = c \quad and \quad b = d$$

Thus two complex numbers are equal if and only if the real terms and the imaginary terms are separately equal.

Writing $\text{Re}(a + bi)$ to indicate the real part of $a + bi$ and $\text{Im}(a + bi)$ to indicate the imaginary part we have:

$$a + bi = c + di \iff \begin{cases} \text{Re}(a + bi) = \text{Re}(c + di) \\ \text{Im}(a + bi) = \text{Im}(c + di) \end{cases}$$

A complex number equation is therefore equivalent to two separate equations. This property provides an alternative method for division by a complex number, e.g. to divide $3 - 2i$ by $5 + i$ we can say,

let $$\frac{3 - 2i}{5 + i} = p + qi$$

hence $$3 - 2i = (p + qi)(5 + i)$$
$$= 5p + pi + 5qi + qi^2$$
$$= (5p - q) + (p + 5q)i$$

Equating real and imaginary parts gives:

$$\begin{cases} 3 = 5p - q \\ -2 = p + 5q \end{cases}$$

Solving these equations simultaneously gives $p = \frac{1}{2}$, $q = -\frac{1}{2}$.

Thus $$(3 - 2i) \div (5 + i) = \frac{1}{2} - \frac{1}{2}i$$

The method of equating real and imaginary parts of a complex equation also provides one way of determining the square root of a complex number,

e.g. to find $\sqrt{15 + 8i}$ we say,

let $\qquad\qquad\qquad \sqrt{15 + 8i} = a + bi \qquad\qquad$ where a and b are real

$\Rightarrow \qquad\qquad\qquad\quad 15 + 8i = (a + bi)^2$

$$= a^2 - b^2 + 2abi$$

Equating real and imaginary parts gives:

$$a^2 - b^2 = 15 \qquad\qquad\qquad\qquad [1]$$

so $\qquad\qquad\qquad\qquad\quad 2ab = 8 \qquad\qquad\qquad\qquad\qquad [2]$

Using $\quad b = \dfrac{4}{a}\quad$ in [1] gives

$$a^2 - \frac{16}{a^2} = 15$$

$\Rightarrow \qquad\qquad\qquad a^4 - 15a^2 - 16 = 0$

$\Rightarrow \qquad\qquad\qquad (a^2 - 16)(a^2 + 1) = 0$

Thus $\qquad\qquad\qquad a^2 - 16 = 0 \quad$ or $\quad a^2 + 1 = 0$

But a is real so $\quad a^2 + 1 = 0 \quad$ gives no suitable values

so $\qquad\qquad\qquad\qquad\qquad a = \pm 4$

Referring to equation [2] we have

a	b
4	1
-4	-1

[**Note.** It is not correct to say $\quad a = \pm 4 \quad$ therefore $\quad b = \pm 1 \quad$ as this offers four different pairs of values for a and b (i.e. $4, 1;\ 4, -1;\ -4, 1;\ -4, -1$) two of which are invalid.]

Hence $\qquad\qquad \sqrt{15 + 8i} = 4 + i \quad$ or $\quad -4 - i$

$$= \pm(4 + i)$$

This result justifies our original assumption that the square root of a complex number is another complex number.

It is sometimes possible, however, to determine the square root of a complex number simply by observation. In the above example, equation [2] shows that the product of a and b is always half the coefficient of i in the original complex number.

Suitable integral values for a and b can then be checked quite quickly, e.g. to find $\sqrt{8-6i}$ we note that $ab = -3$ so possible values for a and b are: $1, -3; 3, -1$.

Checking:
$$(1 - 3i)^2 = -8 - 6i$$
$$(3 - i)^2 = 8 - 6i$$

Hence one square root of $8 - 6i$ is $3 - i$ and the other is $-(3 - i)$

i.e.
$$\sqrt{8 - 6i} = \pm(3 - i)$$

(**Note.** Unless a and b are integers, this method is unlikely to be useful.)

EXERCISE 13a

1) Simplify: $i^7, i^{-3}, i^9, i^{-5}, i^{4n}, i^{4n+1}$.

2) Add the following pairs of complex numbers:
(a) $3 + 5i$ and $7 - i$ (b) $4 - i$ and $3 + 3i$
(c) $2 + 7i$ and $4 - 9i$ (d) $a + bi$ and $c + di$

3) Subtract the second number from the first in each part of Question 2.

4) Simplify:
(a) $(2 + i)(3 - 4i)$ (b) $(5 + 4i)(7 - i)$ (c) $(3 - i)(4 - i)$
(d) $(3 + 4i)(3 - 4i)$ (e) $(2 - i)^2$ (f) $(1 + i)^3$
(g) $i(3 + 4i)$ (h) $(x + yi)(x - yi)$ (i) $i(1 + i)(2 + i)$
(j) $(a + bi)^2$

5) Realise the denominator of each of the following fractions and hence express each in the form $a + bi$.

(a) $\dfrac{2}{1 - i}$ (b) $\dfrac{3 + i}{4 - 3i}$ (c) $\dfrac{4i}{4 + i}$ (d) $\dfrac{1 + i}{1 - i}$

(e) $\dfrac{7 - i}{1 + 7i}$ (f) $\dfrac{x + yi}{x - yi}$ (g) $\dfrac{3 + i}{i}$ (h) $\dfrac{-2 + 3i}{-i}$

6) Solve the following equations for x and y:

(a) $x + yi = (3 + i)(2 - 3i)$ (b) $\dfrac{2 + 5i}{1 - i} = x + yi$

(c) $3 + 4i = (x + yi)(1 + i)$ (d) $x + yi = 2$

(e) $x + yi = (3 + 2i)(3 - 2i)$ (f) $x + yi = (4 + i)^2$

(g) $\dfrac{x + yi}{2 + i} = 5 - i$ (h) $(x + yi)^2 = 3 + 4i$

7) Find the real and imaginary parts of:
(a) $(2 - i)(3 + i)$ (b) $(1 - i)^3$

(c) $\dfrac{3 + 2i}{4 - i}$

(d) $\dfrac{2}{3 + i} + \dfrac{3}{2 + i}$

(e) $\dfrac{1}{x + yi} - \dfrac{1}{x - yi}$

(f) $\left(\cos \dfrac{\pi}{3} + i \sin \dfrac{\pi}{3} \right)^3$

(g) $\left(\cos \dfrac{\pi}{6} + i \sin \dfrac{\pi}{6} \right)^2$

8) Find the square roots of:
(a) $3 - 4i$ (b) $21 - 20i$ (c) $2i$ (d) $15 + 8i$ (e) $-24 + 10i$

COMPLEX ROOTS OF QUADRATIC EQUATIONS

Consider the quadratic equation $x^2 + 2x + 2 = 0$.

Using the formula $x = -\dfrac{b \pm \sqrt{b^2 - 4ac}}{2a}$ we find that $x = \dfrac{-2 \pm \sqrt{-4}}{2}$

Previously we dismissed solutions of this type, in which $b^2 - 4ac < 0$, as not being real. But now, because $\sqrt{-4} = 2i$, we see that the roots of the given equation are the complex numbers $-1 + i$ and $-1 - i$. Further, the roots are conjugate complex numbers.

An examination of the general quadratic equation $ax^2 + bx + c = 0$ shows that, if $b^2 - 4ac < 0$, the roots of the equation are always conjugate complex numbers.

If $\qquad\qquad\qquad ax^2 + bx + c = 0$ [1]

and $\qquad\qquad\qquad b^2 - 4ac < 0$

then $\qquad\qquad\qquad x = \dfrac{-b \pm \sqrt{b^2 - 4ac}}{2a}$

or $\qquad\qquad\qquad x = \dfrac{-b}{2a} \pm i \dfrac{\sqrt{4ac - b^2}}{2a}$

Now using $\qquad p = \dfrac{-b}{2a}$ and $q = \dfrac{\sqrt{4ac - b^2}}{2a}$

the roots of equation [1] are:

$$p + qi \quad \text{and} \quad p - qi$$

and these are conjugate complex numbers.

So if one root of a quadratic equation, with real coefficients, is known to be complex, the other must also be complex and the conjugate of the first.

Similarly a quadratic expression whose discriminant is negative, has conjugate complex factors.

e.g. $\qquad\qquad x^2 + 2x + 2 \equiv (x + 1 - i)(x + 1 + i)$

(the factors are found by solving the equation $x^2 + 2x + 2 = 0$)

CUBIC EQUATIONS WITH COMPLEX ROOTS

Consider a cubic equation $f(x) = 0$ where $f(x)$ has real coefficients.
We have seen that a cubic equation has at least one real root so we can always
find a real value a such that $f(a) = 0$.
Then $(x - a)$ is a factor of $f(x)$ and the remaining factor must be quadratic,
i.e. $(px^2 + qx + r)$

$$\Rightarrow \qquad (x - a)(px^2 + qx + r) = 0$$

so that $\qquad x = a \quad$ or $\quad px^2 + qx + r = 0$

But the roots of $px^2 + qx + r = 0$ are either both real or are a pair of
conjugate complex numbers.
Thus a cubic equation with real coefficients can have either three real roots or
one real root and two conjugate complex roots.
In general, if any polynomial equation with real coefficients has complex roots,
they occur in conjugate pairs.

Cube Roots of Unity

Consider the equation $x^3 - 1 = 0$
An obvious root is $x = 1$, so $x - 1$ is a factor, giving

$$x^3 - 1 \equiv (x - 1)(x^2 + x + 1) = 0$$

Thus $\qquad x = 1,$ or $x^2 + x + 1 = 0$

i.e. $\qquad x = 1 \quad$ or $\quad \dfrac{-1 \pm \sqrt{1 - 4}}{2}$

The roots of $x^3 - 1 = 0$ are therefore

$$1, \quad -\frac{1}{2} + \frac{\sqrt{3}}{2}i, \quad -\frac{1}{2} - \frac{\sqrt{3}}{2}i$$

But if the equation is rewritten in the form

$$x^3 = 1 \quad \text{or} \quad x = \sqrt[3]{1}$$

we see that the three values of x that we have found are the *three cube roots
of unity*.
The following properties should be noted:

(1)
$$\left(-\frac{1}{2} + \frac{\sqrt{3}}{2}i\right)^2 = \frac{1}{4} - \frac{\sqrt{3}}{2}i + \frac{3}{4}i^2 = -\frac{1}{2} - \frac{\sqrt{3}}{2}i$$

also
$$\left(-\frac{1}{2} - \frac{\sqrt{3}}{2}i\right)^2 = \frac{1}{4} + \frac{\sqrt{3}}{2}i + \frac{3}{4}i^2 = -\frac{1}{2} + \frac{\sqrt{3}}{2}i$$

i.e. either complex cube root of 1, when squared, gives the other complex root.
For this reason the three cube roots of 1 are usually denoted by $1, \omega, \omega^2$.

(2) The sum of the three cube roots is zero, since

$$1 + \left(-\frac{1}{2} + \frac{\sqrt{3}}{2}\,i\right) + \left(-\frac{1}{2} - \frac{\sqrt{3}}{2}\,i\right) = 0$$

i.e. $1 + \omega + \omega^2 = 0$

EXAMPLES 13b

1) Find the complex roots of the equation $2x^2 + 3x + 5 = 0$. If these roots are α and β, confirm the relationships $\alpha + \beta = -\dfrac{b}{a}$ and $\alpha\beta = \dfrac{c}{a}$.

If $2x^2 + 3x + 5 = 0$

then $x = \dfrac{-3 \pm \sqrt{9 - 40}}{4}$

$\Rightarrow$ $\alpha = -\dfrac{3}{4} + \dfrac{\sqrt{31}}{4}\,i \qquad \beta = -\dfrac{3}{4} - \dfrac{\sqrt{31}}{4}\,i$

Hence $\alpha + \beta = \left(-\dfrac{3}{4} + \dfrac{\sqrt{31}}{4}\,i\right) + \left(-\dfrac{3}{4} - \dfrac{\sqrt{31}}{4}\,i\right)$

$$= -\frac{3}{2} = -\frac{b}{a}$$

and $\alpha\beta = \left(-\dfrac{3}{4} + \dfrac{\sqrt{31}}{4}\,i\right)\left(-\dfrac{3}{4} - \dfrac{\sqrt{31}}{4}\,i\right)$

$$= \frac{9}{16} - \frac{31}{16}\,i^2$$

$$= \frac{40}{16} = \frac{5}{2} = \frac{c}{a}$$

2) One root of the equation $x^2 + px + q = 0$ is $2 - 3i$. Find the values of p and q.

If one root is $2 - 3i$ the other must be $2 + 3i$.

Then $\alpha + \beta = 4$

and $\alpha\beta = (2 - 3i)(2 + 3i)$

$$= 13$$

Now any quadratic equation can be written in the form

$$x^2 - (\text{sum of roots})x + (\text{product of roots}) = 0$$

So the equation with roots $2 \pm 3i$ is

$$x^2 - 4x + 13 = 0$$

i.e. $p = 4$ and $q = 13$

3) Find the complex factors of $x^2 - 2x + 10$ and hence express $\dfrac{6}{x^2 - 2x + 10}$ in partial fractions with complex linear denominators.

Let $x^2 - 2x + 10 = 0 \Rightarrow x = 1 \pm 3i$.
Thus the factors of $x^2 - 2x + 10$ are $(x - 1 - 3i)$, $(x - 1 + 3i)$

i.e. $x^2 - 2x + 10 \equiv (x - 1 - 3i)(x - 1 + 3i)$

Since the required partial fractions have complex denominators, we must prepare for unknown numerators that are complex.

Let $\dfrac{6}{x^2 - 2x + 10} \equiv \dfrac{A + Bi}{x - 1 - 3i} + \dfrac{C + Di}{x - 1 + 3i}$

$$\equiv \dfrac{(A + Bi)(x - 1 + 3i) + (C + Di)(x - 1 - 3i)}{x^2 - 2x + 10}$$

i.e. $6 \equiv (A + Bi)(x - 1 + 3i) + (C + Di)(x - 1 - 3i)$

When $x = 1 - 3i$

$$6 = (C + Di)(-6i) = 6D - 6Ci$$

Equating real and imaginary parts gives:

$$D = 1 \quad \text{and} \quad C = 0$$

When $x = 1 + 3i$

$$6 = (A + Bi)(6i) = 6Ai - 6B$$

giving $A = 0$ and $B = -1$

Hence $\dfrac{6}{x^2 - 2x + 10} \equiv \dfrac{-i}{x - 1 - 3i} + \dfrac{i}{x - 1 + 3i}$

EXERCISE 13b

1) Solve the following equations:
(a) $x^2 + x + 1 = 0$ (b) $2x^2 + 7x + 1 = 0$
(c) $x^2 + 9 = 0$ (d) $x^2 + x + 3 = 0$
(e) $x^4 - 1 = 0$

2) Form the equation whose roots are:
(a) $i, -i$ (b) $2 + i, 2 - i$ (c) $1 - 3i, 1 + 3i$ (d) $1 + i, 1 - i, 2$

3) Without calculating a, b and c, evaluate $-b/a$ and c/a if one root of the equation $ax^2 + bx + c = 0$ is:

(a) $2 + i$ (b) $3 - 4i$ (c) i (d) $5i - 12$ (e) $-1 - i$

Explain why this question cannot be answered if the given root is 2.

4) Find the complex factors of:

(a) $x^2 + 4x + 5$ (b) $x^2 - 2x + 17$ (c) $x^2 + x + 1$ (d) $x^3 - 8$

5) Express as partial fractions with complex linear denominators:

(a) $\dfrac{1}{x^2 + 1}$ (b) $\dfrac{4}{x^2 - 4x + 5}$

(c) $\dfrac{16}{x^2 + 4x + 8}$ (d) $\dfrac{2}{x^2 + 4}$ (e) $\dfrac{x + 8}{x^2 + 4x + 13}$

6) If $1, \omega, \omega^2$ are the three cube roots of unity, find the value of:

(a) $\dfrac{(1 + \omega)^2}{\omega}$ (b) $(1 + 2\omega + 3\omega^2)(3 + 2\omega + \omega^2)$ (c) $\omega^7 + \omega^8 + \omega^9$

7) By solving the equation $x^3 + 1 = 0$, find the three cube roots of -1. If one of the complex cube roots is λ express the other in terms of λ. Prove that $1 + \lambda^2 = \lambda$.

THE ARGAND DIAGRAM

For a complex number $a + bi$, a and b are both real numbers. Thus $a + bi$ can be represented by the ordered pair $\begin{pmatrix} a \\ b \end{pmatrix}$. This suggests using a and b as the Cartesian coordinates of a point, A say, and using the vector $\overrightarrow{OA}$ as a visual representation of the complex number $a + bi$.

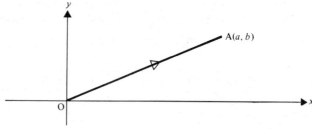

As this idea was introduced by the French mathematician Argand, his name is given to the diagram which demonstrates a complex number in this way. On an Argand diagram real numbers are represented on the x-axis and imaginary numbers on the y-axis. (For this reason the x- and y-axes are often called the real and imaginary axes.)

A general complex number $x + yi$ is represented by the line $\overrightarrow{OP}$ where P is the point (x, y).

In an Argand diagram, the magnitude and direction of a line are used to represent a complex number in the same way that a section of a line can be used to represent a vector quantity. Thus the techniques and operations used in vector problems can be applied equally well to complex number analysis.

A COMPLEX NUMBER AS A VECTOR

On an Argand diagram, the complex number $5 + 3i$ can be represented by the line $\overrightarrow{OA}$ where A is the point $(5, 3)$. But any other line with the same length and direction (e.g. $\overrightarrow{BC}$ or $\overrightarrow{DE}$ as shown) is equally representative.

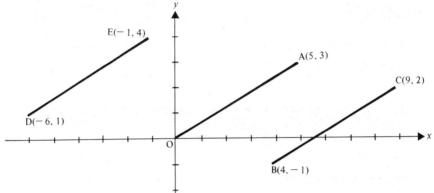

Treated in this way, a complex number behaves as a *free vector*.
If, however, $5 + 3i$ is regarded as a *position vector*, then *only* the line $\overrightarrow{OA}$ represents $5 + 3i$. In this case the *point* $A(5, 3)$ is sometimes taken to represent $5 + 3i$.
The symbol z is often used to denote a complex number, e.g. $z = x + yi$, $z_1 = 5 + 3i$. Used on an Argand diagram, z must be accompanied by an arrow to indicate the direction of the line representing the complex number, e.g.

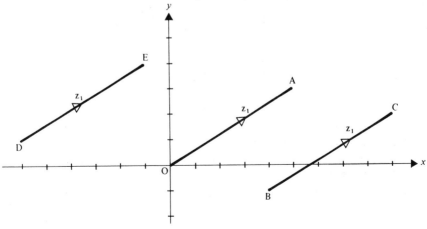

GRAPHICAL ADDITION AND SUBTRACTION

Consider two complex numbers z_1 and z_2 represented on an Argand diagram by $\overrightarrow{OA}$ and $\overrightarrow{OB}$.

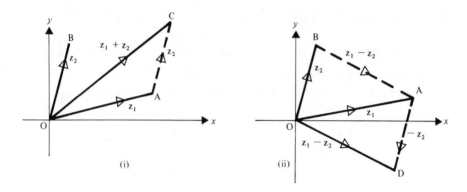

If AC is drawn equal and parallel to OB, then $\overrightarrow{AC}$ also represents z_2.

But the sum of the vectors $\overrightarrow{OA}$ and $\overrightarrow{AC}$ is $\overrightarrow{OC}$.

Therefore, if z_1 and z_2 are represented by $\overrightarrow{OA}$ and $\overrightarrow{OB}$, their sum is represented by the diagonal $\overrightarrow{OC}$ of the parallelogram OACB (see diagram (i)).

If $\overrightarrow{OB}$ represents z_2, $\overrightarrow{AD}$ (equal and parallel to $\overrightarrow{BO}$) represents $-z_2$. Then $\overrightarrow{OD}$ represents $z_1 - z_2$. But $\overrightarrow{OD}$ is equal and parallel to $\overrightarrow{BA}$.

Therefore the line joining B to A $(\overrightarrow{BA})$ represents $z_1 - z_2$.

Similarly the line joining A to B $(\overrightarrow{AB})$ represents $z_2 - z_1$ (see diagram (ii)).

Thus the two diagonals of the parallelogram OACB represent the sum and difference of z_1 and z_2 (see diagram (iii)).

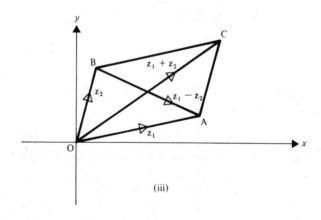

(iii)

Taking $z_1 = x_1 + y_1 i$ and $z_2 = x_2 + y_2 i$ we see that the coordinates of C obtained in this way are $[(x_1 + x_2), (y_1 + y_2)]$. Hence $\overrightarrow{OC}$ represents the complex number

$$(x_1 + x_2) + (y_1 + y_2)i$$

the same result as is obtained by adding separately the real and imaginary parts of z_1 and z_2,

e.g. if $z_1 = 5 + 2i$

and $z_2 = 3 + 4i$

then:

$$z_1 + z_2 = 5 + 2i + 3 + 4i$$
$$= 8 + 6i$$

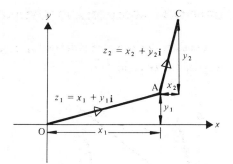

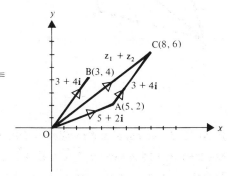

MODULUS AND ARGUMENT

The point $A(a, b)$ can equally well be located using polar coordinates (r, α) where r is the length of $\overrightarrow{OA}$ and α is the angle between the positive x-axis and $\overrightarrow{OA}$.

The length of OA is called the *modulus* of the complex number $a + bi$ and is written $|a + bi|$ so that

$$|a + bi| = r = \sqrt{a^2 + b^2}$$

The angle α is called the argument of $a + bi$ and is written $\text{Arg}(a + bi)$,

thus $$\text{Arg}(a + bi) = \alpha = \text{Arctan}\frac{b}{a}$$

There is an infinite set of angles whose tangent is $\dfrac{b}{a}$ so there is also an infinite set of arguments for $a + bi$. But the position of OA is unique and corresponds to only one value of α in the range $-\pi < \alpha \leqslant \pi$. This value is the *principal argument*, and is written arg $(a + bi)$ whereas Arg $(a + bi)$ represents the infinite set of arguments.

(**Note.** An argument is sometimes given the alternative name *amplitude*.)

When working with arguments it is always wise to draw an Argand diagram. Consider, for example, the complex numbers $4 + 3i$, $-4 + 3i$, $-4 - 3i$, $4 - 3i$.

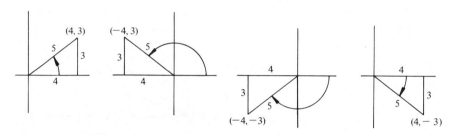

The modulus of each is 5.

The line representing $4 + 3i$ is in the first quadrant, so the principal argument of $4 + 3i$ is positive and acute, and its value is arctan $\frac{3}{4}$,

i.e.   arg $(4 + 3i) = 0.644^c$

The line representing $-4 + 3i$ is in the second quadrant, so the principal argument of $-4 + 3i$ is positive and obtuse and its value is Arctan $(-\frac{3}{4})$,

i.e.   arg $(-4 + 3i) = 2.498^c$

The line representing $-4 - 3i$ is in the third quadrant so the principal argument of $-4 - 3i$ is negative and obtuse and of value Arctan $\frac{3}{4}$

i.e.   arg$(-4 - 3i) = -2.498^c$

$+3.785^c$ is *an* argument of $-4 - 3i$ but not the principal argument.

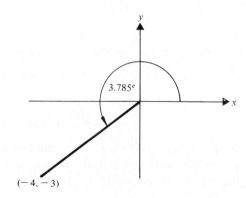

The line representing $4 - 3i$ is in the fourth quadrant so the principal argument of $4 - 3i$ is negative and acute and of value $\arctan\left(-\frac{3}{4}\right)$,

i.e. $$\arg(4 - 3i) = -0.644^c$$

Note. In this example we see why it is not sufficient to say that

$\arg(a + bi) = \arctan\dfrac{b}{a}$.   Both $4 + 3i$ and $-4 - 3i$ have an argument of

value $\arctan\frac{3}{4}$ but their arguments are different because they are in different quadrants. Similarly the principal arguments of $4 - 3i$ and $-4 + 3i$ are both $\text{Arctan}\left(-\frac{3}{4}\right)$ but are different angles. So, in finding the argument of $a + bi$

we use $\text{Arctan}\dfrac{b}{a}$ *together with a quadrant diagram.*

THE MODULUS, ARGUMENT (OR POLAR COORDINATE) FORM FOR A COMPLEX NUMBER

Consider a general complex number $x + yi$ represented on an Argand diagram by $\overrightarrow{OP}$ where P has Cartesian coordinates (x, y) and polar coordinates (r, θ).

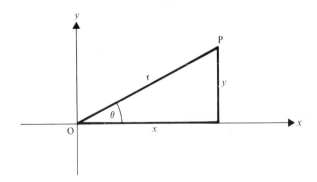

From the diagram we see that $x \equiv r \cos \theta$

$$y \equiv r \sin \theta$$

Hence $$x + yi \equiv r \cos \theta + ir \sin \theta$$

i.e. $$x + yi \equiv r(\cos \theta + i \sin \theta)$$

(**Note.** For clarity we write $\_ \sin \theta$ and not $\sin \theta i$.)
If a complex number is given in the form $x + yi$ it can be converted into the form $r(\cos \theta + i \sin \theta)$ simply by finding the modulus r and the argument θ,

e.g. for the complex number $1 - i$, we find that

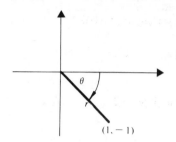

$(1, -1)$

$$r = \sqrt{1^2 + (-1)^2} = \sqrt{2}$$

$$\theta = \arg(1 - i) = -\frac{\pi}{4}$$

Hence

$$1 - i = \sqrt{2}\left[\cos\left(-\frac{\pi}{4}\right) + i\sin\left(-\frac{\pi}{4}\right)\right]$$

Conversely a complex number given in polar form can be expressed directly in Cartesian form,

e.g.

$$4\left(\cos\frac{2\pi}{3} + i\sin\frac{2\pi}{3}\right) = 4\left(-\frac{1}{2}\right) + 4\left(\frac{\sqrt{3}}{2}\right)i$$

$$= -2 + 2i\sqrt{3}$$

EXAMPLES 13c

1) Given that $z_1 = 3 - i$ and $z_2 = -2 + 5i$, represent on an Argand diagram the complex numbers z_1, z_2, $z_1 + z_2$, $z_1 - z_2$. Find the modulus and principal argument of $z_1 + z_2$ and $z_1 - z_2$.

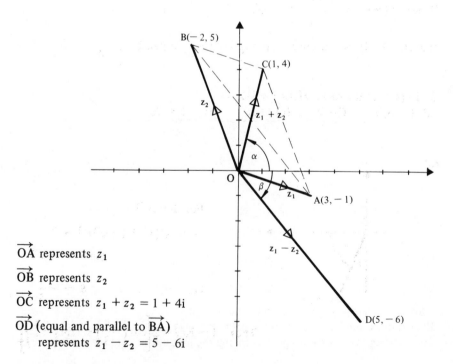

$\overrightarrow{OA}$ represents z_1

$\overrightarrow{OB}$ represents z_2

$\overrightarrow{OC}$ represents $z_1 + z_2 = 1 + 4i$

$\overrightarrow{OD}$ (equal and parallel to $\overrightarrow{BA}$)
    represents $z_1 - z_2 = 5 - 6i$

$$|z_1 + z_2| = \sqrt{1^2 + 4^2} \quad = \sqrt{17} \text{ (length of OC)}$$

$$|z_1 - z_2| = \sqrt{5^2 + (-6)^2} = \sqrt{61} \text{ (length of OD)}$$

$$\arg(z_1 + z_2) = \alpha \quad \text{where} \quad \tan\alpha = \tfrac{4}{1}$$

so $$\arg(z_1 + z_2) = 1.33^c$$

$$\arg(z_1 - z_2) = \beta \quad \text{where} \quad \tan\beta = -\tfrac{6}{5}$$

so $$\arg(z_1 - z_2) = -0.88^c$$

2) Find the modulus and argument of $\dfrac{7-i}{3-4i}$.

First we must express $\dfrac{7-i}{3-4i}$ in the form $a+bi$

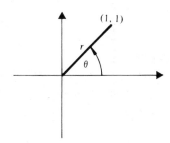

$$\frac{7-i}{3-4i} = \frac{(7-i)(3+4i)}{(3-4i)(3+4i)}$$

$$= \frac{25+25i}{25} = 1+i$$

Hence $|1+i| = \sqrt{1^2 + 1^2} = \sqrt{2}$

and $\arg(1+i)$ is a positive acute angle of value $\arctan 1$, i.e. $\dfrac{\pi}{4}$

3) Express in the form $r(\cos\theta + i\sin\theta)$:
(a) $1 - i\sqrt{3}$ (b) 2   (c) $-5i$ (d) $-2 + 2i$.

(a)

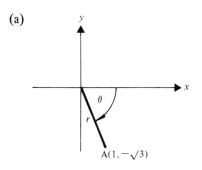

For $1 - i\sqrt{3}$

$$r = \sqrt{[1^2 + (-\sqrt{3})^2]} = 2$$

$$\tan\theta = -\sqrt{3}$$

i.e. $$\theta = -\frac{\pi}{3}$$

Thus $1 - i\sqrt{3} = 2\left[\cos\left(-\dfrac{\pi}{3}\right) + i\sin\left(-\dfrac{\pi}{3}\right)\right]$

(b)

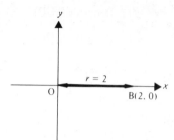

For 2,

$$r = 2 \quad \text{and}$$

$$\theta = 0$$

Thus $2 = 2(\cos 0 + i \sin 0)$

(c)

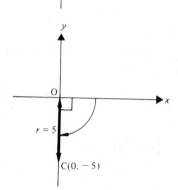

For $-5i$

$$r = 5 \quad \text{and}$$

$$\theta = -\frac{\pi}{2}$$

Thus

$$-5i = 5\left\{\cos\left(-\frac{\pi}{2}\right) + i \sin\left(-\frac{\pi}{2}\right)\right\}$$

(d)

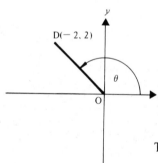

For $-2 + 2i$

$$r = \sqrt{(-2)^2 + 2^2} = 2\sqrt{2}$$

$$\tan \theta = -1$$

i.e. $\theta = \dfrac{3\pi}{4}$

Thus $-2 + 2i = 2\sqrt{2}\left(\cos\dfrac{3\pi}{4} + i \sin\dfrac{3\pi}{4}\right)$

EXERCISE 13c

1) Represent the following complex numbers by lines on Argand diagrams. Determine the modulus and argument of each complex number.

(a) $3 - 2i$ (b) $-4 + i$ (c) $-3 - 4i$ (d) $5 + 12i$

(e) $1 - i$ (f) $-1 + i$ (g) 4 (h) $-2i$

(i) $a + bi$ (j) $1 + i$ (k) $i(1 + i)$ (l) $i^2(1 + i)$

(m) $i^3(1 + i)$ (n) $(3 + i)(4 + i)$ (o) $2\left(\cos\dfrac{\pi}{3} + i \sin\dfrac{\pi}{3}\right)$

(p) $\cos\dfrac{3\pi}{4} + i \sin\dfrac{3\pi}{4}$ (q) $3\left[\cos\left(-\dfrac{5\pi}{6}\right) + i \sin\left(-\dfrac{5\pi}{6}\right)\right]$

2) If $z_1 = 3 - i$, $z_2 = 1 + 4i$, $z_3 = -4 + i$, $z_4 = -2 - 5i$, represent the following by lines on Argand diagrams, showing the direction of each line by an arrow.

(a) $z_1 + z_2$ (b) $z_2 - z_3$ (c) $z_1 - z_3$

(d) $z_2 + z_4$ (e) $z_4 - z_1$ (f) $z_3 - z_4$

(g) z_1 (h) z_4 (i) $z_2 - z_1$ (j) $z_1 + z_3$

3) Express in the form $r(\cos \theta + i \sin \theta)$:

(a) $1 + i$ (b) $\sqrt{3} - i$ (c) $-3 - 4i$ (d) $-5 + 12i$

(e) $2 - i$ (f) 6   (g) -3 (h) $4i$

(i) $-3 - i\sqrt{3}$ (j) $24 + 7i$

4) Express in the form $x + yi$ a complex number represented on an Argand diagram by $\overrightarrow{OP}$ where the polar coordinates of P are:

(a) $\left(2, \dfrac{\pi}{6}\right)$ (b) $\left(3, -\dfrac{\pi}{4}\right)$ (c) $\left(1, \dfrac{2\pi}{3}\right)$ (d) $\left(1, -\dfrac{3\pi}{4}\right)$

(e) $(3, 0)$ (f) $(2, \pi)$ (g) $\left(4, -\dfrac{\pi}{6}\right)$

(h) $\left(1, \dfrac{\pi}{2}\right)$ (i) $\left(3, -\dfrac{\pi}{2}\right)$ (j) $\left(1, -\dfrac{2\pi}{3}\right)$

5) By using $z_1 = x_1 + y_1 i$, $z_2 = x_2 + y_2 i$ show on an Argand diagram the position of the point representing:

(a) $\frac{1}{2}(z_1 + z_2)$ (b) $\frac{1}{3}(2z_1 + z_2)$

6) By solving the equation $z^3 = 1$, find the three cube roots of 1. If $\overrightarrow{OA}$, $\overrightarrow{OB}$ and $\overrightarrow{OC}$ represent the three cube roots on an Argand diagram, show that A, B and C lie equally spaced, on a circle of radius 1 and centre O. Write down each cube root in the form $r(\cos \theta + i \sin \theta)$.

PRODUCTS AND QUOTIENTS

Taking two complex numbers $z_1 = r_1(\cos \theta_1 + i \sin \theta_1)$ and $z_2 = r_2(\cos \theta_2 + i \sin \theta_2)$ we find that:

$z_1 z_2 = r_1 r_2 (\cos \theta_1 + i \sin \theta_1)(\cos \theta_2 + i \sin \theta_2)$

$\qquad = r_1 r_2 [\cos \theta_1 \cos \theta_2 + i \sin \theta_1 \cos \theta_2 + i \sin \theta_2 \cos \theta_1 + i^2 \sin \theta_1 \sin \theta_2]$

$\qquad = r_1 r_2 [\cos \theta_1 \cos \theta_2 - \sin \theta_1 \sin \theta_2 + i(\sin \theta_1 \cos \theta_2 + \cos \theta_1 \sin \theta_2)]$

$\qquad = r_1 r_2 [\cos (\theta_1 + \theta_2) + i \sin (\theta_1 + \theta_2)]$

i.e. $z_1 z_2$ gives a complex number with modulus $r_1 r_2$ and argument $\theta_1 + \theta_2$.

We also find that:

$$\frac{z_1}{z_2} = \frac{r_1(\cos\theta_1 + i\sin\theta_1)}{r_2(\cos\theta_2 + i\sin\theta_2)} = \frac{r_1}{r_2}\left[\frac{(\cos\theta_1 + i\sin\theta_1)(\cos\theta_2 - i\sin\theta_2)}{(\cos\theta_2 + i\sin\theta_2)(\cos\theta_2 - i\sin\theta_2)}\right]$$

which simplifies to $\dfrac{r_1}{r_2}[\cos(\theta_1 - \theta_2) + i\sin(\theta_1 - \theta_2)]$,

i.e. $\dfrac{z_1}{z_2}$ gives a complex number with modulus $\dfrac{r_1}{r_2}$ and argument $\theta_1 - \theta_2$.

These results can be expressed as follows:

$$|z_1 z_2| = |z_1||z_2|$$

$$\left|\frac{z_1}{z_2}\right| = \frac{|z_1|}{|z_2|}$$

$$\mathrm{Arg}(z_1 z_2) = \arg z_1 + \arg z_2$$

$$\mathrm{Arg}\left(\frac{z_1}{z_2}\right) = \arg z_1 - \arg z_2$$

Thus $\qquad |(3 + 4i)(5 - 12i)| = |3 + 4i||5 - 12i|$

$$= 5 \times 13$$

Similarly $\qquad \left|\dfrac{3 + 4i}{5 - 12i}\right| = \dfrac{5}{13}$

Further, a complex equation of the type $\left|\dfrac{z - 3}{z + 5i}\right| = 2$ can be transformed into $|z - 3| = 2|z + 5i|$.

The properties derived above for $z_1 z_2$ and z_1/z_2 can be illustrated on an Argand diagram as shown in the following example.

If $\qquad z_1 = -1 + \sqrt{3}i$ then $\quad r_1 = 2 \qquad$ and $\quad \theta_1 = \frac{2}{3}\pi$

and if $\quad z_2 = 1 + i \qquad$ then $\quad r_2 = \sqrt{2} \qquad$ and $\quad \theta_2 = \frac{1}{4}\pi$

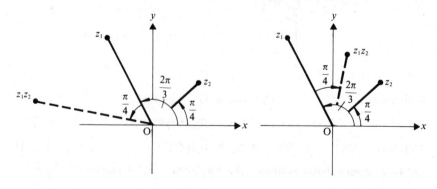

GREATEST AND LEAST VALUES OF $|z_1 + z_2|$

We know that for any two complex numbers z_1 and z_2 represented by $\overrightarrow{OA}$ and $\overrightarrow{OB}$, $z_1 + z_2$ is represented by $\overrightarrow{OC}$ where OACB is a parallelogram.

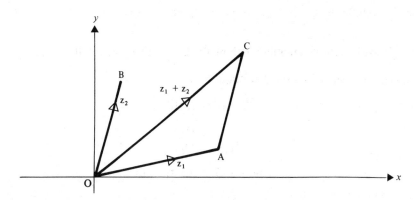

Now $\quad |z_1| = OA, \quad |z_2| = OB = AC \quad$ and $\quad |z_1 + z_2| = OC.$

In any triangle the sum of the lengths of any two sides is greater than the length of the third side.

Thus in triangle OAC

$$OA + AC > OC$$

i.e. $\qquad\qquad |z_1| + |z_2| > |z_1 + z_2|$

In the special case when OB and OA are themselves parallel
(i.e. $\arg z_2 = \arg z_1$) the parallelogram OACB becomes a straight line

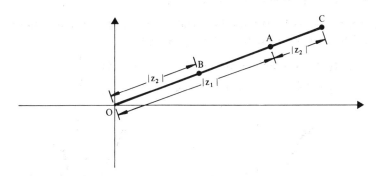

In this case $\qquad\qquad OA + AC = OC$

i.e. $\qquad\qquad |z_1| + |z_2| = |z_1 + z_2|$

Thus $\qquad\qquad |z_1 + z_2| \leqslant |z_1| + |z_2| \qquad\qquad\qquad$ [1]

i.e. the greatest possible value of $\quad |z_1 + z_2| \quad$ is $\quad |z_1| + |z_2|.$

Returning to triangle OAC, two further inequalities can be used:
i.e. $OC + CA > OA$ and $CO + OA > AC$.   These, together with the
extreme case when O, A, B and C are collinear, give the result

$$|z_1 + z_2| \geqslant |z_1| \sim |z_2| \qquad\qquad [2]$$

where $\sim$ means 'the positive difference between'.
Thus the least possible value of $|z_1 + z_2|$ is $|z_1| \sim |z_2|$.
Combining [1] and [2] we have

$$|z_1| \sim |z_2| \leqslant |z_1 + z_2| \leqslant |z_1| + |z_2|$$

Note. $|z_1| \sim |z_2|$ can also be written $\Big| |z_1| - |z_2| \Big|$.

EXAMPLES 13d

1) Write down the moduli and arguments of $-\sqrt3 + i$ and $4 + 4i$. Hence
express in the form $r(\cos\theta + i\sin\theta)$ the complex numbers $(-\sqrt3 + i)(4 + 4i)$
and $\dfrac{(-\sqrt3 + i)}{(4 + 4i)}$.

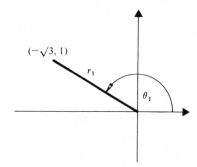

$$r_1 = |-\sqrt3 + i| = 2$$

$$\theta_1 = \arg(-\sqrt3 + i) = \frac{5\pi}{6}$$

$$r_2 = |4 + 4i| = 4\sqrt2$$

$$\theta_2 = \arg(4 + 4i) = \frac{\pi}{4}$$

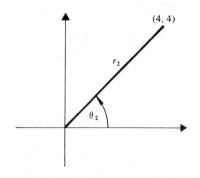

Let $\qquad (-\sqrt3 + i)(4 + 4i) = r_3(\cos\theta_3 + i\sin\theta_3)$

then $\qquad\qquad r_3 = r_1 r_2 = 8\sqrt2$

and
$$\theta_3 = \theta_1 + \theta_2 = \frac{5\pi}{6} + \frac{\pi}{4} = \frac{13\pi}{12}$$

Thus
$$(-\sqrt{3} + i)(4 + 4i) = 8\sqrt{2}\left(\cos\frac{13\pi}{12} + i\sin\frac{13\pi}{12}\right)$$

Note. Although this result is quite correct, the angle $\dfrac{13\pi}{12}$ is not the principal argument since it exceeds π.

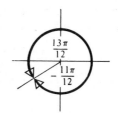

Using the principal argument, which is $-\dfrac{11\pi}{12}$, we can also express $(-\sqrt{3} + i)(4 + 4i)$ in the form $8\sqrt{2}\left[\cos\left(-\dfrac{11\pi}{12}\right) + i\sin\left(-\dfrac{11\pi}{12}\right)\right]$.

Let
$$\frac{(-\sqrt{3} + i)}{(4 + 4i)} = r_4(\cos\theta_4 + i\sin\theta_4)$$

then
$$r_4 = \frac{r_1}{r_2} = \frac{2}{4\sqrt{2}} = \frac{\sqrt{2}}{4}$$

and
$$\theta_4 = \theta_1 - \theta_2 = \frac{5\pi}{6} - \frac{\pi}{4} = \frac{7\pi}{12}$$

(this time the value we have found for θ_4 *is* the principal argument).

Thus
$$\frac{(-\sqrt{3} + i)}{(4 + 4i)} = \frac{\sqrt{2}}{4}\left(\cos\frac{7\pi}{12} + i\sin\frac{7\pi}{12}\right)$$

2) If $z_1 = 3 + 4i$ and $|z_2| = 13$, find the greatest value of $|z_1 + z_2|$. If $|z_1 + z_2|$ has its greatest value and also $0 < \arg z_2 < \dfrac{\pi}{2}$, express z_2 in the form $a + bi$.

Using
$$|z_1 + z_2| \leqslant |z_1| + |z_2|$$

we have
$$|z_1 + z_2| \leqslant \sqrt{3^2 + 4^2} + 13$$

Thus the greatest value of $|z_1 + z_2|$ is 18.

If $\overrightarrow{OA}$ and $\overrightarrow{OB}$ represent z_1 and z_2 where A is the point $(3, 4)$ and B is the point (a, b), both A and B are in the first quadrant $\left(0 < \arg z_2 < \dfrac{\pi}{2}\right)$.

Now if $\overrightarrow{OC}$ represents $z_1 + z_2$ and $|z_1 + z_2|$ has its greatest value, O, A, B and C are collinear.

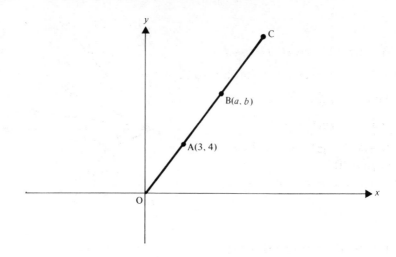

Hence   gradient OB = gradient OA,

i.e.
$$\frac{b}{a} = \frac{4}{3}$$ [1]

also
$$|z_2| = \sqrt{a^2 + b^2} = 13$$

i.e.
$$a^2 + b^2 = 169$$ [2]

Solving [1] and [2] simultaneously gives

$$a^2 + \left(\frac{4a}{3}\right)^2 = 169$$

i.e.
$$\frac{25a^2}{9} = 169$$

$\Rightarrow$
$$a = \pm \frac{39}{5}$$

But a is positive so $a = \frac{39}{5}$ and $b = \frac{52}{5}$

$\Rightarrow$
$$z_2 = \frac{39}{5} + \frac{52}{5}i$$

3) The point P, representing the complex number z on an Argand diagram, is in the first quadrant. Illustrate, on the same diagram, the points Q, R, S, representing $1/z$, z^2, and $z + 1/z$, marking any equal angles. Consider two cases, (a) when $|z| > 1$ (b) when $|z| < 1$.

(a) If $|z| > 1$ then $|1/z| < 1$

and $|z^2| > |z|$

Also $\arg 1/z = -\arg z$

and $\arg z^2 = 2 \arg z$

Hence, if $|z| > 1$ we have
diagram (i)

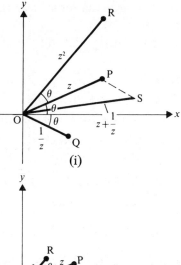

(i)

(b) If $|z| < 1$ then $|1/z| > 1$

and $|z^2| < |z|$

Hence, if $|z| < 1$ we have
diagram (ii)

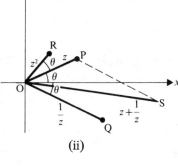

(ii)

EXERCISE 13d

1) Show that $|z_1 - z_2| \leqslant |z_1| + |z_2|$

and $|z_1 - z_2| \geqslant |z_1| \sim |z_2|$

Draw a clear diagram showing the case when

$$|z_1 - z_2| = |z_1| + |z_2|$$

2) $z_1 = 24 + 7i$ and $|z_2| = 6$.   Find the greatest and least values of $|z_1 + z_2|$.

3) Without first expressing them in the form $a + bi$, determine the modulus and argument of the following:

(a) $2(1 + i)$ (b) $(3 - \sqrt{3}i)(1 - i)$ (c) $\dfrac{(-2 - \sqrt{3}i)}{(\sqrt{3}i - 2)}$

In each case illustrate your result on an Argand diagram.

4) If $z = \cos \theta + i \sin \theta$ (i.e. $r = 1$), show that
$z^2 = \cos 2\theta + i \sin 2\theta$ and that $z^3 = \cos 3\theta + i \sin 3\theta$.
(Do not square or cube $\cos \theta + i \sin \theta$.)

5) Using the fact that if $z = r(\cos \theta + i \sin \theta)$ then
$z^2 = r^2(\cos 2\theta + i \sin 2\theta)$, find the square roots of $2\sqrt{3} - 2i$.
(*Hint*. Let $z^2 = 2\sqrt{3} - 2i$.)

6) Illustrate, on an Argand diagram, lines representing z, $1/z$, z^2 and $z-z^2$, if z is:

(a) $2+i$ (b) $\frac{1}{2}-\frac{1}{2}i$ (c) $3+4i$ (d) $\dfrac{\sqrt{3}}{2}+\dfrac{i}{2}$ (e) $5-12i$

SUMMARY

1) $(a+bi)\pm(c+di) \equiv (a\pm c)+(b\pm d)i$

$(a+bi)(c+di) \equiv (ac-bd)+(ad+bc)i$

$\dfrac{(a+bi)}{(c+di)} \equiv \dfrac{(a+bi)(c-di)}{(c+di)(c-di)} \equiv \dfrac{(a+bi)(c-di)}{c^2+d^2}$

2) $x+yi = a+bi \iff x=a$ and $y=b$

3) $|a+bi| = \sqrt{a^2+b^2}$

$\arg(a+bi) = \alpha$ where $\tan\alpha = \dfrac{b}{a}$ and $-\pi<\alpha\leqslant\pi$.

4) $x+yi = r(\cos\theta+i\sin\theta)$ where $r=|x+yi|$ and $\theta=\arg(x+yi)$.

5) If $z_1=x_1+y_1i$ and $z_2=x_2+y_2i$ then:

$$|z_1+z_2|\leqslant|z_1|+|z_2|$$
$$|z_1+z_2|\geqslant|z_1|\sim|z_2|$$

6) If $z=a+bi$, $\bar{z}=a-bi$, where $\bar{z}$ and z are conjugate.

7) If a quadratic or cubic equation has any complex roots, they occur in conjugate pairs.

8) If $z_1=r_1(\cos\theta_1+i\sin\theta_1)$ and $z_2=r_2(\cos\theta_2+i\sin\theta_2)$

then
$$|z_1z_2| = r_1r_2$$
$$\text{Arg } z_1z_2 = \theta_1+\theta_2$$
$$\left|\frac{z_1}{z_2}\right| = \frac{r_1}{r_2}$$
$$\text{Arg }\frac{z_1}{z_2} = \theta_1-\theta_2$$

MULTIPLE CHOICE EXERCISE 13

(Instructions for answering these questions are given on p. xii.)

TYPE I

1) The modulus of $12-5i$ is:

(a) 119   (b) 7   (c) 13   (d) $\sqrt{119}$ (e) $\sqrt{7}$.

2) On an Argand diagram OP represents a complex number z. The conjugate of z is $\bar{z}$.
If P and Q are the points $(3,5)$ and $(5,-3)$ then OQ represents:

(a) $-\bar{z}$ (b) $i\bar{z}$ (c) $-z$ (d) iz (e) $-iz$.

3) $\dfrac{3+2i}{3-2i}$ is equal to:

(a) $\dfrac{5+12i}{13}$ (b) $\dfrac{13+12i}{13}$ (c) $\dfrac{5+6i}{13}$ (d) $\dfrac{5+6i}{5}$ (e) $\dfrac{13+12i}{5}$.

4) When $\sqrt{3}-i$ is divided by $-1-i$ the modulus and argument of the quotient are respectively:

(a) $2\sqrt{2}, \dfrac{7\pi}{12}$ (b) $\sqrt{2}, -\dfrac{11\pi}{12}$ (c) $\sqrt{2}, \dfrac{7\pi}{12}$

(d) $2\sqrt{2}, -\dfrac{11\pi}{12}$ (e) $\sqrt{2}, \dfrac{11\pi}{12}$.

5) The equation $x^2+3x+1=0$ has:
(a) no roots
(b) one real and one complex root
(c) two imaginary roots
(d) two real roots
(e) two complex roots.

6) Expressed in the form $r(\cos\theta + i\sin\theta), -2+2i$ becomes:

(a) $2\left[\cos\left(-\dfrac{\pi}{4}\right)+i\sin\left(-\dfrac{\pi}{4}\right)\right]$ (b) $2\left(\cos\dfrac{3\pi}{4}+i\sin\dfrac{3\pi}{4}\right)$

(c) $2\sqrt{2}\left[\cos\left(-\dfrac{3\pi}{4}\right)+i\sin\left(-\dfrac{3\pi}{4}\right)\right]$ (d) $2\sqrt{2}\left[\cos\left(-\dfrac{\pi}{4}\right)+i\sin\left(-\dfrac{\pi}{4}\right)\right]$

(e) none of these.

7) If $z_1 = 2+i$ and $z_2 = 1+3i$ then z_1z_2 is represented by:

(a)
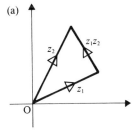

(b)
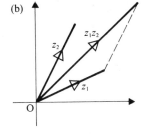

(c)
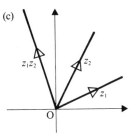

(d)
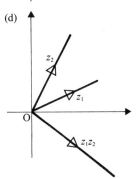

(e)
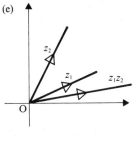

8) $|z_1 - z_2| \leqslant$

(a) $|z_1| - |z_2|$ (b) $|z_2| - |z_1|$ (c) $|z_1| \sim |z_2|$

(d) $|z_1| + |z_2|$ (e) None of these.

TYPE II

9) $-\dfrac{5\pi}{6}$ is an argument of:

(a) $\cos\dfrac{5\pi}{6} - i\sin\dfrac{5\pi}{6}$ (b) $\cos\dfrac{11\pi}{6} + i\sin\dfrac{11\pi}{6}$

(c) $-\sqrt{3} - i$ (d) $\sqrt{3} - i$ (e) $-\sqrt{3} + i$.

10) Which of the following lines can represent z if $z = -2 + 3i$?

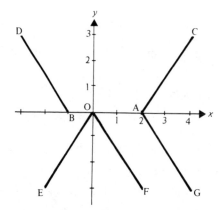

(a) $\overrightarrow{CA}$ (b) $\overrightarrow{BD}$

(c) $\overrightarrow{EO}$ (d) $\overrightarrow{OF}$

(e) $\overrightarrow{GA}$

11) $\bar{z}$ is the conjugate of z:

(a) $|\bar{z}| = |z|$ (b) $\arg z = \arg \bar{z}$ (c) $z\bar{z}$ is real   (d) $z/\bar{z}$ is real

(e) $\bar{z}$ is the mirror image of z in the y-axis.

12) In an Argand diagram, $\overrightarrow{OA}$ represents z, $\overrightarrow{OB}$ represents iz, $\overrightarrow{OC}$ represents i^2z $\overrightarrow{OD}$ represents $-iz$:

(a) ABCD is a straight line,

(b) $\overrightarrow{OC}$ represents $\bar{z}$,

(c) ABCD is a square,

(d) A, B, C and D lie on a circle,

(e) $\overrightarrow{BD}$ represents a real number.

13) If z is any cube root of unity, the value of $1 + z + z^2$ can be:

(a) 0   (b) 1   (c) 2   (d) 3   (e) -1.

TYPE III

14) (a) $z = 1 + i$

(b) $\arg z = \dfrac{\pi}{4}$

15) (a) The sum of the roots of the equation $ax^2 + bx + c = 0$, is $-\dfrac{b}{a}$

(b) The equation $ax^2 + bx + c = 0$ is such that $b^2 < 4ac$

16) (a) $|z_1| = |z_2|$
(b) $\arg z_1 = \arg z_2$

17) (a) $z = 9 - 16i$
(b) $\sqrt{z} = 3 - 4i$

TYPE V

18) A complex number has only one argument.

19) $|z_1 - z_2| \geqslant |z_1| \sim |z_2|$.

20) Any complex number whose modulus is unity can be expressed as $\cos\theta + i\sin\theta$.

21) If any cube root of 1 is squared, the result is another of the cube roots of 1.

22) A complex number $a + bi$ is zero if $a = -b$.

MISCELLANEOUS EXERCISE 13

1) The two complex numbers z_1, z_2 are represented on an Argand diagram. Show that $|z_1 + z_2| \leqslant |z_1| + |z_2|$.
If $|z_1| = 6$ and $z_2 = 4 + 3i$, show that the greatest value of $|z_1 + z_2|$ is 11 and find its least value. (U of L)p

2) If $z_1 = \dfrac{2-i}{2+i}$, $z_2 = \dfrac{2i - 1}{1 - i}$, express z_1 and z_2 in the form $a + ib$.
Sketch an Argand diagram showing points P and Q representing the complex numbers $5z_1 + 2z_2$ and $5z_1 - 2z_2$ respectively. (U of L)p

3) Given that $z = -1 + 3i$, express $z + \dfrac{2}{z}$ in the form $a + ib$, where a and b are real. (U of L)

4) (a) If $z = 4 - 3i$ express $z + \dfrac{1}{z}$ in the form $a + ib$.

(b) Find the two square roots of $4i$ in the form $a + ib$.

(c) If $z_1 = 5 - 5i$ and $z_2 = -1 + 7i$ prove that:

$$|z_1 + z_2| < |z_1 - z_2| < |z_1| + |z_2| \qquad \text{(C)}$$

5) Express the complex number $\dfrac{5 + 12i}{3 + 4i}$ in the form $a + ib$ and in the form $r(\cos\theta + i\sin\theta)$ giving the values of a, b, r, $\cos\theta$, $\sin\theta$. (C)p

6) The complex numbers $z_1 = \dfrac{a}{1+i}$, $z_2 = \dfrac{b}{1 + 2i}$ where a and b are real, are such that $z_1 + z_2 = 1$. Find a and b.
With these values of a and b, find the distance between the points which represent z_1 and z_2 in the Argand diagram. (JMB)

7) Find the modulus and argument of $z_1 = \sqrt{3} + i$. If $z_2 = \sqrt{3} - i$ express $q = z_1/z_2$ in the form $a + bi$ where a and b are real.
Plot z_1, z_2 and q on an Argand diagram and sketch the curve given by the equation $|z_1 - z_2| = |q - z_1|$. (U of L)

8) Given that $(1 + 5i)p - 2q = 3 + 7i$, find p and q when:

(a) p and q are real,

(b) p and q are conjugate complex numbers. (U of L)p

9) If $z_1 = 1 - i$ and $z_2 = 7 + i$, find the modulus of:

(a) $z_1 - z_2$ (b) $z_1 z_2$ (c) $\dfrac{z_1 - z_2}{z_1 z_2}$ (U of L)p

10) If $a = 3 - i$ and $b = 1 + 2i$, find the moduli of:

(a) $2a + 3b$ (b) $\dfrac{a}{2b}$ (C)p

11) The points A and B represent the complex numbers z_1 and z_2 respectively on an Argand diagram, where $0 < \arg z_2 < \arg z_1 < \dfrac{\pi}{2}$.
Give geometrical constructions to find the *points* C and D representing $z_1 + z_2$ and $z_1 - z_2$ respectively. (JMB)p

12) If $z = \cos\theta + i\sin\theta$ where θ is real, show that:

$$\frac{1}{1 + z} = \frac{1}{2}\left(1 - i\tan\frac{\theta}{2}\right)$$

Express: (a) $\dfrac{2z}{1 + z^2}$ (b) $\dfrac{1 - z^2}{1 + z^2}$ in the form $a + ib$ where a and b are real functions of θ. (C)

13) (a) Given that the complex number z and its conjugate $\bar{z}$ satisfy the equation
$$z\bar{z} + 2iz = 12 + 6i$$
find the possible values of z.

(b) Mark in an Argand diagram the points representing the complex numbers $4 + 3i$, $4 - 3i$, $\dfrac{4 + 3i}{4 - 3i}$.   (JMB)p

14) (a) If $z = 3 + 4i$, express $z + \dfrac{25}{z}$ in its simplest form.

(b) If $z = x + yi$, find the real part and the imaginary part of $z + \dfrac{1}{z}$.

(U of L)p

15) (a) Find the square roots of $(5 + 12i)$.

(b) Find the modulus and amplitude of each of the numbers
(i) $(1 - i)$ (ii) $(4 + 3i)$ (iii) $(1 - i)(4 + 3i)$.
If these numbers are represented in an Argand diagram by the points
A, B, C calculate the area of the triangle ABC.   (U of L)p

16) (a) Find the modulus and one value for the argument of $\dfrac{(i + 1)^2}{(i - 1)^4}$.

Find the two square roots of $5 - 12i$ in the form $a + bi$ where a and b are real. Show the points P and Q representing the square roots in an Argand diagram. Find the complex numbers represented by points R_1, R_2 such that the triangles PQR_1, PQR_2 are equilateral. (U of L)

17) (a) Find the modulus and the argument of each root of the equation
$$z^2 + 4z + 8 = 0$$
If the roots are denoted by α and β, simplify the expression
$$(\alpha + \beta + 4i)/(\alpha\beta + 8i)$$

(b) If $z_1 = -1 + i\sqrt{3}$ and $z_2 = \sqrt{3} + i$, show in an Argand diagram points representing the complex numbers z_1, z_2, $(z_1 + z_2)$ and (z_1/z_2).   (U of L)

18) (a) Express the following complex numbers in the form $a + ib$, where a and b are real:

(i) $\dfrac{1 - i}{(3 - i)^2}$ (ii) $(c + i)^4$, where c is real.

(b) If $z = x + iy$ and $z^2 = a + ib$ where x, y, a, b are real, prove that $2x^2 = \sqrt{(a^2 + b^2)} + a$.
By solving the equation $z^4 + 6z^2 + 25 = 0$ for z^2, or otherwise, express each of the four roots of the equation in the form $x + iy$.

(JMB)

CHAPTER 14

PERMUTATIONS AND COMBINATIONS

PERMUTATIONS AND COMBINATIONS

Three pictures are to be hung in line on a wall. Indicating the different pictures by A, B and C, one order in which they can be hung is A, B, C and another is A, C, B.

Each of these arrangements is called a *permutation* of the three pictures (and there are further possible permutations),

i.e. a permutation is an ordered arrangement of a number of items.

Suppose, however, that seven pictures are available for hanging and only three of them can be displayed. This time a choice has first to be made. Representing the seven pictures by A, B, C, D, E, F and G, one possible choice of the three pictures for display is A, B, and C. Regardless of the order in which they are then hung this group of three is just one choice and is called a combination.

thus

$$
\left.
\begin{array}{l}
A, B, C \\
A, C, B \\
B, A, C \\
B, C, A \\
C, A, B \\
C, B, A
\end{array}
\right\}
$$

are *six* different *permutations* but only *one combination*

i.e. a combination is an unordered selection of a number of items from a given set.

In this chapter we will investigate methods for finding the total number of ways of arranging items or choosing groups of items from a given set. But before we do so it is important to be able to distinguish between permutations and combinations.

Consider the following situation.

A street news vendor stocks ten weekly periodicals and has a display stand with five racks in a vertical column. He clearly cannot display all ten of his magazines so first he must *choose a group of five*. The order in which he picks up his chosen five periodicals is irrelevant; the set of five is only *one combination*. Once he has made his choice he is then able to place the five periodicals in various different orders on the display stand. He is now *arranging* them and each arrangement is a *permutation*,

i.e. a particular set of five periodicals is one combination but can be arranged to give several different permutations.

EXERCISE 14a

In each of the following problems determine, *without* working out the answer, whether you are asked to find a number of permutations, or a number of combinations.

1) How many arrangements of the letters A, B, C are there?

2) A team of six members is chosen from a group of eight. How many different teams can be selected?

3) A person can take eight records to a desert island, chosen from his own collection of one hundred records. How many different sets of records could he choose?

4) The first, second and third prizes for a raffle are awarded by drawing tickets from a box of five hundred. In how many ways can the prizes be won?

5) A London telephone number is a seven digit number. How many London telephone lines are available?

6) One red die and one green die are rolled (each numbered one to six). In how many ways can a total score of six be obtained?

PERMUTATIONS

We will now consider methods for finding a number of permutations in different types of problems.

1) How many arrangements of the letters A, B, C are there?

The first (i.e. L.H.) letter can be

$$\left.\begin{array}{c} A \\ \text{or} \\ B \\ \text{or} \\ C \end{array}\right\} \quad \text{i.e. there are three ways of choosing the first letter.}$$

When the first letter has been chosen there are two letters from which to choose the second; and the possible ways of choosing the first two letters are:

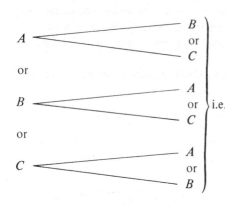

for *each* of the three ways of choosing the first letter there are two ways of i.e. choosing the second letter. Hence there are 3×2 ways of choosing the first two letters.

Having chosen the first two letters there is only one choice for the third letter, i.e. for *each* of the 3×2 ways of choosing the first two letters there is only one possibility for the third letter. Hence there are $3 \times 2 \times 1$ ways of arranging the three letters A, B, C.

2) How many three digit numbers can be made from the integers $2, 3, 4, 5, 6$ if:
(a) each integer is used only once,
(b) there is no restriction on the number of times each integer can be used?

(a) the first digit can be any of the integers $2, 3, 4, 5, 6$, i.e. there are five ways of writing down the first digit. Having written down the first digit there are four integers from which the second digit can be taken. So the possible ways of writing down the first two digits are:

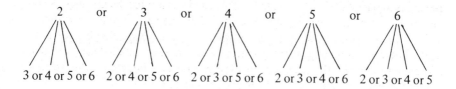

i.e. for *each* of the five ways of writing down the first digit there are four ways of writing down the second digit, i.e. there are 5×4 ways of writing down the first two digits.

The next digit can be taken from the three remaining integers, so for *each* of 5×4 ways of writing down the first two digits there are three ways of writing down the third digit,

i.e. there are 5 × 4 × 3 ways of making the three digit number, i.e. there
are 60 such numbers.

(b) In this problem each integer from the set may be selected up to three times
Thus there are 5 ways of taking the first integer and as the integer used for
the first place can be used again for the second place there are 5 ways of
writing down the second digit. Similarly there are 5 ways of taking the
third digit,
i.e. there are 5 × 5 × 5 ways of making the three digit number, so there are
125 such numbers.

We will now look at some examples where the possible arrangements are
restricted in some way.

3) How many arrangements of the letters of the word BEGIN are there which
start with a vowel?

Starting with the first letter we see that there are only 2 possibilities,
i.e. E or I.

Having taken the first letter there are 4 possibilities for the second letter, so
for *each* of the 2 ways of taking the first letter there are 4 ways of taking the
second letter, so there are 2 × 4 ways of taking the first two letters.

Having removed the first two letters there are 3 possibilities left for the third
letter, and having selected the third letter there are 2 possibilities for the fourth
letter and only 1 possibility for the fifth letter.

Hence there are 2 × 4 × 3 × 2 × 1 ways of arranging the letters B, E, G, I, N
in which a vowel comes first, i.e. there are 48 such arrangements.

4) How many even numbers greater than 2000 can be made from the integers
1, 2, 3, 4, if each integer is used only once?

There are two restrictions on the permutations here:
(a) the number is even so the last digit must be either 2 or 4,
(b) the number exceeds 2000 so the first digit must be either 2, 3 or 4.

Beginning with the last digit, we see from (a) that there are only 2 possibilities.
Having used one of these there are only 2 possibilities left for the first digit. So for
each of the 2 ways of writing down the last digit there are 2 ways of writing
down the first digit, i.e. there are 2 × 2 ways of writing down the first and last
digits. There are now only 2 integers from which the second digit can be taken.
Having written down the second digit this leaves only 1 integer for third place.
Therefore there are 2 × 2 × 2 × 1 = 8 such numbers.
This argument can be seen clearly in the table below.

Number of	1st digit	2nd digit	3rd digit	4th digit
possibilities	2	2	1	2

Note that we start with the item in the arrangement whose choice from the
given set is most restricted.

4) The representatives of five countries attend a conference. In how many ways can they be seated at a round table?

As they are seated at a round table, there is no first or last place to consider. What matters in this arrangement is where each one sits relative to the others, as no one seat is special.

Numbering the chairs 1 to 5, we can say that there are 5 ways of selecting the occupant of the first seat, 4 ways of selecting the occupant of the second seat, ... and so on. Thus there are $5 \times 4 \times 3 \times 2 \times 1$ ways of arranging the representatives in the five seats. But this number includes the arrangements shown below:

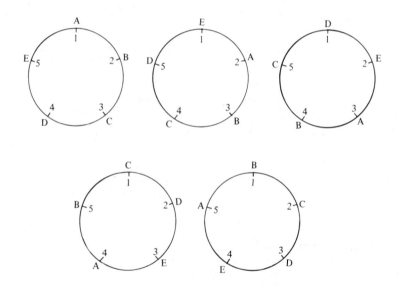

i.e. for any *one* arrangement in the five seats, the representatives can be moved clockwise *five* times and still have the *same* people to the left and to the right of them.

Therefore the $5 \times 4 \times 3 \times 2 \times 1$ ways of arranging the five representatives in numbered seats is *five times* the number of ways of arranging them round a circular table, so there are $\dfrac{5 \times 4 \times 3 \times 2 \times 1}{5}$ ways, or $4 \times 3 \times 2 \times 1$ ways in which the representatives can be seated at the round table.

[This problem does assume that all the chairs are identical and that there are no distinguishing features of the room which might affect the choice of position of a particular delegate (such as not wishing to sit with his back to the window). If considerations such as these do affect a particular arrangement they will be stated.]

5) In how many ways can five beads, chosen from eight different beads be threaded on to a ring?

The number of ways of arranging five beads, taken from eight different beads, in five numbered places is $8 \times 7 \times 6 \times 5 \times 4$.

Thus the number of ways of arranging five (from eight) beads in a circle is

$$\frac{8 \times 7 \times 6 \times 5 \times 4}{5} \quad \text{as:}$$

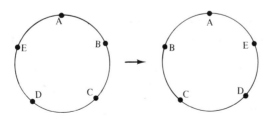

But a ring can be turned over,

i.e.

and these have been counted as two separate arrangements. So the number of circular arrangements is *twice* the number of arrangements on a ring.

Therefore the number of ways of threading five beads, from eight different beads, on a ring is $\dfrac{8 \times 7 \times 6 \times 5 \times 4}{5 \times 2} = 672.$

EXERCISE 14b

1) In how many ways can five different books be arranged on a shelf?

2) How many two digit numbers can be made from the set
$\{2, 3, 4, 5, 6, 7, 8, 9\}$, each number containing two different digits?

3) In how many ways can six different shrubs be planted in a row?

4) How many four digit odd numbers can be made from the set
$\{5, 7, 8, 9\}$, no integer being used more than once?

5) In how many ways can eight people be seated at a round table?

6) How many numbers greater than 4000 can be made from the set
$\{1, 3, 5, 7\}$, if each integer can be used only once?

7) How many arrangements can be made of three letters chosen from PEAT
if the first letter is a vowel and each arrangement contains three different
letters?

8) A telephone dial is numbered 0 to 9. If the 0 is dialled first, the caller
is connected to the STD system. How many local calls (i.e. calls not going
through the STD system) can be rung, if a local number has five digits?

9) How many three digit numbers can be made from the set of integers
$\{1, 2, 3, 4, 5, 6, 7, 8, 9\}$ if:
(a) the three digits are all different,
(b) the three digits are all the same,
(c) the number is greater than 600,
(d) all three digits are the same and the number is odd?

10) Three boxes each contain three identical balls. The first box has red balls
in it, the second blue balls and the third green balls. In how many ways can
three balls be arranged in a row if:
(a) the balls are of different colours,
(b) all three balls are of the same colour?

11) In how many ways can eight cows be placed in a circular milking parlour?

12) In how many ways can six different coloured beads be arranged on a ring?

COMBINATIONS

Consider the number of arrangements of four different books on a shelf.
We have seen that the number of permutations of the four books is

$$4 \times 3 \times 2 \times 1 = 24.$$

But if the order does not matter, there is only one combination of books.
If there are five different books available and only four of them can be placed

on the shelf, the total number of ways in which this can be done is
$$5 \times 4 \times 3 \times 2 = 120.$$
But, for each particular set of four books,
$$4 \times 3 \times 2 \times 1 \text{ permutations} \equiv \text{one combination}$$
i.e. for any one of the 120 permutations there are 23 other permutations of the same combination of four books.

So the *number of different sets* of four books taken from the five available

books is given by $\quad \dfrac{\text{total number of permutations}}{\text{number of permutations of each set}}$

Generalising this argument we see that if we have n objects from which we select groups of r objects, the total number of possible groups (combinations)

is given by $\quad \dfrac{\text{number of permutations of } r \text{ objects from } n \text{ objects}}{\text{number of permutations of } r \text{ objects among themselves}}$

EXAMPLES 14c

1) How many different hands of four cards can be dealt from a pack of fifty two playing cards?

The order in which the cards are dealt is irrelevant, it is the particular set of four cards that matters.

The number of ways of arranging four cards from fifty two is
$$52 \times 51 \times 50 \times 49.$$
The number of ways of arranging any one set of four cards among themselves is $\qquad\qquad 4 \times 3 \times 2 \times 1.$

Therefore the number of combinations of four cards from fifty two cards is

$$\frac{52 \times 51 \times 50 \times 49}{4 \times 3 \times 2 \times 1} = 270\,725$$

In the following examples we investigate problems when the selection is restricted.

2) In how many ways can five boys be chosen from a class of twenty boys if the class captain has to be included?

As one particular boy has to be chosen this leaves four more boys to be chosen from the remaining nineteen boys.

The number of ways of arranging four boys from nineteen is $19 \times 18 \times 17 \times 16$.
The number of ways of arranging the four among themselves is $4 \times 3 \times 2 \times 1$.
Therefore the number of ways of choosing the four boys to join the captain is

$$\frac{19 \times 18 \times 17 \times 16}{4 \times 3 \times 2 \times 1} = 3876$$

3) A bowl of fruit contains a large number of apples, pears, oranges and bananas. How many different groups of three fruits can be chosen if two of them are of the same variety?

The single fruit can be chosen in 4 ways.
The two fruits of the same variety can then be chosen in 3 ways.
(These two must be a variety different from the first, otherwise all three would be the same.)
Hence there are 4 × 3 ways of choosing three fruits, two of which are the same, i.e. the number of combinations is 12.

4) In how many ways can a party of ten children be divided into two groups of five children?

The number of combinations of five children chosen from ten is
$$\frac{10 \times 9 \times 8 \times 7 \times 6}{5 \times 4 \times 3 \times 2 \times 1} = 252$$
Whenever one group of five children is selected, the remaining five children automatically form the other group.
Using the letters A to J to identify the children, two possible selections for the first group are ABCDE and FGHIJ.
The children in the corresponding second group are FGHIJ and ABCDE respectively.
But the division of the ten children into the groups ABCDE/FGHIJ and FGHIJ/ABCDE is the *same* division,
i.e. the 252 combinations of five selected from ten is *twice* the number of divisions into two *equal* groups of five.
Hence there are 126 ways in which the children can be divided into two equal groups.

EXERCISE 14c

1) How many different combinations of six letters can be chosen from the letters A, B, C, D, E, F, G, H, if each letter is chosen only once?

2) In how many ways can the eight letters in Question 1 be divided into two groups of six and two letters?

3) A team of four children is to be selected from a class of twenty children, to compete in a quiz game. In how many ways can the team be chosen if:
(a) any four can be chosen,
(b) the four chosen must include the oldest in the class?

4) A shop stocks ten different varieties of packet soup. In how many ways can a shopper buy three packets of soup if:
(a) each packet is a different variety,
(b) two packets are the same variety?

5) In how many ways can ten different books be divided into two groups of six books and four books?

6) How many different hands of five cards can be dealt from a suit of thirteen cards?

7) In Question 6, if one of the cards dealt is the ace, how many different hands of five cards are there?

8) A large box of biscuits contains nine different varieties. In how many ways can four biscuits be chosen if:
(a) all four are different,
(b) two are the same and the others different,
(c) two each of two varieties are selected,
(d) three are the same and the fourth is different,
(e) all four are the same?

THE FACTORIAL NOTATION

Consider the number of ways of arranging a pack of fifty two playing cards in a row.
This is $52 \times 51 \times 50 \times 49 \times \ldots \times 3 \times 2 \times 1$, which is a *very* large number, is difficult to multiply out, and is cumbersome to write even when left in factor form as above. So we denote this clumsy product by $52!$

In general, $n!$ represents the number $n \times (n-1) \times (n-2) \times \ldots \times 2 \times 1$, i.e. $n!$ means the product of all the integers from 1 to n inclusive and is called 'n factorial'.

EXAMPLES 14d

1) Evaluate $\dfrac{10!}{7!}$.

$$\frac{10!}{7!} = \frac{10 \times 9 \times 8 \times 7 \times 6 \times 5 \times 4 \times 3 \times 2 \times 1}{7 \times 6 \times 5 \times 4 \times 3 \times 2 \times 1} = 10 \times 9 \times 8 = 720$$

Note that we have cancelled by $7!$

2) Write $52 \times 51 \times 50 \times 49 \times 48 \times 47$ in factorial form.

$52 \times 51 \times \ldots \times 47$ is $52!$ with the factors $46 \times 45 \times \ldots \times 3 \times 2 \times 1$ (i.e. $46!$) missing.
Multiplying and dividing by $46!$ we have

$$52 \times 51 \times 50 \times 49 \times 48 \times 47$$

$$= \frac{(52 \times 51 \times 50 \times \ldots \times 47) \times (46 \times 45 \times \ldots \times 2 \times 1)}{46!} = \frac{52!}{46!}$$

3) Evaluate $\dfrac{20!}{15!\,5!}$.

Cancelling by the larger factorial in the denominator gives

$$\frac{20!}{15!\,5!} = \frac{20 \times 19 \times 18 \times 17 \times 16}{5 \times 4 \times 3 \times 2 \times 1} = 15\,504$$

4) Write $\dfrac{9 \times 8 \times 7}{4 \times 3 \times 2}$ in factorial notation.

$$9 \times 8 \times 7 = \frac{9!}{6!}$$

Therefore

$$\frac{9 \times 8 \times 7}{4 \times 3 \times 2} = \frac{9!}{6!\,4!}$$

5) Factorise $8! - 4(7!)$.

As $\quad 8! = 8 \times 7!$

$\qquad\qquad\qquad 8! - 4(7!)$ has a common factor of $7!$

i.e. $\qquad\qquad 8! - 4(7!) = 7!(8 - 4) = 4(7!)$

6) Factorise $\quad (n + 2)! + n^2(n - 1)!$

$$n^2(n - 1)! = [n \times n] \times [(n - 1) \times (n - 2) \times \ldots \times 3 \times 2 \times 1]$$

$$= n \times [n \times (n - 1) \times (n - 2) \times \ldots \times 3 \times 2 \times 1]$$

$$= n \times n!$$

Also $\qquad (n + 2)! = (n + 2) \times (n + 1) \times n!$

Hence $\qquad (n + 2)! + n^2(n - 1)! = (n + 2)(n + 1)n! + n(n!)$

$$= n!\,[(n + 2)(n + 1) + n]$$

$$= n!\,[n^2 + 4n + 2]$$

7) Express as a single fraction $\dfrac{n!}{(n - r)!\,r!} + \dfrac{2(n - 1)!}{(r - 1)!(n - r + 1)!}$.

The numerators of both fractions have a common factor $(n - 1)!$
The denominators of both fractions have a common factor $(n - r)!(r - 1)!$

[**Note** $\quad (n - r + 1)! = (n - r + 1)(n - r)(n - r - 1) \times \ldots \times 3 \times 2 \times 1$

$$= (n - r + 1)(n - r)!$$

and $\qquad\qquad r! = r(r - 1)!]$

Hence

$$\frac{n!}{(n-r)!(r!)} + \frac{2(n-1)!}{(r-1)!(n-r+1)!} = \frac{(n-1)!}{(n-r)!(r-1)!}\left(\frac{n}{r} + \frac{2}{n-r+1}\right)$$

$$= \frac{(n-1)!}{(n-r)!(r-1)!}\left(\frac{n(n-r+1)+2r}{r(n-r+1)}\right)$$

$$= \frac{(n-1)!}{(n-r)!(r-1)!}\left(\frac{n^2-nr+n+2r}{r(n-r+1)}\right)$$

$$= \frac{(n-1)!}{(n-r+1)!\,r!}(n^2-nr+n+2r)$$

EXERCISE 14d

Evaluate:

1) $3!$ 2) $4!$ 3) $5!$ 4) $6!$ 5) $\dfrac{6!}{4!}$ 6) $\dfrac{12!}{10!}$

7) $\dfrac{15!}{12!}$ 8) $\dfrac{7!}{3!}$ 9) $\dfrac{8!}{2!\,6!}$ 10) $\dfrac{20!}{17!\,3!}$ 11) $\dfrac{15!}{6!\,7!}$ 12) $\dfrac{9!}{2!\,3!\,4!}$

13) $\dfrac{8!}{(4!)^2}$ 14) $\dfrac{(3!)^2}{2!\,4!}$

Write in factorial form:

15) $5 \times 4 \times 3$

16) 11×10

17) $39 \times 38 \times 37 \times 36 \times 35$

18) $n(n-1)(n-2)(n-3)$

19) $(n+1)(n)(n-1)$

20) $(n+5)(n+4)(n+3)(n+2)(n+1)$

21) $(n+r)(n+r-1)(n+r-2)$

22) $\dfrac{20 \times 19 \times 18}{3 \times 2 \times 1}$

23) $\dfrac{14 \times 13}{3 \times 2 \times 1}$

24) $\dfrac{8 \times 7 \times 6}{6 \times 5 \times 4}$

25) $\dfrac{n(n-1)(n-2)}{3 \times 2 \times 1}$

26) $\dfrac{(n-2)(n-3)(n-4)}{4 \times 3 \times 2 \times 1}$

Factorise:

27) $8! + 9!$ 28) $7! - 2(5!)$ 29) $3(10!) + 4(8!)$

30) $n! + (n-1)!$ 31) $(n+1)! - (n-1)!$ 32) $n^2(n-1)! + 2n(n-2)!$

33) $n! + (n-1)! + (n-2)!$ 34) $\dfrac{7!}{3!\,4!} + \dfrac{7!}{2!\,5!}$

35) $\dfrac{7!}{3!} + \dfrac{6!}{4!}$

36) $\dfrac{10!}{8!\,2!} + \dfrac{9!}{7!\,2!} + \dfrac{8!}{6!\,2!}$

37) $\dfrac{n!}{r!} + \dfrac{(n-1)!}{(r+1)!}$

38) $\dfrac{2(n+1)!}{r!} - \dfrac{3n!}{(r+1)!}$

39) $\dfrac{n!}{r!(n-r)!} + \dfrac{2(n+1)!}{r!(n-r+1)!}$

40) $\dfrac{n!r^2}{(n-r)!r!} - \dfrac{2(n-1)!}{(r-1)!(n-r)!}$

FURTHER WORK ON PERMUTATIONS AND COMBINATIONS

We have seen that the number of different ways of arranging five objects chosen from eight different objects is $8 \times 7 \times 6 \times 5 \times 4$ or, using factorial notation, $\dfrac{8!}{3!}$.

From this we can generalise to say that:

the number of permutations of r objects chosen from a set of n different objects is $\dfrac{n!}{(n-r)!}$.

Using ^nP_r as a symbol for 'the number of permutations of r objects taken from n different objects' we may write

$$^nP_r = \frac{n!}{(n-r)!}$$

We have also seen that the number of ways of choosing (irrespective of order) five objects from eight is $\dfrac{8 \times 7 \times 6 \times 5 \times 4}{5 \times 4 \times 3 \times 2 \times 1}$ or $\dfrac{8!}{5!\,3!}$.

Generalising again we can say that:

the number of combinations of r objects selected from n different objects is $\dfrac{n!}{(n-r)!\,r!}$.

Using ^nC_r as a symbol for 'the number of combinations of r objects taken from n different objects' we may write

$$^nC_r = \frac{n!}{(n-r)!\,r!}$$

Thus $\qquad ^6P_4 = \dfrac{6!}{2!} = 6 \times 5 \times 4 \times 3 = 360$

and $\qquad ^6C_4 = \dfrac{6!}{2!\,4!} = \dfrac{6 \times 5}{2 \times 1} = 15$

Also $\qquad ^nC_{n-r} = \dfrac{n!}{[n-(n-r)]!(n-r)!} = \dfrac{n!}{r!(n-r)!} = {}^nC_r$

i.e.
$$^nC_{n-r} = {}^nC_r$$

Now ^nC_n means the number of ways of choosing n objects from n objects, and as there is obviously only one way of choosing n objects from n objects, $^nC_n = 1$.

But using the definition

$$^nC_r = \frac{n!}{r!(n-r)!}$$

in the case when $r = n$ we have

$$^nC_n = \frac{n!}{n!0!}$$

and this is equal to unity only if we define $0!$ as having the value 1, i.e.
$$0! = 1$$

We will now consider some more varied problems involving permutations and combinations in which we will use the factorial notation.

Permutations Involving some Identical Objects

Consider the number of possible arrangements of the letters of the word DIGIT.

This word contains two I's which are identical, but which can be distinguished by adding suffixes, i.e. $D\,I_1\,G\,I_2\,T$.

Then the number of permutations of the letters $D\,I_1\,G\,I_2\,T$ is 5!

But this number includes separately the two permutations

$$D\,I_1\,G\,I_2\,T \quad \text{and} \quad D\,I_2\,G\,I_1\,T$$

so that the arrangement DIGIT is counted twice in the 5! permutations.

Because I_1 and I_2 can be arranged in 2! ways, every distinct arrangement of the letters DIGIT is included 2! times in the permutations of $D\,I_1\,G\,I_2\,T$.

Hence the number of arrangements of the letters $D\,I\,G\,I\,T$ is $\dfrac{5!}{2!} = 60$.

Now consider the number of permutations of the letters

$$D\,E\,F\,E\,A\,T\,E\,D$$

There are three E's and two D's.

The number of permutations of $D_1\,E_1\,F\,E_2\,A\,T\,E_3\,D_2$ is 8!

But E_1, E_2, E_3 can be arranged in 3! ways, and D_1, D_2 can be arranged in 2! ways, so

the number of arrangements of $D_1\,E_1\,F\,E_2\,A\,T\,E_3\,D_2$ is $3! \times 2!$ times the number of arrangements of $D\,E\,F\,E\,A\,T\,E\,D$.

Therefore the number of permutations of $D\,E\,F\,E\,A\,T\,E\,D$ is $\dfrac{8!}{3!\,2!} = 3360$.

Generalising this argument we see that,

the number of permutations of n objects, r of which are identical is $\dfrac{n!}{r!}$.

Further, the number of permutations of n objects, p of which are alike and q of which are alike (but different from the set of p objects), is $\dfrac{n!}{p!\,q!}$.

EXAMPLES 14e

1) In how many of the possible permutations of the letters of the word ADDING are the two D's:
(a) together,   (b) separate?

(a) The number of permutations in which the D's are together can be found easily by bracketing the D's and treating them as one item in the arrangements of A, (DD), I, N, G. There are now five different items which can be arranged in 5! ways.

(b) As the D's are either together or separate,

(number of permutations without restriction)

$-$ (number of arrangements with D's together)

$=$ (number of arrangements with D's separate)

Now the number of arrangements without restriction is $\dfrac{6!}{2!}$.

Hence the number of arrangements in which the D's are separated is
$\dfrac{6!}{2!} - 5! = 240$.

Note. The number of permutations in which D and A are next to each other is found in a similar way to (a) above, but these two letters can be written (DA) or (AD). Therefore there are twice as many arrangements of A, D, D, I, N, G in which A and D are adjacent than when the two D's are adjacent.

Independent Permutations or Combinations

Consider the number of ways in which three bottles of wine and ten cans of beer can be selected from a cellar containing thirty different bottles of wine and fifteen different cans of beer.
The choice of the bottles of wine in no way affects the choice of the cans of beer,
i.e. these two combinations are independent of each other.
The three bottles of wine can be chosen in $^{30}C_3$ ways. The ten cans of beer can be chosen in $^{15}C_{10}$ ways. For *each* of the $^{30}C_3$ ways of choosing the bottles of wine there are $^{15}C_{10}$ ways of choosing the cans of beer.

So there are $$^{30}C_3 \times {}^{15}C_{10}$$

ways of choosing the bottles of wine and the cans of beer.

Consider the number of ways in which a code number of two letters followed by three digits can be made if no letter or digit is repeated in any one code. In this problem we are considering the number of permutations of two letters followed by the number of permutations of three digits. These two sets of permutations are independent of each other as the arrangement of the letters has no affect on the arrangement of the digits. There are $^{26}P_2$ ways of arranging the letters and $^{10}P_3$ ways of arranging the digits in the code. For *each* of the $^{26}P_2$ permutations of the letters there are $^{10}P_3$ permutations of the digits.
Therefore the number of permutations of two letters followed by three digits is $^{26}P_2 \times {}^{10}P_3$.
Generalising from these examples we can say that:
the number of permutations, P_1, of objects from one set combined with the number of permutations, P_2, of objects from an *independent* set is $P_1 \times P_2$. This is also true for combinations of objects from two independent sets and can be extended to cover more than two sets.

Mutually Exclusive Permutations or Combinations

Care must be taken to distinguish between problems involving:

(a) *both* one set of objects *and* another set of objects, which involves a product of permutations, or

(b) *either* one set of objects *or* another set of objects.

Consider, for example, the number of ways in which a number greater than 20 can be made from the integers 2, 3, 4, no integer being repeated.
The number may contain *either* two digits *or* it may contain three digits. It *cannot* contain *both* two digits *and* three digits. Such sets of permutations are said to be mutually exclusive.
The number of ways of making a two digit number is $^3P_2$.
The number of ways of making a three digit number is $^3P_3$.
These two cases cover all possible permutations,
so there are $^3P_2 + {}^3P_3$ numbers greater than 20.
Now consider the number of ways in which a class of twenty children can be divided into two groups of six and fourteen respectively, if the class contains one set of twins who are not to be separated.
If we consider the number of ways of choosing six children from twenty, the choice of the six will determine the members of the other group.
However in choosing six children we have to consider two cases, viz,
either six children including the twins
or   six children excluding the twins.

As a combination of six children cannot both include and exclude the twins, the two separate cases are mutually exclusive.

If the twins are included, four more children must be chosen from the remaining eighteen, and this can be done in $^{18}C_4$ ways.

If the twins are excluded, the group can be chosen in $^{18}C_6$ ways.

These two cases cover all possible combinations, so there are

$$^{18}C_4 + {}^{18}C_6$$

ways of dividing the children into two groups as stated.

Generalising from these two examples we can say that if a choice (ordered or not) divides into two cases (A and B) which are mutually exclusive then

(number of choices of *either A or B*)

= (number of choices of A) + (number of choices of B)

This can be extended to cover more than two mutually exclusive cases.

To summarise if P_1 and P_2 are two permutations (or combinations) whose values are p_1 and p_2 then

when P_1 and P_2 are independent

$$P_1 \text{ combined with } P_2 \text{ (i.e. } P_1 \text{ and } P_2) = p_1 \times p_2$$

when P_1 and P_2 are mutually exclusive

$$(\text{either } P_1 \text{ or } P_2) = p_1 + p_2.$$

EXAMPLES 14e (continued)

2) How many permutations of the letters of the word DEFEATED are there in which the E's are separated from each other?

The most direct way of answering this problem is first to remove the E's and consider the number of permutations of the letters D F A T D, of which there are $\dfrac{5!}{2!} = 60$.

If for any one of these permutations (e.g. D F A T D) the E's are reinserted so as to be separated from each other, the three E's can be placed in three of the six positions indicated.

$$\uparrow D \uparrow F \uparrow A \uparrow T \uparrow D \uparrow$$

The number of ways of choosing three positions from six is $^6C_3 = 20$.

Hence there are twenty ways of inserting the E's into *each* of the sixty permutations of DFATD, so there are 20×60, or 1200, permutations of the letters of DEFEATED in which no two E's are adjacent.

(**Note** that the permutations of DFATD and the choice of the possible positions for the E's are independent.)

3) A box contains seven snooker balls, three of which are red, two black, one white and one green. In how many ways can three balls be chosen?

This problem divides itself into three mutually exclusive cases, viz.

(a) Three balls of the same colour;
there is only one way of choosing these (i.e. the three red balls).

(b) Two balls of the same colour, the other being of a different colour.
As there are two ways of choosing the balls of the same colour (two red or two black) and three other colours in each case from which the third ball can be chosen, there are 2×3, i.e. 6 ways of choosing three balls, when two of them are of the same colour.

(c) All three balls of different colours.
There are four colours to choose from, so the three balls can be chosen in $^4C_3$ ways, i.e. 4 ways.

Thus there are $1 + 6 + 4$ ways of choosing the three balls, i.e. 11 ways.

4) Four books are taken from a shelf of eighteen books, of which six are paperback, and twelve are hardback. In how many of the possible combinations of four books is at least one a paperback?

If at least one paperback is to be included then the combinations could

include   one paperback,
or   two paperbacks,
or   three paperbacks,
or   four paperbacks.

As all of these cases are mutually exclusive we could work out the number of combinations in each case and sum them to find the number of combinations which include at least one paperback.

However a more direct approach is to say that the only combination we do *not* want is the one made up entirely of hardbacks. So we need consider only two cases,

i.e.   any four books   (there are $^{18}C_4$ such combinations)

and   four hardbacked books   (there are $^{12}C_4$ such combinations)

Hence $^{18}C_4 = {}^{12}C_4 +$ (number of combinations with at least one paperback).
Therefore the number of combinations including at least one paperback is

$$^{18}C_4 - {}^{12}C_4 = \frac{18!}{4!\,14!} - \frac{12!}{4!\,8!}$$

$$= 2565$$

EXERCISE 14e

1) Find how many numbers between 10 and 300 can be made from the digits $1, 2, 3$, if:
(a) each digit may be used only once,
(b) each digit may be used more than once?

2) How many combinations of three letters taken from the letters A, A, B, B, C, C, D are there?

3) A mixed team of ten players is chosen from a class of thirty, eighteen of whom are boys and twelve of whom are girls. In how many ways can this be done if the team has five boys and five girls?

4) Find the number of permutations of the letters of the word
MATHEMATICS.

5) Find the number of permutations of four letters from the word
MATHEMATICS.

6) How many of the permutations in Question 5 contain two pairs of letters that are the same?

7) Find the number of ways in which twelve children can be divided into two groups of six if two particular boys must be in different groups.

8) In how many of the permutations in Question 4 do all the consonants come together?

9) A team of two pairs, each consisting of a man and a woman, is chosen to represent a club at a tennis match. If these pairs are chosen from five men and four women, in how many ways can the team be selected?

10) A bridge team of four is chosen from six married couples to represent a club at a match. If a husband and wife cannot both be in the team, in how many ways can the team be formed?

11) Two sets of books contain five novels and three reference books respectively. In how many ways can the books be arranged on a shelf if the novels and reference books are not mixed up?

12) A box contains ten bricks, identical except for colour. Three bricks are red, two are white, two are yellow, two are blue and one is black. In how many ways can three distinguishable bricks be
(a) taken from the box   (b) arranged in a row?

13) In how many of the arrangements in a row of all ten bricks in Question 12 are
(a) the three red bricks separated from each other,
(b) just two of the red bricks next to each other?

14) In a multiple choice question there is one correct answer and four wrong answers to each question. For two such questions, in how many ways is it possible to select the wrong answer to both questions?

15) In Question 14, if a correct answer scores one mark and a wrong answer scores zero, in answering three such questions in how many ways is it possible to score:
(a) 0,   (b) 1,   (c) 2?

16) How many even numbers less than 500 can be made from the integers 1, 2, 3, 4, 5, each integer being used only once?

17) Four boxes each contain a large number of identical balls, those in one box are red, those in a second box are blue, those in a third box are yellow and those in the remaining box are green. In how many ways can five balls be chosen if:
(a) there is no restriction   (b) at least one ball is red?

18) How many different hands of four cards can be dealt from a pack of fifty two playing cards if at least one of the cards is an ace?

19) In how many ways can four tins of fruit be chosen from a supermarket offering ten varieties if at least two of the tins are of the same variety?

20) In how many ways can three letters from the word GREEN be arranged in a row if at least one of the letters is E?

SUMMARY

$$n! = n(n-1)(n-2)(n-3) \ldots \ldots (3)(2)(1)$$

$$0! = 1$$

$$^nP_r = \frac{n!}{(n-r)!} \qquad ^nC_r = \frac{n!}{(n-r)!\,r!} \qquad ^nC_r = {}^nC_{n-r}$$

The number of permutations of r objects chosen from n *different* objects is
$$^nP_r = \frac{n!}{(n-r)!}$$

The number of permutations of n objects, r of which are identical, is $\dfrac{n!}{r!}$

The number of circular arrangements of r objects chosen from n *different* objects is
$$\text{objects is } \frac{n!}{(n-r)!\,r}$$

The number of combinations of r objects chosen from n *different* objects is
$$\frac{n!}{(n-r)!\,r!}$$

If P_1 and P_2 are two permutations (or combinations) whose values are p_1 and p_2 then:
when P_1 and P_2 are independent

$$P_1 \text{ combined with } P_2 \text{ (or } P_1 \text{ and } P_2) = p_1 \times p_2$$

when P_1 and P_2 are mutually exclusive

$$(\text{either } P_1 \text{ or } P_2) = p_1 + p_2.$$

MISCELLANEOUS EXERCISE 14

1) Find the number of ways in which a committee of 4 can be chosen from 6 boys and 6 girls
(a) if it must contain 2 boys and 2 girls,
(b) if it must contain at least 1 boy and 1 girl,
(c) if either the oldest boy or the oldest girl must be included but not both.

(U of L)p

2) n boxes are arranged in a straight line and numbered 1 to n. Find:
(a) in how many ways n different articles can be arranged in the boxes, one in each box, so that a particular article A is in box 2;
(b) in how many ways the n articles can be arranged in the boxes so that the article A is in neither box 1 nor box 2 and a given article B is not in box 2.
Deduce the number of ways in which the articles can be arranged so that A is not in box 1 and B is not in box 2.

(JMB)

3) A forecast is to be made of the results of five football matches, each of which can be a win, a draw or a loss for the home team. Find the number of different possible forecasts, and show how this number is divided into forecasts containing $0, 1, 2, 3, 4, 5$ errors respectively.

(U of L)p

4) Find in factor form the number of ways in which 20 boys can be arranged in a line from right to left so that no two of three particular boys will be standing next to each other.

(U of L)p

5) Find how many distinct numbers greater than 5000 and divisible by 3 can be formed from the digits 3, 4, 5, 6 and 0, each digit being used at most once in any number.

(JMB)

6) A certain test consists of seven questions, to each of which a candidate *must* give one of three possible answers. According to the answer that he chooses, the candidate *must* score 1, 2, or 3 marks for each of the seven questions. In how many different ways can a candidate score exactly 18 marks in the test?

(U of L)p

7) A tennis club is to select a team of three pairs, each pair consisting of a man and a woman, for a match. The team is to be chosen from 7 men and 5 women. In how many different ways can the three pairs be selected? (U of L)p

8) n red counters and m green counters are to be placed in a straight line. Find the number of different arrangements of the colours.
A town has n streets running from south to north and m streets running from west to east. A man wishes to go from the extreme south-west intersection to the extreme north-east intersection, always moving either north or east along one of the streets. Find the number of different routes he can take. (JMB)

9) Show that there are 126 ways in which 10 children can be divided into two groups of 5. Find the number of ways in which this can be done
(a) if the two youngest children must be in the same group,
(b) if they must not be in the same group.   (U of L)p

10) A committee of three people is to be chosen from four married couples. Find in how many ways this committee can be chosen
(a) if all are equally eligible,
(b) if the committee must consist of one woman and two men,
(c) if all are equally eligible except that a husband and wife cannot *both* serve on the committee.   (U of L)p

11) Find the number of integers between 1000 and 4000 which can be formed by using the digits 1, 2, 3, 4
(a) if each digit may be used only once,
(b) if each digit may be used more than once.   (U of L)p

12) In how many different ways can the letters of the word MATHEMATICS be arranged? In how many of these arrangements will two A's be adjacent? Find the number of arrangements in which all the vowels come together.
(U of L)p

13) Code numbers, each containing three digits, are to be formed from the nine digits 1, 2, 3, ..., 9. In any number no particular digit may occur more than once.
(a) How many different code numbers may be formed, and in how many of these will 9 be one of the three digits selected?
(b) In how many numbers will the three digits occur in their natural order (i.e. the digits being in ascending order of magnitude reading from left to right, e.g. 359)?   (C)p

CHAPTER 15

SERIES

SEQUENCES

Consider the following sets of numbers

(a) $2, 4, 6, 8, 10, \ldots$

(b) $1, 2, 4, 8, 16, \ldots$

(c) $4, 9, 16, 25, 36, \ldots$

In each set the numbers are in a given order and there is an obvious rule for obtaining the next number and as many subsequent numbers as we wish to find. For example (b) can be continued as follows, $32, 64, 128, 256, \ldots$.

Such sets are called sequences and each member of the set is called a term of the sequence.

Thus *a sequence is a set of terms in a defined order with a rule for obtaining the terms.*

SERIES

When the terms of a sequence are added, e.g. $1 + 2 + 4 + 8 + 16 + \ldots$ a series is formed.

If the series stops after a finite number of terms it is called a finite series. Thus $1 + 2 + 4 + 8 + 16 + 32 + 64$ is a finite series of seven terms.

If the series does not stop but continues indefinitely it is called an infinite series. Thus $1 + \frac{1}{2} + \frac{1}{4} + \frac{1}{8} + \frac{1}{16} + \frac{1}{32} + \ldots + \frac{1}{1024} + \ldots$ is an infinite series.

Consider again the series $1 + 2 + 4 + 8 + 16 + 32 + 64$.

As each term is a power of 2 we can write this series in the form

$$2^0 + 2^1 + 2^2 + 2^3 + 2^4 + 2^5 + 2^6.$$

All terms of this series are of the form 2^r, so that 2^r is a general term. We can then define the series as the sum of terms of the form 2^r where r takes all integral values in order from 0 to 6 inclusive.

THE SIGMA NOTATION

Using Σ as a symbol for 'the sum of terms such as' we can redefine our series more concisely as $\Sigma\, 2^r$, r taking all integral values from 0 to 6 inclusive, or, even more briefly,

$$\sum_{r=0}^{6} 2^r$$

Placing the lowest and highest value that r takes, below and above the sigma symbol respectively, indicates that r also takes all integral values between these extreme values.

Thus $\displaystyle\sum_{r=2}^{10} r^3$ means 'the sum of all terms of the form r^3 where r takes all integral values from 2 to 10 inclusive',

i.e. $$\sum_{r=2}^{10} r^3 = 2^3 + 3^3 + 4^3 + 5^3 + 6^3 + 7^3 + 8^3 + 9^3 + 10^3$$

Note that a finite series, when written out, should always end with the last term even if several intermediate terms are omitted, e.g. $3 + 6 + 9 + \ldots + 99$.

The series $\qquad 1 + \frac{1}{2} + \frac{1}{4} + \frac{1}{8} + \frac{1}{16} + \ldots$

may also be written in the sigma notation. The continuing dots after the last term indicate that the series is infinite (i.e. there is *no* last term).
Each term of series above is a power of $\frac{1}{2}$.
Thus a general term of the series can be written as $(\frac{1}{2})^r$.
The first term is 1 or $(\frac{1}{2})^0$, so the first value that r takes is zero. There is no last term of this series, so there is no upper limit for the value of r.
Therefore $1 + \frac{1}{2} + \frac{1}{4} + \frac{1}{8} + \frac{1}{16} + \ldots$ may be written as

$$\sum_{r=0}^{\infty} (\tfrac{1}{2})^r$$

Note that when a given series is rewritten in the sigma notation it is as well to check that the first few values of r give the correct first few terms of the series.

Writing a series in the sigma notation, apart from the obvious advantage of brevity, allows us to select a particular term of a series without having to write down all the earlier terms.

For example, consider the series $\displaystyle\sum_{r=3}^{10}(2r+5)$.

The first term is the value of $2r+5$ when $r=3$, i.e. $2\times 3+5=11$.
The last term is the value of $2r+5$ when $r=10$, i.e. 25.
The fourth term is the value of $2r+5$ when r takes its fourth value in order
from $r=3$, i.e. when $r=6$.

Thus the fourth term of $\displaystyle\sum_{r=3}^{10}(2r+5)$ is $2\times 6+5=17$.

EXAMPLE 15a

Write the following series in the sigma notation:

(a) $1-x+x^2-x^3+\ldots$

(b) $2-4+8-16+\ldots+128$.

(a) A general term of this series is $\pm x^r$, having a positive sign when r is even
and a negative sign when r is odd.
Because $(-1)^r$ is positive when r is even and negative when r is odd, the
general term can be written $(-1)^r x^r$.
The first term of this series is 1, or x^0.

Hence $\quad 1-x+x^2-x^3+\ldots=\displaystyle\sum_{r=0}^{\infty}(-1)^r x^r$.

(b) $2-4+8-16+\ldots+128=2-(2)^2+(2)^3-(2)^4+\ldots+(2)^7$.
So a general term is of the form $\pm 2^r$, being positive when r is odd and
negative when r is even,
i.e. the general term is $(-1)^{r+1} 2^r$.

Hence $\quad 2-4+8-16+\ldots+128=\displaystyle\sum_{r=1}^{7}(-1)^{r+1} 2^r$.

EXERCISE 15a

1) Write the following series in the sigma notation:

(a) $1+8+27+64+125$

(b) $2+4+6+8+10+\ldots+20$

(c) $3+6+9+12+15+\ldots+99$

(d) $\frac{1}{2}+\frac{1}{3}+\frac{1}{4}+\frac{1}{5}+\ldots+\frac{1}{50}$

(e) $1+\frac{1}{3}+\frac{1}{9}+\frac{1}{27}+\ldots$

(f) $-4-1+2+5+8+\ldots+17$

(g) $8+4+2+1+\frac{1}{2}+\ldots$

2) Write down the first three terms and, where possible, the last term of the
following series:

(a) $\displaystyle\sum_{r=1}^{\infty} \frac{1}{r}$ (b) $\displaystyle\sum_{r=0}^{5} r(r+1)$ (c) $\displaystyle\sum_{r=0}^{20} r!$

(d) $\displaystyle\sum_{r=0}^{\infty} \frac{1}{(r^2+1)}$ (e) $\displaystyle\sum_{r=-1}^{8} r(r+1)(r+2)$ (f) $\displaystyle\sum_{r=0}^{\infty} a^r(-1)^{r+1}$

3) For the following series, write down the term indicated, and the number of terms in the series.

(a) $\displaystyle\sum_{r=1}^{9} 2^r$,   3rd term   (b) $\displaystyle\sum_{r=-1}^{8} (2r+3)$,   5th term

(c) $\displaystyle\sum_{r=-6}^{-1} \frac{1}{r(2r+1)}$,   last term   (d) $\displaystyle\sum_{r=0}^{\infty} \frac{1}{(r+1)(r+2)}$,   20th term

(e) $\displaystyle\sum_{r=2}^{\infty} (-1)^r \frac{2^r}{r!}$,   4th term   (f) $8+4+0-4-8-12\ldots-80.$
                                                                                     15th term

(g) $\displaystyle\sum_{r=1}^{\infty} \left(\frac{1}{2}\right)^r$, nth term

ARITHMETIC PROGRESSIONS

Consider the sequence $5, 8, 11, 14, 17, \ldots, 29$.
Each term of this sequence exceeds the previous term by 3, so the sequence can be written in the form

$$5, (5+3), (5+2\times3), (5+3\times3), (5+4\times3), \ldots, (5+8\times3)$$

This sequence is an example of an arithmetic progression, where an arithmetic progression (A.P.) is a sequence in which any term differs from the proceeding term by a constant, called the common difference. The common difference may be either positive or negative,
e.g. the first 6 terms of an A.P., whose first term is 8 and whose common difference is -3, are $8, 5, 2, -1, -4, -7$,

or $8, [8+1(-3)], [8+2(-3)], [8+3(-3)], [8+4(-3)], [8+5(-3)]$

In general, if an A.P. has a first term a, and a common difference d, the first four terms are $a, (a+d), (a+2d), (a+3d)$,
and the nth term, u_n, is $a+(n-1)d$.

Thus an A.P. with n terms can be written as
$$a, (a+d), (a+2d), \ldots, [a+(n-1)d].$$

EXAMPLES 15b

1) The 8th term of an A.P. is 11 and the 15th term is 21. Find the common difference, the first term of the series, and the nth term.

If the first term of the series is a and the common difference is d, then the 8th term is $a + 7d$,

i.e. $$a + 7d = 11 \qquad [1]$$

and the 15th term is $a + 14d$,

i.e. $$a + 14d = 21 \qquad [2]$$

[2] − [1] gives $\qquad 7d = 10 \implies d = \frac{10}{7}$

and $\qquad\qquad\qquad\qquad a = 1$

i.e. the first term is 1 and the common difference is $\frac{10}{7}$.

Hence the nth term is $\quad a + (n-1)d = 1 + (n-1)\frac{10}{7} = \frac{1}{7}(10n - 3)$.

2) The nth term of an A.P. is $12 - 4n$. Find the first term and the common difference.

If the nth term is $12 - 4n$, the first term $(n = 1)$ is 8.
The second term $(n = 2)$ is 4.

Therefore the common difference is -4.

The Sum of an Arithmetic Progression

Consider the sum of the first ten even numbers.
This series is an A.P. and, writing it in normal and in reverse order, we have

$$S = 2 + 4 + 6 + 8 + \ldots + 18 + 20$$
$$S = 20 + 18 + 16 + 14 + \ldots + 4 + 2$$

Adding gives $\qquad 2S = 22 + 22 + 22 + 22 + \ldots + 22 + 22$

As there are ten terms in this series we have

$$2S = 10 \times 22$$
$$\implies \qquad\qquad S = 110$$

This process is known as finding the sum from first principles. Applying this process to a general A.P., gives formulae for the sum which may be quoted and used, unless a proof from first principles is specifically asked for.

Consider the sum of the first n terms (S_n) of an A.P. whose last term is l,

i.e. $S_n = \quad a \quad + (a+d) + (a+2d) + \ldots + (l-d) + \quad l$

reversing $S_n = \quad l \quad + (l-d) + (l-2d) + \ldots + (a+d) + \quad a$

adding $2S_n = (a+l) + (a+l) + (a+l) + \ldots + (a+l) + (a+l)$

as there are n terms we have

$$2S_n = n(a+l)$$

$\Rightarrow$ $S_n = \dfrac{n}{2}(a+l)$ i.e. (number of terms) $\times$ (average term)

Also, because the nth term, l, is equal to $a+(n-1)d$, we have

$$S_n = \frac{n}{2}[a+a+(n-1)d]$$

or $S_n = \dfrac{n}{2}[2a+(n-1)d]$

Either of these formulae can now be used to find the sum of the first n terms of an A.P.

EXAMPLES 15b (continued)

3) Find the sum of the following series:

(a) An A.P. of eleven terms whose first term is 1 and whose last term is 6.

(b) $\displaystyle\sum_{r=1}^{8}\left(2-\frac{2r}{3}\right)$

(a) We know the first and last terms, and the number of terms so, using

$S_n = \dfrac{n}{2}(a+l)$,   we have

$$S_{11} = \tfrac{11}{2}(1+6) = \tfrac{77}{2}$$

(b) $\displaystyle\sum_{r=1}^{8}\left(2-\frac{2r}{3}\right) = \tfrac{4}{3} + \tfrac{2}{3} + 0 - \tfrac{2}{3} - \ldots - \tfrac{10}{3}.$

This is an A.P. with 8 terms where $a = \tfrac{4}{3}$, $d = -\tfrac{2}{3}$.

Using $S_n = \dfrac{n}{2}[2a+(n-1)d]$ gives

$$S_8 = 4[\tfrac{8}{3} + 7(-\tfrac{2}{3})] = -8$$

4) In an A.P. the sum of the first ten terms is 50 and the 5th term is three times the 2nd term. Find the first term and the sum of the first 20 terms.

If a is the first term and d is the common difference, and there are n terms

using $S_n = \dfrac{n}{2}[2a + (n-1)d]$ gives

$$S_{10} = 50 = 5(2a + 9d) \tag{1}$$

Now using $u_n = a + (n-1)d$ gives

$$u_5 = a + 4d \quad \text{and} \quad u_2 = a + d$$

Therefore $a + 4d = 3(a + d)$ \hfill [2]

From [1] and [2] we get $d = 1$ and $a = \frac{1}{2}$
so the first term is $\frac{1}{2}$.

The sum of the first twenty terms is given by

$$S_{20} = 10(1 + 19 \times 1) = 200$$

5) Show that the terms of $\displaystyle\sum_{r=1}^{n} \ln 2^r$ are in arithmetic progression.

Find the sum of the first 10 terms of this series and the least value of n for which the sum of the first $2n$ terms exceeds 1000.

By putting $r = 1, 2, 3 \ldots$ we have

$$\sum_{r=1}^{n} \ln 2^r = \ln 2 + \ln 2^2 + \ln 2^3 + \ldots + \ln 2^n$$

$$= \ln 2 + 2 \ln 2 + 3 \ln 2 + \ldots + n \ln 2$$

from which we see that there is a common difference of $\ln 2$ between successive terms, so the terms of this series are in arithmetic progression.

Hence $\displaystyle\sum_{r=1}^{10} \ln 2^r = \ln 2 + 2 \ln 2 + 3 \ln 2 + \ldots + 10 \ln 2$

$$= (1 + 2 + 3 + \ldots + 10) \ln 2$$

$$= \tfrac{10}{2}(1 + 10) \ln 2 = 55 \ln 2$$

Now $\displaystyle\sum_{r=1}^{2n} \ln 2^r = (1 + 2 + 3 + \ldots + 2n) \ln 2$

$$= \frac{2n}{2}[1 + 2n] \ln 2$$

So we require the least value of n for which $n(1 + 2n)\ln 2 > 1000$

As $\ln 2$ is positive we can divide both sides by $\ln 2$ giving $2n^2 + n > 1443$

$\Rightarrow$ $n > 26.6$

(26.6 is the positive root of $2n^2 + n - 1443 = 0$)
But n must be an integer so
the least value of n for which $n(2n + 1) \ln 2 > 1000$ is 27.

Note that the sum of the first n natural numbers,

$$\text{i.e. } 1 + 2 + 3 + \ldots + n$$

is an A.P. in which $a = 1$ and $d = 1$ so

$$\sum_{r=1}^{n} r = \frac{n(n+1)}{2}$$

This is a result that may be quoted, unless a proof is specifically asked for.

6) The sum of the first n terms of a series, S_n, is given by $S_n = n(n+3)$. Find the fourth term of the series and show that the terms are in arithmetic progression.

If the terms of the series are $a_1, a_2, a_3, \ldots a_n$

then $\qquad S_n = a_1 + a_2 \ldots + a_n = n(n+3)$

So $\qquad S_4 = a_1 + a_2 + a_3 + a_4 = 28$

and $\qquad S_3 = a_1 + a_2 + a_3 \qquad = 18$

Hence the fourth term of the series, a_4, is 10.

Now $\qquad S_n = a_1 + a_2 + \ldots + a_{n-1} + a_n = n(n+3)$

and $\qquad S_{n-1} = a_1 + a_2 + \ldots + a_{n-1} \qquad = (n-1)(n+2)$

Hence the nth term of the series, a_n, is given by

$$a_n = n(n+3) - (n-1)(n+2) = 2n + 2$$

Replacing n by $n-1$ gives

$$a_{n-1} = 2(n-1) + 2 = 2n$$

Therefore $\qquad a_n - a_{n-1} = (2n+2) - 2n = 2,$

i.e. there is a common difference of 2 between successive terms, showing that the series is an A.P.

EXERCISE 15b

1) Write down the fifth term and the nth term of the following A.P.s:

(a) $1, 5, \ldots$ $\qquad$ (b) $2, 1\frac{1}{2}, \ldots$ $\qquad$ (c) first term 5, common difference 3

(d) $\sum_{r=1}^{n} (2r-1)$ $\quad$ (e) $\sum_{r=1}^{n} 4(r-1)$ $\quad$ (f) first term 6, common difference -2

(g) first term p, common difference q

(h) first term 10, last term 20, 6 terms $\qquad$ (i) $\sum_{r=0}^{n} (3r+3)$

2) Find the sum of the first ten terms of each of the series given in Question (1), parts (a) to (g) inclusive.

3) The 9th term of an A.P. is 8 and the 4th term is 20. Find the first term and the common difference.

4) The 6th term of an A.P. is twice the 3rd term and the first term is 3. Find the common difference and the 10th term.

5) The nth term of an A.P. is $\frac{1}{2}(3 - n)$. Write down the first three terms and the 20th term.

6) Find the sum, to the number of terms indicated, of the following A.P.s:

(a) $1 + 2\frac{1}{2} + \ldots$, 6 terms   (b) $3 + 5 + \ldots$, 8 terms

(c) the first twenty odd integers

(d) $a_1 + a_2 + a_3 + \ldots + a_8$ where $a_n = 2n + 1$

(e) $4 + 6 + 8 + \ldots + 20$ (f) $\sum_{r=1}^{3n} (3 - 4r)$

(g) $S_n = n^2 - 3n$, 8 terms   (h) $S_n = 2n(n + 3)$, m terms

7) The sum of the first n terms of an A.P. is S_n where $S_n = n^2 - 3n$. Write down the fourth term and the nth term.

8) The sum of the first n terms of a series is given by $S_n = n(3n - 4)$. Show that the terms of the series are in arithmetic progression.

9) In an arithmetic progression, the 8th term is twice the 4th term and the 20th term is 40. Find the common difference and the sum of the terms from the 8th to the 20th inclusive.

10) How many terms of the A.P., $1 + 3 + 5 + \ldots$ are required to make a sum of 1521?

11) Find the least number of terms of the A.P., $1 + 3 + 5 + \ldots$ that are required to make a sum exceeding 4000.

12) Find the least number of terms required for the sum of the series $\sum_{r=1}^{n} (3 - 2r)$ to be less than -100.

GEOMETRIC PROGRESSIONS

Consider the sequence

$$12, 6, 3, 1.5, 0.75, 0.375, \ldots$$

Each term of this sequence is half the preceeding term so the sequence may be written

$$12, 12(\tfrac{1}{2}), 12(\tfrac{1}{2})^2, 12(\tfrac{1}{2})^3, 12(\tfrac{1}{2})^4, 12(\tfrac{1}{2})^5, \ldots$$

Such a sequence is called a geometric progression (G.P.) which is a sequence where each term is a constant multiple of the preceding term. This constant multiplying factor is called the common ratio, and it may have any real value.

Hence, if a G.P. has a first term of 3 and a common ratio of -2 the first four terms are

$$3, 3(-2), 3(-2)^2, 3(-2)^3$$

or

$$3, -6, 12, -24$$

In general if a G.P. has a first term a, and a common ratio r, the first four terms are

$$a, ar, ar^2, ar^3$$

and the nth term, u_n, is ar^{n-1}.
Thus a G.P. with n terms can be written

$$a, ar, ar^2, \ldots, ar^{n-1}$$

Sum of a Geometric Progression

Consider the sum of the first eight terms, S_8, of the G.P. with first term 1 and common ratio 3

i.e. $$S_8 = 1 + 1(3) + 1(3)^2 + 1(3)^3 + \ldots + 1(3)^7$$

$\Rightarrow$ $$3S_8 = \quad 3 + 3^2 + 3^3 + \ldots + 3^7 + 3^8$$

Hence $$S_8 - 3S_8 = \quad 1 + 0 \qquad + \ldots + \quad 0 \quad - 3^8$$

So $$S_8(1 - 3) = 1 - 3^8$$

$\Rightarrow$ $$S_8 = \frac{1 - 3^8}{1 - 3} = \frac{3^8 - 1}{2}$$

Applying this process to a general G.P. gives a formula for the sum of the first n terms.
Consider the sum, S_n, of the first n terms of a G.P. whose first term is a and whose common ratio is r,

i.e. $$S_n = a + ar + ar^2 + \ldots + ar^{n-1}$$

Multiplying by r gives

$$rS_n = \quad ar + ar^2 + \ldots + ar^{n-1} + ar^n$$

Hence $$S_n - rS_n = a - ar^n$$

$\Rightarrow$ $$S_n(1 - r) = a(1 - r^n)$$

$\Rightarrow$ $$S_n = \frac{a(1 - r^n)}{1 - r}$$

This formula can now be used to find the sum of the first n terms of any G.P.

If $r > 1$ the formula may be written $\dfrac{a(r^n - 1)}{r - 1}$

EXAMPLES 15c

1) The 5th term of a G.P. is 8, the third term is 4, and the sum of the first ten terms is positive. Find the first term, the common ratio, and the sum of the first ten terms.

If the first term is a and the common ratio is r then the nth term is ar^{n-1}

Thus   we have $ar^4 = 8$ $(n = 5)$

and $ar^2 = 4$ $(n = 3)$

dividing gives $r^2 = 2$

$\Rightarrow$ $r = \pm\sqrt{2}$ and $a = 2$

Using the formula $S_n = \dfrac{a(r^n - 1)}{r - 1}$ gives,

when $r = \sqrt{2}$, $S_{10} = \dfrac{2[(\sqrt{2})^{10} - 1]}{\sqrt{2} - 1} = \dfrac{62}{\sqrt{2} - 1}$

when $r = -\sqrt{2}$, $S_{10} = \dfrac{2[(-\sqrt{2})^{10} - 1]}{-\sqrt{2} - 1} = \dfrac{-62}{\sqrt{2} + 1}$

But we are told that $S_{10} > 0$,   so we deduce that

$$r = \sqrt{2} \quad \text{and} \quad S_{10} = \dfrac{62}{\sqrt{2} - 1}$$

2) The sum of the first n terms of a series is $3^n - 1$. Show that the terms of this series are in geometric progression and find the first term, the common ratio and the sum of the second set of n terms of this series.

If the series is $a_1 + a_2 + \ldots + a_n$

then $S_n = a_1 + a_2 + \ldots + a_{n-1} + a_n = 3^n - 1$

and $S_{n-1} = a_1 + a_2 + \ldots + a_{n-1}$ $= 3^{n-1} - 1$

therefore the nth term, a_n, is given by

$$a_n = (3^n - 1) - (3^{n-1} - 1)$$
$$= 3^n - 3^{n-1}$$
$$= 3^{n-1}(3 - 1)$$
$$= (2)3^{n-1}$$

Similarly $\qquad a_{n-1} = (2)3^{n-2}$

Then $\quad a_n \div a_{n-1} = 3 \quad$ showing that successive terms in the series have a constant ratio of 3.

Hence this series is a G.P. with a first term of 2 and a common ratio of 3.

The sum of the second set of n terms is

(the sum of the first $2n$ terms) $-$ (the sum of the first n terms)

$$
\begin{aligned}
&= S_{2n} - S_n \\
&= (3^{2n} - 1) - (3^n - 1) \\
&= 3^{2n} - 3^n \\
&= 3^n(3^n - 1)
\end{aligned}
$$

3) A prize fund is set up with a single investment of £2000 to provide an annual prize of £150. The fund accrues interest at 5% p.a. paid yearly. If the first prize is awarded one year after the investment, find the number of years for which the full prize can be awarded.

After one year the value of the fund is the initial investment of £2000, plus 5% interest, less one £150 prize.
If £ P_n is the value of the fund after n years we have

$P_1 = 2000 + (0.05)(2000) - 150 = 2000(1.05) - 150$

$P_2 = P_1 + 0.05P_1 - 150 = 1.05P_1 - 150 = 2000(1.05)^2 - 150(1.05) - 150$

$P_3 = 1.05P_2 - 150 = 2000(1.05)^3 - 150(1.05)^2 - 150(1.05) - 150$

....

$P_n = 2000(1.05)^n - 150(1.05)^{n-1} - 150(1.05)^{n-2} - \dots - 150$

$\qquad = 2000(1.05)^n - 150[1 + 1.05 + (1.05)^2 + \dots + (1.05)^{n-1}]$

The expression in the square brackets is a GP of n terms in which $a = 1$, $r = 1.05$, hence

$$
P_n = 2000(1.05)^n - 150\left[\frac{(1.05)^n - 1}{1.05 - 1}\right]
$$

$$
= 2000(1.05)^n - 3000(1.05)^n + 3000
$$

$$
= 3000 - 1000(1.05)^n
$$

The fund can award the full prize while $P_n \geqslant 0$

i.e. $\qquad\qquad 3000 - 1000(1.05)^n \geqslant 0$

$\Rightarrow$ $(1.05)^n \leqslant 3$

$\Rightarrow$ $n \leqslant \dfrac{\lg 3}{\lg 1.05} = 22.5$

But the number of years required must be an integer, so $n = 22$.

Therefore the full prize can be awarded for 22 years.

EXERCISE 15c

1) Write down the fifth term and the nth term of the following G.P.s:
(a) $2, 4, 8, \ldots$ (b) $2, 1, \frac{1}{2}, \ldots$ (c) $3, -6, 12, \ldots$
(d) first term 8, common ratio $-\frac{1}{2}$ (e) first term 3, last term $\frac{1}{81}$, 6 terms

2) Find the sum, to the number of terms given, of the following G.P.s.
(a) $3 + 6 + \ldots$ 6 terms   (b) $3 - 6 + \ldots 8$ terms
(c) $1 + \frac{1}{2} + \frac{1}{4} + \ldots$ 20 terms   (d) first term 5, common ratio $\frac{1}{5}$, 5 terms
(e) first term $\frac{1}{2}$, common ratio $-\frac{1}{2}$, 10 terms
(f) first term 1, common ratio -1, 2001 terms.

3) The 6th term of a G.P. is 16 and the 3rd term is 2. Find the first term and the common ratio.

4) Find the common ratio, given that it is negative, of a G.P. whose first term is 8 and whose 5th term is $\frac{1}{2}$.

5) The nth term of a G.P. is $(-\frac{1}{2})^n$. Write down the first term and the 10th term.

6) Evaluate $\displaystyle\sum_{r=1}^{10} (1.05)^r$.

7) Find the sum to n terms of the following series:

(a) $x + x^2 + x^3 + \ldots$ (b) $x + 1 + \dfrac{1}{x} + \ldots$ (c) $1 - y + y^2 - \ldots$

(d) $x + \dfrac{x^2}{2} + \dfrac{x^3}{4} + \dfrac{x^4}{8} + \ldots$ (e) $1 - 2x + 4x^2 - 8x^3 + \ldots$

8) Find the sum of the first n terms of the G.P. $2 + \frac{1}{2} + \frac{1}{8} + \ldots$ and find the least value of n for which this sum exceeds 2.65.

9) The sum of the first 3 terms of a G.P. is 14. If the first term is 2, find the possible values of the sum of the first 5 terms.

10) Evaluate $\displaystyle\sum_{r=1}^{10} 3(3/4)^r$.

11) A mortgage is taken out for £10 000 and is repaid by annual instalments of £2000. Interest is charged on the outstanding debt at 10%, calculated annually. If the first repayment is made one year after the mortgage is taken out find the number of years it takes for the mortgage to be repaid.

12) A bank loan of £500 is arranged to be repaid in two years by equal monthly instalments. Interest, *calculated monthly*, is charged at 11% p.a. on the remaining debt. Calculate the monthly repayment if the first repayment is one month after the loan is made.

CONVERGENCE OF SERIES

If a piece of string, of length l, is cut up by first cutting it in half and keeping one piece, then cutting the remainder in half and keeping one piece, then cutting the remainder in half and keeping one piece, . . . and so on, the sum of the lengths retained is

$$\frac{l}{2} + \frac{l}{4} + \frac{l}{8} + \frac{l}{16} + \ldots$$

As this process can (in theory) be carried on indefinitely the series formed above is infinite.

After several cuts have been made the remaining part of the string will be very small indeed, so that the sum of the cut lengths will be very nearly equal to the total length, l, of the original piece of string. The more cuts that are made the closer to l this sum becomes,

i.e. if after n cuts, the sum of the cut lengths is

$$\frac{l}{2} + \frac{l}{2^2} + \frac{l}{2^3} + \ldots + \frac{l}{2^n}$$

then as $n \to \infty$,

$$\frac{l}{2} + \frac{l}{2^2} + \ldots + \frac{l}{2^n} \to l$$

or

$$\lim_{n \to \infty} \left[\frac{l}{2} + \frac{l}{2^2} + \ldots + \frac{l}{2^n} \right] = l$$

l is called the sum to infinity of this series.

In general, if S_n is the sum of the first n terms of any series
and if $\lim_{n \to \infty} [S_n]$ exists and is finite, the series is said to be convergent.

In this case the sum to infinity, S_∞, is given by

$$S_\infty = \lim_{n \to \infty} [S_n]$$

Thus, for example, the series $\dfrac{l}{2} + \dfrac{l}{2^2} + \dfrac{l}{2^3} + \ldots$ is convergent, as its sum to infinity is l.

But the series $1 + 2 + 3 + 4 + \ldots + n$ has a sum given by $S_n = \frac{1}{2}n(n + 1)$ (see p. 593) so, as $n \to \infty$ $S_n \to \infty$ and this series does not converge. It is said to be divergent.

For any A.P., $S_n = \dfrac{n}{2}(2a + [n-1]\,d)$, which always approaches infinity as $n \to \infty$ so any A.P. is divergent.

THE SUM TO INFINITY OF A G.P.

The series discussed above, viz. $\dfrac{l}{2} + \dfrac{l}{2^2} + \dfrac{l}{2^3} + \ldots$

is a G.P. whose first term is $\frac{1}{2}l$ and whose common ratio is $\frac{1}{2}$. Its sum to infinity, which we have seen is l, can also be found using the alternative method below.

The sum, S_n, of the first n terms of this series is given by

$$S_n = \frac{\frac{1}{2}l\{1 - (\frac{1}{2})^n\}}{1 - \frac{1}{2}} = l\{1 - (\frac{1}{2})^n\}$$

As n increases, $(\frac{1}{2})^n$ decreases, i.e. as $n \to \infty$, $(\frac{1}{2})^n \to 0$

therefore

$$\lim_{n \to \infty} [l\{1 - (\tfrac{1}{2})^n\}] = l$$

So $S_n \to l$ as $n \to \infty$, i.e. $S_\infty = l$.

This approach can also be applied to the general G.P. $a + ar + ar^2 + \ldots$

where

$$S_n = \frac{a(1 - r^n)}{1 - r}$$

If $|r| < 1$, then $\lim\limits_{n \to \infty} r^n = 0$

and

$$\lim_{n \to \infty} S_n = \lim_{n \to \infty} \left[\frac{a(1 - r^n)}{1 - r}\right] = \frac{a}{1 - r}$$

If $|r| > 1$, $\lim\limits_{n \to \infty} r^n = \infty$ and the series does not converge.

Therefore, provided $|r| < 1$, a G.P. converges to a sum of $\dfrac{a}{1 - r}$

i.e. for a G.P.

$$S_\infty = \frac{a}{1 - r} \iff |r| < 1$$

Arithmetic Mean

If three numbers, p_1, p_2, p_3, are in arithmetic progression then p_2 is called the *arithmetic mean* of p_1 and p_3.

If $p_1 = a$, we may write p_2, p_3 as $a + d, a + 2d$ respectively,

hence $\qquad\qquad p_2 = a + d = \tfrac{1}{2}(2a + 2d) = \tfrac{1}{2}(p_1 + p_3)$

i.e. the arithmetic mean of two numbers x and y is $\tfrac{1}{2}(x + y)$.

Geometric Mean

If p_1, p_2, p_3 are in geometric progression, p_2 is called the *geometric mean* of p_1 and p_3.

If $p_1 = a$, then we may write $p_2 = ar$, $p_3 = ar^2$
thus $p_1 p_3 = a^2 r^2 = p_2^2 \Rightarrow p_2 = \sqrt{p_1 p_3}$,

i.e. the geometric mean of two numbers x and y is $\sqrt{xy}$.

EXAMPLES 15d

1) Determine whether the series given below converge. If they do, give their sum to infinity.

(a) $3 + 5 + 7 + \ldots$ $\qquad$ (b) $1 - \dfrac{1}{4} + \dfrac{1}{16} - \dfrac{1}{64} + \ldots$ $\qquad$ (c) $3 + \dfrac{9}{2} + \dfrac{27}{4} + \ldots$

(a) $3 + 5 + 7 + \ldots$ is an A.P. $(d = 2)$ and so does not converge.

(b) $1 - \dfrac{1}{4} + \dfrac{1}{16} - \dfrac{1}{64} + \ldots = 1 + (-\tfrac{1}{4}) + (-\tfrac{1}{4})^2 + (-\tfrac{1}{4})^3 + \ldots$

which is a G.P. where $r = -\tfrac{1}{4}$, i.e. $|r| < 1$

So this series does converge to $\dfrac{1}{1 - (-\tfrac{1}{4})} = \dfrac{4}{5}$, using $S_\infty = \dfrac{a}{1 - r}$.

(c) $3 + \tfrac{9}{2} + \tfrac{27}{4} + \ldots = 3 + 3(\tfrac{3}{2}) + 3(\tfrac{9}{4}) + \ldots$

$\qquad\qquad\qquad\quad = 3 + 3(\tfrac{3}{2}) + 3(\tfrac{3}{2})^2 + \ldots$

This series is a G.P. where $r = \tfrac{3}{2}$.
As $|r| > 1$, the series does not converge.

2) Find the condition on x so that $\displaystyle\sum_{r=0}^{\infty} \dfrac{(x - 1)^r}{2^r}$ converges.

Evaluate this expression when $x = 1.5$

$$\sum_{r=0}^{\infty} \dfrac{(x - 1)^r}{2^r} = 1 + \dfrac{x - 1}{2} + \left(\dfrac{x - 1}{2}\right)^2 + \ldots$$

This series is a G.P. with common ratio $\dfrac{x - 1}{2}$

so the series converges if $\left|\dfrac{x - 1}{2}\right| < 1$,

i.e. if
$$-1 < \frac{x-1}{2} < 1$$

$\Rightarrow$
$$-2 < x-1 < 2$$

$\Rightarrow$
$$-1 < x < 3$$

When $x = 1.5$,   the series converges

and
$$\sum_{r=0}^{\infty} \frac{(x-1)^r}{2^r} = \sum_{r=0}^{\infty} (\tfrac{1}{4})^r = 1 + \tfrac{1}{4} + (\tfrac{1}{4})^2 + \ldots$$

$$= \frac{1}{1 - \tfrac{1}{4}} = \frac{4}{3}$$

$\left(\text{using } S_\infty = \dfrac{a}{1-r} \quad \text{where} \quad r = \tfrac{1}{4}, \quad a = 1\right)$.

3) Express the recurring decimal $0.1\dot{5}7\dot{6}$ as a fraction in its lowest terms.

$0.1\dot{5}7\dot{6} = 0.1\overline{57}\,\overline{657}\,\overline{657}\,\overline{657}\,6 \ldots$

$= 0.1 + 0.0576 + 0.000\,057\,6 + 0.000\,000\,057\,6 + \ldots$

$= \dfrac{1}{10} + \dfrac{576}{10^4} + \dfrac{576}{10^7} + \dfrac{576}{10^{10}} + \ldots$

$= \dfrac{1}{10} + \dfrac{576}{10^4}\left[1 + \dfrac{1}{10^3} + \dfrac{1}{10^6} + \ldots\right]$

$= \dfrac{1}{10} + \dfrac{576}{10^4}\left[1 + \dfrac{1}{10^3} + \left(\dfrac{1}{10^3}\right)^2 + \ldots\right]$

Now the series in the square bracket is a G.P. whose first term is 1, and whose common ratio is $\dfrac{1}{10^3}$. Hence it has a sum to infinity equal to $\dfrac{1}{1 - 10^{-3}} = \dfrac{10^3}{999}$

Therefore $0.1\dot{5}7\dot{6} = \dfrac{1}{10} + \dfrac{576}{10^4} \times \dfrac{10^3}{999} = \dfrac{1}{10} + \dfrac{576}{9990} = \dfrac{1575}{9990} = \dfrac{35}{222}.$

4) The 3rd term of a convergent G.P. is the arithmetic mean of the 1st and 2nd terms.

Find the common ratio and, if the first term is 1, find the sum to infinity.

If the series is $a + ar + ar^2 + ar^3 + \ldots$

then
$$ar^2 = \tfrac{1}{2}(a + ar)$$

$\Rightarrow$
$$2r^2 - r - 1 = 0$$

$\Rightarrow$
$$(2r + 1)(r - 1) = 0$$

i.e.
$$r = -\tfrac{1}{2} \text{ or } 1$$

As the series is convergent, the common ratio is $-\tfrac{1}{2}$.

When $a = 1$ and $r = -\tfrac{1}{2}$, $S_\infty = \dfrac{1}{1 + \tfrac{1}{2}} = \tfrac{2}{3}.$

EXERCISE 15d

1) Determine whether the series given below converge:

(a) $4 + \dfrac{4}{3} + \dfrac{4}{3^2} + \ldots$

(b) $9 + 7 + 5 + 3 + \ldots$

(c) $20 - 10 + 5 - 2.5 + \ldots$

(d) $\dfrac{5}{10} + \dfrac{5}{100} + \dfrac{5}{1000} + \ldots$

(e) $p + 2p + 3p + \ldots$

(f) $3 - 1 + \frac{1}{3} - \frac{1}{9} + \ldots$

2) Find the range of values of x for which the following series converge:

(a) $1 + x + x^2 + x^3 + \ldots$

(b) $x + 1 + \dfrac{1}{x} + \dfrac{1}{x^2} + \ldots$

(c) $1 + 2x + 4x^2 + 8x^3 + \ldots$

(d) $1 - (1-x) + (1-x)^2 - (1-x)^3 + \ldots$

(e) $(a + x) + (a + x)^2 + (a + x)^3 + \ldots$

(f) $(a + x) - 1 + \dfrac{1}{a+x} - \dfrac{1}{(a+x)^2} + \ldots$

3) Find the sum to infinity of the series in Question 1 that are convergent:

4) Express the following recurring decimals as fractions:

(a) $0.16\dot{2}$ (b) $0.\dot{3}\dot{4}$ (c) $0.0\dot{2}\dot{1}$

5) The sum to infinity of a G.P. is twice the first term. Find the common ratio.

6) If $\ln y$ is the arithmetic mean of $\ln x$ and $\ln z$ show that y is the geometric mean of x and z.

THE BINOMIAL THEOREM

When an expression such as $(1 + x)^4$ is expanded, the coefficients of the terms in the expansion can be obtained from Pascal's Triangle (see Chapter 2, p. 37). The same approach could be used to expand $(1 + x)^{20}$, say, but clearly the construction of the triangular array would be tedious in this case. A more general method known as the Binomial Theorem, will now be developed for expanding powers of $(1 + x)$.

Consider $(1 + x)^6 \equiv (1 + x)(1 + x)(1 + x)(1 + x)(1 + x)(1 + x)$.

When the six brackets are expanded, each term in the expansion is obtained by multiplying together either x or 1 from each of the six brackets.

Taking 1 from each bracket we get 1 as a term in the expansion.

Taking x from only one bracket and 1 from the other five, we get $1 \times x$. But this can be done six times because we can choose to take x from each of the six brackets in turn, (i.e. in $^6C_1$ ways). So the x term in the expansion is $6x$.

Taking x from any two brackets and 1 from all the remaining brackets we get $1 \times x^2$. But this can be done $^6C_2$ times, as the number of ways in which two brackets can be selected from six brackets is $^6C_2$.

So the x^2 term in the expansion is $^6C_2x^2$.

It can similarly be shown that the coefficients of $x^3, x^4 \ldots$ are $^6C_3, ^6C_4 \ldots$

In this context an alternative notation is used for $^6C_2$ and similar quantities. For $^6C_2$ we can write $\binom{6}{2}$, etc.

Thus, arranging the expansion of $(1 + x)^6$ as a series of ascending powers of x,

$$(1+x)^6 \equiv 1 + \binom{6}{1}x + \binom{6}{2}x^2 + \binom{6}{3}x^3 + \binom{6}{4}x^4 + \binom{6}{5}x^5 + \binom{6}{6}x^6$$

$$\equiv 1 + 6x + 15x^2 + 20x^3 + 15x^4 + 6x^5 + x^6$$

The R.H.S. of this identity is called the *series expansion* of $(1 + x)^6$.

Note that the expansion has 7 (i.e. 6 + 1) terms.

This argument can be generalised as follows.

If n is any positive integer, $(1 + x)^n$ can be expanded to give a series of terms in ascending powers of x where the term containing x^r is obtained by multiplying together x's from r brackets and 1's from the remaining brackets. There are ^nC_r ways in which r brackets can be chosen from n brackets, so the term containing x^r is $^nC_rx^r$, or $\binom{n}{r}x^r$.

Hence

$$(1 + x)^n \equiv 1 + \binom{n}{1}x + \binom{n}{2}x^2 + \ldots + \binom{n}{r}x^r + \ldots + \binom{n}{n}x^n \qquad [1]$$

$$\equiv 1 + nx + \frac{n(n-1)}{2!}x^2 + \ldots + \frac{n(n-1)\ldots(n-r+1)}{r!}x^r + \ldots + x^n$$
$$[2]$$

This result is known as the Binomial Theorem and may be written more briefly as

$$(1 + x)^n \equiv \sum_{r=0}^{n} \binom{n}{r}x^r$$

where n is a positive integer.

Note that:

(a) the expansion of $(1 + x)^n$ is a finite series with $n + 1$ terms,

(b) the coefficient of x^r, $\binom{n}{r}$, when written $\dfrac{n(n-1)(n-2)\ldots(n-r+1)}{r!}$ has r factors in the numerator,

(c) the term containing x^2 is the third term, the term in x^3 is the fourth term, and *the term in x^r is the $(r + 1)$th term*, so the rth term is $\binom{n}{r-1}x^{r-1}$,

(d) the form of the expansion given in [1] is useful when the coefficient of a large power of x is required, or when the general term is required. The form of the expansion given in [2] is useful when the first few terms of an expansion are required. Thus *both* forms of the Binomial Theorem should be memorised.

Now consider $(a + x)^n$, where n is a positive integer

$$(a + x)^n \equiv a^n \left(1 + \frac{x}{a}\right)^n$$

Replacing x by $\dfrac{x}{a}$ in the binomial series gives

$$(a + x)^n \equiv a^n \left[{}^nC_0 + {}^nC_1 \left(\frac{x}{a}\right) + {}^nC_2 \left(\frac{x}{a}\right)^2 + \ldots + {}^nC_r \left(\frac{x}{a}\right)^r + \ldots + {}^nC_n \left(\frac{x}{a}\right)^n \right]$$

$$\equiv {}^nC_0 a^n + {}^nC_1 a^{n-1} x + {}^nC_2 a^{n-2} x^2 + \ldots + {}^nC_r a^{n-r} x^r + \ldots + {}^nC_n x^n$$

$$\equiv a^n + na^{n-1}x + \frac{n(n-1)}{2} a^{n-2} x^2 + \ldots + x^n$$

i.e. if n is a positive integer

$$(a + x)^n \equiv \sum_{r=0}^{n} \binom{n}{r} a^{n-r} x^r$$

EXAMPLES 15e

1) Write down the first three terms in the expansion in ascending powers of x of:

(a) $\left(1 - \dfrac{x}{2}\right)^{10}$,   (b) $(3 - 2x)^8$

(a) Using the result [2] above and replacing x by $-\dfrac{x}{2}$ and n by 10 we have

$$\left(1 - \frac{x}{2}\right)^{10} = 1 + (10)\left(-\frac{x}{2}\right) + \frac{10 \times 9}{2!}\left(-\frac{x}{2}\right)^2 + \ldots$$

$$= 1 - 5x + \tfrac{45}{4}x^2 + \ldots$$

(b) Using the general result above and replacing a by 3, x by $-2x$ and n by 8 we have

$$(3 - 2x)^8 \equiv \sum_{r=0}^{8} \binom{8}{r} (3)^{8-r} (-2x)^r$$

Therefore the first three terms of this series $(r = 0, 1, 2)$ are

$$3^8 + 8 \times (3)^7(-2x) + \frac{8 \times 7}{2}(3)^6(-2x)^2$$

i.e. $3^8 - 16 \times 3^7 x + 112 \times 3^6 x^2$

2) In the expansion of $(2a - b)^{20}$ as a series in ascending powers of b find:

(a) the term containing a^3 (b) the sixth term.

$$(2a - b)^{20} \equiv \sum_{r=0}^{20} \binom{20}{r} (2a)^{20-r}(-b)^r$$

(a) The term containing a^3 is the term for which $20 - r = 3$ i.e. $r = 17$.
Therefore the required term is $\binom{20}{17}(2a)^3(-b)^{17}$

$$= \frac{20!}{17!3!}(8a^3)(-b)^{17}$$

$$= -9120a^3b^{17}$$

(b) As the first term of the series is the term for which $r = 0$,   the sixth term
of the series is that for which $r = 5$,

i.e. $\binom{20}{5}(2a)^{15}(-b)^5 = -\dfrac{20!}{15!5!}2^{15}a^{15}b^5$

(Note that the numerical part of this term is so large that it is better left in factor
form rather than multiplied out.)

3) Evaluate $\text{Re}(z^6)$ where $z = 2 + 3i$.

$$(2 + 3i)^6 \equiv \sum_{r=0}^{6} \binom{6}{r}(2)^{6-r}(3i)^r$$

Now $\text{Re}(z^6)$ is the sum of the terms containing even powers of i in the
binomial expansion of z^6,

i.e. $\text{Re}(z^6) = \binom{6}{0}(2)^6(3i)^0 + \binom{6}{2}(2)^4(3i)^2 + \binom{6}{4}(2)^2(3i)^4 + \binom{6}{6}(2)^0(3i)^6$

$$= 2^6 - 15(2^4)(3^2) + 15(2)^2(3)^4 - (3)^6$$

$$= 2035$$

USE OF SERIES FOR APPROXIMATIONS

Consider $(1 + x)^{20}$ and its binomial expansion

$$(1 + x)^{20} \equiv 1 + 20x + \frac{(20)(19)}{2!}x^2 + \frac{(20)(19)(18)}{3!}x^3 + \ldots + x^{20}$$

This identity is valid for all values of x so if, for example, $x = 0.01$ we have

$$(1.01)^{20} = 1 + 20(0.01) + \frac{(20)(19)}{2!}(0.01)^2 + \frac{(20)(19)(18)}{3!}(0.01)^3$$

$$+ \ldots + (0.01)^{20}$$

i.e.

$$(1.01)^{20} = 1 + 0.2 + 0.019 + 0.001\,14 + 0.000\,048\,45 + \ldots + 10^{-40}$$

Because the value of x (i.e. 0.01) is small, we see that adding successive terms
of the series makes progressively smaller contributions to the accuracy of
$(1.01)^{20}$.
In fact, taking only the first four terms gives $(1.01)^{20} \simeq 1.220\,14$

This approximation is correct to three decimal places as the fifth and succeeding terms do not add anything to the first four decimal places.

In general, for any positive integer n,

$$(1 + x)^n \equiv 1 + nx + \frac{n(n-1)}{2!} x^2 + \ldots + x^n$$

and, if x is small (i.e. successive powers of x quickly become negligible in value) the sum of the first few terms of the expansion of $(1 + x)^n$ gives an approximate value for $(1 + x)^n$.

The number of terms required to obtain a good approximation depends on two considerations:

(a) the value of x (the smaller x is, the fewer are the terms needed to obtain a good approximation),

(b) the accuracy required (an answer correct to 3 s.f. needs fewer terms than an answer correct to 6 s.f.).

When finding an approximation, the binomial expansion of $(1 + x)^n$ and *not* $(a + x)^n$ should be used.
The following examples illustrate these points.

EXAMPLES 15e (continued)

4) By substituting 0.001 for x in the expansion of $(1 - x)^7$ find the value of $(1.998)^7$ correct to six significant figures.

$$(1 - x)^7 \equiv 1 - 7x + \frac{(7)(6)}{2!} x^2 + \ldots - x^7$$

and

$$(1.998)^7 = (2 - 0.002)^7 = 2^7 (1 - 0.001)^7$$
$$= 2^7 (1 - x)^7 \quad \text{when} \quad x = 0.001$$

Hence

$$(1.998)^7 = 2^7 \left[1 - 7(0.001) + \frac{(7)(6)}{2!} (0.001)^2 - \frac{(7)(6)(5)}{3!} (0.001)^3 \right.$$
$$\left. + \ldots - (0.001)^7 \right]$$

$$\simeq 128 [1 - 0.007 + 0.000\,021 - 0.000\,000\,035]$$

To give an answer correct to 6 s.f. we must work to 7 s.f. This requires the expression in brackets to be worked to 7 d.p. so only the first three terms need be considered,

i.e. $\qquad (1.998)^7 = 128(1 - 0.007 + 0.000\,021\,0) \quad$ to 7 s.f.
$$= 127.107 \text{ to 6 s.f.}$$

Note that $(1.998)^7$ is rational, i.e. it has an exact numerical value. The use of the series expansion enables us to find a value of $(1.998)^7$ correct to as many significant figures as we wish *without* first having to find the exact value.

5) If x is so small that x^2 and higher powers can be neglected show that

$$(1-x)^5 \left(2 + \frac{x}{2}\right)^{10} \simeq 2^9(2 - 5x)$$

Using the binomial expansion of $(1-x)^5$ and neglecting terms containing x^2 and higher powers of x we have

$$(1-x)^5 \simeq 1 - 5x.$$

Similarly

$$\left(2 + \frac{x}{2}\right)^{10} \equiv 2^{10}\left(1 + \frac{x}{4}\right)^{10}$$

$$\simeq 2^{10}\left[1 + 10\left(\frac{x}{4}\right)\right]$$

Therefore

$$(1-x)^5 \left(2 + \frac{x}{2}\right)^{10} \simeq 2^{10}(1 - 5x)\left(1 + \frac{5x}{2}\right)$$

$$\simeq 2^9(1 - 5x)(2 + 5x)$$

$$\simeq 2^9(2 - 5x)$$

again neglecting the term in x^2.

The graphical significance of this approximation is interesting:

If

$$y = (1-x)^5 \left(2 + \frac{x}{2}\right)^{10}$$

then for values of x close to zero

$$y \simeq 2^9(2 - 5x)$$

which is the equation of a straight line,

i.e. $y = 2^9(2 - 5x)$ is the tangent to $y = (1-x)^5 \left(2 + \frac{x}{2}\right)^{10}$ at the point where $x = 0$.

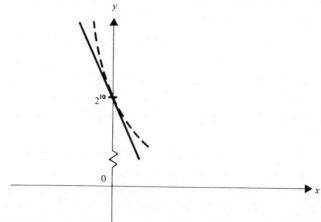

Note that the function $2^9(2 - 5x)$ is called a *linear approximation* for the
function $(1 - x)^5 \left(2 + \dfrac{x}{2}\right)^{10}$ in the region where $x \simeq 0$.

EXERCISE 15e

1) Write down the first four terms in the binomial expansion of:

(a) $(1 + 3x)^{12}$ (b) $(1 - 2x)^9$ (c) $(2 + x)^{10}$

(d) $\left(1 - \dfrac{x}{3}\right)^{20}$ (e) $\left(2 - \dfrac{3}{2}x\right)^7$ (f) $\left(\dfrac{3}{2} + 2x\right)^9$

2) Write down the term indicated in the binomial expansions of the following functions:

(a) $(1 - 4x)^7$, 3rd term (b) $\left(1 - \dfrac{x}{2}\right)^{20}$, 6th term

(c) $(2 - x)^{15}$, 12th term (d) $(2 - 3x)^{30}$, 9th term

(e) $(p - 2q)^{10}$, 5th term (f) $(3a + 2b)^8$, 2nd term

(g) $(1 - 2x)^{12}$, term containing x^4 (h) $\left(2 + \dfrac{x}{2}\right)^9$, term containing x^5

(i) $(a + b)^8$, term containing a^3 (j) $(2a - 3b)^{10}$, term containing b^8

3) Write down the binomial expansions of the following functions as series of ascending powers of x as far as, and including, the term in x^2:

(a) $(1 + x)(1 - x)^9$ (b) $(1 - x)(1 + 2x)^{10}$

(c) $(2 + x)\left(1 - \dfrac{x}{2}\right)^{20}$ (d) $(1 + x)^2(1 - 5x)^{14}$

4) Find Re(z^8) where $z = 2 - i$.

5) Find Im(z^{10}) where $z = 1 + \tfrac{1}{2}i$.

6) For each of the functions given in Question 3, find the term containing x^r and hence express the expansion in the sigma notation.

7) Find the coefficient of x^r in the expansion of $(1 - 2x)(1 + 4x)^8$ as a series of ascending powers of x.

8) Repeat Question 7 for $(1 - x)(2 + x)^{20}$.

9) By substituting 0.01 for x in the binomial expansion of $(1 - 2x)^{10}$, find the value of $(0.98)^{10}$ correct to four decimal places.

10) By substituting 0.05 for x in the binomial expansion of $\left(1 + \dfrac{x}{5}\right)^6$, find the value of $(1.01)^6$ correct to four significant figures.

11) By using the binomial expansion of $(2 + x)^7$, show that $(2.08)^7 = 168.439$ correct to 3 d.p.

12) Show that, if x is small enough for x^2 and higher powers of x to be neglected, the function $(x - 2)(1 + 3x)^8$ has a linear approximation of $-2 - 47x$.

13) If x is so small that x^3 and higher powers of x are negligible, show that $(2x + 3)(1 - 2x)^{10} \simeq 3 - 58x + 500x^2$.

14) By neglecting x^2 and higher powers of x find linear approximations for the following functions in the immediate neighbourhood of $x = 0$.

(a) $(1 - 5x)^{10}$ (b) $(2 - x)^8$ (c) $(1 + x)(1 - x)^{20}$

EXTENDING THE BINOMIAL THEOREM

It has been shown that, for a positive integral value of n

$$(1 + x)^n \equiv 1 + nx + \frac{n(n - 1)}{2!} x^2 + \frac{n(n - 1)(n - 2)}{3!} x^3 + \ldots + x^n \quad [1]$$

Now consider the infinite series $1 - x + x^2 - x^3 + \ldots$

This is a G.P. whose common ratio is $-x$ and, provided that $|x| < 1$, whose sum to infinity is $\dfrac{1}{1 - (-x)}$ i.e. $\dfrac{1}{1 + x}$

so $(1 + x)^{-1} = 1 - x + x^2 - \ldots$ provided $-1 < x < 1$ $\quad [2]$

Now substituting -1 for n in the L.H.S. of identity [1] above we get $(1 + x)^{-1}$ which is the L.H.S. of [2].

Substituting -1 for n in the first few terms of the R.H.S. of [1] we get $1 - x + x^2 - x^3 + \ldots$ which is the R.H.S. of [2].

Thus the form of the binomial expansion given in [1] appears to be valid for a value of n that is not a positive integer except that

(a) the series expansion does not terminate but carries on to infinity, and

(b) x must be in the range $-1 < x < 1$.

(c) It is no longer an identity as the sum of the G.P. is never *exactly* equal to $\dfrac{1}{1 + x}$.

In fact, although it cannot be proved at this stage, the more general series

$$1 + nx + \frac{n(n - 1)}{2!} x^2 + \frac{n(n - 1)(n - 2)}{3!} x^3 + \ldots$$

has a sum to infinity of $(1 + x)^n$ when n is not a positive integer, provided that $-1 < x < 1$.

So for all values of n we can expand $(1 + x)^n$ as a series of ascending powers of x in the form

$$(1 + x)^n = 1 + nx + \frac{n(n - 1)}{2!} x^2 + \frac{n(n - 1)(n - 2)}{3!} x^3 + \ldots$$

provided that $|x| < 1$.

When using this expansion it should be noted that:

(a) The expansion is valid for $(1 + x)^n$ and *not* for $(a + x)^n$.

To expand $(a + x)^n$ it must first be written as $a^n \left(1 + \dfrac{x}{a}\right)^n$.

(b) The expansion is valid for a *restricted range of values of the variable* and this range must *always* be stated.

(c) When $n = -1$,

$$(1 + x)^{-1} = 1 - x + x^2 - x^3 + \ldots + (-1)^r x^r + \ldots, \quad |x| < 1$$

and, replacing x by $-x$,

$$(1 - x)^{-1} = 1 + x + x^2 + x^3 + \ldots + x^r + \ldots, \quad |x| < 1$$

Both of these series are G.P.s with common ratios $-x$ and x respectively. They both occur frequently and are worth memorising.

EXAMPLES 15f

1) Expand each of the following functions as a series of ascending powers of x up to and including the term containing x^4 stating the set of values of x for which each expansion is valid.

(a) $(1 + x)^{\frac{1}{2}}$ (b) $(1 - 2x)^{-3}$ (c) $(2 - x)^{-2}$

$$(1 + x)^n = 1 + nx + \frac{n(n-1)}{2!} x^2 + \ldots \quad |x| < 1 \qquad [1]$$

(a) Replacing n by $\frac{1}{2}$ in [1] gives

$$(1 + x)^{\frac{1}{2}} = 1 + \tfrac{1}{2}x + \frac{(\frac{1}{2})(\frac{1}{2} - 1)}{2!} x^2 + \frac{(\frac{1}{2})(\frac{1}{2} - 1)(\frac{1}{2} - 2)}{3!} x^3$$
$$+ \frac{\frac{1}{2}(\frac{1}{2} - 1)(\frac{1}{2} - 2)(\frac{1}{2} - 3)}{4!} x^4 + \ldots$$

$$= 1 + \frac{x}{2} + \frac{(\frac{1}{2})(-\frac{1}{2})}{2!} x^2 + \frac{(\frac{1}{2})(-\frac{1}{2})(-\frac{3}{2})}{3!} x^3$$
$$+ \frac{(\frac{1}{2})(-\frac{1}{2})(-\frac{3}{2})(-\frac{5}{2})}{4!} x^4 + \ldots$$

$$= 1 + \frac{x}{2} - \frac{x^2}{2^3} + \frac{x^3}{2^4} - \frac{(3)(5)}{(2^4)4!} x^4 + \ldots$$

$$= 1 + \frac{x}{2} - \frac{x^2}{8} + \frac{x^3}{16} - \frac{5x^4}{128} + \ldots \quad \text{provided} \quad |x| < 1$$

(b) Replacing n by -3 and x by $-2x$ in [1] gives

$$(1-2x)^{-3} = 1 + (-3)(-2x) + \frac{(-3)(-4)}{2!}(-2x)^2$$

$$+ \frac{(-3)(-4)(-5)}{3!}(-2x)^3 + \frac{(-3)(-4)(-5)(-6)}{4!}(-2x)^4 + \dots$$

$$= 1 + 6x + 24x^2 + \frac{(4)(5)}{2!}2^3 x^3 + \frac{(5)(6)}{2!}2^4 x^4 + \dots$$

$$= 1 + 6x + 24x^2 + 80x^3 + 240x^4 + \dots$$

provided that $-1 < -2x < 1$, i.e. $\frac{1}{2} > x > -\frac{1}{2}$.

(c) $(2-x)^{-2}$ must be expressed in the form $(1+x)^n$ (see note (a) p. 611).

i.e. $$(2-x)^{-2} \equiv 2^{-2}\left(1 - \frac{x}{2}\right)^{-2}.$$

Consider the expansion of $(1-x)^{-2}$

$$(1-x)^{-2} = 1 + (-2)(-x) + \frac{(-2)(-3)}{(1)(2)}(-x)^2$$

$$+ \frac{(-2)(-3)(-4)}{(1)(2)(3)}(-x)^3 + \dots$$

$$= 1 + 2x + 3x^2 + 4x^3 + \dots$$

Replacing x by $\dfrac{x}{2}$ gives

$$\left(1 - \frac{x}{2}\right)^{-2} = 1 + x + \tfrac{3}{4}x^2 + \tfrac{1}{2}x^3 + \dots$$

So $$(2-x)^{-2} \equiv \tfrac{1}{4}\left(1 - \frac{x}{2}\right)^{-2}$$

$$= \tfrac{1}{4} + \tfrac{1}{4}x + \tfrac{3}{16}x^2 + \tfrac{1}{8}x^3 + \dots$$

provided that $-1 < -\dfrac{x}{2} < 1$, i.e. $2 > x > -2$.

2) Expand $\dfrac{5}{(1+3x)(1-2x)}$ as a series of ascending powers of x giving:

(a) the first four terms

(b) the range of values of x for which the expansion is valid.

Expressing $\dfrac{5}{(1+3x)(1-2x)}$ in partial fractions gives

$$\frac{5}{(1+3x)(1-2x)} \equiv \frac{3}{(1+3x)} + \frac{2}{(1-2x)} \equiv 3(1+3x)^{-1} + 2(1-2x)^{-1}$$

Using $(1+x)^{-1} = 1-x+x^2-x^3+\ldots,$ for $-1<x<1$

and replacing x by $3x$ gives

$$(1+3x)^{-1} = 1-3x+(3x)^2+(3x)^3+\ldots$$
$$= 1-3x+9x^2-27x^3+\ldots$$

for $-1<3x<1.$

Using $(1-x)^{-1} = 1+x+x^2+\ldots$ and replacing x by $2x$ gives

$$(1-2x)^{-1} = 1+(2x)+(2x)^2+(2x)^3+\ldots$$
$$= 1+2x+4x^2+8x^3+\ldots$$

for $-1<-2x<1.$

Hence

$$\frac{5}{(1+3x)(1-2x)} \equiv 3(1+3x)^{-1}+2(1-2x)^{-1}$$
$$= (3+2)+(-9+4)x+(27+8)x^2+(-81+16)x^3+\ldots$$

provided $-\frac{1}{3}<x<\frac{1}{3}$ and $-\frac{1}{2}<x<\frac{1}{2}.$

(a) Therefore the first four terms of the series are $5-5x+35x^2-65x^3.$

(b) The expansion is valid for the range of values of x satisfying $-\frac{1}{3}<x<\frac{1}{3}$ and $-\frac{1}{2}<x<\frac{1}{2}.$

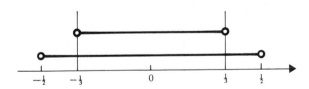

i.e. for $-\frac{1}{3}<x<\frac{1}{3}.$

3) Expand $\sqrt{\left(\dfrac{1+x}{1-2x}\right)}$ as a series of ascending powers of x up to and including the term containing $x^2.$

$$\sqrt{\left(\frac{1+x}{1-2x}\right)} \equiv (1+x)^{\frac{1}{2}}(1-2x)^{-\frac{1}{2}}$$

Now $(1+x)^{\frac{1}{2}} = \left[1+\frac{1}{2}x+\frac{(\frac{1}{2})(-\frac{1}{2})}{2!}x^2+\ldots\right]$

for $-1<x<1$

and $(1-2x)^{-\frac{1}{2}} = \left[1+(-\frac{1}{2})(-2x)+\frac{(-\frac{1}{2})(-\frac{3}{2})}{2!}(-2x)^2+\ldots\right]$

for $-1<2x<1$

Hence $\sqrt{\left(\dfrac{1+x}{1-2x}\right)} \equiv (1+x)^{\frac{1}{2}}(1-2x)^{-\frac{1}{2}}$

$$= (1 + \tfrac{1}{2}x - \tfrac{1}{8}x^2 + \ldots)(1 + x + \tfrac{3}{2}x^2 + \ldots)$$

$$= 1 + (\tfrac{1}{2}x + x) + (\tfrac{1}{2}x^2 - \tfrac{1}{8}x^2 + \tfrac{3}{2}x^2) + \ldots$$

$$= 1 + \tfrac{3}{2}x + \tfrac{15}{8}x^2 + \ldots$$

provided $-1 < x < 1$ *and* $-\tfrac{1}{2} < x < \tfrac{1}{2}$,
i.e. $-\tfrac{1}{2} < x < \tfrac{1}{2}$.

It is interesting to compare the methods used in the last two examples.
In Example 2, the function is expressed as the sum of two binomials and the series is obtained by adding two binomial expansions.
In Example 3 the function is expressed as a product of two binomials and the series is obtained by multiplying two binomial expansions.
The first method has the advantage that it is very much easier to add the terms of two series than it is to multiply them.
Therefore, *whenever possible, a compound function should be expressed as a sum of simpler functions before it is expanded as a series* and, when this is not possible, a compound function should be expressed as a product of simpler functions.

Further Approximations

We saw on p. 606 how a series expansion can be used to find an approximate value of $(1.01)^{20}$ without having to calculate the exact value. The following example shows how to find approximate numerical values (to any degree of accuracy required) for some irrational quantities.

EXAMPLES 15f (continued)

5) By substituting 0.02 for x in $(1-x)^{\frac{1}{2}}$ and its expansion, find $\sqrt{2}$ correct to five decimal places.

$$(1-x)^{\frac{1}{2}} = 1 - \tfrac{1}{2}x + \frac{(\tfrac{1}{2})(-\tfrac{1}{2})(-x)^2}{2!} + \frac{(\tfrac{1}{2})(-\tfrac{1}{2})(-\tfrac{3}{2})(-x)^3}{3!} + \ldots$$

$$= 1 - \frac{x}{2} - \frac{x^2}{8} - \frac{x^3}{16} + \ldots$$

This expansion is valid for $-1 < x < 1$ and is therefore valid when $x = 0.02$
Substituting 0.02 for x gives

$$(0.98)^{\frac{1}{2}} = 1 - 0.01 - 0.000\,05 - 0.000\,000\,5 + \ldots$$

$\Rightarrow$ $\sqrt{\dfrac{98}{100}} = 0.989\,949\,5$ to seven d.p.

$\Rightarrow \qquad \frac{7}{10}\sqrt{2} = 0.989\,949\,5$ to seven d.p.

Hence $\qquad \sqrt{2} = \dfrac{9.899\,495}{7}$ to six d.p.

$\qquad\qquad\qquad = 1.414\,213$ to six d.p.

i.e. $\qquad \sqrt{2} = 1.414\,21$ correct to five d.p.

EXERCISE 15f

Expand the following functions as series of ascending powers of x up to and including the term in x^3. In each case give the range of values of x for which the expansion is valid.

1) $(1 - 2x)^{\frac{1}{2}}$

2) $(3 + x)^{-1}$

3) $\left(1 + \dfrac{x}{2}\right)^{-\frac{1}{2}}$

4) $\dfrac{1}{(1 - x)^2}$

5) $\sqrt{\dfrac{1}{1 + x}}$

6) $(1 + x)\sqrt{(1 - x)}$

7) $\dfrac{x + 2}{x - 1}$

8) $\dfrac{2 - x}{\sqrt{(1 - 3x)}}$

9) $\dfrac{1}{(2 - x)(1 + 2x)}$

10) $\sqrt{\left(\dfrac{1 + x}{1 - x}\right)}$

11) $\left(1 + \dfrac{x^2}{9}\right)^{-1}$

12) $\left(1 + \dfrac{1}{x}\right)^{-1}$ $\quad \left[Hint: \left(1 + \dfrac{1}{x}\right)^{-1} \equiv \left(\dfrac{x + 1}{x}\right)^{-1} \equiv \dfrac{x}{1 + x}\right]$

13) Expand $\left(1 + \dfrac{1}{p}\right)^{-3}$ as a series of descending powers of p, as far as and including the term containing p^{-4}. State the range of values of p for which the expansion is valid. $\left(Hint: \text{ replace } x \text{ by } \dfrac{1}{p} \text{ in } (1 + x)^{-3}\right)$

14) By substituting 0.08 for x in $(1 + x)^{\frac{1}{2}}$ and its expansion find $\sqrt{3}$ correct to four significant figures.

15) By substituting $\dfrac{1}{10}$ for x in $(1 - x)^{-\frac{1}{2}}$ and its expansion find $\sqrt{10}$ correct to six significant figures.

16) Use a suitable binomial expansion to find $\sqrt{1.01}$ correct to five decimal places.

17) Use a suitable binomial expansion to find $\sqrt[3]{8.4}$ correct to seven significant figures.

18) Expand $\sqrt{\dfrac{1 + 2x}{1 - 2x}}$ as a series of ascending powers of x up to and including the term in x^2. By substituting $x = \dfrac{1}{100}$ find an approximation for $\sqrt{51}$, stating the number of significant figures to which your answer is accurate.

19) If x is so small that x^2 and higher powers of x may be neglected show that $\dfrac{1}{(x - 1)(x + 2)} \simeq -\frac{1}{2} - \frac{1}{4}x$.

20) By neglecting x^3 and higher powers of x, find a quadratic function that approximates to the function $\dfrac{1 - 2x}{\sqrt{1 + 2x}}$ in the region close to $x = 0$.

21) Use partial fractions and the binomial series to find a linear approximation for
$$\frac{3}{(1 - 2x)(2 - x)}$$

22) If terms containing x^4 and higher powers of x can be neglected, show that
$$\frac{2}{(x + 1)(x^2 + 1)} \simeq 2(1 - x)$$

23) Show that
$$\frac{12}{(3 + x)(1 - x)^2} \simeq 4 + \tfrac{20}{3}x + \tfrac{88}{9}x^2$$
provided that x is small enough to neglect powers higher than 2.

SUMMATION OF NUMBER SERIES

A series, each of whose terms has a fixed numerical value, is called a number series. For example, $2 + 4 + 8 + 16 + \ldots$ is a number series, whereas $x + x^2 + x^3 + \ldots$ is called a power series as the terms involve powers of a variable quantity.

We are now going to investigate some of the methods available for finding the sum of a number series. The method adopted depends to some extent on the form of the general term of the series. One of the most straightforward ways of summing a given series uses *recognition* of a standard series. For instance it might be recognised as an arithmetic progression (A.P.) or a geometric progression (G.P.).

Consider, for example, the series $\displaystyle\sum_{r=1}^{n} 2^{2r-1}$

Now $\displaystyle\sum_{r=1}^{n} 2^{2r-1} = 2 + 2^3 + 2^5 + \ldots + 2^{2n-1}$

which we *recognise* as a G.P. whose first term is 2, and whose common ratio is 2^2.

Using the formula for the sum of the first n terms of a G.P. gives

$$\sum_{r=1}^{n} 2^{2r-1} = \frac{2[(2^2)^n - 1]}{2^2 - 1} = \tfrac{2}{3}(4^n - 1).$$

Sometimes a given series, although recognised basically as a known series, has slight deviations from the standard series. The sum of the given series can still be deduced from the known series allowing first for any adjustments that may be necessary.

For example $\displaystyle\sum_{r=n}^{\infty} \left(\frac{1}{3}\right)^r \left(\frac{1}{2}\right)^{r-2} = \left(\frac{1}{3}\right)^n \left(\frac{1}{2}\right)^{n-2} + \left(\frac{1}{3}\right)^{n+1} \left(\frac{1}{2}\right)^{n-1} + \ldots$

Each term of this series has a factor $\left(\frac{1}{3}\right)^n \left(\frac{1}{2}\right)^{n-2}$

so we may write

$$\sum_{r=n}^{\infty} \left(\frac{1}{3}\right)^r \left(\frac{1}{2}\right)^{r-2} = \left(\frac{1}{3}\right)^n \left(\frac{1}{2}\right)^{n-2} \left[1 + \left(\frac{1}{3}\right)\left(\frac{1}{2}\right) + \left(\frac{1}{3}\right)^2 \left(\frac{1}{2}\right)^2 + \ldots\right]$$

$$= \left(\frac{1}{3}\right)^n \left(\frac{1}{2}\right)^{n-2} \left[1 + \frac{1}{6} + \left(\frac{1}{6}\right)^2 + \ldots\right]$$

The series in the square bracket is recognised as an infinite G.P. whose sum to infinity is $\displaystyle\frac{1}{1-\frac{1}{6}} = \frac{6}{5}$.

Hence $\displaystyle\sum_{r=n}^{\infty} \left(\frac{1}{3}\right)^r \left(\frac{1}{2}\right)^{r-2} = \frac{6}{5}\left(\frac{1}{3}\right)^n \left(\frac{1}{2}\right)^{n-2} = \frac{4}{5}\left(\frac{1}{6}\right)^{n-1}$

THE NATURAL NUMBER SERIES

The numbers $1, 2, 3, 4, 5, \ldots$ are called the natural numbers.
The series $1 + 2 + 3 + 4 + 5 + \ldots + n$, is called 'the sum of the first n natural numbers' and can be written in the form $\displaystyle\sum_{r=1}^{n} r$.

This series is a simple A.P. in which $\quad a = 1 \quad$ and $\quad d = 1 \quad$ so, using the formula $\quad S_n = \dfrac{n}{2}\{2a + (n-1)d\}, \quad$ we see that

$$\sum_{r=1}^{n} r = \frac{n}{2}(n+1)$$

Now consider the series $1^2 + 2^2 + 3^2 + 4^2 + 5^2 + \ldots + n^2$.
This one is called 'the sum of the squares of the first n natural numbers' and is written $\displaystyle\sum_{r=1}^{n} r^2$.

As this series is neither an A.P. nor a G.P. its sum is not known at this stage so a new method must be devised to find the sum.
This method begins with an identity

$$(r+1)^3 - r^3 \equiv 3r^2 + 3r + 1 \qquad\qquad [1]$$

(The reader may, at this stage, question both the source and the relevance of this identity but these will become clear later.)
Summing both sides of [1] for values of r from 1 to n gives:

$$\sum_{r=1}^{n} \llbracket (r+1)^3 - r^3 \rrbracket \equiv 3\sum_{r=1}^{n} r^2 + 3\sum_{r=1}^{n} r + \sum_{r=1}^{n} 1 \qquad\qquad [2]$$

Considering the L.H.S. and writing down the terms in reverse order (i.e. starting with $r = n$) we have

$$\llbracket (n+1)^3 - n^3 \rrbracket + \llbracket n^3 - (n-1)^3 \rrbracket + \ldots + \llbracket 3^3 - 2^3 \rrbracket + \llbracket 2^3 - 1^3 \rrbracket$$

which simplifies to $(n+1)^3 - 1^3$.
So [2] becomes

$$(n+1)^3 - 1 \equiv 3\sum_{r=1}^{n} r^2 + 3\sum_{r=1}^{n} r + \sum_{r=1}^{n} 1$$

$$\Rightarrow \qquad 3\sum_{r=1}^{n} r^2 \equiv (n+1)^3 - 1 - 3\sum_{r=1}^{n} r - \sum_{r=1}^{n} 1$$

Now $\displaystyle\sum_{r=1}^{n} r = \frac{n}{2}(n+1)$

and $\displaystyle\sum_{r=1}^{n} 1 = n \quad$ (because there are n terms each of value 1)

Hence $\qquad 3\displaystyle\sum_{r=1}^{n} r^2 = (n+1)^3 - 1 - \frac{3n}{2}(n+1) - n$

$$= \frac{n}{2}(n+1)(2n+1)$$

So finally, $$\sum_{r=1}^{n} r^2 = \frac{n}{6}(n+1)(2n+1)$$

This result is quotable when its proof is not specifically asked for.

An essential feature of the above method is obviously the identity used at the outset. There are many other series which can be summed in a similar way and, when this approach is recommended, the reader will be given the appropriate identity.

EXAMPLES 15g

1) Verify that $r(r+1)(r+2)-(r-1)r(r+1) \equiv 3r(r+1)$ and use your

result to find the sum of the series $\sum_{r=1}^{n} r(r+1)$.

$$r(r+1)(r+2)-(r-1)r(r+1) \equiv r(r+1)[(r+2)-(r-1)]$$
$$\equiv 3r(r+1)$$

So the identity is verified.

Summing both sides of the identity from $r=1$ to $r=n$ gives

$$\sum_{r=1}^{n} [r(r+1)(r+2)-(r-1)r(r+1)] \equiv 3\sum_{r=1}^{n} r(r+1)$$

Writing down the terms on the L.H.S. in reverse order gives

$$[n(n+1)(n+2)-(n-1)n(n+1)] + [(n-1)n(n+1)-(n-2)(n-1)n] + \ldots$$
$$+ [(2)(3)(4)-(1)(2)(3)] + [(1)(2)(3)-(0)(1)(2)]$$

So the L.H.S. reduces to $n(n+1)(n+2)$.

Hence $$n(n+1)(n+2) = 3\sum_{r=1}^{n} r(r+1)$$

$\Rightarrow$ $$\sum_{r=1}^{n} r(r+1) = \tfrac{1}{3}n(n+1)(n+2)$$

2) Express $\dfrac{1}{(r-1)r}$ in partial fractions and hence find the sum of the series

$$\frac{1}{(1)(2)} + \frac{1}{(2)(3)} + \frac{1}{(3)(4)} + \ldots + \frac{1}{(n-1)(n)}.$$

The 'cover-up' method gives

$$\frac{1}{(r-1)(r)} \equiv \frac{1}{r-1} - \frac{1}{r} \qquad [1]$$

Now each term in the given series is of the form $\dfrac{1}{(r-1)r}$ where r takes values

from 2 to n, so the series can be expressed in the form $\displaystyle\sum_{r=2}^{n} \dfrac{1}{(r-1)r}$.

Then [1] gives

$$\sum_{r=2}^{n} \frac{1}{(r-1)r} \equiv \sum_{r=2}^{n} \left\{ \frac{1}{r-1} - \frac{1}{r} \right\}$$

This time it is on the R.H.S. that there is a difference of two terms, so we write out the R.H.S. term by term, starting with $r=2$, giving

$$\left(\frac{1}{1} - \frac{1}{2}\right) + \left(\frac{1}{2} - \frac{1}{3}\right) + \left(\frac{1}{3} - \frac{1}{4}\right) + \ldots + \left(\frac{1}{n-1} - \frac{1}{n}\right)$$

and we see that all the terms cancel except 1 and $-\dfrac{1}{n}$,

i.e.

$$\sum_{r=2}^{n} \frac{1}{(r-1)r} = 1 - \frac{1}{n}$$

Hence

$$\frac{1}{(1)(2)} + \frac{1}{(2)(3)} + \ldots + \frac{1}{(n-1)n} = \frac{n-1}{n}$$

METHOD OF DIFFERENCES

The summation of each of the series in the examples above was made possible by using an identity with two special properties.

(a) One of the terms in the identity was the general term of the series to be summed.

(b) One side of the identity comprised a *difference of two terms*.

When summing a series in this way we are using the 'method of differences' which can be generalised as follows:

If the general term, u_r, of a series can be expressed in the form $f(r+1)-f(r)$, then

$$\sum_{r=1}^{n} u_r = \sum_{r=1}^{n} [f(r+1)-f(r)]$$

$$= f(n+1)-f(n)$$
$$+f(n)-f(n-1)$$
$$+f(n-1)-f(n-2)$$
$$\cdots$$
$$+f(3)-f(2)$$
$$+f(2)-f(1) = f(n+1)-f(1)$$

EXAMPLES 15g (continued)

3) If $f(r) \equiv r(r+1)!$ simplify $f(r)-f(r-1)$ and hence sum the series

$$5.2!+10.3!+17.4!+\ldots+(n^2+1)n!$$

$$\begin{aligned} f(r)-f(r-1) &\equiv r(r+1)!-(r-1)r! \\ &\equiv r![r(r+1)-(r-1)] \\ &\equiv r!(r^2+1) \end{aligned}$$

Hence

$$\sum_{r=2}^{n} (r^2+1)r! \equiv \sum_{r=2}^{n} [f(r)-f(r-1)]$$

$$\equiv f(n)-f(n-1)$$
$$+f(n-1)-f(n-2)$$
$$+f(n-2)-f(n-3)$$
$$\cdots$$
$$\cdots$$
$$+f(3)-f(2)$$
$$+f(2)-f(1) = f(n)-f(1)$$

But $f(n) \equiv n(n+1)!$ and $f(1) \equiv 1(1+1)!$

So

$$\sum_{r=2}^{n} (r^2+1)r! = n(n+1)!-2!$$

i.e.

$$5.2!+10.3!+\ldots+(n^2+1)n! = n(n+1)!-2$$

STANDARD RESULTS

Some natural number series have sums that are quotable. These are

(a) the sum of the first n natural numbers,

$$\sum_{r=1}^{n} r = \frac{n}{2}(n+1)$$

(b) the sum of the squares of the first n natural numbers,

$$\sum_{r=1}^{n} r^2 = \frac{n}{6}(n+1)(2n+1)$$

(c) the sum of the cubes of the first n natural numbers,

$$\sum_{r=1}^{n} r^3 = \frac{n^2}{4}(n+1)^2$$

EXAMPLES 15g (continued)

4) Find $\displaystyle\sum_{r=1}^{n} r(r+1)(r+2)$ using the standard results above.

$$\sum_{r=1}^{n} r(r+1)(r+2) = \sum_{r=1}^{n} (r^3 + 3r^2 + 2r)$$

$$= \sum_{r=1}^{n} r^3 + 3\sum_{r=1}^{n} r^2 + 2\sum_{r=1}^{n} r$$

$$= \frac{n^2}{4}(n+1)^2 + 3\left[\frac{n}{6}(n+1)(2n+1)\right] + 2\left[\frac{n}{2}(n+1)\right]$$

$$= \frac{n}{4}(n+1)(n+2)(n+3).$$

Note. This series can also be summed by the method of differences using the identity

$$4r(r+1)(r+2) \equiv r(r+1)(r+2)(r+3) - (r-1)r(r+1)(r+2).$$

5) Find the sum of the squares of the first n odd numbers.

The odd numbers can be represented by $2r-1$ where $r = 1, 2, 3, \ldots$

Hence we require $\displaystyle\sum_{r=1}^{n} (2r-1)^2$.

Now $\displaystyle\sum_{r=1}^{n} (2r-1)^2 = \sum_{r=1}^{n} (4r^2 - 4r + 1)$

$$= 4 \sum_{r=1}^{n} r^2 - 4 \sum_{r=1}^{n} r + \sum_{r=1}^{n} 1$$

$$= 4\left[\frac{n}{6}(n+1)(2n+1)\right] - 4\left[\frac{n}{2}(n+1)\right] + n$$

$$= \frac{n(4n^2-1)}{3}$$

EXERCISE 15g

Find, by recognition, the sum of each of the following series.

1) $1 - \dfrac{1}{2} + \dfrac{1}{4} - \dfrac{1}{8} + \dfrac{1}{16} - \ldots$ 2) $2 - 2.3 + 2.3^2 - 2.3^3 + \ldots + 2.3^{10}$

3) $\displaystyle\sum_{r=2}^{n} ab^{2r}$ 4) $2 + \dfrac{1}{3} + \dfrac{1}{3^2 2} + \dfrac{1}{3^3 2^2} + \ldots$

In Questions 5 to 8, express the general term in partial fractions and hence find the sum of the series.

5) $\displaystyle\sum_{r=1}^{n} \frac{1}{r(r+2)}$ 6) $\displaystyle\sum_{r=3}^{n} \frac{1}{(r+1)(r+2)}$

7) $\displaystyle\sum_{r=n}^{2n} \frac{1}{r(r+1)}$ 8) $\displaystyle\sum_{r=1}^{n} \frac{r}{(2r-1)(2r+1)(2r+3)}$

9) Verify that

$$r(r+1)(r+2)(r+3) - (r-1)(r)(r+1)(r+2) \equiv 4r(r+1)(r+2)$$

Use your result to find the value of $\displaystyle\sum_{r=1}^{n} r(r+1)(r+2)$.

10) Given that $\displaystyle\sum_{r=1}^{n} r^2 = \tfrac{1}{6}n(n+1)(2n+1)$,   use the identity

$$r^4 - (r-1)^4 \equiv 4r^3 - 6r^2 + 4r - 1$$

to find the sum of the cubes of the first n natural numbers, i.e. $\displaystyle\sum_{r=1}^{n} r^3$.

11) Find the sum of the series $(1)(3) + (3)(5) + (5)(7) + \ldots + (2n-1)(2n+1)$ by using the identity

$$(2r-1)(2r+1)(2r+3) - (2r-3)(2r-1)(2r+1) \equiv 6(2r-1)(2r+1)$$

12) If $f(r) \equiv \dfrac{1}{r(r+1)}$,   simplify $f(r+1) - f(r)$.

Hence find $\displaystyle\sum_{r=1}^{n} \dfrac{1'}{r(r+1)(r+2)}$

13) If $f(r) \equiv \dfrac{1}{r^2}$,   simplify $f(r) - f(r+1)$

Hence find the sum of the first n terms of the series

$$\frac{3}{1^2 \cdot 2^2} + \frac{5}{2^2 \cdot 3^2} + \frac{7}{3^2 \cdot 4^2} + \dots$$

14) If $f(r) \equiv r(r+1)(r+2)$,   simplify $f(r+1) - f(r)$

Hence find $\displaystyle\sum_{r=1}^{n} 3(r^2 + 3r + 2)$ and deduce that $\displaystyle\sum_{r=1}^{n} r^2 = \frac{1}{6}n(n+1)(2n+1)$

15) Verify that $4r^3 + r \equiv (r + \frac{1}{2})^4 - (r - \frac{1}{2})^4$.   Hence find $\displaystyle\sum_{r=1}^{n} (4r^3 + r)$

Deduce that $\displaystyle\sum_{r=1}^{n} r^3 = \frac{n^2}{4}(n+1)^2$

16) If $f(r) \equiv r(r+1)(r+2)(r+3)$,   simplify $f(r+1) - f(r)$

and use your result to find $\displaystyle\sum_{r=1}^{n} r^3$

17) If $f(r) \equiv \dfrac{1}{r!}$,   simplify $f(r) - f(r+1)$

and hence find $\displaystyle\sum_{r=1}^{n} \dfrac{r}{(r+1)!}$

18) Given $f(r) \equiv r!$,   find $f(r+1) - f(r)$ and use your result to find the sum of the first $2n$ terms of the series

$$1.1! + 2.2! + 3.3! + 4.4! + \dots$$

19) If $f(r) \equiv \cos 2r\theta$,   simplify $f(r) - f(r+1)$.
Use your result to find the sum of the first n terms of the series

$$\sin 3\theta + \sin 5\theta + \sin 7\theta + \dots$$

Use standard results for $\displaystyle\sum_{r=1}^{n} r, \ \sum_{r=1}^{n} r^2, \ \sum_{r=1}^{n} r^3$ to find

20) $\displaystyle\sum_{r=1}^{n} r(r+1)$ 21) $\displaystyle\sum_{r=1}^{n} r(r+1)(r+2)$

22) The sum of the squares of the first n even numbers.

23) $\displaystyle\sum_{r=n}^{2n} r^2(1+r)$

24) $1.3 + 2.4 + 3.5 + \ldots + (n-1)(n+1)$

25) $1^2 - 2^2 + 3^2 - 4^2 + 5^2 - 6^2 + \ldots - (2n)^2$

(*Hint.* Consider two series, the sum of the squares of even numbers and sum of the squares of odd numbers.)

26) $\displaystyle\sum_{r=10}^{20} r^3$

SEQUENCES

Consider the sequence

$$1.1, 1.01, 1.001, 1.0001, \ldots, 1 + (\tfrac{1}{10})^n, \ldots$$

The nth term of this sequence, u_n, is $1 + (\tfrac{1}{10})^n$
and as $n \to \infty$, $1 + (\tfrac{1}{10})^n \to 1$,

i.e. $$\lim_{n \to \infty} [\![u_n]\!] = \lim_{n \to \infty} [\![1 + (\tfrac{1}{10})^n]\!] = 1$$

and we say that this sequence converges to the value unity.

In general, if u_n is the nth term of a sequence and $\lim\limits_{n \to \infty} [\![u_n]\!]$ exists, the sequence is said to converge and the value of $\lim\limits_{n \to \infty} [\![u_n]\!]$ is called the limiting value, or limit of the sequence.

A Formal Definition of the Limit of $f(n)$ as $n \to \infty$

If $f(n)$ is a function of positive integral values of n, and if a number k exists such that, as n *increases*, $|f(n) - k|$ becomes and *remains* less than an arbitrarily chosen positive quantity ϵ, however small ϵ is, then

$$\lim_{n \to \infty} \{f(n)\} \quad \text{exists and is equal to } k.$$

For example, if $f(n) \equiv 1 + (\tfrac{1}{10})^n$ for $n = 1, 2, \ldots$ and if $k = 1$
then $$|f(n) - k| = |(\tfrac{1}{10})^n|$$

If $\epsilon = 10^{-6}$ then $|f(n) - 1| < 10^{-6}$ for *all* values of $n > 6$.
If $\epsilon = 10^{-20}$ then $|f(n) - 1| < 10^{-20}$ for *all* values of $n > 20$.

Similarly for any given value of ϵ we can find a value of n, N say, such that

$$|f(n) - 1| < \epsilon \quad \text{for all} \quad n > N,$$

i.e.
$$\lim_{n \to \infty} [\![f(n)]\!] = 1$$

From this example we can see that the formal definition above implies that if $\lim_{n \to \infty} \{f(n)\} = k$ then, given a value for ϵ, a value of n, N say, exists such that for all $n > N$, $|f(n) - k| < \epsilon$,

i.e. $\lim_{n \to \infty} \{f(n)\} = k \iff$ $\begin{cases} \text{Given } \epsilon, \text{ there exists a number } N \text{ such} \\ \text{that } |f(n) - k| < \epsilon \text{ for all } n > N \end{cases}$

Note that this definition does not enable the value of a limit to be *found*. All it does is to formalise an intuitive concept of a limit.

The Evaluation of the Limit of $f(n)$ as $n \to \infty$

When evaluating $\lim_{n \to \infty} [f(n)]$, the following assumptions may be made

as $n \to \infty$ $\begin{cases} \dfrac{1}{n} \to 0, \quad \dfrac{1}{n!} \to 0 \\[2mm] a^n \to \infty \text{ if } a > 1, \quad a^n \to 0 \text{ if } 0 < a < 1 \\[2mm] \dfrac{a^n}{n!} \to 0 \end{cases}$

If, as $n \to \infty$, $f(n) \to \dfrac{\infty}{\infty}$ or $\dfrac{0}{0}$, both of which are indeterminate, it may be possible to evaluate $\lim_{n \to \infty} [f(n)]$ by one of the following methods.

(a) Express $f(n)$ as a proper fraction

e.g. if $f(n) \equiv \dfrac{n-1}{n+1}$, then $\lim_{n \to \infty} \dfrac{n-1}{n+1}$ is indeterminate.

But $f(n) \equiv \dfrac{n-1}{n+1} \equiv 1 - \dfrac{2}{n+1}$

so $\lim_{n \to \infty} \dfrac{n-1}{n+1} \equiv \lim_{n \to \infty} 1 - \dfrac{2}{n+1} = 1$

(b) Replace n by $\dfrac{1}{m}$,

e.g. if $f(n) \equiv \dfrac{n^2 - 5n + 6}{2n^2 + 7n - 2}$

replacing n by $\dfrac{1}{m}$ gives

$$f\left(\frac{1}{m}\right) \equiv \frac{\dfrac{1}{m^2} - \dfrac{5}{m} + 6}{\dfrac{2}{m^2} - \dfrac{7}{m} - 2} \equiv \frac{1 - 5m + 6m^2}{2 + 7m - 2m^2}$$

Now as $n \to \infty$, i.e. as $\dfrac{1}{m} \to \infty$, $m \to 0$.

So $\lim\limits_{n \to \infty} [f(n)]$ becomes $\lim\limits_{m \to 0} \left[f\left(\frac{1}{m}\right) \right]$,

i.e. $\qquad \lim\limits_{n \to \infty} \dfrac{n^2 - 5n + 6}{2n^2 + 7n - 2} \equiv \lim\limits_{m \to 0} \dfrac{1 - 5m + 6m^2}{2 + 7m - 2m^2} = \dfrac{1}{2}$

SUM TO INFINITY OF A NUMBER SERIES

Consider the series

$$\frac{1}{1 \cdot 2} + \frac{1}{2 \cdot 3} + \frac{1}{3 \cdot 4} + \frac{1}{4 \cdot 5} + \dots$$

If S_n is the sum of the first n terms of this series then

$$S_n = \sum_{r=1}^{n} \frac{1}{r(r+1)} = \sum_{r=1}^{n} \left(\frac{1}{r} - \frac{1}{r+1} \right) = 1 - \frac{1}{n+1} = \frac{n}{n+1}$$

As $n \to \infty$, $S_n \to 1$,

so we say that the series is *convergent* and that its sum to infinity, S, is unity.
In general, if S_n is the sum of the first n terms of a series, and if $\lim\limits_{n \to \infty} [S_n]$

exists, then the series is said to be convergent with a sum to infinity, S, where

$$S = \lim\limits_{n \to \infty} [S_n]$$

By using the more formal definition of a limit, the sum to infinity of a series
may be defined as follows.
A series is convergent with a sum to infinity S if, given ϵ, there exists a finite
number N such that $|S_n - S| < \epsilon$ for all $n > N$.
Conversely, if $\lim\limits_{n \to \infty} [S_n]$ does not exist, the series is not convergent.

So, to sum an infinite number series which is not recognised as a standard
expansion, we first find S_n and then evaluate $\lim\limits_{n \to \infty} S_n$.

EXAMPLE 15h

Find the sum to infinity, S, of the series

$$\frac{3}{(2)(4)} - \frac{5}{(4)(6)} + \frac{7}{(6)(8)} - \dots + \frac{(-1)^{r+1}(2r+1)}{(2r)(2r+2)} + \dots$$

The form of the general term of this series suggests that a 'difference' identity is likely to arise if partial fractions are used.

$$\frac{(2r+1)}{(2r)(2r+2)} \equiv \frac{\frac{1}{2}}{2r} + \frac{\frac{1}{2}}{2r+2}$$

So $\qquad S_n = \sum_{r=1}^{n} (-1)^{r+1} \frac{(2r+1)}{(2r)(2r+2)} = \sum_{r=1}^{n} (-1)^{r+1} \left(\frac{\frac{1}{2}}{2r} + \frac{\frac{1}{2}}{2r+2} \right)$

$\Rightarrow \qquad 4S_n = \sum_{r=1}^{n} (-1)^{r+1} \left(\frac{1}{r} + \frac{1}{r+1} \right)$

$$= \left(\frac{1}{1} + \frac{1}{2} \right) - \left(\frac{1}{2} + \frac{1}{3} \right) + \left(\frac{1}{3} + \frac{1}{4} \right) - \dots$$

$$+ (-1)^n \left(\frac{1}{n-1} + \frac{1}{n} \right) + (-1)^{n+1} \left(\frac{1}{n} + \frac{1}{n+1} \right)$$

$$= 1 + (-1)^{n+1} \frac{1}{n+1}$$

So S_n, the sum of the first n terms of the series, is given by

$$S_n = \frac{1}{4} \left\{ 1 + (-1)^{n+1} \frac{1}{n+1} \right\}$$

Now, as $n \to \infty$, $\dfrac{1}{n+1} \to 0$, so $S_n \to \frac{1}{4}$

Hence $\qquad\qquad\qquad\qquad S = \lim_{n \to \infty} (S_n) = \frac{1}{4}$

EXERCISE 15h

1) Evaluate the following limits:

(a) $\lim\limits_{n \to \infty} \dfrac{n+1}{n^2}$ (b) $\lim\limits_{n \to \infty} \dfrac{n}{2^n}$ (c) $\lim\limits_{n \to \infty} \dfrac{n^2+2}{n^2+1}$

2) Find the sums to infinity of the following series:

(a) $\dfrac{1}{3.4} + \dfrac{1}{4.5} + \dfrac{1}{5.6} + \dots$

(b) $\dfrac{1}{1.3} + \dfrac{1}{3.5} + \dfrac{1}{5.7} + \dots$ (c) $\dfrac{1}{1.4} + \dfrac{1}{2.5} + \dfrac{1}{3.6} + \dots$

3) Find the least value of n for which $|S_n - S| < 10^{-4}$ where S_n is the sum of the first n terms of a series and S is its sum to infinity and S_n is given by

(a) $\dfrac{n}{n-2}$ (b) $\dfrac{2n+1}{n-1}$ (c) $1 + \dfrac{3}{n^2}$

METHOD OF INDUCTION

All the methods so far used to deal with summation of series have allowed us to *find* the sum by using the general term of the series in some way. We are now going to consider a different case.
Consider, for example, the series

$$S_n = \sum_{r=1}^{n} r(r+1) = (1)(2) + (2)(3) + (3)(4) + \ldots + (n)(n+1)$$

Now $S_1 = (1)(2)$ $= 2 = \frac{1}{3}(1)(2)(3)$

$S_2 = (1)(2) + (2)(3)$ $= 8 = \frac{1}{3}(2)(3)(4)$

$S_3 = (1)(2) + (2)(3) + (3)(4)$ $= 20 = \frac{1}{3}(3)(4)(5)$

These results *suggest*, but do not *prove*, that

$$S_n = \tfrac{1}{3}(n)(n+1)(n+2)$$

We will now prove that this result is true for any number of terms by using a method known as '*proof by induction*'.
Let p_n be the statement $S_n = \frac{1}{3}(n)(n+1)(n+2)$, so that our aim is to prove that p_n is always true when $n \in \mathbb{N}$.
Now *if* the statement is true when $n = k$

i.e. if $S_k = \frac{1}{3}(k)(k+1)(k+2)$

then, adding the $(k+1)$th term to each side, it follows that

$$S_{k+1} = \tfrac{1}{3}(k)(k+1)(k+2) + (k+1)(k+2)$$

$\Rightarrow$ $S_{k+1} = \tfrac{1}{3}(k+1)(k+2)(k+3)$

But this is the statement p_{k+1}.

So, *if* p_k is true, so also is p_{k+1}.   [1]

Note that we have not yet proved either that p_k or that p_{k+1} is true.

Now consider the case when $n = 1$,

$$S_1 = (1)(2) = 2$$
$$\tfrac{1}{3}(1)(2)(3) = 2$$

and   so p_1 *is* true

Then, using [1], we see that

since p_1 is true, so is p_2

since p_2 is true, so is p_3

since p_3 is true, so is p_4

and so on, showing that p_n is true for all positive integer values of n.

It is important to appreciate that the method of induction can never be used to *find* the sum of a series, but only to prove a result that is already suspected. It is also important, for the student who intends to study Mathematics beyond the level of this book, to know that many results apart from the sums of series can be proved by the method of induction.

EXAMPLE 15i

Prove by induction that $\displaystyle\sum_{r=1}^{n} r^2 = \tfrac{1}{6}n(n+1)(2n+1)$.

Let p_n be the statement $\displaystyle\sum_{r=1}^{n} r^2 = \tfrac{1}{6}n(n+1)(2n+1)$.

If p_k is true, i.e. if $\displaystyle\sum_{r=1}^{k} r^2 = \tfrac{1}{6}k(k+1)(2k+1)$

then, adding the $(k+1)$th term to each side, we have

$$\sum_{r=1}^{k+1} r^2 = \tfrac{1}{6}k(k+1)(2k+1) + (k+1)^2$$

$$= \tfrac{1}{6}(k+1)[k(2k+1) + 6(k+1)]$$

$$= \tfrac{1}{6}(k+1)(k+2)(2k+3)$$

$$\Rightarrow \qquad \sum_{r=1}^{k+1} r^2 = \tfrac{1}{6}(k+1)([k+1]+1)(2[k+1]+1)$$

But this is the statement p_{k+1}, so if p_k is true, so is p_{k+1}.

Now, using $n = 1$ in the statement p_n we have,

$$\text{L.H.S.} = \sum_{r=1}^{n} 1^2 = 1$$

$$\text{R.H.S.} = \tfrac{1}{6}(1)(2)(3) = 1$$

$\left.\begin{array}{c} \\ \\ \end{array}\right\} \Rightarrow \quad p_1 \text{ is true}$

Combining the two parts of the argument above,

$$\text{as } p_1 \text{ is true, so is } p_2$$

$$\text{as } p_2 \text{ is true, so is } p_3$$

$$\text{as } p_3 \text{ is true, so is } p_4$$

and so on for all values of n where $n \in \mathbb{N}$,

i.e. $\sum_{r=1}^{n} r^2 = \frac{1}{6}n(n+1)(2n+1)$ for any number of terms.

EXERCISE 15i

Prove by induction that

1) $\sum_{r=1}^{n} r^3 = \frac{n^2}{4}(n+1)^2$

2) $\sum_{r=1}^{n} \frac{1}{r(r+1)} = \frac{n}{n+1}$

3) $\sum_{r=1}^{n} r(3^r) = \frac{3}{4}[1 + 3^n(2n-1)]$

4) $\sum_{r=1}^{n} r^2 = \frac{1}{6}n(n+1)(2n+1)$

5) $\sum_{r=1}^{n} 4^r = \frac{4}{3}(4^n - 1)$

6) $\sum_{r=1}^{n} (2+3r) = \frac{n}{2}(3n+7)$

7) $\sum_{r=1}^{n} \frac{1}{r(r+1)(r+2)} = \frac{n(n+3)}{4(n+1)(n+2)}$

8) $\sum_{r=1}^{n} \frac{r}{2^r} = 2 - (\frac{1}{2})^n(2+n)$

SUMMARY

Arithmetic Progressions

An A.P. with n terms can be written as

$$a + (a+d) + (a+2d) + \ldots + (a+rd) + \ldots + [a+(n-1)d]$$

The sum of the first n terms, S_n, is given by

$$S_n = \frac{n}{2}[2a + (n-1)d]$$

$$= \frac{n}{2}(a+l) \quad \text{where } l \text{ is the last term}$$

The sum of the first n natural numbers is given by

$$\sum_{r=1}^{n} r = \frac{n(n+1)}{2}$$

Geometric Progressions
A G.P. of n terms can be written as

$$a + ar + ar^2 + \ldots + ar^{(n-1)}$$

$$S_n = \frac{a(1-r^n)}{1-r}, \quad S_\infty = \frac{a}{1-r}, \quad |r| < 1$$

Arithmetic mean of two numbers x, y is $\frac{1}{2}(x+y)$

Geometric mean of two numbers x, y is $\sqrt{xy}$

Binomial Theorem

$$(1+x)^n = 1 + nx + \frac{n(n-1)}{2!}x^2 + \frac{n(n-1)(n-2)}{3!}x^3 + \ldots$$

If n is a positive integer the series is finite and valid for all values of x.
If n is not a positive integer the series is infinite and valid only for
$-1 < x < 1$.

MULTIPLE CHOICE EXERCISE 15

(Instructions for answering these questions are given on page xii.)

TYPE I

1) The sum of the series $1 + 5 + 9 + 13 + 17 + 21 + 25 + 29$ is:
(a) 30   (b) 240   (c) 120   (d) 112   (e) 28

2) The sum to infinity of the series $1 + 2x + 4x^2 + 8x^3 + \ldots$, for
$-\frac{1}{2} < x < \frac{1}{2}$ is:

(a) $\dfrac{2x}{1-2x}$ (b) $\dfrac{1}{1-2x}$ (c) $\dfrac{1}{1+2x}$ (d) $\dfrac{2}{1-x}$ (e) $1-2x$

3) The first three terms of the series $\sum\limits_{r=0}^{\infty} (-1)^{r+1} 2^r x^{-r}$ are:

(a) $-1 + \dfrac{2}{x} - \dfrac{4}{x^2}$ (b) $1 + \dfrac{2}{x} - \dfrac{4}{x^2}$ (c) $1 + 2x - 4x^2$

(d) $\dfrac{2}{x} - \dfrac{4}{x^2} + \dfrac{8}{x^3}$ (e) none of these.

4) 3 is the geometric mean of a and b. Possible values of a and b are:

(a) 5, 4   (b) 0, 9   (c) 3, 1   (d) 4, 2   (e) 9, 1

5) The series $1 - x + 2x^2 - 3x^3 + 4x^4 + \ldots$ may be written more briefly as:

(a) $\sum_{r=0}^{\infty} (-1)^r r x^r$ (b) $1 + \sum_{r=0}^{\infty} (-1)^r r x^r$ (c) $\sum_{r=1}^{\infty} r x^r$

(d) $1 - \sum_{r=0}^{\infty} (-1)^r r x^r$ (e) $\sum_{r=1}^{\infty} (-1)^{r+1} r x^r$

6) The coefficient of x^3 in the binomial expansion of $(2 - x)^8$ is:

(a) 1792   (b) 56   (c) -1792 (d) -2000 (e) -448

7) The series $1 - 3x + 9x^2 - 27x^3 + \ldots$ converges to the value:

(a) $\frac{1}{10}$ when $x = 3$ (b) $\frac{1}{2}$ when $x = \frac{1}{3}$ (c) $\frac{2}{3}$ when $x = \frac{1}{6}$

(d) $\frac{3}{2}$ when $x = \frac{1}{9}$ (e) $\frac{1}{4}$ when $x = 1$

TYPE II

8) The sum of the first n terms, S_n, of a given series is given by

$$S_n = \frac{2n^2}{n^2 + 1}.$$

(a) The first two terms of the series are $1, \frac{8}{5}$

(b) The sum of the third and fourth terms is $\frac{24}{85}$

(c) The series converges.

9) $\sum_{r=2}^{12} \frac{2^r}{r}$.

(a) The series has eleven terms.

(b) The series is a G.P.

(c) The third term of the series is $\dfrac{2^3}{3}$.

10) $(1 + x)^3 (1 - x)^{20}$ is expanded as a series of ascending powers of x.

(a) The series is finite.

(b) The first two terms of the series are $1 - 17x$.

(c) The last term of the series is x^{20}.

11) In the expansion of $\dfrac{1}{(1 - x)(1 + x)}$ as a series of ascending powers of x:

(a) there are only even powers of x,

(b) the first three terms are $\frac{1}{2} + \frac{1}{2}x^2 + \frac{1}{2}x^4$,

(c) the expansion is valid for $1 \leqslant x \leqslant 1$.

TYPE III

12) (a) $f(x) \equiv \dfrac{1}{1 - x}$.   (b) $f(x) = \sum_{r=0}^{\infty} x^r$

13) (a) $f(x) \equiv \dfrac{1 - x^n}{1 - x}$.

(b) $f(x) \equiv 1 + x + x^2 + \ldots + x^{n-1}$.

14) (a) The terms of a series are in arithmetic progression.

(b) The sum of the first n terms of a series is $n(3 - 2n)$.

TYPE V

15) The third term in the binomial expansion of $(2 - 3x)^{10}$ is $(45)(2^7)(3^3)(x^3)$.

16) If $2^n - 1$ is the sum of the first n terms of a series, $4^n - 1$ is the sum of the first $2n$ terms.

17) 3 is the arithmetic mean of 4 and 1.

18) The fourth term of the series $\displaystyle\sum_{r=0}^{n} (-1)^{(r+1)} 2^r$ is 16.

19) The geometric mean of $3x$ and $12x$ is $6x$.

MISCELLANEOUS EXERCISE 15

1) Find the sum of the first n terms of the arithmetic progression $2 + 5 + 8 + \ldots$.

Find the value of n for which the sum of the first $2n$ terms will exceed the sum of the first n terms by 224.   (AEB)p '71

2) The sum of n terms of a series is $3n^2$ for $n = 1, 2, 3 \ldots$ Show that the series is an arithmetic progression and find its nth term.   (AEB)p '73

3) The first term of an arithmetic series is $\ln x$ and the rth term is $\ln (xc^{r-1})$.

Show that the sum S_n of the first n terms of the series is $\dfrac{n}{2} \ln (x^2 c^{n-1})$.

(AEB)p '76

4) The first and last terms of an arithmetic progression are a and l respectively.

If the progression has n terms, prove from first principles that its sum is $\frac{1}{2}n(a + l)$.

A circular disc is cut into twelve sectors whose areas are in arithmetic progression. The area of the largest sector is twice that of the smallest. Find the angle (in degrees) between the straight edges of the smallest sector.   (JMB)p

5) Evaluate the following, giving each answer correct to two significant figures:

(a) $\sum_{n=1}^{20} (1.1)^n$ (b) $\sum_{n=1}^{20} \log_{10}(1.1)^n$. (JMB)

6) If S_n denotes the sum of the first n terms of the geometric progression $1 + \frac{1}{2} + (\frac{1}{2})^2 + \ldots + (\frac{1}{2})^{n-1} + \ldots$ and if S denotes the sum to infinity, find the least value of n such that $S - S_n < 0.001$. (O)p

7) A man borrows a sum of money from a building society and agrees to pay the loan (plus interest) over a period of years. If £A is the sum borrowed and $r\%$ the yearly rate of interest charged it can be proved that the amount (£p_n) of each annual instalment which will extinguish the loan in n years is given by the formula

$$P_n = \frac{A(R-1)R^n}{R^n - 1} \quad \text{where} \quad R = 1 + \frac{r}{100}.$$

Assuming this formula calculate P_n (correct to the nearest £) if $A = 1000$, $n = 25$, $r = 6.5$.
Find in its simplest form the ratio $P_{2n} : P_n$. Show that this ratio is always greater than $\frac{1}{2}$ and, if $r = 6.5$, find the least integral number of years for which this ratio is greater than $\frac{3}{5}$. (C)

8) A rod one metre in length is divided into ten pieces whose lengths are in geometric progression. The length of the longest piece is eight times the length of the shortest piece. Find, to the nearest millimetre, the length of the shortest piece. (JMB)

9) (a) By using an infinite geometric progression show that $0.43\overset{..}{2}\overset{.}{1}$,

$$\text{i.e. } 0.432\ 121\ 212\ 1 \ldots \text{ is equal to } \frac{713}{1650}.$$

(b) Write down the term containing p^r in the binomial expansion of $(q + p)^n$ where n is a positive integer. If $p = \frac{1}{6}$, $q = \frac{5}{6}$ and $n = 30$, find the value of r for which this term has the numerically greatest value. (AEB)'74

10) The coefficient of the term in x^n of the series $16 - x - 3x^2 - \ldots$ is $A(\frac{1}{5})^n + B(\frac{2}{3})^n$. Find the values of the constants A and B and hence obtain the next term in the series. Determine the range of values of x for which the series possesses a sum to infinity and find the value of x for which this sum is $\frac{17}{4}$. (AEB)'72

11) Find the sum $S(x)$ of the series $1 + 2x + 3x^2 + \ldots + (n + 1)x^n$,
(a) by first finding $(1 - x)S(x)$,
(b) by regarding $S(x)$ as the derivative with respect to x of another series. (O)

12) The first three terms of a geometric progression are also the first, ninth and eleventh terms, respectively, of an arithmetic progression. Given that the terms of the geometric progression are all different, find the common ratio r.
If the sum to infinity of the geometric progression is 8, find the first term and find the common difference of the arithmetic progression. (JMB)p

13) An infinite geometric progression is such that the sum of all the terms after the nth is equal to twice the nth term. Show that the sum to infinity of the whole progression is three times the first term. (JMB)p

14) Starting from first principles, prove that the sum of the first n terms of a geometric progression whose first term is a and whose common ratio is r

(where $r \neq 1$) is $S_n = \dfrac{a(1 - r^n)}{1 - r}$

Show that $\dfrac{S_{3n} - S_{2n}}{S_n} = r^{2n}$

Given that $r = \frac{1}{2}$, find $\displaystyle\sum_{n=1}^{\infty} \dfrac{S_{3n} - S_{2n}}{S_n}$.

15) The binomial expansion of $(1 + x)^n$, n a positive integer, may be written in the form

$$(1 + x)^n = 1 + c_1 x + c_2 x^2 + c_3 x^3 + \ldots + c_r x^r + \ldots$$

Show that, if c_{s-1}, c_s, and c_{s+1} are in arithmetic progression, then $(n - 2s)^2 = n + 2$.
Find possible values of s when $n = 62$. (AEB) p '71

16) Find the numerical value of the coefficient of x^{11} in the expansion of $\left(x^2 + \dfrac{1}{x}\right)^{10}$ in powers of x. (U of L)

17) Express the function $f(x) \equiv \dfrac{1}{(1 + x)(1 - 3x)}$ in partial fractions.
Find the first four terms in the expansion of $f(x)$ in ascending powers of x. (JMB)p

18) Expand $z = (1 + ic)^5$ in powers of c, and show that there are five real values of c for which z is real. (JMB)p

19) By putting $x = \dfrac{1}{1000}$ in the expansion of $(1 - x)^{\frac{1}{3}}$ in ascending powers of x obtain the cube root of 37 correct to six decimal places. (C)p

20) Prove that when $(1 + x)^2(1 - x)^{-2}$ is expanded in ascending powers of x, the coefficient of x^n (where $n > 0$) is $4n$. (C)p

21) Show that, for $-1 < x < 1$, the expansion in ascending powers of x of
$$\left(\frac{1+x}{1-x}\right)^{\frac{1}{3}} \quad \text{is} \quad 1 + \tfrac{2}{3}x + \tfrac{2}{9}x^2 + \tfrac{22}{81}x^3 + \ldots$$
By taking $x = 0.02$ obtain the cube root of 357, giving five places of decimals in your answer. (C)p

22) Given that $y = \sqrt{\left\{\dfrac{1+x}{2+x}\right\}}$ find the values of a, b, c, in the

approximation $y \simeq a + bx + cx^2$, where it is assumed that x is so small that its cube and higher powers may be neglected. (JMB)p

23) Expand $y = \left(\dfrac{1 + 2px}{1 + 2qx}\right)^{\frac{1}{2}}$ and $z = \left(\dfrac{1 - px}{1 - qx}\right)^{-\frac{1}{2}}$ $(p \neq q)$ in ascending powers of x as far as the terms in x^2.
Given that px and qx are small:
(a) show that if the terms in x^2 and higher powers of x are neglected, the expressions obtained for y and z satisfy the relation $1 + y = 2z$;

(b) find the ratio $p : q$ for which the relation $1 + y = 2z$ still holds if the terms in x^2 are retained but the terms in x^3 and higher powers of x are neglected. (C)

24) (a) The rth term of a series is $\dfrac{1}{r(r + 2)}$. Find the sum of the first n terms of the series and deduce the sum to infinity.

(b) Prove by induction that $\displaystyle\sum_{r=1}^{n} r^2 = \frac{n}{6}(n + 1)(2n + 1)$

Hence, or otherwise, evaluate $\displaystyle\sum_{r=1}^{51} (98 + 2r)^2$. (AEB)'74

25) (a) Simplify $r(r + 1)(r + 2) - (r - 1)r(r + 1)$ and use your result to prove that
$$\sum_{r=1}^{n} r(r + 1) = \tfrac{1}{3}n(n + 1)(n + 2).$$
Deduce that
$$\sum_{r=1}^{n} r^2 = \frac{n}{6}(n + 1)(2n + 1).$$

(b) Find the sum of the series
$$1.2.3 + 3.4.5 + 5.6.7 + \ldots + (2n - 1)(2n)(2n + 1).$$
(U of L)

26) Show that if $f(r) \equiv r(r + 1)(r + 2)$ then
$$f(r) - f(r - 1) \equiv 3r(r + 1)$$

Hence find the sum of the series
$$2 + 6 + 12 + \ldots + r(r + 1) + \ldots + n(n + 1) \qquad \text{(U of L)p}$$

27) If $f(n)$ denotes $\dfrac{1}{n(n + 1)}$, simplify $f(n) - f(n + 1)$ and hence find the
sum
$$S_n = \sum_{r=1}^{n} \frac{1}{r(r + 1)(r + 2)}$$
Find the smallest integer n for which S_n differs from $\frac{1}{4}$ by less than 10^{-4}.

(U of L)p

28) If $f(r) \equiv r(r + 1)(r + 2)(r + 3)$, simplify $f(r) - f(r - 1)$, and hence
find the sum of the first n terms of the series in which the rth term is
$r(r + 1)(r + 2)$.
Hence, or otherwise, show that $\quad 1^3 + 2^3 + 3^3 + \ldots + n^3 = \frac{1}{4}n^2(n + 1)^2$.

(U of L)p

29) Assuming the formulae for $\sum\limits_{r=1}^{n} r$ and $\sum\limits_{r=1}^{n} r^2$, or otherwise, find in
terms of m the value of
$$\sum_{r=m}^{2m} 3(r - 2)(r + 1) \qquad \text{(U of L)p}$$

30) Prove by induction, or otherwise, that
$$1.1! + 2.2! + 3.3! + \ldots + n.n! = (n + 1)! - 1 \qquad \text{(U of L)p}$$

31) Prove, by induction or otherwise, that
$$2.1! + 5.2! + 10.3! + \ldots + (n^2 + 1)n! = n(n + 1)! \qquad \text{(U of L)p}$$

32) Prove, by induction or otherwise, that the sum of the cubes of the first n
positive integers is $\frac{1}{4}n^2(n + 1)^2$. Hence, or otherwise, obtain a formula for the
sum of the cubes of the first n odd positive integers. (JMB)p

33) Prove, by induction or otherwise that
$$\sum_{r=2}^{n} (r - 1)r(r + 2) = \tfrac{1}{12}(n - 1)(n)(n + 1)(3n + 10) \qquad \text{(U of L)p}$$

34) Prove, by induction or otherwise, that
$$\sum_{r=1}^{n} r2^{r-1} = 1 + (n - 1)2^n \qquad \text{(U of L)p}$$

35) If $-1 < r < 1$ and $S_n = \sum_{p=1}^{n} r^p$ find

(a) $\lim_{n \to \infty} S_n$ (b) $\sum_{n=1}^{\infty} nr^n$ (c) $\sum_{q=1}^{n} S_q$ (U of L)

36) If $f(r) \equiv \log \left(1 + \dfrac{1}{r}\right)$,   show that

$$f(1) + f(2) + \ldots + f(n) = f\left(\dfrac{1}{n}\right)$$ (U of L)p

37) (a) Assuming the formula for $\sum_{r=1}^{n} r^2$, write down

(i) the sum of the squares of the first $2n$ positive integers,
(ii) the sum of the squares of the first n even integers.
Hence find $1^2 + 3^2 + 5^2 + \ldots + (2n-1)^2$.

(b) If $S_n = \sum_{r=0}^{n} a^r(1 + a + a^2 + \ldots + a^r)$, $(|a| \neq 1)$ by considering

$(1-a)S_n$, show that

$$S_n = \frac{1 - a^{2n+2}}{(1 - a^2)(1 - a)} - \frac{a^{n+1}(1 + a^{n+1})}{(1 - a)^2}$$

State the set of values of a for which S_n approaches a limit as
$n \to \infty$ and find the sum to infinity of the series for these values of a.
(U of L)

38) The positive integers are bracketed as follows:

$$(1), (2, 3), (4, 5, 6), \ldots$$

where there are r integers in the rth bracket. Find expressions for the first
and last integers in the rth bracket.
Find the sum of all the integers in the first 20 brackets. Prove that the sum of
the integers in the rth bracket is $\frac{1}{2}(r^2 + 1)$.   (C)

39) (a) Prove that $\sum_{r=1}^{n} \dfrac{1}{r(r+1)} = \dfrac{n}{n+1}$

(b) Sum the series $1 + x + x^2 + \ldots + x^n$ for $x \neq 1$.
By differentiation with respect to x, or otherwise, find the value of

$$1 + 2x + 3x^2 + \ldots + nx^{n-1}$$

and deduce the value of

$$1.2 + 2.2^2 + 3.2^3 + \ldots + n.2^n$$ (U of L)

40) Prove that

$$1^2 + 2^2 + 3^2 + \ldots + n^2 = \tfrac{1}{6}n(n+1)(2n+1)$$

Show that

$$a^2 + (a+d)^2 + (a+2d)^2 + \ldots + (a+nd)^2$$
$$= \tfrac{1}{6}(n+1)[6a(a+nd) + d^2 n(2n+1)]$$

Hence, or otherwise, prove that

$$2^2 + 4^2 + 6^2 + \ldots + l^2 = \tfrac{1}{6}l(l+1)(l+2) \qquad (l \text{ even})$$

and $\qquad 1^2 + 3^2 + 5^2 + \ldots + l^2 = \tfrac{1}{6}l(l+1)(l+2). \qquad (l \text{ odd}) \qquad$ (JMB)

41) Find the sum of the series

$$\frac{4}{3} + \frac{9}{8} + \frac{16}{15} + \ldots + \frac{n^2}{n^2 - 1} \qquad\qquad \text{(U of L)p}$$

CHAPTER 16

SOLUTION OF EQUATIONS

POLYNOMIAL EQUATIONS

The real roots of a quadratic equation can be found, either by factorising, or by using the standard formula. So, even when these roots are irrational, they can always be found.

The roots of an equation of higher degree also can sometimes be found by using the factor theorem, but only when the roots are integers or simple rational fractions. The factorisation of polynomial expressions of this type was illustrated in Chapter 2 and students are advised to remind themselves of the work covered on the Remainder and Factor Theorems on pages 32-5. When this approach fails *exact* solutions of such an equation can be found in certain special cases, some of which are demonstrated below.

Equations with Equal Roots

If the equation $f(x) = 0$ has two equal roots of value a then $(x-a)^2$ is a factor of $f(x)$. In this case it can be shown graphically that $(x-a)$ must also be a factor of $f'(x)$.

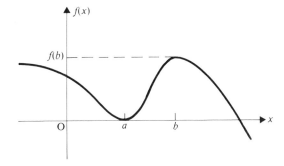

Since $f(x) = 0$ has two equal roots, the graph of $f(x)$ *touches* the x-axis where $x = a$, so $f'(x) = 0$ when $x = a$ i.e. $f'(a) = 0$.

So $f(a) = 0$ *and* $f'(a) = 0$ $\iff$ $f(x)$ has a repeated factor $(x-a)$.

This property can be used to simplify the solution of an awkward polynomial equation which is known to have two equal roots, as shown in the following example.

EXAMPLES 16a

1) Solve the equation $18x^3 - 111x^2 + 224x - 147 = 0$ given that it has two equal roots.

The repeated root of the given equation

$$f(x) \equiv 18x^3 - 111x^2 + 224x - 147 = 0 \qquad\qquad [1]$$

also satisfies the equation

$$f'(x) \equiv 54x^2 - 222x + 224 = 0$$

$\Rightarrow \qquad\qquad\qquad 27x^2 - 111x + 112 = 0$

$\Rightarrow \qquad\qquad\qquad (9x - 16)(3x - 7) = 0$

So *either* $x = \frac{16}{9}$ *or* $x = \frac{7}{3}$ is a solution of the given equation.

To check which of these values is the repeated root of the given equation, each is substituted in turn into equation [1]. We find that

$$f(\tfrac{16}{9}) \neq 0 \quad\text{and}\quad f(\tfrac{7}{3}) = 0$$

Hence $f(x)$ and $f'(x)$ are both zero when $x = \frac{7}{3}$.

So $x = \frac{7}{3}$ is the repeated root of the given equation.

Hence $18x^3 - 111x^2 + 224x - 147 \equiv (3x - 7)^2(ax + b)$

Comparing coefficients of x^3 gives

$$18 = 9a \quad\Rightarrow\quad a = 2$$

Comparing constants gives

$$-147 = 49b \quad\Rightarrow\quad b = -3$$

Therefore $18x^3 - 111x^2 + 224x - 147 \equiv (3x - 7)^2(2x - 3)$

So $(3x - 7)^2(2x - 3) = 0$

$\Rightarrow \qquad\qquad\qquad x = \frac{7}{3}$ or $\frac{3}{2}$

Note that values such as these could not easily be predicted, so the factor theorem would not help in the solution.

Further Equations with Special Properties

It is clearly not possible to demonstrate every type of equation with some particular property that leads to a method of exact solution. Two common varieties are illustrated in the examples that follow but the reader is reminded that adaptations of all basic methods can also be required.

EXAMPLES 16a (continued)

2) Solve the equation

$$3x^4 - 4x^3 - 14x^2 - 4x + 3 = 0.$$

This equation has *symmetrical coefficients*, and can be expressed as follows,

$$3x^2 - 4x - 14 - \frac{4}{x} + \frac{3}{x^2} = 0$$

$\Rightarrow \qquad 3\left(x^2 + \frac{1}{x^2}\right) - 4\left(x + \frac{1}{x}\right) - 14 = 0 \qquad\qquad [1]$

Now we use the substitution

$$y \equiv x + \frac{1}{x} \quad \Rightarrow \quad y^2 \equiv x^2 + 2 + \frac{1}{x^2}$$

So [1] becomes

$$3(y^2 - 2) - 4y - 14 = 0$$

$\Rightarrow \qquad\qquad 3y^2 - 4y - 20 = 0$

$\Rightarrow \qquad\qquad (3y - 10)(y + 2) = 0$

$\Rightarrow \qquad\qquad y = -2 \quad \text{or} \quad \tfrac{10}{3}$

If $\ y = -2, \qquad x + \dfrac{1}{x} = -2 \ \Rightarrow \ x^2 + 2x + 1 = 0$

$\Rightarrow \qquad\qquad\qquad x = -1, -1$

If $\ y = \dfrac{10}{3}, \qquad x + \dfrac{1}{x} = \dfrac{10}{3} \ \Rightarrow \ 3x^2 - 10x + 3 = 0$

$\Rightarrow \qquad\qquad\qquad x = 3, \tfrac{1}{3}$

So the roots of the given equation are $-1, -1, \tfrac{1}{3}, 3$.

Note that the substitution $\ y \equiv x + \dfrac{1}{x}\ $ reduces any quartic equation with symmetrical coefficients, to a quadratic equation.

3) Solve the equation

$$\sqrt{(x + 8)} - \sqrt{(x + 3)} = \sqrt{(2x - 1)}$$

To solve an equation of this type, we must eliminate all square roots.
But whenever both sides of an equation are squared, an extra equation, and hence possible extra solutions, are included,
e.g. if we square both sides of the equation $\ x = 2,\ $ we get $\ x^2 = 4.$
But $\ x^2 = 4\ $ includes *both* $\ x = 2\ $ and $\ x = -2.$

Thus, whenever a solution involves squaring both sides of an equation, all roots must be checked back in the original equation.

Considering the given equation, and squaring both sides, we have

$$x + 8 - 2\sqrt{(x+8)}\sqrt{(x+3)} + x + 3 = 2x - 1$$

$\Rightarrow$
$$12 = 2\sqrt{(x+8)}\sqrt{(x+3)}$$

Squaring both sides again gives

$$36 = (x+8)(x+3)$$

$\Rightarrow$
$$x^2 + 11x - 12 = 0$$

$\Rightarrow$
$$x = 1 \quad \text{or} \quad -12.$$

Returning to the given equation,

when $x = 1$, L.H.S. $= \sqrt{9} - \sqrt{4} = 1$

R.H.S. $= \sqrt{1} \quad = 1.$

So $x = 1$ satisfies the given equation.

When $x = -12$, $\sqrt{(-12+8)}$ is not real,
so $x = -12$ is *not* a solution of the given equation.

Hence the only root is 1.

EXERCISE 16a

Solve the following symmetrical equations.

1) $5x^4 - 16x^3 - 42x^2 - 16x + 5 = 0$ 2) $6x^4 + 5x^3 - 38x^2 + 5x + 6 = 0$

3) $63x^4 - 1024x^3 + 4226x^2 - 1024x + 63 = 0$

4) $4x^4 + 17x^3 + 8x^2 + 17x + 4 = 0$

Solve the following equations, given that they each have a repeated root.

5) $80x^3 + 88x^2 - 3x - 18 = 0$ 6) $45x^3 - 69x^2 + 32x - 4 = 0$

7) $20x^3 - 68x^2 + 69x - 18 = 0$ 8) $112x^3 + 152x^2 + 39x - 9 = 0$

9) $x^4 - 12x^3 - 2x^2 + 36x + 2025 = 0$

Solve the following equations using any suitable method.

10) $1 + \sqrt{x} = \sqrt{(3x-3)}$ 11) $\sqrt{(2x-5)} - \sqrt{(x-2)} = 1$

12) $\sqrt{(3x+1)} + \sqrt{(x-1)} = \sqrt{(7x+1)}$ 13) $x^{\frac{4}{3}} - 5x^{\frac{2}{3}} + 4 = 0$

14) $\dfrac{x^2}{4} + y^2 = 1$ and $xy = 1$

15) $x^2 + y^2 + 4x - 6y = 3$ and $y = x + 1$

16) $x^2 + y^2 + 8x - 4y + 15 = 0$ and $x^2 + y^2 + 6x + 2y - 15 = 0$

17) $x^4 - x^3 - 12x^2 - 4x + 16 = 0$ $\left(\textit{Hint. Use}\quad y \equiv x + \dfrac{4}{x} \right)$

LOCATING THE ROOTS OF A POLYNOMIAL EQUATION

Some polynomial equations of degree higher than two cannot be solved either by using the Remainder Theorem or by one of the special methods previously demonstrated. In such cases we can seek the approximate location of the roots as follows.

Consider, first, a curve with a polynomial equation $y = f(x)$,

e.g.

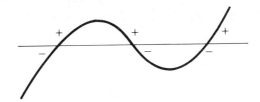

or

etc.

On each occasion that the curve crosses the x-axis, the sign of y changes.

If, therefore, one only of these crossing points lies between two values of x, $x = x_1$ and $x = x_2$, then $f(x_1)$ and $f(x_2)$ are of opposite sign.

Conversely, if we can find two values for x such that $f(x_1)$ and $f(x_2)$ have opposite signs, then we know that the curve $y = f(x)$ has crossed the x-axis between x_1 and x_2.

We cannot assume that there is only one crossing point, however, as is shown by the following diagrams.

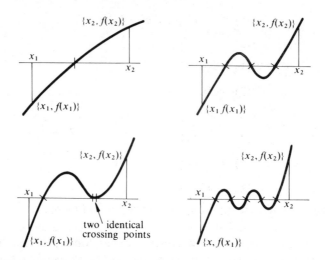

These examples illustrate that

if $f(x_1)f(x_2) < 0$,   then the equation $f(x) = 0$ has an odd number of real roots between x_1 and x_2 (some of these roots may be equal).

Now if $f(x_1)$ and $f(x_2)$ have the same sign then the curve $y = f(x)$ has either not crossed the x-axis at all, or has met it an even number of times between x_1 and x_2, as the following diagrams illustrate.

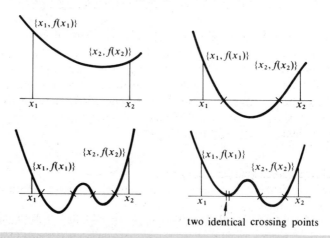

two identical crossing points

So, if $f(x_1)f(x_2) > 0$,   then the equation $f(x) = 0$ has an even number of real roots (distinct or equal) between x_1 and x_2.

EXAMPLES 16b

1) Given that the roots of the equation $12x^3 - 112x^2 + 267x - 77 = 0$ all lie in the range $0 < x < 10$,   find the integral values of x between which each of these roots lies.

If $\qquad$ $f(x) \equiv 12x^3 - 112x^2 + 267x - 77$ $\qquad$ [1]

then $\qquad$ $f(0) = -77$

$\qquad\qquad$ $f(1) = 90$

$\qquad\qquad$ $f(2) = 105$

$\qquad\qquad$ $f(3) = 40$

$\qquad\qquad$ $f(4) = -33$

$\qquad\qquad$ $f(5) = -42$

$\qquad\qquad$ $f(6) = 85$

Because $f(x)$ changes sign between $x = 0$ and 1, $x = 3$ and 4, $x = 5$ and 6, we know that, between each of these pairs of integers, there is an odd number of roots of equation [1]. But a cubic equation cannot have more than three roots, so one root lies in each of these intervals, and no further sign changes can occur.

Hence the roots of the given equation lie between
$x = 0$ and 1,   3 and 4,   5 and 6.

THE NUMBER OF REAL ROOTS OF A POLYNOMIAL EQUATION

Although a polynomial equation of degree n may have, at most, n real roots, the actual number of real roots may be less than n. In this case an attempt to find n intervals within which $f(x)$ changes sign would be a fruitless task. So it is useful to be able to assess the *number* of real roots of a given equation before trying to locate them. It is not always possible to do this, but it is worth trying one or both of the methods given below, each of which can, in some cases, be very helpful.

(1) Consider a cubic equation $f(x) = 0$.

The curve $y = f(x)$ has either two turning points or a point of inflexion (which is not necessarily parallel to the x-axis).

When there are two turning points, the signs of y_{max} and y_{min} can be used to determine the number of real roots of the equation $f(x) = 0$, as follows:

(a)   if $(y_{max})(y_{min}) < 0$ there are three real distinct roots.

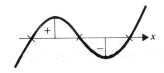

(b) if $(y_{max})(y_{min}) > 0$ there is only one real root.

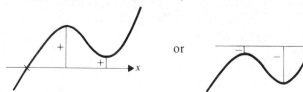

(c) if $(y_{max})(y_{min}) = 0$ there are three real roots one of which is repeated.

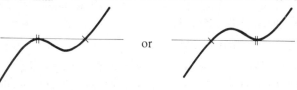

If there is a point of inflexion then either
(d) there are three equal real roots, or

(e) there is only one real root.

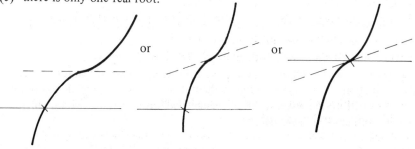

Consider, for example, the equation $2x^3 - 9x^2 + 12x - 1 = 0$.

If $y = 2x^3 - 9x^2 + 12x - 1$,

then $\dfrac{dy}{dx} = 6x^2 - 18x + 12$

At turning points $\dfrac{dy}{dx} = 0$

$\Rightarrow$ $x = 1$ or 2.

When $x = 1$, $y = 4$

When $x = 2$, $y = 3$

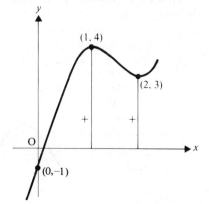

The maximum and minimum values of y have the *same* sign, so the given equation has only one real root.

The coordinates of the turning points, together with a rough sketch of the curve, indicate that this real root is between 1 and zero.

Note. If this method is attempted for equations of degree higher than 3, there is no guarantee that the derived equation, $f'(x) = 0$, can be solved.

(2) Another way of trying to find the number of real roots of a polynomial equation, $f(x) = 0$, is given below. It is offered, without any attempt at proof of its validity, as a useful aid.

First Step. Count the number of times that consecutive coefficients of the terms in $f(x)$ are of opposite sign. If this happens p times then we can say that the equation $f(x) = 0$ has *not more than p positive real roots.*

Second Step. Consider the coefficients of $f(-x)$ in the same way. If the signs of consecutive coefficients change q times we can say that the original equation $f(x) = 0$ *has not more than q negative real roots.*

This process is very quickly completed and, if $p + q < n$, reduces the amount of work subsequently carried out in locating the real roots.

Consider, for example, the equation

$$x^3 - 2x^2 - 7 = 0 \qquad\qquad [1]$$

which, being cubic, may have three real roots.

However, the coefficients $+1, -2, -7$ include only one sign change,

so there is not more than one real positive root

Now examining the equation given by replacing x by $-x$,

i.e. $$-x^3 - 2x^2 - 7 = 0$$

we see that the coefficients have no sign changes,

so there are no real negative roots.

Thus we see that there is only one real root of [1] and that it is positive.

The single root can now be located approximately as before.

The technique described above does not always help however.

For example, if $x^3 + 2x^2 - 7 = 0$

we see from $+1, +2, -7$ that there is not more than one real positive root.

Then, replacing x by $-x$, we see from $-1, +2, -7$ that there are not more than two real negative roots.

So this time the cubic equation can have three real roots (but is not *certain* to have three) and the only useful information we have gained is that only one of the roots can be positive.

Note that the information obtained by this technique is often of value only when used in conjunction with basic knowledge of the graph of the polynomial function $f(x)$.

EXAMPLES 16b (continued)

2) Find the number of real roots of the equation $x^4 + 2x^3 + 8x + 15 = 0$ and find the integral values of x between which each of these roots lies.

If $f(x) \equiv x^4 + 2x^3 + 8x + 15$, the coefficients of $f(x)$, i.e. $1, 2, 8, 15$, include no sign change.

So $f(x) = 0$ has *no* real positive roots.

The coefficients of $f(-x) \equiv x^4 - 2x^3 - 8x + 15$ are $1, -2, -8, 15$ and include two sign changes.

So $f(x) = 0$ has not more than two real negative roots.

This result is not conclusive as the equation $f(x) = 0$ can have either, two real negative roots (distinct or equal)

(a)   (b)

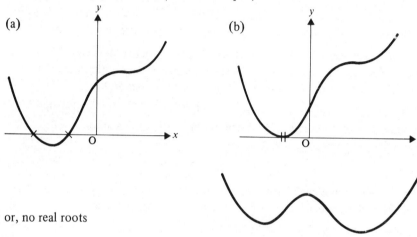

or, no real roots

($f(x) = 0$ *cannot* have only one real root as its complex roots must occur in conjugate *pairs*.)

To determine which is the case we can now look for turning points on the curve $y = f(x)$

$$\frac{dy}{dx} = 4x^3 + 6x^2 + 8 \equiv 2(x + 2)(2x^2 - x + 2)$$

$\Rightarrow \qquad \frac{dy}{dx} = 0$ only when $x = -2$ and $y = -1$.

So there *is* a turning point below the x-axis, showing that the curve crosses the x-axis twice as in diagram (a) above. ($f(x) \to \infty$ when $x \to \pm \infty$)

Thus the equation $f(x) = 0$ has exactly two real negative roots, which lie on opposite sides of $x = -2$.

Then,
$$f(0) = 15$$
$$f(-1) = 6$$
$$f(-2) = -1$$
$$f(-3) = 18.$$

$f(x)$ changes sign between $x = -1$ and -2
and between $x = -2$ and -3.
So one real root lies in each of these intervals.

EXERCISE 16b

Find the maximum number of real roots of each of the following equations.

1) $x^3 + x + 2 = 0$

2) $3x^4 - 4x^3 + x^2 + 7 = 0$

3) $2x^5 + x^3 = 1$

4) $x^4 + 4x - 9 = 0$

Determine the *exact* number of real roots of each of the following equations.

5) $x^3 - 3x^2 - 4 = 0$

6) $3x^4 - 4x^3 + 6x^2 - 12x + 5 = 0$

7) $3x^4 + 8x^3 + 24x^2 + 96x - 10 = 0$

8) $x^3 - 3x^2 + 1 = 0$

9) In Questions 5–8, find the consecutive integers between which each real root lies.

10) If $x^3 + \lambda x^2 + 2 = 0$ explain why, if $\lambda > 0$, the equation must have two complex roots.

11) Find the range of values of λ for which the equation $2x^3 + 9x^2 + 12x + \lambda = 0$ has three real distinct roots.

APPROXIMATE SOLUTIONS

There are many equations whose roots cannot be evaluated exactly by any method. The approximate values of the roots of such equations can be found either by a graphical approach or by one of a number of methods using successive numerical approximations, or by a combination of these two processes.

GRAPHICAL SOLUTIONS

Consider two functions $f(x)$ and $g(x)$. We have seen that when these functions are plotted on the same axes and their graphs intersect at a point where $x = a$ say, then $f(a) = g(a)$.

Conversely if when $x = b$ the graphs do not intersect then $f(b) \neq g(b)$. Thus the values of x at which the two graphs intersect is the set of values of x satisfying the equation $f(x) = g(x)$.

Now consider the two functions $f(x) \equiv \sin x$ and $g(x) \equiv x/2$.

When plotted on Cartesian axes, the curve with equation $y = f(x)$ is a sine wave, while the graph of $y = g(x)$ is a straight line through the origin.

The equation $\sin x = x/2$ can therefore be solved by finding the x coordinates of the points of intersection of the graphs $y = \sin x$ and $y = x/2$.

There are three such values of x and no more, as can be seen by *sketching* the two graphs over a wide range.

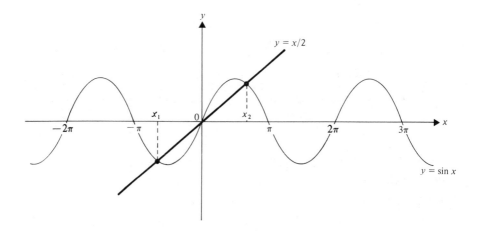

Therefore the equation $\sin x = \dfrac{x}{2}$ has three solutions (roots)

$$x = x_2, \quad x = 0 \quad \text{and} \quad x = x_1$$

However the actual values of x_2 and x_1 cannot be found from a *sketch*. By *plotting* the curves between $-\pi$ and π (the sketch shows that all the roots are within this range), approximate values for x_2 and x_1 can be found. Great accuracy at this stage is not necessary as it is used for a first approximation; a more accurate stage comes later. The following table is quite adequate.

x (radians)	0	± 1	± 2	± 3
$\sin x$	0	± 0.84	± 0.91	± 0.14
$\dfrac{x}{2}$	0	$\pm \frac{1}{2}$	no more points are needed for a line	

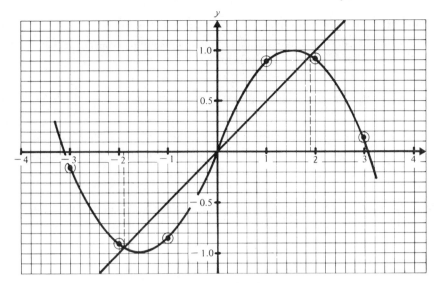

Now it is clear that the roots x_2 and x_1 are approximately ± 1.9. An accurate plot of the graphs in the region of each of these values then gives a very good approximation to the value of each root, e.g. for the positive root we could work within the range $1.7 \leqslant x \leqslant 2.1$, calculating values of $\sin x$ at intervals of 0.05^c $\left(\text{the graph of } y = \dfrac{x}{2} \text{ requires only two points in any case}\right)$.

So the graphical method of solution of an equation $f(x) = g(x)$ usually proceeds as follows:

1) A *sketch* of the two graphs on the same axes indicates the number and rough positions of the roots of the equation.

2) Choosing a range indicated by the sketch, a *plot* of the two graphs on the same axes gives a fair approximation for the value of each root within that range.

3) Using a larger scale and a smaller range near to each required root, the abscissa of a point of intersection of the two graphs now gives a very good approximation for that root.

Note: It is sometimes possible to combine two of the steps listed above, but in all cases a sketch should be made first. This is because deductions can be made from a sketch, whose range is unlimited, which cannot be observed from a plot of limited range. This graphical solution of an equation is invaluable when no analytical method exists, e.g. for $\sin x = \dfrac{x}{2}$ there are no trig identities which provide a solution. Roots found graphically are approximations however, the degree of accuracy depending upon the quality of the graph plotting and drawing.

Some Notes on Drawing Graphs

For those readers who have not had much practice in plotting and using graphs, the following points may be helpful.

1) Use a sharp, but not too hard, pencil.

2) Mark points either with a dot, ringed for visibility ⊙, or with a 'straight' cross (+), rather than a 'diagonal' cross (×) whose centre point can easily be misplaced.

3) Choose a scale for each axis which spreads the available points over as much of the length of the axis as possible. This avoids 'crowding' the points.

e.g.

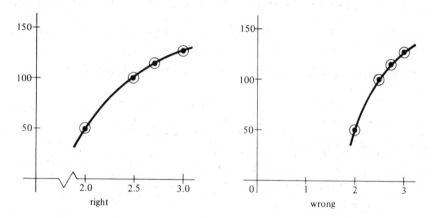

It is not necessary to use the same scales on the two axes, nor is it necessary to begin each scale from zero.
Avoid scales which are inherently difficult to use, e.g. making one square represent three units.

4) When hand drawing a curve, always make sure that you draw from the concave side,

e.g.

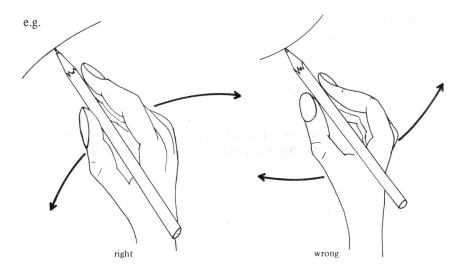

right   wrong

EXAMPLES 16c

1) Show that the equation $x3^x = 1$ has only one real root and find an approximate value for this root using a graphical method.

First we rearrange the given equation so that it contains two simple functions, one on each side, i.e.

$$x = 3^{-x}$$

The solutions of this equation are the values of x for which the graph of $y_1 = x$ meets the graph of $y_2 = 3^{-x}$.

Sketching y_1 and y_2 we have,

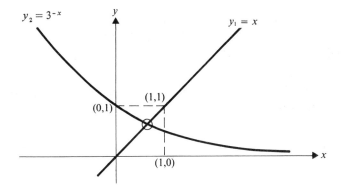

This sketch shows clearly that the graphs cross only once and that, at this point, x is between 0 and 1.

So now we plot accurately the portions of the two graphs between $x = 0$ and $x = 1$.

x	0	0.2	0.4	0.6	0.8	1.0
y_1	0					1.0
y_2	1.0	0.80	0.65	0.52	0.42	0.33

From the graph below we see that $y_1 = y_2$ only when $x \simeq 0.55$. So $x \simeq 0.55$ is the only real solution of the equation $x3^x = 1$.

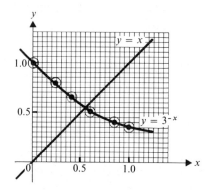

2) A chord AB of a circle subtends an angle θ at the centre. AB divides the area of the circle into two parts in the ratio $1:5$. Find the value of θ.

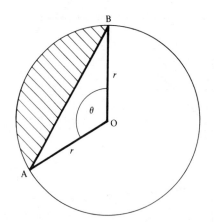

Let the radius of the circle be r.

Area of sector $AOB = \frac{1}{2}r^2\theta$

Area of triangle $AOB = \frac{1}{2}r^2 \sin\theta$

Hence the area of the minor segment (shaded) is $\frac{1}{2}r^2 (\theta - \sin\theta)$.

If AB divides the area in the ratio 1:5 then the area of the minor segment
is $\frac{1}{6}$ of the area of the circle

i.e.
$$\tfrac{1}{2}r^2(\theta - \sin\theta) = \tfrac{1}{6}\pi r^2$$

$\Rightarrow$
$$\theta - \sin\theta = \frac{\pi}{3}$$

This equation cannot be solved by using standard identities so it is rearranged
giving

$$\sin\theta = \theta - \frac{\pi}{3} = \theta - 1.047^c$$

Now we sketch $y = \sin\theta$ and $y = \theta - 1.047$ to find their point(s) of
intersection.
Clearly $0 < \theta < \pi$ (θ is an angle in a minor segment) so only this range need
be considered.
The line $y = \theta - 1.047$ is easily drawn using the points of intersection with
the axes, i.e. $(0, -1.047)$ and $(1.047, 0)$.

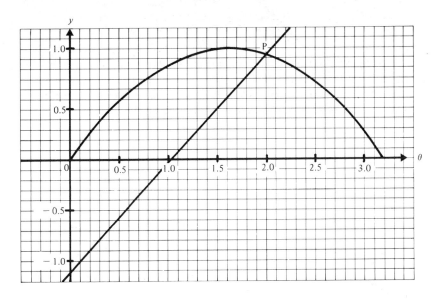

At P, $\theta \simeq 1.9^c$
A more accurate graph is now drawn using $1.7 \leqslant \theta \leqslant 2.1$.

θ	1.7	1.8	1.9	2.0	2.1
$\sin\theta$	0.992	0.974	0.946	0.909	0.863
$\theta - 1.047$	0.653				1.053

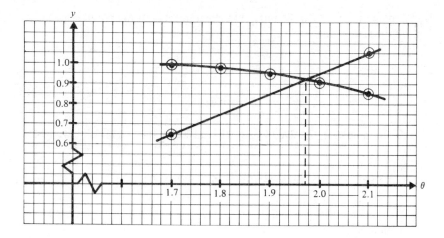

The graphs intersect where $\theta \simeq 1.97^c$ so this is the angle subtended by AB at the centre of the circle.

Note that although this answer is given to three significant figures it is not necessarily correct to three significant figures. In fact it is not possible to specify the degree of accuracy of a graphical solution because curve drawing is not an exact art.

Note also that when one term in an equation contains an angle and another term involves a trig ratio of that angle, it is understood that the angle is measured in radians unless it is specifically indicated that degrees are to be used (a rare situation). Normally, therefore, the graphical solution of such equations requires a *horizontal axis graduated in radians*.

EXERCISE 16c

1) Sketch the pairs of graphs you could use to solve the following equations (do not carry out the solution).

(a) $\sin x = \dfrac{1}{x}$ (b) $\cos x = x^2 - 1$ (c) $2^x = \tan x$

(d) $2^x \sin x = 1$ (e) $(x^2 - 4) = \dfrac{1}{x}$ (f) $x2^x = 1$

(g) $x \ln x = 1$ (h) $\sin x = x^2$ (i) $\ln x + 2^x = 0$

2) Use the sketches in Question (1) to estimate, in each case, the *number* of roots of the given equation (some may have an infinite set of solutions).

3) Find graphically the specified roots of the following equations, giving your answer to three significant figures.

(a) $\tan \theta = 2\theta$;   the root between 0 and $\dfrac{\pi}{2}$.

(b) $2^x = \frac{1}{2}(x+3)$;   the positive root.

(c) $x^2 - 1 = x$;   the negative root.

4) Show by a suitable sketch that there are no positive roots of the equation
$$3^{-x} = x^2 + 2.$$
How many negative roots are there?

5) Show by means of a sketch that the equation $\theta = 1 + \cos\theta$ has only one root. Find this root graphically.

6) Find the length of an arc AB of a circle of radius 0.5 m, if the chord AB divides the area of the circle in the ratio 3 : 1.

NUMERICAL SOLUTIONS

There is a variety of ways in which successive numerical approximations can be used to find an approximate root of an equation. Those given below are some of the commonest of these iterative methods.
Whatever method is used, however, we must first find an interval in which the required root lies. This can be done graphically or by investigation of the sign of $f(x)$ as already shown. We must also be certain that within this interval the graph of $y = f(x)$ is continuous (i.e. the curve has no breaks) and that there is only *one* root in this interval.

METHOD 1   INTERVAL BISECTION

Suppose that we are asked to find the largest root of the equation $e^x - 3x = 0$. First we write the equation in the form $e^x = 3x$ and draw a sketch of the curves $y = e^x$ and $y = 3x$.

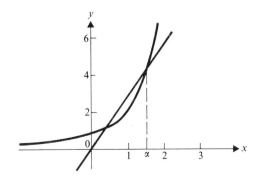

This shows that there are two roots of the equation $e^x - 3x = 0$ and that the larger root, α, is in the interval $(1,2)$.

Also if $f(x) \equiv e^x - 3x$ then it can be shown by calculation or by deduction from the sketch that

$$f(1) < 0 \quad \text{and} \quad f(2) > 0$$

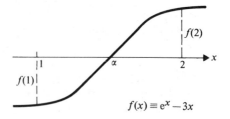

$f(x) \equiv e^x - 3x$

We can now find a better approximation to this root by reducing the interval in which it is known to lie. This can be done by choosing any number between 1 and 2, c say, and finding the sign of $f(c)$.

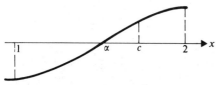

If $f(c) > 0$ then we know that α is in the interval $(1,c)$,

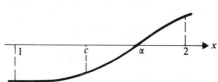

and if $f(c) < 0$ then α is in the interval $(c,2)$.

A fairly obvious choice for c is 1.5, i.e. we bisect the interval $(1,2)$.

Now $\qquad f(1) < 0, \quad f(1.5) < 0, \quad f(2) > 0 \quad \Rightarrow \quad 1.5 < \alpha < 2$

i.e. $\quad \alpha = 2 \quad$ to the nearest integer.

We can now repeat this operation, bisecting the interval $(1.5,2)$ by taking $c = 1.75$

$\Rightarrow \qquad\qquad f(1.5) < 0, \quad f(1.75) > 0, \qquad \Rightarrow \qquad 1.5 < \alpha < 1.75$

This process can now be continued until the required degree of accuracy is achieved. If we want a value for α correct to one decimal place we must continue until α is located in an interval whose length is less than 0.05

$$f(1.5) < 0, \quad f(1.625) > 0, \qquad \Rightarrow \qquad 1.5 < \alpha < 1.625$$

$$f(1.5) < 0, \quad f(1.5625) > 0, \qquad \Rightarrow \qquad 1.5 < \alpha < 1.5625$$

$$f(1.5) < 0, \quad f(1.531\,25) > 0, \qquad \Rightarrow \qquad 1.5 < \alpha < 1.531\,25$$

As this interval is less than 0.05 we now know that $\alpha = 1.5$ correct to one decimal place. If α is required correct to two decimal places then further steps are necessary until the interval is less than 0.005.

All iterative methods follow the same basic principles as this method, i.e. once a first approximation is found, it is used as an input to a predetermined operation which will give a better approximation as output. This is then fed into the procedure again to give an even better approximation. The procedure is repeated until the required degree of accuracy is achieved. Each such operation is called a step, or iteration, in the calculation. The number of iterations necessary to achieve a given degree of accuracy depends upon many factors. If a large number of steps is necessary, the method is said to give a slow rate of convergence whereas a small number of steps gives a rapid rate of convergence. It sometimes happens that an iterative process will give answers that get less accurate rather than more accurate in which case the method fails and is said to diverge. The cause of this varies according to the method adopted, and some of the reasons are discussed in the following work.

Returning to the interval bisection method described above, we can see that, starting with an interval of one unit, five steps were necessary to give α correct to one decimal place. To get α correct to two decimal places would require a further three steps, i.e. eight steps in total. This is a large number of iterations to achieve a fairly low degree of accuracy so the rate of convergence for this method is slow. However the method does have some advantages

(a) the number of steps necessary to achieve a given degree of accuracy is known before embarking on the calculations and

(b) the method will not fail if the initial conditions are met, i.e. that there is only one root of $f(x) = 0$ in the initial chosen interval and that $f(x)$ is continuous in this interval.

METHOD 2   LINEAR INTERPOLATION

This method is similar to the interval bisection method, but uses linear proportion to find the value of c rather than just bisecting the interval. Consider a function $f(x)$ which changes sign between $x = a$ and $x = b$ so that the equation $f(x) = 0$ has a root, α, in the interval (a, b).

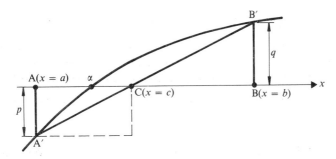

In the diagram, the line joining $A'(a, f(a))$ and $B'(b, f(b))$ crosses the x-axis at C where C divides AB in the ratio $p : q$ (from the similar triangles A'AC and B'BC).

So, at C, $x = \dfrac{aq + bp}{p + q}$ $(= c)$ where $p = |f(a)|$ and $q = |f(b)|$.

As before, by finding the sign of $f(c)$ we can determine if α is in the interval (a,c) or in the interval (c,b). Linear proportion can then be used again to reduce further the length of the interval in which α is known to lie, and so on until the required degree of accuracy is achieved.

Also, if $b-a$ is small, $c-\alpha$ is even smaller, so the value of x at C is an approximation for α.

Suppose, for example, we are required to find the larger root of the equation $\ln x = x - 2$ correct to three decimal places.

Before we can begin using linear interpolation, we must find the approximate location of the specified root by plotting the graphs $y = \ln x$ and $y = x - 2$ on the same axes,

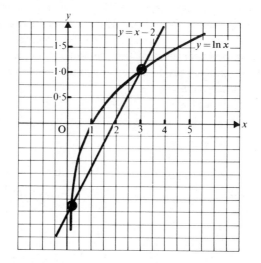

From the graph we see that there are two real roots of the equation $\ln x = x - 2$ and that the larger root, α, is slightly larger than 3.

As it appears that α is much nearer to 3 than 4 we will try $(3, 3.5)$ as our first interval.

If $f(x) \equiv \ln x - x + 2$ then $f(3) = 0.0986$ and $f(3.5) = -0.2472$ confirming that $3 < \alpha < 3.5$.

Using linear interpolation on this interval gives

$$c = \frac{(3)(0.2472) + (3.5)(0.0986)}{0.0986 + 0.2472} = 3.1426$$

(working to four decimal places).

So our first approximation for α is 3.1426.
Now $f(3.1426) = 0.0025$ so $3.1426 < \alpha < 3.5$.
Using linear interpolation on the interval $(3.1426, 3.5)$ gives

$$c = \frac{(3.1426)(0.2472) + (3.5)(0.0025)}{0.0025 + 0.2472} = 3.1462$$

So our second approximation for α is 3.1462.
Now $f(3.1462) = -0.000\,005$ (to 6 d.p.).
As $f(3.1462)$ is so small it is now worth checking to see if our last
approximation is correct to three decimal places.
Evaluating $f(3.1455)$ and $f(3.1464)$ we find that $f(3.1455) > 0$ and
$f(3.1464) < 0$ so that $3.1455 < \alpha < 3.1464$,
i.e. $\alpha = 3.146$ correct to three decimal places.

We have calculated the required root of the equation $\ln x = x - 2$, correct
to three decimal places, in just two steps. This is a very rapid rate of convergence
so linear interpolation has an obvious advantage over the interval bisection
method in this case.
However the method of linear interpolation will not always give an answer so
quickly. The rate of convergence depends on the rate at which the gradient of
$f(x)$ is changing in the interval (a, b),

e.g.

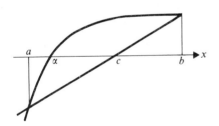

$f'(x)$ changes considerably and
c is not very close to α so the
rate of convergence is slow.

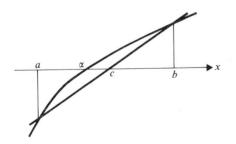

There is not much change in $f'(x)$,
c is close to α and there is a rapid
rate of convergence.

The reader is recommended to vary the shape of the curve between a and b,
including the case when more than one root is in the interval (a, b) and to
investigate the effect that these changes have on the rate of convergence.

METHOD 3 $x = g(x)$

This method can often be used to find a root of an equation $f(x) = 0$ which can be written in the form $x = g(x)$.

The roots of the equation $x = g(x)$ are the values of x at the points of intersection of the line $y = x$ and the curve $y = g(x)$.

Taking x_1 as a first approximation to a root α then

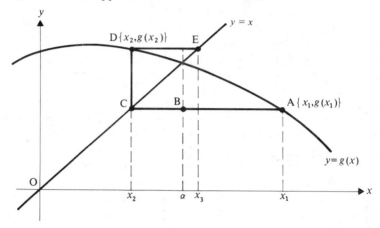

in the diagram,

> A is the point on the *curve* where $x = x_1$, $y = g(x_1)$
> B is the point where $x = \alpha$, $y = g(x_1)$
> C is the point on the *line* where $x = x_2$, $y = g(x_1)$

If, in the region of α, the slope of $y = g(x)$ is less steep than that of the line $y = x$

i.e. provided that $\qquad\qquad |g'(x)| < 1$

then $\qquad\qquad\qquad$ CB < BA

so x_2 is closer to α than is x_1

i.e. x_2 is a better approximation to α.

But C is on the line $y = x$

therefore $\qquad\qquad\qquad x_2 = g(x_1)$

Now taking the point D on the curve where $x = x_2$, $y = g(x_2)$ and repeating the argument above we find that x_3 is a better approximation to α than is x_2

where $\qquad\qquad\qquad x_3 = g(x_2)$

This process can be repeated as often as necessary to achieve the required degree of accuracy.

The rate at which these approximations converge to α depends on the value of $|g'(x)|$ near α. The smaller $|g'(x)|$ is, the more rapid is the convergence.

It should be noted that this method fails if $|g'(x)| > 1$ near α.
The following diagrams illustrate some of the factors which determine the
success, or otherwise, of this method.

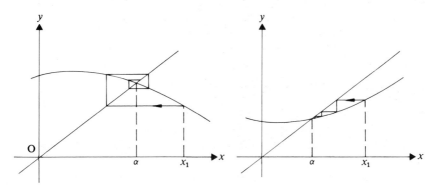

Rapid rate of convergence   ($|g'(x)|$ small).

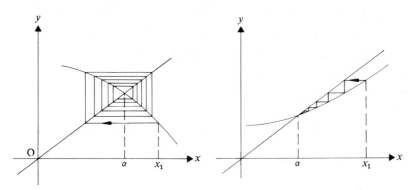

Slow rate of convergence   ($|g'(x)| < 1$ but close to 1)

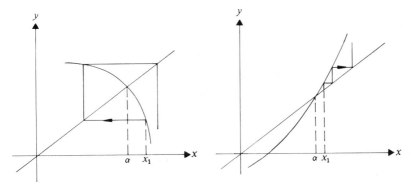

Divergence, i.e. failure,   ($|g'(x)| > 1$)

The reader is recommended to vary the shape of the curve and the position of x_1 to investigate the effect of these changes on the convergence or otherwise of approximations.

As an example consider the equation

$$x^3 + 2x^2 + 5x - 1 = 0$$

This equation can be written in the form

$$x = g(x)$$

where
$$g(x) \equiv -\tfrac{1}{5}(x^3 + 2x^2 - 1)$$

Suppose we want the largest root of this equation.
From the sign changes we see that there is at most one positive root.
Now $f(0) = -1$ and $f(1) = 7$ so there is a root in the interval $(0, 1)$ which is probably nearer 0 than 1.
So we will take $x_1 = 0$ as our first approximation.

A better approximation, x_2, is found from

$$x_2 = g(x_1) = -\tfrac{1}{5}[\![0^3 + 2(0)^2 - 1]\!] = 0.2$$

Further improvements are obtained by repeating this step,

i.e.
$$x_3 = g(x_2) = -\tfrac{1}{5}[\![(0.2)^3 + 2(0.2)^2 - 1]\!]$$
$$= 0.1824$$
$$x_4 = g(x_3) = -\tfrac{1}{5}[\![(0.1824)^3 + 2(0.1824)^2 - 1]\!]$$
$$= 0.1855 \qquad \text{(to 4 d.p.)}$$
$$x_5 = g(x_4) = -\tfrac{1}{5}[\![(0.1855)^3 + 2(0.1855)^2 - 1]\!]$$
$$= 0.1850 \qquad \text{(to 4 d.p.)}$$

and so on.

The degree of accuracy at any stage can be checked by determining the sign of $f(x)$ on either side of the value so far obtained for the root, e.g. taking $x \simeq 0.1850$ we find that $f(0.1846)$ is negative and $f(0.1854)$ is positive, so $x = 0.185$ correct to 3 d.p.
Note. This does *not* show that $x = 0.1850$ to 4 d.p.

METHOD 4   NEWTON'S METHOD (OR THE NEWTON–RAPHSON METHOD)
This method is based on determining a linear approximation for a function. Suppose that the equation $f(x) = 0$ has a root α and that a is an approximation for α.
The curve $y = f(x)$ cuts the x-axis where $x = \alpha$. If we consider tangent to $y = f(x)$ at the point where $x = a$ then the point B where this tangent cuts the x-axis will, in most circumstances, be nearer to the point $x = \alpha$,

i.e.

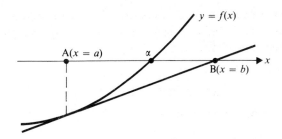

So if this tangent cuts the x-axis at $B(x = b)$,   then b is a better
approximation to α than a is.
The gradient of $y = f(x)$ at the point A is $f'(a)$.
The coordinates of A are $(a, f(a))$.
So the equation of the tangent at A is

$$y - f(a) = f'(a)(x - a)$$

This line cuts the x-axis where $y = 0$,   i.e. where $x = a - \dfrac{f(a)}{f'(a)}$.

So if a is an approximation for a root of the equation $f(x) = 0$,

$$b = a - \frac{f(a)}{f'(a)}$$

is a better approximation.

As an example we will use Newton's Method to find the root of $xe^x = 3$
correct to three decimal places.
A first approximation to the root is found by drawing the graphs of
$y = \dfrac{3}{x}$ and $y = e^x$.

From these graphs we see that $xe^x = 3$ has a root α which is
approximately   1.

Now $$f(x) = xe^x - 3$$

$\Rightarrow$ $$f'(x) = (x+1)e^x$$

Using $b = a - \dfrac{f(a)}{f'(a)}$,   and taking $a = 1$ as our first approximation to α,
the second approximation is

$$1 - \frac{e-3}{2e} = 1.0518 \quad \text{(to 4 d.p.)}$$

Taking $a = 1.0518$ and repeating the procedure, the third approximation
is

$$1.0518 - \frac{(1.0518)e^{1.0518} - 3}{(2.0518)e^{1.0518}} = 1.0499 \quad \text{(to 4 d.p.)}$$

So, to three decimal places, the root is likely to be   1.050   and we can check this
by calculating $f(1.0495)$ and $f(1.0504)$.
Now $f(1.0495)$ is negative and $f(1.0504)$ is positive. Thus the root lies
between these values and, correct to   3 d.p., is   1.050.

The rate of convergence using Newton's Method depends on the crudeness, or
otherwise, of the first approximation and on the shape of the curve in the
neighbourhood of the root. In extreme cases these factors may lead to failure.
Some of these cases are illustrated by the following graphs.

In the following diagrams, α is a root of $f(x) = 0$. a is the first
approximation for α and b is the second approximation for α given by
Newton's Method.

(a) $f'(a)$ is too small.

(b) $f'(x)$ increases too rapidly.

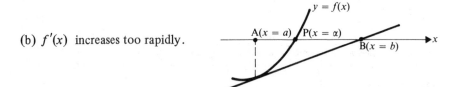

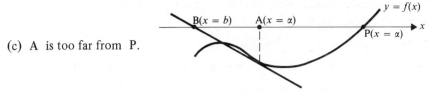

(c) A is too far from P.

Show that each of the following equations has a root between $x = 0$ and
$x = 1$ and find this root correct to 2 decimal places.

1) $x^3 - x^2 + 10x - 2 = 0$

2) $3x^3 - 2x^2 - 9x + 2 = 0$

3) $2x^3 + x^2 + 6x - 1 = 0$

Find, correct to 3 significant figures, the smallest positive root of each of the
following equations.

4) $x^3 + x - 11 = 0$

5) $x^4 - 4x^3 - x^2 + 4x - 10 = 0$

6) $4 + 5x^2 - x^3 = 0$

Use a graphical method to find a first approximation to the root(s) of the
following six equations. Then apply two stages of Newton's Method to give a
better approximation. State the accuracy of each of your results.

7) $\tan x = 2x$ (the positive root)

8) $x^2 = \ln (x + 1)$

9) $e^x = 2x + 1$

10) $\sin x = 1 - x$

11) $x^3 - 6x + 3 = 0$ (the negative root)

12) $e^x(1 + x) = 2$

Use linear interpolation to find the root(s) of the following equations, correct
to 3 decimal places.

13) $e^x = 3x + 1$

14) $x = 1 + \ln x$

15) $3 + x - 2x^2 = e^x$

16) $e^x = 2 \cos x$ $\left(\text{the roots between } -\frac{\pi}{2} \text{ and } \frac{\pi}{2}\right)$

17) $x^3 - 3x^2 - 1 = 0$

SUMMARY

For a polynomial equation $f(x) = 0$,

(a) $f(a) = 0 \iff x = a$ is a solution.

(b) $f(a) = 0$ and $f'(a) = 0 \iff x = a$ is a repeated solution.

A polynomial equation of degree n has
(a) not more than n real roots
(b) at least one real root if n is odd.

If there are p occasions when there is a change in the sign of consecutive coefficients of the terms in a polynomial equation $f(x) = 0$, then the equation has

not more than p real positive roots

and if there are q sign changes in the coefficients of $f(-x)$, then the equation $f(x) = 0$ has

not more than q real negative roots.

If a polynomial function $f(x)$ is such that $f(a)f(a + 1) < 0$ then there is at least one real root of the equation $f(x) = 0$ between $x = a$ and $x = a + 1$.

The roots of the equation $f(x) = g(x)$ are the values of x for which the graph of $f(x)$ meets the graph of $g(x)$ when the two graphs are drawn on the same axes.

If $x = a$ is a first approximation to a solution of the equation $f(x) = 0$, then

$$x = a - \frac{f(a)}{f'(a)}$$

is a better approximation (the Newton–Raphson Method).

MISCELLANEOUS EXERCISE 16

1) Plot on the same axes the graphs whose equations are $y = 2 \sin x$ and $y = \cos 2x$ for values of x from 0 to π. Hence give approximate values of x which satisfy the equation $2 \sin x = \cos 2x$. Check your answers by solving the equation exactly.

2) Using a graphical method, find good approximations to the roots of the following equations, within the range $0 \leqslant x \leqslant \pi$.

(a) $x = \cos x$ (b) $\tan x = \cos x$ (c) $1 + \cos x = 2 \sin x$

(d) $\dfrac{1}{x} = 2 \tan x$ (e) $\cos\left(x - \dfrac{\pi}{4}\right) = x - 1$

3) By drawing suitable sketches, state the number of (i) positive, (ii) negative roots of the following equations:

(a) $\sin x = x^2$ (b) $x + 1 = \cos x$ (c) $2^x = \dfrac{1}{x}$

(d) $3^x = \tan x$ (e) $3^x = \tan 3x$ (f) $\sin x + x^2 = 1$

4) Two equal circles of radius a intersect and the common chord subtends an angle of 2θ radians at either centre. Find an expression for the area of the region common to the two circles.
If this area is equal to half the area of either circle show that $\sin 2\theta = 2\theta - \tfrac{1}{2}\pi$.
Using a graphical method, or otherwise, estimate the value of θ which satisfies this equation.   (C)

5) By using sketch graphs, or otherwise, show that the equation

$$x = 3 \sin x$$

where the angle is measured in radians, has only one positive root.
Verify, with the use of tables, that this root lies between 2.2 and 2.3.
Determine the root correct to two places of decimals.   (C)

6) The chord AB of a circle divides the circle into two portions whose areas are in the ratio $3:1$. If AB makes an angle θ with the diameter passing through A show that θ satisfies the equation

$$\sin 2\theta = \frac{\pi}{2} - 2\theta$$

Solve this equation graphically.   (AEB)'66

7) By using a graphical method, find approximate solutions to the equation
$2 \sin \theta + \cos 3\theta = 1$ for values of θ in the range $0 \leqslant \theta \leqslant 90°$.   (AEB)'72 p

8) By investigating the turning values of

$$f(x) = x^3 + 3x^2 + 6x - 38$$

or otherwise, show that the equation $f(x) = 0$ has only one real root.
Find two consecutive integers, n and $n + 1$, which enclose the root.
Describe a method by which successive approximations to the root can be obtained. Starting with the value of n as a first approximation, calculate two further successive approximations to the root.
Give your answers correct to 3 significant figures.   (C)

9) Show that the equation $x^3 - 5x - 3 = 0$ has a root between 2 and 3.
Use linear interpolation to find an approximation to this root.
Determine the root correct to two places of decimals.   (C)

10) Show that the equation $x = \frac{1}{5}(x^4 + 2)$ has two real roots, both of which are positive.

Evaluate the smaller root correct to 3 places of decimals, showing that your answer has the required degree of accuracy. (O)

11) Find the positive root of the equation $x = \ln(1 + x + x^2)$, correct to three significant figures, showing that your answer has the required degree of accuracy. (O)

12) (a) Show that the equation

$$x^4 - 2x - 1 = 0$$

has exactly two real roots. Find integers a and b such that one of these roots lies between a and $a + 1$ and another between b and $b + 1$.

(b) Using the substitution $y \equiv x + x^{-1}$, obtain the four roots of the equation

$$6x^4 - 25x^3 + 37x^2 - 25x + 6 = 0. \qquad \text{(U of L)}$$

13) By drawing appropriate graphs show that the equation

$$x(\pi - x) = 4 \sin 2x$$

has four real roots. Show that the root between $x = 0$ and $x = \pi$ occurs at approximately $x = 1.2$ and by means of a single application of Newton's Method determine a more accurate estimate of its value. (AEB '72)

14) Prove that, if $x = a$ is an approximation to one root of the equation

$f(x) = 0$, then $x = a - \dfrac{f(a)}{f'(a)}$ is a closer approximation.

Hence

(a) establish the formula $x_{r+1} = x_r(2 - Nx_r)$ as a method of successive approximation to the reciprocal of N,

(b) show that $x^4 + x^2 - 80 = 0$ has a root near $x = 3$.

Taking $x = 3$ as the first approximation use Newton's Method twice to obtain a better approximation. (AEB '74)

15) A student, asked to find a root of the equation

$$f(x) \equiv x^3 - 14x^2 + 49x - 8 = 0$$

did not notice the solution $x = 8$ but chose, instead, to use Newton's Method taking $x = 7.2$ as the first approximation. He then calculated, correctly, $f(7.2) = -7.712$, $f'(7.2) = 2.92$ and deduced (again correctly) the second approximation 9.84. By means of a graph, or otherwise, explain why Newton's Method failed to give a better approximation in this case. Prove that using Newton's Method, a first approximation α, for a value of α in the interval $7.2 < \alpha < 8$, would give a second approximation which is closer to the root $x = 8$ provided that

$$2(8 - \alpha)f'(\alpha) + f(\alpha) > 0$$

and deduce that any value of α in the above range exceeding $5 + \sqrt{(5.6)}$ would in fact give improvement.

[You may, if you wish, assume without proof that $f'(x)$ is positive and increasing for $x > 7.2$; also that

$$2(8 - \alpha)f'(\alpha) + f(\alpha) = (8 - \alpha)(5\alpha^2 - 50\alpha + 97).]$$ (C)

16) An equation can be written in the form $x = F(x)$ and it is known that the equation has only one real root and that this root is near $x = x_1$.

Explain, with the aid of a diagram, how, if $|F'(x)| < 1$, the iterative formula

$$x_{r+1} = F(x_r), \ (r = 1, 2, \ldots)$$

will give the root to whatever degree of accuracy is required.

Show that the cubic equation $x^3 + 3x - 15 = 0$ has only one real root and that this root is near $x = 2$.

This cubic equation can be written in any one of the forms

(a) $\qquad\qquad\qquad x = \frac{1}{3}(15 - x^3)$

(b) $\qquad\qquad\qquad x = 15/(x^2 + 3)$

(c) $\qquad\qquad\qquad x = (15 - 3x)^{\frac{1}{3}}$

Determine which of these forms would be suitable for the use of the previous iterative formula.   (U of L)

CHAPTER 17

NUMERICAL APPLICATIONS

In the major part of this book we have concentrated on the general analysis of functions and in this Chapter we examine a variety of numerical applications of this work.

REDUCTION OF A RELATIONSHIP TO LINEAR FORM

If it is thought that a certain relationship exists between two variable quantities, this hypothesis can be tested by experiment, i.e. by giving one variable certain values and measuring the corresponding values of the other variable. The experimental data collected can then be displayed graphically. If the graph shows points that lie approximately on a straight line (allowing for experimental error) then a linear relationship between the variables (i.e. a relationship of the form $Y = mX + c$) is indicated. Further, the gradient of the line (m) and the vertical axis intercept (c) provide the values of the constants.

On the other hand, if the relationship is not of a linear form, the points on the graph will lie on a section of a curve. It is very difficult to identify the equation of a curve from a section of it, so the form of a non-linear relationship can rarely be verified in this way. Non-linear relationships, however, can often be reduced to a linear form. The following examples illustrate some of the relationships which can be verified by plotting experimental data in a form which gives a straight line.

Linear Relationships

An elastic string is fixed at one end and a variable weight is hung on the other end. It is believed that the length of the string is related to the weight by a linear law. Use the following experimental data to confirm this belief and find the particular relationship between the length of the string and the weight.

Weight (W) in newtons	1	2	3	4	5	6	7	8
Length (l) in metres	0.33	0.37	0.4	0.45	0.5	0.53	0.56	0.6

If l and W are related by a linear law, then $l = aW + b$ where a and b are constants. Plotting l against W gives

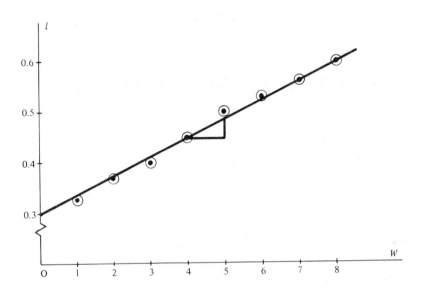

We see that the points do all lie on a straight line (allowing for experimental error) so that l and W are connected by a relationship of the form
$$l = aW + b.$$
Now we draw the line of 'closest fit' (i.e. the line which has the points distributed above and below it as evenly as possible, which is not necessarily the line through the most points).

By measurement from the graph

the gradient $= 0.04$

the intercept on the vertical axis $= 0.3$

So comparing $\left.\begin{array}{l} l = aW + b \\ \\ Y = mX + c \end{array}\right\}$ we have $a = 0.04, \quad b = 0.3$

with

i.e. within the limits of experimental accuracy

$$l = 0.04W + 0.3$$

Note that in finding the gradient, the *increase in l is taken from the vertical scale used* and the *increase in W is taken from the horizontal scale used* and these two scales are not necessarily the same.

Relationships of the Form $y = ax^n$

The following data, collected from an experiment, is believed to obey a law of the form $p = aq^n$ where a and n are constants.

q	1	2	3	4	5	6
p	0.5	0.63	0.72	0.8	0.85	0.9

If the relationship $p = aq^n$ is correct

then $$\log p = n \log q + \log a$$

Comparing $$\log p = n \log q + \log a$$

with $$Y = mX + c$$

[1]

we see that $\log p$ and $\log q$ are related by a linear law. So if $\log p$ is plotted against $\log q$ we expect a straight line.

$\log q$	0	0.30	0.48	0.60	0.70	0.78
$\log p$	-0.30	-0.20	-0.14	-0.10	-0.07	-0.05

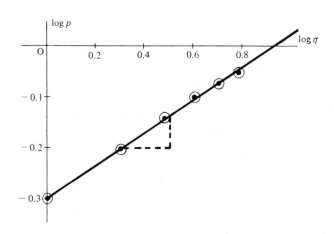

This straight line graph confirms that there is a linear relationship between $\log q$ and $\log p$.

By measurement, the gradient of the line is 0.33.
Reading from the graph, the intercept on the vertical axis is -0.3.

From [1], $n = 0.33$ and $\log a = -0.3 \Rightarrow a = 0.5$

(**Note** that these values of a and n are approximate. Apart from experimental errors in the data, selecting the line of best fit is a personal judgement and so is subject to slight variations which affect the values obtained for a and n.)
The data given confirms that p and q are related by the law $p = aq^n$ where $a \simeq 0.5$ and $n \simeq 0.33$.

Relationships of the Form $y = ab^x$

A relationship of the form $y = ab^x$ where a and b are constant can be reduced to a linear relationship by taking logs, since

$$y = ab^x \iff \log y = x \log b + \log a$$

Comparing $$\log y = x \log b + \log a$$

with $$Y = mX + c$$

we see that plotting values of $\log y$ against corresponding values of x gives a straight line whose gradient is $\log b$ and whose intercept on the vertical axis is $\log a$.

Relationships of the Form $\dfrac{1}{y} + \dfrac{1}{x} = \dfrac{1}{a}$

If a is a constant, $\dfrac{1}{y} + \dfrac{1}{x} = \dfrac{1}{a}$ is a linear relationship between $\left(\dfrac{1}{y}\right)$ and $\left(\dfrac{1}{x}\right)$

i.e. if values of $\left(\dfrac{1}{y}\right)$ are plotted against corresponding values of $\left(\dfrac{1}{x}\right)$, a straight line graph will result.

By comparing $$\left(\frac{1}{y}\right) = -\left(\frac{1}{x}\right) + \left(\frac{1}{a}\right)$$

with $$Y = mX + c$$

we note that the gradient of the graph should be -1 and the intercept on the $\left(\dfrac{1}{y}\right)$ axis gives the value of $\dfrac{1}{a}$.

EXAMPLES 17a

1) In an experiment, values of a variable y were measured for selected values of a variable x.

The results are shown in the table below. It is believed that x and y are related by a law of the form $2y + 10 = ab^{(x-3)}$.   Confirm this graphically and find approximate values for a and b.

x	10	12	15	20	21
y	37.5	90	320	2440	3700

If $2y + 10 = ab^{(x-3)}$,   taking logs of both sides gives

$$\log (2y + 10) = (x - 3) \log b + \log a$$

which is of the form $Y = mX + c$

where $\left. \begin{array}{l} Y \equiv \log (2y + 10) \\ X \equiv x - 3 \end{array} \right\}$ and $\left\{ \begin{array}{l} m = \log b \\ c = \log a \end{array} \right.$

i.e. $\log (2y + 10)$ and $(x - 3)$ obey a linear law.
So we tabulate corresponding values of $x - 3$ and $\log (2y + 10)$ from the given values of x and y:

$x - 3$	7	9	12	17	18
$\log (2y + 10)$	1.9	2.3	2.8	3.7	3.9

Then plotting $\log (2y + 10)$ against $x - 3$ gives the graph below.

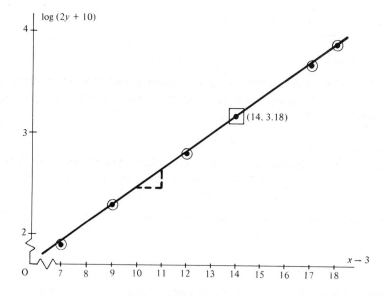

The straight line shows that there is a linear relationship between $\log (2y + 10)$ and $(x - 3)$, confirming that $2y + 10 = ab^{x-3}$.

Measurement from the graph gives the gradient $\simeq 0.175$

i.e. $\qquad\qquad\qquad\qquad\qquad \log b \simeq 0.175$

$\Rightarrow \qquad\qquad\qquad\qquad\qquad\qquad b \simeq 1.49$

In all graphical work the scales should be chosen to give the greatest possible accuracy, i.e. the range of values given in the table should have as much spread as possible. This sometimes means that the horizontal scale does not include zero and the value of c cannot then be read from the graph. In these circumstances, which arise in this example, we find c by using the equation $Y = mX + c$ together with the measured value of m (i.e. 0.175) and the co-ordinates of any point P on the graph (*not* a pair of values of X and Y from the table).

Thus $\qquad\qquad\qquad 3.18 = (0.175)(14) + c$

$\Rightarrow \qquad\qquad\qquad\qquad c = 0.73$

i.e. $\qquad\qquad\qquad\qquad \log a \simeq 0.73$

$\qquad\qquad\qquad\qquad\qquad a \simeq 5.37$

2) It is known that two variables x and y are related by the law

(a) $ae^y = x^2 - bx$

(b) $\quad y = \dfrac{1}{(x-a)(x-b)}$

In each case state how you would reduce the law to a linear form so that a straight line graph could be drawn from experimental data.

When attempting to reduce a relationship between x and y to a form from which a straight line graph can be drawn, the given equation must be expressed in the form

$$Y = mX + c$$

where X and Y are variable terms, values for which must be calculable from the given data, i.e. X and Y must not contain unknown constants. On the other hand m and c must be constants, but may be unknown.

Now X and Y may be functions of x and/or y, as

$$f(xy) = mg(xy) + c$$

is a linear relationship between $f(xy)$ and $g(xy)$.

So to reduce a non-linear relationship to a linear form we:

(1) try to express it in a form containing three terms,

(2) make one of those terms constant,

(3) remove unknown constants from the coefficient of one of the variable terms.

These objectives will now be applied to the problem in hand.

(a) $ae^y = x^2 - bx$.

This equation has three terms, one of which becomes constant when we divide by x, giving

$$a\frac{e^y}{x} = x - b$$

We also have a variable term (x) whose coefficient is a known constant. This equation may now be written as

$$x = a\frac{e^y}{x} + b$$

Comparing with $\qquad Y = mX + c$

we see that if values of x are plotted against corresponding values of $\dfrac{e^y}{x}$,

a straight line will result whose gradient is a and whose intercept on the vertical axis is b.

Note that the original equation can be arranged in linear form in a variety of ways

e.g. $\qquad \left(\dfrac{e^y}{x^2}\right) = -\dfrac{b}{a}\left(\dfrac{1}{x}\right) + \dfrac{1}{a} \quad$ or $\quad \left(\dfrac{x^2}{e^y}\right) = b\left(\dfrac{x}{e^y}\right) + a$

(b) $y = \dfrac{1}{(x-a)(x-b)}$.

Although this form suggests the use of partial fractions, this approach increases the number of times the unknown constants appear.

It is better to invert the equation giving

$$\frac{1}{y} = (x-a)(x-b)$$

$\Rightarrow \qquad\qquad \dfrac{1}{y} = x^2 - x(a+b) + ab$

$\Rightarrow \qquad\qquad x^2 - \dfrac{1}{y} = (a+b)x - ab$

We can now compare this form with $\quad Y = mX + c$.

Thus plotting values of $x^2 - \dfrac{1}{y}$ against corresponding values of x will give

a straight line (whose gradient is $a + b$ and whose intercept on the vertical axis is $-ab$).

EXERCISE 17a

1) Reduce each of the given relationships to the form $Y = mX + c$. In each case give the functions equivalent to X and Y and the constants equivalent to m and c.

(a) $\dfrac{1}{y} = ax + b$ (b) $y(y - b) = x - a$ (c) $ae^x = y(y - b)$

In each of the following questions, the table gives sets of values for the related variables and the law which relates the variables. By drawing a straight line graph find approximate values for a and b.

2) $y = ax + ab$.

x	3	5	7	10
y	-2	2	6	12

3) $s = ab^{-t}$.

t	1	2	3	4
s	1.5	0.4	0.1	0.02

4) $r^2 = a\theta - b$.

θ	1	4	10	25	40
r	1.6	2	2.6	3.8	4.7

5) $ay = b^x$.

x	5	6	7	8
y	1.07	2.13	4.27	8.53

6) $\dfrac{a}{V} + \dfrac{b}{L} = 1$.

V	2.5	3	5.5	7	12
L	2.5	1.5	0.79	0.7	0.6

7) $y = (x - a)(x - b)$.

x	1	2	3	4	5	6
y	-6	-4	0	6	14	24

INTEGRATION USED AS A PROCESS OF SUMMATION

We have seen in Chapter 9 that

$$\lim_{\delta x \to 0} \sum_{x=a}^{b} y\,\delta x = \int_a^b y\,dx$$

i.e. integration is a process of summation.

Thus if Q is a quantity made up of elements δQ where $Q = \Sigma\, \delta Q$ and if δQ can be expressed approximately in the form $f(x)\,\delta x$, i.e. $\delta Q \simeq f(x)\,\delta x$ then $Q \simeq \Sigma\, f(x)\,\delta x$.

Further, if as $\delta x \to 0$ the approximation becomes more accurate,

then
$$Q = \lim_{\delta x \to 0} \Sigma\, f(x)\,\delta x = \int f(x)\, dx$$

AREAS BOUNDED BY CURVES

We have seen in Chapter 9 how this process of summation is used to find areas. The area is divided into elements which are strips. Each element is approximately rectangular. For example, to find the area bounded by the x-axis, and the curve $y = x^2 - 1$, we divide it into vertical strips.

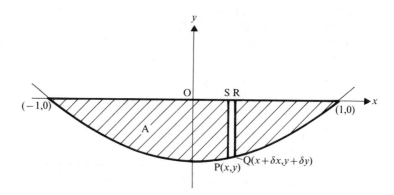

Taking PQRS as a typical element, where $P(x, y)$, $Q(x + \delta x, y + \delta y)$ are points on $y = x^2 - 1$, then the area of PQRS $\simeq y\,\delta x$ (taking the element as approximately rectangular),

i.e.
$$A \simeq \sum_{x=-1}^{1} y\,\delta x$$

Therefore

$$A = \lim_{\delta x \to 0} \sum_{x=-1}^{1} y\,\delta x = \int_{-1}^{1} y\, dx = \int_{-1}^{1}(x^2 - 1)\, dx$$

$$= [\tfrac{1}{3}x^3 - x]_{-1}^{1} = [\tfrac{1}{3} - 1] - [-\tfrac{1}{3} + 1] = -\tfrac{4}{3}$$

We expect $\Sigma\, y\, \delta x$ to give a negative result for $-1 \leqslant x \leqslant 1$ as y is negative in this range.

Therefore the area is $\frac{4}{3}$ square units.

Note that the area A is symmetrical about the y-axis, i.e. about $x = 0$. Hence A can also be found by writing

$$A \simeq 2 \sum_{x=0}^{1} y\, \delta x$$

i.e.
$$A = 2 \int_{0}^{1} (x^2 - 1)\, dx$$

We will now look further at problems in which areas are to be found and will adopt the following procedure:

(a) Draw a diagram clearly showing the area to be found.

(b) Take a vertical or horizontal strip for the element of area.
The choice should be made so that:
(i) the element has the same format throughout the area. For example, if the area shown in the diagram below is divided into vertical strips, some of the elements are bounded at the top by the curve and others by the line. On the other hand, horizontal strips are all bounded on the left by the y-axis and on the right by the curve.
Thus horizontal strips should be chosen in this case.

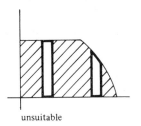 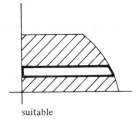

unsuitable suitable

(ii) the element does *not* have both ends on the *same* curve.
e.g.

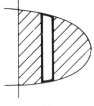 

unsuitable suitable

(c) Taking the element as being approximately rectangular, write down its approximate area.

(d) Sum, by integrating, the areas of the elements over the specified range.

EXAMPLES 17b

1) Find the area of the region of the xy plane which satisfies the following inequalities:

$$y \geqslant x^2, \qquad y \leqslant 4$$

$y \geqslant x^2$ defines the curve $y = x^2$ and the region above the curve.

$y \leqslant 4$ defines the line $y = 4$ and the region below the line.

$\left. \begin{array}{l} y = x^2 \\ y = 4 \end{array} \right\}$ intersect at $(2, 4)$ and $(-2, 4)$.

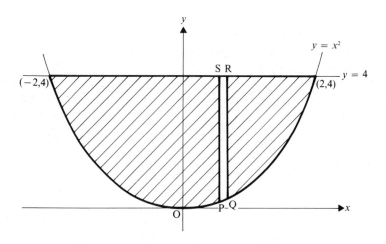

Hence the area to be found is the shaded region in the diagram. Taking PQRS as a typical element of area, and treating it as approximately a rectangle of height SP and width SR we have,

$$SP = y_S - y_P \quad \text{and} \quad SR = \delta x$$

But

$$y_P = x^2, \quad y_S = 4$$

$\Rightarrow$

$$SP = 4 - x^2$$

So the area of   PQRS $\simeq (4 - x^2) \, \delta x$.

Therefore the   total area $\simeq \displaystyle\sum_{x=-2}^{2} (4 - x^2) \, \delta x$.

or, as the area is symmetrical about $x = 0$,   the total area is also given approximately by $2 \displaystyle\sum_{x=0}^{2} (4 - x^2) \delta x$.

i.e.   the total area $= 2 \int_{0}^{2} (4 - x^2) \, dx$

$$= 2 \left[4x - \frac{x^3}{3} \right]_{0}^{2} = 2(8 - \tfrac{8}{3}) = 10\tfrac{2}{3} \text{ square units}$$

2) Find the area of the region of the xy plane defined by the inequalities

$$y^2 \leqslant 1 - x \quad \text{and} \quad y \leqslant x + 1$$

$y^2 = 1 - x$ is the parabola shown
in the diagram.

Writing $y^2 \leqslant 1 - x$

as $x \leqslant 1 - y^2$

we see that this inequality is
satisfied by all points on, and to the
left of, $x = 1 - y^2$.

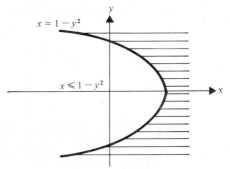

$y \leqslant x + 1$ defines the line $y = x + 1$ and the region below the line.

$\left. \begin{array}{l} y^2 = 1 - x \\ y = 1 + x \end{array} \right\}$ intersect where $y^2 + y - 2 = 0$,

i.e. at $(-3, -2)$ and $(0, 1)$.

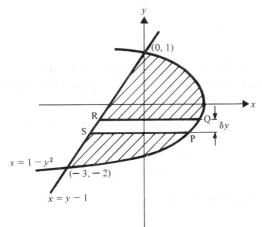

Taking a horizontal strip PQRS as a typical element of area (a vertical strip is
unsuitable because it would change its structure at $x = 0$ and would have
both ends on the parabola for $x > 0$)   we have

$$\text{SP} = x_P - x_S$$

where $x_P = 1 - y^2$ and $x_S = y - 1$

Hence $SP = (1 - y^2) - (y - 1) = 2 - y^2 - y$

The area of the element $\simeq (2 - y^2 - y)\,\delta y$

Therefore the total area $\simeq \displaystyle\sum_{y=-2}^{1} (2 - y^2 - y)\,\delta y$

i.e.   the total area $= \displaystyle\int_{-2}^{1} (2 - y^2 - y)\,dy$

$$= \left[2y - \frac{y^3}{3} - \frac{y^2}{2} \right]_{-2}^{1}$$

$$= \left(2 - \frac{1}{3} - \frac{1}{2} \right) - \left(-4 + \frac{8}{3} - \frac{4}{2} \right) = \frac{9}{2}$$

Therefore the area is $4\frac{1}{2}$ square units.

Note that although it might appear from the diagram that for the range under consideration, SP should be the sum of the x coordinates of S and P, the length of a horizontal line is *always* the difference of x coordinates, e.g.

x $AB = x_B - x_A = 3 - (-2)$

$A(-2, 0)$ O $B(3, 0)$

EXERCISE 17b

1) Find the area of the region of the xy plane bounded by the curve $y = x^2 - 4$ and the line:
(a) $y = 0$ (b) $y = -\frac{7}{4}$ (c) $y = \frac{9}{4}$.

2) Find the area of the region of the xy plane bounded by the curve $y^2 = x + 1$ and the line:
(a) $x = 0$ (b) $x = 3$.

3) Find the area of the region of the xy plane defined by the following inequalities:

(a) $y \geqslant e^x$, $x \geqslant 0$, $y \leqslant e$ (b) $y \geqslant \dfrac{1}{x}, 0 < y \leqslant 2, 0 < x \leqslant 2$.

4) Find the area of the region of the xy plane defined by the following inequalities:
(a) $y \leqslant x(1 - x)$, $y \geqslant x - 1$
(b) $x \leqslant y(1 - y)$, $x \geqslant y - 1$.

5) Find the area of the region of the xy plane defined by the following inequalities:

$$y \geqslant (x + 1)(x - 2), \quad y \leqslant x$$

VOLUMES OF REVOLUTION

If part of a curve is rotated about a straight line, the solid formed is called a solid of revolution.

Note that such a solid is always symmetrical about the axis of rotation.

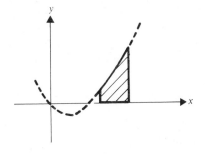

Consider the part of the curve

$$y = x(2x - 1)$$

between $x = 1$ and $x = 2$.

If the area enclosed by this section of the curve, the x-axis and the ordinates $x = 1$, $x = 2$, is rotated about the x-axis, a solid of revolution is formed.

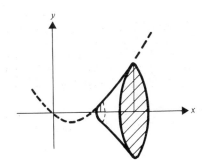

Note that any cross-section perpendicular to the axis of rotation is circular.
To calculate the volume of this solid we can divide it up into sections by making cuts perpendicular to the axis of rotation.

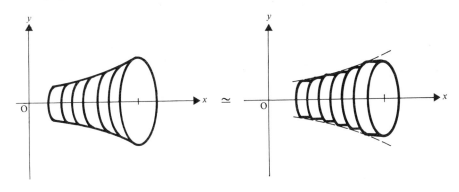

If the cuts are reasonably close, each section obtained is approximately cylindrical. Hence the volume of the solid is approximately equal to the sum of the volumes of these cylinders.

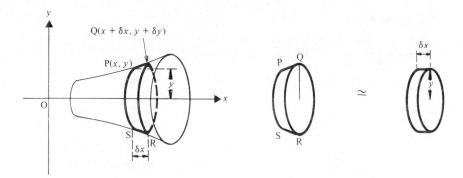

Taking PQRS as a typical element where $P(x,y)$ is a point on the curve $y = x(2x-1)$ then the section PQRS is approximately a cylinder whose radius is y and whose 'height' is δx.

Hence the volume of PQRS $\qquad \simeq \left| \pi y^2 \; \delta x \right.$

Therefore the volume, V, of the solid $\simeq \displaystyle\sum_{x=1}^{2} \pi y^2 \; \delta x$

The smaller δx is, the closer this approximation is to V

i.e. $\qquad V = \displaystyle\lim_{\delta x \to 0} \sum_{x=1}^{2} \pi y^2 \; \delta x$

$$= \int_1^2 \pi \, y^2 \; \mathrm{d}x$$

$$= \pi \int_1^2 x^2 (2x-1)^2 \; \mathrm{d}x = \pi \int_1^2 (4x^4 - 4x^3 + x^2) \; \mathrm{d}x$$

$$= \pi \left[\frac{4x^5}{5} - x^4 + \frac{x^3}{3} \right]_1^2$$

$$= \frac{182\pi}{15}$$

i.e. the volume is $182\pi/15$ cubic units.

Most volumes of revolution can be found in the same way,
i.e. by taking sections perpendicular to the axis of rotation, to give elements of volume, δV, which are approximately cylindrical,

i.e. $\qquad\qquad \delta V \simeq \pi (\text{radius})^2 \times \text{thickness}$

Then $\qquad\qquad V \simeq \Sigma \, \delta V$

EXAMPLES 17c

1) Find the volume generated when the area between $y = e^x$, the x-axis, the y-axis and $x = 1$ is rotated through one revolution about the x-axis.

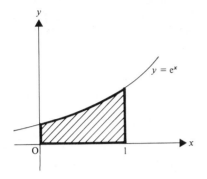

The area described is the shaded region in the diagram on the left.

When this area is rotated about the x-axis the solid of revolution formed is shown in the diagram below.

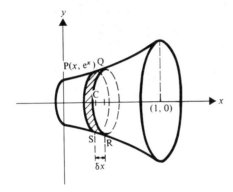

Taking the section PQRS as a typical element of volume then the volume, δV, of PQRS is approximately that of a cylinder of radius CP and thickness δx,

i.e. $$V \simeq \sum_{x=0}^{1} \pi (e^x)^2 \, \delta x$$

Therefore $$V = \pi \int_0^1 e^{2x} \, dx = \left[\frac{\pi}{2} e^{2x} \right]_0^1 = \frac{\pi}{2} (e^2 - 1)$$

Hence the volume generated is $\frac{\pi}{2} (e^2 - 1)$ cubic units.

2) The area defined by the inequalities

$$y \geqslant x^2 + 1, \quad x \geqslant 0, \quad y \leqslant 2$$

is rotated completely about the y-axis. Find the volume of the solid generated.

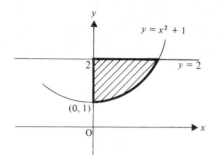

The shaded region in the diagram on the left is that defined by the given inequalities.

Rotating this area about the y-axis gives the solid shown below

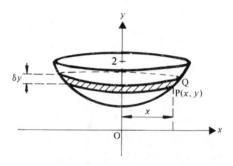

If $P(x, y)$ is any point on the curve $y = x^2 + 1$, sections through P and Q perpendicular to Oy give an element of volume δV, where

$$\delta V \simeq \pi x^2 \, \delta y$$

Therefore

$$V = \int_1^2 \pi x^2 \, \mathrm{d}y$$

As $y = x^2 + 1$,

$$x^2 = y - 1$$

Hence

$$V = \pi \int_1^2 (y - 1) \, \mathrm{d}y$$

$$= \pi \left[\frac{y^2}{2} - y \right]_1^2$$

$$= \frac{\pi}{2}$$

3) The area enclosed by the curve $y = 4x - x^2$ and the line $y = 3$ is rotated about the line $y = 3$. Find the volume of the solid generated.

The curve $\left. \begin{array}{l} y = 4x - x^2 \\ \text{and the line} \quad y = 3 \end{array} \right\}$ intersect at $(1, 3)$ and $(3, 3)$.

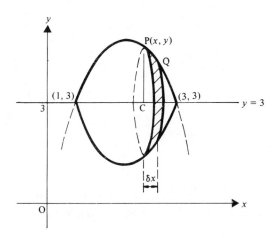

If P and Q are points on $y = 4x - x^2$, sections through P and Q, perpendicular to $y = 3$, give an element of volume δV, where $\delta V \simeq \pi (PC)^2 \, \delta x$.

Now
$$PC = (y_P) - 3$$
$$= (4x - x^2) - 3$$
$$= -x^2 + 4x - 3$$

Hence
$$V = \int_1^3 \pi(-x^2 + 4x - 3)^2 \, dx$$

$$= \pi \int_1^3 (x^4 - 8x^3 + 22x^2 - 24x + 9) \, dx$$

$$= \pi \left[\frac{x^5}{5} - 2x^4 + \frac{22}{3}x^3 - 12x^2 + 9x \right]_1^3$$

$$= \frac{16\pi}{15}$$

4) The area of the region defined by the inequalities

$$y^2 \leqslant x, \quad y \geqslant x$$

is rotated about the x-axis.
Find the volume of the solid generated.

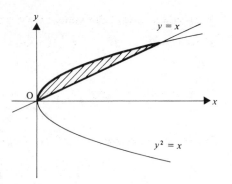

The area defined by

$$y^2 \leqslant x, \quad y \geqslant x$$

is the shaded region in the diagram.

The points of intersection of $\left.\begin{array}{l} y^2 = x \\ y = x \end{array}\right\}$ are $(0,0), (1,1)$

When this area is rotated about Ox the solid generated is bowl shaped on the outside with a conical hole inside. The cross-section this time is not a simple circle but is an annulus, i.e. the area between two concentric circles.

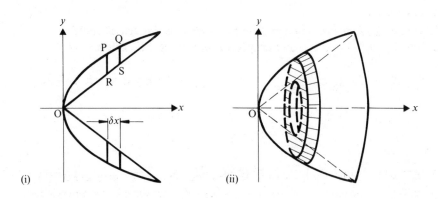

(i)   (ii)

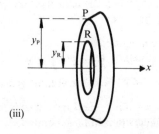

(iii)

A typical element of this volume is shown in diagram (iii),

where $\qquad$ P is a point on $\quad y^2 = x$

and $\qquad$ R is a point on $\quad y = x$

The cross-section through P and R is an annulus of area $\pi y_P^2 - \pi y_R^2$

$$= \pi[(\sqrt{x})^2 - x^2]$$

Therefore the volume, δV, of the element is given by

$$\delta V \simeq \pi(x - x^2)\,\delta x$$

Hence $\qquad\qquad\qquad V = \int_0^1 \pi(x - x^2)\,dx$

$$= \pi\left[\frac{x^2}{2} - \frac{x^3}{3}\right]_0^1 = \frac{\pi}{6}$$

i.e. the volume generated is $\dfrac{\pi}{6}$ cubic units.

EXERCISE 17c

In each of the following questions, find the volume generated when the area defined by the following sets of inequalities is rotated completely about the x-axis.

1) $0 \leqslant y \leqslant x(4 - x)$.

2) $0 \leqslant y \leqslant e^x$, $\quad 0 \leqslant x \leqslant 3$.

3) $0 \leqslant y \leqslant \dfrac{1}{x}$, $\quad 1 \leqslant x \leqslant 2$.

4) $0 \leqslant y \leqslant x^2$, $\quad -2 \leqslant x \leqslant 2$.

5) $y^2 \leqslant x$, $\quad x \leqslant 2$.

In each of the following questions, the area bounded by the curve and line(s) given is rotated about the y-axis to form a solid. Find the volume generated.

6) $y = x^2$, $\quad y = 4$.

7) $y = 4 - x^2$, $\quad y = 0$.

8) $y = x^3$, $\quad y = 1$, $\quad y = 2$, $\quad$ for $\quad x \geqslant 0$.

9) $y = \ln x$, $\quad x = 0$, $\quad y = 0$, $\quad y = 1$.

10) Find the volume generated when the area enclosed between $y^2 = x$ and $x = 1$ is rotated about the line $x = 1$.

11) The area defined by the inequalities

$$y \geqslant x^2 - 2x + 4, \quad y \leqslant 4$$

is rotated about the line $y = 4$. Find the volume generated.

12) The area enclosed by $y = \sin x$ and the x-axis for $0 \leqslant x \leqslant \pi$ is rotated about the x-axis. Find the volume generated.

13) The area enclosed by $y = x^2$ and $y^2 = x$ is rotated about the x-axis. Find the volume generated.

14) The area defined by the inequalities

$$0 \leqslant y \leqslant x^2, \quad 0 \leqslant x \leqslant 2$$

is rotated about the y-axis. Find the volume generated.

15) Find the volume generated when the area in the first quadrant enclosed by $y = |x^2 - 1|$, and the line $y = 1$ is rotated about the line $y = 1$.

CENTROID OF AREA

The centroid of an area is the *point* about which the area is evenly distributed,

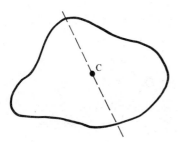

From this it follows that the centroid lies on any line of symmetry.
Hence, the centroid of a rectangle is the point of intersection of its diagonals.
The centroid of a circle is the centre. The centroid of a triangle is the point of intersection of the medians.

First Moment of Area

If the point C is the centroid of an area A, the first moment of A about an axis is defined as

$$A \times (\text{distance of C from the axis}).$$

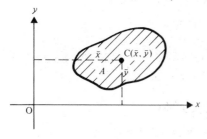

In particular, for an area A in the xy plane, whose centroid is the point $C(\bar{x}, \bar{y})$.

The first moment of A about Ox is $A\bar{y}$.
The first moment of A about Oy is $A\bar{x}$.

To Find the Coordinates of the Centroid of an Area

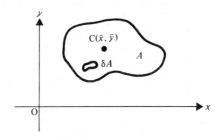

If $C(\bar{x},\bar{y})$ is the centroid of an area A and if δA is the area of a small element of A, then first moment of A about any axis is given by

$$\sum(\text{first moment of } \delta A \text{ about the same axis})$$

Hence $A\bar{x} = \sum(\text{first moment of } \delta A \text{ about } Oy)$

and $A\bar{y} = \sum(\text{first moment of } \delta A \text{ about } Ox)$

Consider the area between the curve $y = x(2-x)$ and the x-axis.

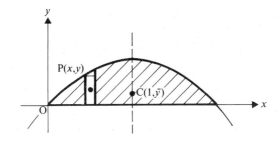

From symmetry the centroid C lies on $x = 1$,   i.e. $\bar{x} = 1$.

To find $\bar{y}$ we consider the first moment about Ox.
Taking a vertical strip through P as shown we have $\delta A \simeq y\,\delta x$.
The centroid of the element is distant approximately $y/2$ from Ox. So the first moment of δA about Ox is approximately $(y\,\delta x)\tfrac{1}{2}y$.

Hence $A\bar{y} \simeq \displaystyle\sum_{x=0}^{2} \frac{1}{2}y^2\,\delta x$

But $A = \displaystyle\int_0^2 y\,dx$

Therefore
$$\bar{y}\int_0^2 y \, dx = \int_0^2 \frac{1}{2}y^2 \, dx$$

$$\bar{y}\int_0^2 x(2-x) \, dx = \frac{1}{2}\int_0^2 x^2(2-x)^2 \, dx$$

$$\bar{y}\left[x^2 - \frac{x^3}{3}\right]_0^2 = \frac{1}{2}\left[\frac{4x^3}{3} - x^4 + \frac{x^5}{5}\right]_0^2$$

$$\frac{4\bar{y}}{3} = \frac{8}{15}$$

$$\bar{y} = \tfrac{2}{5}$$

i.e. the centroid of this area is the point $(1, 2/5)$ and the first moment of this area about Ox is $8/15$.

CENTROID OF VOLUME

The centroid of a volume is the *point* about which the volume is evenly distributed.
Hence *the centroid of a volume of revolution lies on the axis of rotation*.

First Moment of Volume

If the point C is the centroid of a volume V, the first moment of V about an axis is defined as

$$V \times \text{(distance of C from the axis)}.$$

In particular, for a volume V obtained by rotating an area about the x-axis, and whose centroid is the point $C(\bar{x}, 0)$,

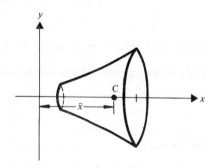

the first moment of V about Ox is zero and
the first moment of V about Oy is $V\bar{x}$.
Further, if δV is the volume of an element of V,

$$V\bar{x} = \sum \text{(first moment of } \delta V \text{ about } Oy)$$

For example, the area between $y^2 = x$ and $x = 4$ is rotated about the x-axis, giving the solid of revolution shown below

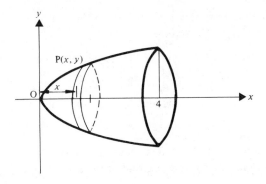

Taking a section of thickness δx as shown, gives

$$\delta V \simeq \pi y^2 \, \delta x$$

Hence $$V = \int_0^4 \pi y^2 \, dx = \pi \int_0^4 x \, dx = \pi \left[\frac{x^2}{2} \right]_0^4 = 8\pi$$

Also, the distance from Oy of the centroid of the element is approximately x. So the first moment of δV about Oy is approximately $(\pi y^2 \delta x)x$. Using $V\bar{x} = \Sigma$ (first moment of δV about Oy) we have

$$8\pi\bar{x} \simeq \sum_{x=0}^{4} (\pi y^2 x \, \delta x)$$

i.e. $$8\pi\bar{x} = \int_0^4 \pi x^2 \, dx = \pi \left[\frac{x^3}{3} \right]_0^4 = \frac{64\pi}{3}$$

$\Rightarrow$ $$\bar{x} = \tfrac{8}{3}$$

i.e. the centroid of this volume is the point $(\tfrac{8}{3}, 0)$

and the first moment of this volume about Oy is $\dfrac{64\pi}{3}$.

EXERCISE 17d

1) A region of the xy plane is defined by the inequalities
$$0 \leqslant y \leqslant \sin x, \quad 0 \leqslant x \leqslant \pi$$
Find: (a) the area of the region,
      (b) the first moment of this area about the x-axis,
      (c) the coordinates of the centroid of this area.

Find also:

    (d) the volume obtained when this area is rotated completely about the x-axis,

    (e) the first moment of this volume about the y-axis,

    (f) the centroid of this volume.

2) Repeat Question 1 for the region of the xy plane defined by the inequalities

$$0 \geqslant y \geqslant x(x-1)$$

3) A region of the xy plane is bounded by the curve $y = x^2$ and the line $y = 4$.

Find: (a) the area of this region,

    (b) the first moment of this area about the x-axis,

    (c) the y coordinate of the centroid of the area.

Find also:

    (d) the volume obtained when this area is rotated about the y-axis,

    (e) the first moment of the volume about the x-axis,

    (f) the centroid of this volume.

4) Repeat Question 3 for the area bounded by $y = \ln x$, $x = 0$, $y = 0$, $y = 1$.

5) Find the x coordinate of the centroid of the area bounded by

$$y = e^x, \ y = 0, \ x = 0, \ x = 2.$$

APPROXIMATE METHODS FOR EVALUATING A DEFINITE INTEGRAL

To determine the indefinite integral $\int f(x)\,\mathrm{d}x$ we must find a function whose differential is $f(x)$, and this is not always possible.

But the definite integral $\int_a^b f(x)\,\mathrm{d}x$ is a number which represents the area between $y = f(x)$, the x-axis and the ordinates $x = a$ and $x = b$.

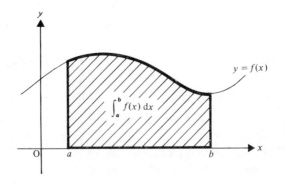

So, even if $\int f(x)\,dx$ cannot be found, an approximate value for $\int_a^b f(x)\,dx$ can be found by evaluating the appropriate area using another method.

We will now look at two methods for finding the approximate value of an area bounded partly by a curve and the x-axis.

1. The Trapezium Rule

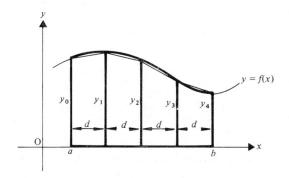

If the area represented by $\int_a^b f(x)\,dx$ is divided into strips, each of width d, as shown above, then each such strip is approximately a trapezium.

The area of a trapezium

$$= \text{(half sum of } \| \text{ sides)} \times \text{(distance between them)}.$$

Using the sum of the areas of these strips as an approximation for the actual value of the area we have

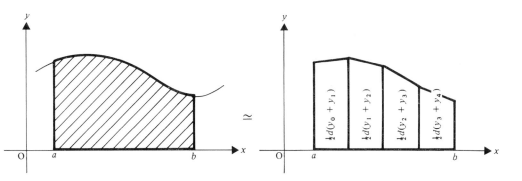

i.e. $\int_a^b f(x)\, dx \simeq \frac{1}{2}d(y_0 + y_1) + \frac{1}{2}d(y_1 + y_2) + \frac{1}{2}d(y_2 + y_3) + \frac{1}{2}d(y_3 + y_4)$

$$\simeq \frac{1}{2}d\left[y_0 + 2y_1 + 2y_2 + 2y_3 + y_4 \right]$$

This method is known as the *trapezium rule* and here we used it with five ordinates (i.e. y_0 to y_4).

Note that the ordinates must be evenly spaced (i.e. the widths of all strips must be the same).

Generalising, used with n ordinates the trapezium rule is

$$\int_a^b f(x)\, dx \simeq \frac{1}{2}d\left[y_0 + 2y_1 + \ldots + 2y_{n-2} + y_{n-1} \right]$$

EXAMPLES 17e

1) Use the trapezium rule, with five ordinates, to evaluate $\int_0^{0.8} e^{x^2}\, dx$.

For five ordinates, $(y_0, y_1, \ldots, y_4)$ evenly spaced in the range $0 \leqslant x \leqslant 0.8$, we need to divide this range into four equal parts.

i.e.

Thus $d = 0.2$

and if $y = e^{x^2}$

$y_0 = 1$

$y_1 = e^{0.04} = 1.0408$

$y_2 = e^{0.16} = 1.1735$

$y_3 = e^{0.36} = 1.4333$

$y_4 = e^{0.64} = 1.8965$

Hence the trapezium rule gives

$$\int_0^{0.8} e^{x^2}\, dx \simeq \frac{1}{2}d\left[y_0 + 2y_1 + 2y_2 + 2y_3 + y_4 \right]$$

$$\simeq \frac{1}{2}(0.2)[1 + 2(1.0408) + 2(1.1735) + 2(1.4333) + 1.8965]$$

$$\simeq 1.0192$$

2. Simpson's Rule

Suppose that the area represented by $\int_a^b f(x)\,dx$ is divided by the ordinates y_0, y_1, y_2 into two strips each of width d as shown below. A particular parabola can be found passing through the three points with the same coordinates.

Simpson's rule uses the area under that parabola as an approximation for the value of the area under the curve $y = f(x)$.

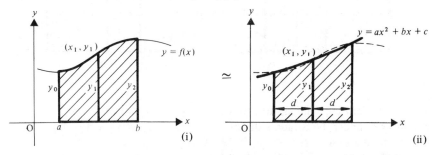

i.e.
$$\int_c^b f(x)\,dx \simeq \int_{x_1-d}^{x_1+d} (ax^2 + bx + c)\,dx \qquad [1]$$

If $y = ax^2 + bx + c$ is the parabola through the ordinates as shown, then $(x_1 - d, y_0), (x_1, y_1), (x_1 + d, y_2)$ are on this parabola,

i.e.
$$y_0 = a(x_1 - d)^2 + b(x_1 - d) + c \qquad [2]$$

$$y_1 = ax_1^2 + bx_1 + c \qquad [3]$$

$$y_2 = a(x_1 + d)^2 + b(x_1 + d) + c \qquad [4]$$

Now the area in diagram (ii) is

$$\int_{x_1-d}^{x_1+d} (ax^2 + bx + c)\,dx$$

$$= \frac{a}{3}\{(x_1 + d)^3 - (x_1 - d)^3\} + \frac{b}{2}\{(x_1 + d)^2 - (x_1 - d)^2\}$$

$$+ c\{(x_1 + d) - (x_1 - d)\}$$

which simplifies to

$$\frac{d}{3}\left[2a\{(x_1 + d)^2 + (x_1 + d)(x_1 - d) + (x_1 - d)^2\} + 3b(2x_1) + 6c \right]$$

Then using [2], [3] and [4] we find that

$$\int_{x_1-d}^{x_1+d} (ax^2 + bx + c)\,dx = \frac{d}{3}(y_0 + 4y_1 + y_2)$$

From [1]
$$\int_a^b f(x)\,dx \simeq \tfrac{1}{3}d[y_0 + 4y_1 + y_2]$$

This argument can be extended to another two strips, also of width d.

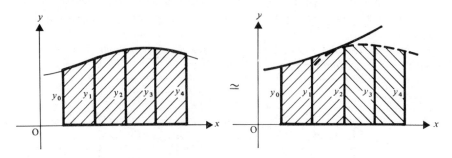

The area under the parabola through y_0, y_1, y_2 is $\tfrac{1}{3}d[y_0 + 4y_1 + y_2]$
and the area under the parabola through y_2, y_3, y_4 is $\tfrac{1}{3}d[y_2 + 4y_3 + y_4]$
giving a total area of $\tfrac{1}{3}d[y_0 + 4y_1 + 2y_2 + 4y_3 + y_4]$
i.e. with five ordinates, Simpson's rule is

$$\int_a^b f(x)\,dx \simeq \tfrac{1}{3}d[y_0 + 4y_1 + 2y_2 + 4y_3 + y_4]$$

This argument can be extended to cover any even number of strips, i.e. any *odd* number of ordinates.
Hence Simpson's rule, with $(2n+1)$ ordinates, is

$$\int_a^b f(x)\,dx \simeq \tfrac{1}{3}d[y_0 + 4y_1 + 2y_2 + 4y_3 + 2y_4 + \ldots + 4y_{2n-1} + y_{2n}]$$

Note that the use of Simpson's rule requires an *odd* number of ordinates.
For ease of computation, the ordinates used can be arranged in the form

$$\frac{d}{3}\Big[(1\text{st} + \text{last}) + 4(2\text{nd} + 4\text{th} + \ldots) + 2(3\text{rd} + 5\text{th} + \ldots)\Big]$$

or
$$\frac{d}{3}\Big[y_0 + y_{2n} + 4\sum_{r=1}^{2n} y_{2r-1} + 2\sum_{r=1}^{2n-2} y_{2r}\Big]$$

EXAMPLES 17e (continued)

2) Use Simpson's rule with five ordinates to find an approximate value for

$$\int_0^\pi \sqrt{\sin\theta}\,d\theta$$

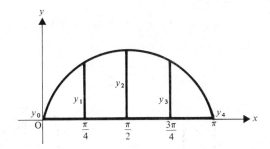

Taking five ordinates from $\theta = 0$ to $\theta = \pi$ gives four strips each of width $\pi/4$

$$y = \sqrt{\sin \theta}$$

$\Rightarrow$

$$y_0 = \sqrt{\sin 0} \quad = 0$$

$$y_1 = \sqrt{\sin \pi/4} \quad = 2^{-\frac{1}{4}} = 0.8409$$

$$y_2 = \sqrt{\sin \pi/2} \quad = 1$$

$$y_3 = \sqrt{\sin (3\pi/4)} = 2^{-\frac{1}{4}} = 0.8409$$

$$y_4 = \sqrt{\sin \pi} \quad = 0$$

Hence, using Simpson's rule,

$$\int_0^\pi \sqrt{\sin \theta} \, d\theta \simeq \tfrac{1}{3}d[y_0 + y_4 + 4(y_1 + y_3) + 2y_2]$$

$$\simeq \left(\frac{1}{3}\right)\left(\frac{\pi}{4}\right)\left[0 + 8(0.8409) + 2\right]$$

$$\simeq 2.2848$$

3) Estimate, to 4 decimal places, $\int_0^1 \dfrac{1}{1 + x^2} \, dx$, using five ordinates and applying

(a) the trapezium rule   (b) Simpson's rule.

Using the substitution $x = \tan \theta$, or otherwise, evaluate $\int_0^1 \dfrac{1}{1+x^2} \, dx$

and hence determine the accuracy of your estimated values. (Take π as 3.1416.)

Taking five ordinates from $x = 0$ to $x = 1$ gives four strips, each of width 0.25.

So

$$y = \frac{1}{1 + x^2}$$

$\Rightarrow$

$$y_0 = \frac{1}{1 + (0)^2} = 1$$

$$y_1 = \frac{1}{1 + (0.25)^2} = 0.9412$$

$$y_2 = \frac{1}{1 + (0.5)^2} = 0.8$$

$$y_3 = \frac{1}{1 + (0.75)^2} = 0.64$$

$$y_4 = \frac{1}{1 + (1)^2} = 0.5$$

(a) Using the trapezium rule

$$\int_0^1 \frac{1}{1 + x^2}\, dx \simeq \frac{0.25}{2}\left[1 + 0.5 + 2(0.9412 + 0.8 + 0.64)\right]$$

$$\simeq 0.7828$$

(b) Using Simpson's rule

$$\int_0^1 \frac{1}{1 + x^2}\, dx \simeq \frac{0.25}{3}\left[1 + 0.5 + 4(0.9412 + 0.64) + 2(0.8)\right]$$

$$\simeq 0.7854$$

If $x \equiv \tan \theta$, ... $dx \equiv \ldots \sec^2 \theta\, d\theta$

then $$\int_0^1 \frac{1}{1 + x^2}\, dx \simeq \int_0^{\frac{\pi}{4}} \frac{1}{\sec^2 \theta} \sec^2 \theta\, d\theta$$

$$= \int_0^{\frac{\pi}{4}} d\theta = \left[\theta\right]_0^{\frac{\pi}{4}} = \frac{\pi}{4}$$

Taking $\pi = 3.1416$, $\dfrac{\pi}{4} = 0.7854$.

Hence, the trapezium rule gave an estimate correct to 2 d.p. and Simpson's rule gave an estimate correct to 4 d.p. (at least).

Approximate Integration using a Series Expansion

Consider $\displaystyle\int_0^{0.1} \frac{1}{\sqrt{(1 + x^2)}}\, dx$.

Since the range of values of x for which the integral is required is $0 \leqslant x \leqslant 0.1$, x is small for the whole of the range.

Using the binomial theorem,

$$\frac{1}{\sqrt{(1+x^2)}} \equiv (1+x^2)^{-\frac{1}{2}} = 1 - \tfrac{1}{2}x^2 + \tfrac{3}{8}x^4 \ldots$$

Because x is small, terms in x^5 and higher powers do not affect the fourth decimal place and can be neglected.

Hence
$$\int_0^{0.1} \frac{1}{\sqrt{(1+x^2)}} \, dx \simeq \int_0^{0.1} (1 - \tfrac{1}{2}x^2 + \tfrac{3}{8}x^4) \, dx$$

$$\simeq \left[x - \tfrac{1}{6}x^3 + \tfrac{3}{40}x^5 \right]_0^{0.1}$$

$$\simeq 0.0998 \text{ to 4 d.p.}$$

Note that we are assuming that the integral of a function is equal to the sum of the integrals of terms of a series expansion of that function.

EXERCISE 17e

Estimate the values of the following definite integrals, taking the number of ordinates given in each case, using (a) the trapezium rule, (b) Simpson's rule.

1) $\int_0^{\frac{\pi}{2}} \dfrac{1}{1+\cos x} \, dx$ 3 ordinates   2) $\int_0^{0.4} \sqrt{(1-x^2)} \, dx$ 5 ordinates

3) $\int_0^{\pi} (\sin^4 \theta) \, d\theta$,   5 ordinates   4) $\int_0^{0.6} x e^x \, dx$,   7 ordinates

5) $\int_1^{e} \ln x \, dx$,   5 ordinates   6) $\int_0^{1} \sqrt{(1+x^2)} \, dx$,   11 ordinates

Use the first three terms of the appropriate series expansion as an approximation for each of the functions to be integrated. Hence estimate the values of the following definite integrals.

7) $\int_0^{0.1} \sqrt{(1+x)} \, dx$ 8) $\int_{0.1}^{0.2} \sqrt[3]{(1-x^2)} \, dx$

SMALL INCREMENTS

If $P(x, y)$ is a general point on the curve with equation $y = f(x)$, and $Q(x + \delta x, y + \delta y)$ is a point on the curve close to P,

then δx is a small increase in x

and δy is the corresponding increase in y.

We saw in Chapter 5 that

$$\lim_{\delta x \to 0} \left(\frac{\delta y}{\delta x} \right) = \frac{dy}{dx}$$

so when δx is small we can say

$$\frac{\delta y}{\delta x} \simeq \frac{dy}{dx}$$

or

$$\delta y \simeq \left(\frac{dy}{dx} \right)(\delta x)$$

This approximation can be used to estimate the value of a function close to a known value, i.e. $y + \delta y$ can be estimated if y is known.

For example, an approximate value of $\ln 1.1$ can be found as follows

$$y = \ln x \quad \Rightarrow \quad \frac{dy}{dx} = \frac{1}{x}$$

Hence

$$\delta y \simeq \left(\frac{1}{x} \right)(\delta x)$$

Taking P as the point $(1, \ln 1)$ and Q as $(1.1, \ln 1.1)$ we have

$$\delta y \simeq \left(\frac{1}{1} \right)(0.1)$$

Hence

$$\ln 1.1 = y + \delta y \simeq \ln 1 + 0.1$$

$\Rightarrow$

$$\ln 1.1 \simeq 0.1$$

EXAMPLES 17f

1) Using $y = \sqrt{x}$, estimate $\sqrt{101}$.

$$y = \sqrt{x} \quad \Rightarrow \quad \frac{dy}{dx} = \frac{1}{2} x^{-\frac{1}{2}}$$

Using $\delta y \simeq \dfrac{dy}{dx} \delta x$ we have $\delta y \simeq \dfrac{1}{2\sqrt{x}} \delta x$

Taking $x = 100$, $y = \sqrt{100}$, and $\delta x = 1$ gives

$$\delta y \simeq \left(\frac{1}{2\sqrt{100}} \right) \quad (1)$$

Then

$$\sqrt{101} = y + \delta y \simeq \sqrt{100} + \tfrac{1}{20}$$

$\Rightarrow$

$$\sqrt{101} \simeq 10.05$$

2) Use $f(\theta) \equiv \sin\theta$ to find an approximate value for
(a) $\sin 31°$ (b) $\sin 29°$. (Take $1° = 0.0175^c$.)

If $f(\theta) \equiv \sin\theta$, $f'(\theta) \equiv \cos\theta$.

Using $\delta y \simeq \dfrac{dy}{dx}\delta x$ in the form $\delta(f(\theta)) \simeq f'(\theta)\delta\theta$,

we have $$\delta(\sin\theta) \simeq (\cos\theta)\,\delta\theta$$

(a) Taking $\theta = \dfrac{\pi}{6}$, $\sin\theta = \sin\dfrac{\pi}{6} = \tfrac{1}{2}$, $\delta\theta = 0.0175$

we have $$\delta(\sin\theta) \simeq \left(\cos\dfrac{\pi}{6}\right)(0.0175)$$

Then $$\sin 31° \simeq \sin 30° + \delta(\sin\theta)$$
$$\simeq 0.5 + (0.866)(0.0175)$$
$\Rightarrow$ $$\sin 31° \simeq 0.5152$$

(b) Again $\theta = \dfrac{\pi}{6}$ but $\delta\theta = -0.0175$ (as the angle *decreases* from $30°$
    to $29°$)

so $$\delta(\sin\theta) \simeq \left(\cos\dfrac{\pi}{6}\right)(-0.0175)$$

Then $$\sin 29° \simeq \sin 30° + \delta(\sin\theta)$$
$$\simeq 0.5 + (0.866)(-0.0175)$$
$\Rightarrow$ $$\sin 29° \simeq 0.4848$$

SMALL PERCENTAGE INCREMENTS

The method used above requires only minor adaptation in order to estimate the percentage change in a dependent variable due to a small change in the independent variable. In this case we use the additional fact that if x increases by $r\%$ then δx can be found using $\delta x = \dfrac{r}{100} \times x$.

Then the corresponding percentage increase in y is $\dfrac{\delta y}{y} \times 100\%$.

EXAMPLES 17f (continued)

3) The period, T, of a simple pendulum is calculated from the formula $T = 2\pi\sqrt{l/g}$ where l is the length of the pendulum and g is the constant gravitational acceleration.

Find the percentage change in the period caused by lengthening the pendulum by 2%.

$$T = 2\pi\sqrt{l/g} \quad \Rightarrow \quad \frac{\mathrm{d}T}{\mathrm{d}l} = \pi/\sqrt{lg}$$

The small increment in the length is given by

$$\delta l = \frac{2}{100} \times l = \frac{l}{50}$$

So, using $\quad \delta T \simeq \delta l(\mathrm{d}T/\mathrm{d}l), \quad$ we have

$$\delta T \simeq \left(\frac{l}{50}\right)\left(\frac{\pi}{\sqrt{lg}}\right)$$

i.e.
$$\delta T \simeq \frac{\pi}{50}\sqrt{\frac{l}{g}}$$

The percentage change in the period is given by $\dfrac{\delta T}{T} \times 100\,\%$,

i.e.
$$\left(\frac{\pi}{50}\sqrt{\frac{l}{g}}\right) \div \left(2\pi\sqrt{\frac{l}{g}}\right) \times 100\,\% = 1\,\%$$

As this is a positive quantity, we see that

the period *increases* by 1%

RATE OF CHANGE PROBLEMS

The identity $\quad \dfrac{\mathrm{d}y}{\mathrm{d}x} \equiv \dfrac{\mathrm{d}y}{\mathrm{d}t} \times \dfrac{\mathrm{d}t}{\mathrm{d}x} \quad$ is useful when solving certain rate of change problems. The way in which it can be applied is demonstrated in the following examples.

EXAMPLES 17f (continued)

4) A spherical balloon is blown up so that its volume increases at a constant rate of $2\,\mathrm{cm}^3/\mathrm{s}$.

Find the rate of increase of the radius when the volume of the balloon is $50\,\mathrm{cm}^3$.

If, at time t, the volume of the balloon is V and the radius is r then

$$V = \tfrac{4}{3}\pi r^3 \quad \Rightarrow \quad \frac{\mathrm{d}V}{\mathrm{d}r} = 4\pi r^2 \qquad\qquad [1]$$

We wish to find $\dfrac{dr}{dt}$, we are given $\dfrac{dV}{dt}(=2)$ and we know that $\dfrac{dV}{dr} = 4\pi r^2$.

Using

$$\frac{dr}{dt} = \frac{dr}{dV} \times \frac{dV}{dt} = \frac{dV}{dt} \bigg/ \frac{dV}{dr}$$

gives

$$\frac{dr}{dt} = \frac{2}{4\pi r^2} = \frac{1}{2\pi r^2}$$

From [1], when $V = 50$, $r = 2.29$

$$\Rightarrow \qquad \frac{dr}{dt} = \frac{1}{2\pi(2.29)^2} = 0.03$$

So the radius is increasing at 0.03 cm/s when the volume is 50 cm^3.

5) A vessel containing water is in the form of an inverted hollow cone with a semi-vertical angle of $30°$. There is a small hole at the vertex of the cone and the water is running out at a rate of 3 cm^3/s.
Find the rate at which the surface area in contact with the water is changing when there are 81π cm^3 of water remaining in the cone.

At any time, the volume, V, of water in the vessel is given by

$$V = \tfrac{1}{3}\pi r^2 h$$

Now both r and h are variable so we cannot yet find dV/dr.
Using $r = h \tan 30° = h/\sqrt{3}$ gives

$$V = \frac{\pi h^3}{9} \qquad \Rightarrow \qquad \frac{dV}{dh} = \frac{\pi h^2}{3}$$

The wet surface, S, at any time, t, is given by

$$S = \pi r l = \pi h^2 \sec 30° = 2\pi h^2/3 \qquad \Rightarrow \qquad \frac{dS}{dh} = 4\pi h/3$$

Now $dV/dt = -3$ (the volume is decreasing), and dS/dt is required. In this problem we have no direct relationship between S and V so we use a three step link as follows,

$$\frac{dS}{dt} = \frac{dS}{dh} \times \frac{dh}{dV} \times \frac{dV}{dt}$$

$$\Rightarrow \qquad \frac{dS}{dt} = \left(\frac{4\pi h}{3}\right)\left(\frac{3}{\pi h^3}\right)(-3) = -\frac{12}{h^2}$$

At the time that the value of dS/dt is required, the remaining volume of water is $81\pi\,\text{cm}^3$,

i.e. $$\tfrac{1}{9}\pi h^3 = 81\pi \quad \Rightarrow \quad h = 9$$

Then, when $V = 81\pi$,

$$\frac{dS}{dt} = -\frac{12}{9} = -\frac{4}{3}$$

So the wet surface area is decreasing at $\dfrac{4}{3}\,\text{cm}^2/\text{s}$

Note that in all problems of this type, all the necessary differentiation must be done *before* particular values are introduced.

EXERCISE 17f

1) Using $y = \sqrt[3]{x}$, find approximate values for:
(a) $\sqrt[3]{9}$ (b) $\sqrt[3]{28}$ (c) $\sqrt[3]{126}$.

2) Using $f(\theta) \equiv \cos\theta$, find approximate values for:
(a) $\cos 29°$ (b) $\cos 61°$ (c) $\cos 44°$.
(Take $1° = 0.0175^c$.)

3) If $f(x) \equiv x \ln(1 + x)$, find an approximation for the increase in $f(x)$ when x increases by δx.
Hence estimate the value of $\ln(2.1)$, given that $\ln 2 = 0.6931$.

4) If $y = \tan x$, find an approximation for δy when x is increased by δx.
Use your approximation to estimate the value of $\tan\dfrac{9\pi}{32}$.

5) Using $f(x) \equiv x^{\frac{1}{5}}$, find an approximate value for $\sqrt[5]{33}$.

6) Ink is dropped on to blotting paper forming a circular stain which increases in area at the rate of $5\,\text{cm}^2/\text{s}$. Find the rate of change of the radius when the area is $30\,\text{cm}^2$.

7) A container in the shape of a right circular cone of height $20\,\text{cm}$ and radius $5\,\text{cm}$ is held vertex downward and filled with water which then drips out from the vertex at the rate of $5\,\text{cm}^3/\text{s}$. Find the rate of change of the height of water in the cone when it is half empty (measured by volume).

8) The surface area of a cube is increasing at the rate of $10\,\text{cm}^2/\text{s}$. Find the rate of increase of the volume of the cube when the edge is of length $12\,\text{cm}$.

9) The surface area of a sphere is decreasing at a rate of $0.9\,\text{m}^2/\text{s}$ when the radius is $0.6\,\text{m}$. Calculate the rate of change of the volume of the sphere at this instant.

10) Sand falls on to horizontal ground at the rate of $9\,m^3$ per minute and forms a heap in the shape of a right circular cone with vertical angle $60°$. Show that 10 seconds after the sand begins to fall, the rate at which the radius of the base of the pile is increasing is $3^{1/2}(4/\pi)^{1/3}\,m$ per minute.

MISCELLANEOUS EXERCISE 17

1) The table below shows the values of y obtained experimentally for the given values of x. Show graphically that, allowing for small errors of observation, there is a relation of the form $y - 2 = k(1 + x)^n$ and find approximate values of k and n.

x	4	8	15	19	24
y	2.45	2.60	2.80	2.89	3.00

(AEB)p'74

2) The variables x and y are believed to satisfy a relationship of the form $y = ab^x$, where a and b are constants. Show graphically that the values obtained in an experiment and shown in the table below do verify the relationship. From your graph calculate approximate values of a and b.

x	1	2	3	4	5
y	14.1	15.8	17.8	19.9	22.4

(AEB)p'76

3) The variables x and y below are believed to be related by a law of the form $y = \ln(ax^2 + bx)$, where a and b are constants. Plot a suitable graph to show that this is so and determine the probable values of a and b.

x	1	2	3	4	5	6
y	-1.897	0.588	1.599	2.262	2.757	3.153

(AEB)'71

4) In an experiment, sets of values of the related variables (x, y) are obtained. State how you would determine whether x and y were related by a law of the form:

(a) $y = a^{x+b}$,

(b) $ay^2 = (x + b)\ln x$,

where in each case a and b are unknown constants. State briefly how you would be able to determine the values of a and b for each law. (AEB)p'72

5) The following values of x and y are believed to obey a law of the form

$$y = \frac{a}{bx + c}$$ where a, b and c are constants. Show that they do obey this law and hence estimate the value of the ratios $a:b:c$.

x	0	1	2	3	4
y	1.00	0.67	0.50	0.40	0.33

(AEB)p'73

6) It is thought that two variables s and t satisfy a relation of the form $(s/t)^a = be^{-t}$, where the constants a and b are positive integers. Show by drawing a linear graph that this is supported by the following table of measured values, and find the values of a and b.

t	0.2	0.4	0.6	0.8	1.0	
s	1.09	1.96	2.67	3.22	3.64	(U of L)p

7) It is known that two variables x and y are connected by a relation of the form $y = kx^n$, where k and n are constant integers.
From the given table of approximate values of x and y, find graphically the values of k and n.

x	1.34	3.58	7.60	12.1	14.8	
y	208.0	10.9	1.14	0.283	0.155	(U of L)p

8) The variables x and y are related by a law of the form $y = ax^n$, where a, n are integers. Approximate values for y, corresponding to the given values for x, are tabulated below.

x	2	3	4	5	6	7
y	50	250	775	1875	3900	7200

Plot $\lg y$ against $\lg x$ and use your graph to obtain the integer values of a and n.
Using the relation between y and x so obtained, calculate the approximate percentage change in the value of y corresponding to a 1% change in the value of x.

$$[\lg x \text{ denotes } \log_{10} x.] \qquad \text{(U of L)}$$

9)

x	0.5	1.5	3.0	4.5	5.5	6.0	7.5
$\ln y$	0.6	1.6	3.6	5.4	6.8	7.2	9.0

The table gives a set of corresponding values for x and $\ln y$, where the variables x and y are known to satisfy a relation of the form $y = ab^x$, where a, b are constants. Show, by plotting $\ln y$ against x, or otherwise, that the values are not consistent with this relation, but that they are consistent if two of the $\ln y$ entries are assumed to be incorrect. Estimate the corrected values of $\ln y$ and calculate the values of a and b, correct to one decimal place.

(U of L)

10) Sketch the curve $2y = x(x-4)(2x-5)$.
The line $y = x$ cuts this curve at the origin O and at A and B where A is between O and B. Find the area bounded by the arc OA of the curve and the line $y = x$.

(U of L)p

11) Find the area of the region bounded by the curve $y = e^x$, the tangent at the point $(1, e)$ and the y-axis.

Find the volume of the solid formed by rotating this region through a complete revolution about (a) the x-axis, (b) the y-axis.   (O)

12) C is the curve given by $y = \sin x$ for values of x between 0 and π.
Find:
(a) the area of the region enclosed between C and the x-axis;
(b) the volume of the solid formed by rotating this region through four right angles about the x-axis.   (C)

13) Find the x coordinate of the turning point of the curve whose equation is

$$y = \frac{a}{x} + \ln x$$

where $x > 0$ and $a > 0$,   and determine whether this turning point is a maximum or a minimum. Deduce the range of values of the constant a for which $y \geq 0$ for all $x > 0$.
In the case when $a = 1$,   find the area and the x coordinate of the centroid of the region bounded by the curve, the x-axis and the ordinates $x = 1$ and $x = 2$.   Express both answers in terms of $\ln 2$.   (JMB)

14) A curve joining the points $(0, 1)$ and $(0, -1)$ is represented parametrically by the equations

$$x = \sin \theta, \quad y = (1 + \sin \theta) \cos \theta,$$

where $0 \leq \theta \leq \pi$.   Find dy/dx in terms of θ, and determine the x, y coordinates of the points on the curve at which the tangents are parallel to the x-axis and of the point at which the tangent is perpendicular to the x-axis. Sketch the curve.
The region in the quadrant $x \geq 0$, $y \geq 0$ bounded by the curve and the coordinate axes is rotated about the x-axis through an angle of 2π. Show that the volume swept out is given by

$$V = \pi \int_0^1 (1 + x)^2 (1 - x^2) \, dx$$

Evaluate V, leaving your result in terms of π.   (JMB)

15) Use the substitution $x = a \sin \theta$ to evaluate $\int_0^a \sqrt{a^2 - x^2} \, dx$.

The area A included between the parts of the two curves $x^2 + y^2 = 4$ and $4x^2 + y^2 = 16$ for which $x \geq 0$ and $y \geq 0$ is rotated completely about the x-axis. Find the volume of the solid thus formed.

Calculate the first moment of the area A about the y-axis and the x coordinate of the centroid of this area.   (AEB)'74

16) Show that the curves whose equations are

$$y - 1 = x^3 \quad \text{and} \quad y + 3 = 3x^2$$

intersect at a point on the x-axis, and find the coordinates of this point. Show also that the only other point at which the curves meet is $(2, 9)$ and that the curves have a common tangent there. Sketch the two curves on the same diagram.

Show that the area of the finite region bounded by the curves is $27/4$, and find the x coordinate of the centroid of this region.   (JMB)

17) A region R of the plane is defined by $y^2 - 4ax \leqslant 0$, $x^2 - 4ay \leqslant 0$, $x + y - 3a \leqslant 0$.   Find:
(a) the area of R,
(b) the volume obtained by rotating R through 4 right angles about the
x-axis.   (O)

18) The curves $cy^2 = x^3$ and $y^2 = ax$ (where $a > 0$ and $c > 0$) intersect at the origin O and at a point P in the first quadrant. The areas of the regions enclosed by the arcs OP, the x-axis and the ordinate through P are A_1 and A_2 for the two curves; the volumes of the two solids formed by rotating these regions through four right angles about the x-axis are V_1 and V_2 respectively Prove that $A_1/A_2 = \frac{3}{5}$ and $V_1/V_2 = \frac{1}{2}$.   (O)

19) Sketch the curve whose equation is

$$y = 1 - \frac{1}{x + 2},$$

indicating any asymptotes which the curve possesses.
The region bounded by the curve, the x-axis and the ordinates $x = 0$ and $x = 2$ is denoted by R. Find:

(a) the area of R;
(b) the x coordinate of the centroid of R;
(c) the volume swept out when R is rotated about the x-axis through an angle of 2π.   (JMB)

20) Sketch the curves $y = e^x$ and $y = e^{-x}$ for $-2 \leqslant x \leqslant 2$.
The interior of a wine glass is formed by rotating the curve $y = e^x$ from $x = 0$ to $x = 2$ about the y-axis. If the units are centimetres, find, correct to two significant figures, the volume of liquid that the glass contains when full.
                                                                           (C)

21) Find the area of the region in the first quadrant enclosed between the curve $y = x^2 + 4$, the line $y = 8$ and the y-axis.
The region is rotated through four right angles about Oy to form a uniform solid. Find the volume and the coordinates of the centre of mass of the solid. (C)

22) (a) Integrate $\dfrac{1}{x(1-x^2)}$ with respect to x.

    (b) Find the mean value of $\sin^5 x$ over the range $x = 0$ to $x = \pi/2$.

    (c) Obtain an approximate value of $\displaystyle\int_{\frac{\pi}{6}}^{\frac{\pi}{2}} \sqrt{\sin\theta}\ d\theta$ by the trapezium rule

    using five ordinates. Work as accurately as your tables will allow.

                                                  (AEB)'74

23) Apply Simpson's rule using five ordinates to find an approximate value of

$\displaystyle\int_0^\pi \sin^{\frac{3}{2}} x\ dx.$ (AEB)p'75

24) The table below gives values of $f(x) = \sqrt{1 - x^2/25}$ correct to three decimal places for values of x between 0 and 5, at intervals of 0.5.

x	0	0.5	1.0	1.5	2.0	2.5	3.0	3.5	4.0	4.5	5.0
$f(x)$	1.000	0.995	0.980	0.954	0.917	0.866	0.800	0.714	0.600	0.436	0.000

Use Simpson's rule with eleven ordinates to calculate an approximate value of

$\displaystyle\int_0^5 \sqrt{1 - x^2/25}\ dx.$ Hence find an approximate value, to one decimal place, for

the area of the ellipse $x^2/25 + y^2/4 = 1$.
Using an appropriate substitution, evaluate the above integral and hence find an approximate value for π, to two decimal places.   (AEB)'74

25) R is the region in the first quadrant bounded by the y-axis, the x-axis from 0 to $\frac{1}{2}\pi$, the line $x = \frac{1}{2}\pi$ and part of the curve $y = (1 + \sin x)^{\frac{1}{2}}$.

    (a) Show that, when R is rotated about the x-axis through four right angles, the volume of the solid formed is $\frac{1}{2}\pi(\pi + 2)$.

    (b) Use the trapezium rule with three ordinates to show that the area of R is approximately 0.63π.   (C)p

26) Tabulate values of $f(x) = \sqrt{\{27 + (x-3)^3\}}$ for integral values of x from 0 to 6 inclusive and sketch the graph of $y = f(x)$ for the interval $0 \leqslant x \leqslant 6$.

Given that $F(t) \equiv \displaystyle\int_0^t f(x)\ dx$, use Simpson's rule and the calculated values

of $f(x)$ to estimate $F(2)$ and $F(6)$.   (U of L)

27) Let $y = \sqrt[3]{x}$. Write down $\dfrac{dy}{dx}$, and hence find an expression for the approximate small change in y when x changes by a small amount δx. Use your result to estimate the cube root of 1001. (C)p

28) Tabulate, to three decimal places, the values of the function
$f(x) = \sqrt{(1 + x^2)}$ for values of x from 0 to 0.8 at intervals of 0.1.

Use these values to estimate $\displaystyle\int_0^{0.8} f(x)\, dx$:

(a) by the trapezium rule, using all the ordinates,

(b) by Simpson's rule, using only ordinates at intervals of 0.2. (U of L)

29) Find the coordinates of the centroid of the finite region bounded by the curve $y = x^2$, the lines $x = 2$, $x = 3$ and the x-axis. (U of L)p

30) Estimate $\displaystyle\int_{\frac{\pi}{6}}^{\frac{\pi}{2}} (\sin x)\, dx$ by Simpson's rule, using five ordinates and giving your answer to two decimal places. (U of L)p

31) Given that $y = \sqrt{\left(\dfrac{1 + x}{2 + x}\right)}$ determine the value of $\dfrac{dy}{dx}$ when $x = 2$,

and deduce the approximate increase in the value of y when x increases in value from 2 to $2 + \epsilon$ (ϵ small). (JMB)p

32) The height h and the base radius r of a right circular cone vary in such a way that the volume remains constant. Find the rate of change of h with respect to r at the instant when h and r are equal. (U of L)

33) A hemispherical bowl of radius a cm with its axis vertical is being filled with water at a steady rate of $5\pi a^3$ cm^3 per min. Find in cm per min the rate at which the level is rising when the depth of water is $\frac{1}{3}a$ cm.
[The volume of a cap of height h of a sphere of radius r is $\frac{1}{3}\pi h^2(3r - h)$.]
(U of L)p

34) A bowl is formed by rotation about the y-axis of the arc of the curve $ay = x^2$ from $x = 0$ to $x = a$. Initially, the bowl is full of water.

(a) Find the height of the centre of mass of the water above the lowest point of the bowl.

(b) The water evaporates from the surface so that, at any instant, the rate of decrease in its volume is proportional to the surface area. Show that the depth of the water decreases at a uniform rate, giving this rate in terms of a and T, where T is the time taken for the bowl to empty. (U of L)

35) A bowl is formed by rotating about the y-axis that part of the x-axis between $x = 0$ and $x = 2$ and that part of the curve $y = (x-2)^2$ between $x = 2$ and $x = 4$. Calculate the volume inside the bowl. Water is poured into the bowl at a constant rate of 20 cubic units per second. Find the rate at which the depth of water is increasing when the water level is 0.25 units above the base of the bowl.

ANSWERS

Exercise 1a – p. 4

1) a) equation   b) identity   c) identity
    d) equation   e) equation   f) identity
    g) equation   h) identity

2) a) $x > 5$ b) $x \leqslant -2$
    c) $x \leqslant 4$ d) $x < 1$

3) a) $0, -2, 10, 10$
    b) $-\frac{1}{2}, -1, -\frac{1}{4}, \frac{1}{3}$
    c) $-14, -18, 0, -14$
    d) $1\frac{2}{3}, 2\frac{1}{2}, -\frac{5}{7}, -\frac{5}{42}$

Exercise 1b – p. 9

1) $\dfrac{3}{2(x-1)} - \dfrac{3}{2(x+1)}$

2) $\dfrac{4}{3(x-4)} - \dfrac{1}{3(x-1)}$

3) $\dfrac{3}{4(x+2)} + \dfrac{1}{4(x-2)}$

4) $\dfrac{2}{3(x-2)} - \dfrac{4}{3(2x-1)}$

5) $\dfrac{3}{x} - \dfrac{2}{x+1}$

6) $\dfrac{3}{x+1} - \dfrac{7}{3x+2}$

7) $\dfrac{3}{2(x-1)} - \dfrac{6}{x-2} + \dfrac{9}{2(x-3)}$

8) $\dfrac{7}{24(x-3)} + \dfrac{7}{8(x+1)} - \dfrac{2}{3x}$

9) $\dfrac{4}{5(2x+1)} - \dfrac{1}{5(3x-1)}$

10) $\dfrac{4}{3(x-1)} - \dfrac{11}{3(2x+1)}$

11) $-\dfrac{1}{5(2x-1)} - \dfrac{7}{5(x+2)}$

12) $1 - \dfrac{2}{x-1} + \dfrac{6}{x-2}$

13) $\dfrac{3}{2(x+1)} + \dfrac{5}{6(x+3)} - \dfrac{4}{3x}$

14) $-\dfrac{3}{25(x+1)} + \dfrac{1}{3(x-1)} - \dfrac{16}{75(x-4)} + \dfrac{2}{5(x-4)^2}$

15) $\dfrac{1}{6(x-3)} + \dfrac{5}{24(x+3)} - \dfrac{1}{8(x-1)} - \dfrac{1}{4(x+1)}$

16) $\dfrac{5}{8(x-1)} + \dfrac{3}{4(x-3)} + \dfrac{5}{8(x+3)}$

17) $\dfrac{1}{x-1} - \dfrac{x+1}{x^2+1}$

18) $\dfrac{x-2}{2(x^2-2)} - \dfrac{1}{2(x+4)}$

19) $\dfrac{3}{2x} - \dfrac{x}{2(x^2+2)}$

20) $\dfrac{22}{19(x-3)} + \dfrac{1-6x}{19(2x^2+1)}$

21) $\dfrac{3}{5(x+2)} - \dfrac{3}{5(2x+1)} + \dfrac{x-1}{5(x^2+1)}$

22) $\dfrac{1}{x} - \dfrac{6x+3}{2x^2-1} + \dfrac{2}{x-1}$

23) $\dfrac{1}{x-1} - \dfrac{1}{x-2} + \dfrac{2}{(x-2)^2}$

24) $\dfrac{2}{x} - \dfrac{1}{x^2} - \dfrac{3}{2x+1}$

25) $\dfrac{3}{x} - \dfrac{9}{3x-1} + \dfrac{9}{(3x-1)^2}$

26) $\dfrac{3}{4(x-1)} - \dfrac{1}{4(x+1)} - \dfrac{x}{2(x^2+1)}$

27) $1 + \dfrac{1}{2(x-1)} - \dfrac{1}{2(x+1)}$

28) $1 - \dfrac{7}{4(x+3)} - \dfrac{1}{4(x-1)}$

29) $x + \dfrac{2}{x-1} - \dfrac{1}{x+1}$

Exercise 1c − p. 11

1) $-6, 1$ 2) $0, \frac{7}{3}$ 3) $1, 1$
4) $2, \infty$ 5) $1, -\frac{3}{2}$ 6) a, a
7) $1, \infty$ 8) $0, \frac{2}{3}$ 9) $2, \frac{1}{2}$
10) $\sqrt{2}, -\sqrt{2}$

Exercise 1d − p. 13

1) $2, 1$ 2) $1.46, -5.46$
3) $-\frac{1}{2}, -3$ 4) $1 \pm \sqrt{(1-a)}$
5) $a \pm \sqrt{(a^2 - b)}$ 6) $\dfrac{-b \pm \sqrt{(b^2 - 4ac)}}{2a}$
7) equal 8) not real
9) real and distinct 10) not real
11) equal 12) real and distinct
13) ± 24 14) ± 12
15) $6\frac{1}{4}$ 18) $q^2 - 4p = 0$

Exercise 1e − p. 17

1) a) $3, 2$ b) $-\frac{7}{4}, -\frac{3}{4}$
   c) $4, -4$ d) $-1, -8$
   e) k, k^2 f) $\dfrac{a+2}{a}, -1$

2) a) $x^2 - 3x + 4 = 0$
   b) $2x^2 + 4x + 1 = 0$
   c) $15x^2 - 5x - 6 = 0$
   d) $4x^2 + x = 0$
   e) $x^2 - ax + a^2 = 0$
   f) $x^2 + (k + 1)x + k^2 - 3 = 0$
   g) $abx^2 - b^2 x + ac^2 = 0$

3) a) $\frac{4}{5}$ b) $5\frac{1}{2}$ c) -1 d) 5
   e) -6 f) $-\frac{2}{5}$ g) $\frac{8}{11}$ h) $\frac{4}{7}$ i) $\frac{4}{25}$

4) a) $x^2 - 6x + 11 = 0$
   b) $3x^2 - 2x + 1 = 0$
   c) $x^2 + 2x + 9 = 0$
   d) $3x^2 + 2x + 3 = 0$
   e) $x^2 + 8 = 0$

5) $x^2 - 2x - 3 = 0$
6) $2x^2 - 5x - 4 = 0$
7) $5x^2 + 13x + 7 = 0$
8) $2x^2 + 3x - 1 = 0$
9) $\frac{11}{12}$ 10) -6
11) $cx^2 + bx + a = 0$
12) $a = c$ 14) 1

Multiple Choice Exercise 1 − p. 19

1) e 2) e 3) a 4) b
5) d 6) b 7) d 8) b
9) b, c 10) c 11) a, b, c 12) a
13) E 14) D 15) B 16) C
17) I 18) A 19) a 20) F
21) F 22) T 23) F
24) T 25) F

Miscellaneous Exercise 1 − p. 21

1) $\dfrac{1}{1-x} + \dfrac{2}{2x+1} - \dfrac{3}{(2x+1)^2}$

2) $\dfrac{1}{x-1} - \dfrac{1}{2(x-1)^2} - \dfrac{2x+1}{2(x^2+1)}$

3) $\dfrac{1}{x-3} - \dfrac{1}{x+2} + \dfrac{2}{(x+2)^2}$

4) $acx^2 - b(a + c)x + a^2 + b^2 + c^2$
   $- 2ac = 0$

5) $qy^2 + 2py + 4 = 0$
7) $x^2 + pqx + q^3 = 0$
8) $3y^2 - (p + 8)y + 3 = 0$,
   $p = -2$ or -14

9) $\dfrac{2x+3}{1+x^2} + \dfrac{2}{1-x}$

10) $A = 3, B = -1$

11) $\dfrac{1}{a}(a - b + c)$

13) $\dfrac{2}{x+1} - \dfrac{1}{x}$

Exercise 2a − p. 24

1) 4 2) 2 3) $\frac{1}{3}$ 4) $\frac{4}{7}$
5) $\frac{1}{5}$ 6) 1331 7) $\frac{1}{32}$ 8) 27
9) $\frac{27}{1000}$ 10) $\frac{9}{4}$ 11) 49 12) 0.6
13) 625 14) $\frac{5}{8}$ 15) $\frac{125}{27}$ 16) 1
17) $\frac{4}{3}$ 18) 6 19) 3 20) 4
21) 6 22) 1 23) 16 24) 1
25) $y^{-\frac{3}{4}}$ 26) 1 27) x^5 28) t
29) $x^{5/2} + x^3$ 30) $y^{5/4} + y^{1/2}$
31) $\dfrac{x}{x-1}$ 32) $m - 1$

Exercise 2b − p. 26

1) a) $2\sqrt{2}$ b) $2\sqrt{3}$ c) $5\sqrt{2}$
   d) $3\sqrt{2}$ e) $10\sqrt{2}$ f) $6\sqrt{2}$
   g) $5\sqrt{5}$ h) $12\sqrt{2}$ i) $15\sqrt{2}$
   j) $20\sqrt{5}$
2) a) $3\sqrt{2} - 2$ b) $3\sqrt{2} - 4$
   c) $9 - \sqrt{3}$ d) 1
   e) $5 - 3\sqrt{3}$ f) $2 - 3\sqrt{2}$
   g) 23 h) $24 + 5\sqrt{5}$
   i) $\sqrt{6} + \sqrt{3} - \sqrt{2} - 1$
   j) $33 - 12\sqrt{6}$

3) a) $\dfrac{\sqrt{3}}{3}$ b) $\dfrac{\sqrt{2}}{4}$

c) $\dfrac{\sqrt{2}}{4}$ d) $\sqrt{2}-1$

e) $\dfrac{\sqrt{3}+1}{2}$ f) $3(2+\sqrt{3})$

g) $5(\sqrt{5}-2)$ h) $\frac{2}{3}(2\sqrt{3}+3)$

i) $\dfrac{2\sqrt{5}-3}{11}$ j) $\sqrt{6}+\sqrt{5}$

k) $\dfrac{2\sqrt{3}-\sqrt{2}}{10}$ l) $2\sqrt{2}$

Exercise 2c – p. 28

1) a) $\log_5 125 = 3$
b) $\log_7 49 = 2$
c) $\log_8 4096 = 4$
d) $\log_4 32 = \frac{5}{2}$
e) $\log_{121} 1331 = \frac{3}{2}$
f) $\log 0.01 = -2$
g) $\log_5 1 = 0$
h) $\log_8 \frac{1}{2} = -\frac{1}{3}$
i) $\log_{1/5} 125 = -3$
j) $\log_9 \frac{1}{27} = -\frac{3}{2}$
k) $\log_a 1 = 0$
l) $\log_\pi 9.8696 = 2$
m) $\log_q p = 2$
n) $\log_a b = c$
o) $\log_x 2 = y$

2) a) $5^4 = 625$ b) $10^3 = 1000$
c) $3^3 = 27$ d) $10^0 = 1$
e) $(\frac{1}{2})^{-2} = 4$ f) $25^{1/2} = 5$
g) $a^0 = 1$ h) $x^2 = y$
i) $4^q = p$ j) $a^b = 5$
k) $x^z = y$ l) $q^p = r$

3) a) 3 b) 4 c) $\frac{1}{2}$ d) -2
e) -1 f) $\frac{2}{3}$ g) $\frac{1}{2}$ h) $\frac{1}{4}$
i) -1 j) $-\frac{1}{3}$ k) $\frac{1}{2}$ l) 3
m) 0 n) 0 o) 3 p) b
q) 10 r) 25

Exercise 2d – p. 31

1) a) $\log a + \log b$
b) $\log a + \log b + \log c$
c) $\log a - \log b$
d) $\log a + \log b - \log c$
e) $\log a - \log b - \log c$
f) $-\log a$
g) $2 \log a + \log b$
h) $\frac{1}{2} \log a - \frac{1}{2} \log b$
i) $2 \log a - \log b$

j) $3 \log a - 1$
k) $-1 - \log a$
l) $2 - \frac{1}{2} \log b$

2) a) $\log 12$ b) $\log 3$ c) $\log 3$
d) $\log 18$ e) $\log \frac{1}{3}$ f) $\log 4$
g) $\log 30$ h) $\log 20$ i) 2
j) $-\log(x-1)$ k) 3 l) 0

3) a) 1.6309 b) 0.8614
c) 1.1610 d) 2.7712
e) -0.1038 f) 1.2153
4) a) 3.3219 b) 0.8271
c) 0.2314 d) -1.3368
5) a) $-1, 1$ b) 2 c) 3
6) 2 7) $x = 2, y = 4$
8) $y = 1, x = \frac{1}{2}$ 9) $x = 1, y = 0$
10) 3 12) $x = 1, y = 0$

Exercise 2e – p. 35

1) a) 3 b) 18 c) 47 d) $\frac{35}{16}$ e) $-\frac{16}{27}$
f) $a^3 - 2a^2 + 6$ g) $c^2 - ac + b$
2) a) yes b) no c) no d) yes
e) yes f) yes
3) a) $(x-1)(x+2)(x+1)$
b) $(x-2)(x^2 + x + 1)$
c) $(x-1)(x+1)(x^2 + 1)$
d) $(x-2)(x+2)(x^2 + x + 1)$
e) $(2x-1)(x^2 + 1)$
f) $(3x-1)(9x^2 + 3x + 1)$
g) $(x+a)(x^2 - ax + a^2)$
h) $(x-y)(x^2 + xy + y^2)$
4) -7 5) 5 6) $a = -3$ 7) 1

Exercise 2f – p. 38

1) $1 + 8x + 24x^2 + 32x^3 + 16x^4$
2) $1 - 3x + 3x^2 - x^3$
3) $x^3 - 3x^2 + 3x - 1$
4) $1 - 12y + 54y^2 - 108y^3 + 81y^4$
5) $1 + 7x + 21x^2 + 35x^3 + 35x^4$
  $+ 21x^5 + 7x^6 + x^7$
6) $8x^3 - 12x^2 + 6x - 1$
7) $8x^3 + 12x^2 y + 6xy^2 + y^3$
8) $x^4 + 4x^2 + 6 + \dfrac{4}{x^2} + \dfrac{1}{x^4}$
9) $p^3 - 6p^2 q + 12pq^2 - 8q^3$
10) $x^{10} - 5x^8 y + 10x^6 y^2 - 10x^4 y^3$
  $+ 5x^2 y^4 - y^5$
11) $x^3 - 3x + \dfrac{3}{x} - \dfrac{1}{x^3}$
12) $a^6 - 3a^4 b^2 + 3a^2 b^4 - b^6$
13) $7 + 5\sqrt{2}$ 14) $49 - 20\sqrt{6}$
15) 20 16) 1.030 301
17) 19.4481

Multiple Choice Exercise 2 — p. 39

1) c   2) a   3) a
4) a   5) b   6) e
7) c   8) a, b, c   9) a
10) b, c   11) a, b   12) a, c
13) C 14) A 15) C
16) D 17) T 18) T
19) F 20) T

Miscellaneous Exercise 2 — p. 41

1) $9, \frac{1}{3}$ 2) a) 0.59   b) 1.6
3) a) $\frac{7}{4}$ b) 6   4) $b = a^2$
5) $nm + m - 1$ 6) -1
7) $x = 4, y = 2$ or $x = 9, y = 3$
8) $x = 4, y = 2$
9) $x = 10, y = 4; x = 4, y = 2$
10) $4^8, \frac{1}{4}$
11) $h = 4, g = -1, (x + 2), (x + 3)$
12) $14, x + 1$
14) a) 125^- or $\frac{1}{125}$ b) $\dfrac{\sqrt{2}}{2}$
15) $3x^3 - 7x^2 - 2x + 8$
16) $x = 3^6, y = \frac{1}{3}$ or $x = \frac{1}{3}, y = 3^6$

Exercise 3a — p. 48

1) The range of f is the set of positive real numbers including zero
2) a) and c) are functions
3) Yes. The domain is the set of rational numbers. The range is the set of positive rational numbers and zero.
4) $T = 0, I \in \mathbb{Q}, I \leqslant 10\,000$
$T = \frac{5}{100}I - 500, I \in \mathbb{Q},$
$10\,000 < I \leqslant 20\,000$
$T = \frac{1}{10}I - 1500, I \in \mathbb{Q}, I > 20\,000$

Exercise 3b — p. 51

1) $\frac{11}{4}$ 2) 3   3) 4
4) $\frac{29}{4}$ 5) -2 6) 1
7) least value 4 at $x = 1$
8) least value -12 at $x = -2$
9) least value $-\frac{3}{2}$ at $x = \frac{3}{2}$
10) greatest value $\frac{65}{4}$ at $x = -\frac{7}{2}$
11) least value -10 at $x = 0$
12) greatest value $\frac{49}{12}$ at $x = -\frac{5}{6}$

Exercise 3c — p. 53

1) $x > 1$ 2) $x < -3$
3) $x < \frac{1}{2}$ 4) $x < 3$
5) $x < 1$ 6) all values
7) $x < 1, x > 2$ 8) $x < -1, x > 2$
9) $3 < x < 5$ 10) $-1 < x < \frac{1}{2}$

11) $x < -1, x > 5$ 12) $-\frac{1}{2} < x < \frac{1}{2}$
13) $x < -\frac{2}{5}, x > 1$ 14) $x < -4, x > 2$
15) $-5 < x < \frac{3}{2}$ 16) $1 < x < 2$
17) $-1 < x < \frac{1}{3}$ 18) $-\frac{1}{2} < x < 2$

Exercise 3d — p. 56

1) $x > \frac{11}{5}$
2) $x > \frac{11}{4}$
3) i) $k < 1, k > 9$ ii) $0 < k < 1, k > 9$
4) $-4 < k < 0$
5) $c \leqslant -2, c \geqslant 1$
7) $k > 1$
8) $-1 \leqslant x < 1$ and $4 < x \leqslant 5$
9) a) $\alpha = 1$ or -3, other root 4 or $\frac{4}{9}$
   b) $\alpha > \frac{5}{2}(\sqrt{2} - 1), \alpha < -\frac{5}{2}(\sqrt{2} + 1)$
10) a) $x < -5.16$ or $x > 1.16$
    b) $-8.24 < x < 0.24$
11) a) $-4 < x < \frac{3}{2}$
    b) real roots, $a = 1$
12) $(m + 1)^2 = ma, a \leqslant 0, a \geqslant 4$
13) $-1\frac{1}{2} < p < \frac{1}{2}$
14) a) $-4 < x < 2, a = 3$
    b) $-2x^2 - 8x + 10,$
      $-50x^2 + 40x + 10$
15) $k < -3\frac{1}{12}, k = -2\frac{1}{12}$
16) $k = -\frac{1}{48}$ a) $k > 1$ b) $k \leqslant \frac{1}{2}$
18) $x < -2, x > \frac{1}{2}$
19) b) $k \leqslant 0, k \geqslant 3$ (c) $k \geqslant 3$
20) $f \in \mathbb{R}, f \geqslant -\frac{25}{4}; x = -1, 4$

Exercise 3e — p. 64

1) $\frac{1}{4}, \frac{1}{16}, \frac{1}{64}, 4, 16, 64$ 3) $x = 1$
8) By reflection in the line at $45°$ to the x-axis
9) By reflection in the line at $45°$ to the x-axis

Exercise 3f — p. 68

1) a) $f^{-1}: x \to \frac{1}{3}x$
   b) $f^{-1}: x \to \frac{1}{2}(x - 1)$
   d) $f^{-1}: x \to 5^x$
3) a) $f^{-1}: x \to 2x + 6, \quad x \in \mathbb{R}$
   b) $f^{-1}: x \to \frac{1}{3}x, x \in \{-5, 0, 5\}$
   c) $f^{-1}: x \to \frac{1}{3}x, x = 5n$ where $n \in \mathbf{N}$
   d) $f^{-1}: x \to 2 + \sqrt{(x + 1)}, x \in \mathbb{R},$
     $x \geqslant -1$
   e) $f^{-1}: x \to \sqrt{(x + 4)}, x \in \mathbb{R}, x > -4$
   f) $f^{-1}: x \to \lg x, x \in \mathbb{R}, x > 1$

Exercise 4b — p. 76

1) a) 5   b) $\sqrt{2}$ c) $\sqrt{13}$ d) $\sqrt{13}$
   e) $\sqrt{5}$ f) $\sqrt{8}$

2) a) $(\frac{5}{2}, 4)$ b) $(\frac{5}{2}, \frac{1}{2})$ c) $(3, \frac{7}{2})$
   d) $(\frac{1}{2}, 5)$ e) $(-\frac{1}{2}, -1)$
   f) $(-2, -3)$
3) $\sqrt{65}$ 6) $(-\frac{7}{2}, -\frac{1}{2}), \frac{35}{2}$
8) $(0, \frac{9}{2}), 2\frac{1}{2}$ 9) $(-5, -3)$

Exercise 4c – p. 82

1) a) 3   b) $\frac{3}{2}$ c) $\frac{1}{3}$ d) $\frac{3}{4}$
   e) -4 f) 6   g) $-\frac{7}{3}$ h) $-\frac{3}{2}$
   i) k/h
2) a) 1   b) -1 c) 0   d) ∞
3) a) yes   b) no   c) yes
4) a) parallel   b) perpendicular
   c) perpendicular   d) parallel
5) $a = 0, b = 4$ 6) $\frac{45}{2}$
7) 8   8) $b(d - b) = ac$
10) $a^2 + b^2 = 4$ 11) $3k = 2 - h$
12) 5, 35   13) $b^2 = 8a - 16$

Exercise 4d – p. 88

5) no   6) yes   7) no   8) no
9) yes   10) yes

Exercise 4e – p. 94

1) a) $y = 2x$ b) $y = -x$
   c) $3y = x$ d) $4y + x = 0$
   e) $y = 0$ f) $x = 0$
2) a) $2y = x + 2$ b) $2y = x$
   c) $2y = x - 7$
3) a) $2y = x$ b) $2x + y - 6 = 0$
   c) $y + 2x - 4 = 0$ d) $y = 2x + 5$
4) a) $y > x$ b) $y < 4 - 2x$
   c) $y > 2 - x$ d) $y < 2x + 4$
5) a) $3y - 2x = 0$ b) $y + 2x = 0$
6) a) $3y = x + 1$ b) $2y + 3x + 7 = 0$
7) a) $5x - y - 17 = 0$
   b) $x + 7y + 11 = 0$

Exercise 4f – p. 99

1) $2x - y - 8 = 0$ 2) $x - y = 0$
3) a) $3x - 2y + 2 = 0$
   b) $3x - 2y + 7 = 0$
   c) $x = 3$
4) a), c), d) are perpendicular
5) a) $(\frac{2}{5}, \frac{11}{5})$ b) $(-\frac{3}{2}, \frac{5}{2})$
   c) $(-\frac{2}{5}, \frac{4}{5})$ d) $(-\frac{6}{5}, \frac{7}{5})$
6) a) $x + 2y - 5 = 0$
   b) $6y - 16x - 19 = 0$
   c) $16y - 10x = 23$
7) $5y + 4x = 0$ 8) $4y + 5x = 0$
9) $(-5, 0), (\frac{7}{2}, 0), (\frac{16}{7}, \frac{17}{7})$

10) $x + 2y - 11 = 0, \dfrac{11\sqrt{5}}{5}$

Multiple Choice Exercise 4 – p. 101

1) c   2) d   3) b   4) a
5) a   6) c   7) a   8) d
9) b   10) a   11) a, c   12) b
13) a, b   14) c   15) a, c   16) D
17) B   18) C   19) c   20) I
21) A   22) I   23) A   24) T
25) T   26) F   27) T   28) F

Miscellaneous Exercise 4 – p. 104

2) $\frac{1}{4}$
3) $(-5, 0), (6, 0), (-\frac{4}{3}, \frac{11}{3})$
4) $x - y - 4 = 0, 8$
5) $(3, 0), (-2, -5), 5\sqrt{2}$
6) $x + 3y - 11 = 0, (2.6, 2.8)$
7) $y = 2x - 3$
8) $8by = 2ax - 8b^2 - 3a^2$
9) $(x - 2)^2 + y^2 < 16$
10) $(x - 2)^2 + y^2 = 16$
11) $(0, 0), (2, 4)$
12) $A(1 + \sqrt{2}, 1 - \sqrt{2}), B(1 - \sqrt{2}, 1 + \sqrt{2})$
13) $M(1, 1)$, AB is a diameter
14) $(0, 4), (12 + 4\sqrt{2}, 8\sqrt{2}),$
    $(4\sqrt{2}, 4 + 8\sqrt{2})$
15) $(4, 3\frac{1}{2}), (7, 5)$ or $(\frac{2}{5}, -\frac{19}{10}), (2\frac{1}{5}, 9\frac{4}{5})$
16) $(-\frac{2}{11}, \frac{3}{11}), (1, 4), (4\frac{2}{11}, 1\frac{8}{11})$
    $13y - 41x = 11, 7y + 5x = 33,$
    $13y - 41x + 149 = 0$

Exercise 5a – p. 111

1) 6   2) 4   3) -8
4) -5 5) 3   6) -2

Exercise 5b – p. 113

1) $2x$ 2) $3x^2$ 3) $4x^3$
4) 1   5) $6x$ 6) $15x^2$
7) $-\dfrac{2}{x^3}$ 8) $2x + 3$ 9) $2x - 2$

Exercise 5c – p. 116

1) $5x^4$ 2) $10x^9$ 3) $-3x^{-4}$
4) $\frac{1}{3}x^{-2/3}$ 5) $-x^{-2}$ 6) $-\frac{1}{2}x^{-3/2}$
7) $-4x^{-5}$ 8) $\frac{3}{2}x^{1/2}$ 9) 2
10) $-5x^{-6}$ 11) $-\frac{1}{2}x^{-3/2}$ 12) $\frac{1}{4}x^{-3/4}$
13) 0   14) -4 15) $\frac{1}{2}x^{-1/2}$
16) $\frac{2}{3}x^{-1/3}$

Exercise 5d — p. 118

1) $6x - 7$ 2) $4x^3 - 27x^2$
3) $4x - 1$ 4) $2x - 8$

5) $147x^2 - 98x - 1$ 6) $2x - \dfrac{2}{x^2}$

7) $-2x^{-3} - x^{-2}$ 8) $-\dfrac{1}{x^2} + \dfrac{14}{x^3} - \dfrac{12}{x^4}$

9) $\frac{3}{2}\sqrt{x} - \dfrac{1}{2\sqrt{x}}$ 10) $\dfrac{1}{2x^{1/2}} + \dfrac{1}{2x^{3/2}}$

11) $24x^2 + 24x + 6$ 12) $-\dfrac{1}{x^2} - \dfrac{2}{x^3} - \dfrac{3}{x^4}$

13) 2   14) 18   15) $\dfrac{\sqrt{2}}{4}$

16) -1 17) $-\frac{1}{9}$ 18) -1
19) $\frac{1}{9}$ 20) 34   21) 4
22) $-\frac{3}{2}$ 23) -5 24) $\frac{1}{54}$
25) $(1, 3)$ 26) $(\frac{3}{2}, 5)$
27) $(1, 0)$ 28) $(1, -16)$
29) $(-1, 0), (\frac{1}{3}, \frac{4}{27})$
30) $(-4, -\frac{1}{3}), (-2, -3\frac{2}{3})$

31) no points   32) $\left(\dfrac{1}{16}, \dfrac{1}{4}\right)$

33) $(\frac{1}{2}, -1), (-\frac{1}{2}, 3)$ 34) $(-2, \frac{1}{4})$
35) $(1, 1), (-1, -1)$ 36) no points

Exercise 5e — p. 121

1) $y = 2x - 3$ 2) $3x - y - 1 = 0$
3) $x + y + 2 = 0$ 4) $y = 8x + 2$
5) $x - y - 7 = 0$ 6) $y = -2$
7) $x + 2y + 1 = 0, x + 3y + 3 = 0,$
   $x - y = 0, x + 8y + 49 = 0$
   $x + y + 1 = 0, x = 0$
8) $x + 3y + 6 = 0$ 9) $13x - y - 18 = 0$
10) $2y = 6x - 3, 3x + y + 3 = 0,$
   $(-\frac{1}{4}, -\frac{9}{4})$
11) $x + y - 3 = 0, x - y - 2 = 0$
12) $y = 5x - 1, 9x + 3y + 19 = 0$
13) $(1, 1), y = 2x - 1$
14) $(\frac{3}{2}, -\frac{11}{4}), -\frac{29}{4}$
15) $2y = x + 1$ 16) $y = x - 2$
17) $-\frac{87}{32}$ 18) $y = -\frac{121}{8}$

Exercise 5f — p. 123

1) 0   2) $\frac{5}{2}$ 3) ± 2
4) $0, 1, -\frac{1}{4}$ 5) -6 6) $2, -2$
7) $\frac{2}{9}, \frac{34}{9}$ 8) $(-\frac{1}{2}, -\frac{25}{4})$
9) $(\frac{1}{3}, -\frac{41}{27}), (3, -11)$ 10) no points

Exercise 5g — p. 131

1) -2 min, 2 max   2) 4 min, -4 max

3) -3 max   4) 0 min, 4 max
5) $-\frac{49}{8}$ min   6) 0 inflexion
7) 0 min   8) -1 min, 0 max
9) 0 min, $\frac{5}{16}$ max, 2 max
10) $(0, 0), (\frac{2}{3}, \frac{4}{27})$ 11) $(0, 1)$
12) $(1, 2), (-1, -2)$
13) $(0, -4)$ 14) $(\frac{1}{2}, 2\frac{3}{4})$
15) $(-4, -173), (\frac{1}{2}, 9\frac{1}{4})$
16) none   17) $(0, 0)$
18) $(1, -4), (-1, 4)$

Multiple Choice Exercise 5 — p. 132

1) b   2) d   3) a   4) a
5) e   6) c   7) b   8) a, b
9) b, c   10) b, c   11) A 12) C
13) C 14) E 15) I 16) a
17) A 18) I

Miscellaneous Exercise 5 — p. 134

1) 24   2) $-\frac{9}{7}$
3) $2t - 5, 2y - 2, -4v$

4) $1 - \dfrac{1}{x^2}$ 5) 0.289

6) $(\frac{9}{2}, -\frac{33}{4}), x + 2y + 12 = 0$
7) $y = 15x - 16, y = 15x + 16$
8) $4x + 12y + 15 = 0$
9) $(0, 0), (1, -5), (-2, -32)$
10) $2, -2$ 12) $(200\pi)^{-1/3}$
13) $a = 2, b = 2, c = \frac{3}{2}$ 14) $\frac{1}{108}$ m³
15) $\dfrac{500\pi\sqrt{3}}{9}$ 17) 200 ms⁻¹
18) $\frac{17}{4}$ m   19) 3.5°/min
20) 1.5 cms⁻¹
21) $y = 8px - 4p^2$
22) $t^2 y + x - 2t = 0; 2$

Exercise 6a — p. 139

1) $\frac{1}{6}\pi; \frac{3}{2}\pi; \frac{2}{3}\pi; \frac{1}{4}\pi; \frac{4}{3}\pi; \frac{5}{6}\pi; \frac{1}{9}\pi; \frac{5}{3}\pi; \frac{1}{8}\pi; \frac{4}{9}\pi$
2) $135°; 150°; 18°; 270°; 330°; 60°;$
   $240°; 15°; 22.5°; 80°$
3) $0.62^c; 1.36^c; 0.96^c; 1.22^c; 0.28^c$
4) $163.9°; 241.2°; 57.3°; 198.8°; 171.9°$

Exercise 6b — p. 142

1) $d = 1.57; r = 4.3; \frac{1}{2}^c; \dfrac{\pi x}{90}; 3.85; 8.26$

2) 3445 km
3) $112° 45'$

4) $4.19; 3.69; 1.78^c; \dfrac{49\pi x}{360}; 2.37; 14.1$

5) $16°55'$

6) 7.5 cm^2

7) 1.83 cm^2 ; 32.38 cm^2

8) a) 6.29 cm　b) 9.44 cm^2
   c) 5.54 cm^2

9) a) 18.5 cm　b) 118.9 cm^2

10) 8.94 cm

Exercise 6c – p. 151

1) $87°; \frac{1}{4}\pi; 52°; \frac{1}{3}\pi; 22°; 36°; \frac{1}{6}\pi; \frac{1}{4}\pi;$
   $\frac{1}{4}\pi; 4°$

2) a) $-0.9205, 23°, 157°$
   b) $-0.7071, 45°, 225°$
   c) $0.5, -0.866, -0.5774, \frac{1}{6}\pi$
   d) $-0.866, 0.5, -1.7321, 60°$
   e) $0.1736, -0.9848, -0.1763, 10°$
   f) $-0.5774, 30°, 150°$
   g) $\pm 0.866, \pm 1.7321, 60°, \pm 60°$
   h) $\pm 0.7071, \pm 0.7071, \pm 1, \pm 45°$ or
   $\pm 135°$

3) a) $23.6°, 156.4°, -203.6°, -336.4°$
   b) $\pm 120°, \pm 240°$
   c) $50.2°, 230.2°, -129.8°, -309.8°$
   d) $-41.8°, -138.2°, 221.8°, 318.2°$
   e) $\pm 66.4°, \pm 293.6°$
   f) $-206.6°, -26.6°, 153.4°, 333.4°$

4) a) $36.9°$　b) $-36.9°$　c) $26.6°$

Exercise 6d – p. 155

1) $17.7°$　　2) 8.26 cm

3) $91.8°$ or $38.2°$　　4) 6.4 cm

5) $B = 28.4°, C = 88.6°, c = 10.23$ cm

6) $A = 99.4°, B = 61.3°, a = 6.45$ cm or
   $A = 42°, B = 118.7°, a = 4.37$ cm

7) $A = 53.1°$　$b = 24.7$ cm, $c = 17.2$ cm

10) 18.7 m

Exercise 6e – p. 157

1) $A = 34°, B = 101.5°, C = 44.5°$

2) $a = 30.3$ cm, $B = 61°, C = 40°$

3) $b = 34$ cm, $A = 23.4°, C = 35.6°$

4) $B = 26°, A = 121.2°, C = 32.8°$

6) $29°, 46.6°, 104.4°$

Exercise 6f – p. 172

1) a) $\dfrac{\pi}{6}$　b) $\dfrac{\pi}{4}$　c) π　d) $\dfrac{\pi}{6}$　e) 0　f) $\dfrac{\pi}{2}$

g) $\dfrac{2\pi}{3}$　h) $\dfrac{\pi}{2}$　i) $\dfrac{\pi}{4}$　j) 0.675^c　k) $-\dfrac{\pi}{3}$

l) 1.104^c　m) -1.204^c　n) $-\dfrac{\pi}{4}$　o) $\dfrac{5\pi}{6}$

2) a) $\frac{1}{3}\pi$　　b) $\frac{3}{4}\pi$　　c) $\frac{1}{3}\pi$
   d) $\frac{1}{6}\pi$　　e) $-\frac{1}{2}\pi$　f) 0
   g) $-\frac{1}{4}\pi$　h) $\frac{1}{2}\pi$　　i) 0
   j) π　　　k) $-\frac{1}{6}\pi$　l) $\frac{1}{2}\pi$
   m) $39.1°$　n) $63.9°$　p) $70.6°$

3) a) $\frac{1}{4}\pi, \frac{3}{4}\pi$　　b) $\frac{5}{6}\pi, -\frac{5}{6}\pi$
   c) $-\frac{1}{3}\pi, \frac{2}{3}\pi$　d) $-32.7°, -147.3°$
   e) $50.9°, -50.9°$
   f) $56.3°, -123.7°$

4) $\sin \theta = 1, \sec \theta = 1$

Exercise 6g – p. 181

1) a) $\frac{1}{3}\pi, \frac{5}{3}\pi$　b) $\frac{5}{6}\pi, \frac{11}{6}\pi$　c) $\frac{1}{4}\pi, \frac{3}{4}\pi$
   d) $109.5°, 250.5°$　e) $26.6°, 206.6°$
   f) $14.5°, 165.5°$　　g) $32.9°, 327.1°$
   h) $203.6°, 336.4°$　i) $36.9°, 216.9°$
   j) $0, \pi, 2\pi, \frac{1}{6}\pi, \frac{11}{6}\pi$
   k) $0, \pi, 2\pi, \frac{2}{3}\pi, \frac{4}{3}\pi$
   l) $14.5°, 165.5°, 90°, 270°$

2) a) $\begin{cases} \frac{1}{3}\pi + 2n\pi \\ \frac{2}{3}\pi + 2n\pi \end{cases}$　b) $\frac{1}{2}(2n+1)\pi$
   c) $-\frac{1}{3}\pi + n\pi$
   d) $\begin{cases} -14.5° + 360n° \\ -165.5° + 360n° \end{cases}$
   e) $\pm 68.2° + 360n°$
   f) $63.4° + 180n°$
   g) $\begin{cases} 19.5° + 360n° \\ 160.5° + 360n° \end{cases}$
   h) $2n\pi$
   i) $\begin{cases} \frac{1}{6}\pi + 2n\pi \\ \frac{5}{6}\pi + 2n\pi \end{cases}$
   and $\begin{cases} -\frac{1}{6}\pi + 2n\pi \\ -\frac{5}{6}\pi + 2n\pi \end{cases} \Rightarrow \pm \frac{1}{6}\pi + n\pi$

Exercise 6h – p. 184

1) a) $22\frac{1}{2}°, 112\frac{1}{2}°, 202\frac{1}{2}°, 292\frac{1}{2}°$
   b) $14.8°, 45.2°, 134.7°, 165.3°,$
   $254.7°, 285.3°$
   c) $63.6°$
   d) $12°, 60°, 84°, 132°, 156°, 204°,$
   $228°, 276°, 300°, 348°$

2) a) $\begin{cases} 1080n° + 125.4° \\ 1080n° + 414.6° \end{cases}$
   b) $4.6° + 45n°$
   c) $\pm 25.5° + 180n°$
   d) $\begin{cases} \frac{1}{2}\pi + 6n\pi \\ \frac{5}{2}\pi + 6n\pi \end{cases}$

Exercise 6i – p. 186

1) $\frac{2}{7}n\pi$　　2) $\frac{1}{5}n\pi$

3) $\frac{1}{2}(4n-1)\pi, \frac{1}{10}(4n+1)\pi$

4) $\frac{1}{18}(2n+1)\pi$　　5) $2n\pi, \frac{1}{7}(2n+1)\pi$

6) $\frac{1}{2}n\pi, \frac{1}{3}n\pi$　　7) $\frac{2}{3}n\pi - \frac{1}{6}\pi, \frac{2}{5}n\pi$

8) $\frac{1}{6}(2n + 1)\pi$

9) $0, \frac{1}{5}\pi, \frac{2}{5}\pi, \frac{1}{2}\pi, \frac{3}{5}\pi, \frac{4}{5}\pi, \pi$

10) $\frac{1}{10}\pi, \frac{3}{10}\pi, \frac{1}{2}\pi, \frac{7}{10}\pi, \frac{9}{10}\pi$

11) $0, \frac{1}{9}\pi, \frac{2}{9}\pi, \frac{1}{3}\pi, \frac{5}{9}\pi, \frac{7}{9}\pi, \frac{8}{9}\pi, \pi$

12) $0, \frac{2}{11}\pi, \frac{4}{11}\pi, \frac{6}{11}\pi, \frac{8}{11}\pi, \frac{10}{11}\pi$

13) $-140°, -60°, -20°, 100°$

14) $-115°, -25°, 65°, 155°$

15) $-150°, -110°, 10°, 130°$

Exercise 6j – p. 192

1) $\frac{\pi}{2}, 4$ 　　2) $\pi, 2$ 　　3) $4\pi, \frac{1}{4}$

4) $\pi, 2$ 　　5) $\frac{\pi}{2}, 2$ 　　6) $2\pi, 1$

7) $\frac{2\pi}{3}, 3$ 　　8) $\frac{2\pi}{3}, 3$ 　　9) $8\pi, \frac{1}{4}$

10) $\pi, 2$ 　　11) $\pi, 1$

12) $\begin{cases} 2n\pi - \frac{7}{12}\pi \\ 2n\pi + \frac{1}{12}\pi \end{cases}$ 　　13) $\frac{1}{2}n\pi + \frac{1}{24}\pi$

14) $\frac{2}{3}n\pi + \frac{1}{9}\pi \pm \frac{5}{18}\pi$ 　　15) $\begin{cases} n\pi \\ n\pi + \frac{1}{3}\pi \end{cases}$

Multiple Choice Exercise 6 – p. 193

1) d 　　　2) d 　　　3) c
4) c 　　　5) d 　　　6) c
7) b 　　　8) b, c, d 　9) b, d
10) a, c 　11) b, c 　12) b
13) a, d 　14) b, d 　15) B
16) D 　　17) C 　　18) A
19) C 　　20) C 　　21) B
22) d 　　23) d 　　24) I
25) F 　　26) F 　　27) F
28) T 　　29) T 　　30) F

Miscellaneous Exercise 6 – p. 197

2) $\frac{1}{4}a^2(2\sqrt{3} - \pi)$ 　3) $78.79a^2$

4) 13.32 cm

6) a) $2n\pi - \frac{1}{12}\pi, 2n\pi + \frac{1}{12}\pi$
　b) $2n\pi - \frac{1}{12}\pi, 2n\pi + \frac{7}{12}\pi$

7) $\pm \frac{1}{6}\pi, \frac{5}{6}\pi$ 　　8) $\frac{1}{3}n\pi + \frac{1}{18}\pi$

9) $n\pi - \frac{1}{4}\pi \pm \frac{1}{6}\pi$ 　10) $\frac{1}{14}(2n + 1)\pi$

12) $\frac{2\pi}{k}; k = 2, \frac{1}{4}(2n + 1)\pi, n\pi, \frac{1}{2}(2n + 1)\pi$

　　$k = \frac{1}{2}, (2n + 1)\pi, 4n\pi, (4n + 2)\pi$

13) a) $7.5°, 45°, 97.5°$
　b) $17.5°, 107.5°, 145°$

14) $\frac{1}{14}\pi + \frac{1}{7}n\pi, \frac{1}{2}(4n - 1)\pi$

15) a) $\frac{2}{5}\pi, \frac{4}{5}\pi, \frac{6}{5}\pi, \frac{8}{5}\pi, 2\pi$

　b) $k\theta = \begin{cases} 2n\pi + \theta \\ (2n + 1)\pi - \theta \end{cases}$; (i) 6　(ii) 2

Exercise 7a – p. 204

1) $\pm 57.7°, \pm 122.3°$
2) $-10.1°, -169.9°$
3) $38.2°, 141.8°$
4) $-135°, 45°$
5) $30°, 150°$
6) $\pm 131.8°, 0°$
7) $\pm 41.4° + 360n°$
8) $-18° + 360n°, -162° + 360n°$
9) $45° + 180n°, -14° + 180n°$
10) $360n° \pm 60°, (2n + 1) 180°$

20) a) $\dfrac{x^2}{16} - \dfrac{y^2}{25} = 1$

　b) $\dfrac{x^2}{a^2} - \dfrac{y^2}{b^2} = 1$

　c) $y^2(x^2 + 4) = 36$

　d) $(x - 1)^2 + (y - 1)^2 = 1$

　e) $(x - 2)^2 + 1 = \left(\dfrac{2}{y}\right)^2$

　f) $\left(\dfrac{a}{x}\right)^2 + \left(\dfrac{y}{b}\right)^2 = 1$

21) a) $\tan^4 A$ 　b) 1 　c) $\operatorname{cosec} \theta \sec \theta$
　d) $\sec^2 \theta$ 　e) $\tan \theta$

22) a) $-\frac{12}{13}, \frac{12}{5}$ 　b) $-\frac{4}{5}, -\frac{3}{4}$
　c) $\frac{7}{25}, \frac{24}{25}$ 　d) $0, 0,$ straight line

Exercise 7c – p. 209

1) a) $\frac{1}{2}$ 　b) $\frac{1}{2}$ 　c) $\frac{1}{4}(\sqrt{6} + \sqrt{2})$
　d) $\frac{1}{4}(\sqrt{6} - \sqrt{2})$ 　e) $\frac{1}{4}(\sqrt{6} - \sqrt{2})$
　f) $2 + \sqrt{3}$

2) a) $\frac{1}{2}\sqrt{3}, \frac{1}{2}, \sqrt{3}; \frac{1}{2}, -\frac{1}{2}\sqrt{3}, -\frac{1}{3}\sqrt{3};$
　　$-\frac{1}{2}, -\frac{1}{2}\sqrt{3}, \pm \infty$

　b) $\frac{1}{2}\sqrt{2}, -\frac{1}{2}\sqrt{2}, -1; -\frac{1}{2}, -\frac{1}{2}\sqrt{3},$
　　$\frac{1}{3}\sqrt{3}; -\frac{1}{4}(\sqrt{6} - \sqrt{2}), \frac{1}{4}(\sqrt{6} + \sqrt{2}),$
　　$-2 - \sqrt{3}$

　c) $\frac{24}{25}, \frac{7}{25}; -\frac{3}{5}, -\frac{4}{5}; \frac{3}{5}, -\frac{4}{5}, \frac{117}{44}$

　d) $-\frac{4}{5}, -\frac{3}{4}, \frac{12}{13}; -\frac{5}{13}, -\frac{63}{65}, -\frac{16}{65}, \frac{33}{56}$

　e) $\frac{7}{25}, \frac{24}{25}; \frac{7}{24}, \frac{7}{7}; \frac{336}{625}, \frac{527}{625}, 0$

　f) Negative acute, $-\frac{5}{12}$; acute, $\frac{3}{5}, \frac{3}{4},$
　　$\frac{63}{65}, -\frac{56}{63}$

9) $52.5°, 232.5°$ 　10) $7.4°, 187.4°$

11) $37.9°, 217.9°$ 　12) $15°, 195°$

Exercise 7d – p. 214

1) a) $\sin 28°$ 　　b) $\tan 70°$
　c) $\cos 8\theta$ 　　d) $\tan 6\theta$
　e) $\sqrt{2} \cos 3\theta$ 　f) $\cos 52°$
　g) $\frac{1}{2} \sin 2\theta$ 　h) $\cos 68°$
　i) $\tan (x + 45°)$

2) a) (i) $\frac{24}{25}, -\frac{7}{25}$ (ii) $-\frac{24}{25}, -\frac{7}{25}$

   b) (i) $\frac{336}{625}, \frac{527}{625}$ (ii) $-\frac{336}{625}, \frac{527}{625}$

   c) (i) $\frac{120}{169}, -\frac{119}{169}$ (ii) $\frac{120}{169}, -\frac{119}{169}$

3) $7, \frac{7}{10}\sqrt{2}, \frac{1}{10}\sqrt{2}; \frac{7}{25}, -\frac{24}{25}$

4) a) $2y = x(1 - y^2)$

   b) $x = 2y^2 - 1$

   c) $y^2(1 - x) = 2$

   d) $y(1 - 2x^2) = 1$

13) $30°, 150°, 270°; 30° + 360n°,$
    $150° + 360n°, 270° + 360n°$

14) $90°, 210°, 270°, 330°; 90° + 180n°,$
    $210° + 360n°, 330° + 360n°$

15) $60°, 300°; \pm 60° + 360n°$

16) $35.26°, 144.74°, 215.26°, 324.74°;$
    $\pm 35.26° + 180n°$

17) $90°, 270°, 45°, 225°; 90° + 180n°,$
    $45° + 180n°$

18) $90°, 270°, 23.58°, 156.42°;$
    $90° + 180n°, 23.58° + 360n°,$
    $156.42° + 360n°$

Exercise 7e – p. 217

6) $\dfrac{2x}{1 + x^2}$ 7) $\frac{1}{2}\pi$ 8) $\frac{7}{11}$

9) $\frac{2}{9}$ 10) ± 1 11) 0

Exercise 7f – p. 224

1) a) $\frac{24}{25}$ b) $\frac{1}{2}$ c) $-\frac{7}{24}$

2) a) t^2 b) $\dfrac{1}{t}$ c) $\dfrac{1 - t^2}{2t^2}$

   d) $\dfrac{1}{3 + 6t - 5t^2}$ e) $\dfrac{1 - 4t + t^2}{3 - t^2}$

5) a) $0, 67.4°$

   b) $-153°, 130.4°$

   c) $36.9°, 126.9°$

   d) $-90°, 36.8°$

6) a) $2, \frac{1}{6}\pi$ b) $\sqrt{10}, 71.57°$

   c) $5, 36.87°$ d) $\sqrt{2}, \frac{1}{4}\pi$

   e) $\sqrt{29}, 21.80°$

7) a) $2, -2; 300°, 120°$

   b) $28, -22; 286.26°, 106.26°$

   c) $-\dfrac{1}{\sqrt{2}}, \dfrac{1}{\sqrt{2}}; 112.5°, 292.5°, 22.5°,$
      $202.5°$

   d) $-\sqrt{\frac{2}{3}}, \sqrt{\frac{2}{3}}; 125.26°, 305.26°$

   e) $25; 53.13°, 233.13°$

8) a) $2n\pi + \frac{1}{4}\pi$

   b) $360n° + 118.1°$ or $360n° - 36.9°$

   c) $360n°$ or $360n° - 143.2°$

   d) $360n°$ or $360n° - 53.2°$

Exercise 7g – p. 232

1) a) $2 \sin 2A \cos A$

   b) $2 \cos 4A \cos A$

   c) $2 \cos 3A \sin A$

   d) $-2 \sin 4A \sin 3A$

   e) $-2 \cos 4A \sin A$

   f) $2 \sin 3A \sin 2A$

   g) $2 \sin 45° \cos 15°$

   h) $2 \cos 60° \cos 10°$

   i) $2 \sin (A + 45°) \cos (A - 45°)$

   j) $2 \cos^2 2A$

2) a) $\sin 3\theta + \sin \theta$

   b) $\cos 5\theta + \cos \theta$

   c) $\sin 5\theta + \sin 3\theta$

   d) $\cos 4\theta - \cos 2\theta$

   e) $\cos 2\theta - \cos 6\theta$

   f) $\frac{1}{2}(\cos 5\theta + \cos 3\theta)$

   g) $1 - \sin 30°$

   h) $\frac{1}{2} + \cos 20°$

4) a) $\cos \theta$ b) 0

6) $30°, 90°, 150°, 210°, 270°, 330°$

7) $0°, 45°, 135°, 180°, 225°, 315°, 360°$

8) $0°, 60°, 90°, 120°, 180°, 240°, 270°,$
   $300°, 360°$

9) $20°, 90°, 100°, 140°, 220°, 260°,$
   $270°, 340°$

10) $22\frac{1}{2}°, 90°, 112\frac{1}{2}°, 202\frac{1}{2}°, 270°, 292\frac{1}{2}°$

11) $0°, 30°, 60°, \ldots 330°, 360°$ and
    $40°, 80°, 160°, 200°, 280°, 320°$

12) $30°, 45°, 135°, 150°, 225°, 315°$

13) $0°, 60°, 105°, 120°, 165°, 180°,$
    $240°, 285°, 300°, 345°, 360°$

14) $10°, 130°, 250°, 330°$

Exercise 7h – p. 237

1) a) $\frac{1}{2}$ b) $\dfrac{\theta^2}{1 - 2\theta^2}$

   c) $\dfrac{\theta}{2 - \theta^2}$ d) 2

   e) $\frac{1}{2}\theta(\sin \alpha + \frac{1}{2}\theta \cos \alpha)$

   f) 1 g) 1

   h) $-\frac{1}{8}(\theta \cos \alpha + 4 \sin \alpha)$

3) a) -2.129 b) 0.819

   c) -0.423 d) 0.136

   e) -0.087 f) -0.398

Exercise 7i – p. 243

1) a) 20.1 b) 55 c) 32.8

2) 4.32 m; 6.13 m

5) $28.5°$

6) $13.6, 10.1; 2.5$

Multiple Choice Exercise 7 — p. 245

1) c 2) b 3) b
4) a 5) b 6) c
7) b, c 8) d, e 9) c, d
10) b, c 11) a, b, e 12) d
13) d 14) b, c, d 15) e
16) D 17) B 18) B
19) D 20) B 21) E
22) D 23) A 24) C
25) D 26) I 27) d
28) F 29) T 30) F
31) F

Miscellaneous Exercise 7 — p. 249

1) a) $\pm\dfrac{8\sqrt{5}}{21},\pm\dfrac{4\sqrt{5}}{21}$ b) $10.9°,-129°$

2) a) (i) $336.3°,143.7°$
 (ii) $226.3°,346.3°$
 b) $\frac{1}{8}$

3) a) $-\frac{63}{65},\frac{56}{33}$ b) $\frac{1}{6}\pi,\frac{5}{6}\pi$

4) a) $0,126°52'$ b) $0,90°,180°$

5) a) $60°,109°28'$ b) $113°42',345°6'$

6) a) $45°,161°34',225°,341°34'$
 b) $\frac{1}{2}(2n+1)\pi,\begin{cases}2n\pi+\frac{1}{6}\pi\\2n\pi+\frac{5}{6}\pi\end{cases}$

7) a) $\tan\theta=\left(\dfrac{1+k}{1-k}\right)\tan\alpha;120°,300°$
 b) $\sin\dfrac{\pi}{10}=\dfrac{\sqrt{5}-1}{4}$

8) a) $\frac{1}{12}\pi,\frac{5}{12}\pi,\frac{13}{12}\pi,\frac{17}{12}\pi$

9) $\frac{1}{8}\pi,\frac{1}{2}\pi,\frac{5}{8}\pi,\frac{9}{8}\pi,\frac{3}{2}\pi,\frac{13}{8}\pi$

10) $36°52';-36°52';143°8';$
 0 and $-73°44'$

11) b) $72°24',220°12'$

12) a) $-29°33',103°17'$
 b) $0,180°,\pm75°31'$

13) $\tan\left(\dfrac{\pi}{4}-\dfrac{x}{2}\right);-(2+\sqrt{3}),2-\sqrt{3}$

15) a) $n\pi,\frac{2}{3}n\pi$ b) $26.6°,-153.4°$

16) $2\sin(\theta-30°);70°,190°,330°,310°$

17) a) $32.8°,147.2°$
 b) $n\pi+\frac{1}{2}\pi,\frac{4}{3}n\pi\pm\frac{1}{3}\pi,\frac{2}{3}n\pi$

19) b) $0,30°,60°,120°,150°,180°$
 c) $360n°+119.6°,360n°-13.3°$

20) a) $1+\sqrt{2},2+\sqrt{3}$
 b) $\cot\frac{1}{2}\theta-\cot4\theta$

21) $\frac{5}{12}\pi$

22) a) $110°36',216°52'$ b) $x=y=60°$

23) a) $n\pi,\frac{1}{4}\pi(4n-1)$
 b) $\frac{1}{8}\pi(1+8n),\frac{1}{24}\pi(8n-1)$

24) a) $-120°,-90°,-60°,0°,60°,$
 $90°,120°$

25) a) $\begin{cases}n\pi+\frac{1}{12}\pi\\n\pi+\frac{5}{12}\pi\end{cases}$

26) a) $0,180°,360°;54.7°,125.3°,$
 $234.7°,305.3°$
 b) $257.6°,349.8°$

27) a) $(6n\pm1)+\dfrac{3}{\pi}\arctan\frac{5}{12}$
 b) $\frac{1}{2}\pi,\frac{3}{2}\pi,0.67^c,2.47^c$

28) a) $n\pi,2n\pi\pm\frac{1}{6}\pi$
 b) (i) $73.7°,180°$ (ii) $1,\frac{1}{11}$

29) a) $\sin nt+\sqrt{3}\cos nt,$
 $2\sqrt{3}\sin nt+2\cos nt;$
 $4x^2+y^2-2\sqrt{3}xy=4$
 b) $199.5°,340.5°$

30) b) $51.3°,128.7°$

31) a) $51.7°,14.8°$
 b) $\pm90°,\pm41.4°,180°$

32) a) (i) $90°,210°,330°$
 (ii) $60°,90°,180°,270°,300°$
 c) $\dfrac{1}{2n+1};\dfrac{2n^2-1}{2n^2+4n+1}$

33) a) $1+\theta-\frac{1}{2}\theta^2$ b) $\frac{1}{2}$ c) $\frac{1}{2}\theta$
 d) $\dfrac{2}{\sqrt{2+\theta}}$

34) $0,\frac{1}{2}r^2,\dfrac{\sqrt{3}}{2}r^2$

35) $0,\frac{1}{4}\pi$

37) $\frac{1}{10}\sqrt{2}$

38) $\pm\frac{1}{3}\sqrt{3}$

Exercise 8a — p. 256

1) $-\sin x$

2) a) $\cos x-\sin x$ b) $\sin x$
 c) $2\cos\theta$ d) $-4\sin\theta$
 e) $2\cos\theta+3\sin\theta$ f) $4\cos t$

3) a) $-\sin x$ b) $\dfrac{\sqrt{2}}{2}(\cos x-\sin x)$
 c) $-\frac{1}{2}(\cos x+\sqrt{3}\sin x)$

4) a) -1 b) -1 c) $\dfrac{5\sqrt{2}}{2}$ d) $\dfrac{2\pi}{3}-\dfrac{1}{2}$

5) a) $\dfrac{2\pi}{3}$ b) 2.5^c c) 2.82^c

6) a) $2n\pi,2n\pi+1.107^c,2n\pi-0.927^c$
 b) $(2n+1)\pi,2n\pi+4.25^c$
 $2n\pi+2.215^c$

7) a) $5x+\sqrt{2}y-1-\dfrac{5\pi}{4}=0$
 b) $18y=(12\pi-27)x+9\pi$
 $-2\pi^2-27\sqrt{3}$

Exercise 8b – p. 261

1) 20.1, 0.135, 5.47, 0.819
2) a) $2e^x$ b) $2x - e^x$ c) $e^x - \sin x$
3) a) 0   b) -0.69
    c) no stationary values

14) a) $\dfrac{1}{\cos x}$ b) $\ln\left(\dfrac{1}{x}\right)$ c) $\cos(\ln x)$

d) $\cos\left(\dfrac{1}{\ln x}\right)$

Exercise 8c – p. 262

1) 2.0568, -0.7277, 3.0680
2) a) $\ln x - \ln(x + 1)$
    b) $\ln 3 + 2\ln x$
    c) $\ln(x + 2) + \ln(x - 2)$
    d) $\frac{1}{2}\ln(x - 1) - \frac{1}{2}\ln(x + 1)$
    e) $\ln \cos x - \ln \sin x$
    f) $\ln 4 + 2\ln \cos x$

3) a) $\ln\left[\dfrac{x}{(1 - x)^2}\right]$ b) $\ln\dfrac{e}{x}$

    c) $\ln \sin x \cos x$ d) $\ln\left[\dfrac{x^3}{\sqrt{(x - 1)}}\right]$

4) a) 2.10   b) 0   c) 1.05   d) 0

Exercise 8d – p. 264

1) a) $\dfrac{2}{x}$ b) $\dfrac{5}{x}$ c) $\dfrac{2}{x}$

    d) $-\dfrac{3}{2x}$ e) $-\dfrac{1}{2x}$ f) $\dfrac{1}{2x}$

    g) $-\dfrac{5}{2x}$ h) $\dfrac{-1}{4x}$ i) $\dfrac{1}{x \ln 10}$

2) a) $(1, 1)$ b) $(1, -1), (-1, -1)$

Exercise 8e – p. 266

1) $uv, u = e^x, v = x^2 - 1$
2) $e^u, u = 3x^2$
3) $\sin u, u = x^2 - 2$
4) $uv, u = x^3, v = \cos x$
5) $\dfrac{u}{v}, u = \sqrt{(x - 1)}, v = \sqrt{(x + 1)}$
6) $u^4, u = x + 1$
7) $\ln u, u = x^2 - 1$
8) $uv, u = x^2 - 1, v = (x - 2)^5$
9) $\tan u, u = x^2$
10) $\ln u - \ln v, u = x + 1, v = x - 1$
11) $u^2, u = \cos x$
12) $u^{\frac{1}{2}}, u = \ln x$
13) $f[g(x)] \equiv (x + 1)^2, g[f(x)] \equiv x^2 + 1$

Exercise 8f – p. 268

1) $3e^{3x}$
2) $\dfrac{2}{x - 1}$
3) $5(x + 1)^4$
4) $-3\sin\left(3\theta - \dfrac{\pi}{4}\right)$
5) $10x(x^2 + 1)^4$
6) $2xe^{x^2}$
7) $\dfrac{8x}{x^2 + 1}$
8) $3x(3x^2 + 4)^{-\frac{1}{2}}$
9) $10e^{2x}$
10) $2\theta \cos \theta^2$
11) $2\sin \theta \cos \theta$
12) $3e^{3x}$
13) $-(x + 1)^{-2}$
14) $\dfrac{-2x}{(x^2 + 1)^2}$
15) $-\dfrac{1}{2(x - 1)^{3/2}}$
16) $6(x + 1)^5$
17) $6(2x - 4)^2$
18) $2x \, e^{(x^2 + 2)}$
19) $3\cos\left(3\theta + \dfrac{\pi}{4}\right)$
20) $-2\cos \theta \sin \theta$
21) $\dfrac{2x}{x^2 + 2}$
22) $-4(2x + 3)^{-3}$
23) $-(2x - 1)^{-\frac{3}{2}}$
24) $e^x + e^{-x}$
25) $-e^{-x} + 2e^{-2x}$
26) $6\cos\left(2\theta - \dfrac{\pi}{4}\right)$
27) $-\dfrac{\cos \theta}{\sin^2 \theta}$
28) $\dfrac{\sin \theta}{\cos^2 \theta}$
29) $-6e^{-3x} + 4e^{4x}$
30) $\cot x$

31) $-k\,e^{-kt}$

32) $\dfrac{3}{x-1}$

33) $\cos x\,e^{\sin x}$

34) $\dfrac{-2\cos x}{\sin^3 x}$

35) $\dfrac{2\sin t}{\cos^3 t}$

36) $6\cos 3\theta\,e^{\sin 3\theta}$

37) $\dfrac{2}{x-1}$

38) $\dfrac{2\theta}{\sqrt{(\theta^2+4)}}\cos(\theta^2+4)^{\frac{1}{2}}$

39) $e^x\,e^{e^x}$

40) $-8x\cos(x^2+1)\sin(x^2+1)$

41) $3e^x\cos e^x$

42) $\dfrac{\sin\theta\,\cos\theta}{\sqrt{(3-\cos^2\theta)}}$

43) $2\cot 2x$

44) $\dfrac{-e^{\sqrt{1-x}}}{2\sqrt{1-x}}$

45) $4x\sin(x^2+1)\cos(x^2+1)$

46) $-\dfrac{2}{x^2}\,e^{1/x}$

47) $-\operatorname{cosec} x$

48) $-(e^x+e^{-x})(e^x-e^{-x})^{-2}$

Exercise 8g – p. 272

1) $1+\ln x$

2) $\dfrac{5x^2-4x}{2\sqrt{(x-1)}}$

3) $\dfrac{\ln x-1}{(\ln x)^2}$

4) $3\cos x\cos 2x-6\sin x\sin 2x$

5) $\sec^2 x$

6) $\dfrac{e^x(x-2)}{(x-1)^2}$

7) $1+\ln(x-1)$

8) $\cos x\,\ln x+\dfrac{1}{x}\sin x$

9) $\sec x(1+x\tan x)$

10) $-\dfrac{1}{\sin^2 x}$

11) $\dfrac{2}{(x+1)^2}$

12) $e^x(\sin x+\cos x)$

13) $\dfrac{e^x(\sin x-\cos x)}{\sin^2 x}$

14) $\dfrac{1+\ln x}{\ln 10}$

15) $\dfrac{-2}{(e^x-e^{-x})^2}$

16) $\dfrac{1-x^2}{(x^2+1)^2}$

17) $\dfrac{x^2-1}{x^2}$

18) $\dfrac{2-x}{2x^2\sqrt{(x-1)}}$

19) $\dfrac{2x\cos x-\sin x}{2x\sqrt{x}}$

20) $\dfrac{(x-1)\ln(x-1)-x\ln x}{x(x-1)\ln^2(x-1)}$

21) $-\operatorname{cosec}^2 x$

22) $\dfrac{-2x}{(x-1)^2(x+1)^2}$

23) $-\dfrac{2}{3(x-2)^2}-\dfrac{1}{3(x+1)^2}$

24) $\dfrac{4}{5(2x-1)^2}-\dfrac{6}{5(x-3)^2}$

25) $-\dfrac{7}{5(x+2)^2}-\dfrac{2}{5(2x-1)^2}$

26) $-\dfrac{5}{2(x-1)^2}+\dfrac{10}{(x-2)^2}-\dfrac{15}{2(x-3)^2}$

27) $\dfrac{10}{3(2x-1)^2}-\dfrac{1}{(x-1)^2}-\dfrac{1}{3(x+1)^2}$

28) $-\dfrac{9}{10(x+3)^2}-\dfrac{x^2-6x-1}{10(x^2+1)^2}$

29) $-\dfrac{1}{4(x+1)^2}+\dfrac{1}{4(x-1)^2}-\dfrac{1}{(x-1)^3}$

30) $\dfrac{9}{10(x+3)^2}-\dfrac{9x^2+6x-9}{10(x^2+1)^2}$

Exercise 8h – p. 273

1) $3\cos x\,e^{\sin x}$

2) $-e^{-x}\sin x+e^{-x}\cos x$

3) $\dfrac{-(1+2x)}{(1+x+x^2)^2}$

4) $3\sin^2 x-4\sin^4 x$

5) $\tan x$

6) $7 \cos 7t - \cos t$

7) $2e^{x^2} \left(x \ln 2x^2 + \dfrac{1}{x} \right)$

8) $4 \sec^2 2x \tan 2x$

9) $\dfrac{-1 - \sin x - \cos x}{(1 + \cos x)(1 + \sin x)}$

10) 4

12) $\dfrac{4}{(1-x)^3}$

13) -1

14) $\dfrac{e^x}{2x\sqrt{(x-1)}} [(2x^2 - x) \ln x + 2x - 2]$

15) $\dfrac{1}{\sqrt{(1-x^2)}}$

16) $-\dfrac{1}{2(\frac{1}{2}\pi - 1)^{3/2}}$

17) ± 1 18) 0.8

19) $\dfrac{7\pi}{4}$ 20) $-\dfrac{1}{e}$

15) $\pm \dfrac{1}{2\sqrt{2}}$

17) $\dfrac{dy}{dx} = -y \ln 2, -\frac{1}{4} \ln 2$

18) $\frac{1}{3}(\ln 3)^2$

19) $\frac{1}{2}$

22) a) $a^x \ln a + b^x \ln b$

b) $\left(\dfrac{1}{x} \sin x + \cos x \ln x \right) x^{\sin x}$

c) $(\sin x)^x (\ln \sin x + x \cot x)$

d) $(x + x^2)^x \left[\ln (x + x^2) + \dfrac{1 + 2x}{1 + x} \right]$

e) $\dfrac{x}{(x-1)(x+3)^2(x^2-1)}$

$\times \left(\dfrac{1}{x} - \dfrac{1}{x-1} - \dfrac{2}{x+3} - \dfrac{2x}{x^2-1} \right)$

f) $\dfrac{6 + 4x - 9x^2}{(x^2-1)^{3/2}(3x-4)^3}$

23) a) $\ln y + \dfrac{x}{y} \dfrac{dy}{dx} = \dfrac{1}{x}$

b) $\ln x \dfrac{dy}{dx} + \dfrac{y}{x} = \cot x$

c) $\dfrac{1}{y} \dfrac{dy}{dx} + \dfrac{1}{1+x} = \frac{1}{2} \cot x$

d) $\dfrac{\sin x}{y} \dfrac{dy}{dx} + \cos x \ln y = \dfrac{1}{2x}$

24) a) $xx_1 - 3yy_1 = 2(y + y_1)$

b) $x(2x_1 + y_1) + y(2y_1 + x_1) = 6$

26) $3x + 12y - 7 = 0$

Exercise 8i – p. 281

1) $2x + 2y \dfrac{dy}{dx} = 0$

2) $2x + y + (x + 2y) \dfrac{dy}{dx} = 0$

3) $2x + x \dfrac{dy}{dx} + y = 2y \dfrac{dy}{dx}$

4) $-\dfrac{1}{x^2} - \dfrac{1}{y^2} \dfrac{dy}{dx} = e^y \dfrac{dy}{dx}$

5) $-\dfrac{2}{x^3} - \dfrac{2}{y^3} \dfrac{dy}{dx} = 0$

6) $\dfrac{x}{2} - \dfrac{2y}{9} \dfrac{dy}{dx} = 0$

7) $\cos x + \cos y \dfrac{dy}{dx} = 0$

8) $\cos x \cos y - \sin x \sin y \dfrac{dy}{dx} = 0$

9) $e^y + xe^y \dfrac{dy}{dx} = 1$

10) $(1 + x) \dfrac{dy}{dx} = 2x - 1 - y$

11) $\dfrac{dy}{dx} = \pm \dfrac{1}{\sqrt{(2x + 1)}}$

14) $\dfrac{x}{(2 - x^2)^{3/2}}$

Exercise 8j – p. 288

1) a) $y^2 = 4x$ b) $x^2 + y^2 = 1$

c) $xy = 2$

3) a) $\dfrac{1}{t}$ b) $-\cot \theta$ c) $-\dfrac{1}{2t^2}$

4) a) $-\dfrac{1}{2t^3}$ b) $-\text{cosec}^3 \theta$ c) $\dfrac{1}{2t^3}$

5) $\left(\dfrac{\sqrt{3}}{3}, -\dfrac{2\sqrt{3}}{9} \right) \text{min}, \left(-\dfrac{\sqrt{3}}{3}, \dfrac{2\sqrt{3}}{9} \right) \text{max}$

6) $\left(2n\pi + \dfrac{\pi}{2}, 1 \right)$

7) $2x + y + 2 = 0$

8) $xt^3 - yt + 1 - t^4 = 0$

9) $2x - yt + 2t^2 = 0$

10) $x = y$, $-\dfrac{\sqrt{2}}{2}, -\dfrac{\sqrt{2}}{2}$

11) $e^{-t}\cos t, -e^{-2t}(\cos t + \sin t)$

12) $x + t^2 y - 2t = 0, (2t, 0), \left(0, \dfrac{2}{t}\right)$

13) $tx + 2y - 8t - t^3 = 0$,
$(8 + t^2, 0), (0, \tfrac{1}{2}t\{8 + t^2\})$

Multiple Choice Exercise 8 — p. 290

1) b	2) c	3) b	4) a
5) e	6) d	7) a	8) a
9) b	10) b	11) d	12) a,b
13) b	14) b,c	15) b,c	16) a
17) C	18) C	19) A	20) D
21) E	22) c	23) c	24) A
25) I	26) F	27) T	28) T
29) F	30) F	31) F	32) F
33) T			

Miscellaneous Exercise 8 — p. 294

1) a) $A = a^2 - b^2, B = 2ab$
   b) $16/(1 - 2x)^3$
   c) $\dfrac{x}{y}, (y^2 - x^2)/y^3$

2) $7x + 11y = 18$

3) a) $-\dfrac{(x^3 + 4)}{2x^3(1 + x^3)^{1/2}}$,
   $\dfrac{1 + 2\cos x - 3\sin x}{(2 + \cos x)(3 - \sin x)}$

4) a) (i) $2\sec 2x$ (ii) $2x/(1 + x^2)^2$
   b) $\ln(3\pi/4)$

5) $x^x(1 + \ln x), -1/(1 + x^2)$

6) a) $\left(\dfrac{1}{e}, -\dfrac{1}{e}\right)$ min
   b) $(0, 2), (0, -2), 3x - 5y + 8 = 0$

9) a) $\dfrac{\sin 2x}{x^3 e^{3x}} - \dfrac{\cos 2x}{x^2 e^{3x}} - 3y$ b) 1

10) -2 11) -1.15

13) $1 + \dfrac{1}{3(2x - 1)} - \dfrac{2}{3(x + 1)}, \dfrac{5}{2}$

14) a) $\tan x + a^x \ln a$ b) $\tfrac{1}{9}$ max, 1 min

15) b) (iii) $(3t^2 + 3)/2t$

16) a) (i) $1/\sqrt{(x^2 + 1)}$
    (ii) $4\sec^2 2x \tan 2x$
    (iii) $(3\ln 10) 10^{3x}$
    b) $\dfrac{dy}{dx} = \dfrac{x}{1 - y}$

17) a) $-2e^{-2t}(1 - 2t)^2$ b) $2/(1 + 4t^2)$

19) $-\tfrac{1}{2}$

20) a) (i) $2x\sin 3x + 3x^2\cos 3x$
    (ii) $\dfrac{2}{x^2} e^{-2/x}$ (iii) $\dfrac{2(x - 1)}{(2 - x)^3}$
    b) max

21) $\left\{\begin{array}{l} (x - 1)\ln(1 - x) \\ \quad - (x + 1)\ln(x + 1) - 2x \end{array}\right\}$
    $\div (1 - x^2)(2 - 2\ln[1 - x])^2$

22) $\left(a\left\{\dfrac{\pi}{2} - 1\right\}, a\right), \left(a\left\{\dfrac{3\pi}{2} + 1\right\}, a\right)$,
    $y = x - a\left(\dfrac{\pi}{2} - 2\right)$,
    $y = -x + a\left(\dfrac{3\pi}{2} + 2\right)$

23) $-2\sqrt{(1 - x^2)}$

24) f does not have an inverse as it is
    defines a one–two mapping but g does
    have an inverse as it defines a one–one
    mapping. $\pi/6$.

Note. A constant of integration should be
added to every indefinite integral in
Chapter 9

Exercise 9a — p. 301

1) $\tfrac{1}{7}x^7; \tfrac{3}{4}x^{\frac{4}{3}}; -\tfrac{1}{3}x^{-3}; -2x^{-\frac{1}{2}}; \tfrac{4}{5}x^{\frac{5}{4}}; -\tfrac{1}{6}x^{-6}$

2) $\dfrac{2}{3}x^3 + \dfrac{1}{x} + \dfrac{x^2}{2}$

3) $\tfrac{2}{5}x^{\frac{5}{2}} + \tfrac{3}{2}x^{\frac{2}{3}}$

4) $\tfrac{3}{4}x^4 - \tfrac{1}{2}x^{-2} + 3x$

5) $-\dfrac{2}{x^2} + \dfrac{1}{x} - \dfrac{x^3}{3}$

6) $\tfrac{4}{7}x^{\frac{7}{2}} - \tfrac{5}{3}x^{\frac{3}{5}}$

7) $x^5 - x^3 + 7x$

8) $-2x^{-2} - \tfrac{1}{3}x^{-3} + x$

9) $6x^{\frac{1}{2}} + 2x^{-\frac{1}{2}}$

10) $\tfrac{1}{4}x^2 - 4\sqrt{x} - x$

11) $-\tfrac{1}{3}x^{-3} + \tfrac{4}{3}x^{\frac{3}{4}} - 4x$

12) $4x^{\frac{3}{2}} - \tfrac{3}{4}x^4 - x^{-1} + 2x$

Exercise 9b — p. 302

1) $3x; \tfrac{1}{2}x^6; \tfrac{81}{2}x^6; \tfrac{1}{2}(x + 1)^6$

2) $x^4 - \tfrac{5}{2}x^2 + 6x$

3) $\tfrac{1}{6}(2x - 1)^3$

4) $\frac{1}{28}(2 + 7x)^4$

5) $-\frac{8}{3}(1 - x)^{\frac{3}{2}}$

6) $-\frac{2}{5}\sqrt{(2 - 5x)^3}$

7) $-\dfrac{1}{8(4x + 5)^2}$

8) $-\sqrt{1 - 2x}$

9) $-\frac{3}{14}(3 - 7x)^{\frac{2}{3}}$

10) $-\dfrac{3}{4(2x + 1)^2} - \dfrac{\sqrt{(1 - 2x)^3}}{3}$

11) $\frac{8}{3}\sqrt{x^3} + \frac{1}{6}\sqrt{(4x + 1)^3} + \frac{1}{3}(1 - 3x)^4$

12) $\dfrac{(px + q)^{r+1}}{p(r + 1)}$

13) $\frac{4}{15}(3 + 5x)^{\frac{3}{4}}$

14) $-\frac{9}{10}(4 - 5x)^{\frac{2}{3}}$

15) $-\frac{2}{3}\sqrt{(1 - x)^3} - 2\sqrt{(1 - x)} - \dfrac{1}{1 - x}$

Exercise 9c – p. 303

1) $-\frac{1}{2}\cos 2x$

2) $\frac{3}{4}\sin(4x - \frac{1}{2}\pi)$

3) $-\frac{1}{2}\tan(\frac{1}{3}\pi - 2x)$

4) $-\frac{1}{3}\cot(3x + \frac{1}{6}\pi)$

5) $-5\sec(\frac{1}{4}\pi - x)$

6) $-\frac{2}{3}\operatorname{cosec} 3x$

7) $-\frac{2}{3}\cos(3x + \alpha)$

8) $-10\sin(\alpha - \frac{1}{2}x)$

9) $\frac{1}{3}\sin 3x - 3\cos x$

10) $\frac{1}{2}\tan 2x + \frac{1}{4}\cot 4x$

Exercise 9d – p. 305

1) $\frac{1}{2}e^{2x}$

2) $-3e^{-x}$

3) $\frac{1}{4}e^{4x+1}$

4) $-\frac{4}{3}e^{5-3x}$

5) $\dfrac{2^x}{\ln 2}$

6) $\dfrac{3^{2x}}{2\ln 3}$

7) $-\dfrac{3^{1-x}}{\ln 3}$

8) $2\sqrt{e^x}$

9) $\frac{1}{2}(e^{2x} - e^{-2x})$

10) $-e^{-3x} - \frac{1}{4}e^{2x}$

Exercise 9e – p. 306

1) $\frac{1}{2}\ln|x|$

2) $2\ln|x|$

3) $\frac{1}{3}\ln|3x + 1|$

4) $-\frac{1}{3}\ln|1 - 3x|$

5) $2\ln|1 + 2x|$

6) $-\frac{3}{2}\ln|4 - 2x|$

7) $6\ln|x - 2| - 3\ln|x - 1|$

8) $\dfrac{1}{6(2 - 3x)^2}$

9) $-\frac{2}{3}\sqrt{2 - 3x}$

10) $-\frac{1}{3}\ln|2 - 3x|$

11) $\frac{1}{3}\cos(\frac{1}{2}\pi - 3x)$

12) $\frac{1}{12}(4x + 1)^3$

13) $\frac{1}{3}x^3$

14) $\dfrac{4^x}{\ln 4}$

15) $-\frac{1}{5}e^{4 - 5x}$

16) $\frac{1}{4}\tan 4x$

17) $-4\ln|1 - x|$

18) $\frac{1}{3}\sqrt{(2x + 3)^3}$

19) $\frac{1}{6}e^{6x}$

20) $-\frac{5}{7}\ln|6 - 7x|$

Exercise 9f – p. 307

1) $-\arcsin x$

2) $\arctan 2x$

3) $\arcsin 3x$

4) $3\arctan x$

5) $\frac{1}{4}\arctan 4x$

6) $3\arcsin 2x$

7) $\frac{1}{2}\arctan \frac{1}{2}x$

8) $\arcsin \frac{1}{2}x$

Exercise 9g – p. 310

1) e^{x^4}

2) $-e^{\cos x}$

3) $e^{\tan x}$

4) $e^{x^2 + x}$

5) $e^{1 - \cot x}$

6) $\frac{1}{10}(x^2 - 3)^5$

7) $-\frac{1}{3}\sqrt{(1 - x^2)^3}$

8) $\frac{1}{6}(\sin 2x + 3)^3$

9) $-\frac{1}{6}(1 - x^3)^2$

10) $\frac{2}{3}\sqrt{(1 + e^x)^3}$

11) $\frac{1}{5}\sin^5 x$

12) $\frac{1}{4}\tan^4 x$

13) $\dfrac{1}{3(n + 1)}(1 + x^{n+1})^3$

14) $-\frac{1}{3}\cot^3 x$

15) $\frac{4}{9}\sqrt{(1 + x^{\frac{3}{2}})^3}$

16) $\frac{1}{12}(x^4 + 4)^3$

17) $-\frac{1}{4}(1 - e^x)^4$

18) $\frac{2}{3}\sqrt{(1 - \cos\theta)^3}$

19) $\frac{1}{3}\sqrt{(x^2 + 2x + 3)^3}$

20) $\frac{1}{2}e^{x^2 + 1}$

Exercise 9h – p. 316

1) $x\sin x + \cos x$

2) $e^x(x^2 - 2x + 2)$

3) $\frac{1}{16}x^4(4 \ln 3x - 1)$

4) $-e^{-x}(x + 1)$

5) $3(\sin x - x \cos x)$

6) $\frac{1}{5}e^x(\sin 2x - 2 \cos 2x)$

7) $\frac{1}{5}e^{2x}(\sin x + 2 \cos x)$

8) $\frac{1}{32}e^{4x}(8x^2 - 4x + 1)$

9) $-\frac{1}{2}e^{-x}(\cos x + \sin x)$

10) $x(\ln 2x - 1)$

11) $x\, e^x$

12) $\frac{1}{72}(8x - 1)(x + 1)^8$

13) $\sin\left(x + \frac{\pi}{6}\right) - x \cos\left(x + \frac{\pi}{6}\right)$

14) $\dfrac{1}{n^2}(\cos nx + nx \sin nx)$

15) $\dfrac{x^{n+1}}{(n + 1)^2}[(n + 1) \ln |x| - 1]$

16) $\frac{3}{4}(2x \sin 2x + \cos 2x)$

17) $\frac{1}{5}e^x(\sin 2x - 2 \cos 2x)$

18) $(2 - x^2) \cos x + 2x \sin x$

19) $\dfrac{e^{ax}}{a^2 + b^2}(a \sin bx - b \cos bx)$

20) $\frac{1}{3} \sin \theta\, (3 \cos^2 \theta + 2 \sin^2 \theta)$

21) $\frac{1}{2}e^{x^2 - 2x + 4}$

22) $(x^2 + 1)\, e^x$

23) $-\frac{1}{4}(4 + \cos x)^4$

24) $e^{\sin x}$

25) $\frac{2}{15}\sqrt{(1 + x^5)^3}$

26) $\frac{1}{5}(e^x + 2)^5$

27) $\cos\left(\frac{\pi}{4} - x\right) - x \sin\left(\frac{\pi}{4,} - x\right)$

28) $\frac{1}{4}e^{2x - 1}(2x - 1)$

29) $-\frac{1}{20}(1 - x^2)^{10}$

30) $\frac{1}{6} \sin^6 x$

Exercise 9i – p. 320

1) $\ln |3 + \sin x|$

2) $-\ln |1 - e^x|$

3) $\ln (1 + x^2)$

4) $-\frac{1}{3} \ln |1 - 3 \tan x|$

5) $-\ln (e^{-x} + 1)$

6) $\frac{1}{4} \ln (1 + x^4)$

7) $\frac{1}{4} \ln |2 \sin 2x + 1|$

8) $\ln |\sec x + 1|$

9) $\ln |\ln x|$

10) $\ln |x^2 + 3x + 4|$

11) $\ln |\sin x|$

12) $\ln |\tan x + \sec x|$

13) $-\ln |\cot x + \operatorname{cosec} x|$

14) $\sqrt{x^2 + 1}$

15) $-\dfrac{1}{5 \sin^5 x}$

16) $2\sqrt{1 + \sin x}$

17) $-\dfrac{1}{e^x + 4}$

18) $-\dfrac{1}{2 \tan^2 x}$

19) $\dfrac{1}{(n - 1) \cos^{n-1} x}$

20) $\dfrac{-1}{(n - 1) \sin^{n-1} x}$

21) $2\sqrt{1 + e^x}$

22) $-\ln |3 - \sec x|$

23) $\dfrac{1}{3(2 + \cot x)^3}$

24) $\frac{1}{6} \ln |3x^2 - 6x + 1|$

25) $x \arcsin x + \sqrt{1 - x^2}$

26) $x \arctan x - \frac{1}{2} \ln (1 + x^2)$

27) $x \arccos x - \sqrt{1 - x^2}$

28) $x - \ln |x + 1|$

29) $x - \frac{1}{2} \ln \left|\dfrac{x - 1}{x + 1}\right|$

30) $x + \ln \left|\dfrac{x + 1}{(x + 2)^4}\right|$

31) $x + 4 \ln |x|$

32) $x + 3 \ln |x + 1|$

33) $\ln |x^2 - 4|$

34) $\ln \left|\dfrac{u^4}{u + 1}\right|$

35) $x + 5 \ln \left|\dfrac{x}{x + 1}\right|$

36) $\ln \left|\dfrac{y - 2}{(y - 1)^2}\right|$

37) $\ln |z^2 - 5z + 6|$

38) $12 \ln \left|\dfrac{(3 - x)^3}{(2 - x)(4 - x)^2}\right|$

39) $\frac{1}{2} \ln \left|\dfrac{(x - 1)^7 (x + 1)}{(2x - 1)^7}\right|$

40) $2u - \ln |u + 1| - \frac{5}{2} \ln |2u + 3|$

Exercise 9j – p. 324

1) a) $\frac{3}{8}\theta - \frac{1}{4} \sin 2\theta + \frac{1}{32} \sin 4\theta$

b) $\sin \theta - \frac{1}{3} \sin^3 \theta$

c) $\frac{1}{3} \tan^3 \theta - \tan \theta + \theta$

d) $\frac{1}{3} \cos^3 \theta - \cos \theta$

734   Mathematics – The Core Course for A-level

e) $\frac{3}{8}\theta + \frac{1}{4}\sin 2\theta + \frac{1}{32}\sin 4\theta$

f) $\frac{1}{4}\tan^4\theta - \frac{1}{2}\tan^2\theta - \ln|\cos\theta|$

2) a) $-(\frac{1}{7}\cos 7\theta + \cos\theta)$

b) $\frac{1}{21}(3\sin 7\theta + 7\sin 3\theta)$

c) $\frac{1}{16}(2\cos 4\theta - \cos 8\theta)$

d) $\frac{1}{8}(2\sin 2\theta - \sin 4\theta)$

e) $-\dfrac{1}{n+m}\cos(n+m)x$

$-\dfrac{1}{n-m}\cos(n-m)x$

f) $\dfrac{6}{5}\left(5\sin\dfrac{u}{6} + \sin\dfrac{5u}{6}\right)$

g) $\dfrac{1}{2(n+m)}\sin(n+m)x$

$+\dfrac{1}{2(n-m)}\sin(n-m)x$

3) a) $\frac{1}{3}\sin^3 x - \frac{1}{5}\sin^5 x$

b) $\frac{1}{11}\sin^{11}x - \frac{1}{13}\sin^{13}x$

c) $\frac{1}{32}(4x - \sin 4x)$

d) $\frac{1}{5}\tan^5 x + \frac{1}{3}\tan^3 x$

e) $\frac{1}{10}\cos^5 2x - \frac{1}{6}\cos^3 2x$

f) $\dfrac{1}{n+1}\sin^{n+1}x - \dfrac{1}{n+3}\sin^{n+3}x$

g) $\frac{1}{5}\cos^5 x - \frac{1}{3}\cos^3 x$

h) $\frac{1}{4}\tan^4 x$

Exercise 9k – p. 328

1) $\frac{1}{21}(x+3)^6(3x+2)$

2) $\frac{1}{2}\arctan\frac{x}{2}$

3) $-\frac{2}{3}(x+6)\sqrt{3-x}$

4) $\frac{2}{15}(3x-2)(x+1)^{\frac{3}{2}}$

5) $\dfrac{1-5x}{10(x-3)^5}$

6) $\ln|x + \sqrt{1+x^2}|$

7) $\frac{4}{135}(9x+8)(3x-4)^{\frac{3}{2}}$

8) $\frac{3}{10}\arctan\frac{2}{5}x$

9) $\ln|x+2 + \sqrt{(1-\{x+2\}^2)}|$

10) $-\frac{1}{36}(8x+1)(1-x)^8$

11) $\arcsin\dfrac{x}{3}$

12) $\dfrac{5+4x}{12(4-x)^4}$

13) $\frac{1}{2}e^{2x+3}$

14) $\frac{1}{6}(2x^2-5)^{\frac{3}{2}}$

15) $\frac{1}{12}(6x - \sin 6x)$

16) $-\frac{1}{2}e^{-x^2}$

17) $-\frac{1}{8}(\cos 4\theta + 2\cos 2\theta)$

18) $\frac{1}{110}(10u - 7)(u+7)^{10}$

19) $-\dfrac{1}{12(x^3+9)^4}$

20) $\frac{1}{2}\ln|1 - \cos 2y|$

21) $\frac{1}{2}\ln|2x+7|$

22) $\arcsin u$

23) $-\frac{2}{9}(1+\cos 3x)^{\frac{3}{2}}$

24) $\frac{1}{16}(\sin 4x - 4x\cos 4x)$

25) $\frac{1}{2}\ln|x^2+4x-5|$

26) $\frac{1}{3}\ln|(x+5)^2(x-1)|$

27) $-\frac{1}{4}(x^2+4x-5)^{-2}$

28) $-(9-y^2)^{\frac{3}{2}}$

29) $\frac{1}{13}e^{2x}(2\cos 3x + 3\sin 3x)$

30) $x(\ln|5x| - 1)$

31) $\frac{1}{6}\sin 2x(3 - \sin^2 2x)$

32) $-e^{\cot x}$

33) $-2\sqrt{7+\cos y}$

34) $e^x(x^2 - 2x + 2)$

35) $\frac{1}{2}\ln|x^2-4|$

36) $x + \ln\left|\dfrac{x-2}{x+2}\right|$

37) $\frac{1}{4}\ln\left|\dfrac{x-2}{x+2}\right|$

38) $\frac{1}{30}(3\sin 5x + 5\sin 3x)$

39) $-\frac{1}{30}\cos 2\theta(15 - 10\cos^2 2\theta + 3\cos^4 2\theta)$

40) $\frac{1}{15}\cos^3 u(3\cos^2 u - 5)$

41) $\frac{1}{3}\tan^3\theta - \tan\theta + \theta$

42) $\frac{1}{4}\tan^4\theta - \frac{1}{2}\tan^2\theta - \ln\cos\theta$

43) $\arcsin x + 2\sqrt{1-x^2}$

44) $\ln|\ln u|$

45) $\frac{1}{27}(9y^2\sin 3y + 6y\cos 3y - 2\sin 3y)$

46) $-\ln|1 - \tan x|$

47) $\frac{1}{3}(7+x^2)^{\frac{3}{2}}$

48) $-\frac{1}{5}\cos\left(5\theta - \dfrac{\pi}{4}\right)$

49) $\sin\theta\{\ln|\sin\theta| - 1\}$

50) $e^{\tan u}$

51) $\dfrac{2x-1}{10(3-x)^6}$

52) $\frac{1}{3}\tan^3 x$

Exercise 9l – p. 331

1) $2y = 3x^2 - 8x + 9$

2) $2y = 2x^3 - 5x^2 + 2x + 6$

3) $y = 3e^{2x} - 1$

4) $15(1-y) = (7-5x)^3$

5) $3y = \sin 3x + 4$
6) $y = x(x - 1)$
7) $\left(0, \dfrac{7 - e^3}{3}\right)$

Exercise 9m – p. 336

1) $y^2 = A - 2\cos x$
2) $x + \dfrac{1}{y} = A$
3) $\ln(x^2 + 1) = 2y + A$
4) $y = A(x - 3)$
5) $x + A = 4\ln|\sin y|$
6) $u^2 = v^2 + 4v + A$
7) $16y^3 = 12x^4 \ln|x| - 3x^4 + A$
8) $y^2 + 2(x + 1)e^{-x} = A$
9) $\sin x = A - e^{-y}$
10) $2r^2 = 2\theta - \sin 2\theta + A$
11) $u + 2 = A(v + 1)$
12) $y^2 = A + (\ln|x|)^2$
13) $y^2 = Ax(x + 2)$
14) $4v^3 = 3(2 + t)^4 + A$
15) $x^2 = A(1 + y^2)$
16) $e^{-x} = e^{1-y} + A$
17) $A - \dfrac{1}{y} = 2\ln|\tan x|$
18) $Ar = e^{\tan\theta}$
19) $y^2 = A - \text{cosec}^2 x$
20) $v^2 + A = 2u - 2\ln|u|$
21) $2y^2 = x^2 - 1$
22) $e^t(5 - 2\sqrt{s}) = 1$
23) $y^3 = x^3 + 3x - 13$
24) $(6 - y)x = 4$
25) $2(x - 1) = (x + 1)(y + 1)^2$
26) $y = e^x - 2$
27) $s\dfrac{ds}{dt} = k$ 28) $\dfrac{dh}{dt} = k\ln|H - h|$
29) a) $\dfrac{dn}{dt} = 0$ b) $\dfrac{dn}{dt} = k\sqrt{n}$
30) a) $\dfrac{dn}{dt} = K_1 n$ b) $\dfrac{dn}{dt} = \dfrac{K_2}{n}$
c) $\dfrac{dn}{dt} = -K_3$

Exercise 9n – p. 343

1) $10\frac{1}{3}$
2) 0
3) $\frac{1}{4}(e^4 - 1)$
4) $\ln\frac{5}{4}$
5) $12\frac{2}{3}$
6) $-\frac{1}{3}$
7) $\frac{1}{24}(2\pi - 3\sqrt{3})$
8) $\frac{1}{2}e(e^3 - 1)$
9) $\frac{31}{160}$
10) $\ln 2$
11) 1
12) $\ln\frac{32}{27}$

13) $\ln\sqrt{\frac{5}{2}}$
14) $\frac{1}{4}$
15) $\frac{31}{15}$
16) $\ln 16 - \frac{15}{16}$
17) $\frac{2}{3}$
18) 8
19) 5.14
20) $\frac{1}{2}\left(e^{\frac{\pi}{2}} + 1\right)$
21) 1
22) $\frac{35}{72}$
23) $\frac{1}{3}$
24) $\frac{1}{27}(8\sqrt{3} - 9)$
25) $\frac{1}{3}\pi$
26) $\frac{1}{48}\pi$
27) $-\frac{2}{3}$
28) $\dfrac{16\sqrt{2}}{3}$
29) $4 - 2\sqrt{2}$
30) $\frac{1}{10}\ln\frac{9}{4}$

Exercise 9p – p. 348

1) 12
2) $2\ln 4$
3) $2 - 3\ln 4$
4) $\frac{4}{3}$
5) 36
6) 1
7) $\ln\sqrt{2}$
8) $\frac{2}{3}$
9) $e - 1$
10) 36
11) a) -2 b) 2 c) 0
12) 2

Multiple Choice Exercise 9 – p. 349

1) d   2) d   3) b   4) c
5) b   6) c   7) d   8) d
9) b   10) a, b   11) b, c   12) a, d
13) b   14) a, c   15) b   16) e
17) B   18) D   19) C   20) B
21) A   22) a or c   23) a   24) F
25) F   26) T   27) F

Miscellaneous Exercise 9 – p. 354

1) a) $2 + \ln 2$ b) $\frac{1}{5}$
2) a) $\frac{1}{6}$ b) $\frac{1}{2}\ln\frac{3}{2}$
3) a) $\frac{1}{12}(1 + x^2)^6$ b) $-\frac{1}{4}e^{-2x}(2x + 1)$
c) $\frac{1}{2}\ln\frac{32}{27}$ d) $\frac{2}{15}$
4) a) (i) $\frac{1}{4}e^{4x} - e^{2x} + x$ (ii) $2\sqrt{\sin\theta}$
b) (i) $\frac{2}{3}$ (ii) $\ln\frac{12}{7}$
5) a) $\frac{1}{2}\ln 2$ b) $\frac{4}{15}$
6) a) 78 b) $\frac{1}{3}\ln 4$ c) $\ln 2$
7) a) 0.13 b) $\frac{1}{12}\ln\left|\dfrac{2 + 3\tan x}{2 - 3\tan x}\right|$
8) a) $\frac{15}{8} - \ln 4$ b) $\frac{7}{3} - \sqrt{2}$ c) $\frac{1}{2}\ln\frac{12}{5}$
9) a) $\frac{3}{4}\ln|x - 3| + \frac{1}{4}\ln|x + 1|$
b) $\frac{16}{105}$ c) $y = 1 - e^{-x}$
10) a) 0.275 b) 0.158 c) 0.693
11) a) $\ln 6$ b) $\arccos\dfrac{1}{\sqrt{5}} - \dfrac{1}{4}\pi - \dfrac{1}{10}$
12) a) $2\ln|x + 1| + \dfrac{3}{x + 1}$;
$\frac{1}{2}(e^{2x} - 4x - e^{-2x})$
b) $1 - \frac{1}{4}\pi$

13) a) $2 \ln \frac{7}{3}$ b) $4(\sqrt{7} - \sqrt{3})$
    c) $2 - \frac{3}{2} \ln \frac{7}{3}$

14) b) (i) $-\ln (1 - \sqrt{x})^2$

    (ii) $\frac{2}{9}(1 - 3 \cos^2 x)^{3/2}$

15) a) 0.068   b) $\frac{1}{4}(1 + \sqrt{3})$ c) $\frac{62}{5}, \frac{116}{15}$

16) a) $\frac{1}{2} \cos 2x - \frac{1}{4} \cos 4x;$

    $\frac{2}{3}(x - 1)^{\frac{3}{2}} - 2(x - 1)^{\frac{1}{2}}$

    b) $\frac{1}{18}\pi$

17) a) $\ln \dfrac{x}{\sqrt{(2x + 1)}}$

    b) $x^2 - x + \ln |x + 1|$

18) a) 2   b) $-\dfrac{1}{n}, 0, \dfrac{1}{n}$ c) 0.752

19) $\dfrac{3}{5(x + 1)} - \dfrac{1}{5(2x - 3)}, \frac{3}{5} \ln 4 - \frac{7}{10} \ln 3$

20) a) $\frac{3}{2} \ln |2x + 1| - \ln |x - 2|$ b) $\frac{5}{24}$

21) a) $\frac{1}{4}$ b) $\frac{1}{9}(1 - 4e^{-3})$

22) a) $-\frac{2}{5}$ b) $\frac{1}{16}\pi$ c) 1   d) $\ln \frac{32}{27}$

23) a) $\frac{1}{4}$ b) $\frac{1}{3} \sin^3 \theta - \frac{1}{5} \sin^5 \theta$

    c) $\frac{2}{3}(3 + x)^{\frac{1}{2}}(x - 6)$

24) a) $\ln 6$ b) $\frac{1}{9}\pi$ c) 6

25) $y = x e^{1 - x^2}$

26) a) $\ln \frac{2}{3}$ b) $\frac{1}{4}(2x - 1) e^{2x}$

    c) $2 \ln \left| \dfrac{y}{2} \right| = x^2$

27) $1 + e^{-y} = 2\sqrt{2} \cos x$

28) $\dfrac{8x}{3x + 5} = 20 \arctan 2y + 5 \ln (1 + 4y^2)$

29) a) (i) $\frac{1}{2} + \frac{1}{4} \ln 3$ (ii) $\frac{1}{3}$ b) $\frac{4}{3}$

30) $2y = (y + 1)\sqrt{1 + x^2}$

31) $y^2 = x^2 + A$ 32) $y^2 = 4x^2 + 5$

33) a) $y = (1 + x)^2 e^{-x}$
    b) $y^2 = 2x$

34) $y = \dfrac{4x}{3 - x}$

35) a) (i) $\frac{1}{2}x^2 - 2x + 4 \ln |x + 2|$
       (ii) $-\frac{1}{10}(\cos 5x + 5 \cos x)$

    b) $(\ln 3, 3); 2 \ln 3$

36) $2(\sqrt{3} - \frac{1}{3}\pi)$ 37) $\frac{1}{4}\pi$

38) $1, 4; \frac{45}{4}$

40) a) $y = (2e^{-x^2/2} - 1)^{-1}$

    b) $\dfrac{dy}{dx} = 1 - \dfrac{dz}{dx}; y = e^{-x} + x - 1$

41) $a = x + y; \dfrac{2}{ka} \ln 9$

1) a) $(1, 5)$ b) $(6, \frac{19}{2})$ c) $(5, 1)$
   d) $(11, -18)$

3) a) $\dfrac{\sqrt{5}}{5}$ b) $\frac{1}{5}$ c) $\left| \dfrac{ma - b + c}{\sqrt{(m^2 + 1)}} \right|$

   d) $\dfrac{3\sqrt{2}}{2}$ e) $\dfrac{ax + by + c}{\sqrt{(a^2 + b^2)}}$

   f) $\left| \dfrac{c}{\sqrt{(a^2 + b^2)}} \right|$

4) a) opposite   b) same   c) same

5) a) $\frac{15}{16}$ b) $\frac{5}{31}$ c) $\left| \dfrac{a_1 b_2 - a_2 b_1}{a_1 a_2 + b_1 b_2} \right|$

6) $\dfrac{3}{14}, \dfrac{9}{83}, \dfrac{3}{29}; \dfrac{9}{\sqrt{170}}, \dfrac{9}{\sqrt{41}}, \dfrac{3}{\sqrt{5}}$

7) a) $(-1, 8)$ b) $(-11, 10)$
   c) $(-3, -14)$

8) $5y = x, y + 5x = 0$

9) $4X^2 + Y^2 - 14X + 12Y - 4XY + 11 = 0$

10) no

11) $14X + 112Y = 93$ and $64X - 8Y = 63$

12) $x + 2y = 7$

1) $y^2 = 2y + 12x - 13$
2) $x + y = 4$
3) $8x^2 + 9y^2 - 20x = 28$
4) $11y = 3x$ and $99x + 27y + 130 = 0$
5) $x^2 + y^2 - 6x - 10y + 30 = 0$
6) $4x - 3y = 26$ and $4x - 3y + 24 = 0$
7) $x^2 + y^2 = 1$

1) a) $x^2 + y^2 - 2x - 4y = 4$
   b) $x^2 + y^2 - 8y + 15 = 0$
   c) $x^2 + y^2 + 6x + 14y + 54 = 0$
   d) $x^2 + y^2 - 8x - 10y + 32 = 0$

2) a) $(-4, 1); 5$

   b) $(-\frac{1}{2}, -\frac{3}{2}); \dfrac{3\sqrt{2}}{2}$

   c) $(-3, 0); \sqrt{14}$ d) $(\frac{3}{4}, -\frac{1}{2}); \dfrac{\sqrt{5}}{4}$

   e) $(0, 0); 2$ f) $(2, -3); 3$

   g) $(1, 3); 3$ h) $(-1, \frac{1}{2}); \dfrac{\sqrt{69}}{6}$

3) a) yes   b) yes   c) no   d) yes

4) a) $3y + 4x = 23$ b) $4y = x + 22$
   c) $3y = 2x + 25$

5) a) touch ext.   b) orthog.
   c) touch int.   d) neither
   e) orthog.
6) $9 \pm 6\sqrt{5}$
7) $2mac = a^2 - c^2$
8) $x^2 + y^2 + 4x + 4y + 4 = 0$ or
   $196(x^2 + y^2) + 84(x + y) + 9 = 0$
9) $2x^2 + 2y^2 - 15x - 24y + 77 = 0$
10) $x^2 + y^2 - 8x + 6y = 0$
11) $x^2 + y^2 - 4ax - 6by + 3a^2 + 5b^2 = 0$
12) $x^2 + y^2 - 4x - 6y + 9 = 0$
13) $x^2 + y^2 - 4x - 14y = 116$
14) $x^2 + y^2 - 6y = 0$
15) $y = 0, 8y = 15x$
16) $2y + x = 0$
17) $2x + y = 11, 2x + y + 9 = 0$
18) $(-\frac{4}{3}, \frac{2}{3}); \dfrac{4\sqrt{5}}{3}$

Exercise 10d p. 383

1) a) $x^2 + y^2 - 10x - 10y - 2xy = 5$
   b) $x^2 - 6x - 8y + 25 = 0$
   c) $y^2 + 4ax = 0$
2) a) $y = x + 1; y + x = 3$
   b) $x = 0; y = 1$
   c) $y = 8x - 2; 16y + 2x = 33$
   d) $2y = x - 4; 2x + y + 12 = 0$

Exercise 10e – p. 391

1) touch at $(1, 2)$
2) $(1, 1)$ and $(-5, 4)$
3) no intersection
4) $(0, 2)$ and $(\frac{8}{5}, \frac{14}{5})$
5) $(1, 0)$ and $(2, 0)$
6) $(0, 0), (-3, 0)$ and $(-2, 0)$
7) touch at $(1, 0)$ and $(2, 0)$
8) touch (inflexion) at $(-3, 0)$;
   meet at $(-2, 0)$
9) touch at $(0, 0)$
10) $k = 8$ 11) $k = \frac{1}{4}$
12) $k = \pm 2\sqrt{19}$ 13) $(-1, 1)$
14) $(0, 0)$ 15) $(-\frac{7}{2}, -\frac{7}{6})$
16) $(\frac{1}{2}, 2)$
17) $(1, 2)$ and $(1, -2)$
18) touch at $(-1, -2)$; cut at $(1, 2)$
    and $(4, \frac{1}{2})$
19) $(0, 0)$ and $(\frac{2}{3}, 2)$
20) a) $\pm 2\sqrt{3}$ b) $-2\sqrt{3} < \lambda < 2\sqrt{3}$
    c) $\lambda < -2\sqrt{3}, \lambda > 2\sqrt{3}$
22) a) (i) $y = \pm(x - 1)$
    (ii) $2y + x = 4$
    b) (i) $y = 0, y = 36(1 - x)$
    (ii) $2y + x = \pm 6\sqrt{2}$

c) (i) $3y = 2(x - 1), y = 0$
   (ii) $8y + 4x + 3 = 0$

Exercise 10g – p. 403

1) a) $y^2 - 6y + 8 = x$
   b) $x^2 = y^3$
   c) $xy = 9$
   d) $y = x^2 - 3x + 2$
   e) $\dfrac{x^2}{4} + \dfrac{y^2}{9} = 1$
   f) $y^2 = 4(x^2 - 1)$
   g) $2y^2 = a(x + a)$
3) $y^2 = 8x$ 4) $2y = x + 8$
5) $(p + q)y = 2x + 4pq$
6) $(18, 12); (2, -4)$ 7) $k = 2$
8) $(2, 4); (2, -4)$ 9) $(8, 8)$
10) $t_1 t_2 y + x = c(t_1 + t_2)$
11) $(p + q)y = 2x + 2apq$
12) $(T^2 + 2T + 4)y = (T + 2)x + 4T^2$
13) a) $at\sqrt{t^2 + 4}$ b) $\dfrac{c}{t}\sqrt{t^4 + 1}$
    c) $(t^4 + 3t^2 - 2t + 2)^{\frac{1}{2}}$
    d) $(a^2 \cos^2\theta + b^2 \sin^2\theta)^{\frac{1}{2}}$
14) $y + t_1 x = 2at_1 + at_1^3$;
    $\left[a\left(\dfrac{2}{t_1} + t_1\right)^2, -2a\left(\dfrac{2}{t_1} + t_1\right) \right]$
15) $py = p^3 x + c(1 - p^4); \left(-\dfrac{c}{p^3}, -cp^3 \right)$
16) $y = px + b(p^3 + p - 4);$
    $\left[-\dfrac{b}{p^2}(p^4 + 3p^2 + 4), \right.$
    $\left. -\dfrac{2b}{p}(p^2 + 2p + 2) \right]$
17) $(\frac{11}{2}, -4)$ 18) $(\frac{7}{4}, \frac{7}{2})$ 19) $(\frac{7}{2}, 0)$

Exercise 10h – p. 409

1) $xy = 1$
2) $9y^2 = 4(3x - 8)$
3) $y^2 = 6x$
4) $y^4 = c^2(c^2 - 2xy)$
5) $x^2 = 4(1 + y)$
6) $4x^2 + 9y^2 = 144$
7) $y^2 = 8(x + 1)$
8) the directrix
9) $xy = 2$

Multiple Choice Exercise 10 – p. 410

1) c   2) d   3) c
4) c   5) b   6) a

7) e   8) c   9) b

10) e   11) b, c   12) a, b

13) a, d   14) C 15) A

16) B 17) A 18) C

19) A 20) D 21) A

22) a or b or c   23) a, c, d   24) b or c

25) I 26) F 27) T

28) T 29) T 30) F

34) $2y = 3tx - at^3$

35) a) $y^2 = x^2(x + 3)$

36) $e^{-\frac{1}{2}}$

38) $(0, \pm \frac{32}{15})$

39) $x = 4, y = x + 5; (4, 0), (-\frac{16}{5}, \frac{9}{5})$

41) $\left(-ct, -\dfrac{c}{t}\right)$

Exercise 11a – p. 425

1) a, c, d, f, g

2) a) a b) b, e, h c) a, b, e

3) a) i, ii, iii, v   b) i, ii, viii

c) v, vi   d) ii, iii, iv

Miscellaneous Exercise 10 – p. 414

1) a) $x^2 + y^2 - 2x = 24$

b) $(\frac{8}{3}, 0)$

c) $x^2 + y^2 - 2x + 5y = 24$

2) a) $(1, 3), (-7, -1)$

b) $x^2 + y^2 + 6x - 2y = 0$

c) $1\frac{2}{3}$

3) $x^2 + y^2 + 2y = 1; x^2 + y^2 - 14y = 1$

4) $(\frac{12}{5}, \frac{9}{5}); x(5x - 6) + y(5y - 17) = 0$

5) a) $x^2 + y^2 - 6x - 6y + 5 = 0$

b) $3x^2 + 3y^2 - 6x - 26y + 3 = 0$

6) $\frac{3}{5}\sqrt{5}$

7) $x^2 + y^2 - 6x = 16;$

$2(5 + 5\sqrt{5}) = 10(1 + \sqrt{5})$

8) $x + y = 1; (\frac{7}{3}, -\frac{4}{3}), (-\frac{1}{3}, \frac{4}{3})$

11) a) $(-14, -10)$ b) $(2, -\frac{10}{3})$

12) $x^2 + y^2 - 20x - 10y + 100 = 0,$

$x^2 + y^2 - 40x - 20y + 400 = 0$

13) $2, 5; a = \frac{3}{10}; b = \pm\frac{2}{5}$

14) $x^2 + y^2 - 8x - 10y + 16 = 0;$

$4y + 3x = 7, 4y + 3x = 57$

15) $(x - 3)^2 + (y - 2)^2 = 4$

16) $(\frac{1}{5}, -\frac{3}{5})$

17) $\left[\dfrac{3}{\sqrt{2}}, \; 2 + \dfrac{3}{\sqrt{2}}\right], \; \left[-\dfrac{3}{\sqrt{2}}, \; 2 - \dfrac{3}{\sqrt{2}}\right]$

19) $ty = x + t^2; y = 3; 2x + 3y = 0$

21) $y^2 = 4a(x - a)$

23) a) $y^2 = a(x - a)$ c) $(3a, \pm 2\sqrt{3}a)$

24) $y = \pm(\frac{3}{4}x + 5)$

25) $y = \dfrac{2a}{m}, y(y + a) = 2ax$

27) $(6, \pm 4\sqrt{2})$

29) $x \cos \theta + y \sin \theta = 2 + 2 \cos \theta;$

$(-2, 0); x^2 + y^2 = 2x + 2$

30) $y \sin T + x = 1 + 2 \sin^2 T; \frac{1}{4}\pi$

31) $3 \pm 2\sqrt{2}$

32) $y \sin 2\theta + x \cos 2\theta$

$= a(5 \cos \theta - \cos 3\theta);$

points where $\theta = \frac{1}{8}\pi, \frac{1}{4}\pi, \frac{3}{8}\pi$

33) $y = (\frac{5}{4}x_0^2 - \frac{13}{9})x - \frac{5}{6}x_0^3;$

$\pm\frac{1}{3}\sqrt{5}, \pm\frac{2}{3}\sqrt{2}$

Exercise 11b – p. 434

1) a) $x = 1, y = 0$

b) $x = 1, y = -1$

c) $x = 0, y = -1$

d) $x = 0, y = 0$

e) $x = 4, y = 1$

f) $x = -\frac{1}{2}, y = \frac{1}{2}$

g) $x = 4, y = 3$

h) $x = -\frac{5}{2}, y = 2$

i) $x = -\frac{3}{2}, y = \frac{1}{2}$

2) a) $(0, 1)$ b) $(0, 0)$ c) $(1, 0)$

d) $(-1, 0)$ e) $(0, \frac{1}{2}), (2, 0)$

f) $(0, 3), (-3, 0)$

g) $(0, 0)$

h) $(\frac{3}{4}, 0), (0, -\frac{3}{5})$

i) $(0, 0)$

3) a) $y \to 0; y \to 0$

b) $y \to \pm \infty$ as $x \to \infty$; $x \not\to -\infty$

c) $y \to 0; y \to \infty$

d) $y \to \infty$ as $x \to \infty; x \not\to -\infty$

e) $y \to \infty; y \to -\infty$

f) $y \to \infty; y \to -\infty$

4) a) $x < -6, x > 1$

b) $1 < x < 2, x > 3$

c) $x > 1$

d) $0 < x < 1$

e) $x > 0$

f) $-1 < x < 1, x > 2$

g) $x < -2 -\sqrt{3}, x > -2 + \sqrt{3}$

h) $x < 2, x > 4$

Exercise 11d – p. 442

1) $7 < x < 11\frac{1}{2}$

2) $x < \frac{1}{4}, x > 1$

3) $4 < x < 4.8$

4) $-0.8 < x < \frac{1}{7}$

5) $x < \frac{3}{2}, x > \frac{5}{3}$

6) $x < 0, 1 < x < 2$

7) $-5 < x < -3, x > -1$

8) $x > 1$

9) $1 < x < 2, x > 3$

10) $-5 < x < 1, 2 < x < 4$

11) $-1 \leqslant \dfrac{4(x-2)}{4x^2+9} \leqslant \dfrac{1}{9}$

12) $-\dfrac{1}{2} \leqslant \dfrac{2x+1}{x^2+2} \leqslant 1$

13) $-3 < x < 3, x > 5$

14) $f(x) \in \mathbb{R}$

15) $-\frac{1}{2} \leqslant f(x) \leqslant \frac{1}{2}$

16) $f(x) < -2, f(x) > 2$

17) $f(x) \leqslant 1 - \dfrac{\sqrt{3}}{2}, f(x) \geqslant 1 + \dfrac{\sqrt{3}}{2}$

18) $f(x) \geqslant -\frac{1}{4}$

19) $f(x) \leqslant -4, f(x) \geqslant 0$

20) $f(x) \in \mathbb{R}$

21) $f(x) \leqslant -1, f(x) \geqslant 1$

22) $f(x) \leqslant 9 - 4\sqrt{5}, f(x) \geqslant 9 + 4\sqrt{5}$

24) $p \leqslant 6$

25) a) $k \leqslant 6$ b) $2 < x < 3$

Exercise 11e – p. 449

15) $1, 2$

16) $\frac{2}{3}, \frac{4}{5}$

17) $\frac{1}{2}(\sqrt{17} - 3), 3$

18) $\sqrt{3}, 2 - \sqrt{7}$

22) d, b, e, f

Miscellaneous Exercise 11 – p. 451

3) both continuous

5) $4 - 2\sqrt{3} \leqslant y \leqslant 4 + 2\sqrt{3}$

8) a) $(2, \frac{1}{4})$,   b) $(1, 0)$

9) $g(x) \in \mathbb{R}, g(x) \neq 2,$

$g^{-1} : x \to \dfrac{x+1}{x-2}$ $(x \in \mathbb{R}, x \neq 2)$

10) $|K| \leqslant 1$

11) $-289/24; -\frac{3}{2}, \frac{4}{3}; -1.56, -1.44, 1.27,$
    1.39

12) (i) $x = 2, y = 1$ (ii) $x + 2y = 0$

14) a) no,   b) yes, π

16) a) yes, m defines a many–one mapping
    b) $m(x) \in \mathbb{R}, m(x) \geqslant 0$

18) $\frac{5}{3} < x < 3$

19) $2 < x < 3, 4 < x < 6$

20) $x < -3, -1 < x < 2, x > 4$

21) $-3 < p < 0$

22) $a \leqslant -14, a \geqslant 6, a \leqslant -4$

23) $0 < x < 7, x > 8$

24) $\lambda \leqslant -1, y \leqslant -2, y \geqslant 6$

25) a) $f(x)$ is even, $g(x)$ is even,
    $h(x)$ is neither
    b) p^{-1} exists and is odd,
    q^{-1} does not exist

26) $f(x) \in \mathbb{Z}, 0 \leqslant g(x) < 1$, a) 0, b) $\mathbb{R}$

Exercise 12a – p. 460

1) a) $\overrightarrow{AC}$ b) $\overrightarrow{BD}$ c) $\overrightarrow{AD}$ d) $\overrightarrow{DB}$

2) $a, b - a, 2b - 2a$

5) $b - a, a + b, a - 3b, 2b, 2a - 2b$

6) $b - a, -a, -b, a - b$

7) $\overrightarrow{DE} = \overrightarrow{CH} = \overrightarrow{BG} = c,$
   $\overrightarrow{DC} = \overrightarrow{EH} = \overrightarrow{FG} = a,$
   $\overrightarrow{FE} = \overrightarrow{GH} = \overrightarrow{BC} = b$

9) $c - a, b - a, b - c$

10) $b + c - a, b - c - a, a + b + c,$
    $a + b - c$

11) $\frac{1}{2}b - c, \frac{1}{2}(c + b) - a$

Exercise 12b – p. 468

1) a) $3i - j$ b) $i + 3j$
   c) $5i$ d) $-2i - 11j$

2) a) $\sqrt{65}$ b) $\frac{1}{2}(3i + 2j)$
   c) $\frac{1}{4}(5i - 4j)$

3) a) $2i + j$ b) $\sqrt{73}$ c) $-3i - 7j$

4) $i + 2j$ 6) $\arctan(-2/3)$

Exercise 12c – p. 474

1) a) $3i + 6j + 4k$,   b) $i - 2j - 7k$,
   c) $i - 3k$

2) a) $(5, -7, 2)$ b) $(1, 4, 0)$ c) $(0, 1, -1)$

3) a) $\sqrt{21}$ b) 5 c) 3

4) a) 6 b) 7 c) $\sqrt{206}$

5) a) $3i + 4k$ b) $2i - 2j + 2k$
   c) $2i + 3j + 3k$ d) $-6i + 12j - 8k$

6) a) $-2i - 3j + 7k$ b) $-j + 3k$
   c) $-2i - 2j + 4k$

7) $\sqrt{62}, \sqrt{10}, 2\sqrt{6},$

8) $\sqrt{5}$

9) $\overrightarrow{AB} = -i + 3j, \overrightarrow{BD} = -j - 4k,$
   $\overrightarrow{CD} = -2i + 2j - 6k, \overrightarrow{AD} = -i + 2j - 4k$

Exercise 12d – p. 481

1) a) $\frac{2}{3}, \frac{2}{3}, -\frac{1}{3}$ b) $\frac{6}{7}, -\frac{2}{7}, -\frac{3}{7}$ c) $\frac{3}{5}, 0, \frac{4}{5}$
   d) $\frac{1}{9}, \frac{8}{9}, \frac{4}{9}$

2) a) $\frac{1}{3}(2i + 2j - k)$ b) $\frac{1}{7}(6i - 2j - 3k)$
   c) $\frac{1}{5}(3i + 4k)$ d) $\frac{1}{9}(i + 8j + 4k)$

3) a) $54.74°$ b) $90°$ c) $45°$
   d) $131.81°$

4) a) $(\frac{1}{3}, \frac{2}{3}, -\frac{2}{3})$ b) $(\frac{3}{7}, \frac{2}{7}, \frac{6}{7})$

5) a) $54.74°$ to Ox, $125.26°$ to Oy,
   $54.74°$ to Oz
   b) $63.61°$ to Ox, $27.27°$ to Oy,
   $83.62°$ to Oz
   c) $90°$ to Ox, $45°$ to Oy, $135°$ to Oz

6) a) $(\frac{12}{7}, -\frac{18}{7}, \frac{36}{7})$ b) $(\frac{16}{9}, \frac{2}{9}, -\frac{8}{9})$
   c) $\left(\dfrac{4}{\sqrt{3}}, \dfrac{4}{\sqrt{3}}, \dfrac{4}{\sqrt{3}} \right)$

7) a) $4j + 5k$ b) $16i - 16j - 8k$
   c) $\pm 4\sqrt{2}i + \frac{4}{?}j + \frac{4}{?}k$
   d) $\pm (\frac{8}{3}i + \frac{1}{3}j + \frac{4}{3}k)$

8) a) $5\sqrt{2}$; $64°54'$, $55°34'$, $45°$
   b) $\sqrt{3}$; $125°16'$, $54°44'$, $125°16'$
   c) $\sqrt{42}$; $39.5°$, $81.1°$, $51.9°$

9) a) $\dfrac{1}{\sqrt{3}}i - \dfrac{1}{\sqrt{3}}j + \dfrac{1}{\sqrt{3}}k$
   b) $\frac{3}{7}i + \frac{2}{7}j - \frac{6}{7}k$

10) $2\sqrt{3}(i + j + k)$

11) $\sqrt{22}$, $3/\sqrt{22}$, $2/\sqrt{22}$, $3/\sqrt{22}$; $3\sqrt{2}$,
    $1/3\sqrt{2}$, $-4/3\sqrt{2}$, $-1/3\sqrt{2}$

Exercise 12e – p. 486

1) $\sqrt{13}$; 3

2) $2/\sqrt{5}, 0, -1/\sqrt{5}$; $-1/\sqrt{5}, 0, 2/\sqrt{5}$

3) a) $\frac{1}{5}(2i + j + 7k)$ b) $-2i + j + 5k$

4) a) no b) no c) no

Exercise 12f – p. 492

1) a) $x - 2 = y - 3 = z + 1$
   b) $\dfrac{x}{3} = \dfrac{z}{5}$ and $y = 4$ c) $\dfrac{x}{2} = \dfrac{y}{3} = \dfrac{z}{4}$

2) a) $r = 3i + j + 7k + \lambda(2i + j + 3k)$
   $r = 2i - 5j + k + \lambda(3i + j + 4k)$
   c) $r = i + \lambda(-3i + 5j + k)$

3) a) $r = i - 3j + 2k + \lambda(5i + 4j - k)$;
   $\dfrac{x - 1}{5} = \dfrac{y + 3}{4} = \dfrac{z - 2}{-1}$
   b) $r = 2i + j + \lambda(3j - k)$;
   $x = 2, \dfrac{y - 1}{3} = \dfrac{z}{-1}$
   c) $r = \lambda(i - j - k)$; $x = -y = -z$.

4) a) no b) yes c) yes d) yes
   e) no

5) $r = 4i + 5j + 10k + \lambda(i + j + 3k)$,
   $4 - x = 5 - y = \dfrac{10 - z}{3}$
   $r = 2i + 3j + 4k + \lambda(i + j + 5k)$,
   $2 - x = 3 - y = \dfrac{4 - z}{5}$; $3i + 4j + 5k$.

6) a) $r = 3i + j - 4k + \lambda(i - 3j - 6k)$;
   $(\frac{7}{3}, 3, 0)$; $(0, 10, 14)$; $(\frac{10}{3}, 0, -6)$
   b) $r = i + j + 7k + \lambda(2i + 3j - 6k)$;
   $(\frac{10}{3}, \frac{9}{2}, 0)$; $(0, -\frac{1}{2}, 10)$; $(\frac{1}{3}, 0, 9)$

7) $r = 5i - 2j - 4k + \lambda(3i + 4j + 5k)$;
   $(\frac{13}{2}, 0, -\frac{3}{2})$

8) $6 : 3 : 2$; $r = 2i - j - k + \lambda(6i + 3j + 2k)$

Exercise 12g – p. 495

1) a) parallel b) intersecting; $r = i + 2j$
   c) skew

2) $a = -3$; $r = -i + 3j + 4k$

3) $(\frac{1}{3}, -\frac{2}{3}, \frac{2}{3})$; $(\frac{2}{7}, \frac{3}{7}, -\frac{6}{7})$; arccos $-\frac{16}{21}$

Exercise 12h – p. 502

1) a) 30
   b) 0, a and b are perpendicular
   c) -1

2) a) 7, $\sqrt{7}/3$ b) 14, $\sqrt{7/19}$ c) 3, $3/\sqrt{58}$

3) a) $a^2 - a.b$ b) $a^2 + b^2 - 2a.b$
   c) $b^2 - a^2$ d) $2b.c$

4) a) 0 b) $-b^2$ c) a^2 d) $2a^2 - b^2$

5) 4

6) a) -4 b) 15 c) 11
   d) -12 e) 29

7) $-\sqrt{(\frac{7}{34})}$; arccos $\sqrt{(\frac{7}{10})}$

11) a) arccos $\frac{19}{21}$ b) arccos $\frac{2}{3}$
    c) arccos $8\sqrt{3}/15$

12) a) $90°$ b) Parallel lines

15) a) $(i + 9j - 5k)/\sqrt{107}$
    b) $(-i - 2j + k)/\sqrt{6}$

Exercise 12i – p. 510

1) a) $r = i + 2j - k + \lambda(j + 3k)$
   $+ \mu(i - 2k)$,
   $r. (2i - 3j + k) = -5$
   b) $r = i + j - 2k + \lambda(j - 3k)$
   $+ \mu(3i + k)$,
   $r. (-i + 9j + 3k) = 2$

2) a) $r. (5i - 2j - 3k) = 7$
   b) $r. (-j + 2k) = -2$
   c) $r. (2i + k) = 5$

3) a) $x + y - z = 2$
   b) $2x + 3y - 4z = 1$
   c) $2x - 5y - z = -15$

4) $r. (i - j - k) = 0$

5) $r = -3i - 2j + \lambda(i - 2j + k)$
   $+ \mu(2i - j + 2k)$

6) $r = i + j + \lambda(i + k) + \mu(i - j + k)$;
   the second line is contained in the
   plane.

7) $\mathbf{r}.(7\mathbf{i}+2\mathbf{j}-3\mathbf{k})=3,\ 3/\sqrt{62}$

8) $3/\sqrt{5}$ 9) 2 10) $\mathbf{r}=\lambda(\mathbf{i}-2\mathbf{j}+\mathbf{k})$

11) $\mathbf{r}=2\mathbf{i}+\mathbf{j}+\mathbf{k}+\lambda(\mathbf{i}+2\mathbf{j}-3\mathbf{k})$

12) $\mathbf{r}.(\mathbf{i}-2\mathbf{j})=-2$

13) $\mathbf{r}.(\mathbf{j}-\mathbf{k})=0$ 14) $(\frac{5}{2},-\frac{1}{2},-\frac{1}{2})$

15) $(1,-1,4)$ 16) $7\sqrt{3}/3$

17) a) intersecting b) intersecting
   c) parallel d) contained in the plane.

Exercise 12j – p. 516

1) a) $3/\sqrt{11}$ b) $1/\sqrt{1102}$
   c) $20/\sqrt{442}$

2) a) $\sqrt{3}/9$ b) $\sqrt{15}/5$ c) $\frac{8}{21}$

3) a) 2 b) $\frac{22}{9}$ c) $9/\sqrt{5}$

4) a) $\mathbf{r}=-\frac{1}{3}(10\mathbf{j}+11\mathbf{k})+\lambda(3\mathbf{i}+5\mathbf{j}+7\mathbf{k})$
   b) $\mathbf{r}=\mathbf{j}+4\mathbf{k}+\lambda(\mathbf{i}-\mathbf{j}-3\mathbf{k})$

Exercise 12k – p. 523

1) 1 2) $\frac{1}{5}(13\mathbf{i}+10\mathbf{j}-7\mathbf{k})$

3) $4/\sqrt{259}$

4) $(4,4,2);\ \mathbf{r}.(\mathbf{i}+4\mathbf{j}-\mathbf{k})=18$

5) $(0,0,0);\ (2,0,0);\ (0,-4,0);$
   $(0,0,4);\ \frac{16}{3}$

6) $\frac{16}{3};\ \sqrt{50}/6;\ 200\pi/27$ 7) $4,\ 16\pi$

8) $\mathbf{r}=2\mathbf{i}+\mathbf{k}+\lambda(\mathbf{i}-2\mathbf{j}+2\mathbf{k});$
   $(\frac{11}{3},-\frac{10}{3},\frac{13}{3});\ (-\frac{4}{3},-\frac{25}{3},-\frac{17}{3});$
   $625\pi/3$

9) $(\mathbf{j}+\mathbf{k})/\sqrt{2};\ \mathbf{r}=(3\mathbf{i}+2\mathbf{j})+\lambda(\mathbf{j}+\mathbf{k});$
   $(3,2+3\sqrt{2},3\sqrt{2})$

10) $\mathbf{r}=\mathbf{i}+\mathbf{j}+\lambda(3\mathbf{i}-\mathbf{j}+5\mathbf{k})+\mu(\mathbf{i}-2\mathbf{j})$

11) $\mathbf{r}=\lambda(2\mathbf{i}+\mathbf{j}-5\mathbf{k});$
   $\mathbf{r}=\mathbf{i}+2\mathbf{j}-\mathbf{k}+\mu(2\mathbf{i}+\mathbf{j}-5\mathbf{k});$
   $\mathbf{r}.(2\mathbf{i}+\mathbf{j}-5\mathbf{k})=0;\ \sqrt{30}/5$

Multiple Choice Exercise 12 – p. 526

1) b 2) c 3) b 4) b
5) d 6) a 7) d 8) a
9) c 10) c,d 11) b,d 12) b
13) D 14) C 15) D 16) B
17) b,d 18) A 19) I 20) c,d
21) F 22) T 23) F 24) F

Miscellaneous Exercise 12 – p. 528

8) $\mathbf{c}=\mathbf{a}+2\mathbf{b};\ 2:5$

9) $\mathbf{r}.(\mathbf{i}+2\mathbf{j}+5\mathbf{k})=9;\ \frac{1}{2}\sqrt{30},\ 1/\sqrt{30}$

11) a) $\mathbf{p}.\mathbf{q}=|\mathbf{p}||\mathbf{q}|$ b) $\mathbf{p}.\mathbf{q}=0;90°,\frac{220}{3}$,
   $-7\mathbf{i}+22\mathbf{j}-\mathbf{k}$

12) b) $\frac{4}{3},3$

13) b) $(1,1,2);70.53°$

14) $\mathbf{r}.(2\mathbf{i}+\mathbf{k})=0,\sqrt{5}$

16) $\dfrac{x+1}{2}=\dfrac{y-2}{-3}=\dfrac{z-3}{4};\ (-\frac{53}{29},\frac{94}{29},\frac{39}{29})$

17) a) $\frac{2}{27}$ b) $(0,-5,19)$ c) $(-\frac{1}{9},\frac{1}{9},-\frac{22}{9})$

18) $x=y=2z,\ 4x-3y-2z=0,\ 90°$

19) $\dfrac{x-14}{2}=\dfrac{y-5}{2}=\dfrac{z-2}{-1},\ \cos\theta=\frac{2}{3}$,
   $x-2y-2z=0$

22) $\mathbf{a}-\mathbf{c}$

23) a) $\lambda(\mathbf{a}+\mathbf{b})$ b) $\mathbf{b}+\mu(\frac{1}{2}\mathbf{a}-\mathbf{b})$

Exercise 13a – p. 537

1) $-\mathbf{i},\mathbf{i},\mathbf{i},-\mathbf{i},1,\mathbf{i}$

2) a) $10+4i$ b) $7+2i$
   c) $6-2i$ d) $(a+c)+(b+d)i$

3) a) $-4+6i$ b) $1-4i$
   c) $-2+16i$ d) $(a-c)+(b-d)i$

4) a) $10-5i$ b) $39+23i$
   c) $11-7i$ d) 25
   e) $3-4i$ f) $-2+2i$
   g) $-4+3i$ h) x^2+y^2
   i) $-3+i$ j) $(a^2-b^2)+2abi$

5) a) $1+i$ b) $\frac{9}{25}+\frac{13}{25}i$ c) $\frac{4}{17}+\frac{16}{7}i$
   d) i e) $-i$
   f) $\dfrac{x^2-y^2}{x^2+y^2}+\dfrac{2xy}{x^2+y^2}i$
   g) $1-3i$ h) $-3-2i$

6) a) $x=9,y=-7$
   b) $x=-\frac{3}{2},y=\frac{7}{2}$
   c) $x=\frac{7}{2},y=\frac{1}{2}$
   d) $x=2,y=0$
   e) $x=13,y=0$
   f) $x=15,y=8$
   g) $x=11,y=3$
   h) $x=2,y=1$ or $x=-2,y=-1$

7) a) $7,-1$ b) $-2,-2$
   c) $\frac{10}{17},\frac{11}{17}$ d) $\frac{9}{5},-\frac{4}{5}$
   e) $0,\dfrac{-2y}{x^2+y^2}$ f) $-1,0$
   g) $\frac{1}{2},\frac{1}{2}\sqrt{3}$

8) a) $\pm(2-i)$ b) $\pm(5-2i)$
   c) $\pm(1+i)$ d) $\pm(4+i)$
   e) $\pm(1+5i)$

Exercise 13b – p. 541

1) a) $-\frac{1}{2}\pm\frac{1}{2}\sqrt{3}i$ b) $-\frac{7}{4}\pm\frac{1}{4}\sqrt{41}$
   c) $\pm3i$ d) $-\frac{1}{2}\pm\frac{1}{2}\sqrt{11}i$
   e) $\pm1,\pm i$

2) a) $x^2+1=0$
   b) $x^2-4x+5=0$
   c) $x^2-2x+10=0$
   d) $x^3-4x^2+6x-4=0$

3) a) $4, 5$ b) $6, 25$ c) $0, 1$
   d) $-24, 169$ e) $-2, 2$
4) a) $(x + 2 - i)(x + 2 + i)$
   b) $(x - 1 - 4i)(x - 1 + 4i)$
   c) $(x + \frac{1}{2} - \frac{1}{2}\sqrt{3}i)(x + \frac{1}{2} + \frac{1}{2}\sqrt{3}i)$
   d) $(x + 1 - \sqrt{3}i)(x + 1 + \sqrt{3}i)$
5) a) $\dfrac{i}{2(x + i)} - \dfrac{i}{2(x - i)}$

   b) $\dfrac{-2i}{x - 2 - i} + \dfrac{2i}{x - 2 + i}$

   c) $\dfrac{4i}{x + 2 + 2i} - \dfrac{4i}{x + 2 - 2i}$

   d) $\dfrac{i}{2(x + 2i)} - \dfrac{i}{2(x - 2i)}$

   e) $\dfrac{1 - 2i}{2(x + 2 - 3i)} + \dfrac{1 + 2i}{2(x + 2 + 3i)}$
6) a) 1 b) $3\omega^2$ c) 0
7) $-1, \frac{1}{2} \pm \frac{1}{2}\sqrt{3}i; -\lambda^2$

Exercise 13c – p. 550

1) a) $\sqrt{13}$ b) $\sqrt{17}$
   c) 5 d) 13
   e) $\sqrt{2}, -\frac{1}{4}\pi$ f) $\sqrt{2}, \frac{3}{4}\pi$
   g) $4, 0$ h) $2, -\frac{1}{2}\pi$
   i) $\sqrt{a^2 + b^2}, \arctan\dfrac{b}{a}$
   j) $\sqrt{2}, \frac{1}{4}\pi$ k) $\sqrt{2}, \frac{3}{4}\pi$
   l) $\sqrt{2}, -\frac{3}{4}\pi$ m) $\sqrt{2}, -\frac{1}{4}\pi$
   n) $\sqrt{170}$ o) $2, \frac{1}{3}\pi$
   p) $1, \frac{3}{4}\pi$ q) $3, -\frac{5}{6}\pi$

3) a) $\sqrt{2}\left(\cos\dfrac{\pi}{4} + i\sin\dfrac{\pi}{4}\right)$

   b) $2\left\{\cos\left(-\dfrac{\pi}{6}\right) + i\sin\left(-\dfrac{\pi}{6}\right)\right\}$

   c) $5\{\cos(-2.214^c) + i\sin(-2.214^c)\}$
   d) $13(\cos 1.966^c + i\sin 1.966^c)$
   e) $\sqrt{5}\{\cos(-0.464^c)$
      $+ i\sin(-0.464^c)\}$
   f) $6(\cos 0 + i\sin 0)$
   g) $3(\cos \pi + i\sin \pi)$

   h) $4\left(\cos\dfrac{\pi}{2} + i\sin\dfrac{\pi}{2}\right)$

   i) $2\sqrt{3}\left\{\cos\left(-\dfrac{5\pi}{6}\right) + i\sin\left(-\dfrac{5\pi}{6}\right)\right\}$
   j) $25(\cos 0.284^c + i\sin 0.284^c)$
4) a) $\sqrt{3} + i$ b) $\frac{3}{2}\sqrt{2} - \frac{3}{2}\sqrt{2}i$
   c) $-\frac{1}{2} + \frac{1}{2}\sqrt{3}i$ d) $-\frac{1}{2}\sqrt{2} - \frac{1}{2}\sqrt{2}i$
   e) 3 f) -2

g) $2\sqrt{3} - 2i$ h) i
i) $-3i$ j) $-\frac{1}{2} - \frac{1}{2}\sqrt{3}i)$
6) $1, -\frac{1}{2} \pm \frac{1}{2}\sqrt{3}i; 1(\cos 0 + i\sin 0),$
   $1\left(\cos\dfrac{2\pi}{3} + i\sin\dfrac{2\pi}{3}\right),$
   $1\left\{\cos\left(-\dfrac{2\pi}{3}\right) + i\sin\left(-\dfrac{2\pi}{3}\right)\right\}$

Exercise 13d – p. 557

2) $31; 19$
3) a) $2\sqrt{2}, \frac{1}{4}\pi$ b) $2\sqrt{6}, -\frac{5}{12}\pi$
   c) $1, 2\arctan\sqrt{3}/2$
5) $\pm 2\left\{\cos\left(-\dfrac{\pi}{12}\right) + i\sin\left(-\dfrac{\pi}{12}\right)\right\}$

Multiple Choice Exercise 13 – p. 558

1) c 2) e 3) a 4) c
5) d 6) e 7) c 8) d
9) a,c 10) b,e 11) a,c 12) c,d
13) a,d 14) A 15) E 16) E
17) D 18) F 19) T 20) T
21) F 22) F

Miscellaneous Exercise 13 – p. 561

1) 1 2) $\frac{1}{5}(3 - 4i), \frac{1}{2}(-3 + i)$
3) $-\frac{6}{5} + \frac{12}{5}i$
4) a) $\frac{104}{25} - \frac{72}{25}i$ b) $\pm \sqrt{2}(1 + i)$
5) $\frac{63}{25}, \frac{16}{25}, \frac{13}{5}, \frac{63}{65}, \frac{16}{65}$
6) $4, -5; 5$
7) $2, \frac{1}{6}\pi; \frac{1}{2}(1 + i\sqrt{3})$
8) a) $p = \frac{7}{5}, q = -\frac{4}{5}$
   b) $p = 2 - i, q = 2 + i$
9) (i) $2\sqrt{10}$ (ii) 10 (iii) $\frac{1}{5}\sqrt{10}$
10) (i) $\sqrt{97}$ (ii) $\frac{1}{2}\sqrt{2}$
12) a) $\sec\theta$ b) $-i\tan\theta$
13) a) $3 - i, 3 + 3i$
14) a) 6
    b) $\dfrac{x(x^2 + y^2 + 1)}{x^2 + y^2}, \dfrac{y(x^2 + y^2 - 1)}{x^2 + y^2};$
15) a) $\pm(3 + 2i)$
    b) (i) $\sqrt{2}, -\frac{1}{4}\pi$ (ii) $5, 0.643^c$
    (iii) $5\sqrt{2}, -0.142^c; 12$
16) a) $\frac{1}{2}, -\frac{1}{2}\pi$
    b) $3 - 2i, -3 + 2i; \pm(2 + 3i)\sqrt{3}$
17) a) $2\sqrt{2}, \frac{3}{4}\pi; 2\sqrt{2}, -\frac{3}{4}\pi; \frac{1}{2}i$
18) a) (i) $\frac{7}{50} - \frac{1}{50}i$
    (ii) $(c^4 - 6c^2 + 1) + (4c^3 - 4c)i$
    b) $\pm(1 + 2i); \pm(1 - 2i)$

Exercise 14a — p. 565

1) permutation
2) combination
3) combination
4) permutation
5) permutation
6) permutation

Exercise 14b — p. 570

1) 120
2) 56
3) 720
4) 18
5) 5040
6) 12
7) 12
8) 9×10^4
9) a) 504   b) 9   c) 324   d) 5
10) a) 6   b) 3   11) 5040   12) 60

Exercise 14c — p. 572

1) 28
2) 28
3) a) 4845   b) 969
4) a) 120   b) 90
5) 210
6) 1287
7) 495
8) a) 126   b) 252   c) 36   d) 72   e) 9

Exercise 14d — p. 575

1) 6
2) 24
3) 120
4) 720
5) 30
6) 132
7) 2730
8) 840
9) 28
10) 1140
11) 360 360
12) 1260
13) 70
14) $\frac{3}{4}$
15) $\frac{5!}{2!}$
16) $\frac{11!}{9!}$
17) $\frac{39!}{34!}$
18) $\frac{n!}{(n-4)!}$
19) $\frac{(n+1)!}{(n-2)!}$
20) $\frac{(n+5)!}{n!}$
21) $\frac{(n+r)!}{(n+r-3)!}$
22) $\frac{20!}{17!\,3!}$
23) $\frac{14!}{12!\,3!}$
24) $\frac{8!\,3!}{5!\,6!}$
25) $\frac{n!}{(n-3)!\,3!}$
26) $\frac{(n-2)!}{4!\,(n-5)!}$
27) 10(8!)   28) 40(5!)   29) 274(8!)
30) $(n+1)(n-1)!$
31) $(n^2 + n - 1)(n - 1)!$
32) $n(n^2 - n + 2)(n - 2)!$
33) $n^2(n-2)!$ 34) $\frac{8(7!)}{15(2!\,4!)}$
35) $\frac{29(6!)}{4!}$ 36) $\frac{109(8!)}{28(6!\,2!)}$
37) $\frac{(n-1)!}{(r+1)!}(nr + n + 1)$
38) $\frac{n!}{(r+1)!}(2nr + 2r + 2n - 1)$
39) $\frac{n!}{r!\,(n-r+1)!}(3n - r + 3)$
40) $\frac{(nr-2)(n-1)!}{(n-r)!\,(r-1)!}$

Exercise 14e — p. 581

1) a) 10   b) 27   2) 13
3) $\frac{18!}{13(5!)^2\,(7!)}$ 4) $\frac{11!}{(2!)^3}$
5) 2454
6) 18
7) 252
8) 75 600
9) 120
10) 240
11) 1440
12) a) 27   b) 109
13) a) $7 \times 7!$ b) $7 \times 7!$
14) 16   15) a) 64   b) 48   c) 12
16) 28   17) a) 56   b) 35
18) 76 145   19) 505
20) 27

Miscellaneous Exercise 14 — p. 584

1) a) 225   b) 465   c) 240
2) a) $(n-1)!$ b) $(n-2)^2\,(n-2)!$
   $(n^2 - 3n + 3)(n - 2)!$
3) $1 + 10 + 40 + 80 + 80 + 32 = 243$
4) $16 \times 17 \times 18!$ 5) 126
6) 77   7) 2100
8) $\frac{(n+m)!}{n!m!}, \frac{(n+m-2)}{(n-1)!(m-1)!}$
9) a) 56   b) 70
10) a) 56   b) 24   c) 32
11) a) 18   b) 192
12) $\frac{11!}{8}, \frac{10!}{4}, 4!\,7!$
13) a) 504; 168   b) 84

Exercise 15a — p. 588

1) a) $\displaystyle\sum_{r=1}^{5} r^3$ b) $\displaystyle\sum_{r=1}^{10} 2r$

c) $\displaystyle\sum_{r=1}^{33} 3r$ d) $\displaystyle\sum_{r=2}^{50} \frac{1}{r}$

e) $\displaystyle\sum_{r=0}^{\infty} \frac{1}{3^r}$ f) $\displaystyle\sum_{r=0}^{7} (-4 + 3r)$

g) $\displaystyle\sum_{r=0}^{\infty} \left(\frac{8}{2^r}\right) = \sum_{r=0}^{\infty} 2^{3-r}$

2) a) $1 + \frac{1}{2} + \frac{1}{3} + \ldots$

b) $0 + 2 + 6 + 12 + \ldots + 30$

c) $1 + 1 + 2 + \ldots + 20!$

d) $1 + \frac{1}{2} + \frac{1}{5} + \ldots$

e) $0 + 0 + 6 + 24 + 60 + \ldots + 720$

f) $-1 + a - a^2 + \ldots$

3) a) 8, 9   b) 9, 10

c) 1, 6   d) $\frac{1}{420}, \infty$

e) $-\frac{4}{15}, \infty$ f) $-48, 23$

g) $(\frac{1}{2})^n, \infty$

Exercise 15b – p. 593

1) a) $17, 4n - 3$ b) $0, \frac{1}{2}(5 - n)$

c) $17, 3n + 2$ d) $9, 2n - 1$

e) $16, 4(n - 1)$ f) $-2, 8 - 2n$

g) $p + 4q, p + (n - 1)q$

h) $18, 8 + 2n$ i) $15, 3n$

2) a) 190   b) $-\frac{5}{2}$ c) 185

d) 100   e) 180   f) -30

g) $5(2p + 9q)$

3) $a = 27\frac{1}{5}, d = -2\frac{2}{5}$ 4) $d = 3; 30$

5) $1, \frac{1}{2}, 0; -8\frac{1}{2}$

6) a) $28\frac{1}{2}$ b) 80   c) 400

d) 80   e) 108   f) $3n(1 - 6n)$

g) 40   h) $2m(m + 3)$

7) $4, 2n - 4$ 9) $2, 364$

10) 39   11) 64   12) 12

Exercise 15c – p. 598

1) a) $32, 2^n$ b) $\frac{1}{8}, \dfrac{1}{2^{n-2}}$

c) $48, 3(-2)^{n-1}$

d) $\frac{1}{2}, (-1)^{n-1}(\frac{1}{2})^{n-4}$

e) $\frac{1}{27}, (\frac{1}{3})^{n-2}$

2) a) 189   b) -255 c) $2 - (\frac{1}{2})^{19}$

d) $\frac{781}{125}$ e) $\frac{341}{1024}$ f) 1

3) $\frac{1}{2}, 2$ 4) $-\frac{1}{2}$

5) $-\frac{1}{2}, \frac{1}{1024}$ 6) 13.21

7) a) $\dfrac{x - x^{n+1}}{1 - x}$ b) $\dfrac{x^n - 1}{x^{n-2}(x - 1)}$

c) $\dfrac{1 + (-1)^{n+1}y^n}{1 + y}$

d) $\dfrac{x(2^n - x^n)}{2^{n-1}(2 - x)}$ e) $\dfrac{1 - (-2)^n x^n}{1 + 2x}$

8) $\frac{8}{3}(1 - (\frac{1}{4})^n), 4$ 9) 62 or 122

10) 8.49   11) 8   12) £23.31

Exercise 15d – p. 603

1) a) yes   b) no   c) yes

d) yes   e) no   f) yes

2) a) $-1 < x < 1$

b) $x < -1, x > 1$

c) $-\frac{1}{2} < x < \frac{1}{2}$

d) $0 < x < 2$

e) $-1 - a < x < 1 - a$

f) $x < -1 - a, x > 1 - a$

3) a) 6   c) $13\frac{1}{3}$ d) $\frac{5}{9}$ f) $\frac{9}{4}$

4) a) $\frac{161}{990}$ b) $\frac{34}{99}$ c) $\frac{7}{330}$

5) $\frac{1}{2}$

Exercise 15e – p. 609

1) a) $1 + 36x + 594x^2 + 5940x^3$

b) $1 - 18x + 144x^2 - 672x^3$

c) $1024 + 5120x + 11\,520x^2$
$+ 15\,360x^3$

d) $1 - \frac{20}{3}x + \frac{190}{9}x^2 - \frac{380}{9}x^3$

e) $128 - 672x + 1512x^2 - 1890x^3$

f) $\left(\dfrac{3}{2}\right)^9 + \dfrac{3^{10}}{2^7}x + \dfrac{3^9}{8}x^2 + \dfrac{7}{2}(3^7)x^3$

2) a) $336x^2$ b) $-\frac{969}{2}x^5$

c) $-\dfrac{2^3 15!}{12!}x^{11}$ d) $2^{22} 3^8 \,^{30}C_8 x^8$

e) $3360p^6q^4$ f) $16(3a)^7 b$

g) $7920x^4$ h) $63x^5$

i) $56a^3 b^5$ j) $20(3^{10})a^2 b^2$

3) a) $1 - 8x + 27x^2$

b) $1 + 19x + 160x^2$

c) $2 - 19x + 85x^2$

d) $1 - 68x + 2136x^2$

4) -527 5) $-\frac{779}{256}$ i

6) a) $\displaystyle\sum_{r=0}^{9} (-1)^r \dfrac{9! x^r (10 - 2r)}{r!(10 - r)!}$

b) $\displaystyle\sum_{r=0}^{10} \dfrac{2^{r-1}(22 - 3r)10! x^r}{r!(11 - r)!}$

c) $\displaystyle\sum_{r=0}^{20} \dfrac{(-1)^r(21 - 2r)20! x^r}{2^{r-1} r!(21 - r)!}$

d) $\displaystyle\sum_{r=0}^{14} \dfrac{(-1)^r 14!(36r^2 - 936r + 6000)5^{r-2} x^r}{r!(16 - r)!}$

7) $\dfrac{4^{r-1}(36 - 6r)8!}{r!(9 - r)!}$

8) $\dfrac{2^{20-r} 20!(21 - 3r)}{r!(21 - r)!}$

9) 0.8171

10) 1.062

14) a) $1 - 50x$ b) $256 - 1024x$
    c) $1 - 19x$

Exercise 15f — p. 615

1) $1 - x - \dfrac{x^2}{2} - \dfrac{x^3}{2}, -\frac{1}{2} < x < \frac{1}{2}$

2) $\dfrac{1}{3} - \dfrac{x}{9} + \dfrac{x^2}{27} - \dfrac{x^3}{81}, -3 < x < 3$

3) $1 - \dfrac{x}{4} + \dfrac{3x^2}{32} - \dfrac{5}{128}x^3, -2 < x < 2$

4) $1 + 2x + 3x^2 + 4x^3, -1 < x < 1$

5) $1 - \frac{1}{2}x + \frac{3}{8}x^2 - \frac{5}{16}x^3, -1 < x < 1$

6) $1 + \frac{1}{2}x - \frac{5}{8}x^2 - \frac{3}{16}x^3, -1 < x < 1$

7) $-2 - 3x - 3x^2 - 3x^3, -1 < x < 1$

8) $2 + 2x + \frac{21}{4}x^2 + \frac{27}{8}x^3, -\frac{1}{3} < x < \frac{1}{3}$

9) $\frac{1}{2} - \frac{3}{4}x + \frac{13}{8}x^2 - \frac{51}{16}x^3, -\frac{1}{2} < x < \frac{1}{2}$

10) $1 + x + \frac{1}{2}x^2 + \frac{1}{2}x^3, -1 < x < 1$

11) $1 - \dfrac{x^2}{9}, -3 < x < 3$

12) $x - x^2 + x^3, -1 < x < 1$

13) $1 - \dfrac{3}{p} + \dfrac{6}{p^2} - \dfrac{10}{p^3} + \dfrac{15}{p^4}, |p| > 1$

14) 1.732

15) 3.162 28

16) 1.004 99

17) 2.032 793

18) $1 + 2x + 2x^2, \sqrt{51} = 7.1414$

20) $1 - 3x + \frac{7}{2}x^2$

21) $\frac{3}{2} + \frac{15}{4}x$

Exercise 15g — p. 623

1) $\frac{2}{3}$

2) $\frac{1}{2}(1 + 3^{11}) = 88\,574$

3) $\dfrac{ab^2(b^2 - b^{2n})}{1 - b^2}$

4) $2\frac{2}{5}$

5) $\dfrac{n(3n + 5)}{4(n + 1)(n + 2)}$

6) $\dfrac{n - 2}{4(n + 2)}$

7) $\dfrac{n + 1}{n(2n + 1)}$

8) $\dfrac{n(n + 1)}{2(2n + 1)(2n + 3)}$

9) $\dfrac{n}{4}(n + 1)(n + 2)(n + 3)$

10) $\dfrac{n^2}{4}(n + 1)^2$

11) $\frac{1}{2} + \frac{1}{5}(2n - 1)(2n + 1)(2n + 3)$

12) $\dfrac{n(n + 3)}{4(n + 1)(n + 2)}$

13) $\dfrac{n(n + 2)}{(n + 1)^2}$

14) $n(n^2 + 6n + 11)$

15) $\dfrac{n}{2}(2n^3 + 4n^2 + 6n + 1)$

16) $4(r + 1)(r + 2)(r + 3)$

17) $1 - \dfrac{1}{(n + 1)!}$

18) $(2n + 1)! - 1$

19) $\dfrac{\cos 2\theta - \cos(2n + 2)\theta}{2\sin\theta}$

20) $\dfrac{n}{3}(n + 1)(n + 2)$

21) $\dfrac{n}{4}(n + 1)(n + 2)(n + 3)$

22) $\dfrac{2n}{3}(n + 1)(2n + 1)$

23) $\dfrac{n}{12}(n + 1)(45n^2 + 37n + 2)$

24) $\dfrac{n}{6}(2n - 1)(n - 1)$

25) $-2n^2 - n$

26) 42 075

Exercise 15h — p. 628

1) a) 0   b) 0   c) 1

2) a) $\frac{1}{3}$ b) $\frac{1}{2}$ c) $\frac{11}{18}$

3) a) 20 003   b) 30 002   c) 174

Multiple Choice Exercise 15 — p. 632

1) c   2) b   3) a   4) e
5) b   6) c   7) c   8) b, c
9) a   10) a, b   11) a   12) E
13) C 14) B 15) F 16) T
17) F 18) F 19) T

Miscellaneous Exercise 15 — p. 634

1) $S_n = \dfrac{n(3n + 1)}{2}, n = 7$

2) $6n - 3$ 4) $20°$

5) a) 63   b) 8.7   6) 11

7) £82, 7 years   8) 29 mm

9) b) 5

10) $A = 25, B = -9, -\frac{37}{15}x^3, |x| < \frac{3}{2}$,

11) $\dfrac{[1 - (n+2)x^{n+1} + (n+1)x^{n+2}]}{(1-x)^2}$

12) $\frac{1}{4}, 6, -\frac{9}{16}$ 14) $\frac{1}{3}$

15) 35, 27   16) 120

17) $1 + 2x + 7x^2 + 20x^3$

18) $z = (1 - 10c^2 + 5c^4)$
  $+ i(5c - 10c^3 + c^5)$

19) 3.332 222   21) 7.093 97

22) $a = \dfrac{1}{\sqrt{2}}, b = \dfrac{1}{4\sqrt{2}}, c = -\dfrac{5}{32\sqrt{2}}$

23) a) $1 + (p-q)x + (\frac{3}{2}q^2 - pq - \frac{1}{2}p^2)x^2$,
  $1 + \frac{1}{2}(p-q)x$
  $+ (\frac{3}{8}p^2 - \frac{1}{4}pq - \frac{1}{8}q^2)x^2$
  b) $-\frac{7}{5}$

24) a) $\frac{1}{2}\left(\dfrac{3}{2} - \dfrac{1}{n+1} - \dfrac{1}{n+2}\right), \frac{3}{4}$
  b) 1 191 700

25) b) $n(n+1)(2n^2 + 2n - 1)$

26) $\frac{1}{3}n(n+1)(n+2)$

27) 70

28) $\dfrac{n}{4}(n+1)(n+2)(n+3)$

29) $7m^3 + 3m^2 - 10m - 6$

32) $n^2(n^2 - 2n - 2)$

35) a) $\dfrac{r}{1-r}$ b) $\dfrac{r}{(1-r)^2}$
  c) $\dfrac{r[n - (n+1)r + r^{n+1}]}{(1-r)^2}$

37) a) $\dfrac{n}{3}(4n^2 - 1)$ b) $|a| < 1$,
  $\dfrac{1}{(1-a^2)(1-a)}$

38) $\dfrac{r^2 - r + 2}{2}, \dfrac{r^2 + r}{2}$, 22 155

39) b) $\dfrac{1 - x^{n+1}}{1-x}, \dfrac{1 - (n+1)x^n + nx^{n+1}}{(1-x)^2}$

41) $\dfrac{2 + (n-1)2^{n+1}}{4}, \dfrac{4n-1}{4} - \dfrac{2n+1}{2n(n+1)}$

Exercise 16a – p. 644

1) $-1, -1, \frac{1}{5}, 5$

2) $-3, -\frac{1}{3}, \frac{1}{2}, 2$

3) $\frac{1}{9}, \frac{1}{7}, 7, 9$

4) $-\frac{1}{4}, -4, \pm i$

5) $-\frac{3}{4}, -\frac{3}{4}, \frac{2}{5}$

6) $\frac{2}{3}, \frac{2}{3}, \frac{1}{5}$

7) $\frac{3}{2}, \frac{3}{2}, \frac{2}{5}$

8) $-\frac{3}{4}, -\frac{3}{4}, \frac{1}{7}$

9) $9, 9, -3, \pm 4i$

10) 4

11) $6 \pm 2\sqrt{3}$

12) 5

13) $1, 8, -8$

14) $(\sqrt{2}, \frac{1}{2}\sqrt{2}), (-\sqrt{2}, -\frac{1}{2}\sqrt{2})$

15) $(2, 3), (-2, -1)$

16) $(-3, 4), (-6, 3)$

17) $4, 1, -2, -2$

Exercise 16b – p. 651

1) 1   2) 2   3) 1   4) 2
5) 1   6) 2   7) 2   8) 3
9) 3 and 4; 1 and 2, 0 and 1; −4 and −3,
  0 and 1; −1 and 0, 0 and 1, 2 and 3
11) $4 < \lambda < 5$

Exercise 16c – p. 658

2) a) infinite set   b) 2
  c) infinite set
  d) infinite set (all +ve)
  e) 3   f) 1   g) 1   h) 2   i) 1
3) a) 1.17^c b) 1 (exact) c) −0.618
4) 1
5) 1.28^c
6) 1.16 m

Exercise 16d – p. 669

1) 0.20   2) 0.22   3) 0.16
4) 2.07   5) 4.15   6) 5.15

In Questions 7–12 the degree of accuracy
depends upon the accuracy of the reader's
first graphical approximation.

7) 1.1656 (correct to 4 d.p.)
8) 0 (exact), 0.746 85 (correct to 5 d.p.)
9) 0 (exact), 1.2564 (correct to 4 d.p.)
10) 0.510 97 (correct to 5 d.p.)
11) −2.6691 (correct to 4 d.p.)
12) 0.374 82 (correct to 5 d.p.)
13) 0 (exact), 1.904
14) 1 (exact)
15) 0
16) 0.540, −1.454
17) 3.104

Miscellaneous Exercise 16 – p. 670

1) 0.38^c, 2.77^c

2) a) 0.74^c b) 0.67^c, 2.48^c
   c) 0.93^c, 3.14^c d) 0.65^c
   e) 1.65^c

3) a) (i) 1 (ii) 0   b) (i) 0 (ii) 0
   c) (i) 1 (ii) 0   d) (i) ∞ (ii) ∞
   e) (i) ∞ (ii) ∞ f) (i) 1 (ii) 1

4) $a^2(2\theta - \sin 2\theta)$, 1.16^c

5) 2.28^c

6) 0.42^c

7) 0, $30°$, $70°$

8) $n = 2$, $a_2 = 2.20$, $a_3 = 2.19$

9) 2.4, 2.49

10) 0.405

11) 1.79

12) a) $-1, 1$ b) $\frac{1}{2}$, 2, $\frac{1}{6}(5 \pm i\sqrt{11})$

13) 1.26

14) 2.908

15) The point where $x = 7.2$ is too near to the tangent at the point where $x = 7$

16) c) is suitable

Exercise 17a – p. 681

1) a) $Y = aX + b$, $Y = \dfrac{1}{y}$, $X = x$, $m = a$,
      $c = b$
   b) $Y = y^2 - x$, $X = y$, $m = b$, $c = -a$
   c) $X = e^x/y$, $Y = y$, $m = a$, $c = b$

2) $a = 2$, $b = -4$ 3) $a = 6$, $b = 4$

4) $a = 0.5$, $b = -2$ 5) $a = 30$, $b = 2$

6) $a = 2$, $b = \frac{1}{2}$ 7) $a = 3$, $b = -2$

Exercise 17b – p. 686

1) a) $\frac{32}{3}$ b) $4\frac{1}{2}$ c) $\frac{125}{6}$

2) a) $\frac{4}{3}$ b) $\frac{32}{3}$

3) a) 1   b) $3 - \ln 4$

4) a) $\frac{4}{3}$ b) $\frac{4}{3}$

5) $4\sqrt{3}$

Exercise 17c – p. 693

1) $\dfrac{512\pi}{15}$ 2) $\dfrac{\pi}{2}(e^6 - 1)$

3) $\dfrac{\pi}{2}$ 4) $\dfrac{64\pi}{5}$

5) 2π 6) 8π

7) 8π 8) $\dfrac{3\pi}{5}(\sqrt[3]{32} - 1)$

9) $\dfrac{\pi}{2}(e^2 - 1)$ 10) $\dfrac{16\pi}{15}$

11) $\dfrac{16\pi}{15}$ 12) $\dfrac{\pi^2}{2}$

13) $\dfrac{3\pi}{10}$ 14) 8π

15) $\pi(32\sqrt{2} - 40)/15$

Exercise 17d – p. 697

1) a) 2   b) $\dfrac{\pi}{4}$ c) $\left(\dfrac{\pi}{2}, \dfrac{\pi}{8}\right)$
   d) $\dfrac{\pi^2}{2}$ e) $\dfrac{\pi^3}{4}$ f) $\left(\dfrac{\pi}{2}, 0\right)$

2) a) $\frac{1}{6}$ b) $\frac{1}{60}$ c) $(\frac{1}{2}, -\frac{1}{10})$
   d) $\dfrac{\pi}{30}$ e) $\dfrac{\pi}{60}$ f) $(\frac{1}{2}, 0)$

3) a) $\frac{32}{3}$ b) $\frac{128}{5}$ c) $\frac{12}{5}$
   d) 8π e) $\dfrac{64\pi}{3}$ f) $(0, \frac{8}{3})$

4) a) $e - 1$ b) 1   c) $\dfrac{1}{e - 1}$
   d) $\dfrac{\pi}{2}(e^2 - 1)$ e) $\dfrac{\pi}{4}(e^2 + 1)$
   f) $\left(0, \dfrac{e^2 + 1}{2(e^2 - 1)}\right)$

5) $\dfrac{e^2 + 1}{e^2 - 1}$

Exercise 17e – p. 705

1) a) 1.049   b) 1.006

2) a) 0.3887   b) 0.3891

3) a) 1.178   b) 1.047

4) a) 0.2727   b) 0.271 15

5) a) 0.990 33   b) 0.9996

6) a) 1.148 38   b) 1.147 79

7) 0.102 460

8) 0.099 215 56

Exercise 17f – p. 710

1) a) 2.0833   b) 3.037
   c) 5.0133

2) a) 0.874 77   b) 0.4848
   c) 0.719 48

3) $\left[\dfrac{x}{1 + x} + \ln(1 + x)\right]\delta x$, 0.7385

4) $(\sec^2 x)\,\delta x$, $1 + \dfrac{\pi}{16}$

5) 2.0125　　　　6) 0.2575 cm/s

7) − 0.1011 cm³/s　　8) 30 cm³/s

9) −0.27 m³/s

15) $\dfrac{a^2\pi}{4}$, 16π, $\dfrac{8}{3}$, $\dfrac{8}{3\pi}$

16) $(-1, 0)$, $\frac{1}{5}$

17) $\frac{13}{6}a^2$, $\frac{59}{15}\pi a^3$

Miscellaneous Exercise 17 – p. 711

1) $k = 0.2, n = 0.5$

2) $a = 12.6, b = 1.12$

3) $a = 0.75, b = -0.6$

5) $a:b:c = 2:1:2$

6) $a = 2, b = 36$

7) $501, -3$

8) $n \simeq 4, a \simeq 3.125, 4\%$

9) $1.8, 6.6, a = 1, b = 3.3$

10) $4\frac{2}{3}$

11) $\frac{1}{2}e - 1$, $\pi(\frac{1}{6}e^2 - \frac{1}{2})$, $2\pi(1 - \frac{1}{3}e)$

12) $2, \frac{1}{2}\pi^2$

13) a, min, $a \geqslant \dfrac{1}{e}$, $-1 + 3\ln 2$,

$(1 + 8\ln 2)/(12\ln 2 - 4)$

14) $(\cos 2\theta - \sin\theta)/\cos\theta$, $\left(\dfrac{1}{2}, \dfrac{3\sqrt{3}}{4}\right)$,

$\left(\dfrac{1}{2}, -\dfrac{3\sqrt{3}}{4}\right)$, $(1, 0)$, $\dfrac{13\pi}{10}$

19) a) $2 - \ln 2$　　b) $\dfrac{2\ln 2}{2 - \ln 2}$

c) $\pi(\frac{9}{4} - 2\ln 2)$

20) 40

21) $\frac{16}{3}$, 8π, $(0, \frac{20}{3})$

22) a) $\ln\dfrac{kx}{\sqrt{(1 - x^2)}}$　　b) $\dfrac{16}{15\pi}$　　c) 0.945

23) 1.77

24) $3.91, 31.3, \pi \simeq 3.13$

26) $7.512, 30.005$

27) $\dfrac{1}{3x^{2/3}}$, $\dfrac{\delta x}{3x^{2/3}}$, 10.003

28) a) 0.879,　　b) 0.879

29) $(\frac{195}{76}, \frac{633}{190})$　　　30) 0.95

31) $\sqrt{3}\,e/30$　　　32) -2

33) $9a$ cm per min

34) a) $\frac{2}{3}a$　　b) $-\dfrac{a}{T}$

35) $136\pi/3, 16/5\pi$

INDEX

Abscissa 71
Altitude of a triangle 241
Ambiguous case 153
Amplitude 157, 545
Angle
    associated acute 145
    between a line and a plane 512
    between two lines 363, 499
    between two planes 512
    between two vectors 463
    general 144
    negative 148
    small 234
Approximate solution of equations 651
    graphical 652
Arbitrary constant 333
Arc, length of 140
Area of a triangle 241
    bounded by a curve 344, 682
Argand diagram 542
Argument 545
Arithmetic mean 600
Arithmetic progression 589
Asymptote 59, 428

Base vectors 469, 471
Binomial expression 36
    theorem 603, 610

Cartesian coordinates 17, 70
    base vectors 471
Centroid
    of area 694
    of a triangle 241, 362, 468
    of volume 696
Change of variable 309, 317
Circles 370
    orthogonal 373
    tangent to 371, 372
    touching 374
Circular functions 143
    graphs of 157
    inverse 163
Complementary angles 166
Complete primitive 333

Complex number 533
    argument of 545
    conjugate 534
    imaginary part of 535
    modulus of 545
    polar form of 547
    real part of 535
    vector representation of 543
Complex roots of an equation 538
Compound angles 186, 205
Conjugate 534
Constant of integration 299, 330
Continuous 60, 422
Convergence 599
Coordinates 69
    Cartesian 70
    parametric 283, 397
    polar 70
Cover-up method 271
Cosine rule 155
Cotangent formula 238
Cube roots of unity 539
Cubic curve 392
    function 393
Curve sketching 427

Definite integration 340
Degree 5, 137
Derivative 112
    second 127, 286
Derived function 112
Difference method 620
Differential
    coefficient 112
    equation 331
        linear 331
        variables separable 332
Differentiation 106
    logarithmic 280
    parametric 264
Differentiation of
    compound functions 265
    exponential functions 258, 280
    function of a function 266
    implicit functions 275
    inverse trig functions 278

logarithmic functions 263
polynomials 115
products 269
quotients 270
trig functions 255
Direction cosines 475, 482
ratios 475, 482
vector 479
Directrix 380
Discontinuity 60
Discriminant 12
Distance
of plane from origin 505
of point from line 363
of point from plane 513
Divergent 600
Division of a line 361, 483
Domain 46

Ellipse 388
Empty region 428
Equation 2
approximate solution of 651
graphical solution of 652
meaning of 2, 83
of a circle 370
of a curve 89
of a parabola 382
of a plane 504
of a straight line 90, 486
polynomial 641
quadratic 10
trigonometric 167, 173, 177
Equations
conjugate complex roots of 538
number of real roots of 647
with repeated roots 641
Even function 420
Exponential equations 30
function 58

Factor theorem 34
Factorial 573
First moment of area 694
of volume 696
Focal length 381
Focus 380
Fractions
partial 5, 311
proper 5
Frame of reference 59, 470
Free vector 462
Function 4, 43
continuous 60, 422
even 420
graphical representation of 58
implicit 269
inverse 65
logarithmic 61

odd 421
periodic 422
quadratic 48
rational 59

General angle 144
trig ratios of 144
General solution of trig equations 167
Geometric mean 601
Geometric progression 594
Gradient 76
function 111
of a curve 106
Graphical representation of functions 58
solution of equations 652

Identity 3
Image set 46
Imaginary axis 542
number 532
Implicit functions 269
Increment, small 745
Incentre of a triangle 241
Independent arrangements 578
Indices (laws of) 23
Induction 629
Inequality 3, 52, 439
modulus 445
quadratic 52
Infinite series 610
Inflexion 124
Integral
definite 340
Integration 299, 325
approximate 698, 704
as the reverse of differentiation 299
as the limit of a sum 337, 681
by parts 312
by substitution 309, 317
constant of 299, 330
of exponential functions 304
of fractions 316
of products 308
of trig functions 302, 322
Intersection 96, 384, 400, 445, 493
Inverse circular function 163
function 65
Iterative methods 651

Line of intersection of two planes 515
Linear differential equation 331
interpolation 661
Locus 368, 404
Logarithm 27
base of 27
change of base of 29
laws of 28
Naperian 261
natural 262

Logarithmic differentiation 280
Logarithmic function 61
 differentiation of 263

Mapping 43
Maximum value 124
Median of a triangle 241
Method of differences 620
 of induction 629
Minimum value 124
Modulus
 equation 445
 of a complex number 545
 of a function 443
 of a vector 456, 473
Moment
 first, of area 694
 first, of volume 696
Multiple angles 181
Mutually exclusive 579

Naperian logarithm 261
Natural logarithm 262
 number series 617
Negative angles 148
Newton–Raphson method 666
Normal 106, 119, 285
Number series 616
Numerical solution of equations 659
 by interval bisection 659
 by linear interpolation 661
 by Newton's method 666

Odd function 421
Ordinate 71
Orthocentre 241
Orthogonal circles 373

Parabola 380
 vertex of 380
Parameter 283
Parametric equations 283, 397
 differentiation 264
Partial fractions 5
 cover-up method 271
Pascal's triangle 36
Periodic function 422
Perpendicular distance of a
  point from a line 363
Planes
 angle between 512
 equation of 504
 line of intersection of two 515
Point of inflexion 124
Polar coordinates 70
Polynomial 5
 equation 641
Position vector 461

Primitive, complete 333
Principle argument 545
 value 168
Proper fraction 5

Quadratic
 equation 10
  roots of 12, 14
 function 48
 inequality 52

Radian 138
Range 46
Rational function 59
Rationalise 26
Rate of change 708
Real axis 542
 number 532
Realise 536
Reciprocal function 435
Rectangular hyperbola 389
Remainder theorem 32
Resultant vector 457, 473
Roots of a quadratic equation 12, 14
Roots of a polynomial equation
 location of 645
 number of real 647

Scalar 456
 product 456
Secondary value 170
Sector, area of 140
Segment 141
Separating the variables 332
Sequence 586, 625
Series 586
Sigma notation 587
Simpson's rule 701
Sine rule 152
Small increments 705
Stationary values 122
Substitution, integration by 309, 317
Sum
 by integration 344, 681
 of an A.P. 590
 of a G.P. 600
 of a number series 616
 to infinity 627
Surds 25
System of reference 59

Tangent 106, 119, 277, 285, 383
Trapezium rule 699
Triangle
 altitude of 241
 area of 241
 centroid of 241
 circumcentre of 240
 incentre of 241

law for vector sum 457
orthocentre of 241
properties of 152, 155, 237
Trigonometric equations 167, 173, 177
identities 199, 207, 211, 218, 228
Trigonometric functions
differentiation of 255
integration of 302, 322
inverse 163
Trigonometric ratios 144
Turning points 123

Unit vector 478
Unity, cube roots of 539

Variables separable 332
Vector 455
angle between 463
base 469, 471
Cartesian components of 464
direction 479
free 462
modulus of 456, 473
position 461
resultant 457, 473
unit 478
Vectors
applications of 524
Volume of revolution 687